CLASSICAL NOVA EXPLOSIONS

Related Titles from the AIP Conference Proceedings Subseries on Astronomy and Astrophysics

599 X-Ray Astronomy: Stellar Endpoints, AGN, and the Diffuse X-Ray Background
Edited by Nicholas E. White, December 2001, 0-7354-0043-1

598 Solar and Galactic Composition: A Joint SOHO/ACE Workshop
Edited by Robert F. Wimmer-Schweingruber, December 2001,CD-ROM included, 0-7354-0042-3

587 Gamma 2001: Gamma-Ray Astrophysics 2001
Edited by Steven Ritz, Neil Gehrels, and Chris R. Shrader, October 2001, 0-7354-0027-X
CD-ROM: 0-7354-0030-X

565 Young Supernova Remnants: Eleventh Astrophysics Conference
Edited by Stephen S. Holt and Una Hwang, May 2001, 0-7354-0001-6

556 Explosive Phenomena in Astrophysical Compact Objects: First KIAS Astrophysics Workshop
Edited by Heon-Young Chang, Chang-Hwan Lee, Mannque Rho, and Insu Yi, March 2001, 1-56396-987-4

526 Gamma-Ray Bursts: 5th Huntsville Symposium
Edited by R. Marc Kippen, Robert S. Mallozzi, and Gerald J. Fishman, June 2000, CD-ROM included, 1-56396-947-5

522 Cosmic Explosions: Tenth Astrophysics Conference
Edited by Stephen S. Holt and William W. Zhang, June 2000, 1-56396-943-2

510 The Fifth Compton Symposium
Edited by Mark L. McConnell and James M. Ryan, March 2000, 1-56396-932-7

499 Small Missions for Energetic Astrophysics: Ultraviolet to Gamma-Ray
Edited by Steven P. Brumby, December 1999, 1-56396-912-2

To learn more about these titles, or the AIP Conference Proceedings Series, please visit the webpage **http://proceedings.aip.org**

CLASSICAL NOVA EXPLOSIONS

International Conference on Classical Nova Explosions

Sitges, Spain   20–24 May 2002

EDITORS
Margarita Hernanz
Jordi José
*Institut d'Estudis Espacials de Catalunya
(IEEC/CSIC/UPC)
Barcelona, Spain*

SPONSORING ORGANIZATIONS
IEEC (Institute for Space Research of Catalonia)
CSIC (National Research Council of Spain)
UPC (Polytechnic University of Catalonia)
MCYT (Spanish Ministry of Science and Technology)
DURSI (Department of Universities and Research of the
       Generalitat de Catalunya)

Melville, New York, 2002
AIP CONFERENCE PROCEEDINGS ■ VOLUME 637

Editors:

Margarita Hernanz
Jordi José

Institut d'Estudis Espacials de Catalunya (IEEC)
Ed. Nexus 201
c./ Gran Capità, 2-4
E-08034 Barcelona
SPAIN

E-mail: hernanz@ieec.fcr.es
         jjose@ieec.fcr.es

Authorization to photocopy items for internal or personal use, beyond the free copying permitted under the 1978 U.S. Copyright Law (see statement below), is granted by the American Institute of Physics for users registered with the Copyright Clearance Center (CCC) Transactional Reporting Service, provided that the base fee of $19.00 per copy is paid directly to CCC, 222 Rosewood Drive, Danvers, MA 01923. For those organizations that have been granted a photocopy license by CCC, a separate system of payment has been arranged. The fee code for users of the Transactional Reporting Service is: 0-7354-0092-X/02/$19.00.

© 2002 American Institute of Physics

Individual readers of this volume and nonprofit libraries, acting for them, are permitted to make fair use of the material in it, such as copying an article for use in teaching or research. Permission is granted to quote from this volume in scientific work with the customary acknowledgment of the source. To reprint a figure, table, or other excerpt requires the consent of one of the original authors and notification to AIP. Republication or systematic or multiple reproduction of any material in this volume is permitted only under license from AIP. Address inquiries to Office of Rights and Permissions, Suite 1NO1, 2 Huntington Quadrangle, Melville, N.Y. 11747-4502; phone: 516-576-2268; fax: 516-576-2450; e-mail: rights@aip.org.

L.C. Catalog Card No. 2002113020
ISBN 0-7354-0092-X
ISSN 0094-243X
Printed in the United States of America

Contents

Preface... xiii
Organizing Committees xv
Conference Photo xvii

SCENARIO

General Properties of Quiescent Novae 3
    B. Warner
Alternatives to Hibernation 16
    T. Naylor
Spectroscopic Evidence from HST and Archival IUE Studies of
Dwarf Novae as Past Classical Novae 21
    E. M. Sion
More on Peculiar Abundances in Dwarf Novae and NLs: Implications
for Past Novae Explosions 28
    P. Szkody
The Recurrent Novae and Their Relation with Classical Novae 32
    G. C. Anupama
Evolution of the Symbiotic Nova RX Puppis 42
    J. Mikołajewska, E. Brandi, L. Garcia, O. Ferrer, C. Quiroga,
    and G. C. Anupama
An Evolutionary Scenario for the U Scorpii 47
    M. J. Sarna, E. Ergma, and J. Gerškevitš
V838 Mon and the New Class of Stars Erupting into
Cool Supergiants (SECS) 52
    U. Munari, A. Henden, R. M. L. Corradi, and T. Zwitter
On the Maximum Mass of C-O White Dwarfs 57
    I. Domínguez, O. Straniero, J. Isern, and A. Tornambé
The Evolution of Intermediate Mass Close Binary Systems: Scenarios
Leading to Novae 62
    E. García-Berro, P. Gil-Pons, and J. W. Truran
Abnormal CNO Abundances in Magnetic Cataclysmic Variables 67
    M. Mouchet, J. M. Bonnet-Bidaud, M. Abada-Simon, K. Beuermann,
    D. de Martino, R. Ferlet, R. Fried, B. Gänsicke, S. Howell, A. Lecavelier,
    K. Mukai, D. Porquet, E. Roueff, and P. Szkody
Did EY Cyg Go through a Nova Explosion? 72
    G. Tovmassian, M. Orio, S. Zharikov, J. Echevarría, R. Costero,
    and R. Michel
The 2001 Superoutburst of WZ Sge: Is There Any Connection with
Nova Outbursts? 77
    J. Echevarría, R. Costero, G. Tovmassian, S. Zharikov, R. Michel,
    M. Richer, and A. Arellano-Ferro

The Influence of Mass Loss on Orbital Elements of Binary Systems
by Periastron Effect 82
    M. Andrade and J. A. Docobo

EXPLOSION MECHANISM AND MASS LOSS

Studies of Novae in the 20th Century 89
    S. Starrfield
H-Accreting CO WDs: Accretion Regimes and Final Outcomes 99
    L. Piersanti, S. Cassisi, I. Iben Jr., and A. Tornambé
Nuclear Ashes: Reviewing Thirty Years of Nucleosynthesis in
Classical Novae 104
    J. José
The Effects of Thermonuclear Reaction Rate Variations on
Nova Nucleosynthesis 114
    C. Iliadis, A. Champagne, J. José, S. Starrfield, and P. Tupper
Nuclear Astrophysics at ISAC with DRAGON: Initial Studies 119
    A. Olin, S. Bishop, L. Buchmann, M. L. Chatterjee, A. Chen,
    J. M. D'Auria, S. Engel, D. Gigliotti, U. Greife, D. Hunter, A. Hussein,
    D. Hutcheon, C. Jewett, J. King, S. Kubono, M. Lamey, A. M. Laird,
    R. Lewis, W. Liu, S. Michimasa, D. Ottewell, P. Parker, J. Rogers,
    F. Strieder, M. Wiescher, and C. Wrede
Multidimensional Nova Simulations 124
    A. Glasner and E. Livne
Mixing by Nonlinear Gravity Wave Breaking on a
White Dwarf Surface 134
    A. C. Calder, A. Alexakis, L. J. Dursi, R. Rosner, J. W. Truran, B. Fryxell,
    P. Ricker, M. Zingale, K. Olson, F. X. Timmes, and P. MacNeice
Onset of Convection on a Pre-runaway White Dwarf 139
    L. J. Dursi, A. C. Calder, A. Alexakis, J. W. Truran, M. Zingale, B. Fryxell,
    P. Ricker, F. X. Timmes, and K. Olson
The Role of Novae in Galactic Chemical Evolution 144
    D. Romano and F. Matteucci
The Electrostatic Screening of Nuclear Reactions in Dense Plasma 150
    N. J. Shaviv and G. Shaviv
Movies of Novae Explosions: Restricted Three-Body Dynamics and
Geometry of Novae Shells for Purely Gravitational Development 155
    D. K. Lynch, S. Mazuk, E. Campbell, and C. C. Venturini
Nova Nucleosynthesis Calculations: Robust Uncertainties,
Sensitivities, and Radioactive Ion Beam Measurements 161
    M. S. Smith, W. R. Hix, S. Parete-Koon, L. Dessieux, M. W. Guidry,
    D. W. Bardayan, S. Starrfield, D. L. Smith, and A. Mezzacappa
The Imprint of Nova Nucleosynthesis in Presolar Grains 167
    J. José, M. Hernanz, S. Amari, and E. Zinner

OBSERVATIONS: OPTICAL, UV, IR, AND RADIO SPECTRA

Panchromatic Study of Novae in Outburst: Phenomenology and Physics 175
    S. N. Shore
Some Clues to the Understanding of the Ultraviolet Spectra of Novae 188
    A. Cassatella and R. González-Riestra
Spectroscopic Monitoring of Classical and Recurrent Novae at Asiago Observatory 193
    T. Iijima
Infrared and Radio Observations of Classical Novae: Physical Parameters and Abundances in the Ejecta 198
    R. D. Gehrz
0.8–2.5 μm Spectroscopy of Novae 208
    D. K. Lynch, R. J. Rudy, C. C. Venturini, S. Mazuk, W. L. Dimpfl, J. C. Wilson, N. A. Miller, and R. Puetter
Spectral Evolution of Galactic Novae 214
    E. Mason, M. Della Valle, and A. Bianchini
A Photoionization Model Analysis of the ONeMg Nova QU Vul 219
    G. J. Schwarz
Elemental Abundance Analysis of Nova Cygni 1992 224
    K. M. Vanlandingham, S. Starrfield, S. N. Shore, and R. M. Wagner
The INES Guide for Classical Novae 228
    R. González-Riestra and A. Cassatella
The Spectral Evolution of Nova Velorum 1999 (V382 Vel) 233
    A. Augusto and M. Diaz
WSO/UV: World Space Observatory/Ultraviolet 238
    M. Hernanz, R. González-Riestra, W. Wamsteker, B. Shustov, M. Barstow, N. Brosch, C. Fu-Zhen, M. Dennefeld, M. Dopita, A. I. Gómez de Castro, N. Kappelmann, I. Pagano, J. Sahade, H. Haubold, J.-E. Solheim, and P. Martínez
Radio Emission from V723 Cas 242
    I. Heywood, T. J. O'Brien, S. P. S. Eyres, M. F. Bode, and R. J. Davis

MODEL ATMOSPHERES AND LIGHT CURVES

Nova Model Atmospheres 249
    P. H. Hauschildt, G. Schwarz, C. I. Short, E. Baron, and S. Starrfield
Classical Novae as Super-Eddington Objects 259
    N. J. Shaviv
A Few Comments on Nova Models 266
    M. Friedjung
Formation and Evolution of Dust in Novae 270
    J. M. C. Rawlings and A. Evans
The Properties of the Dust around Nova V705 Cas. 275
    A. Evans, O. Smith, V. H. Tyne, J. M. C. Rawlings, T. R. Geballe, and S. P. S. Eyres

A Solution to the Transition Phase in Classical Novae 279
   A. Retter
Recurrent Novae as a Progenitor System of Type Ia Supernovae 284
   I. Hachisu and M. Kato
IM Normae and N Sgr 2002: CCD Spectroscopy and Photometry 289
   W. Liller
The Recurrent Nova IM Nor: A Representative Example of the Use of
a New Classification of Novae by the Shape of Their Light Curves 294
   A. E. Rosenbush
Spectroscopic and Photometric Observations of the Recurrent
Nova IM Normae 299
   H. W. Duerbeck, C. Sterken, R. Baptista, M. P. Diaz, C. M. Dutra,
   L. Freyhammer, H. Hensberge, and A. F. Jones
Nova Monocerotis 2002 (V838 Mon) in the Early Stages
of Its Outburst 303
   E. A. Barsukova, N. V. Borisov, V. P. Goranskij, A. V. Kusakin,
   N. V. Metlova, and S. Y. Shugarov
V723 Cas a Borderline Classical Nova 308
   M. Friedjung and T. Iijima
Radial Pulsation of the Cooling White Dwarf in the Decay of Nova
Cassiopeiae 1995 (V723 Cas)... 311
   V. P. Goranskij, N. V. Metlova, and S. Y. Shugarov
Photometric Observations of Two Novae in Cygnus 315
   I. Voloshina, H. Rovithis-Livaniou, and N. Metlova
BVRI Photometry of Extremely Slow Nova Aql=V1548 Aql............... 319
   N. V. Primak, E. P. Pavlenko, S. Y. Shugarov, and V. P. Goranskij
The Photometry of V1974 Cyg = N Cyg 1992 323
   S. Y. Shugarov, V. P. Goranskij, and E. P. Pavlenko
The Problem of the Flickering Activity of the Recurrent Nova T CrB 328
   L. Hric, K. Petrík, A. Dobrotka, and R. Gális
Activity of the Supersoft X-ray Source V Sge 333
   V. Šimon and J. A. Mattei
The Colors and Luminosities of the Supersoft X-Ray Sources and
Classical Novae 338
   V. Šimon

X-RAYS FROM NOVAE

X-Ray Observations of Novae... 345
   J. Krautter
A XMM-Newton Observation of Nova LMC 1995........................ 355
   M. Orio, J. Greiner, W. Hartmann, and M. Still
The Search for Extended Ionization and Reflection Nebulae................ 360
   M. Orio and G. Tovmassian
On the Possibility of Detecting Remnants of Novae in the X-Rays and
Recovering the Remains of an Explosion after a Century 365
   Ş. Balman

Chandra Observations of Old Novae..372
    K. Mukai, M. Orio, F. Ringwald, and M. Still
Chandra ACIS-I and LETGS X-Ray Observations of
Nova 1999 Velorum (V382 Vel)..377
    V. Burwitz, S. Starrfield, J. Krautter, and J.-U. Ness
XMM-Newton Observations of Classical Novae...........................381
    M. Hernanz and G. Sala
Photoionization as a Source of X-Ray Emission from Classical Novae........386
    G. Sala and M. Hernanz
Novae in M31: How Many of Them Turn into Supersoft
X-Ray Sources?...391
    M. Orio, A. Dalmazzo, and P. Nedialkov

GAMMA RAYS FROM NOVAE

Gamma-Ray Emission from Classical Novae 399
    M. Hernanz
On the Formation of CVs with ONeMg White Dwarfs and Their
Contribution to the ^{26}Al Production in the Galaxy 409
    M. Politano
Global Galactic Distribution of Classical Novae 415
    A. F. Iyudin, V. Schönfelder, K. Bennett, R. Diehl, W. Hermsen,
    G. G. Lichti, and J. Ryan
A New Experiment for the Determination of the ^{18}F(p,α) Reaction
Rate at Nova Temperatures..420
    N. de Séréville, A. Coc, C. Angulo, M. Assunção, D. Beaumel, B. Bouzid,
    S. Cherubini, M. Couder, P. Demaret, F. de Oliveira Santos, P. Figuera,
    S. Fortier, M. Gaelens, F. Hammache, J. Kiener, D. Labar, A. Lefebvre,
    P. Leleux, A. Ninane, M. Loiselet, S. Ouichaoui, G. Ryckewaert,
    N. Smirnova, V. Tatischeff, and J.-P. Thibaud
Study of the ^{18}F(p,α)^{15}O Reaction at Energies Relevant for ^{18}F
Nucleosynthesis in Novae..425
    D. W. Bardayan, J. C. Batchelder, J. C. Blackmon, A. E. Champagne,
    T. Davinson, R. Fitzgerald, W. R. Hix, C. Iliadis, R. L. Kozub, Z. Ma,
    S. Parete-Koon, P. D. Parker, N. Shu, M. S. Smith, and P. J. Woods
The Diffuse 1.275 MeV Emission from Galactic ONe Novae..................430
    P. Jean, M. Hernanz, and J. José
Future INTEGRAL Observations of Classical Novae 435
    M. Hernanz, P. Jean, J. José, A. Coc, S. Starrfield, J. Truran, J. Isern,
    G. Sala, and A. Giménez

NOVA POPULATIONS/NOVAE IN EXTERNAL GALAXIES

Nova Populations 443
    M. Della Valle

400 Novae in M87 457
    M. M. Shara and D. R. Zurek

The Galactic Nova Rate 462
    A. W. Shafter

The Mysterious Eruption of V2434-LMC 472
    W. Liller and M. Morel

The Parent Population of Novae in the Large Magellanic Cloud 476
    A. Subramaniam and G. C. Anupama

Novae in External Galaxies from the POINT-AGAPE Survey and the Liverpool Telescope 481
    M. J. Darnley, M. F. Bode, E. J. Kerins, and T. J. O'Brien

Early Decline Spectra of Nova SMC 2001 and Nova LMC 2002 486
    E. Mason, M. Della Valle, R. Gilmozzi, R. E. Williams, and G. Lo Curto

A Survey for Novae in M33: Preliminary Results 491
    S. J. Williams and A. W. Shafter

OLD NOVA SHELLS AND REMNANTS

The Evolution of Nova Remnants 497
    M. F. Bode

The Structure of the Shell of HR Del 509
    T. J. O'Brien, D. J. Harman, and M. F. Bode

Nova Outburst Luminosities, Postnova Magnitude Behaviour, and Long-Term Evolution of Nova Shell Luminosities 514
    H. W. Duerbeck

Multicolour Studies of the Old Novae Behavior 519
    E. P. Pavlenko, S. Y. Shugarov, V. P. Goranskij, and N. V. Primak

Photometric and Spectroscopic Properties of Four Old Novae 527
    L. Schmidtobreick, C. Tappert, A. Bianchini, and R. Mennickent

The UCT Survey of Old Novae 532
    P. A. Woudt and B. Warner

Optical Spectroscopy of GK Persei during Outburst and Quiescence 534
    U. S. Kamath and G. C. Anupama

Physical and Chemical Diagnostic of Structured Nova Shells 538
    M. P. Diaz

Nova V Persei—A classical Nova in the "Period Gap" 543
    N. A. Katysheva, E. P. Pavlenko, and S. Y. Shugarov

A Possible Detected Faint Shell of the Classical Nova QZ Aurigae 548
    H. H. Esenoglu

The Active Quiescence of the Ex-Nova HR Del 553
    P. Selvelli and M. Friedjung

Emission Line Flaring in the SW Sex Old Nova V533 Herculis 558
    P. Rodríguez-Gil and I. G. Martínez-Pais

CONCLUDING REMARKS

Summary of the Meeting 565
    S. Starrfield

APPENDICES

Conference Program 573
List of Participants 581
Author Index 587

Preface

The *International Conference on Classical Nova Explosions* was held at the *sala Gaudí* of *Hotel Antemare*, in the charming town of Sitges (Barcelona, Spain), from May 20-24, 2002. Since the conference "Physics of Classical Novae", held in Madrid (Spain) in June 1989, there had not been any specialized workshop dedicated exclusively to this topic. In 1976, there was a previous Conference about "Novae and Related Stars" in Paris, France. Therefore, it seems that these conferences follow somewhat a 13-year period. Although sessions devoted to classical novae are always included in the workshops on "Cataclysmic Variables", they can't cover in depth all the topics involved in the field of classical novae. This fact, together with the encouragement of some local and non local colleagues, led us to organize a specific meeting about these fascinating stellar explosions, including all the recent advances from both the theoretical and the observational points of view. The aim of the conference was to foster contacts within the different experts in the field, including observers and theoreticians, in order to address the current knowledge, outstanding problems and future perspectives in the field of classical novae. We all enjoyed the pleasant and relaxing atmosphere of Sitges that contributed to a lively Conference.

The Scientific Programme of the Conference was established, with the help of the Scientific Organizing Committee, according to eight sessions: scenario; explosion mechanism and mass loss; observations: optical, UV, IR and radio spectra; model atmospheres and light curves; x-rays from novae; gamma-rays from novae; nova populations/novae in external galaxies; old nova shells and remnants. The Conference convincingly demonstrated the health of this discipline through a number of recent advances, from pioneering multidimensional simulations of the outburst to remarkable results from HST, showing hundreds of novae discovered in some external galaxies, or wonderful spectra from recent X-ray satellites, to quote only a few. 14 invited reviews, 38 contributed talks, and 50 posters, that fuelled lively discussions among the 97 participants from 23 countries worldwide, were presented during the Conference. Such contributions, with only a few remarkable absences, form the main body of the present volume that not only summarizes the contents of the Conference, but also provides an overview of the current understanding of classical novae. We thank all the authors for their effort to write their contributions within a very short deadline and, in particular, the participants who made the effort to keep their contributions according to the expected length.

It is with great pleasure that we thank the Scientific Organizing Committee for providing advice and enthusiasm already from the early outline of the Conference format. Nevertheless, since the success of a Conference relies not only on its scientific content, we would also like to acknowledge support from a number of institutions and individuals that contributed to its organization. First, we would like to mention the Sitges local authorities for the visit and reception held at the Palau Maricel. We also thank Caixa del Penedès for kindly providing folders to the

Conference attendees. We, as well, acknowledge partial funding from the following institutions: the MCYT (Spanish Ministry of Science and Technology), the DURSI (Department of Universities and Research of the *Generalitat de Catalunya*), and our home institutions, the CSIC (National Research Council of Spain), the UPC (Polytechnic University of Catalonia), and the IEEC (Institute for Space Research of Catalonia). And last, but not least, we would like to express our gratitude to the numerous colleagues from our home institutions, and, in particular, to Joan Bausells, for his continuous and efficient help and encouragement, and to the other members of the Local Organizing Committee: Jordi Isern, for many stimulating ideas, suggestions and remarks; Glòria Sala, for providing us with her enthusiasm and invaluable help; Eva Notario, for her devoted and efficient job as the Conference Secretary; and Josep Guerrero, who deserves probably more than a couple of lines for his extraordinary task before, during, and after the Conference. His unselfish investment of time and helpful assistance has not only contributed to the success of the Conference, but also to many aspects concerning the edition of this book.

We hope that less than 13 years will pass before the next Conference on Classical Novae, where we would like to meet all the participants (at least) again.

Barcelona, August 2002
Margarita Hernanz and Jordi José

Scientific Organizing Committee

Michael Bode (Liverpool John Moores University, United Kingdom)
Angelo Cassatella (Istituto di Astrofisica Spaziale, Rome, Italy)
Hilmar Duerbeck (Brussels Free University, Belgium)
Michael Friedjung (Institut d'Astrophysique, CNRS, Paris, France)
Robert Gerhz (University of Minnesota, Minneapolis, USA)
Margarita Hernanz (IEEC/CSIC, Barcelona, Spain, Chair)
Jordi José (IEEC/UPC, Barcelona, Spain, Co-Chair)
Mariko Kato (Keio University, Japan)
Joachim Krautter (Landessternwarte, Heidelberg, Germany)
Dina Prialnik (Tel Aviv University, Israel)
Giora Shaviv (Technion, Haifa, Israel)
Sumner Starrfield (Arizona State University, Tempe, USA)
James Truran (University of Chicago, USA)
Brian Warner (University of Cape Town, South Africa)

Local Organizing Committee

Margarita Hernanz (IEEC/CSIC, Barcelona, Chair)
Jordi José (IEEC/UPC, Barcelona, Co-Chair)
Jordi Isern (IEEC/CSIC, Barcelona)
Josep Guerrero (IEEC, Barcelona)
Glòria Sala (IEEC/CSIC, Barcelona)
Eva Notario (IEEC, Barcelona, Conference Secretary)

SCENARIO

General Properties of Quiescent Novae

Brian Warner

Department of Astronomy, University of Cape Town, Rondebosch 7700, South Africa

Abstract. The observed properties of novae before and after eruption are discussed. The distribution of orbital periods of novae shows a concentration near 3.2 h, which resembles that of magnetic cataclysmic variables, and there is some evidence that many of the novae themselves are magnetic near that orbital period. Desynchronisation of polars by nova eruptions can lead to an estimate ($\sim 2 \times 10^3$ y) for the time between eruptions for the strongly magnetic systems; this is much shorter than that found from other methods. The similarity of pre- and post-nova luminosities, at high rates of mass transfer, is ascribed to irradiation of the secondary producing a self-sustained high $\dot{M}$ state. This slows cooling of the white dwarf after eruption, delays the onset of full scale dwarf nova outbursts in most systems, and delays any descent into a hibernation state of low rate of mass transfer.

INTRODUCTION

This conference is mostly about the high luminosity state, and the transitions into and out of it, but full understanding of the nova process must include the nature of the low luminosity white dwarf and its accretion environment, in which it spends most of its life.

CYCLIC EVOLUTION AND HIBERNATION

As is well known, the mechanisms of orbital angular momentum loss generally invoked to drive CV evolution are magnetic braking (MB) for orbital periods longer than about 2 h and gravitational radiation (GR) for shorter periods (Note that it has long been pointed out that GR alone is not sufficient to account for the observed luminosities of many of the short period systems (Warner 1987), though these may not be representative of the long-term mean brightness). Basic models of MB (e.g. Verbunt & Zwaan 1981) are single valued, giving a unique relationship between mass transfer rate ($\dot{M}$) and orbital period P_{orb}. The large range (factors of 1000 or more) of $\dot{M}$ that is observed at most values of P_{orb} (Patterson 1984; Warner 1987, 1995a) shows that at least one other parameter is determining the instantaneous $\dot{M}$. Whether standard MB and GR set the long-term average, or whether additional mechanisms are required, is not yet certain. Large temporary excursions of $\dot{M}$ above the average set by MB can be generated by the effect of irradiation of the secondary by the primary and inner disc, which increases the scale height of the atmosphere of the secondary. If the response of the entire secondary to irradiative heating at the surface is calculated, cyclical evolution is found which alternates between a high $\dot{M}$ state in which the secondary expands and a low $\dot{M}$ state in which it contracts (King et al. 1995). Time scales for the transitions are $\sim 10^4 - 10^5$

y, and durations in the high states are $\sim 10^7$ y. Such cycles do not develop for secondary masses below ~ 0.65 M$_\odot$ because of the large thermal inertia in the convective envelope (King et al. 1996).

In addition, the observed high $\dot{M}$ after nova eruption, and the subsequent steady reduction, which is in quantitative agreement (e.g., for V1500 Cyg: Somers & Naylor 1999) with the prediction (Prialnik 1986) of the effects of irradiation of the disc and the secondary by post-nova cooling of the white dwarf (Schreiber & Gänsicke 2001), show that variation about the mean certainly occurs through the intervention of novae. The additional mass transfer caused by heating of the secondary during (and for a century or more after) a nova eruption is, in theory, followed by a phase of lowered accretion rate (Kovetz, Prialnik & Shara 1988) which is needed to maintain the secular average (but allowing for the fact that the eruption itself alters the orbital separation and therefore changes the radius of the Roche lobe of the secondary).

The question remains as to whether this is sufficient to send CVs into deep and sustained hibernation. Some evidence that this does not happen for at least 200 y has been given by Somers, Mukai & Naylor (1996) and Somers, Ringwald & Naylor (1997) for WY Sge (Nova Sagittae 1783), which is the oldest definitely recovered nova and still shows enhanced $\dot{M}$. But T Sco, the nova of 1860 that occurred in the globular cluster M 80, has been recovered and appears about a factor of ten fainter than normal nova remnants and does not appear (from absence of orbital modulation) to be a high inclination system (Shara & Drissen 1995). In addition, not all novae that occurred early in the twentieth century have been recovered, and until they are it cannot be claimed that there is no evidence for hibernation within the first century or so of nova eruption. The most important fact continues to be that no stellar remnants of the bright novae noted in Oriental records of one to two millennia ago are observable today (Shara 1989) – instead of being the nova-like systems at magnitudes 12 – 15 that modern bright naked eye novae become, they have faded beyond easy identification. This is the most persuasive case for eventual extended very low $\dot{M}$ states for post novae.

Another case, and certainly easier to observe (at $V \sim 12$), is AE Aqr, with its magnetic primary rotating at $P_{rot} = 33$ s, which must have been established as an equilibrium rotation period at high $\dot{M}$, but which has currently a low $\dot{M}$ suggestive of quite deep hibernation. Systems like AE Aqr would be quite difficult to discover at great distances (AE Aqr is one of the closest of CVs, at a distance of 86 pc), so they may be more common than realised.

It should also be kept in mind that at the shortest orbital periods there are dwarf novae like WZ Sge for which the estimated values of $\dot{M}$ are an order of magnitude or more below that set by GR, and so these systems are in at least partial hibernation. WZ Sge's spin period of ~ 28 s (Warner & Woudt 2002) also indicates a high $\dot{M}$ in the past.

PRE- AND POST-ERUPTION BEHAVIOUR

The survey of pre-eruptive behaviour of novae, made by Robinson (1975), disclosed only one (V446 Her) that could be thought to have had dwarf nova (DN)-like outbursts, with a range of nearly 4 mag. The variations were largely irregular, but the 'flares' or 'outbursts' had such slow rises to maxima ($\sim$ 10 d) that they are incompatible with a CV having P_{orb} = 4.97 h and (as concluded by Robinson on other grounds) are probably not normal DN outbursts. A similar remark may be made about the pre-outburst 1.6 mag variations in V3890 Sgr (Nova Sgr 1962: Dinerstein 1973). There is therefore no authenticated case of a nova eruption having taken place in a normal DN.

Robinson also found that pre- and post-eruptive magnitudes are very similar (an apparent exception, BT Mon, was later found to conform: Schaefer (1983)). This in general continues to be true, with definite exceptions of the three very fast novae GQ Mus (N 1983), CP Pup (N 1942) and V1500 Cyg (N 1975), all of which rose from exceptionally faint magnitudes (to which they have not returned), though V1500 Cyg was in a brighter state for a week before eruption. Their pre-eruptive luminosities were so low that they were certainly in states of low $\dot{M}$ at those times.

The identity of pre- and post-eruptive luminosities is commonly used as evidence that a nova eruption does not seriously change the state of a CV. Yet the mass and angular momentum ejected in a nova eruption will certainly perturb the long-term orbital evolution – the question is: How long will it take for the effects to show themselves? This is relevant to the evidence or otherwise for hibernation, discussed above. Here I want to examine the implications of pre- and post-novae being, both spectroscopically and in luminosity, indistinguishable from nova-like variables (e.g. Chapter 4 of Warner 1995a).

The interrelationships between nova-likes, VY Scl stars and DN, as functions of P_{orb}, can be summarised as follows:

- The VY Scl stars (which are nova-likes showing randomly distributed states of low $\dot{M}$), all lie roughly in the P_{orb} range 3.0 – 4.0 h.
- There are very few DN in the P_{orb} range 3.0 – 4.0 h.
- The values of $\dot{M}$ that appear among the nova-likes in the region of P_{orb} = 3 – 4 h are more than an order of magnitude greater than that predicted by the theory of magnetic braking (Warner 1987).

These strong correlations have been quantitatively explained by Wu, Wickramasinghe & Warner (1995a,b, hereafter WWW; see also Warner 1995a) in the following way: For P_{orb} < 4 h the separation between the secondary and primary is small enough to produce significant irradiative heating of the secondary by the hot central regions of the disc and the primary which results in greatly enhanced $\dot{M}$. The surface temperature T_{eff} of the primary is largely governed by $\dot{M}$ – in particular, the large values of T_{eff} found in nova-likes (up to 50 000 K, Sion 1999) are what are expected for accretion at $\dot{M} \sim 10^{-8}$ M$_\odot$ y^{-1} (Section 9.4.4 of Warner 1995a). WWW found that, as the thickness of the accretion disc increases with $\dot{M}$ and shields the secondary, this negative feedback automatically leads to an upper limit of $\dot{M} \sim$ few $\times 10^{-8}$ M$_\odot$ y^{-1}, and that the non-linearity of the situation leads to short lived high/low $\dot{M}$ states as observed in the VY Scl stars (in which the important time scales are the response time of the outer envelope of the secondary, and the cooling time of the outer parts of the primary) and/or to long-lived ($10^4 - 10^6$ y) high and low $\dot{M}$ excursions. The upper limit on $\dot{M}$ corresponds to $M_V \sim 3.3$ in the 3.0 < P_{orb} < 4.0 h region, only slightly brighter than what is actually observed (Warner 1987).

To summarise: the predominance of nova-likes, and in particular the VY Scl stars, in the P_{orb} = 3 – 4 h range is probably due to irradiation-enhanced $\dot{M}$; the near absence of DN is due to the relative rapidity of passage through states of intermediate $\dot{M}$. Below

$P_{orb} \sim 3$ h the mechanism causing the orbital period gap dominates.

The relevance of this theory to novae is clear: almost all novae are observed to erupt from CVs of the nova-like subtype, where $\dot{M}$ is high (naturally high for $P_{orb} > 4$ h, enhanced by irradiation to a high value for $P_{orb} < 4$ h) and where a primary with T_{eff} up to $\sim 50\,000$ K resides. After eruption, the primary has an even higher temperature, and $\dot{M}$ is thereby even more enhanced (Prialnik 1986), but both decrease as the primary cools. However, for $P_{orb} \lesssim 4$ h, the primary is prevented from cooling below $T_{eff} \sim 50\,000$ K because the irradiation-enhanced high $\dot{M}$ equilibrium is re-established after eruption. It is this effect that results (at least for the shorter P_{orb} systems) in equality of luminosity before and after eruption. The equilibrium value of T_{eff} is sensitive to the mass of the primary (higher masses lead to more gravitational potential energy release but a smaller area to radiate it away); using equations 2.83b and 9.55 of Warner (1995a) we find $T_{eff} \propto M^3$ for $M > 1.0$ M$_\odot$, where a large primary mass is adopted because there is higher probability for nova eruptions to be observed in high mass systems. As the selection for higher masses does not apply to nova-likes we should find that T_{eff}, $\dot{M}$ (and hence M$_V$) in nova-likes is on average smaller (fainter) than in old post-novae. Comparison of Figures 4.16 and 4.20 of Warner (1995a) shows this is in fact the case.

For non-magnetic CVs with $P_{orb} < 4$ h, therefore, the primary's cooling curve as computed by Prialnik (1986) applies only until the irradiation-enhanced pre-nova equilibrium is regained. How long the latter phase will last is not yet known. Clearly, the effects of mass and angular momentum lost during eruption eventually must take their toll – in order to maintain the long-term average angular momentum drain from the orbit. For $P_{orb} < 4$ h the irradiative equilibrium phase acts to delay the onset of the necessary low $\dot{M}$ state – an example of deferred compensation.

These remarks are also relevant to the nature of the pre-eruption outbursts seen in V446 Her and V3890 Sgr, as mentioned above. The pre-eruption range of V446 Her was $m_{pg} \sim 14.9 - 18.4$ (Robinson 1975) – and is larger than the range seen in its post-eruption variations, which are clearly standard DN outbursts (Honeycutt, Robertson & Turner 1995, Honeycutt et al. 1998). This and the slow rises appear more like VY Scl behaviour, but the large P_{orb} argues against this and it is likely that what is seen is a combination of normal DN outbursts (greatly under-sampled by the photographic archive plates) and some variations in $\dot{M}$, implying that $\dot{M}$ was low enough before (as after) eruption for the accretion disc to be thermally unstable. We note that at $P_{orb} \sim 5$ h the irradiation in V446 Her will not be sufficient to hold the system in a high $\dot{M}$ nova-like state if its natural $\dot{M}$ is below the critical value.

This last effect is seen as responsible for the non-appearance of DN outbursts in the majority of post-novae. The only systems for which authenticated standard DN outbursts have been seen are V446 Her ($P_{orb} = 4.97$ h), GK Per ($P_{orb} = 47.9$ h) and V1017 Sgr ($P_{orb} = 137$ h), for all of which irradiation of the secondary is not important because of the large separations implied by the long orbital periods.

On the other hand, 'stunted' DN outbursts in nova-likes and old post-novae are commonly observed (Honeycutt, Robertson & Turner 1998). In these the time scales of typical DN outbursts are seen, but the amplitudes are only ~ 0.6 mag. These can be understood as arising from outbursts in only the outer parts of the accretion discs, with the inner parts kept permanently in a high temperature state through irradiation by the hot primary (Warner 1995b; Schreiber, Gänsicke & Cannizzo 2000).

It should be mentioned that, in addition to the variations in brightness on DN time scales (weeks or months), there are low amplitude (typically 0.1 – 0.2 mag) variations on time scales of years believed to be caused by $\dot{M}$ variations resulting from magnetic cycling within the secondary (see Table 9.3 of Warner 1995a).

To return to pre-eruption light curves, Robinson (1975) found that 5 out of 11 well observed systems showed slow increases of brightness of 0.25 – 1.5 mag during 1 – 15 y before eruption. With the possible exception of the largest value (which is for V533 Her) the M_V increases are modest and may be the result of increases in $\dot{M}$ from the secondaries as seen in the decadal cycles mentioned above. (Note, however, that an increase of only 0.33 mag in V for a high $\dot{M}$ disc implies an increase in $\dot{M}$ by a factor ~ 2 (Smak 1989) – this may not apply to V533 Her, which is an intermediate polar (see below) and has a truncated disc). Such an increase in $\dot{M}$, with its concomitant increase of compressional heating in the surface layers of the primary, could well trigger a nova eruption. The observations imply that about half of nova-likes that have accreted almost a critical mass are triggered during a high part of a decadal cycle. Of course, this is what would be expected randomly anyway, so it is not evidence for such a triggering mechanism!

Finally, high $\dot{M}$ discs in CVs with $P_{orb} \lesssim 4$ h commonly show superhumps arising from precessing elliptical accretion discs (e.g. Patterson 1999), and post-novae are no exception. V603 Aql, V1974 Cyg, CP Pup and probably V4633 Sgr, V2214 Oph and GQ Mus are examples. Superhumps are a diagnostic for high $\dot{M}$ discs, and as such can be used to identify high luminosity discs at short P_{orb} – where almost all CVs have very low $\dot{M}$ (e.g. Fig. 9.8 of Warner 1995a). The only one so far found (that is not a known recent nova) is BK Lyn ($P_{orb} = 0.075$ d; Skillman & Patterson 1993), which may possibly be the remnant of Nova Lyn 101 AD (Hertzog 1986). Such systems have very low amplitude photometric modulations and are difficult to find. The only known reason for high $\dot{M}$ at such short P_{orb} is connected with eruption, and such systems therefore are strongly indicative of prehistoric novae. If the identification of BK Lyn with the nova of 101 AD could be proven, it would show that at very short P_{orb} a high $\dot{M}$ might be (self-)sustained for at least a millennium.

ORBITAL PERIODS OF NOVAE

In their discussion five years ago of the orbital period distribution of novae, and their progenitor population, Diaz and Bruch (1997) listed 30 objects with known $P_{orb} < 24$ h (six of which are considered by Downes et al. to be unreliable period determinations). The number of classical novae (omitting recurrent novae, all of which seem to be different from the 'non-recurrent' ones) with known periods < 24 h is now the 50 listed in Table I (which also omits some uncertain determinations - and note that a few may be superhump periods, which are a few percent different from P_{orb}), and demonstrates considerable observational progress in the past 5 years.

The frequency distribution of nova orbital periods is shown in Figure 1.

To the list of novae that have been observed to erupt it should eventually be possible to

TABLE 1. Orbital Periods of Novae

Star	Date	Magn. range	P_{orb}	Star	Date	Magn. range	P_{orb}
RW UMi	1956	6 – 18.5	1.418	WY Sge	1783	5.4 – 20.7	3.687
GQ Mus	1983	7.2 – 18.3	1.425	OY Ara	1910	6.0 – 17.5	3.731
CP Pup	1942	0.5 – 15.2	1.474	V1493 Aql	1999	10.4 – >21	3.74
V1974 Cyg	1992	4.2 – 16.1	1.950	V4077 Sgr	1982	8.0 – 22	3.84
RS Car	1895	7.0 – 18.5	1.980	DO Aql	1925	8.7 – 16.5	4.026
DD Cir	1999	7.7 – 20.2	2.340*	V849 Oph	1919	7.3 – 17	4.146
V Per	1887	9.2 – 18.5	2.571	V697 Sco	1941	10.2 – 19.7	4.53*
QU Vul	1984	5.6 – 17.5	2.682	DQ Her	1934	1.3 – 14.6	4.647
V2214 Oph	1988	8.5 – 20.5	2.804	CT Ser	1948	7.9 – 16.6	4.68
V630 Sgr	1936	1.6 – 17.6	2.831	T Aur	1891	4.2 – 15.2	4.906
V351 Pup	1991	6.4 – 19.0	2.837	V446 Her	1960	3.0 – 17.8	4.97
V4633 Sgr	1998	7.4 – >20	3.014	HZ Pup	1963	7.7 – 17.0	5.11
DN Gem	1912	3.5 – 16.0	3.068	AP Cru	1936	10.7 – 18.0	5.12
V1494 Aql	1999	4.0 – >16	3.232	HR Del	1967	3.5 – 12.3	5.140
V1668 Cyg	1978	6.7 – 19.8	3.322	V1425 Aql	1995	7.5 – >19	5.419
V603 Aql	1918	−1.1 – 11.8	3.324	BY Cir	1995	7.2 – 17.9	6.76*
DY Pup	1902	7.0 – 19.6	3.336	V838 Her	1991	5.4 – 15.4	7.143
V1500 Cyg	1975	2.2 – 18.0	3.351	BT Mon	1939	8.5 – 16.1	8.012
V909 Sgr	1941	6.8 – 20	3.36	V368 Aql	1936	5.0 – 15.4	8.285
RR Cha	1953	7.1 – 18.4	3.370	QZ Aur	1964	6.0 – 17.5	8.580
RR Pic	1925	1.0 – 12.1	3.481	CP Cru	1996	9.2 – 19.6	11.3*
V500 Aql	1943	6.6 – 17.8	3.485	DI Lac	1910	4.6 – 15.0	13.050
V382 Vel	1999	2.7 – 16.6	3.508	V841 Oph	1848	4.2 – 13.5	14.50
V533 Her	1963	3.0 – 14.8	3.53	V723 Cas	1995	7.1 – >18	16.638
V992 Sco	1992	8.3 – 17.2	3.683*				

* Woudt & Warner, unpublished

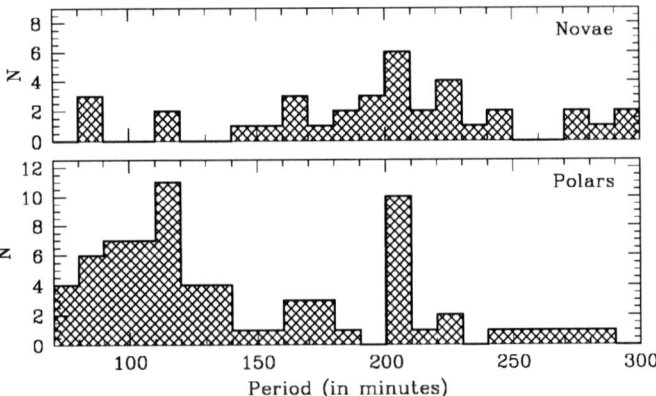

FIGURE 1. The frequency distribution of the orbital periods of Novae (upper panel) and Polars (lower panel).

add ones that are demonstrably novae that were overlooked in the relatively recent past. As a class, the desynchronised polars, discussed below, provide four probable examples with periods 1.85, 3.35, 3.35 and 3.37 h. The X-Ray source RX J1039.7-0507, with P_{orb} = 1.574 h, is probably another (see poster by Woudt & Warner at this conference).

A FAQ (Frequently Asked Question) is whether the P_{orb} distribution for classical novae shows a gap (or, at least, a greatly lowered space density) in the range 2 – 3 h in the same way as dwarf novae (see, e.g., Warner 1995). Using the range 2.11 < P_{orb}(h) < 3.20 for the empirically determined period gap (for CVs of all types: Diaz & Bruch, 1997) there are eight novae in the 'gap', which does not support the presence of a gap – though a reduction of population relative to the number of novae immediately above 3.2 h is undeniable – but so is the rapid fall in numbers for P_{orb} below 3 h.

A FUQ (Frequently Unasked Question) is what we should learn from the answer to the above FAQ. A point to be considered is that all white dwarfs accreting at the rates commonly seen in CVs must eventually undergo nova explosions (we exclude here the highest accretion rates which lead to Ultra Soft X-Ray Sources steadily burning hydrogen near their surfaces). The population of novae is therefore drawn from all of those CV subtypes that have high enough accretion rates to produce novae. If there is a population of detached systems in the period gap, which is the conventional explanation for the 'missing' CVs in the gap, they obviously do not contribute to the census of novae. If there is in fact a lower space density of novae at the position of the traditional period gap, it may simply be representing the gap we already see in some other subtypes.

There are few known short orbital period novae, and the reason for this is not immediately obvious. The space density of CVs in the range $P_{orb} - P_{orb} + dP_{orb}$ should be inversely proportional to dP_{orb}/dt, and therefore proportional to $<\dot{M}>^{-1}$, and the frequency of nova eruptions should be proportional to $<\dot{M}>$ (where <> denotes the long-term secular mean), so the fraction of CVs that become novae should be roughly independent of $<\dot{M}>$. The observed large pile-up of CVs at short P_{orb} (which is even larger in distributions predicted by population syntheses) – is not represented by a similarly large number of novae. Furthermore, the novae are sampled from a larger volume of space than the other CVs, so the short P_{orb} novae have an even lower relative space density than expected.

Even if there is no obvious period gap, Table 1 and Fig. 1 do have one distinctive feature: there is a concentration of novae in the range 2.8 – 4.1 h (44% of all known orbits lie in this 1.3 h range). The remainder are spread widely: another 48% cover the 6.3 h of ranges 1.4 – 2.8 h and 4.1 – 9 h. In some respects this is a mirror image of what happens in the DN, where there is a large number with P_{orb} < 2 h and very few in the range 3 – 4 h. We can partly understand this through the fact that the average mass transfer rate is very much lower below the period gap than above, which ensures that most CVs below the gap lie below the critical transfer rate for stable accretion discs, and at the same time they accrete mass so slowly that the frequency of nova eruption *per star* is low. But above the gap the anticorrelation of classical and dwarf novae populations must be more subtle. We shall see below that there is some evidence that many of the novae in the central part of the 3 – 4 h range have magnetic primaries; these would have to be subtracted from any comparison with the dwarf novae (which at most are weakly magnetic), but this still leaves the 3 – 4 h range clearly favoured by novae.

A possible explanation has already been given above in terms of the effects of irradiation of the secondary (WWW), which begin to be very important for $P_{orb} < 4$ h. If mean values of $\dot{M}$ over times of $10^3 - 10^4$ y are either very high or very low in the 3 – 4 h range, with transitions between taking place relatively rapidly (hundreds of years), then dwarf novae (which would only exist in the transition region) will be rare. The only evidence for possible secular change in $\dot{M}$ due to such transition is the difference in mean outburst intervals in U Gem (P_{orb} = 4.25 h), which are 96.5 d for 1855–1905 and 107.6 d for 1905–1955 (Warner 1987).

MAGNETIC NOVAE

At least a quarter of CVs have primaries with magnetic fields strong enough to affect the accretion flow. The strongest fields, in the polars, prevent the formation of accretion discs; the intermediate polars and DQ Her stars, which have progressively lower field strengths, have accretion discs truncated at their inner edge by the magnetosphere of the primary.

Eruptions on magnetic white dwarfs have occurred in RR Cha, GK Per, HZ Pup and V697 Sco, which are intermediate polars, and in DQ Her and V533 Her, but Nova Cygni 1975 (V1500 Cyg) remains unique as the only observed nova eruption that has been proven to have arisen on a strongly magnetic white dwarf, i.e. a polar. Photometry 12 years after the eruption showed it to have a light curve like that of a polar (Kaluzny & Semeniuk 1987) and subsequent polarimetric observations revealed the characteristics of a polar (Stockman, Schmidt & Lamb 1988) with a field ~ 25 MG. An unexpected feature, however, was that the rotation period of the white dwarf is shorter by 1.8% than the orbital period – which is interpreted as being the result of coupling between the expanded atmosphere of the primary and secondary during eruption, with subsequent spin-up as the atmosphere collapsed back later. Observations have shown that the spin period P_{spin} is increasing at a rate that implies resynchronisation in ~ 185 y (Schmidt, Liebert & Stockman 1995).

Nova Puppis 1991 (V351 Pup) has recently been found to have a light curve very similar to that observed for V1500 Cyg a decade after its eruption; although not yet detectably polarized, it may well turn out to be another example of eruption of a strongly magnetic system (Woudt & Warner 2001).

Indirect evidence for eruptions that occurred in magnetic systems in the past, but went unrecorded as novae, is given by the occurrence of three other desynchronised polars: BY Cam (Mason et al. 1998), V1432 Aql (RX J1940.1-1025) (Geckeler & Staubert 1997) and CD Ind (RX J2115-5840) (Ramsay et al. 2000). Their properties are listed in Table 2. The fact that $P_{spin} > P_{orb}$ in V1432 Aql may be explained as coupling of the magnetic field of the primary with the dense wind during eruption (this is in competition with the purely dynamical transfer of angular momentum mentioned above for V1500 Cyg). The measured resynchronisation time scales, listed in Table 2, range over an order of magnitude: $\sim 100 - 1000$ y, not correlated with the amount of asynchronism (as measured by the beat period P_{beat} between the orbital and white dwarf spin periods).

V1432 Aql provides another form of indirect evidence for historical nova eruption.

TABLE 2. Desynchronised Polars

Star	P_{orb} (mins)	P_{spin} (mins)	P_{beat} (d)	T_{syn} (y)
V1432 Aql	201.94	202.51	49.5	110
BY Cam	201.26	199.33	14.5	1107
V1500 Cyg	201.04	197.50	7.8	185
CD Ind	110.89	109.55	6.3	

Schmidt & Stockman (2001) measure an effective temperature of 35 000 K for the primary, which is much greater than the typical 8000 – 20 000 K found for other polars (other than V1500 Cyg, which has 90 000 K from its recent eruption) and is out of equilibrium with the present rate of accretion heating. From the cooling calculations made by Prialnik (1986) this indicates a nova eruption in the past 75 – 150 y.

AE Aqr, already mentioned above, has an unknown T_{eff} because the dominant measurable UV flux comes from spots with $T \sim 26000$ K produced by heating by the accretion columns (Eracleous et al. 1994). If the general non-heated surface temperature of the primary could be measured more accurately than the currently estimated 10 000 – 16 000 K this would provide a measure of how long ago the high $\dot{M}$ phase (characteristic of the large P_{orb} of AE Aqr) was interrupted, probably by a nova eruption.

As has been pointed out before (Warner 1995a), the fraction f of polars that are desynchronised may provide a means of estimating observationally the otherwise inaccessible average time T_R between nova eruptions, at least for the magnetic systems. If $<T_{syn}>$ is the average resynchronisation time then $T_R = <T_{syn}>/f$. There are about 68 known polars, so with 4 observed to be desynchronised and a mean $<T_{syn}> \sim 300$ y we have $T_R \sim 5000$ y. This crude figure can be refined in several ways but, as we see below, it produces a serious conflict with what is known about T_R from other directions.

First, T_{syn} depends on a number of system parameters, including the mass of the white dwarf, the mass ejected and how well angular momentum was exchanged with it, the time since the eruption, etc. A global average of these effects could be obtained by appropriate theoretical modelling, but this may not be justified at present because of the small number of systems included in the statistics.

Second, f is undoubtedly underestimated because of insufficient observational coverage of many of the fainter polars. Perhaps not even all the polars with $m_v < 16.0$ have been studied enough to detect asynchronism, but assuming that they have we note that V1432 Aql, BY Cam and CD Ind (we exclude V1500 Cyg as having been discovered in a non-standard way) constitute 3 out of about 20 systems with high state magnitudes brighter than 16. If this fraction applies also to the total population of polars, then there are another 7 de-synchronised systems among the fainter members (one of which is already known: V1500 Cyg, at $m_v = 18.0$). This gives $f \sim 0.15$ and reduces T_R to 2000 y.

Third, there are difficulties in detecting asynchronism for the older magnetic novae where synchronism has been nearly re-established. In essence, when the beat period P_{beat} becomes $\sim$ months there is an observational bias against finding such systems. Superficially, it might be thought that, as the observed values of P_{beat} lie in the range 6 – 50 d, leaving the range 50 – ∞ d unexplored, there could be a large fraction of currently undetected desynchronised polars. However, the following reasoning suggests that the loss is not great.

The key aspect is that the synchronization torque is independent of the amount of de-synchronisation and is constant with time. The general form of the torque N_{syn} is $N_{syn} \sim \mu_1 \mu_2 / a^3$, where μ is the magnetic moment (indigenous or induced in the case of the secondary) and a is the separation of stellar components (e.g. Hameury, King & Lasota 1987). The synchronisation time is $T_{syn} = (P_{orb} - P_{spin})/\dot{P}$, where $\dot{P} = dP_{spin}/dt$ is essentially constant with time because of the constant torque. At any time after the nova eruption we have the relationship

$$P_{beat} = \frac{P_{orb}^2}{|\dot{P}|T_{syn}}, \tag{1}$$

where T_{syn} is the time remaining until synchronisation. Suppose that the initial de-synchronisation produces a beat period $P_{beat}(\text{init})$ (typically a few days) and that current observational techniques make it difficult to detect asynchronism for beat periods longer than some limit $P_{beat}(\text{limit})$ (typically a few weeks). Then the fraction F of asynchronous systems that is detectable is

$$F = \frac{T_{syn}(\text{init}) - T_{syn}(\text{limit})}{T_{syn}(\text{init})} = 1 - \frac{P_{beat}(\text{init})}{P_{beat}(\text{limit})}. \tag{2}$$

The result is that with $P_{beat}(\text{init}) \sim$ week and $P_{beat}(\text{limit}) \sim$ months, very few of the desynchronised systems are overlooked, so the difficulty of detecting differences between P_{orb} and P_{spin} when they are very small has little effect on the estimate of f.

The deduced value of $T_R \sim 2000$ y is distressingly incompatible with the estimated mass $\sim 2 \times 10^{-4}$ M$_\odot$ of ejecta in V1500 Cyg (Hjellming 1990) and the average mass transfer rate $\sim 10^{-9}$ M$_\odot$ y^{-1} estimated for longer period polars (Beuermann & Burwitz 1995), which give $T_R \sim 2 \times 10^5$ y. It could be that nova eruption is not the only mechanism that is capable of desynchronising polars, or that the ejecta masses of magnetic novae are grossly overestimated. Weakening of the synchronising torque with time does not help because it puts a larger fraction into systems with large P_{beat} that would currently be overlooked, necessitating an even larger correction to obtain the true number of desynchronised polars.

Another FAQ is whether polars show any period gap. The latter subject is one set about with controversy (Beuermann & Burwitz 1995; Wickramasinghe & Ferrario 2000). The list of polars shows that no absolutely empty gap is present. Beuermann & Burwitz note that comparing the number of systems in the period range of the gap with that just outside shows that gap polars are relatively twice as frequent as non-magnetic gap systems. They say that nevertheless the orbital distributions of magnetic and non-magnetic systems are not statistically different. Another point of view could be that the 'period gap' would probably not be noticed in polars if it were not so prominent in the non-magnetic CVs – and that it is probably equally true that the polar distribution is not statistically different from one that has no gap at all. However, here it is necessary to distinguish between two kinds of polars – several of the polars in the gap are strong cyclotron line emitters and are interpreted as having extremely low accretion rates $\sim 10^{-13}$ M$_\odot$ y^{-1} (Reimers & Hagen

2000). Such rates are $< 10^{-3}$ of what is normally seen in polars and constitute systems in which accretion has nearly shut down, as predicted by some explanations of the period gap. For present purposes these systems should therefore be removed from the census of polars in the gap. These particular gap-filling polars are not going to contribute to the census of novae – they will take at least 10^8 y to accumulate sufficient mass to trigger a nova eruption.

Novae are drawn from all those CV subtypes that have sufficiently high long-term average accretion rates. In the case of the magnetic systems we might look for similarities in the distributions of polars and novae. The frequency distribution of polars has spikes near 114 min (1.90 h) and 202 min (3.37 h) (Wickramasinghe & Ferrario 2001); it happens that all four of the desynchronised polars listed in Table 2 are members of these two groups.

Two novae (V1974 Cyg and RS Car) out of the five that have $P_{orb} < 2.0$ h coincide with the 114 min peak. RS Car is strongly modulated in brightness but has not been observed well enough to detect any periods other than P_{orb} (or a superhump period), but V1974 Cyg could be a desynchronised polar – its multiple periodicities are interpreted as such by Semeniuk et al. (1995), but an explanation in terms of superhumps is preferred by Skillman et al. (1997) and Retter, Leibowitz & Ofek (1997). On the other hand, Shore et al. (1997) find that the flux distribution in V1974 Cyg can be accounted for entirely by a hot white dwarf, leaving no evidence for the presence of the disc required to generate superhumps, and thus indirectly supporting the desynchronised polar model.

This may be pure coincidence, but it is interesting that in addition to the 114 min peak, the concentration of novae around 3.3 h is centred roughly on 202 min and includes the magnetic nova type specimen V1500 Cyg. This suggests looking for possible evidence of magnetism in other novae within the 3.3 h cluster – the results are shown in Table 3.

Although partly indirect, this evidence for several magnetic systems within the 3.2 h cluster is strong, though most are not desynchronised polars.

A comparison between the P_{orb} distributions for polars and novae is shown in Figure 1. The clustering near 3.3 h is seen in both subclasses, but is much tighter and more pronounced in the polars. Adding the few intermediate polars to the polars does not change the distribution noticeably (especially as the former also have a weak preference for periods around 3.3 h, 5 of about 15 with $P_{orb} < 5$ h being close to that value). There are selection effects, depending on period, that distort these period histograms (Diaz & Bruch 1997), but for comparisons between the subclasses these are probably not important.

Direct and indirect evidence for magnetic primaries also exists outside of the two period spikes. DQ Her and V533 Her are classic DQ Her stars. Confusion between the effects of asynchronism and those of superhumps leaves the statuses of V2214 Oph (Baptista et al. 1993) and GQ Mus (Diaz & Steiner 1989) uncertain. On the other hand, a high mass transfer disc with the centre missing is a characteristic of an intermediate polar, and such evidence exists for V Per (Shafter & Abbot 1989).

A possible prehistoric magnetic nova is the polar RX J1313.2-3259 (Gänsicke et al 2000), which, with the relatively long orbital period of 4.19 h, would be expected to have a moderately high accretion rate and, from comparison with other CVs, a primary accretion-heated to about 30 000 K. In fact, RXJ1313 has a measured temperature of 15 000 K, which is compatible with heating at the estimated accretion rate of $\sim 10^{-10}$

TABLE 3. Properties of Novae with Orbital Periods near 3.3 hours

Star	Remarks
V351 Pup	Possibly magnetic like V1500 Cyg (Woudt & Warner 2001).
V4633 Sgr	Possible asynchronous polar (but also could simply be a superhump modulation) (Lipkin et al. 2001).
DN Gem	No evidence for magnetism.
V1494 Aql	41.7min X-Ray period attributed to pulsations (Starrfield & Drake 2001). No direct evidence for magnetism.
V1668 Cyg	Well observed but no magnetic signature at quiescence *.
V603 Aql	Possible intermediate polar, P_{spin} = 62.9 min (Schwarzenberg-Czerny, Udalski & Monier 1992).
DY Pup	Observations too sparse to check for magnetic signature.
V1500 Cyg	Polar.
V909 Sgr	Not well observed.
RR Cha	Intermediate polar with P_{spin} = 32.50 min (Woudt & Warner 2002).
RR Pic	At one time thought to be an intermediate polar but not confirmed by later observations (Haefner & Schoembs 1985).
V500 Aql	Well observed but no magnetic signature.

* However, the apparently well-defined photometric modulation at a period of 10.54 h, seen during early decline (Campolonghi et al. 1980), could be a spin beat with the orbital period. A cycle of this modulation was detected by Di Paolantonio et al. (1981) but not by some other observers (Piccioni et al. 1984; Kaluzny 1990).

$M_\odot$ y^{-1}. To cool and reach equilibrium at 15 000 K after a nova eruption takes $\sim 10^4$ y, so Gänsicke et al. suggest that the low accretion rate and temperature may be the result of an extended hibernation state that has lasted since a nova eruption $\sim 10^4$ y ago.

ACKNOWLEDGMENTS

Steve Potter kindly communicated orbital periods for some recently recognised polars. Patrick Woudt assisted with preparation of the paper and with helpful conversations. The author's research is funded by the University of Cape Town.

REFERENCES

1. Baptista, R, Jablonski, F.J., Cieslinski, D., and Steiner, J.E., ApJL, **406**, 67 (1993).
2. Beuermann, K., and Burwitz, V. 1995. *ASP Conf. Ser.*, **85**, 99 (1995).
3. Campolonghi, F., et al., A&A, **85**, L4 (1980).
4. Diaz, M.P., and Bruch, A. 1997. A&A, **322**, 807 (1997).
5. Diaz, M.P., and Steiner, J.E., ApJL, **339**, 41 (1989).
6. Dinerstein, H., IBVS No. 845 (1973).
7. Di Paolantonio, A., Patriarca, R., and Tempesti, P., IBVS No. 1913 (1981).
8. Eracleous, M., et al., ApJ, **433**, 313 (1994).
9. Gänsicke, B.T., Beuermann, K., de Martino, D., and Thomas, H.-C., A&A, **354**, 605 (2000).
10. Geckeler, R.D., and Staubert, R., A&A, **235**, 1070 (1997).
11. Haefner, R., and Schoembs, R., A&A, **150**, 325 (1985).
12. Hameury, J.-M., King, A.R., and Lasota, J.-P., A&A, **171**, 140 (1987).

13. Hertzog, K.P., Observatory, **106**, 38 (1986).
14. Hjellming, R.M., Lect Notes Phys., **369**, 169 (1990).
15. Honeycutt, R.K., Robertson, J., and Turner, G.W., ApJ, **446**, 838 (1995).
16. Honeycutt, R.K., Robertson, J., and Turner, G.W., AJ, **115**, 2527 (1998).
17. Honeycutt, R.K., Robertson, J., Turner, G.W., and Henden, A.A., ApJ **495**, 933 (1998).
18. Kaluzny, J., MNRAS, **245**, 547 (1990).
19. Kaluzny, J., and Semeniuk I., Acta Astr., **37**, 349 (1987).
20. King, A.R., Frank, J., Kolb, U., and Ritter, H., ApJ, **444**, L37 (1995).
21. King, A.R., Frank, J., Kolb, U., and Ritter, H., ApJ, **467**, 761 (1996).
22. Kovetz, A., Prialnik, D., and Shara, M.M., ApJ, **325**, 828 (1988).
23. Lipkin, Y., Leibowitz, E.M., Retter, A., and Shemmer, O., ApJ, **328**, 1169 (2001).
24. Mason, P.A., et al., MNRAS, **295**, 511 (1998).
25. Patterson, J., ApJS, **54**, 443 (1984).
26. Patterson, J., in *Disc Instabilities in Close Binary Systems*, edited by S. Minishege & J.C. Wheeler, Front. Sci. Ser. **26**, 61 (1999).
27. Piccioni, A., et al., Acta Astr., **34**, 473 (1984).
28. Prialnik, D., ApJ, **310**, 222 (1986).
29. Ramsay, G., et al., MNRAS, **316**, 225 (2000).
30. Reimers, D., and Hagen, H.-J., A&A, **358**, L45 (2000).
31. Retter, A., Leibowitz, E.M., and Ofek, E.O., MNRAS, **286**, 745 (1997).
32. Robinson, E.L., AJ, **80**, 515 (1975).
33. Schaefer, B., ApJ, **268**, 710 (1983).
34. Schmidt, G.D., Liebert, J., and Stockman, H.S., ApJ, **441**, 414 (1995).
35. Schreiber, M.R., and Gänsicke, B.T., A&A, **375**, 937 (2001).
36. Schreiber, M.R., Gänsicke, B.T., and Cannizzo, J.K., A&A, **362**, 268 (2000).
37. Schwarzenberg-Czerny, A., Udalski, A., and Monier, R., ApJL, **401**, 19 (1992).
38. Semeniuk, I., De Young, J.A., Pych, W., Olech, A., Ruszkowski, M., and Schmidt, R.E., Acta Astr., **45**, 365 (1995).
39. Shafter, A.W., and Abbott, T.M.C., ApJL, **339**, 75 (1989).
40. Shara, M.M., PASP, **101**, 5 (1989).
41. Shara, M.M., and Drissen, L., ApJ, **448**, 203, (1995).
42. Shore, S., Starrfield, S., Ake, T.B., and Hauschildt, P.H., ApJ, **490**, 393 (1997).
43. Sion, E.M., PASP, **111**, 532 (1999).
44. Skillman, D.R., and Patterson, J., ApJ, **417**, 298 (1993).
45. Skillman, D.R., Harvey, D., Patterson, J., and Vanmunster, T., PASP **109**, 114 (1997).
46. Smak, J., Acta Astr., **39**, 317 (1989).
47. Somers, M.W., and Naylor, T., A&A, **352**, 563 (1999),
48. Starrfield, S., and Drake, J., in *Two Years of Science with Chandra*, Washington Symposium, September 2001, 50 (2001).
49. Stockman, H.S., Schmidt, G.D., and Lamb, D.Q., ApJ, **332**, 282 (1988).
50. Verbunt, F., and Zwaan C., A&A, **100**, L7 (1981).
51. Warner, B., MNRAS, **227**, 23 (1987).
52. Warner, B., *Cataclysmic Variable Stars*, Cambridge University Press (1995a).
53. Warner, B., Ap.Sp.Sci., **230**, 83 (1995b).
54. Warner, B., and Woudt, P.A., MNRAS, in press (2002).
55. Wickramasinghe, D.T., and Ferrario, L., PASP, **112**, 873 (2000).
56. Woudt, P.A., and Warner, B., MNRAS, **328**, 159 (2001).
57. Woudt, P.A., and Warner, B., MNRAS, in press (2002a).
58. Woudt, P.A., and Warner, B., MNRAS, submitted (2002b).
59. Wu, K., Wickramasinghe, D.T., and Warner, B., (WWW), PASA, **12**, 60 (1995)
60. Wu, K., Wickramasinghe, D.T., and Warner, B., (WWW), Astrophys. Sp. Libr. **205**, 315 (1995).

Alternatives to Hibernation

T. Naylor

School of Physics, University of Exeter, Stocker Road, Exeter, EX4 4QL, U.K.

Abstract. I outline the evidence pertinent to the connection between the nova explosion and mass transfer rates in CVs. I conclude that there is still insufficient evidence to decide whether or not such a connection exists.

THE PROBLEM

Systems in which classical nova explosions have been observed are structurally indistinguishable from other cataclysmic variables (CVs). They all consist of a low-mass late-type, normally main-sequence star losing mass via Roche-lobe overflow to a white dwarf. The magnetic field of the white dwarf can cause differences in the detail of how the accretion flow reaches the white dwarf. A strong magnetic field will channel the accretion flow straight onto the white dwarf (a "polar" or AM Her star) whilst a weak field will allow an accretion disc to form, with intermediate cases occurring when the inner disc alone is disrupted by the field (an "intermediate polar" or DQ Her star). However, it should be emphasized that all these sub-classes occur amongst both classical novae and other CVs, again arguing for a close relationship between the two. The obvious conclusion to draw is that all CVs eventually undergo a nova explosion, and that those we classify as old novae, happen to have had an explosion in the recent past.

One of the outstanding problems in understanding CVs is the large range of mass transfer rates they have. If we concentrate on the non-magnetic systems, where the problem is best understood, we find a range of about a factor of 100, even if we restrict ourselves to systems with similar orbital periods. Part of the non-magnetic CV classification system is based on this difference, with the high mass transfer rate systems being classified as UX UMa stars, the intermediate ones as VY Scl and Z Cam stars, and the low mass transfer rate systems as dwarf novae.

THE SOLUTIONS

Shara and collaborators [1] suggested that nova explosions may hold the key to explaining the differences in mass transfer rate. In essence their idea is that the explosion co-incides with a period of high mass transfer rate, and centuries after the explosion the mass transfer decreases, and finally ceases altogether. The detached binary then "hibernates", until angular momentum loss brings the system back into contact, mass transfer begins again, laying down on the white the material for the next nova explosion.

The attraction of what became known as the "hibernation scenario" was twofold. First, at the date it was proposed, it appeared that the observed mass transfer rates in CVs were much higher than those which the models required for a nova explosion to occur [2]. However, whilst the models of Starrfield, Sparks & Truran still require low mass transfer rates [3], the Prialnik & Kovetz models [e.g. 4] do undergo explosions for mass transfer rates similar to those observed in CVs. The second attractive feature was that a host of evidence seemed to imply that old novae continued to fade for at least two hundred years after outburst, as though their mass transfer rates were declining and the binary heading towards "hibernation". The evidence for this decline was questioned by myself and others [5], leading us to suggest that the evidence is perfectly consistent with the idea that the nova explosion is unconnected with the mass transfer rate in the binary [6]. Thus if a system was a dwarf nova before the classical nova outburst, it will be so again immediately afterwards, and continue to be so for many nova explosion cycles. For the purposes of contrast, I will refer to this as the "constant mass transfer model" or simply CMT.

That two apparently orthogonal theories can be consistent with the observational evidence is, perhaps, surprising. In this paper I aim to review the current evidence for links between mass transfer rate and the time of nova outburst. I shall begin by discussing the evidence for the systems with long orbital periods, *i.e.* greater than 0.2 days.

ARE ALL NOVAE UX UMA STARS?

A few years after the nova explosion, virtually all classical novae appear to be high mass transfer rate cataclysmic variables. In the case of the non-magnetic systems, this means they are UX UMa stars. Furthermore, their very similar magnitudes before and after the explosion [7], implies that they were also at high mass transfer rates before the explosion. Whilst this may, at first sight, appear to support the idea that there is a mass transfer cycle, at the peak of which the system explodes as a classical nova, in fact it probably simply tells us that low mass transfer rate systems only rarely have nova outbursts. The reason is that it will take them longer to build up the layer of material required for the runaway [e.g. 8]. Further evidence that this is the correct interpretation was provided by the discovery that Nova Her 1960 (V446 Her) is a dwarf nova [9], and the data of [7] imply it was probably a dwarf nova before its nova outburst.

THE POST NOVA DECLINE

In the first hundred years after the nova outburst the system luminosity declines by around 2 magnitudes. This was first shown by correlating the age of each nova with its current brightness [10], but later also by following individual systems [11]. Whilst such a decline could be caused by a decline in mass transfer rate, it could also be due to irradiation of the disc and secondary star by the white dwarf, which is hot as a result of the nova explosion, and is cooling. Support for this idea came first from observations of Nova Cyg 1975 (V1500 Cyg) [12]. The observations show that the inner face of the secondary star is heated by the white dwarf, and that the degree of heating is declining with time. The data can be modeled to deduce the temperature of the white dwarf, which is found to be falling at the rate expected by theory. This suggests that it is not actually light from the white dwarf itself which is responsible for the post-nova decline, but the decline in flux which is reprocessed by the secondary star and accretion disc. There are calculations of the reprocessing from the disc alone, which suggest this is correct [13].

If white dwarf cooling is the reason for the post-nova decline, more observations are explained. First, there is evidence that the decline in magnitude ceases after about 100 years, as the white dwarf cooling models predict. Nova Sge 1783 (WY Sge) is now 200 years old, but the binary still has the magnitude expected for high mass transfer rate system [14, 15]. Secondly, old novae sometimes seem to to undergo low amplitude dwarf nova-like outbursts. It seems these can be explained by the white dwarf irradiation maintaining the inner disc in the viscous state, whilst an outer annulus undergoes dwarf nova outbursts [16]. The small region of the disc participating in the outburst explains its low amplitude.

CONCLUSIONS FOR LONG PERIOD SYSTEMS

The real problem here is that the predictions of hibernation and CMT have converged. The hibernation model predicts that the decline in mass transfer will drive a decline in luminosity after the nova outburst. Presumably this is already happening in Nova Her 1960, and will happen for Nova Sge 1783, if we wait long enough. A CMT model accepts there is a decline in luminosity caused by the falling irradiation from the white dwarf, but asserts this stops after about a hundred years. In the CMT view, those old novae which show mass transfer rates below the mean are the few low mass transfer rate systems we expect to find. Thus the discovery by [17] that Nova Sco 1860 (T Sco) lies at a brightness level suggesting a long period dwarf nova, is consistent with either model.

SHORT PERIOD SYSTEMS

Whatever view one takes of the post-nova decline, until very recently all the evidence supported the idea that mass transfer rate is broadly the same before and after the explosion. Again both theories accept this, as the hibernation models bring the CV up to the high mass transfer state before the explosion, and the CMT model requires it. Such a picture, though, was built up when all old novae with reliably determined orbital periods had periods longer than about 0.2 days. This has now begun to change, and as new systems are being discovered, there are signs that the picture presented above may not apply to short period systems.

Perhaps the most important point to make first is that high mass transfer rate, short orbital period systems simply should not exist. Figure 1 shows the usual way of classifying non-magnetic CVs, in a plot of mass transfer rate against period [19]. For short periods the only angular momentum loss mechanism available is gravitational radiation, which cannot support (at least in the long term) the high mass transfer rates observed

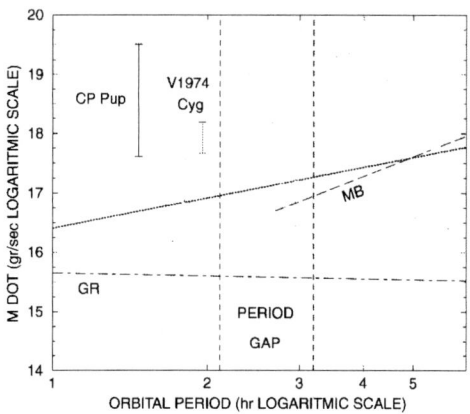

FIGURE 1. The orbital period vs mass transfer rate plane for cataclysmic variables. The two dotted vertical lines mark the approximate limits of the period gap. The lines marked MB and GR show the expected mass transfer rates for magnetic braking and gravitational radiation respectively, and the unlabeled line divides high mass accretion rate (stable) discs from low mass accretion rate ones. The points for CP Pup and V1794 Cyg are marked. Adapted from [18].

in the old novae CP Pup and V1974 Cyg. Retter and I [18] pointed out that these two systems (which are the best studied ones below the period gap), appear brighter after the nova outburst than they were before it – in contrast to the behaviour of the long period systems. Duerbeck [20] suggests that this effect may be more widespread, including systems with periods above the period gap, but less than 0.2 days.

Retter argues that these results suggest that mass transfer rates below the period gap are driven by the nova explosion. After the nova explosion the mass transfer rate is high, before dropping decades or centuries later. Although this is broadly a "weak hibernation scenario", one should note a crucial difference – the system is faint immediately before the nova outburst, whilst in the normal hibernation models it is bright. If mass transfer rates do vary in this way, it would explain why there are some systems well above the mass transfer rate allowed by gravitational radiation; they are only there for a short while as a result of the nova event, and will eventually fall back to their original level. I have my own reservations. First, there is no good physical theory of how the cycles would work since simple mass loss cannot drive them [21], nor can irradiation [22]. Secondly, due to their small orbital separation, these short period systems are the ones we would expect to show the greatest effects of irradiation. Thus systems which were probably dwarf novae before their nova explosions, like Nova Cyg 1992 (V1974 Cyg), will find their discs being held in the bright state by irradiation for a relatively long time by the mechanism outlined in [16]. Clearly one solution to this debate is to wait, and see if as the irradiation declines, V1974 Cyg begins to show dwarf nova outbursts. However, as the white dwarf takes decades to cool, quicker resolutions would clearly be preferable.

CONCLUSIONS

For the long orbital period systems (>0.2 days), it is clear we see considerable irradiation from the white dwarf in the decades after outburst, which certainly raises the overall luminosity. Whether when this phase is over, the binary simply returns to its pre-outburst state, or mass transfer then declines into "hibernation" remains an open question. For the systems with orbital periods below 0.2 days, there is emerging evidence that they are fainter before outburst than afterwards. This may simply be because most of them are low mass transfer rate systems (dwarf novae) whose post outburst luminosity is held high for a few decades by intense irradiation. Alternatively short period systems may have mass transfer cycles driven by the explosion. Discovering which of these scenarios is correct is crucial to our understanding not only of classical novae, but also of CVs in general.

ACKNOWLEDGMENTS

My thanks are due to Koji Mukai, who not only first kindled my interest in classical novae, but has also collaborated in developing many of the ideas presented in our papers reviewed here. Nye Evans nurtured that interest, whilst Fred Ringwald, Mark Somers and Alon Retter not only contributed ideas, but did much of the hard work. Finally, without Mike Shara to argue with, the field would have been much duller.

REFERENCES

1. Shara, M. M., *PASP*, **101**, 5–31 (1989).
2. Shara, M. M., Livio, M., Moffat, A. F. J., and Orio, M., *ApJ*, **311**, 163–171 (1986).
3. Starrfield, S., Sparks, W. M., Truran, J. W., and Wiescher, M. C., *ApJS*, **127**, 485–495 (2000).
4. Kovetz, A., and Prialnik, D., *ApJ*, **477**, 356–367 (1997).
5. Naylor, T., Charles, P. A., Mukai, K., and Evans, A., *MNRAS*, **258**, 449–456 (1992).
6. Mukai, K., and Naylor, T., "Hibernation- Problems and Alternatives," in *ASSL Vol. 205: Cataclysmic Variables*, 1995, p. 517.
7. Robinson, E. L., *AJ*, **80**, 515–524 (1975).
8. Iben, I. J., Fujimoto, M. Y., and MacDonald, J., *ApJ*, **384**, 580–586 (1992).
9. Honeycutt, R. K., Robertson, J. W., Turner, G. W., and Henden, A. A., *ApJ*, **495**, 933 (1998).
10. Vogt, N., *ApJ*, **356**, 609–612 (1990).
11. Duerbeck, H. W., *MNRAS*, **258**, 629–638 (1992).
12. Somers, M. W., and Naylor, T., *A&A*, **352**, 563–566 (1999).
13. Schreiber, M. R., and Gänsicke, B. T., *A&A*, **375**, 937–943 (2001).
14. Somers, M. W., Mukai, K., and Naylor, T., *MNRAS*, **278**, 845–853 (1996).
15. Somers, M. W., Ringwald, F. A., and Naylor, T., *MNRAS*, **284**, 359–364 (1997).
16. Schreiber, M. R., Gänsicke, B. T., and Cannizzo, J. K., *A&A*, **362**, 268–272 (2000).
17. Shara, M. M., and Drissen, L., *ApJ*, **448**, 203 (1995).
18. Retter, A., and Naylor, T., *MNRAS*, **319**, 510–516 (2000).
19. Osaki, Y., *PASP*, **108**, 39 (1996).
20. Duerbeck, H., "Longterm evolution of nova shell luminosities and postnova magnitudes," in *These proceedings*, 2003, p. ???
21. Kolb, U., Rappaport, S., Schenker, K., and Howell, S., *ApJ*, **563**, 958 (2001).
22. King, A. R., Frank, J., Kolb, U., and Ritter, H., *ApJ*, **467**, 761 (1996).

Spectroscopic Evidence from HST and Archival IUE Studies of Dwarf Novae as Past Classical Novae

Edward M. Sion

Department of Astronomy and Astrophysics, Villanova University, Villanova, PA 19085, USA

Abstract. I present recent spectroscopic evidence suggesting the past occurences of thermonuclear runaways on the accreting white dwarfs in certain dwarf novae both above and below the orbital period gap. This information is based upon synthetic spectral analyses of HST and IUE archival spectra obtained when the underlying white dwarf photosphere is detected spectroscopically or when peculiar N V/C IV emission line ratios, associated with the accretion disk or accretion column, appear. Hubble STIS observations of VW Hydri 2 days and 7 days after the end of a superoutburst reveal a heated white dwarf with deep broad Lymanα, narrow metallic absorption features and evidence of a hotter Keplerian-broadened component. These spectra confirm a large N/C ratio and the existence of enhanced abundances of odd-numbered nuclear species P, Mn and Al as well as a N/C ratio indicative of proton capture by even-numbered nuclear species outside the CNO by-cycle during the CNO H-burning thermonuclear processing. These STIS observations confirm that a past (pre-historic?) thermonuclear runaway has occured on the white dwarf in VW Hyi. It is expected that the thermonuclear runaway would be strong enough to produce a nova outburst. Therefore, these two classes of close binaries, namely dwarf novae and classical novae, are linked and can overlap. Additional evidence for such a link to novae is also discussed for the dwarf nova U Gem, nova-like variables in low optical brightness states of little or no accretion and in dwarf novae with very peculiar N V/C IV emission line ratios.

INTRODUCTION

Due to their enormous surface gravities and hence very short gravitational diffusion timescales, single white dwarfs display spectra that are essentially mono-elemental with the lightest element in the envelope always appearing at the surface. All heavier elements have diffused downward. For a typical DA white dwarf with log g = 8 and T_{eff} = 15,000 K (the highest T_{eff} H$\sim$-rich case tabulated by Paquette et al. 1986), the diffusion timescale for metals should be shorter than 0.011 to 0.007 year (i.e., $< 3 - 4$ days). Hence, in an accreting white dwarf any metals detected in the photosphere cannot have remained there more than a few days (unless radiative levitation is operative). For white dwarfs in cataclysmic variables, the presence of heavy elements in the atmosphere is continually being replenished by disk accretion and presumably dredge-up from deeper layers where nucleosynthetic products have settled. A likely candidate mechanism for the dredge-up may be the dwarf nova explosion itself and its associated tangential accretion with shear mixing which could stir up deeper layers of the envelope (see below).

TABLE 1.

Obs. #	V	Si	C	N	O	Al	Fe
1	400	0.3	0.3	3	3	2	0.5
2	500	0.4	0.4	4	4	2	0.05
	Mg	Mn	Ni	P	Ti	χ^2	Scale
1	3	50	0.3	15	0.1	1.7110	3.9916×10^{-2}
2	5	50	0.3	20	0.4	4.9370	4.0693×10^{-1}

The accretion of material from the secondary during the dwarf nova with no mixing would tend to cover over the TNR–processed material. The secondary is normally expected to consist of solar composition. It may, however, be contaminated by captured processed material during the nova ejection and during the common envelope stage following the TNR. Since it would be mixed with secondary solar material, the original nova would be even more enhanced than observed now. It is with this scenario in mind that I discuss surface abundances of accreting white dwarfs in cataclysmic variables below.

SPECTROSCOPIC CLUES FOR DWARF NOVAE AS PAST CLASSICAL NOVAE

Dwarf novae and nova-like variables contain accreting white dwarfs which may have undergone numerous thermonuclear runaways as classical novae. In order to demonstrate their connection with novae however, attempts have been made to detect ejected shells without success. However, a new approach has recently emerged for systems in which the white dwarf photosphere has been detected spectroscopically. Sion et al. (1997) showed that the surface abundances of the white dwarf in VW Hydri during its quiescence manifests a direct evolutionary to a past thermonuclear event. This conclusion is based upon the presence of a large abundance ratio of nitrogen to carbon and the spectroscopic presence of odd-numbered proton capture nuclei in abundances greatly elevated above solar. Both of these spectroscopic characteristics point to hot CNO processing as the source of the abundances. In addition to VW Hyi, the dwarf nova U Gem has also been studied extensively with HST and manifests some of the same surface abundance characteristics as VW Hyi. I will discuss both of these cases below.

In figure 1 (left) we display the HST STIS E140M spectra of VW Hyi obtained 2 days and 7 days following a superoutburst, compared with the best-fitting single temperature white dwarf and white dwarf plus equatorial accretion belt. The resulting parameters and chemical abundances are summarized in Table 1 (see also Sion et al. 2001).

For the single temperature fits, the same T_{eff} was derived for both observations. However, spectrum 1 had V sin i = 400 km/s while spectrum 2 has 500 km/s. The difference is probably not significant. Two metal abundance differences could be statistically significant. The Fe and P reveal the greatest difference in abundance between the two observations. The odd-numbered element phosphorus has an abundance of 15 times solar in spectrum 1 compared to 20 times solar in spectrum 2. The odd-numbered element Man-

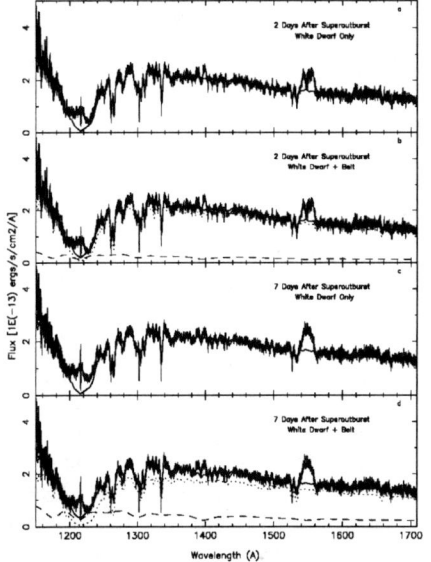
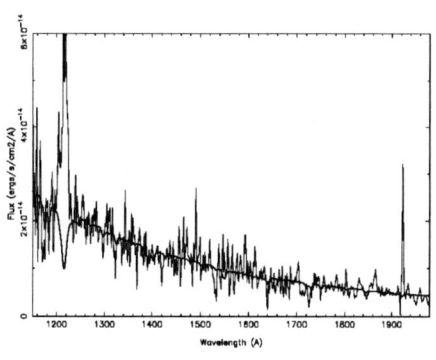

FIGURE 1. LEFT. (a). The best fit synthetic spectrum of a white dwarf alone, to spectrum 1 2 days post-superoutburst.($T_{eff} = 22,500$K, log g = 8, $V_{sin} = 400$ km/s); (b). The best fit two-temperature component model (white dwarf plus accretion belt) to spectrum 1 (2 days post-superoutburst). The white dwarf has $T_{eff} = 22,000$, log g = 8.0, $V \sin i = 400$ km/s and contributes 89% of the flux, plus an accretion belt with solar abundance, T = 32,000 K, V = 3350 km/s, log g = 6, emitting area and flux contribution 3% and 11% respectively. The dashed line is the contribution of the accretion belt alone, the dotted line represents the flux of the white dwarf and the solid line is the flux combining the white dwarf plus the belt; (c). The best-fit synthetic spectrum of a white dwarf alone, to spectrum 2 of VW Hydri obtained 7 days post-superoutburst. ($T_{eff} = 22,500$K, log g = 8, $V_{sin} = 500$ km/s); (d). The best fit two-temperature component model (white dwarf plus accretion belt) to spectrum 2 (7 days post-superoutburst). The white dwarf has $T_{eff} = 21,000$, log g = 8.0, $V \sin i = 500$ km/s and contributes 89% of the flux, plus an accretion belt with solar abundance, T = 32,000 K, V = 3350 km/s, log g = 6.0, emitting area and flux contribution 5% and 20% respectively. The dashed line is the contribution of the accretion belt alone, the dotted line represents the flux of the white dwarf and the solid line is the flux combining the white dwarf plus the belt.
RIGHT. The best-fitting single temperature white dwarf model ($T_{eff} = 49,000$K, log g = 8) to the IUE spectrum of MV Lyra, obtained during a low state.

ganese is elevated to 50 times solar in both observations. However, the difference in the Fe abundance between the two observations is extraordinary. In obs. 2, the Fe abundance has declined by a factor of 10. We have no explanation for this other than diffusion. All of the other metal species show slight increases in spectrum 2 but the differences may not be significant. The two-temperature model provides a large improvement in the χ^2 indicating that the white dwarf plus accretion belt models are better fits than the single temperature models.

We believe that our two STIS observations strengthen the evidence that TNR induced by the accretion of material from the secondary has occured in the past on the white dwarf of VW Hyi for the following reasons. (1) Confirmation of the elevated abundances of the odd-numbered nuclear species P, Al and Mn relative to their even-numbered nuclear neighbors. (2) The absolute abundance of nitrogen is larger than carbon. Solar composition contains 4 times as much carbon as nitrogen. During a TNR the carbon will capture protons and become nitrogen (Starrfield et al. 1978). Even a TNR on a ONeMg white dwarf will end up with more N than C as the O proton captures, ejects He, and becomes N (Politano et al. 1995). While the CNO nuclei are capturing protons during a TNR, the heavier nuclei will also be capturing protons. Because of the very short time that the temperature is high during a TNR, most nuclei can only capture a few protons. Thus, the even-numbered nuclei peaks in a solar abundance distribution will tend to be leveled out by filling in the valleys occupied by the odd-numbered nuclei (Na, Al, P, Mn). Since the even-numbered nuclei are 10 to 100 times more abundant than their odd-numbered neighbor nuclei, a smoothing of these peaks and valleys would lead to ~ 0.5 solar abundances for the even-numbered nuclei and 5 to 50 times solar abundances for the odd-numbered nuclei. The actual value also depends on the proton capture rates and the number of stable states. This simple picture explains approximately the observed abundances in both of the STIS spectra. Hydrodynamics simulations with nucleosynthesis are required for a more accurate comparison.

In U Gem, Sion et al.(1998) proposed that sufficient contamination of the secondary star by the brief common envelope phases of numerous classical nova explosions would result in the secondary transferring material back to the white dwarf that was enriched by the products of CNO processing. They found the white dwarf in U Gem to have a mass $M_{wd} = 1.12$ $M_\odot$ and $K_1 = 107$ km/s. They predicted that the N abundance in the white dwarf photosphere would be several times solar. All these results, including the white dwarf mass, were confirmed in the HST G140L study of Long & Gilliland (1998). Of greatest interest here are the abundances: C: 0.05 to 1 solar, N: 4.0 solar, Si: 0.4 to 1.3 solar, He: 1.0 solar and all other metals 1.0 solar. Thus, the white dwarf in U Gem presents another solid example of a large N/C ratio, well above the solar value. Since the secondary stars in these dwarf novae are of very low mass, it is unlikely that the large N/C ratio is associated with CNO processing intrinsic to the M dwarf components.

SPECTROSCOPIC EVIDENCE FROM IUE STUDIES OF DWARF NOVAE

While the number of cataclysmics with exposed white dwarf photospheres (and hence the possibility of abundance analyses) is rather limited, the signature of CNO processing may also found in the emission line spectra originating in the accretion disk or, in the case of magnetic CVs, the accretion column. The most prominent emission line in the far UV spectra of dwarf novae is virtually always C IV (1548, 1550). However, a recent examination of the IUE archive (Winter & Sion 2002) revealed two dwarf novae which appear to be exceptions to this rule. These systems are EY Cyg and BZ UMa. Could the depleted or absent carbon be pointing to a different thermonuclear history for these dwarf novae, related to thermonuclear burning and contamination of the mass-transferring secondary by many past novae? the accreting white dwarf or nucleosynthetic processing in the secondary? The observed properties of these two systems are found in Warner (1995) and reference therein. EY Cyg is classified as a U Gem system, the most common type of dwarf novae system. EY Cyg is known to have a period of 0.2185 days with mV(min)= 15.5 and mV(max)= 11.5. The secondary spectral type is determined as K7 ± 2. BZ UMa however is below the period gap with a period of 0.0679 days and is possibly an SU system. BZ UMa is a more studied system with mV(min)=15.2 and mV(max)=10.2. The secondary spectral type is M5.5 ± 0.5. Unfortunately, the noisy IUE spectra and presence of strong disk emission do not allow a determination of surface abundances for the white dwarfs inn these two systems. However, the fact that N V emission is very strong while C IV emission is extremely weak or absent suggests that the secondary is contaminated by the products of CNO burning.

The results of combined accretion disk and photospheric model analyses of EY Cyg and BZ UMa will be reported elsewhere (Winter and Sion 2002). However, we note that in this volume there are two contributed poster papers which are relevant to the topic of this paper. Tomvassian et al. (this volume) suggest the possibility that EY Cygni was indeed a past classical novae while Mouchet et al. (this volume) present evidence of peculiar NV to C IV emission line ratios in magnetic (AM Her) systems.

EVIDENCE FROM NOVA-LIKE VARIABLES DURING LOW STATES

The nova-like variables of subclasses VY Scl and UX UMa undergo low optical brightness states of little or no accretion when a luminous accretion disk no longer hides the underlying white dwarf. Due to their higher time-averaged accretion rates, these systems should contain hotter white dwarfs (since they mostly remain in their high states).

Preliminary evidence (Sion & Clancy 2002) suggest that three such systems, MV Lyrae, TT Ari and DW UMa have white dwarf surface abundances determined fromm their low state spectra which are depleted in carbon and enhanced in nitrogen. Preliminary results for MV Lyra indicate $T_{eff} = 49,000K$, $\log g = 8$, Si abundance = 2.0 solar, C abundance = 0.5 solar and the N abundance = 5-10 solar. These results will discussed in detail elsewhere (Sion & Clancy 2002). The IUE archival spectrum of MV Lyra during its low state is shown in figure 1 (right) together with the best-fitting single-temperature white dwarf model atmosphere.

CONCLUSIONS

There are at least three processes which can modify the abundances of the material left on the white dwarf after the TNR to that which we are now observing: diffusion, accretion of material from the secondary, and mixing during accretion. For the slow rotating polar regions which we are considering, one can visualize that only diffusion occurs, diffusion with accreting material spreading out from the fast rotating belt, or all three processes. We will investigate each of these. Recall that the diffusion timescale for a 20,000K H-rich white dwarf is extremely short as discussed in the introduction. Thus, only if a nova outburst was very recent, could observed metal abundances represent directly the remains of the TNR. The metal we are detecting (e.g., Al, Mn, P) almost certainly cannot have remained there since the last nova explosion of VW Hydri. Instead it seems more likely that the Al, Mn and P have been dredged-up from deeper layers where nucleosynthetic products have settled. A likely candidate mechanism for the dredge-up may be the dwarf nova explosion itself and its associated tangential accretion with shear mixing which could stir up deeper layers of the envelope (see the discussion of accretion belt C and Si abundance below).

The accretion of material from the secondary during the dwarf nova with no mixing would tend to cover over the TNR–processed material. The secondary is normally expected to consist of solar composition. It may, however, be contaminated by captured processed material during the nova ejection and during the common envelope stage following the TNR. Since it would be mixed with secondary solar material, the original nova material original nova material would be even more enhanced than observed now.

For the case of all three processes, there would be mixing of the accreted material with the underlying nova remnant material and possibly with the original white dwarf material in the belt region. In fact for the only two elements that we have abundances (C $10\times$ solar and Si $15\times$ solar, see Sion et al. 1996), their values are far above solar and above their values in the polar regions. The large overabundance of carbon in the belt is expected for a 0.86 $M_\odot$ white dwarf and is a strong indicator of shear mixing. The overabundance of Si is not expected. A TNR on a 0.8 $M_\odot$ white dwarf is not expected to reach high enough temperatures for the CNO nuclei to break out of the CNO bi-cycle or to α-capture to produce heavier elements. Multiple proton captures on solar Ne and Mg could in principal increase the Si to six times solar but this is unlikely. Other possibilities include a ONeMg white dwarf (Politano et al. 1995, Starrfield et al. 1978) or a ONeMG mantle built up during He thermal pulses during the pre-CV phase (Shara and Prialnik 1994). Both of these seem unlikely since these studies find that all of the even-numbered nuclei are strongly enhanced. In addition an initially massive ONeMg white dwarf demands an implausibly large amount of core erosion.

It is therefore compelling that accretion and mixing will not by themselves produce the observed N to C ratio nor the high abundance of P, Mn and Al relative to the even-numbered nuclei. In fact these processes will tend to produce the opposite effect. It is also doubtful that the even-numbered nuclei would be strongly preferentially diffused downward relative to their odd-numbered neighbor nuclei. Therefore, in spite of the complications these processes cause, these observations still reveal that a TNR has occured on the white dwarf of VW Hyi.

While it would seem that only a past thermonuclear event could be responsible for the peculiar abundances, there is some question as to whether the contamination of the secondary ring of the many brief common envelope transient stages of a nova would be sufficient to account for the observed abundances. Although it may seem far-fetched, the possibility exists that thermonuclear processing occurs during the high accretion episodes of dwarf novae in outbursts and nova-like variables in high states. At the present time, there are too many uncertainties to rule this out completely.

For the nova-like variables, only a handful of systems have been observed in the low state of little or no accretion when the white dwarf can be directly observed. However, three objects have been analyzed and contain very hot white dwarfs; TT Ari, MV Lyra and DW UMa. The higher temperatures are presumably due to the higher time-averaged accretion rates. In all three cases, there are preliminary indications that the nitrogen abundances are elevated above solar and the C abundance is depleted (Sion & Clancy 2002).

ACKNOWLEDGMENTS

This research was supported by HST grant GO-8139 and by NSF grant AST99-09155.

REFERENCES

1. Long, K., & Gilliland, R., *ApJ* **511**, 916 (1998).
2. Mouchet, M., et al., this volume, 2002.
3. Paquette, C., Pelletier, C., Fontaine, G., & Michaud, G., *ApJS*, **61**, 197 (1986).
4. Politano,M., Starrfield, S.G., Truran, J., Weiss, A., & Sparks, W.M., *ApJ* **448**, 807 (1995).
5. Shara, M., Prialnik, D., & Kovetz, A., *ApJ* **406**, 220 (1993).
6. Sion, E.M., Szkody, P. Cheng, F.H., & Huang, M., *ApJ* **444**, L97 (1995a).
7. Sion, E.M., Huang, M., Szkody, P., & Cheng, F.H., *ApJ* **445**, L31 (1995b).
8. Sion, E.M, Cheng, F., Sparks, W.M., Szkody, P., Huang, M., & Hubeny, I., *ApJL* **480**, L17 (1997).
9. Sion, E. M., Cheng, F., Szkody, P., Gänsicke, B., Sparks, W., & Hubeny, I., *ApJ* **54**, 127 (2001).
10. Sion, E.M., & Clancy, K., *AJ*, in preparation (2002).
11. Sion et al., *ApJ* **496**, 449 (1998).
12. Starrfield, S.G., Truran, J., & Sparks, W.M., *ApJ* **226** 186 (1978).
13. Tovmassian, G., et al., this volume, 2002.
14. Warner, B., *Cataclysmic Variable Stars*, Cambridge University Press, Cambridge, GB, 1995.
15. Winter, L., & Sion, E.M., in preparation (2002).

More on Peculiar Abundances in Dwarf Novae and NLs: Implications for Past Novae Explosions

Paula Szkody

Department of Astronomy, Box 351580, University of Washington, Seattle, WA 98195, USA

Abstract. Enhanced abundances of P, Mn and Al and a general enhancement of N/C ratios provide the first direct evidence that past thermonuclear runaways have occurred on the white dwarfs in dwarf novae and novalikes and indicate that these systems share a common history. Four different types of observations (elevated Al in HST spectra, unusual N/C ratios in FUSE and IUE spectra, the CN blend in optical spectra and low CO in IR spectra) show that a wide variety of systems exhibit peculiar abundances. Since these abundances share a common pattern, they provide clues to the past history of these systems, yet also provoke unsolved questions.

INTRODUCTION

The previous paper (Sion, this volume) presented evidence from HST studies that dwarf novae have undergone prior nova events. This paper extends this result by offering four further clues that also point to a past nova: 1) the presence of Al absorption lines in the HST spectra of some dwarf novae, 2) the presence of inverted N/C ratio in the UV and FUSE emission lines of several magnetic and dwarf novae, 3) the presence of a CN emission blend in the optical spectra of novalikes and 4) the weakness of CO absorption in the IR spectra of dwarf novae. In each of these cases, the overabundance of nitrogen, the depletion of carbon and the presence of odd-numbered nuclei all implicate a past thermonuclear runaway event.

AL IN HST SPECTRA OF DN

A medium Cycle 8 program using STIS on HST obtained UV spectra of 14 low mass transfer dwarf novae at quiescence and determined the temperature, composition and rotation of the underlying white dwarfs (Szkody et al. 2002a) by fitting the spectra to Hubeny white dwarf models. Among those with sufficient S/N to accomplish detailed fitting of the AlII 1671Å absorption line, two dwarf novae required an enhanced Al abundance. SW UMa required an aluminum abundance of 0.8 times solar and BC UMa required 1.5 times solar. Together with VW Hyi, which also shows enhanced aluminum (along with decreased C and increased N abundances), these 3 short period dwarf novae are prime candidates for a past TNR. Unfortunately, the HST spectrum of the known old nova DI Lac (Nova Lac 1910) does not show these abundance peculiarities, but its spectrum is completely dominated by a thick accretion disk so that the white dwarf itself is not visible (Moyer et al. 2002). It is unfortunate for the study of abundance anomolies

TABLE 1. Objects with N/C > 1

Object	Type	P (hr)
AE Aqr	IP	9.9
V1309 Ori	Polar	8.0
TX Col	IP	5.7
EY Cyg	DN	5.2
BY Cam	Polar	3.4
BZ UMa	DN	1.6

due to TNRs on the white dwarf that most novae have high accretion rates following outburst, so that the white dwarf cannot be directly observed.

UNUSUAL N/C RATIOS IN UV

IUE first revealed 4 magnetic systems with completely inverted N/C emission line ratios compared to normal Polars (AM Hers) and Intermediate Polars (IPs or DQ Hers) (Bonnet-Bidaud & Mouchet 1987, de Martino 1995, Szkody & Silber 1996). While this effect first seemed related to an origin in long period magnetic systems, the recent addition (Winter & Sion 2001) of two dwarf novae (including one below the period gap), destroyed this association. Table 1 summarizes the known systems. To explore the abundance question further, using the wide range of line transitions available with FUSE, observations of 3 of the 4 magnetics (BY Cam, V1309 Ori and AE Aqr) were undertaken (Mouchet et al. this volume). The presence of strong NIII (990Å) and NIV (1718Å) as well as the weakness of the CIII (977,1175Å) lines confirms the carbon depletion and nitrogen enhancement that was evident in the IUE spectra. The N/C ratios cannot be matched with photoionization models and are most consistent with those expected from core dredgeup during a TNR. However, it is discouraging that the recent nova V1500 Cyg does not show this inverted ratio (Schmidt, Liebert & Stockman 1995).

CN BLEND IN NLS

The optical portion of the spectrum of some cataclysmic variables reveals a strong emission component consisting of a blend of NIII 4640 and CIII 4650. Novae typically show this feature (Ringwald, Taylor & Mukai 1996), including even the oldest novae such as WY Sge which had its outburst in 1783. The other cataclysmic variables (CVs) that always show this feature are the SW Sex stars, the group of novalikes with high accretion rates and orbital periods between 3-4 hrs that are recently identified with magnetic IPs (Rodriguez-Gil et al. 2001). The strength of the CN feature shows a large range, with the CN/HeII 4686 ratio anywhere from ~0.25 to greater than 1 (Hoard 1998). The Sloan Digital Sky Survey (SDSS), which is finding many CVs missed in previous surveys (Szkody et al. 2002b), shows about 4% of CVs with a strong CN blend. While this feature is almost always present together with HeII, Williams & Ferguson (1983) showed that it cannot be explained by a Bowen fluorescence mechanism, and at least in some cases, is due to enhanced nitrogen. In addition, it is conspicuously absent in Polars, even though they have very strong HeII (e.g. see RXJ101.5+0904 in Szkody, Armstrong & Fried 2000). If the SW Sex stars were recent novae, it could explain both their high accretion and their strong CN blend.

CO IN DN SECONDARIES

The above 3 clues have been found in material associated with the white dwarf. However, there is also evidence of thermonuclear processing available from spectra of the atmospheres of the secondaries in CVs. As shown in Harrison et al. (2000), IR spectroscopy of the M star in U Gem and the K star in SS Cyg reveal that the CO features are weak compared to normal dwarfs, implying an underabundance of carbon and/or oxygen. Since the secondary is not massive enough to undergo CNO processing in its interior, a possible scenario is that it has accreted its peculiarities from the common envelope phase of past novae explosions. Then, in turn, it transfers this peculiar abundance material to the white dwarf to account for the enhanced nitrogen and decreased carbon (as observed in the white dwarf in U Gem; Sion et al. 1998; Long & Gilliland 1999). Further IR observations of long and short orbital period systems are needed to exclude other evolutionary scenarios.

CONCLUSIONS

While the evidence for abundance peculiarities is clear, the explanations for the cause could be due to several sources:

- TNR events which produce P, Mn, and Al and result in C less than solar abundance and N greater than solar
- CNO processing (on the white dwarf or the secondary)
- Photoionization effects

Item one is the most direct link to a past nova, as P, Mn and Al cannot be formed by other means. Item 2 is indirectly linked to a TNR, for if the settling time for heavy elements in the atmospheres of white dwarfs is very short, the CNO effects observed would have to be due to dredgup occurring during the nova TNR on the white dwarf or to matter ejected during the TNR and then accreted by the secondary and transferred to the white dwarf. Alternatively, if the secondary starts out as a massive star and then loses much of its mass, it could supply the CNO processed material which is then transferred to the white dwarf. As photoionization models (Mauche, Lee & Kallman 1997; Mouchet et al. this volume) cannot produce the observed inverted N/C ratios, the last item does not seem viable. Thus, the common thread of past novae explosions can explain most of the abundance peculiarities that are evident. However, it cannot explain why these affects are not apparent in past known novae, nor why they are not more common throughout dwarf novae and novalikes. Being able to study the abundances of a white dwarf that has returned to a low accretion state following a known nova event will help to clarify the connection. Unfortunately, most known novae remain at high accretion rates for long periods of time, so that the white dwarf cannot be directly observed as its light is overwhelmed by the accretion disk.

ACKNOWLEDGMENTS

It has been my great pleasure to work with many colleagues on data dealing with this abundance issue, including Ed Sion, Boris Gänsicke and Steve Howell on HST spectra, Martine Mouchet and her collaborators in France on FUSE spectra, the SDSS collaboration on SDSS data and Tom Harrison and collaborators in New Mexico on IR spectra. This work was partially supported by NASA FUSE grants NAG5-8981,10400 and HST grant GO-0813-97A.

REFERENCES

1. Bonnet-Bidaud, J. M. and Mouchet, M., *AAp* **188**, 89 (1987)
2. de Martino, D., *ASP Conf. Ser* **85**, 238 (1995)
3. Harrison, T. E., McNamara, B. J., Szkody, P. and Gilliland, R. L., *AJ* **120**, 2649 (2000)
4. Hoard, D. W., *PhD Thesis*, U of Washington (1998)
5. Long, K. S. and Gilliland, R. L., *ApJ* **511**, 916 (1999)
6. Mauche, C.W., Lee, Y. P. and Kallman, T. R., *ApJ* **477**, 832 (1997)
7. Moyer, E., Sion, E. M., Szkody, P., Gänsicke, B., Howell, S. and Starrfield, S., *AJ*, in press (2002)
8. Ringwald, F., Taylor, A. and Mukai, K., *MNRAS* **281**, 192 (1996)
9. Rodriguez-Gil, P., Casares, J., Martinez-Pais, I. G., Hakala, P. and Steeghs, D., *ApJ* **548**, 49L (2001)
10. Schmidt, G., Liebert, J. and Stockman, H. S., *ApJ* **441**, 414 (1995)
11. Sion, E. M., Cheng, F. H., Szkody, P., Sparks, W., Huang, M. and Mattei, J., *ApJ* **496**, 449 (1998)
12. Szkody, P. and Silber, A., *AJ* **112**, 289 (1996)
13. Szkody, P., Armstrong, J. and Fried, R., *PASP* **112**, 228 (2000)
14. Szkody, P., Sion, E. M., Gänsicke, B. T. and Howell, S. B., *ASP Conf. Ser* **261**, 21 (2002a)
15. Szkody, P. et al. *AJ* **123**, 430 (2002b)
16. Williams, R. E. and Ferguson, D., *IAU Coll.* **72**, 97 (1983)
17. Winter, L. M. and Sion, E. M., *BAAS* **199**, 6201 (2001)

The Recurrent Novae and Their Relation with Classical Novae

G.C. Anupama

Indian Institute of Astrophysics, II Block Koramangala, Bangalore 560 034, India

Abstract. Recurrent novae (RNe) are a subclass of cataclysmic variables classified based on recorded observations of more than one nova outburst. They form a small, but heterogeneous group of objects, with 10 known members, that undergo classical nova-like outbursts, with the significant differences being in the amplitude and the frequency of the outbursts. RNe outbursts are of smaller amplitude with lesser energy released as compared to the classical novae. The outbursts occur with intervals of $\sim 10 - 100$ yr. Some RNe systems are short period binaries similar to the classical novae, while other members of this group are long period binaries, with orbital periods of the order of several hundred days and consist of a hot white dwarf and a red giant similar to the symbiotic stars. It is widely believed that the white dwarfs in RNe are massive and close to the Chandrasekhar limit. The mass transfer rate from the companion is high, of the order of $\sim 10^{-7} M_\odot$ yr^{-1}. The massive white dwarf and the high mass-transfer rates, and the nature of the companion (in some systems) are responsible for the differences in the observational properties between RNe and the classical novae.

INTRODUCTION

Recurrent novae (RNe) form a small, heterogeneous group of cataclysmic variables and are classified based on more than one recorded observations of a nova outburst. The outburst luminosities of RNe ($M_V \leq -5.5$) lie at the lower end of the range of outburst luminosities observed in classical novae. Similar to the classical novae, RNe outbursts are also accompanied by the ejection of matter with velocities $V_{\exp} \geq 300$ km s^{-1}. The recurrence periods of the outbursts are of the order of decades. The outburst is the result of thermonuclear reaction in the accreted layer formed on the surface of the white dwarf, widely believed to be massive and close to the Chandrasekhar limit (in most systems), accreting at rates of $\sim 10^{-8} - 10^{-7}$ M$_\odot$ yr^{-1}(e.g. [1], [2], [3]). Theoretical models indicate that RNe are possible progenitors of sypernovae of type Ia, if the accreting white dwarf is a carbon-oxygen one (e.g. [3]). With the recent outbursts of CI Aql in 2000 and IM Nor in 2002, the number of known RNe now stands at ten.

RNe can be grouped as (a) short period binaries and (b) long period binaries. The short period binaries, U Sco, V394 CrA, LMC 1990#2, T Pyx, CI Aql and IM Nor, consist of an evolved main sequence or subgiant secondary, with orbital periods ranging from 1.8 hr (T Pyx) to 1.7 day (U Sco), similar to the classical nova systems. The outburst and quiescence properties of U Sco, V394 CrA and LMC 1990#2 are very similar, while the outburst properties of T Pyx, CI Aql and IM Nor are similar. The long period binaries have orbital periods of $\sim$ several hundred days, and consist of a red giant secondary, similar to the symbiotic binary systems. All four systems belonging to this group, T CrB, RS Oph, V3890 Sgr and V745 Sco, have very similar outburst and quiescence behaviour.

We discuss here some of the recent developments in the study of recurrent novae.

SHORT PERIOD SYSTEMS

CI Aquilae

CI Aquilae was recorded as a 'doubtful' nova based on an outburst in 1917 with $\Delta m \approx 4.6$ mag [4], [5]. The quiescence spectrum of CI Aql was found to be very different from that of other cataclysmic variables with Balmer lines in absorption and very weak emission lines due to N III/C III 4640, He II 4686 and O I 6300 [6], [7]. It was also discovered to be an eclipsing binary with an orbital period of 0.618355 days [6]. The recent brightening discovered in April 2000 [8] indicated a nova-like outburst with $\Delta m \approx 7.5$ mag. A re-examination of the Harvard Patrol Plates obtained in 1917 [9] indicated the 1917 outburst was much brighter than previously reported, and also similar to the 2000 one. A probable outburst in 1941 with a late decline light curve similar to the 2000 outburst is also reported [10]. This places CI Aql amongst those recurrent novae belonging to the class of the short period binaries. The quiescence spectra suggest a K–M type star, while the orbital period implies a subgiant luminosity class similar to the U Sco systems.

The 2000 Outburst

The light curve indicates CI Aql is a moderately fast nova with $t_2 = 30 \pm 1$ days and $t_3 = 36 \pm 1$ days [11]. A plateau was seen in the light curve beginning around 20 days since maximum, which was not evident in the 1917 outburst. Near-IR photometry showed indications of dust formation ≈ 50 days after optical maximum [12].

Early spectra showed lines due to H I, Fe II and N III [11], [13]. P-Cyg absorption features were detected at -1100, -1700 and -2200 km s^{-1}. About a month later, the Fe II lines disappeared and the He I and He II lines were strong. Other strong lines due to N II and N III were present and the spectrum appeared very similar to the 'He/N' class of novae.

Near-IR spectrum obtained at day +11 showed lines due to hydrogen Paschen and Brackett series, as well as N I and O I. He I line was self-absorbed, while the Lymanβ fluoroscence O I lines were very strong [14]. At day +74, the C I and N I lines in the near-IR spectrum disappeared, as also the O I lines. The He I absorption was gone and He II lines were prominent, similar to the development in the optical. [Si VI] line at 1.9629 μm was present, although the nova had not yet entered its coronal phase [15].

The Balmer lines had broad wings with initial FWZI $\sim 8000 - 9000$ km s^{-1}, which decreased to 3000–4000 km s^{-1} about 20 days after discovery. The Hα profile changed from a broad, Gaussian profile in the early phases to a saddle shaped profile with two maxima at ± 1100 km s^{-1} about 60 days later [11].

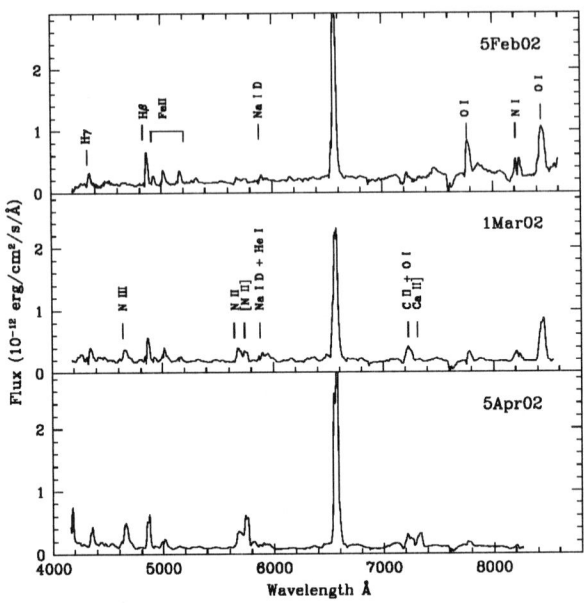

FIGURE 1. Temporal development of the outburst spectra of IM Nor during 2002 February – April. Based on spectra obtained from the Vainu Bappu Observatory, India.

The light curve and spectral development of the 2000 outburst of CI Aql was found to be very similar to that of T Pyx. Theoretical models of the outburst indicate that the white dwarf mass in CI Aql is 1.2 $M_\odot$, similar to T Pyx [16].

IM Normae - 2002 Outburst

IM Nor was discovered as a possible nova in 1920, with a light curve that suggested similarity with slow novae such as DQ Her and T Pyx [17]. A second outburst was discovered in January 2002, by W. Liller, when IM Nor was found to be at a red magnitude of 8.3 on photographs taken during Jan 10.343-10.360 [18]. Subsequent observations confirmed the nova outburst, adding to the group of recurrent novae. The light curve of the recent outburst indicates it to be a moderately fast – slow nova with $t_3 \gtrsim 50$ days [19], [20]. The light curve is very similar to that of CI Aql.

Early spectrum obtained on Jan 12.351 showed a strong, narrow Hα with FWHM of 950 km s^{-1}. Other Balmer lines, Fe II (42, 74) and O I 7774 lines were clearly present [18]. Spectra obtained during Jan 14.4–17.4 and on Jan 26.75 showed emission lines with P-Cyg profiles [21], [22]. Apart from the Balmer lines, the spectrum in the blue regions was dominated by the Fe II (27, 38, 42, 49) lines, while the O I (7774, 8446) and Ca II lines dominated the red region. The emission lines had an average FWHM of

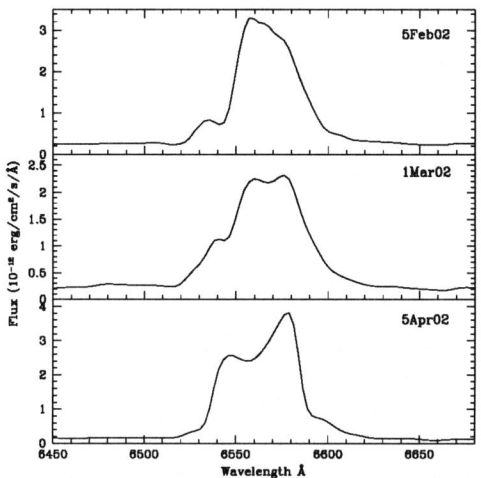

FIGURE 2. The Hα line profile IM Nor during 2002 February – April. Based on spectra obtained from the Vainu Bappu Observatory, India.

1150 km s^{-1}. The weak P-Cyg absorption had a velocity of -860 km^{-1}. Early spectra indicated the nova belonged to the 'Fe II' class.

Not much evolution in the spectrum was detected even in early February, about 20 days since maximum (Figure 1). Fe II and O I lines were still prominent, and the weak P-Cyg absorption still present. The FWHM of emission lines had however increased to 1600 ± 50 km s^{-1}. As seen in Figure 1, a change in the spectrum occurred around +45 days since maximum. The Fe II and O I 7774 line strengths decreased, and N III 4640, N II 5680 and other N II lines were present and strong in the blue region of the spectrum. He I 5876 was also weakly present. The P-Cyg absorption was however still visible. About +80 days since maximum, in early April, the spectrum resembled that of the 'He/N' class of novae. The spectrum had changed from Fe II to He/N type and the emission line widths had marginally decreased.

The Hα profile which had an asymmetric single-peaked profile in the early phases had changed to a saddle shaped profile during the later phase (Figure 2).

Both CI Aql and IM Nor are moderately fast novae with a light curve development similar to those of moderately fast classical novae. The spectral development resembles those of the 'hybrid' novae [23] which have "FeII" type spectra in the early phases that evolve to the "He/N" type. The "FeII" spectra probably originate from discrete massive shells, which are optically thick. As the expanding shells become optically thin, coinciding with the transition phase in the light curve, the spectrum changes to the "He/N" type.

The light curve and spectral development of both CI Aql and IM Nor resemble that of T Pyx rather than the U Sco systems with subgiant secondaries.

The 1999 Outburst of U Sco

The 1999 outburst of U Sco, provided, for the first time, and opportunity to study its development during the early phases of the outburst. Optical spectra were obtained as early as 0.45 days [24], 0.64 days [25] and 0.67 [26] days since maximum. U Sco was observed, for the first time, in the near-IR, 2.34 – 27.28 days since maximum [27].

Early spectra showed strong, broad emission lines due to hydrogen Balmer, N III, C III and He I, with initial FWZI $\sim 10,000$ km s^{-1}. High resolution spectra obtained on day 0.67 showed weak, narrow absorption features associated with the broad emission lines, with a mean blue shifted velocity of ~ 4500 km s^{-1} [26]. He II lines developed rapidly by day 2 since maximum [27] The width of the Hα line decreased, almost linearly, with time [25]. The spectrum was similar to the 'He/N' class of novae. HST spectra provided evidence for the presence of an accelerated clump in the nova ejecta [28], and were consistent with the existence of a moderately collimated bipolar outflow, where the acclerated clump may be located.

U Sco was discovered as a supersoft X-ray source about 19–20 days after outburst maximum [29], as predicted by the theoretical models of the outburst, consistent with a TNR outburst on a massive white dwarf. A hard X-ray component was also detected, due to emission from a postoutburst strong shocked wind from the white dwarf.

The helium abundance estimated, based on spectra obtained at different phases is inconsistent and range from a normal abundance of 0.16 to an overabundance of 4 [24], [26], [27].

The binary system

The spectrum of U Sco obtained at late decline phases clearly show Mg Ib and Fe I + Ca I absorption lines from the secondary. The indices of these absorption features indicate the secondary to be a K2 subgiant [24], [29].

U Sco is an eclipsing binary, with an orbital period of 1.2305 days [30]. Spectroscopic observations obtained during the late decline phases of the 1999 outburst, covering one complete oribital period including an eclipse [31], give the following system parameters: $i = 82.7° \pm 2.9°$; $K_s = 170 \pm 10$ km s^{-1}; $M_s = 0.88 \pm 0.17$ M$_\odot$; $R_s = 2.1 \pm 0.2 R_\odot$; $K_w = 93 \pm 10$ km s^{-1}; $M_w = 1.55 \pm 0.24$ M$_\odot$. The radius estimated for the secondary confirms its luminosity class estimated based on the absorption indices.

The estimates for the mass ejected during the outbursts indicate that only 30–60% of the total matter accreted during outbursts is ejected. This implies that the white dwarf could increase its mass at an average rate of $\dot{M}_{av} \sim 10^{-7}$ M$_\odot$ yr^{-1} and can evolve to a Type Ia supernova in 700,000 years [31], [3].

T Pyxidis

T Pyx is the only recurrent nova that is known to have a Roche-lobe filling main sequence secondary. It has an orbital period of 0.0762 days [32]. A transient signal has also been detected at 0.1098 days, suggesting the white dwarf could be magnetic [32].

The outburst characteristics, such as the light curve and the spectral development, are very similar to the slow classical novae. The spectral development of the 1966 outburst indicates T Pyx also belongs to the class of 'hybrid' novae. Theoretical models to explain the outburst light curve indicate a white dwarf mass of $1.2 M_\odot$ [33].

T Pyx is the only recurrent nova that shows a discernable shell. The structure of the shell is similar to that of GK Per, and is composed of thousands of unresolved knots, some of which are variable, typically on a time scale of months [34]. The observations support the model in which the rapidly moving, initial ejecta from a nova eruption catches up and collides with the slow moving ejecta from a previous eruption [34], [35]. This collision produces the observed clumping, observed emission lines and the knot variability.

T Pyx also exhibits extremely blue colours at quiescence: $B - V = -0.3$, $U - B = -1.3$, with a corresponding bolometric luminosity $L_{bol} \gtrsim 10^{36}$ erg s^{-1} [32]. This luminosity is much higher than expected for short-period CVs, but similar to those seen in supersoft X-ray sources. It is hence suggested that T Pyx is a supersoft X-ray source. The observed luminosity can be explained as being due to (a) quasi-steady nuclear burning of accreted gas on the white dwarf [32], or, (b) a strong, radiation-induced wind excited from the secondary; it is a wind-driven supersoft X-ray source [36]. The latter case could lead to a complete evaporation of the secondary star.

LONG PERIOD BINARIES

The outburst and quiescence properties of all the four members of this group are similar. They are fast novae and eject matter with extremely high velocities. High excitation coronal lines and narrowing of emission lines are seen in the outburst spectrum as a result of shock heating as the fast moving nova ejecta interacts with the slow moving pre-outburst circumstellar material from the stellar wind of the giant companion [37], [38]. This shock interaction also gives rise to the non-thermal X-ray and radio radiation detected during the 1985 outburst of RS Oph [39], [40].

The optical spectrum at quiescence is dominated by that of the giant secondary, with emission lines due to H I and He I superimposed. Fe II and Ca II lines are also present (except in T CrB). He II lines are extremely weak or absent (see e.g. [41]). The UV spectrum of T CrB shows a relatively hot continuum, with emission lines and shell-like absorption features. The UV continuum as well as the emission lines fluxes are variable and strongly correlated. The UV spectrum of RS Oph shows a flat continuum with a few weak emission lines [42]. Variability is also detected in the UBV/visual magnitudes and the optical emission line fluxes in all the four systems [41]. As can be seen in Figure 2, the emission lines fluxes are strongly correlated with the variability in the V/visual magnitude. These variations are found to correlate with the activity of the hot component [41].

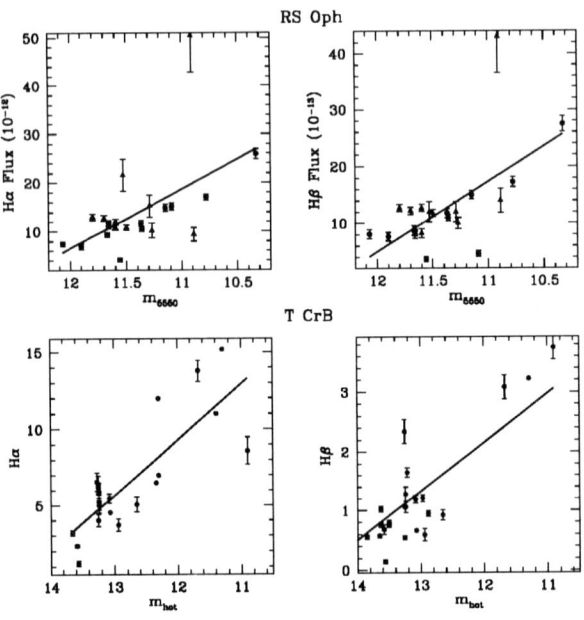

FIGURE 3. Bottom panel: correlation of the Hβ and Hα emission line fluxes in T CrB with the V magnitude of the hot component $m_{\rm hot}$. Fluxes are in 10^{-12} erg cm^{-2} s^{-1}. Top panel: correlation of the Hβ and Hα emission line fluxes in RS Oph with the monochromatic magnitude at 5500 Å. All fluxes are in erg cm^{-2} s^{-1}. Based on [41]

The Hot Component

Although sufficient observational evidence existed for the presence of an accreting white dwarf as the hot component, and the outbursts being powered by thermonuclear runaway reactions [43], the major problem for the TNR model arose in the inconsistency of the hot component's luminosity and effective temperature with standard massive white dwarf tracks in the HR diagram. For instance, a spectral decomposition of the quiescence spectrum in the optical indicates the hot component mimics the spectrum of A–F stars in these systems, while the $L_{\rm EUV}$ obtained using the observed H, He I and He II fluxes is higher, by atleast an order of magnitude, than expected for normal stars corresponding to the estimated spectral types. Further, the O I 8446/7774 line ratios imply a mean Lyβ radiation density $\gtrsim 1.3 \times 10^3$ erg cm^{-2} s^{-1} sr^{-1}. However, there is a general lack of He II and other high excitation lines normally associated with the presence of a hot ionizing source. Also the observed X-ray luminosity of RS Oph at quiescence is found to be much lower than expected for a hot white dwarf [44].

Based on a detailed study of these systems at quiescence, [41] interpret the lack of He II and other high excitation lines, and the flat UV continuum (in RS Oph), despite the high EUV luminosity as being due to the fact that the hot white dwarf is embedded in

TABLE 1. Recurrent Novae: Parameters

Name	m_{max}	m_{min}	t_3 days	$<t_{rec}>$ yrs	dist kpc	Secondary	P_{orb} days	Ref
T CrB	2.0p	10.2v	6.8	80	1.3	M3 III	227.67	[41]
RS Oph	5.0v	11.5v	9.5	22	1.6	M0/2 III	455.72	[41], [46]
V3890 Sgr	8.2v	17.0:	17.0	28	5.2	M5 III		[41]
V745 Sco	9.6v	19.0:	14.9	52	4.6	M6 III		[41]
U Sco	7.6v	18-19v	4.3	27	14	K2 IV	1.2305	[30], [24]
V394 CrA	7.0v	18.8B	10,5.5	38	5	K	0.7577	[47]
LMC 1990#2	10.9v	20:	5.26	22	49.4			[47]
T Pyx	6.5v	15.2v	88	19	4.5:		0.0762	[32], [47]
CI Aql	8.9v	17-17.8v	36	42		K-M IV:	0.6184	[11], [6]
IM Nor	7.7v	19.5:	≳ 50	82	1.9-2.4			[19], [21]

the optically thick red giant wind. This interpretation is similar to the model invoked for the symbiotic systems [45], wherein the UV continuum and emission lines are shown to be suppressed by differential extinction due to absorption lines in the red giant wind produced by neutral and singly ionized iron peak elements. This effect is seen for column densities $\geq 5 \times 10^{22}$ cm^{-2}.

The observed Fe II and Ca II (IR triplet) line ratios in RS Oph indicate electron densities $\sim 10^{11} - 10^{12}$ cm^{-3} and column density $\geq 10^{23}$ cm^{-2}. Further, the presence of Fe II emission lines in the optical indicate that line blanketing is indeed possible in these systems. The red giant wind could thus absorb the direct photons from the white dwarf+accretion disc and soften by reradiation. This model explains the observed UV spectra and the lack of high excitation lines. It also explains the observed X-ray luminosity of RS Oph at quiescence.

All observed variations in the emission lines and the UV total fluxes, and in the X-ray luminosity are related to the radiation flux from the ionizing source and are caused by fluctuations in the mass accretion rates and changes in the absorption column density [41].

SUMMARY

The basic properties of all the known RNe are summarised in Table 1.

The speed classes of RNe span from the slow to the very fast range.

RNe may be grouped as systems with (a) long period binaries, or (b) systems with short period binaries.

The long period binaries are a homogeneous group with (almost) similar outburst and quiescence properties. The binary components are a red giant secondary and a hot white dwarf that is embedded in the red giant wind, similar to symbiotic systems.

The white dwarf mass is believed to be close to the Chandrasekhar limit.

The short period binaries are a heterogeneous group.

U Sco, V394 CrA and LMC 1990#2 are very similar in their outburst and quiescence properties. The secondary in U Sco is a K2 subgiant. A similar secondary is indicated by the quiescence spectra and the orbital period in V394 CrA. Observations indicate the white dwarf mass in U Sco could be close to the Chandrasekhar limit, and also growing. This implies that U Sco is the most likely candidate as a Type Ia supernova progenitor.

CI Aql, IM Nor and T Pyx have very similar outburst properties. These objects eject more matter during outburst than the other RNe systems. The outbursts of T Pyx and CI Aql are modelled to have occurred on ~ 1.2 $M_\odot$ white dwarfs.

The orbital period of CI Aql indicates a subgiant secondary, with a spectral type K–M. T Pyx has an orbital period very similar to classical novae and other cataclysmic variables. The bolometric luminosity at quiescence is however similar to the supersoft X-ray sources.

T Pyx is the only recurrent nova that has an extended, clumpy shell.

RELATION TO CLASSICAL NOVAE

The overall development of the spectra of recurrent novae following the nova outburst can be classified under the general classification scheme used for the classical novae [23], [46].

The U Sco systems follow the spectral development of He/N novae.

CI Aql, IM Nor and T Pyx are similar to the 'hybrid' novae, evolving from an initial Fe IIb class to He/N class.

The shell around T Pyx is very similar to that of GK Per.

The outburst properties imply an outburst mechanism similar to the classical novae. The theoretical models used for classical nova outbursts reproduce the recurrent nova outbursts also.

The major differences are in the recurrence timescales and the outburst luminosities.

These differences arise due to the nature of the white dwarf, which is more massive (1.2–1.34 $M_\odot$), and accretion rates ($10^{-8} - 10^{-7}$ $M_\odot$ yr^{-1}) that are higher in the RNe.

Another difference observed with respect to classical nova outbursts is that not all the accreted matter is ejected during a recurrent nova outburst. Some fraction is left behind, which can lead to an increase in the mass of the white dwarf. This can, in some systems, eventually lead to a Type Ia supernova event.

REFERENCES

1. Starrfield, S., Sparks, W.M., Truran, J.W., *ApJ*, **291**, 136-146 (1985).
2. Kato, M., *ApJ*, **369**, 471-474 (1991).
3. Hachisu, I., Kato, M., Kato, T., Matsumoto, K., *ApJL*, **528**, L97-L100 (2000).
4. Reinmuth, K., *AN*, **225**, 385 (1925).
5. Duerbeck, H.W., *Space Sci., Rev.*, **45**, (1987).
6. Mennickent, R.E., Honneycut, R.K., *IBVS*, No. 4232 (1995).
7. Greiner, J., Alcala, J.M., *IBVS*, No. 4338 (1996).
8. Takamizawa, K., Kato, T., Yamamoto, M., et al., *IAUC*, No. 7409 (2000).
9. Williams, D.B., *IBVS*, No. 4904 (2000).
10. Schefer, B.E., *IAUC*, No. 7750 (2001).
11. Kiss, L.L., Thomson, J.R., Ogloza, W., Fürész, G., Sziládi, K., *A&A*, **366**, 858-864 (2001).
12. Schmeja S., Armsdorfer, B., Kimeswenger, S., *IBVS*, No. 4957 (2000).
13. Matsumoto, K., Uemura, M., Kato, T., et al., *A&A*, **378**, 487-494 (2001).
14. Wilson, J.C., Dunscombe, K.R., *IAUC*, No. 7426 (2000).
15. Mazuk,S., Rudy, R.J., Lynch, D.K., et al., *IAUC*, No. 7490 (2000).
16. Hachisu, I., Kato, M., in *The Physics of Cataclysmic Variables and Related Objects*, Edited by B. T. Gänsicke, et al., ASP Conference Proceedings 261, Astronomical Society of the Pacific, San Francisco, 2002, p. 627.
17. Elliot, J. L., Liller, W., *ApJ*, **175**, L69 (1972).
18. Liller, W., *IAUC*, No. 7789, (2002).
19. Kato, T., Yamaoka, H., Liller, W., Monard, B., *preprint: astro-ph/0204354*, (2002)
20. Duerbeck, H.W., Baptista, R., Dutra, C.M., Sterken, C., *IAUC*, No. 7799 (2002).
21. Duerbeck, H.W., Baptista, R., Dutra, C.M., et al. *These proceedings* (2002).
22. Retter, A., O'Toole, S.J., Starrfield, S., *IAUC*, No. 7818 (2002).
23. Williams, R.E., *AJ*, **104**, 725-733 (1992).
24. Anupama, G.C., Dewangan, G.C., *AJ*, **119**, 1359-1364 (2000).
25. Munari, U., Zwitter, T., Tomov, T., et al. *A&A*, **347**, L39-L42 (1999).
26. Iijima, T., *A&A*, **387**, 1013-1021 (2002).
27. Evans, A., Krautter, J., Vanzi, L., Starrfield, S., *A&A*, **378**, 132-141 (2001).
28. Lépine, S., Shara, M., Livio, M., Zurek, D., *ApJL*, **522**, 121-124 (1999).
29. Kahabka, P., Hartmann, H.W., Parmar, A.N., Negueruela, I., *A&A*, **347**, L43-L46 (1999).
30. Schaefer, B.E., Ringwald, F.A., *ApJL*, **447**, L45-L48 (1995).
31. Thoroughgood, T.D., Dhillon, V., Littlefair, S.P., et al. *MNRAS*, **327**, 1323-1333 (2001).
32. Patterson, J., Kemp, J., Shambrook, A., et al. *PASP*, **110**, 380-395 (1998).
33. Hachisu, I., *These proceedings*, (2002).
34. Shara, M., Zurek, D.R., Williams, R.E., et al. *AJ*, **114**, 258-264 (1997).
35. Contini, M., Prialnik, D., *ApJ*, **475**, 803-811 (1997).
36. Knigge, Ch., King, A.R., Patterson, J., *A&A*, **364**, L75-79 (2000).
37. Bode M.F., Kahn, F.D., *MNRAS*, **217**, 205-215 (1985).
38. O'Brien, T., Bode, M.F., Kahn, F.D., *MNRAS*, **255**, 683-693 (1992).
39. Mason, K.O., Córdova, F.A., Bode, M.F., Barr, P., in *RS Ophiuchi (1985) and the Recurrent Nova Phenomenon*, Edited by M.F. Bode, VNU Science Press, Utercht, 1987, p. 167.
40. Hjellming, R.M., van Gorkom, J.H., Seaquist, E.R., et al. *ApJL*, **305**, L71-L75 (1986).
41. Anupama, G.C., Mikołajewska, J., *A&A*, **344**, 177-187 (1999).
42. Dobrzycka, D., Kenyon, S.J., Proga, D., et al. *AJ*, **111**, 2090-2098 (1996).
43. Anupama, G.C., in *Cataclysmic Variables*, Edited by A. Bianchini et al., Kluwer Academic, Netherlands, 1995, pp.49-57.
44. Orio, M., Covington, J., Ögelman, H., *A&A*, **373**, 542-554 (2001).
45. Shore, S.N., Aufdenberg, J.P., *ApJ*, **416**, 355-367 (1993)
46. Fekel, F.C., Joyce, R.R., Hinkle, K.H., Skrutskie, M.F., *AJ*, **119**, 1375-1388 (2000).
47. Warner, B., *Cataclysmic Variable Stars*, Cambridge Astrophysics Series, 28, Cmabridge University Press, Cambridge, 1995.
46. Williams, R.E., Hamuy, M., Phillips, M.M., et al., *ApJ*, **376**, 721-737 (1991).

Evolution of the symbiotic nova RX Puppis

J. Mikołajewska*, E. Brandi†, L. Garcia†, O. Ferrer†, C. Quiroga† and G.C. Anupama**

*N. Copernicus Astronomical Center, Bartycka 18, 00716 Warsaw, Poland
mikolaj@camk.edu.pl
†Facultad de Ciencias Astronómicas y Geofísicas, UNLP - CIC - CONICET, 1900 La Plata, Argentina
**Indian Institute of Astrophysics, Bangalore 560034, India

Abstract. We present and discuss a hundred year history of activity of the hot component of RX Pup based on optical photometry and spectroscopy. The outburst evolution of RX Pup resembles that of other symbiotic novae whereas at quiescence the hot component shows activity (high and low activity states) resembling that of symbiotic recurrent novae T CrB and RS Oph.

INTRODUCTION

RX Pup is a long-period interacting binary system consisting of a Mira variable pulsating with P=578 days, surrounded by a thick dust shell, and a hot white dwarf companion accreting material from the Mira's wind. The binary separation could be as large as $a \geq 50$ a.u. (corresponding to $P_{orb} \geq 200$ yr) as suggested by the permanent presence of a dust shell around the Mira component ([1]).

The analysis of multifrequency observations by [1] has shown that most, if not all, photometric and spectroscopic activity of RX Pup in the UV, optical and radio range is due to activity of the hot component, while the Mira variable and its circumstellar environment is responsible for practically all changes in the IR range (Fig.1). In particular, RX Pup underwent a nova-like eruption during the last three decades. The evolution of the hot component in the HR diagram (Fig.11 of [1]) as well as evolution of the nebular emission in 1970−1993 is consistent with a symbiotic nova eruption, with the luminosity plateau reached in 1972/75 and a turnover in 1988/89. The hot component contracted in radius at roughly constant luminosity from c. 1972 to 1986; during this phase it was the source of a strong stellar wind and therefore could not accrete any further material. By 1991 the luminosity of the nova remnant had decreased to a few per cent of the maximum (plateau) luminosity, and the hot wind had practically ceased. By 1995 the hot component start to accrete material from the Mira wind, as indicated by a general increase of the optical continuum and Balmer H I emission. The quiescent optical spectrum of RX Pup resembles the quiescent spectra of symbiotic recurrent novae, while the hot component luminosity is consistent with variable wind-accretion at a high rate, $\dot{M}_{acc} \sim 10^{-7} M_\odot \, yr^{-1}$ (≈ 1 per cent of $\dot{M}_{cool}$). RX Pup may be a recurrent nova; there is some evidence that a previous eruption occurred around 1894.

In the following we discuss results of optical and red spectroscopic observations of

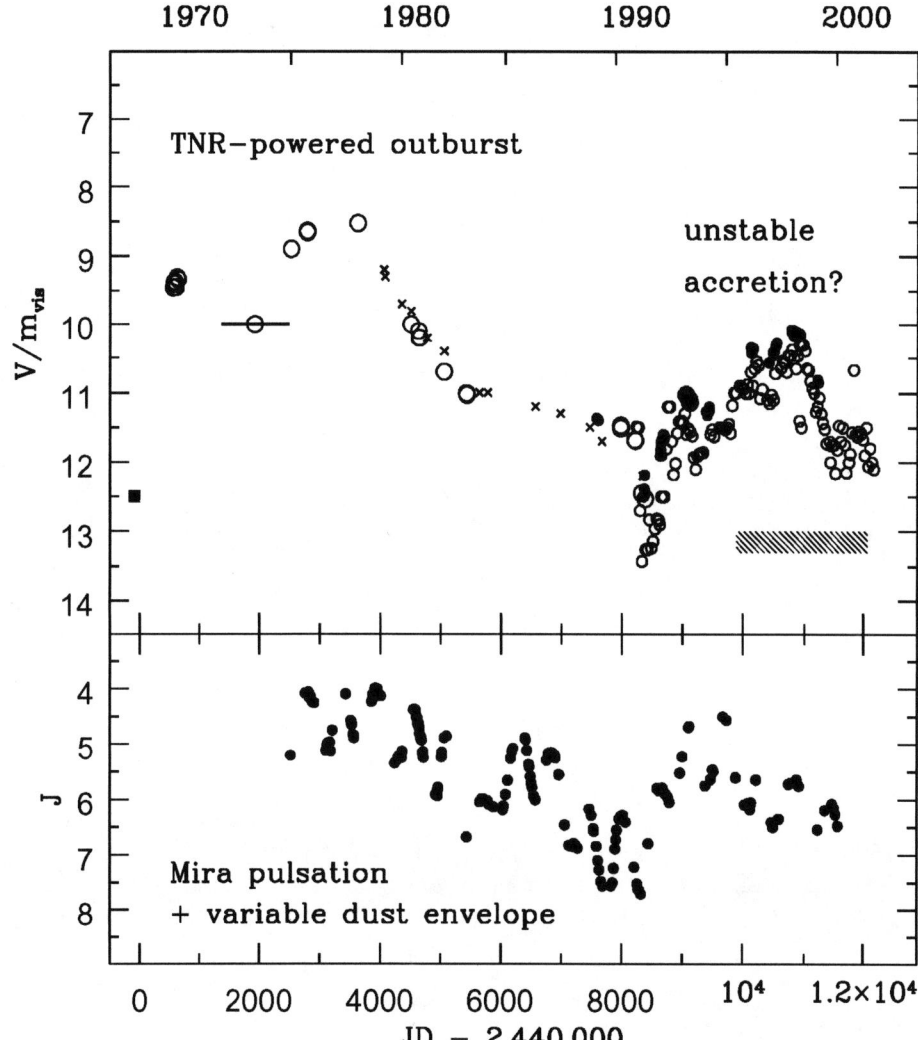

FIGURE 1. Optical and near- IR (J) light curves of RX Pup ([1]). In the V/m_{vis} light curve, small open circles represent RASNZ observations; large open circles and dots published V data; crosses FES magnitudes. The optical light curve is dominated by the hot component activity whereas the J light curve is dominated by the Mira pulsations and variable obscuration of the Mira by circumstellar dust. The shaded area indicates the period of our spectroscopic observations discussed in Sec. 2.

RX Pup obtained during 1995–2001 with the REOSC echelle spectrograph at the 2.15-m CASLEO telescope at San Juan, Argentina (see [1] for details), and low resolution CCD spectra obtained from the Vainu Bappu Observatory, India (see e.g. [2] for details).

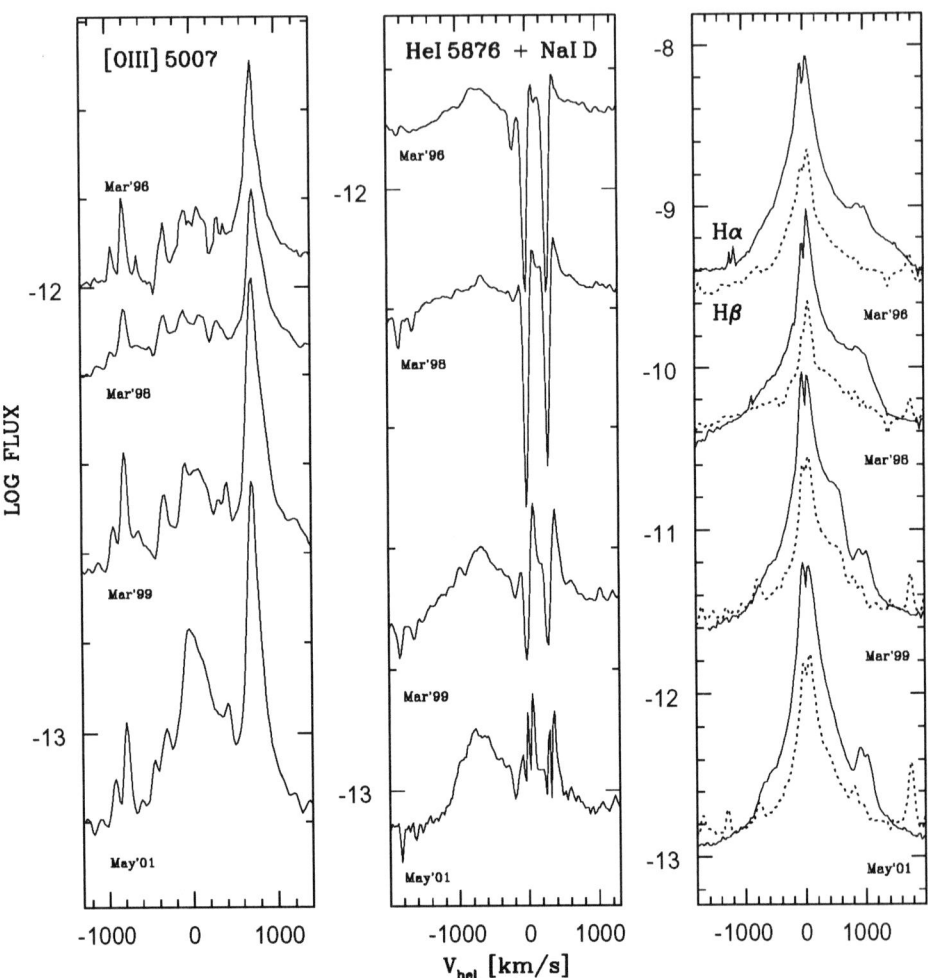

FIGURE 2. Evolution of emission line profiles in RX Pup in 1996–2001. The profiles are shifted vertically for better display.

RESULTS AND DISCUSSION

In the period covered by our spectroscopic observations, both components of RX Pup were active, although their changes were generally not correlated (Fig. 1). In particular, the near-IR flux was gradually decreasing in 1995–2000, suggesting that the Mira had entered a new obscuration phase ([3]). At the same time, the visual light curve revealed a small eruption with maximum in 1998. The lack of any molecular absorption bands in our spectra ($\lambda \leq 8600$), as well as the absence of any Mira pulsations in the optical and red light curves (Fig.1; [1]; [3]) implies that the hot component is responsible for all changes in the optical and red spectral range.

The optical brightening was accompanied by significant spectroscopic changes. The changes were the most remarkable in Balmer H I lines: they were strong and narrow in 1988 – 91 ([1]) whereas in 1995 – 2001 they developed a prominent flattened red wing (Fig. 2). The broad asymmetric wings were also present in He I and near infrared Ca II triplet. The ionization level remained low, although the decline from the visual maximum in 1998 was associated with increase in [O III] and He I emission line fluxes. Our spectra also showed the presence of a variable blue continuum with A/F-type shell absorption lines. The colour temperature of the continuum is low, $\sim 6000 - 9000$ K for $E_{B-V} \sim 0.8$, and it has a luminosity of $\sim 500 - 1000\,(d/1.8\,\text{kpc})^2\,L_\odot$. On the other hand, the optical emission lines indicate a higher, $\sim 2.5 - 4 \times 10^4$ K, temperature source with a roughly comparable luminosity. [3] also detected large changes in the degree of polarization in the optical and red spectral range, and found that the polarized component is radiation from the blue continuum source scattered in the dust envelope surrounding the Mira.

We find here a striking similarity between RX Pup and the quiescent symbiotic recurrent novae RS Oph and T CrB, as well as the hot component of CH Cyg during bright phases. In all these systems, the optical data indicate the presence of relatively cool F/A- (CH Cyg) or A/B-type source (RS Oph and T CrB), while the optical emission lines indicate a higher temperature source, with a roughly comparable luminosity ([4]; [5]; [2]). For example, in RS Oph the IR colours indicate the presence of an additional warm, ≥ 7000 K, source whereas the IUE and optical spectra show an A-B type shell source with $L \sim 100 - 600 L_\odot$, accompanied by strong H I and moderate He I emission lines ([5], and references therein). Similar variable UV/optical continuum with $L \sim 40 - 100 L_\odot$ was observed in T CrB ([2] and references therein). Although both in the symbiotic recurrent novae, T CrB and RS Oph, and in RX Pup, the average luminosity of the B/A/F-type shell source is consistent with the accretion rate, $\dot M \geq 10^{-8} M_\odot\,\text{yr}^{-1}$, required by the theoretical models, the effective temperatures places the hot components far from the standard massive white dwarf tracks in the HR diagram. Simultaneously, the X-ray data suggest $1-2$ orders of magnitude lower accretion rates ([1], and references therein). All three systems show similar quiescent behaviour: their hot components have highly variable luminosity and occasionally display blue-shifted absorption features, and broad asymmetric emission line wings (see also [1]; [2]; [5]).

We note here that similar A/F-type shell absorption spectrum was detected in AR Pav and other Z And-type symbiotic stars with multiple outburst activity(e.g. [6]). The blue absorption system in these systems traces the orbit of the hot component, and it is probably formed in a geometrically and optically thick accretion disc seen nearly edge-on and in a gas stream ([7]; [6], and references therein). Parenthetically, the timescales and amplitudes for their eruptions are very similar to the timescales and amplitudes of the hot component luminosity variations (high and low states) in RX Pup and other symbiotic recurrent novae between their nova outbursts, as well as in accretion-powered systems CH Cyg and MWC 560.

It is also interesting that the shell spectrum appears only during the late decline from the nova outburst. In particular, the optical/visual light curves from the outbursts of T CrB, RS Oph and RX Pup show more or less pronounced minima followed by a stand-still (or secondary maximum) associated with the appearance of the shell spectrum. We suggest the minima are due to a decline in the hot component's luminosity after it passes the turnover in the HR diagram. The strong hot component wind during the plateau phase prevents accretion onto the hot component. Following the decline in luminosity the wind also ceases, and accretion of the material from the cool giant can be restored. We believe that the shell-type features and variable "false atmosphere" observed at quiescence together with the complex and broad emission lines originate from the accretion flow. The observed variability could be due to fluctuations in the mass-loss rate from the cool giant, the shell spectrum becomes stronger as $\dot{M}_{cool}$ increases and as a result $\dot{M}_{acc}$ increases. In the recurrent novae T CrB, RS Oph and the accretion- powered systems of CH Cyg and MWC 560, the brightening of the hot component (or high activity stage) is associated with the presence of flickering (e.g. [2]). Such flickering, however, has not yet been detected in RX Pup.

Summarizing, the spectral development of RX Pup during the large outburst in 1969–1990 resembles that of symbiotic novae AG Peg, HM Sge and V1329 Cyg as well as of the symbiotic recurrent nova RS Oph ([1]), although these other systems seem to be evolving on very different timescales, possibly reflecting differences in the mass of the white dwarf component. At quiescence the hot component shows activity (high and low activity states) resembling that of symbiotic recurrent novae T CrB and RS Oph, which is probably due to relatively high mass transfer/accretion rate, of order of $\sim 10^{-7} M_\odot \, yr^{-1}$. A high accretion rate, $\geq 10^{-8} M_\odot \, yr^{-1}$, is also required to account for the recurrence time, $\sim 80, 80$ and 22 yr, for RX Pup, T CrB, and RS Oph, respectively.

ACKNOWLEDGMENTS

This study was partly supported by KBN research grant No. 5P03D 019 20.

REFERENCES

1. Mikołajewska, J., Brandi, E., Hack, W., Whitelock, P.A., Barba, R., Garcia, L., Marang, F., 1999, MNRAS, 305, 190
2. Anupama, G.C., Mikołajewska, J., 1999, A&A, 344, 177
3. Mikołajewska, J., Brandi, E., Garcia, L., Ferrer, O., W., Whitelock, P.A., Marang, F., 2001, in Szczerba R. et al., eds, Post-AGB Objects as a Phase of Stellar Evolution, Kluwer, 227
4. Mikołajewska, J., Selvelli, P.L., Hack, M., 1988, A&A, 198, 150
5. Dobrzycka, D., Kenyon, S.J., Proga, D., Mikołajewska, J., Wade, R., 1996, AJ, 111, 2090
6. Quiroga, C., Mikołajewska, J., Brandi, , Ferrer, O., Garcia, L., 2002, A&A, 387, 139
7. Mikołajewska, J., Kenyon, S.J., 1992, AJ, 103, 579

An evolutionary scenario for the U Scorpii

Marek J. Sarna*, Ene Ergma† and Jelena Gerškevitš†*

*N. Copernicus Astronomical Center, Polish Academy of Sciences, ul. Bartycka 18, 00–716 Warsaw, Poland
†Physics Department, Tartu University, Ülikooli 18, 50510 Tartu, Estonia

Abstract. We perform evolutionary calculations of binary stars to find progenitors of systems with parameters similar to the recurrent novae U Sco. We show that a U Sco type–system may be formed starting with an initial binary system which has a low–mass carbon-oxygen white dwarf as an accretor. Since the evolutionary stage of the secondary is not well known, we calculate sequences with hydrogen rich and helium rich secondaries. The evolution of the binary may be devided into several observable stages as: classical nova, supersoft X–ray source with stable hydrogen shell burning and strong wind phases, ending up with the formation of a massive white dwarf near the Chandrasekhar mass limit. We follow the chemical evolution of the secondary as well as of the matter lost from the system, and we show that observed $^{12}C/^{13}C$ and N/C ratios may give some information about the nature of the binary.

INTRODUCTION

Recurrent novae are a small class of objects which bear many similarities to other cataclysmic variable systems. They experience recurrent outbursts at intervals of 20–80 yrs. Webbink et al. (1987) discussed the nature of the recurrent novae, and they concluded that according to outburst mechanisms there are two subclasses of these systems: (a) powered by thermonuclear runaway on the surface of the white dwarf (e.g. U Sco), and (b) powered by the transfer of a burst of matter from the red giant to the main–sequence companion.

Recurrent novae have been proposed as possible progenitors of type Ia supernovae (Starrfield, Sparks & Truran, 1985). Although they experience recurrent outbursts it is thought that the white dwarf mass would grow to finally exeed Chandrasekhar mass limit.

OBSERVATIONAL CHARACTERISTIC

U Sco is one of the best observed recurrent novae. Historically, its outbursts were observed in 1863, 1906, 1936, 1979, 1987 and in 1999. Determination of the system visual luminosity at maximum and minimum, indicate a range $\Delta m_V \sim 9$. Schaefer (1990) and Schaefer & Ringwald (1995) observed eclipses of U Sco in the quiescent phase, and determined the orbital period P_{orb}=1.23056 d. Recent spectroscopic analysis of U Sco was presented by Thoroughgood et al. (2001). Determination of radial velocity semiamplitude for primary and secondary stars allows to obtain M_{wd}= 1.55±0.24$M\odot$ for the white dwarf and of M_2=0.88±0.17 $M\odot$ for a secondary star. For the 1979 outburst ejecta abundances have been estimated from optical and UV studies by Williams et al. (1981) and Barlow et al. (1981). They derived extremely helium rich ejecta He/H~ 2 (by number), while the CNO abundance was solar with an enhanced N/C ratio. From the analysis of the 1999 outburst Anupama & Dewangan (2000) obtained an average helium abundance of He/H~0.4±0.06. The estimated mass of the ejected shell for 1979 and 1999 outbursts is ~ 10^{-7} $M_\odot$ (Williams et al. 1981; Anupama & Dewangan 2000).

THE MAJOR PHASES OF SEMIDETACHED BINARY EVOLUTION.

We can identify three different phases during semidetached evolution of the close binary system with white dwarf as accretor:

- the nova phase (nova outbursts);
- the stable hydrogen shell burning phase on a surface of white dwarf and
- a strong, optically thick wind phase powered by steady hydrogen burning.

While calculating evolutionary models of binary stars, we must take into account mass transfer and associated physical mechanisms which lead to mass and angular momentum loss.

The novae phase

In the novae phase besides magnetic stellar wind braking (MSW) the angular momentum loss that accompanies mass loss due to novae outbursts is taken into account. During novae outburst the matter which is re–accreted by secondary changes significantly its chemical composition.

The stable hydrogen shell burning phase

During the stable hydrogen shell burning phase only angular momentum loss due to MSW will be important.

The strong optically thick wind phase

In this phase besides MSW three more processes changing total angular momentum and mass of the system are important:

- strong optically thick wind (Hachisu, Kato & Namoto, 1996);
- frictional deposition of orbital energy into wind (Livio, Govarie & Ritter, 1991);
- re–accretion by the secondary of material from strong wind.

Table 1 Results for computed sequences

model	M_{sg} [$M_\odot$]	M_{wd} [$M_\odot$]	$\log T_{eff}$ [K]	$\log L/L\odot$	$\dot{M}_{sg}$ [$M_\odot$ yr^{-1}]
A	0.545	0.983	3.669	0.153	2.02×10^{-8}
B	0.936	1.304	3.693	0.416	3.58×10^{-8}
C	0.581	1.263	3.719	0.370	4.07×10^{-8}
D	1.696	0.852	3.957	1.707	1.14×10^{-7}

EVOLUTIONARY CALCULATIONS

Our calculations were carried out under the following assumptions:

- All evolutionary calculations were carried out using a standard Henyey–type code which utilizes the results of the theoretical novae calculations of Prialnik & Kovetz (1995) and Kovetz & Prialnik (1997). By interpolation from the data of Kovetz & Prialnik, we use the white dwarf mass and the mass transfer rate to determine the nova characteristic: the amounts of matter ejected in a nova outburst, amplitude of the outburst, recurrence period and chemical composition of the ejected material.
- To understand the evolution of close binary system consisting of a C–O white dwarf and a main sequence or "turn–off main sequence" stars we computed various evolutionary sequences for: three chemical compositions (Population II – X=0.756, Z = 0.001; Population I – X=0.68, Z=0.02 and helium rich – X=0.5, Z=0.02); four initial mass ratios $q_i = M_{sg,i}/M_{wd,i}$=1.5, 2.0, 2.5, 3.0; five white dwarf masses: 0.7, 0.85, 1.0, 1.15, 1.3 $M_\odot$, and initial orbital period ranging from 0.35 to 1.5 days
- For radiative transport, we used the opacity tables of Iglesias & Rogers (1996); for temperature less than 6000 K we used the opacity given by Alexander & Ferguson (1994).

THE RESULTS

We found that only a few evolutionary sequences are able to produce system parameters similar to U Sco. Since the evolutionary stage of the secondary (hydrogen or helium rich) is not observationally determined, we selected from our grid of models four different sequences with hydrogen (A, B) and helium (C, D) rich secondaries.

The sequences C and D represent the helium rich channel proposed by Hachisu et al. (1999). In Table 1 results for computed sequences are shown. The masses of the subgiant and the white dwarf are given at the moment when orbital period of the system is equal to 1.23 d. The effective temperature and luminosity are for the subgiant star.

We infer that sequences B (hydrogen rich) and C (helium rich) may evolve into systems like U Sco. After stable hydrogen burning both sequences enter into recurrent nova phase.

We find from grid of models (Kovetz & Prialnik) the value of the amplitude of the recurrent novae outburst A=7.6 mag, and P_{rec} equal to 23 and 54 yrs for sequences B and C, respectively. The mean recurrent interval implied by known outbursts is $P_{rec} = 23$ yrs.

Table 2 Time–scales for evolutionary phases

model	Δt_{n1}	Δt_{s1}	Δt_w	Δt_{s2}	Δt_{rn}	Δt_{n2} *
			$[\log(\Delta t/\text{yr})]$			
A	4.00	6.57	–	–	–	8.17
B	5.97	6.46	–	–	7.51	6.86
C	4.76	5.41	5.39	6.17	7.17	6.72
D	4.89	5.02	5.60	5.58	–	–

* n1 – first novea phase, s1 – first stable hydrogen shell burning phase, w – wind phase, s2 – second stable hydrogen shall burning phase, rn – recurrent novea phase, n2 – second novae phase

Chemical composition of the subgiant and ejected matter

From the grid of models calculated by Kovetz & Prialnik, we interpolate chemical composition during semidetached evolution (for novae phase) of each model.

If we compare the helium rich model C with the hydrogen rich model B we see that at P_{orb}= 1.23 d the N/C ratio in the envelope of the subgiant is higher for helium rich model (11.3) than hydrogen rich one (1.8), but the isotopic ratio $^{12}C/^{13}C$ is higher for the hydrogen rich model (4.6 vs. 2.9). In the ejected matter both ratios are similar.

Our calculations show that during the evolution He/H ratio of the matter lost from the system changes from 0.56 to 1.26. For sequences B and C at P_{orb}= 1.23 d the He/H ratio is 0.75 and 0.65, respectively, which is inside limit (0.4 – 2) obtained from observations.

OBSERVATIONAL TESTS.

Chemical composition and isotopic analysis may give more information about the evolutionary stage of U Sco. We think that blue and red domain spectra of U Sco could show some absorption structure in the region of 4216Å and 7920Å in the CN sequence, similar to that observed in DQ Her (Chanan, Nelson & Margon 1978, Bauschlicher, Langhoff & Taylor 1988). Analysis of the blue region of the spectra is more complicated because the structure of the CN violet system can be affected by absorption features from the disc. We suggest that the red CN band is more useful for observations. We propose to observe spectral region near 7920–7940Å to identify two ^{13}CN lines at 7921.13Å and 7935,67Å which are important for both $^{12}C/^{13}C$ isotopic ratio and N/C ratio determination.

In the case of $^{12}C/^{13}C < 10$ these lines are clearly recognized (Fujita 1985), whereas, in stars with $^{12}C/^{13}C > 20$ both lines are undetectable.

CONCLUSIONS

We calculated several evolutionary sequences to reproduce orbital and physical parameters of the recurrent novae of U Sco–type. The results of calculations can be summarized as follows:

- We find a new evolutionary channel for the formation of the U Sco–type systems. We show that U Sco–type systems may form from binaries with a low–mass C–O white dwarf as accretor if $M_{wd,i} < 0.85\ M_\odot$ and $1.5 < q_i < 2.5$. Such a system evolves through several observable stages: supersoft X-ray source with stable hydrogen burning, strong wind phase and long term recurrent novae phase (longer than 10^7 years, Table 2). In final phase of evolution a massive white dwarf near the Chandrasekhar mass limit is formed, but it never exceeds this limit.
- We propose that the evolutionary sequence B is able to produce binary system with parameters similar to U Sco. Our best fitting model has initial parameters: $M_{sg,i} = 1.7\ M_\odot$, $M_{wd,i} = 0.85\ M_\odot$ and $P_i(RLOF) = 1.61\mathrm{d}$. Based on evolutionary model we found that the best fit parameters for U Sco are: $M_{sg} = 0.94\ M_\odot$, $M_{wd} = 1.31\ M_\odot$, $\log L_{sg}/L_\odot = 0.42$, $\dot{M}_{sg} = 3.58 \times 10^{-8} M_\odot\ yr^{-1}$ for $P_{orb} = 1.23056\ d$.

ACKNOWLEDGMENTS

This work is partly supported through grants 2–P03D–005–16 of the Polish National Committee for Scientific Research and Collaborative Linkage Grant PST.CLG.977383. JG and EE acknowledge support through Estonian SF grant 4338.

REFERENCES

1. Alexander, D. R., and Ferguson, J. W., 1994, ApJ, 437, 879
2. Anupama, G. C., and Dewangan, G. C., 2000, AJ, 119, 1359
3. Barlow, M. J. et al., 1981, MNRAS, 195, 61
4. Bauschlicher, C. W., Langhoff, S. R., and Taylor P. R., 1988, ApJ, 332, 531
5. Chanan, G. A., Nelson, J. E., and Margon, B., 1978, ApJ, 226, 963
6. Fujita, Y., it Cool Stars with Excesses of Haevy Elements, eds. M. Jaschek and P. C. Keenan, Dordrecht, Reidel, 1985, p. 31
7. Hachisu, I., Kato, M., and Nomoto, K., 1996, ApJ, 470, L97
8. Hachisu, I., Kato, M., Nomoto, K., and Umeda H., 1999, ApJ, 519,314
9. Iglesias, C. A., and Rogers, F. J., 1996, ApJ, 464, 943
10. Kovetz, A., and Prialnik, D., 1997, ApJ, 477, 356
11. Livio, M., Govarie, A., and Ritter, H., 1991, A&A, 246, 84
12. Prialnik, D., and Kovetz, A., 1995, ApJ, 445, 789
13. Schaefer, B.E., 1990, ApJ.,335,L39
14. Schaefer, B. E., and Ringwald, F. A., 1995, ApJ, 447, L45
15. Starrfield, S., Sparks, W. M., and Truran, J. W., 1985, ApJ, 291, 136
16. Thoroughood, T. D., Dhillon, V. S., Littlefair, S. P., Marsh, T. R., and Smith, D. A., 2001, MNRAS, 327, 1323
17. Webbink, R. F., Livio, M., Truran, J. W., and Orio, M., 1987, ApJ, 314, 653
18. Williams, R. E., Sparks, W. M., Gallagher, J. S., and Ney, E. P., Starrfield S. G., Truran J. W., 1981, ApJ, 251, 221

V838 Mon and the new class of stars erupting into cool supergiants (SECS)

U. Munari*, A.Henden†, R.M.L.Corradi** and T.Zwitter‡

*Osservatorio Astronomico di Padova – INAF, Sede di Asiago, I-36012 Asiago (VI), Italy
†Univ. Space Research Ass./U. S. Naval Obs., P. O. Box 1149, Flagstaff AZ 86002-1149, USA
**ING, Apartado de Correos 321, 38700 Santa Cruz de La Palma, Canarias, Spain
‡University of Ljubljana, Dept. of Physics, Jadranska 19, 1000 Ljubljana, Slovenia

Abstract. V838 Mon has undergone one of the most mysterious stellar outbursts on record. The spectrum at maximum closely resembled a cool AGB star, evolving toward cooler temperatures with time, never reaching optically thin conditions or showing increasing ionization and a nebular stage. The latest spectral type recorded is M8-9. The amplitude peaked at $\Delta V=9$ mag, with the outburst evolution being characterized by a fast rise, three maxima over four months, and a fast decay (possibly driven by dust condensation). BaII, LiI and s–element lines were prominent in the outburst spectra. Strong and wide (500 km/sec) P-Cyg profiles affected low ionization species, while Balmer lines emerged to modest emission only during the central phase of the outburst. A light-echo discovered expanding around the object constrains its distance to 790±30 pc, providing $M_V = +4.45$ in quiescence and $M_V = -4.35$ at optical maximum (dependent on the still uncertain E_{B-V}=0.5 reddening). The visible progenitor resembles a somewhat under-luminous F0 main sequence star, that did not show detectable variability over the last half century.

V838 Mon together with M31-RedVar and V4332 Sgr seems to define a new class of astronomical objects, *Stars that Erupt into Cool Supergiants* (SECS). They do not develop optically thin or nebular phases, and deep P-Cyg profiles denounce large mass loss at least in the early outburst phases. Their progenitors are photometrically located close to the Main Sequence, away from the post-AGB region. After the outburst, the remnants still closely resemble the precursors (same brightness, same spectral type). Many more similar objects could be buried among poorly studied variable stars that have been classified as Miras or SemiRegulars on the base of a single spectrum at maximum brightness.

THE OUTBURST OF V838 MON

A detailed description of the outburst of V838 Mon is given by [1], to which the reader is referred. In this note only the main features are summarized with some updates on the late photometric and spectroscopic evolution of the outburst.

An updated lightcurve of the eruption of V838 Mon is presented in Figure 1. A first maximum was reached by $+10^d$ (see abscissae scale on Figure 1) when the continuum energy distribution was characterized by a temperature of 4150 K. A second maximum at $+37^d$ peaked around 5200 K and a third one at $+68^d$ reached 4600 K. Each decline from maxima was accompanied by a monotonic cooling, with the last one taking V838 Mon to 3500 K by $+90^d$. From $+90^d$ to $+120^d$ the color temperature in the region of the V, R_c, I_c bands decreased to that of an M8-9 supergiant, or 2600 K. The retracing of the $U-B$ and $B-V$ color indexes when the spectrum developed the coolest temperatures is a real effect (cf. absolute spectrophotometry in Figure 2 at $+119^d$), and it is normally seen

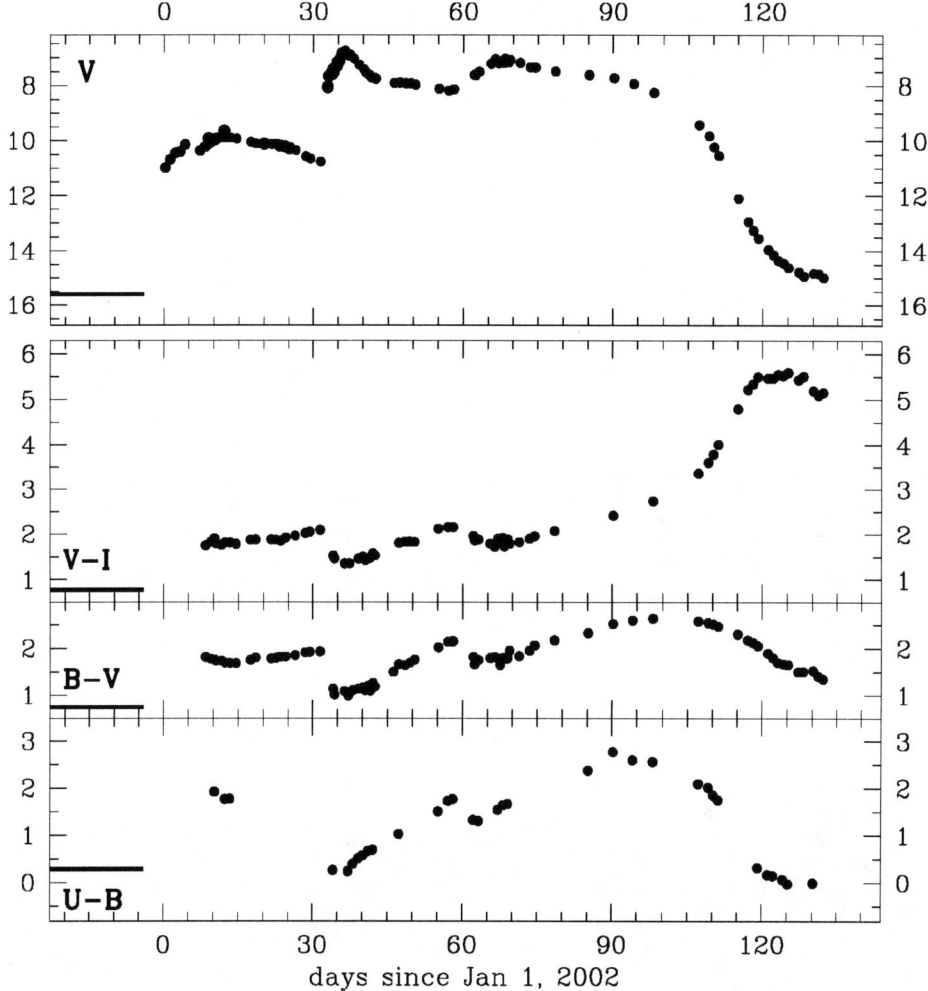

FIGURE 1. V, $B-V$ and $V-I_C$ lightcurves of the outburst of V838 Mon. Dots mark NOFS data, open circles Tsukuba data. Crosses and open triangles are values from various IAUC and VSNET circulars (mainly from SAAO, D.West, P.Sobotka, L.Smelcer, F.Lomoz and J.Bedient). The solid line indicates the quiescence brightness.

in M-type stars of the corresponding spectral types due to progressive disappearance of TiO absorptions at the shortest wavelengths.

The spectral evolution well followed the $V-I$ color temperature evolution. During the first three months the spectrum closely resembled a K giant, slowing progressing toward later spectral types and reaching K5 by $+90^d$. In the following month the spectrum rapidly entered the M-type realm and reached M8-9 by $+119^d$. The spectrum of V838 Mon for this date is shown in Figure 2. The spectral evolution of V838 Mon has been

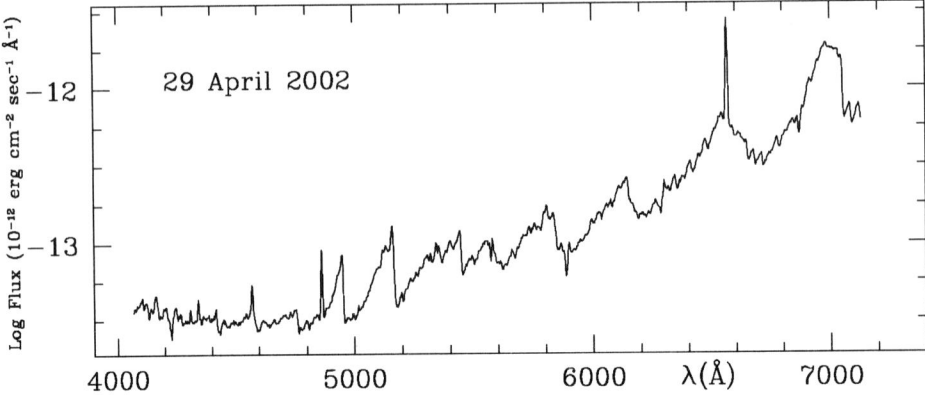

FIGURE 2. The spectrum of V838 Mon for April 29, 2002 obtained with WHT 4.2m in La Palma.

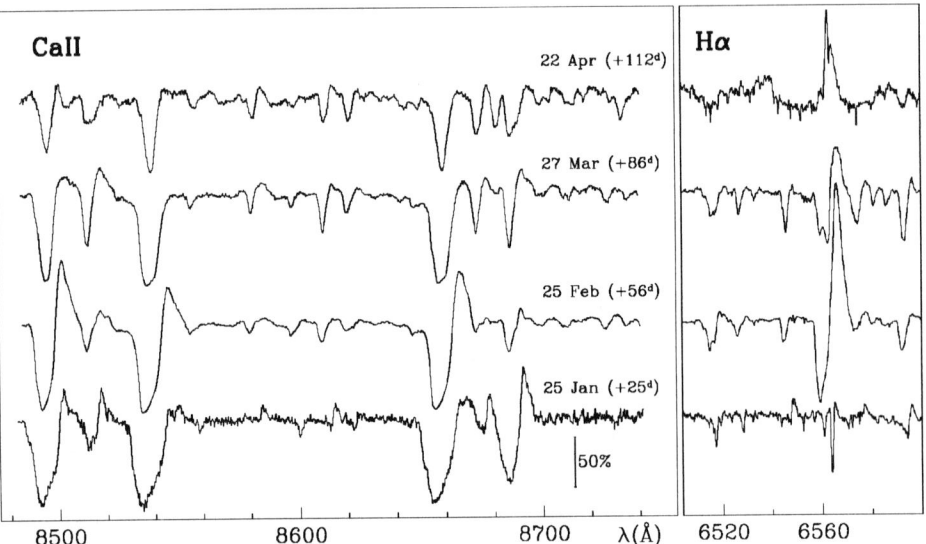

FIGURE 3. Small sections of sample of Asiago Echelle spectra to document the evolution around the far-red Calcium triplet and Hα of the V838 Mon outburst.

exciting also on a much finer scale, as Figure 3 indicates, where Echelle spectra around the CaII far-red triplet and Hα are presented for $+25^d$, $+56^d$, $+86^d$ and $+112^d$. P Cyg line profiles for low-excitation species have a terminal velocity which monotonically decreased with time from the initial value of -500 km/sec, with Balmer lines appearing in emission with their own P-Cyg profiles only after the second maximum. BaII, LiI and $s-$elements are present in the V838 Mon spectra.

In mid-February, [2] discovered the formation of a light-echo around V838 Mon, when the light from the second maximum began illuminating pre-existing circumstel-

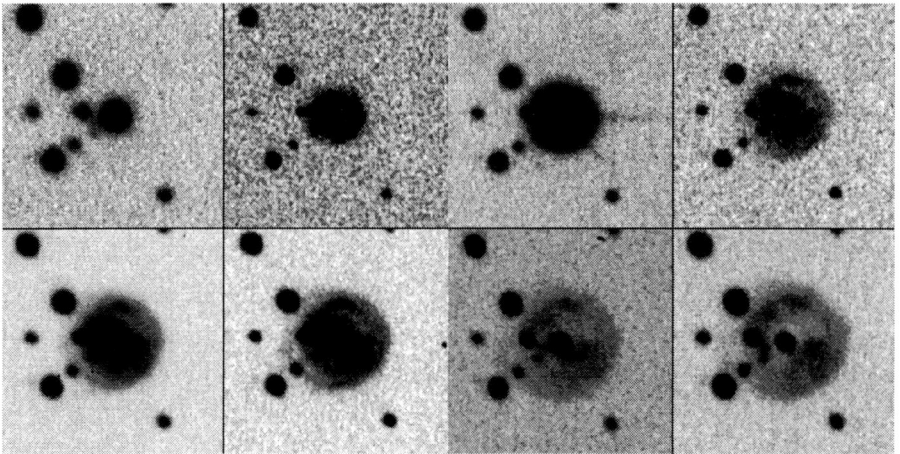

FIGURE 4. Expansion of the light-echo around V838 Mon, revealing a previously invisible ring of circumstellar material. U band 67×67 arcsec images obtained with the USNO 1m telescope. Dates from left to right and top to bottom: January 13, February 27, March 10, March 27, March 31, April 4, April 20 and April 30.

lar material responsible for the IRAS detection of the precursor. This light-echo was followed as it expanded to a maximum diameter of 35 arcsec, a size that has remained essentially constant during the following months, as Figure 4 shows. The light-echo expansion rate of 0.44 ± 0.017 arcsec day^{-1} sets the distance of V838 Mon to 790 ± 30 pc for a spherical distribution of the scattering material. The outburst light sweeping through the circumstellar material allows us to read the recent mass loss history of the progenitor: assuming 15 km sec^{-1} velocity for its wind (typical for an AGB), the light-echo has reached by April 1 material lost $\sim$4900 years ago. High resolution imaging with HST by [3] confirms the spherical symmetric dust distribution around V838 Mon and reveals multiple circularly-symmetric rings, along with a central void. This void was also visible on ground-based images after V838 Mon faded. The void is the likely reason why the light-echo was not seen until some time after the rise to second maximum. The angular separation of the concentric rings in the HST images indicate a $\sim$500 year recurrence time in the enhanced mass loss events.

Using the 790pc distance estimate and the peak brightness, we can derive $M_V = +4.45$ for V838 Mon in quiescence and $M_V = -4.35$ at peak outburst. The precise values depend on the exact amount of reddening, here estimated to be E_{B-V}=0.5. At galactic coordinates $l = 217.80$ $b = +1.05$, the height over the Galactic plane is just $z = 13$ pc. It is relevant to note that the progenitor of V838 Mon was not detected by Hα emission-line surveys in the region (these surveys discovered several faint emission line stars close to V838 Mon), and inspection of Palomar and SERC plates as well as results from many archival plates presented at this conference by [4] reveal absence of photometric variability in quiescence. Both Hα emission and variability would have supported an interactive binary nature of the precursor.

V838 MON AND THE STARS ERUPTING INTO COOL SUPERGIANTS (SECS)

In 1989 an erupting star in the Andromeda Galaxy (M31) developed a M-type cool supergiant spectrum at maximum, with pronounced P-Cyg profiles and Balmer lines in emission ([5], [6]). The progenitor was too faint to be identified and the event has been modeled by [7] in terms of a cool WD accreting at a very low rate from a companion and under such circumstances the entire WD could experience a thermonuclear runaway. The similarities with V838 Mon are the cool supergiant spectrum at maximum, the emission in the Balmer lines, the variable P-Cyg profiles and the faintness of the progenitor. However, the event in M31 peaked to $M_V = -9.95$, much brighter that the $M_V = -4.45$ reached by V838 Mon. Nevertheless, the light curve presented by [8] is quite similar to V838 Mon.

Another close match is Nova Sgr 1994 V4332 Sgr (Nova Sgr 1994 #1). As described in [9], V4332 Sgr displayed a flat outburst, dramatic increase in $(V - I)$ during its fade from maximum, an M-type spectrum at maximum and P-Cyg profiles. The progenitor was a K star close to the Main Sequence. The outburst duration was only 20 days, however, and perhaps this variable was not discovered until some time after the beginning of the outburst. Imagery during outburst as well as more recently does not show any signs of a light-echo surrounding V4332 Sgr.

There have been several possible novae reported over the past century that have been ascribed to Mira variability on the base of the spectrum in outburst. Perhaps some of these are similar objects to M31 RedVar, V4332 Sgr and V838 Mon. The latter may represent a new class of astronomical objects: stars erupting into cool supergiants but that never develop an optically thin phase (increasing excitation temperature and development of a nebular spectrum) like in classical novae. Their progenitors lies away from the post-AGB stars and appear close to the cool main sequence on the HR diagram. These stars, characterized by havy mass loss at least in the early outburst phases, could be binaries even if no evidence of this has been found (in V838 Mon the circumstellar dust envelope could be the relic of an AGB phase of an hypotetical companion to the F0 V star seen in quiescence). We suggest for this new class of astronomical objects the name of *stars erupting into cool supergiants* (SECS).

REFERENCES

1. Munari, U., Henden, A., Kiyota, S., et al., A&ALett 389, L51 (2002)
2. Henden, A., Munari, U., Schwartz, M.B., IAUC 7859 (2002)
3. Bond, H.E., Panagia, N., Sparks, W.B., Starrfield, S.G., Wagner, R.M., IAUC 7892 (2002)
4. Barsukova, E.A., Borisov, N.V., Goranskij, V.P., et al., this volume (2002)
5. Rich, R.M., Mould, J., Picard, A., et al., ApJ 341, L51 (1989)
6. Mould, J., Cohen, J., Graham, J.R., et al., ApJ 353, L35 (1990)
7. Iben, I.Jr, Tutukov, A.V., ApJ 389, 369 (1992)
8. Sharov, A.S., Sov. Astron. Lett. 19, 33 (1993)
9. Martini, P., et al., AJ 118, 1034 (1999)

On the Maximum Mass of C-O White Dwarfs

Inma Domínguez*, Oscar Straniero†, Jordi Isern** and Amedeo Tornambé†

*Dpto. Física Teórica y del Cosmos, Universidad de Granada, 18071 Granada, Spain
†Osservatorio Astronomico di Collurania, 64100 Teramo, Italy
**Institut de Estudis Espacials de Catalonia, Edif. Nexus-104, Gran Capitá, 08034 Barcelona, Spain

Abstract. The formation of massive C-O white dwarfs in the range 1.1 to 1.4 $M_\odot$ opens interesting channels for novae, supernovae and accretion induced collapse events. We will review the present situation concerning the maximum mass attained by the C-O core inside an AGB star, both following standard and non-standard evolution, and taking into account the observational constraints. We find that to form C-O cores with masses greater than 1.1 $M_\odot$, the 2nd dredge-up should be delayed and that the only obvious mechanism for that is rotation.

INTRODUCTION

As it is well known low and intermediate mass stars, $M \leq 8 M_\odot$, end their lives as C-O WDs [1], [2] and [3]. Stars with higher masses form, after the exhaustion of He in their centers, C-O cores greater than around 1.1 $M_\odot$ and this happens to be the critical mass for C ignition in a non-degenerate C-O core; the corresponding Main Sequence mass for that limit C-O core mass is called M_{up}.

To understand the final mass of the C-O core we have to analyze what happens during the Asymptotic Giant Branch phase (AGB). The structure, starting from the center, is composed by (1) a C-O core that cools down and contracts, increasing the level of degeneracy, (2) the He shell that advances outward in mass increasing the mass of the C-O core and (3) a more external H shell. During the first part of the AGB, the E-AGB, both burning shells are active shells, but as the C-O core contracts the temperature in the He shell increases, pushes the overlying layers and nearly extinguishes the H shell. At this point, the convective envelope penetrates inward and H is reignited starting the TP-AGB phase, in which the H and He shells are alternatively active. In the more massive models $M \geq 4 M_\odot$, the convective envelope penetrates the He core (2nd dredge-up) with the practical consequence of decreasing the final C-O core mass. As it is shown in Figure 1 the main growth of the C-O core occurs during the E-AGB phase, during the subsequent phase, the thermal pulses, the C-O core still increases but at a smaller rate.

MODELS, OBSERVATIONS AND RESULTS

All the calculations presented in this work have been done with the Evolutionary Code FRANEC, this is a 1D hydrostatic code, completely updated [4] and [5]. This code has

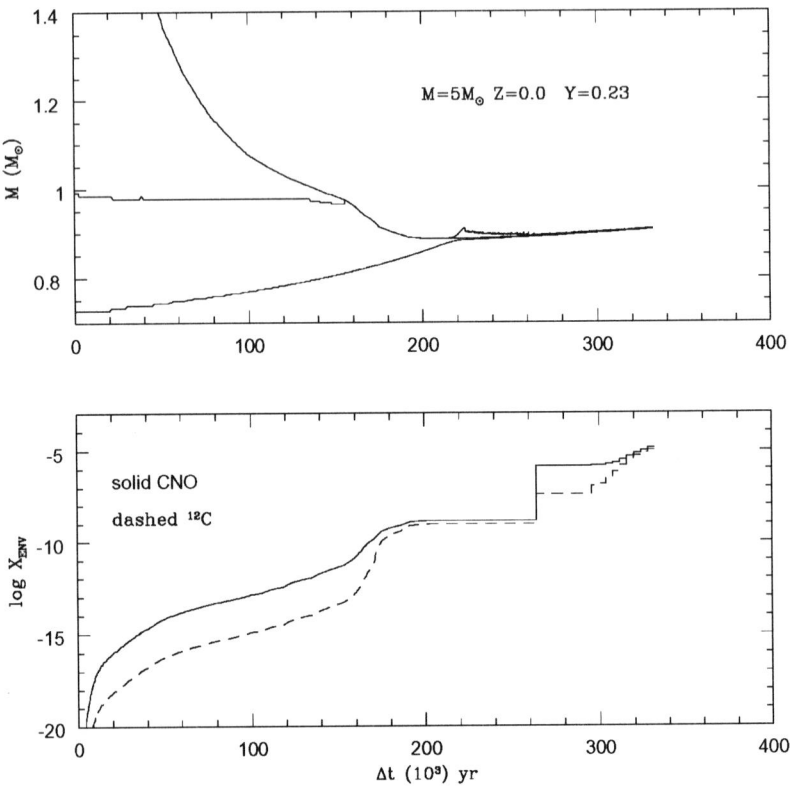

FIGURE 1. Upper panel: location, in mass coordinate, of the He and H shells and the base of the convective envelope for a 5 $M_\odot$, Z=0 model during the AGB (E-AGB and TP-AGB) phase. Lower panel: evolution of the surface abundances (mass fraction) during the AGB phase (2nd and 3rd dredge-up).

been used for a wide range of applications, from globular clusters to SN progenitors.

During the TP-AGB the mass of the C-O core still increases. We see in Figure 2 the evolution in time of the location of the He and H shells along the first 300 thousands years during the TP phase. The average rate at which the C-O core mass increases in this model is 10^{-7} $M_\odot$/yr. At this rate in 5 million years the C-O core would attain the Chandrasekhar mass; however, it is believed, based on observations, that the star losses the envelope before many pulses occurs.

The relevant observations are: (1) the maximum luminosity of AGB stars observed in the Galaxy and in the Magallanic Clouds combined with the fact that the luminosity is related to the C-O core mass and (2) the high mass loss rates observed. An increasing amount of IR (IRAS and ISO) and radio (submilimiter) observational data combined with models, including dust, give mass loss rates during the AGB phase, increasing from the 10^{-7} to 10^{-3}, with a mean value centered around 10^{-5} $M_\odot$/yr. Related with

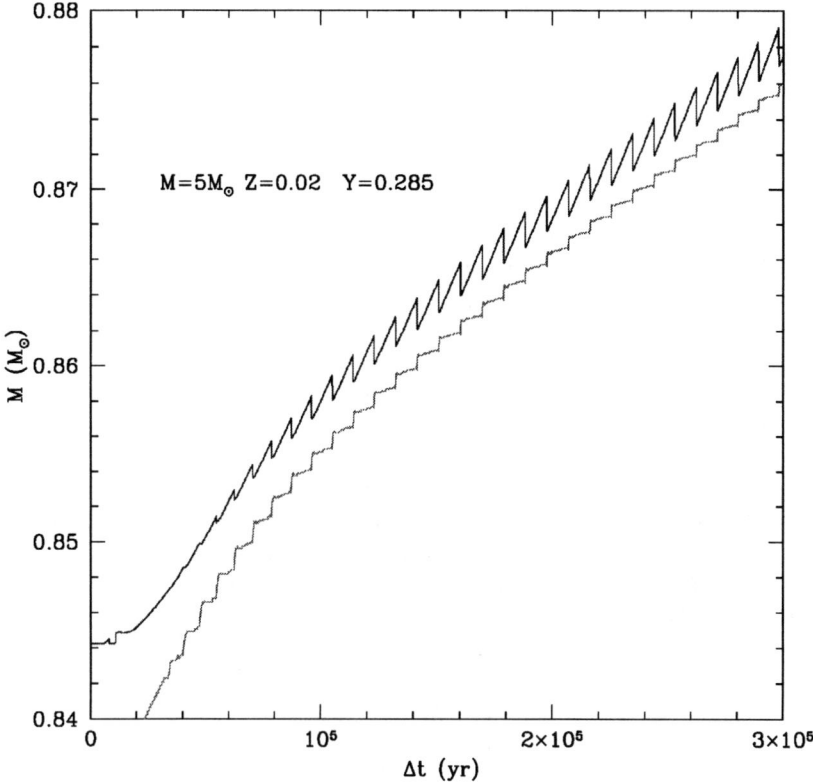

FIGURE 2. Location, in mass coordinate, of the He and H shells for a 5 $M_\odot$, Z=0.02 during the TP-AGB phase.

this, a longstanding problem is the physical mechanism responsible of the ejection of the envelope. I would like to note that [6] and [7] have found, in their computations based on hydrostatic evolutionary codes, evidence for the ejection of the envelope; after around 30-40 TPs, in mass losing models, a minimum in the ratio of the gas to radiation pressure occurs [7].

The relation between the final WD mass and the MS mass, the final to initial mass relation, is obtained both, from the observations in a semiempirical way and from the models. The C-O core masses attained in our computations [9] before all the envelope is lost and the AGB phase ends adopting the mass loss rate of [10] are in excellent agreement with the new relation given by [8]. I would like to remark that based on the observations there are no reasons to search for a relation different from a single-valued one and that for more massive stars the corresponding WD masses are higher. So, and this is important, few pulses are expected to occur.

We can obtain also the observed mass distribution of WDs. In recent samples of

TABLE 1. Rotating models at the end of the E-AGB (6.5 $M_\odot$ Z=0.02)

	f_c/f_{gr}	$M_{co}(M_\odot)$
Case 0	0.00	0.999
Case 1	0.10	1.120
Case 2	0.25	1.423

extreme-UV WDs, WDs with masses higher than $1.1 M_\odot$ have been identified: 2 WDs in the sample of [11] and 15 in the sample of [12]. These WDs could be O-Ne WDs, as [8] suggested or, alternatively, be the results of WD-merging. It would be very interesting to identify the composition and scenario of those WDs.

From the previous considerations it seems that to have a massive CO WD, we need a massive CO core at the 1st TP, the standard evolution gives a maximum C-O core mass of 1.1 $M_\odot$. For a further increase of the C-O core mass, mass-accretion from a companion would be necessary; however there are problems for this (see Piersanti et al. this volume and [13]) or to have a merger, but also in this case it is not clear at all that a massive C-O WD would form, the result could be a collapse (accretion induced collapse) or a thick disk around the C-O WD [14]. A long TP phase, increasing the mass pulse by pulse, would do it, but this seems to be excluded by the observations. This means that we may have to consider non-standard evolution to increase that limit of 1.1 $M_\odot$, for example avoiding C-ignition by cooling down the C-O core or delaying the inward penetration of the envelope, extending E-AGB phase in which, as we saw, the C-O core increases markedly.

With the first possibility in mind we include an extra sink of energy, axions [15], that, as the neutrinos carry out energy, the result was interesting but opposite to the expectations, the 2nd dredge-up is anticipated because the He shell has to supply for the energy losses, increasing its temperature, and the final C-O core mass is smaller than in the standard case. In another numerical experiment we just include the energy sink within the C-O core but also in this case the TP phase is anticipated because the C-O contracts faster leading a faster increase of the He shell temperature. The energy loss is proportional to the square of the axion mass.

Finally, we perform models including rotation, this was done in a very approximate way [16], just as a centrifugal force in the hydrostatic equation and so, the results should be considered as *qualitative*. In anycase the trend is well founded and clear, rotation delays the 2nd dredge-up due to the *lifting effect*, the contraction of the core is slower and slower is the increase in temperature of the He-shell. In Table 1 we show the mass of the C-O cores at the end of the 2nd dredge-up for the standard case (Case 0) and 2 models including rotation (Case 1 and Case 2), f_c/f_g is the ratio of the centrifugal to gravitational force. The 2nd dredge-up occurs later in the model in which the degree of rotation is higher and the final C-O core mass is higher. We think that rotation is probably the only way to obtain massive C-O cores.

SUMMARY

We search for mechanisms able to produce massive C-O WDs, we do not know if they have been observed for sure. We conclude that the best possibility is to delay the 2nd dredge-up as rotation does. Another possibility would be a *long* TP phase but observations favour a *short* TP phase, due to the observed high mass loss rates, the maximum luminosity of AGB stars and the initial to final mass relation. We may think that the C-O core mass could increase later by accretion, but this mechanism is not clearly succesfull, mergers are also candidates but merger simulations obtain different outcomes.

Other *ingredients* in the models, like convection, composition, microphysics (opacities, equation of state, nuclear reaction rates and the like) would change the initial to final mass relation but not the maximum mass of the C-O core.

We conclude that the only obvious way to achieve C-O WDs with masses greater than 1.1 $M_\odot$ is rotation.

REFERENCES

1. Becker, S.A., Iben, I.J. ApJ **232**, 831 (1979).
2. Reimers, D., Koester, D., A&A **116**, 341 (1982).
3. Weidemann, V., Koester, D. A$A **121**, 77, (1983).
4. Chieffi, A., Limongi, M., Straniero, O. ApJ **502**, 737 (1998).
5. Straniero, O., Chieffi, A., Limongi, M. ApJ **490**, 425 (1997).
6. Sweigart, A.V. IAUS **91**, 533 (1997).
7. Straniero, O., Limongi, M., Chieffi, A., Domínguez, I., Busso, M., Gallino, R., Mem. SAIt. **71**, 719 (2000).
8. Weidemann, V., A&A **363**, 647 (2000).
9. Domínguez, I., Chieffi, A., Limongi, M., Straniero, O., ApJ **524**, 226 (1999).
10. Groenewegen, M.A.T., de Jong, T. A&A **283**, 463 (1994).
11. Napiwotski, R., Green P.J., Saffer, R.A. ApJ **517**, 399 (1999).
12. Vennes, S., ApJ **525**, 995 (1999).
13. Piersanti, L., Cassisi, S., Icko, I.Jr., Tornambé, A., ApJ **535**, 932 (2000).
14. Guerrero, J., Ph.D. thesis (2001).
15. Domínguez, I., Straniero, O., Isern, J., MNRAS **306**, L1 (1999).
16. Domínguez, I., Straniero, O., Tornambé, A., Isern, J. ApJ **472**, 783 (1996).

The evolution of intermediate mass close binary systems: scenarios leading to novae

Enrique García–Berro*, Pilar Gil–Pons† and James W. Truran**

*Institute for Space Studies of Catalonia, c/Gran Capitá 2–4, Edif. Nexus 104, 08034 Barcelona, Spain, (e-mail: garcia@fa.upc.es)
†Departament de Física Aplicada, Universitat Politècnica de Catalunya, c/Jordi Girona s/n, Mòdul B-4, Campus Nord, 08034 Barcelona, Spain, (e-mail: pilar@fa.upc.es)
**Department of Astronomy and Astrophysics, 5640 S. Ellis Ave, Chicago, IL 60637, USA (e-mail: truran@nova.uchicago.edu)

Abstract. We revisit the problem of the determination of the frequency of occurrence of ONe white dwarfs in galactic novae. In doing that we use our evolutionary scenarios for intermediate mass close binary systems. We also consider observational selection effects. Our results indicate that approximately 1/3 of the observed novae should host an ONe white dwarf.

INTRODUCTION

The determination of the frequency of occurrence of ONe white dwarfs in Galactic classical novae systems has been already considered by several authors — see, for instance, Truran & Livio (1986), and Ritter et al. (1991). If the evolution of the primary in the binary system were equivalent to that of single stars, and the observational selection effects were negligible, one should expect a peak in the mass distribution of white dwarfs surrounding the canonical value of $0.6 M_\odot$. Hence, we would expect to find a small number of oxygen-neon white dwarfs in novae systems. However, there is a wealth of observations showing that the actual mass distribution of white dwarfs in novae is skewed towards higher masses and, in fact, is peaked around $1.0 M_\odot$ (Webbink, 1990). Consequently, an important fraction of novae systems could host an ONe white dwarf. Livio & Soker (1984), Ritter & Burkert (1986) and Ritter et al. (1991) found that observational selection effects in novae systems play a key role. On the other hand, another aspect which has an influence on the mass distribution in novae is the evolutionary history of the binary system previous to the formation of the nova. Here we recompute the frequency of ONe white dwarfs in novae by taking into account both the observational selection effects and our recent evolutionary sequences for binary systems with intermediate–massive primaries (Gil–Pons & García–Berro, 2001).

SCENARIOS LEADING TO NOVAE

Our systems undergo the so called case BB' of mass transfer. The first Roche lobe overflow (RLOF) from the primary component occurs when hydrogen burning has set

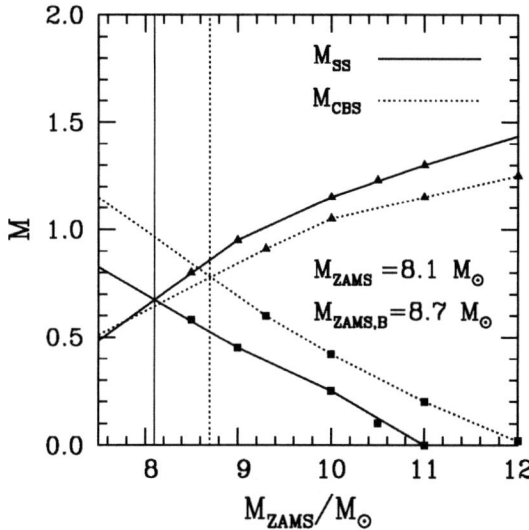

FIGURE 1. Size of the CO cores at the begining of carbon burning and mass point at which carbon is ignited.

in a shell and a deep convective envelope surrounds its helium core. Mass loss in the presence of convection induces unstable mass transfer between the components and leads to the formation of a common envelope. This, in turn, induces orbital shrinkage. The primary is deprived of its hydrogen-rich envelope during the first RLOF, and most of the mass is lost from the system. The second RLOF episode leaves a remnant which is composed by an ONe core surrounded by a CO buffer and a thin helium layer. Reversal mass transfer occurs when the secondary reaches giant dimensions, whether during the H- or He-burning shell phase or during its carbon burning phase. This mass transfer process is also unstable and is accompanied by another common envelope phase that allows the orbital period to decrease down to a value of the order of hours.

In Figure 1 we show the size of the CO core just before carbon burning sets in and the mass point at which carbon is ignited, both for the single case (solid line) and for close binary evolution (dotted line). Carbon burning starts off-center for the low-mass range and, as the initial mass of the star increases the mass point at which carbon is ignited approaches the center. The intersection of both curves determines the minimum mass for carbon ignition to take place. However, this mass cannot be taken as the limiting mass separating CO from ONe white dwarfs as the fact that carbon ignition is able to proceed does not necessarily imply that burning is going to be extended enough in the CO core to

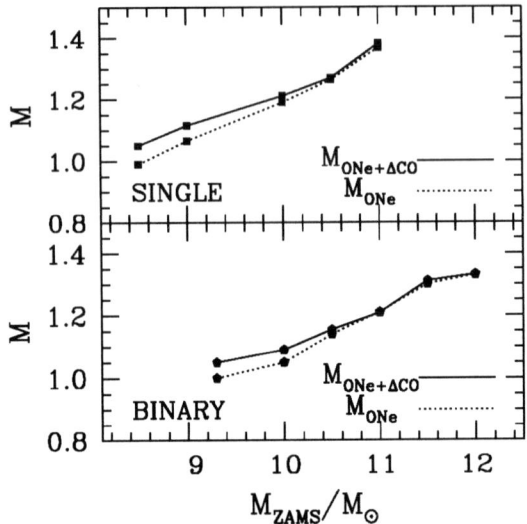

FIGURE 2. Size of the final cores as a function of the ZAMS mass for single and binary star evolution.

TABLE 1. White dwarf masses as a function of ZAMS masses.

M_{ZAMS}	M_{ONe}	$M_{ONe+\Delta CO}$
9.3	1.00	1.07
10.0	1.05	1.09
10.5	1.14	1.15
11.0	1.21	1.22
11.5	1.30	1.31
12.0	1.33	1.33

change drastically the composition of the core and form an ONe white dwarf. Primary stars of close binary systems yield remnants whose mass is lower than that resulting from single star evolution. Figure 2 shows the relationship between the initial mass of the primary and the mass of the white dwarf (solid line) and the size of the ONe core (dotted line) for both the single and the binary cases. These two quantities, for the sake of completeness, are also given in Table 1.

OXYGEN–NEON NOVA RATES

The observed fraction of nova outbursts in which an ONe white dwarf is involved depends on the relative frequency of ONe white dwarfs in binary systems and on the recurrence time between the outbursts. A white dwarf undergoes a thermonuclear flash when a certain critical pressure, P_{crit}, is reached. A good approximation to this quantity is

$$P_{crit} = \frac{GM_{WD}\Delta m}{4\pi R_{WD}^4} \qquad (1)$$

where M_{WD} and R_{WD} are the white dwarf mass and radius, respectively, and Δm is the mass of the accreted layer (Truran & Livio, 1986). The amount of accreted mass necessary for the ignition to take place can be obtained from a fit to full numerical simulations (Fujimoto, 1982; Prialnik et al., 1982):

$$\Delta m_{ign} = \left(\frac{M_{WD}^4}{R_{WD}}\right)^{-\alpha} \dot{M}_{WD}^{-\beta} \qquad (2)$$

where α and β are parameters to be fitted. The calculations show only a weak dependence in the mass accretion rate, $\dot{M}_{WD}$ and, therefore, we will assume $\beta = 0$. On its hand, α is typically $\simeq 1.0$ (Truran & Livio, 1986). However, numerical calculations which include a detailed treatment of diffusion (Kovetz & Prialnik, 1985) show that $\alpha = 0.7$ may be a more appropriate value. Therefore, we will perform calculations for both $\alpha = 1$ and $\alpha = 0.7$.

The frequency of recurrence between outbursts can be computed as the ratio between the mass accretion rate and the mass of the layer required to reach P_{crit}. Following Ritter et al. (1991) and Ritter & Burkert (1986), a selection function is defined as:

$$S_N(M_{WD}) = L^n(M_{WD})\left(\frac{M_{WD}}{R_{WD}^4}\right)^{-\alpha} \qquad (3)$$

where $L(M_{WD})$ is the luminosity of the hydrogen burning shell (Kippenhahn 1981), and $R(M_{WD})$ is determined using the mass–radius relationship of Eggleton (1982). The exponent n in Eq. 3 accounts for the possible spatial distribution of novae. For a disklike distribution the exponent is $n = 1/2$, whereas for a volume-limited sample we have $n = 0$ and for an isotropic distribution $n = 3/4$ must be adopted (Ritter et al. 1991). We will perform calculations for two nova distributions corresponding to $n = 0$ and $n = 0.5$.

Finally, the probability of observing an outburst from a nova of a certain mass will be given by

$$P(M_{WD}) = S_N(M_{WD}) \times M_{ZAMS}^{-2.35} \qquad (4)$$

where we have assumed a Salpeter-like IMF. A more precise treatment of the problem should include a statistical study of the orbital parameters of the binaries leading to novae. We have not done such a study, but work in this direction is in progress. With all these inputs our results for the frequencies of occurrence of novae of the different types are shown in table 2.

TABLE 2. Frequency of occurrence of novae of different types

		P_{CO}	P_{ONe}
$\alpha = 1$	$n = 0$	0.65	0.35
	$n = 0.5$	0.70	0.30
$\alpha = 0.7$	$n = 0$	0.65	0.35
	$n = 0.5$	0.72	0.28

CONCLUSIONS

We have computed detailed evolutionary scenarios for the formation of ONe novae in intermediate mass close binary systems. We have found that the minimum mass allowing for extensive carbon burning in the CO core is $\simeq 9.3\,M_\odot$. The mass of the resulting white dwarf is $\simeq 1.1\,M_\odot$. We have also derived the initial to final mass relationship and used our evolutionary sequences to compute the relative fraction of ONe novae. In doing that we have followed closely the treatment of Ritter et al. (1991) and we have found that about 35% of the observed novae are expected to host ONe white dwarfs for a volume limited sample. This fraction decreases down to 30% for a disklike distribution. We have not performed a detailed statistical study of the evolution of the orbital parameters of the systems that evolve into novae, but work in this direction is under way. Finally, one of our most important findings is the presence of a CO buffer of $\simeq 5 \times 10^{-2}\,M_\odot$ surrounding the ONe core. This buffer is almost deprived of ^{22}Ne. Hence, although the occurence of ONe novae is high, the first nova outburst will not show an enhancement in Ne. Instead, since this CO buffer is rich in ^{25}Mg we expect that important amounts of magnesium should be found in the ejecta during the early phases. We estimate that the number of nova outbursts required to remove the CO buffer is typically $\sim 4 \times 10^3$. Once the CO buffer is removed the nova outburst will, indeed, show an enhancement in Ne.

ACKNOWLEDGMENTS

This work has been supported by the MCYT grant AYA2000–1785.

REFERENCES

1. Gil–Pons, P., & García–Berro, E., A&A, 2001, **375**, 87.
2. Kippenhahn, R., 1981, A&A, **102**, 293
3. Kovetz, A., & Prialnik, D., 1995, ApJ, **291**, 812.
4. Livio, M., & Soker, N., 1984, MNRAS, **208**, 783.
5. Prialnik, D., Livio, M., Shaviv, G., & Kovetz, A., 1982, ApJ, **257**, 312.
6. Ritter, H., & Burkert, A., 1986, A&A, **158**, 161.
7. Ritter, H., Politano, M., & Livio, M., 1991, ApJ, **376**, 177.
8. Truran, J., & Livio, M., 1986, ApJ, **308**, 721.
9. Webbink, R.F., 1990, in IAU Colloq. 122, *The Physics of Classical Novae*, ed. A. Casatella (Berlin: Springer)

Abnormal CNO abundances in magnetic cataclysmic variables

M. Mouchet*, J.M. Bonnet-Bidaud[†], M. Abada-Simon*, K. Beuermann**,
D. de Martino[‡], R. Ferlet[§], R. Fried[¶], B. Gänsicke**, S. Howell[∥], A.
Lecavelier[§], K. Mukai[††], D. Porquet[†], E. Roueff* and P. Szkody[‡‡]

* *Observatoire de Paris, Meudon, F*
[†] *Service d'Astrophysique CEN, Saclay, F*
** *Universtäts-Sternwarte, Göttingen, D*
[‡] *Astronomical Observatory Capodimonte, Naples, I*
[§] *Institut d'Astrophysique de Paris, F*
[¶] *Braeside Observatory, Flagstaff, AZ, USA*
[∥] *Astrophysics Group, Planetary Science Institute, Tucson, USA*
[††] *Laboratory for High Energy Astrophysics, NASA/GSFC, USA*
[‡‡] *University of Washington, Seattle, USA*

Abstract. We present far-UV observations of three peculiar magnetic cataclysmic variables (MCVs), BY Cam, V1309 Ori and AE Aqr, obtained with the FUSE satellite. Previous IUE spectra of these three objects revealed quite unusual resonance lines compared to other MCVs: an intense NV line and a weak CIV line. The FUSE spectra of these sources exhibit broad OVI lines as well as a strong NIII line at 991Å, while the CIII 1175Å line is nearly absent. Photoionisation models fail to produce the observed line intensities. This confirms non-solar CNO abundances. We discuss possible origins for these peculiar abundances, including the signature of nova thermonuclear runaways (TNR) at the surface of the magnetized white dwarf.

THREE PECULIAR MEMBERS AMONG MCVS

Four MCVs have revealed strong NV and weak CIV resonance emission lines. Three of them are synchronized systems (polars) : BY Cam [1], V1309 Ori [29] and MN Hya [23]. The fourth source, AE Aqr [9], is an intermediate polar. The pathological UV emission lines can result from peculiar ionization conditions [1] or from abnormal chemical abundances, following an unrecorded nova explosion or revealing the composition of an evolved secondary [19]. We obtained FUSE spectra of BY Cam, V1309 Ori and AE Aqr which give access to the OVI resonance line for the first time.

In addition to their pathological UV lines, all of the three observed sources have peculiar properties. The polar BY Cam is slightly desynchronized [24], and emits a hard X-ray spectrum (kT $\sim$ 20 keV, [10]). V1309 Ori is a long orbital period (8h) system with a slightly evolved secondary [4], and it has an extremely high soft-to-hard X-ray ratio luminosity [15]. AE Aqr is also a long orbital period (10h) binary but its white dwarf (WD) rotates very fast (33s). This prevents an efficient accretion onto the white dwarf, which explains the low observed X-ray luminosity (see review in [30]).

FAR-UV FUSE AVERAGE SPECTRA

The FUSE spectra were obtained in the range 910-1180Å, at a resolution of 0.032Å, using the time-tag mode and the LWRS aperture (30"x30") [18]. The total exposure times were of 20 ks, 12 ks and 40 ks respectively for BY Cam, V1309 Ori and AE Aqr with a nearly complete coverage of the orbital phases (P_{orb} = 3.35 h, 7.98 h, 9.88 h respectively). The average spectra shown in Fig.1 reveal broad emission lines of OVI, HeII, NIII and SIV. Interstellar lines are also present in BY Cam (Ar, FeII, SiII) and in AE Aqr (HI, CII, OI, NII). In addition, strong H_2 molecular lines (Lyman and Werner series) are seen in BY Cam. Their intensities correspond to a column density $N_{H2} \sim 10^{19}$ cm^2 (Mouchet et al. in preparation).

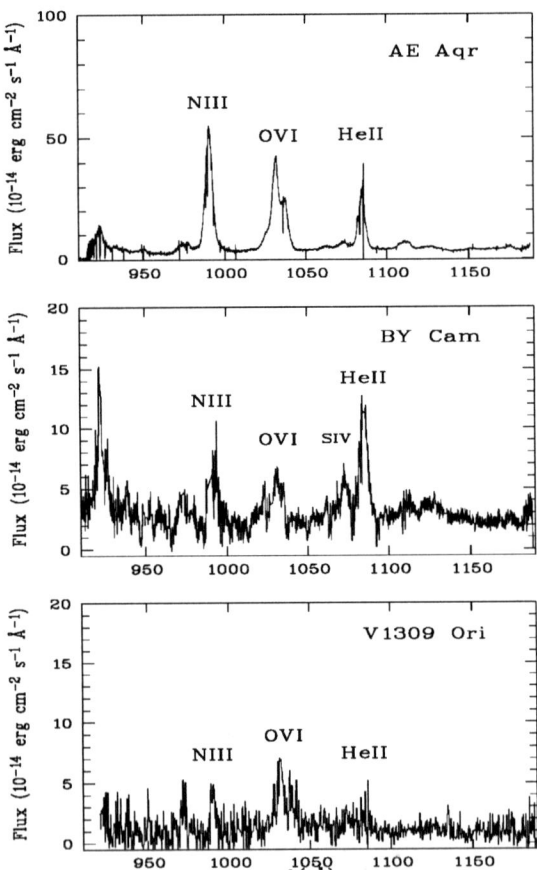

FIGURE 1. Average FUSE spectra of AE Aqr, BY Cam and V1309 Ori and identification of the strongest lines. Note the absorption H_2 bands in BY Cam and interstellar lines in AE Aqr.

In Table 1, measurements of the resonance lines are reported for the three targets as well as for AM Her, the prototype of the polar class. AM Her was observed in the far-UV with Orfeus [17], HUT [5], and recently with FUSE [8]. The intensities of the broad lines in AM Her are those derived from HUT and Orfeus spectra. We note that no narrow components are detected in BY Cam and AE Aqr contrary to AM Her. The OVI/CIV, NV/CIV and NV/OVI ratios are stronger in all three sources than in AM Her.

TABLE 1. Line intensities in 10^{-13} erg s^{-1}cm^{-2} and line ratios

source	C IV	N V	O VI	NV/CIV	OVI/CIV	OVI/NV
BY Cam (FUSE + IUE)	2.9	15.0	2.0	5.2	0.70	0.13
V1309 Ori (FUSE + HST)	2.4	17.4	3.7	7.3	1.5	0.21
AE Aqr (FUSE + IUE)	26	162	27	6.2	1.0	0.17
AM Her (Orfeus + HUT broad)	59	7.3	20	0.12	0.34	2.7

MODELISATION OF THE BROAD OVI LINE

Broad emission resonance lines are thought to be produced in the accretion flow which is photoionised by the X-rays. The photoionisation state is related to the ionisation parameter: $U = \frac{Q(1Ryd - 100MeV)}{4\pi R^2} \frac{1}{n_H c}$, where Q is the number of ionising photons and R is the distance of the medium, of density n_H, to the ionisation source. The line intensities of CIV, NV and OVI have been computed using the photoionisation code Cloudy [3] for two typical X-ray spectra emitted by MCVs, a black body (Tbb = 30 eV) and a bremsstrahlung (Tbr = 10 keV), for a density of $n_H = 10^{12}$ cm^{-3} and a column density of $N_H = 10^{21}$ cm^{-2}, assuming solar abundances. The line intensity ratios of AM Her can be reproduced within a factor three for both types of ionising spectrum and for a U parameter in the range $10^{-3} - 3\,10^{-2}$, but these models cannot reproduce the ratios for the three other sources.

More detailed computation of the resonance lines has been done for BY Cam by taking into account the peculiar ionisation spectrum of BY Cam (sum of a 20 keV bremsstrahlung and a 50 eV black body with a bolometric luminosity ratio of 0.1), the density gradient in the column computed in the framework of a dipole geometry and assuming free-fall velocities in an homogeneous column extending up to 50 R_{WD}. The contribution of the lowest parts of the column to the line intensities is negligible compared to the outer regions (Fig.2). Neither the observed luminosities nor the line ratios can be reproduced in the case of solar composition. For illustration, we have reported results obtained for non-solar abundances (C depletion of 25, N enhancement of 8 and O depletion of 4) which give agreement within a factor two for the luminosities (Fig. 2). Corresponding N/C, O/C and O/N ratios of 5.7, 0.54 and 0.09 are very close to the observed ones. An overabundance of nitrogen is consistent with the presence of strong NIII (990Å), and NIV (1718Å) lines as well as the detection of the NVI-NVII X-ray lines in the XMM spectrum of BY Cam [21]. Similarly a depletion of carbon is compatible with the weakness of the CIII (977Å, 1175Å) lines.

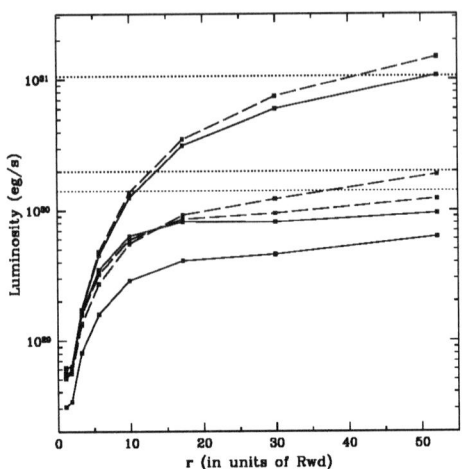

FIGURE 2. Computed luminosities (erg/s) of NV, CIV and OVI formed in eight slabs of the accretion column, up to a distance of 50 R_{WD}, for ionisation parameters corresponding to BY Cam and for two cases: solar abundances (dashed lines for CIV, OVI and NV from top to bottom) and C:25 - Nx8 - O:4 abundances (full lines for NV, CIV and OVI from top to bottom). Horizontal dotted lines are observed luminosity values of NV, CIV and OVI from top to bottom. The cumulated luminosities are roughly in agreement with BY Cam measurements for the non-solar case.

POSSIBLE ORIGINS OF NON-SOLAR ABUNDANCES

The peculiar chemical composition found in four MCVs can either reflect the composition of the secondary or result from an unnoticed TNR at the white dwarf surface.

The first possibility requires an evolved secondary. Due to strong mass loss, stripping of the upper layers of the secondary might reveal the CNO processed core. However recent computations of core evolution indicate that only few secondaries in CVs evolve prior to contact [7], although computed evolutionary paths for long orbital periods, as applied to the soft X-ray transient XTE J1118+480 also showing a high NV/CIV ratio, can produce partial nuclear evolved secondaries [2,6,7]. Such paths might be relevant for V1309 Ori and AE Aqr, for which observational clues of slightly evolved secondaries have been found.

The second possibility which attributes the peculiar abundances to a nova explosion is appealing, although most of classical novae exhibit CNO redistribution, with global metal enrichment ($Z/Z_{\odot}$ up to 45) [14], indicating dredge-up of matter from the underlying CO core [27]. Recent 2D calculations of turbulent nuclear burning indeed predict self-enrichment of the envelope composition [11], as do shear flow instabilities [22]. Multicycle models show that a large quantity of C is reprocessed into N while O might be non affected or almost completely destroyed [13]. In order to produce the depletion of C, as observed in recurrent novae (f.i. U Sco), high WD mass and high accretion rates are required leading to very fast novae [20]. Since no obvious nova eruption was recorded for any of the three sources, an only very weak TNR could have happened

[25] or steady nuclear burning phases, as they are observed in supersoft X-ray sources, including one intermediate polar [12]. Note that the WD spectrum of U Gem reveals C depletion, explained in terms of weak TNR contaminating the secondary during common envelope phase with depleted carbon re-accreted during dwarf nova (DN) events [26]. Several other DN show signatures of CNO processing in the WD spectra and interestingly two DN also exhibit strong NV and very weak CIV lines (see texts by Sion and by Szkody in these proceedings).

Enhancement of N and depletion of C can also result from evolutionary models of semi-detached systems in which the secondary surface is polluted by re-accretion of material ejected during nova outbursts [16]. However this change of metallicity is less sensitive for Pop. I CVs [28].

To reinforce the hypothesis of peculiar abundances in these objects, the chemical composition of the underlying WD atmosphere should be inferred from high resolution spectra obtained during low brightness states. Similarly, knowledge of the surface chemical composition of the secondary requires spectra in the near IR (CN bands) and in the IR (CO bands) with a possible isotopic determination.

REFERENCES

1. Bonnet-Bidaud J.M., Mouchet M., *A&A*, 188, 89 (1987)
2. Ergma E., Sarna M.J., *A&A*, 374, 195 (2001)
3. Ferland G.J., Korista K.T., Verner D.A., et al., *PASP*, 110, 761 (1998)
4. Garnavich P.M., Szkody P., Robb R.M et al. *ApJ*, 435, L141 (1994)
5. Greeley B., Blair W., Long K., Raymond J., *ApJ*, 513, 491 (1999)
6. Haswell C.A., Hynes R.I., King A.R., Schenker K., *MNRAS*, 332, 928 (2002)
7. Howell S.B., *PASJ*, 53, 675 (2001)
8. Hutchings J.B., Fullerton A.W., Cowley A.P., Schmidtke P.C., *AJ*, 123, 2841 (2002)
9. Jameson R.F., King A.R., Sherrington M.R., *MNRAS*, 191,559 (1980)
10. Kallman T.R., Mukai K., Schlegel E.M., Paerels F.B., *ApJ*, 466, 973 (1996)
11. Kercek A., Hillebrandt W., Truran J.W., *A&A* 337, 379 (1998)
12. King A.R., Osborne J.P., Schenker K., *MNRAS*, 329, L43 (2002)
13. Kovetz, A., Prialnik, D., *ApJ*, 477, 356 (1997)
14. Livio M., Truran J.W., *ApJ*, 425, 797 (1994)
15. de Martino D., Barcaroli R., Matt G. et al., *A&A*, 332, 904 (1998)
16. Marks P.B., Sarna M.J., *MNRAS*, 301, 699 (1998)
17. Mauche C.W., Raymond J.C., *ApJ*, 505, 869 (1998)
18. Moos H.W., Cash W.C., Cowie L.L. et al., *ApJ*, 538, L1 (2000)
19. Mouchet M., Bonnet-Bidaud J.M., Hameury J.M., in "Accretion-Powered Compact Binaries", ed C. Mauche (CUP), 247 (1990)
20. Politano M., Starrfield S., Truran J.W. et al., *ApJ*, 448, 807 (1995)
21. Ramsay G., Cropper M., astro-ph/0203444 (2002)
22. Rosner R., Alexakis A., Young Y.N. et al., *ApJ*, 562, L177 (2002)
23. Schmidt G.D., Stockman H.S., *ApJ*, 548, 410 (2001)
24. Silber A., Bradt H.V., Ishida M. et al., *ApJ* 389, 704 (1992)
25. Sion E.M., *PASP*, 111, 532 (1999)
26. Sion E.M., Cheng F.H., Szkody P. et al., *ApJ*, 496, 449 (1998)
27. Starrfield S., Truran J.W. Sparks, *New Ast. Rev.*, 44, 81 (2000)
28. Stehle R., Ritter H., *MNRAS*, 309, 245 (1999)
29. Szkody P., Silber A., *AJ*, 112, 239 (1996)
30. Welsh W.F., in Proc. of "Annapolis workshop on MCVs", eds C. Hellier & K. Mukai, ASP Conf. Series 157, 357 (1999)

Did EY Cyg go through a Nova explosion?

G. Tovmassian*, M. Orio†**, S. Zharikov*, J. Echevarría‡, R. Costero‡ and R. Michel*

*Observatorio Astronómico Nacional, Instituto de Astronomía, UNAM,
P.O. Box 439027, San Diego, CA 92143-9027, USA
†INAF, Osservatorio Astronomico di Torino, Strada Osservatorio 20,
Pino Torinese (TO), I-10025 Italy
**Astronomy Department, University of Wisconsin, 475 N. Charter Str., Madison WI 53706, USA
‡Instituto de Astronomía, UNAM, Apartado Postal 70-264, Ciudad Universitaria, Mexico City,
D.F., Mexico 04510

Abstract. We studied the environment of the Cataclysmic Variable EY Cyg with direct images obtained in Hα. We were motivated by the very unusual light curve of this extremely long orbital period and extremely low inclination CV, which could be an indication of a common envelope around the object. The complex Hα emission at the background complicates the identification of nebular emission associated with EY Cyg.

INTRODUCTION

EY Cyg is a relatively bright Dwarf Nova (DN), studied back in 1962 by Kraft [8]. However, since then not much has been done about this object, probably owing to the low inclination angle of its orbital plane. AAVSO archives indicate that it undergoes fairly regular outbursts with average cycle duration of 2000 days (10 times longer than it was previously reported by Piening [11]).

It has been detected by *ROSAT* in soft X-Rays with energies between 0.4 and 2.2 KeV [10] and proved to be one of the brightest DN on the sky.

The first orbital period determination was made by Hacke and Andronov [7] from photographic observations, with a period of 0.181228 days. Later Sarna et al. [12], based on R and I photometric observations carried out during two nights, found a larger orbital period with probable values of 0.2630 or 0.2185 days. The authors argue in favor of the latter value by claiming a spectral type of dM2-dM3 from the spectroscopy by Smith et al. [13].

But recently high resolution Echelle spectroscopy by Costero et al. [4] and Echevarria et al. [6] showed that the radial velocity curve of the emission and absorption components, yield semiamplitudes $K_{em} = 33 \pm 2$ km s^{-1} and $K_{abs} = 54 \pm 1$ km s^{-1}. The orbital period is $P_{orb} = 0.45932590 \pm 0.00000002$ days, a much larger value than previously reported. We also found a K0 to K5 phase dependent spectral type for the secondary star. The observed rotational velocity of the secondary is $V_{rot} \sin i = 32 \pm 13$ km s^{-1}, consistent with the predicted rotational velocity assuming that EY Cyg B fills its Roche-Lobe and co-rotates with the binary. The masses and the binary separation of the components are $M_W \sin^3 i = 0.01770 \pm 0.0016 M_\odot$, $M_R \sin^3 i = 0.00950 \pm 0.0014 M_\odot$, and

$a \sin i = 0.7872 \pm 0.0067 R_\odot$. They also conducted CCD VRI photometry at the same time as the spectroscopic runs. From the Chandrasekhar limit, the observed spectral type of the secondary, as well as the light curves, very tight limits can be imposed on the inclination angle of this cataclysmic variable, with values between 13 to 17 degrees. The most likely value is $i = 16°$, consistent with an analysis of the mass and radius of the secondary star. For this value $M_R = 0.59 M_\odot$, $M_W = 1.26 M_\odot$ and $a = 3.03 R_\odot$. These results imply that EY Cyg B is in fact a K0 star with a mean radius 1.6 times greater than a normal main sequence star for the same mass, with its surface probably heated by the accretion disk and/or the white dwarf.

RATIONALE

The newly determined system parameters of EY Cyg immediately place it at the very edge of any statistical sampling of DNe and Cataclysmic Variables (CV) in general. Among 300+ CVs [5], only a dozen are known to have periods similar to or exceeding that of EY Cyg, most of them being novalikes. Among them GK Per, a magnetic CV that had a Classical Nova (CN) outburst in the past. The 2000+ day outburst frequency also places the object among heavyweights, that according to the well known Kukarkin and Parenago [9] relation should produce outbursts with amplitudes rather similar to recurrent Novae (RN) than DN. The estimated mass of the White Dwarf(WD), $M_W \geq 1.2 M_\odot$, is also typical of a CN candidate or even a type Ia SN progenitor.

Finally the highly variable and unusual light-curves that are difficult to explain in terms of the usual DN configuration prompted us to seek an explanation in the frame of a common envelope scenario. The lightcurves obtained at the 1.5m telescope of SPM (México) at various epochs are presented in Figure 1. They are separated by colors (V & I) and also by shapes of the curve in different panels. The lightcurves are folded according to the spectroscopic phases. There are a few patterns that remain constant in the light curves. The highest maximum occurs always at phase 0.2. And there is a shallow and narrow minimum at phase 0.0 in all observations prior to the end of August 2001. There is a broader and deeper minimum at phase 0.5, which disappears in the latest data obtained some 70 days before the outburst.

Generally, the light curve seems to be shifted by 0.5 phase in comparison to the usual DN light curve with similar parameters. One would expect a maximum at the phase 0.7-0.9 when the hot spot is viewed under an optimal angle. Also, the minimum at phase 0.0 is usually deeper than at phase 0.5 due to the ellipsoidal variations of a Roche-lobe distorted and irradiated secondary. We would even assume that some strange error occurs in our calculation of phases if the photometric observations of July 99 were not done *simultaneously* with the spectroscopy.

However it is possible to find a plausible solution for ellipsoidal variations of the non-irradiated secondary in the models of Avni and Bahcall [1] and Bochkarev et al. [2], using the tables and graphs of the latter, where the minimum at phase 0.5 exceeds in depth the minimum at 0.0. The gravity darkening effect is more pronounced in the direction of inner Lagrangian point L_1, and in low inclination systems the observer sees a larger projected area at phase 0.0 than at 0.5. Thus, the depth of minima at the latter

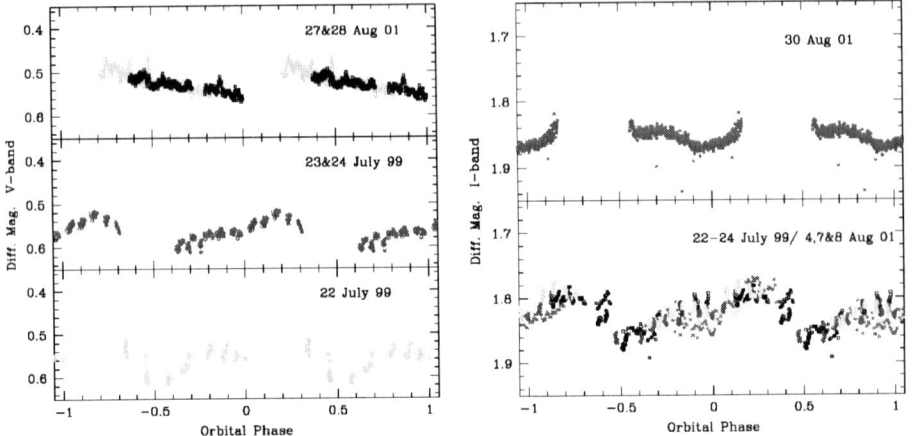

FIGURE 1. The light curves of EY Cyg grouped by the shape of the lightcurve and filters are presented in various panels. Corresponding dates and filters are marked.

phase becomes somehow deeper. The strange inclination of the ligthcurves of EY Cyg and their strong variability does not allow the determination of the inclination from these ellipsoidal variations, as is often done in various LMXB and CVs.

We may explain the disappearance of the minima at phase 0.5 in the latest observations (28-30 August 2001), if we assume that in quiescence the secondary of EY Cyg was not irradiated until immediately before the outburst. This, however, contradicts our findings derived from the spectral observations, that the face of the secondary shows earlier spectral type than the back. Nevertheless, even assuming absence of irradiation, we are left with an additional source of light causing disparity of maxima at phases 0.25 and 0.75.

DIRECT IMAGING

We observed EY Cyg in narrow filters ($H\alpha$ and the corresponding continuum) with the 3.5 m WYIN telescope at Kitt Peak. We obtained a series of short exposure so as not to saturate the relatively bright object. The images were co added to study the fainter nebular features. The seeing was not bad and with the superb adaptive optics of the telescope we reached FWHM of stellar objects of 0.6-0.7 arcsec. The images were reduced (bias, flatfield, combined) using MIDAS routines. The seeing in continuum images was slightly worse (about 0.1 arc sec) than in $H\alpha$. We had to do a Gaussian filtering to degrade the $H\alpha$ images to the level of the continuum in order to do subtraction.

The surroundings of EY Cyg show complex $H\alpha$ clouds common to Cygnus. In Figure 2 the large image of a mosaic covering an 8×8 arcmin area after continuum subtraction is presented. The zoomed in images of the immediate surroundings of EY Cyg are presented in Figure 3, where the location of the object is marked. The star is situated at the rim of a nearly circular ring of Hα emission. We checked the possibility that EY Cyg has the proper motion required to move it to the center of that ring in a reasonable time stretch, but could not find any indication of a significant positional shift. It is then very unlikely that this ring is associated with our object.

The other suspicious structure that can be detected in the immediate surrounding of EY Cyg is in the form of faint arch. It is not homogeneous. The size is about 25 arcsec. Based on the spectral type of the secondary, the minimum distance to EY Cyg is 250 pc. The upper limit is probably 700 pc, otherwise the object would be embedded in the background Cygnus super-bubble and would hardly sustain its X-ray luminosity [3]. Thus, a reasonable distance estimate to EY Cyg is around 400–500 pc. In this case the arch would be just slightly larger and older than the well known Nova shells of DQ Her and/or GK Per.

The morphology of this formation is a bit different from the larger scale strips crossing the region, and EY Cyg is close to the center of the arch. So it may well be associated with the object. However, we lack any real evidence to prove this. Flux calibrated multi-line observations are required to check if the excitation conditions in the arch are different than in the background and if EY Cyg can be the source of it.

CONCLUSIONS

From high-resolution spectroscopy we have revealed that the EY Cyg is an extremely long period Dwarf Nova with an very low inclination angle. The estimate of the mass of the primary white dwarf and the unusual shape of the light curves lead us to believe that EY Cyg might have undergone nova explosions in the past or is a probable CN candidate. Our direct images with a narrow Hα filter of the object and its surrounding were inconclusive whether any of the nebular emission around the object is associated with it.

REFERENCES

1. Avni, Y. & Bahcall, J.N., Ap.J., 1975, 197, 675
2. Bochkarev, N.G., Karitskaya, E.A., Shakura, N.I., Sov. Astron., 1977, 23, 8
3. Bochkarev, N.G. & Sitnik, T.G., ApSS, 1985, 108, 237
4. Costero, R., Echevarria, J., Pineda, L., BAAS, 1998, 30, 1156
5. Downes, R.A., etal., PASP, 2001, 113, 764
6. Echevarria, J., Costero, R., Tovmassian, G., etal, A&A, 2002, in preparation
7. Hacke, G., & Andronov, I.L., Mitt. Ver. Sterne, 1988, 11, 74
8. Kraft, R.P., Ap.J., 1964, 135, 408
9. Kukarkin, B.V., & Parenago, P.P., Var. Star Bull., 1934, 4, 44
10. Orio, M., & Ogelman, H., IAU Circ., 1992, 5680
11. Piening A.T., AAVSO J., 1978, 6, 60
12. Sarna, M.J., etal., IBVS, 1995, 4165
13. Smith, R.C., etal., MNRAS, 1997, 287, 271

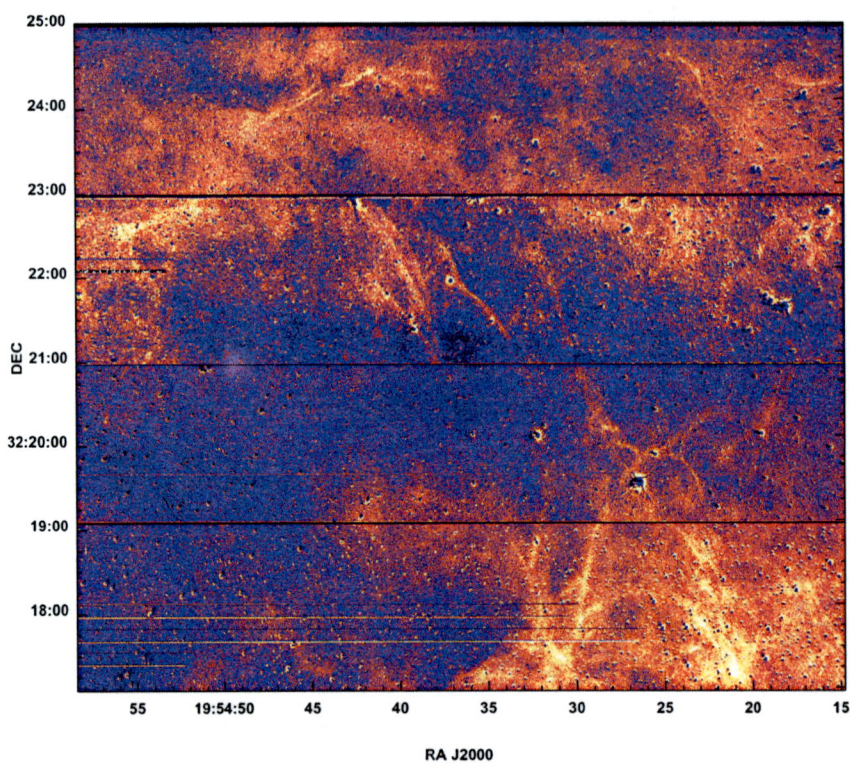

FIGURE 2. A mosaic of direct images of EY Cyg field obtained at the WYIN telescope with a narrow Hα filter. The continuum has been subtracted.

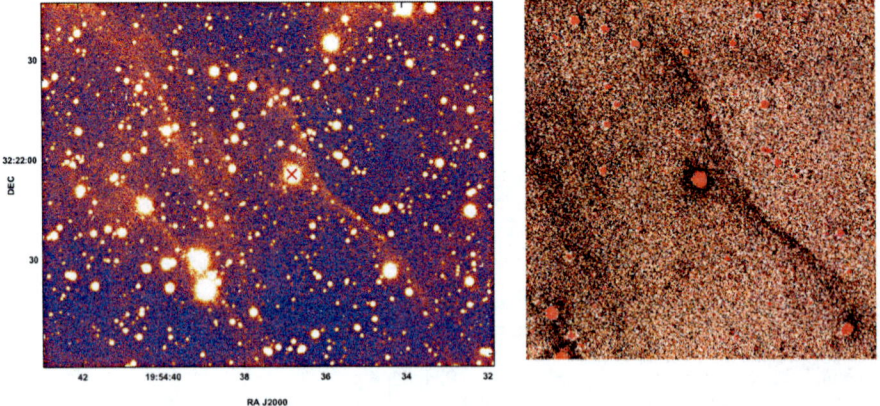

FIGURE 3. The blowup of continuum subtracted image of EY Cyg (right panel). A faint arc crossing the larger scale strips of nebular emission can be seen to the north-east of the the star. In the left panel a larger area than on the right is presented without continuum subtraction showing the surroundings of EY Cyg.

The 2001 superoutburst of WZ Sge: Is there any connection with Nova Outbursts?

Juan Echevarría*, Rafael Costero*, Gaghik Tovmassian[†], Sergei Zharikov[†], Raúl Michel[†], Michael Richer[†] and Armando Arellano-Ferro*

*Instituto de Astronomía, Universidad Nacional Autónoma de México, Ciudad Universitaria, Apartado Postal 70-264, México, D.F. 04510
[†]Instituto de Astronomía, Universidad Nacional Autónoma de México, Observatorio Astronómico Nacional, Apartado Postal 877, Ensenada, B.C. México, 228000

Abstract. We have observed WZ Sge with high resolution *echelle* spectroscopy during the 2001 superoutburst. The hot-spot arising near the secondary star is again visible 15 days after its first detection. We have been able to measure its position as a function of orbital phase directly from individual spectra. We derive a a semiamplitude of $K_{hs} = 457 \pm 16 \text{kms}^{-1}$ and a systemic velocity of $\gamma = 0 \pm 14 \text{km s}^{-1}$. We also show that the accretion disc has an ellipsoidal shape with a value around $e = 0.75$. Based on these facts we derive a phase shift of $\Delta\Phi = -0.12$ between the eclipse and the inferior conjunction ephemeris. If this is the case, then the hot-spot appears located at the trailing side of the red dwarf and must be the result of in-falling material back to the donor star.

INTRODUCTION

WZ Sge, first considered a recurrent nova, is now a well known dwarf nova which undergoes a superoutburst stage on an 33 year cycle, gaining as much as 8 magnitudes from its quiescent magnitude V~15.5. For an excellent review see [1]. Unexpectedly, after only 23 yr in quiescence the object went into a new superoutburst. It was first detected by T. Oshima on July 23.565 UT, 2001 at a magnitude of V=9.7 [2]. As the night reached all longitudes of the planet, telescopes from many observatories around the globe pointed to WZ Sge, and shortly became the best observed dwarf nova during outburst. Due to the amount of data collected both photometric and spectroscopic, results have been pouring very slowly into the literature. On August 13, 2001, Steeghs et al. [3] has detected Balmer emission arising from the secondary and revealing a large velocity amplitude, as well the presence of an asymmetric accretion disc. Other works include first detection of *early superhumps* [4]; results from multi-wavelength observations [5]; spiral arms detection from Doppler tomography [6]; and the results of a worldwide photometric campaign, where the authors conclude that the disc must be very eccentric [7]. The original idea of this contribution was to compare nova outbursts and the eight magnitude superoutburst in WZ Sge. However, after we submitted this idea to the Nova Conference, events have superseded this proposal, and we devote this contribution to show new evidence of the hot-spot associated with the donor star and calculate its K_{hs} value. Further results are under way from photometric and spectroscopy data during early decline [8], and from spectroscopy during the echo outbursts [9].

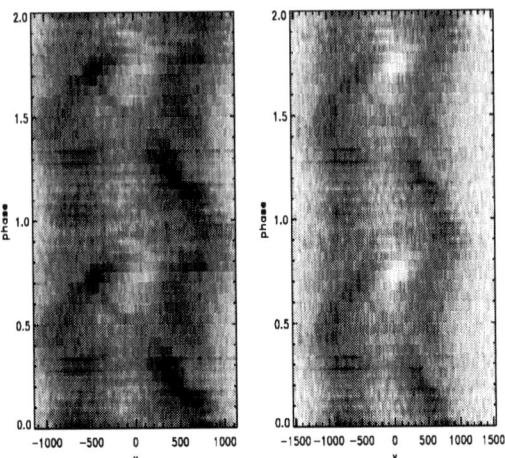

FIGURE 1. Hα (*left*) and Hβ (*right*) trailed spectra as a function of spectroscopic orbit during the night of 29 August, 2001. The ephemeris are taken from [13].

OBSERVATIONS

Observations of WZ Sge were obtained with the *echelle* Spectrograph attached to the 2.1m Telescope at the Observatorio Astronómico Nacional during August 29, 2001. The spectral coverge is from 3700 to 7000 Å. We used a Site CCD with a 24 μm pixel size and obtained a spectral resolution of 25 km s^{-1}. 66 exposures were adquired, 180 sec each, starting at HJD 2452150.640071 and ending at HJD 2452150.788307. This period covers 2.61 orbital cycles.

RADIAL VELOCITY OF THE SECONDARY

In Figure 1 we show the trailed spectra of Hα and Hβ as a function of spectroscopic phase [13]. It is evident that, apart prom the typical double peaks and the presence of a strong S-wave arising from the bright spot usully seen in the disc during quiesence we see, at many orbital phases, other features are seen, including a narrow emission which is in phase with the secondary star. This narrow hot-spot was first detected by Steehgs et al. [3] from spectra taken on August 13, 2001,i.e. sixteen days earlier than our observations. This narrow feature is not present most of the time during this stage of the so called echo outbursts [7]. Particulary We have taken spectra one day before and one day after the observations reported here, and we found no evidence of this bright spot. The discussion of this fact is done elsewhere [9].

The narrow emission line was sufficiently strong on August 29 to enable us to measure its position directly from the *echelle* spectra and to calculate its radial velocity curve.

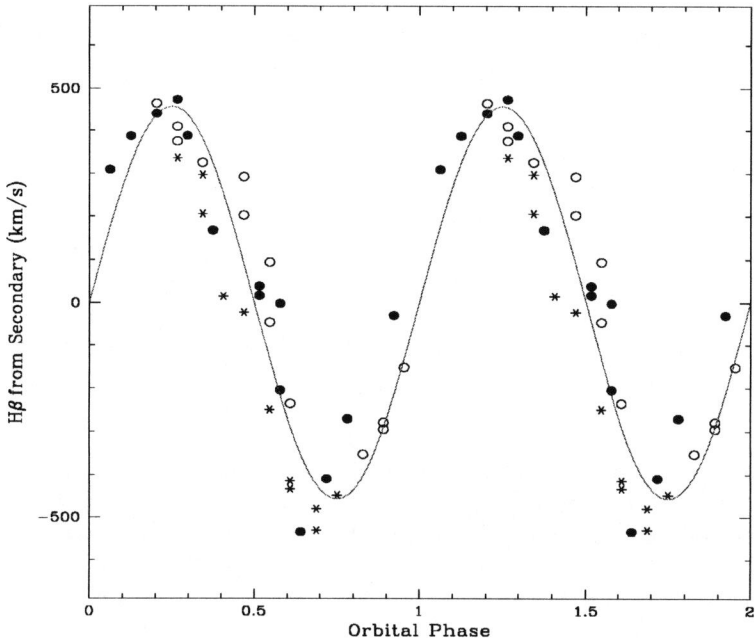

FIGURE 2. Radial Velocity curve of the Hα hot-spot arising from the secondary star. Again the ephemeris are taken from [13].

The results are shown in Figure 2. The lack of points near phase zero is due to the fact that the hot-spot and the secondary star are behind the disc which at this stage must be very thick. We have taken great care in isolating these measurements from contamination from the disc spot, as they cross each other twice during an orbital period. Its radial velocities were measured with gaussian fits and have been corrected to an heliocentric reference frame. We have applied a sinusoidal rms fit and found: $\gamma = 0 \pm 14$ km s^{-1}; $K_2 = 457 \pm 15$ km s^{-1}; and HJD$_\odot$ = 2452150.6282 days for a fixed $P_{orb} = 0.05668784603$ [11],

DOPPLER TOMOGRAPHY AT Hα AND Hβ

We have applied Doppler tomography to our spectra using the Maximum-entropy programs developed by Spruit [14]. Since our orbital phase resolution is not optimal, we have adjusted the velocity resolution of the Doppler map by reducing the velocity bins roughly by a factor of two. This is done within the map reconstruction program, without much loss in quality [10]. In the upper left part of Plate 1, we show the Hα Doppler map using the eclipse ephemeris [11] corrected to the inferior conjunction of the secondary star by a phase shift of $\Delta\Phi = -0.041$ [13]. We adopt $\gamma = -72$ km s^{-1} [12], [13] and [3]. We note in this map, that the large stream hot-spot lies outside the stream and the Kepler-velocity along-stream paths, derived from the system parameters by [13] and shown at the top of the maps. The fact that the γ velocity for the narrow spot is nearly zero and also that the broad hot-spot appears shifted from its expected position makes us question the validity of the adopted γ and $\Delta\Phi$ values. If the disc is not symmetric and has a high eccentricity [3], [7], we could expect that the eclipse timings between the blue and red broad peaks may not not be symmetric with respect to the inferior conjunction of the secondary. There is also suggesting evidence from our Doppler map that there might be disc material reaching back the secondary star, in which case the position of the narrow hot-spot is also incorrect.

We can evaluate $\Delta\Phi$ from two methods. First we have measured the disc eccentricity from the Hα map, taking the mayor axis of the ellipse almost perpendicular to the rotational axis of the binary, we find $e = 0.75$. This would results in a phase shift of $\Delta\Phi = -0.10$. On the other, if we simply take the γ velocity of the secondary hot-spot as the correct velocity for the system then, given that the total radial velocity amplitude of the narrow spot is 914 km s^{-1}, then a shift from -72 to 0 km s^{-1} implies $\Delta\Phi = -0.12$. Using this shift and a zero velocity for the binary gives the results shown in the upper right section of Plate 1. Now the broad hot-spot lies within the stream and Kepler-velocity along-stream paths. The narrow spot now lies now on the trailing side of the donor star. The same effect is seen in the Hβ Balmer line as shown at the bottom left section in Plate 1. Here the broad hot-spot is very large and almost blends with the accretion disc which, at phase 0.9, seems to reach back at the secondary star. We have made an Hβ Doppler map calculation with higher resolution for lower velocities only. The results are shown in the lower right in Plate 1. Note the high definition for the position of the narrow hot-spot and the disc material falling back into the donor star.

REFERENCES

1. Smak, J. 1993. Acta Astron. **43**, 101
2. Ishioka, R., Uemura, M., Matsumoto, K., Kato, T.,Ayani, K. & Yamaoka, H. 2001. IAU Circ. 7669
3. Steeghs, D., Marsh, T., Knigge, C., Maxted, P.F.L., Kuulkers, E. & Skidmore, W. 2001, Ap.J. Lett. **562** L145
4. Ishioka, R. et al. 2002. A.A. **382** L41
5. Kuulkers, E. et al. 2002. *The Physics of Cataclysmic Variables and Related Objects* ASP Conference Series, Vol.xxx, 2002. eds. B.T. Hänsicke, K. Beuermann, K. Reinssch.
6. Baba, J. et al., 2002, PASJ **54** L7
7. Patterson, J. et al., 2002, PASP **114** 721
8. Echevarría, Costero, R., J., Michel, R., Zharikov, S.V., Tovmassian, G.H., Arellano-Ferro, A. & Richer, M. 2002, Ap.J.Lett. sumbitted
9. Costero, R., Echevarría, J., Michel, R., Zharikov, S.V., Tovmassian, G.H., Arellano-Ferro, A. & Richer, M. 2002, in preparation
10. Spruit, H.C. 2001, "Dopmap: Doppler mapping of disks in binaries" Version 2.3.1. Documentation, 22 Jan,2001
11. Patterson, J., Richman, H.R., Kemp, J. & Mukai, K. 1998, PASP **110** 403
12. Gilliland,R.L., Kemper, E. & Suntzeff, N. 1986, ApJ **301** 252
13. Spruit, H.C. & Rutten, G.M. 1998, MNRAS **299** 768
14. Spruit, H.C. 1994, AA **289** 441

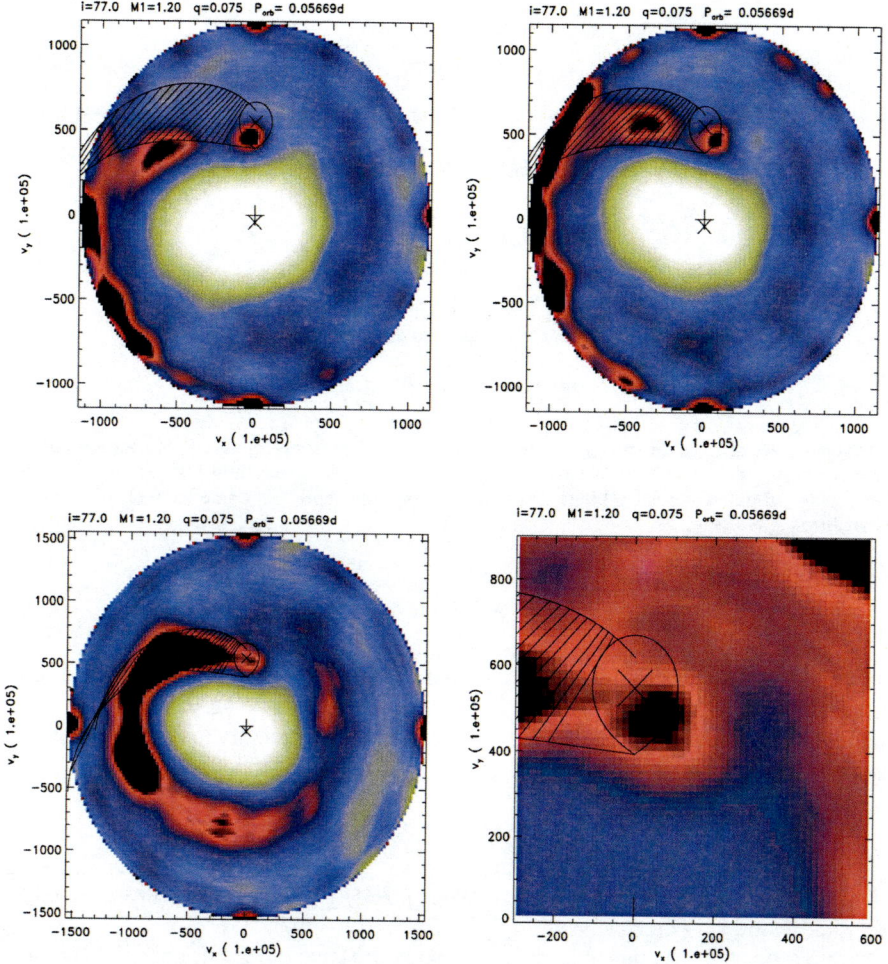

FIGURE 3. Plate 1. *Upper Left*: Doppler Tomography of Hα using [13] ephemeris; *Upper Right*: Doppler Tomography of Hα using our ephemeris (see text); *Bottom Left*: Doppler Tomography of Hβ using our ephemeris; *Bottom Right*: Blown-up region of Hβ around the secondary and the narrow spot.

The influence of mass loss on orbital elements of binary systems by periastron effect

M. Andrade* and J.A. Docobo*

Astronomical Observatory Ramón María Aller
University of Santiago de Compostela
P.O. Box 197
Santiago de Compostela, Spain

Abstract. We analyse the enhanced mass loss rate during periastron passage considering several reasonable physical processes which may contribute to that. Moreover, mass loss laws are proposed to explain orbital variations in binary systems. We also show how they can effectively describe the *periastron effect*.

INTRODUCTION

It is well—known that the star mass decreases throughout the time. In the case of binary systems we can suppose that this variation has some influence on the system evolution. Under traditional assumption that the mass is being isotropically lost, the motion of the system can be described by the equation

$$\ddot{\vec{r}} = -\frac{\mu(t)}{r^3}\vec{r} \qquad (1)$$

This task is known as *Gylden—Meščerskij [G-M] problem*, which has been exhaustively addressed [1]-[8].

In this scenario, it is possible to introduce successive perturbations with the aim to take into account new phenomena. In the present study we examine the more complex case of an enhanced mass loss rate during periastron passages which will hereafter be named as *periastron effect*.

Periastron effect

In recent papers [9-10] we have already analysed this problem confirming the difficulties that appear when one tries to approach the analytical integration of the canonical equations of motion by means of perturbation methods based on Lie transform techniques [11].

This problem is very interesting and can possibly explain [12] the relatively high eccentricity found in some tidally strongly interacting binary systems, since the periastron effect counteracts the tidal circularization predicted by Zahn's theory [13].

The periastron effect gives rise to new variations in contrast with the non−perturbed Gylden−Meščerskij problem, and our purpose is to analyse and to quantify such variations.

New equation of motion will be

$$\ddot{\vec{r}} = -\frac{\mu(t,r)}{r^3}\vec{r} \tag{2}$$

where $\mu(t,r) = Gm(t,r)$ is a function which has to explain an enhanced mass loss rate close to the periastron.

METHODOLOGY

To treat this type of dynamical problems, two different feasible approaches do exist: first one by means of perturbation analytical methods based on Lie transform techniques and another, which we shall use here, by means of numerical integration of the differential equations system with the aid of an algebraic calculator.

We get a mass loss law simultaneously depending on time and distance. By considering the dependence on distance as a perturbation to the G-M problem we obtain the appropriate perturbation potential. By its substitution in the Gauss equations of perturbed motion, we obtain the timely variation of orbital elements due to periastron effect.

After achieving this first order equations system, named as Lagrange equations, we are going to integrate it. To this end, we shall use a fourth order Runge−Kutta method for first order systems of equations by means of the algebraic calculator Mathematica®.

We are always assuming that α and β are two small parameters close to zero and that $\beta < \alpha$, since we are considering periastron effect as a perturbation to the G-M problem.

For a better visualization of the periastron effect contribution to the time−dependence of orbital elements we shall make a graphical overlapping. We plot the time−dependence of the argument of periastron, ω. Dashed line represents non−perturbed G-M problem and the solid line is used when we have in addition periastron effect.

RESULTS

The hamiltonian governing the G-M problem is given by

$$H = \frac{1}{2}\left(p_r + \frac{p_\theta^2}{r^2}\right) - \frac{\mu(t)}{r} \tag{3}$$

now, instead of $\mu(t)$, we take

$$\mu(t,r) = \mu(t) - \frac{\beta}{r} \tag{4}$$

This is equivalent to add to (3) the perturbation depending on distance $V'(r) = \frac{\beta}{r^2}$. Therefore, the hamiltonian of the G-M problem with periastron effect will be

$$H = \frac{1}{2}\left(p_r + \frac{p_\theta^2}{r^2}\right) - \frac{\mu(t)}{r} + \frac{\beta}{r^2} \tag{5}$$

where $\mu(t)$ is given by the Eddington–Jeans law $\dot{\mu}(t) = -\alpha\mu^n$, and μ is a function of time, uniformly decreasing, continuous and differentiable, and α and n are two real numbers that in practice take concrete values, α close to zero and n between 0.4 and 4.4.

The respective potential $V'(r)$ gives rise to a repulsive radial perturbing force. By considering the Gaussian equations of perturbed motion we obtain the Lagrange equations

$$\frac{de}{dt} = -\frac{\dot{\mu}(t)}{\mu(t)}(e + \cos f) + 2\beta\sqrt{\frac{1}{\mu(t)[a(1-e^2)]^5}}(1 + e\cos f)^3 \sin f \tag{6}$$

$$\frac{d\omega}{dt} = -\frac{\dot{\mu}(t)}{\mu(t)}\frac{\sin f}{e} - 2\beta\sqrt{\frac{1}{\mu(t)[a(1-e^2)]^5}}(1 + e\cos f)^3 \frac{\cos f}{e} \tag{7}$$

$$\frac{da}{dt} = -\frac{\dot{\mu}(t)}{\mu(t)}a\frac{1 + 2e\cos f + e^2}{1 - e^2} + 4\beta e\sqrt{\frac{1}{\mu(t)a^3(1-e^2)^7}}(1 + e\cos f)^3 \sin f \tag{8}$$

$$\frac{df}{dt} = \sqrt{\frac{\mu(t)}{[a(1-e^2)]^3}}(1 + e\cos f)^2 + \frac{\dot{\mu}(t)}{\mu(t)}\frac{\sin f}{e} + 2\beta\sqrt{\frac{1}{\mu(t)[a(1-e^2)]^5}}(1 + e\cos f)^3 \frac{\cos f}{e} \tag{9}$$

Putting in this differential equations system $\beta = 0$, we recover the Lagrange equations with the isotropic mass loss depending on time without periastron effect. After its numerical integration we already can plot the time–dependence of each orbital elements.

When considering the G-M problem without periastron effect, we can see that the eccentricity and the argument of periastron show periodic variations, while the major semi–axis and the true anomaly demonstrate secular variations. However, the periastron effect with the previous potential (the perturbed G-M problem), gives rise to a regression of the argument of periastron (see Table 1).

It can be demonstrated that a more general potential

$$V'(t,r) = \frac{\beta}{r^n} \tag{10}$$

not depending on power n, gives rise to the precession of the argument of periastron too. Even so, the eccentricity undergoes merely periodic variations (see Table 1).

APPLICATIONS

Let us consider the nova *GK Persei*, a classical nova system that erupted in 1901. It contains a white dwarf primary with an evolved K2 sub–giant secondary. Moreover, it is a special case because it has the longest known orbital period for a classical nova.

TABLE 1. Secular variations

$\Delta\lambda_i$	$[G-M]$	$[G-M]+V'(t,r)$
Δe	0	0
$\Delta\omega$	0	$-\frac{2\pi\beta}{\mu a(1-e^2)}$
Δa	$2\pi\alpha a^{\frac{5}{2}}\mu^{n-\frac{3}{2}}$	0
Δf	$\frac{\pi(2+e^2)}{\sqrt{-(-1+e^2)^3}}$	$\frac{2\pi\beta}{\mu a(1-e^2)}$

For this system we have a set of circular orbital elements [14]. However, now we shall consider that the orbit is slightly eccentric. In this way, we take the values for the eccentriticy (compatible with the observations) and the argument of periastron that appear in Table 2.

TABLE 2. Initial orbital elements and parameters

Orbital elements		Parameters*
$e_0 = 0.01$	$f_0 = 0.0$	$\alpha = 10^{-7}$
$\omega_0 = 0.0$	$P_0[days] = 1.997$	$\beta = 2 \cdot 10^{-8}$
$a_0[AU] = 0.0325$	$\mu_0[M_\odot] = 1.15$	

* Assumed values for the mass loss parameters.

Below we plot the time−dependence of ω (Fig. 1) with and without periastron effect for the given initial conditions (see Table 2) over one year.

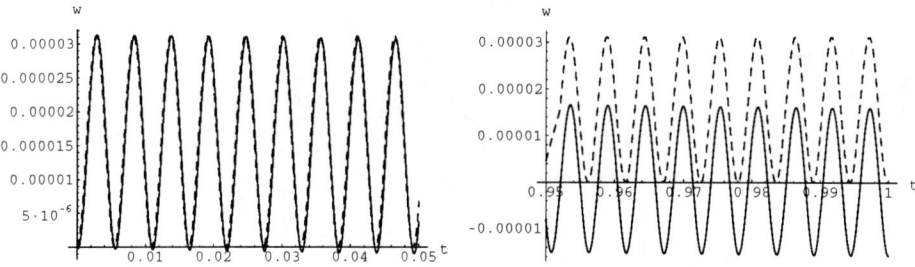

FIGURE 1. Argument of periastron

At the end of one year the following values of the orbital elements are obtained (see Table 3).

These results indicate that periastron regression is, for the considered $V'(r)$, the most important consequence of mass loss by periastron effect.

CONCLUSIONS

The present analysis clearly indicates that periastron effect may have a considerable influence on the variation of orbital elements and, therefore, on the dynamics of binary systems too. For the family of the studied potential (10), this influence gives rise to

TABLE 3. Values of the orbital elements

$Elements(\Delta t = 1\ year)$	$[G-M]$	$[G-M]+V'(t,r)$
e	0.0100000243413	0.010000024676
ω	$1.88457929002 \cdot 10^{-7}$	-0.0000158255680388
$a[UA]$	0.032505828498	0.0325058284982
f	1149.98122246	1149.98117425
$\mu(t,r)[M_\odot]$	1.14979379838	1.14979317697

the regression of the argument of periastron. The smaller major semi−axis the larger regression, and the larger eccentricity the bigger regression.

Other mass loss laws with periastron effect may lead to different phenomena such as a secular increase of eccentricity. It would be useful to link such results with the statistical studies of the orbital elements distribution and the dynamical and physical characteristics of binary systems.

Our purpose is to carry out an exhaustive examination of other mass loss laws with periastron effect by means of analytical methods of integration. Then, we shall add successive perturbations, such as those due to mass exchange between the components, influence of accretion disks, existence of electromagnetic fields, etc.

ACKNOWLEDGMENTS

This work was supported by the Project Grant AYA 2001-3073 of the *Ministerio de Ciencia y Tecnología* (Spain).

REFERENCES

1. Hadjidemetriou, J.D., *Icarus* **2**, 440–451 (1963).
2. Hadjidemetriou, J.D., *Icarus* **5**, 34–46 (1966).
3. Verhulst, F., *Bull. Astr. Netherlands* **20**, 215–221 (1969).
4. Deprit, A., *Celest. Mech.* **31**, 1–22 (1983).
5. Floría, L. 'Perturbed Gylden Systems and Time−Dependent Delaunay−Like Transformations', in *Visual Double Stars: Formation, Dynamics and Evolutionary Tracks*, edited by J.A. Docobo, A. Elipe and H. McAlister, Astrophys. & Space Sci. Lib. 223, Dordrecht, 1997, pp. 347–356.
6. Prieto, C., and Docobo, J.A., *Astron. Astrophys.* **318**, 657–661 (1997).
7. Prieto, C., and Docobo, J.A., *Celest. Mech. & Dyn. Astron.* **68**, 53–62 (1997).
8. Docobo, J.A., Prieto, C., and Ling, J.F., *Astrophys. & Space Sci.* **261**, 205–207 (1999).
9. Andrade, M., and Docobo, J.A., 'The influence of decreasing mass on the orbits of wide binaries: an approach to the problem', in *Highlights of Spanish Astrophysics II*, edited by J. Zamorano, J. Gorgas and J. Gallego, Kluwer Academic Publishers, Dordrecht, 2001, pp. 273–276
10. Docobo, J.A., and Andrade, M., 'El problema de dos cuerpos con masa decreciente en función del tiempo y de la distancia. Aplicación de un método triparamétrico de perturbaciones.', Pending of publication on *Monografías del Seminario Matemático "García de Galdeano"*, Zaragoza, 2002.
11. Deprit, A., *Celest. Mech.* **1**, 12–30 (1969).
12. Soker, N., *Astron. Astrophys.* **357**, 557–560 (2000).
13. Zahn, J-P., *Astron. Astrophys.* **57**, 383–394 (1977).
14. Crampton, D., Cowley, A.P., and Fisher, W.A., *Astrophys. J.* **300**, 788–793 (1986).

EXPLOSION MECHANISM AND MASS LOSS

Studies of Novae in the 20th Century

Sumner Starrfield

*Department of Physics and Astronomy, Arizona State University,
P. O. Box 871504 Tempe, Arizona 85287-1504
email: starrfield@asu.edu*

Abstract. A nova outburst is not only the third most violent explosion that can occur in a galaxy, it is the largest hydrogen bomb in the Universe. The last meeting devoted purely to studies of novae was held in Madrid in 1989. At that time IUE spectra of novae were just beginning to be analyzed in detail, the EXOSAT x-ray results had just been published, and ROSAT had yet to be launched. Since that time, the IUE studies of novae in the LMC (known distance and small reddening) showed that all fast novae exceed the Eddington luminosity at maximum which was not predicted by numerical hydrodynamic simulations. A number of novae have now been observed in x-rays with ROSAT, BEPPOSAX, ASCA, CHANDRA, and XMM (plus others). The x-ray studies determined the turn off times for several classical novae, showed that at maximum x-ray light they resemble the Super Soft Sources, and found rapid variations in the x-ray light curves of two novae. There have been abundance analyses of V838 Her (oxygen poor, sulfur rich), LMC 1991, and other classical and recurrent novae. Infrared studies have resulted in a number of puzzling results such as the same nova producing both carbon (first) and silicate (later) dust at different times during its outburst. There have been advances in theoretical modeling and most studies now include large nuclear reaction networks combined with the hydrodynamic calculations. One perplexing problem produced by the simulations is that they predict that much less mass is ejected in the outburst than is observed, and the problem has worsened as the input physics has improved. Nevertheless we have made significant progress in the past 100 years and I will present some of the more important developments that have occurred over the past century. I will end with a list of questions to be answered.

INTRODUCTION

In the first half of the 20th century, the observations of novae in outburst reached a point where it was possible to determine the features that were typical of the outburst. For example, it was clear from the spectroscopic observations done early in the past century that a large amount of matter was ejected at high velocities and an explosion had occurred (Payne-Gaposchkin 1957; McLaughlin 1960; Gallagher and Starrfield 1978). It was also noted that emission lines of abundant elements would appear and gradually come to dominate the spectrum. At later times, emission lines from more highly ionized species would appear suggesting that the underlying central engine had become extremely hot (Gallagher and Starrfield 1976, 1978; Krautter and Williams 1989). Some years after the outburst, the ejected material would become spatially resolved, the physical conditions measured, and elemental abundances obtained (Gehrz et al. 1998). In all cases, it was found that the ejecta were enriched in helium implying that hydrogen fusion had powered the explosion (Starrfield 1989; Krautter et al. 1996).

In this review I will present the developments in our understanding of the cause and evolution of the nova outburst over the past 100 years. I have sufficient space, however, to describe only the major steps in this evolution.

OBSERVATIONS AT MINIMUM

The pioneering observations of Cataclysmic Variable systems (CV's), which led to the basic understanding of their binary nature, occurred when Joy (1943) discovered that AE Aqr was a short period spectroscopic binary, and that SS Cyg and RU Peg showed similar features (Joy 1960, and references therein). Somewhat later, Walker (1954) reported that DQ Her was a short period eclipsing binary, and discovered the 71 second oscillation in its light curve. The presence of this oscillation indicated that there was a compact object in the system. It was originally thought to be caused by pulsation which implied a low mass white dwarf. Later it became clear that it was caused by rotation and, therefore, that the white dwarf could be massive. Based on these discoveries, Struve (1955) suggested that all CV's were close binary systems, and Crawford and Kraft (1956) proposed a model for AE Aqr consisting of a close binary stellar system with one star filling its Roche Lobe and the other star small and hot. They also suggested that mass was being lost by the Roche Lobe filling star and this mass eventually reached the compact object.

Over the next few years, there were studies of a number of systems culminating in the extremely important paper by Kraft (1964) in which he studied 10 "old" novae and showed that they satisfied the close binary model. By this time, it was also clear that the compact object was a white dwarf. However, Kraft argued that the cause of the nova explosion was not a thermonuclear runaway (TNR) in the accreted material but a pulsation driven outburst as suggested by Schatzman (1965, and references therein). Kraft based his arguments on the high degeneracy and, thereby, the high electron conductivity expected for material that had been accreted onto a white dwarf. Heat would be transferred away from the nuclear burning region as fast as it was produced. Theoretical modeling, however, showed that this gas was partially degenerate and sufficient material could be accreted to produce a TNR and an explosion (Giannone and Weigert 1967; Starrfield 1971). Further discussion of the early history of CV's can be found in Warner (1995).

THEORETICAL MODELING

Although there were a number of theoretical suggestions prior to 1950 to explain the outbursts of novae (see Payne-Gaposchkin 1957; McLaughlin 1960; Schatzman 1965; Gallagher and Starrfield 1978; Starrfield 1989 and references therein), it was not until the 1950's that Evry Schatzman proposed that the explosion was caused by runaway nuclear reactions in the interior of a star. A few years later, however, he proposed that the explosion was caused, instead, by a pulsational instability (Schatzman 1965, and references therein).

After it was realized that the outburst occurred on the white dwarf, Giannone and Weigert (1967) and Starrfield (1971) numerically followed the growth of TNR's in accreting white dwarf envelopes. At the same time, Sparks (1969) studied the propagation of a shock wave through a stellar envelope and his simulated light curve resembled that of a nova. The first one-dimensional hydrodynamic study, which incorporated a nuclear reaction network that included the β^+-unstable nuclei, was done by Starrfield et al. (1972). They discovered, in their initial tests of the TNR hypothesis, that at temperatures above 10^8K, most of the CNO nuclei in the envelope were transformed into the β^+-unstable nuclei. As a consequence, they reported that one could not get a rapid outburst, mass ejected, and large ejection velocities unless the CNO nuclei were enriched in the envelope of the white dwarf. Starrfield et al. (1972, 1974) also discovered that not all the accreted plus core material was ejected in the initial stages of the outburst but some fraction (depending on a number of factors) remained on the surface of the white dwarf burning hydrostatically in a thin shell. However, their studies of the evolution of this layer showed that nuclear burning could last for hundreds of years which strongly disagreed with the observations. Because the luminosity of the white dwarf remained virtually constant during this phase, it was called the constant bolometric luminosity phase of the outburst.

This paper established the predictive nature of the TNR scenario for the nova outburst since both enriched CNO nuclei and the constant bolometric luminosity phase were later confirmed by observations (Gehrz et al. 1998). A possible explanation for the discrepancy between theory and observations of the length of the outburst was proposed by Starrfield (1979) and MacDonald et al. (1986), who calculated that mass lost both through a wind and through dynamical friction with the secondary could shorten the post outburst evolution time of the rekindled white dwarf. In later papers it was shown that the white dwarf had to be massive for an extremely fast nova or recurrent nova outburst (Starrfield et al. 1974, 1978, 1985, 1988).

In the late 1970's a group in Israel led by Giora Shaviv began their own 1-D hydrodynamic studies of TNR's on white dwarfs (Prialnik et al. 1978, 1979; Shara et al. 1980). They were the first group to include accretion in a hydrodynamic study (Prialnik et al. 1982), chemical diffusion in the accreting material (Prialnik & Kovetz 1984; Kovetz & Prialnik 1985), and follow the evolution of the white dwarf through multiple outbursts (Prialnik 1986). Interestingly enough, they predicted the existence of Super Soft Sources long before their discovery in the ROSAT all sky survey (Shara et al. 1977). More recently, the organizers of this meeting, Hernanz and José, have also been doing theoretical studies of TNR's on white dwarfs, but concentrating on the nucleosynthesis and γ-ray emission associated with the explosions (Hernanz et al. 1996, 1999; José et al. 1997, 1999; José & Hernanz 1998; Gomez-Gomar et al. 1998). They have also been involved in the study of pre-solar grains from novae (Amari et al. 2001).

NOVAE IN OUTBURST

The prediction that a fast nova outburst required enriched CNO abundances was not verified until the late 1970's when Williams and his collaborators began a study of spatially resolved novae ejecta and reported that they were rich in carbon, nitrogen and oxygen (Williams et al. 1978; Williams and Gallagher 1979; Gallagher et al. 1980). However, it is important to point out that Cecilia Payne-Gaposchkin and Lawrence Aller analyzed the shell of DQ Her around 1940 and realized that the ejecta abundances were non-solar and probably enriched in the CNO nuclei (Aller 1994; Aller 1993, private communication). They were not allowed to publish this result, however, since it flew in the face of the accepted dogma of the time that all objects had Solar abundances (Aller 1993, private communication).

INFRARED STUDIES The pioneering observations in the UV and IR showed the importance of multiwavelength observations of novae during their outbursts (Gehrz 1988; Starrfield and Snijders 1987; Starrfield 1988; Gehrz et al. 1998). The first nova detected in the IR was FH Ser (Geisel et al. 1970; Hyland & Neugebauer 1970). V1500 Cyg was the brightest nova in the second half of 20th Century and its evolution was well studied in the IR (Gallagher & Ney 1976). Another well studied nova was NQ Vul (Ney & Hatfield 1978). These investigators found a rich variety in the IR behavior of novae and verified that the transition phase was the signature of dust formation in the nova outburst. First observed in the IR by Hyland and Neugebauer (1970) for FH Ser, it has been studied in detail for a number of novae with more sensitive detectors providing more information about the dust and the ejected shell (Gehrz 1988; Gehrz et al. 1998). For many years it was assumed that carbon-oxygen novae formed only amorphous carbon dust (even in oxygen rich environments) while oxygen- neon- magnesium novae formed silicate dust. Recently, observations of those novae for which the optical drops by about 5 magnitudes when the dust forms show that the same nova can form multiple kinds of dust ranging from amorphous carbon grains to silicates (Gehrz et al 1998). QV Vul and V705 Cas are two examples. In addition, V705 Cas was caught by IUE just as it was forming grains and they were found to be larger than the typical ISM grain (Shore et al. 1994). Presumably, the explanation of this phenomenon will require non-equilibrium grain condensation calculations (Rawlings & Evans 2002).

ULTRAVIOLET STUDIES Because the rekindled white dwarf is small and luminous, one expects it to be hot and emit a major fraction of its energy outside the optical. This was first realized by the detection of FH Ser by the Orbiting Astronomical Observatory-A2 satellite (Code 1972; Gallagher and Code 1974). An extremely important set of UV data were later obtained by the International Explorer Satellite (IUE) which observed novae from V1668 Cyg to LMC 1995 (Starrfield and Snijders 1987; Shore et al. 1997). From these spectral data, it was finally possible to verify the occurance of the constant bolometric luminosity phase, and show that it lasted for some time after optical maximum (Shore et al. 1993). The high quality, fluxed, UV spectra provided important information on the evolution of the nebular emission lines with time. In addition, there were lines in the UV from elements with no lines in the optical and, thus, the abundances for a large number of elements could be obtained. As a result, a second compositional class of novae was identified. One class was an outburst that occurred on a carbon-oxygen (CO) white dwarf. The second, was discovered when the IUE observations of V693 CrA were analyzed (Williams et al 1985). The enrichments of neon, magnesium, aluminum, and silicon implied that the outburst had occurred on a white dwarf with a core composition of oxygen, neon, and magnesium (Starrfield et al. 1986). In addition, the IUE studies provided confirmation that material from the underlying white dwarf was dredged up into the accreted layers.

The large number of novae studied with the IUE has required that two new analysis methods be developed, which have broad applicability in astronomy, and so I briefly mention them here. The first is a versatile, Non-LTE, spherical, expanding stellar atmosphere code (Hauschildt et al. 1992, 1997, 2002) that has now been used for studies of hot stars, cool stars, and brown dwarfs. The stellar atmosphere studies of novae showed that the emission "lines" in the UV were not actually spectral lines but regions of transparency between overlapping absorption lines from the iron group elements (the iron curtain: Hauschildt et al. 1992). It was also realized that, because of the expansion, a large fraction of the entire ejected shell was visible at the same time. This result explained why so many different ionization stages were simultaneously present in nova ejecta. The Hauschildt code was recently used to study the extraordinary nova LMC 1991 (Schwarz et al. 2001). LMC 1991 reached V$\sim$9 mag, ejected material with velocities exceeding 3000 km s^{-1}, and was super-Eddington for more than 10 days. Nevertheless, LMC 1991 appears to be a CO nova, but the underlying metallicity of the material (the iron group abundances) are at least a factor of 3 lower than typical LMC material (Z=0.1Z$_\odot$). The existence of this nova has led to theoretical studies of nova outbursts that occur in low metallicity gas (Starrfield et al. 1999a,b). Another important result of the studies of the LMC novae is that, because they are at a known distance, we can unambiguously determine their peak luminosities. In all cases, LMC novae are Super-Eddington at maximum (Della Valle et al. 1994).

The second new method is the use of an optimization technique in combination with CLOUDY (Ferland 1998). With this method, it is possible to analyze novae ejecta in which the shell is not spatially resolved, the densities are sufficiently large so that collisional processes are important, and there is an underlying hot, time dependent, luminous source (Vanlandingham et al. 1996, 1999, 2002, and references therein). One interesting result of her studies is that the abundances for LMC 1990 #1 are nearly identical to those of V693 CrA, yet these novae occurred in two different galaxies. Another result is that V838 Her ejected gases which were enriched in sulfur and strongly depleted in oxygen (Vanlandingham et al. 1996). These abundances imply that nuclear breakout has occurred from the CNO cycle, but it has not been possible to find initial conditions that result in breakout (white dwarf mass, luminosity, composition, mass accretion rate).

X-RAY STUDIES X-ray observations have also provided important information on the outburst. Data obtained with EXOSAT for GQ Mus and QU Vul initiated the x-ray studies of classical novae in outburst (Ögelman et al. 1984, 1987). They showed that a nova became hot and luminous in x-rays at some time after the outburst. The most complete set of data have been obtained by ROSAT for V1974 Cyg (Krautter et al. 1996). This satellite observed the complete x-ray light curve for this nova and showed that its outburst lasted $\sim$18 months. At the peak this nova became the brightest Super Soft Source in the x-ray sky. Analysis of the x-ray results suggested that the rise was caused by the clearing of the ejected hydrogen shell, and the decline was caused by cessation of nuclear burning on the rekindled white dwarf. The length of the outburst suggested that the explosion had occurred on a massive white dwarf, and the rate of decline implied that a shell of about 10^{-5}M$_\odot$ of helium rich material was left on the white dwarf after the outburst (Krautter et al. 1996). In contrast, Orio and collaborators have examined the ROSAT data on novae in outburst and find that a typical nova does not stay bright in x-rays for more than a few months (Orio et al. 2001a). Two mechanisms have been proposed to explain the rapid loss of this material after the explosive phase of the outburst: common envelope evolution and radiation pressure driven mass loss

(Starrfield 1979; MacDonald et al. 1985; Starrfield et al. 1991).

More recently, there have been ASCA, BEPPOSAX, CHANDRA, and XMM studies of V382 Vel and CHANDRA studies of V1494 Aql and IM Nor. Observations of V382 Vel were obtained by ASCA and BEPPOSAX early in the outburst (Orio et al. 2001b; Mukai & Ishida 2001) and they show that most of the early emission was at energies around 1 keV, the so-called "hard" component. By November 1999, however, BEPPOSAX observations show that it had evolved to the Super Soft phase (Orio et al. 2002). A CHANDRA spectrum was obtained with the ACIS low resolution instrument on December 30, 1999, and it was still in the Super Soft phase (Burwitz et al. 2002). However, a high resolution spectrum was obtained by CHANDRA on 14 February 2000, using the Low Energy Grating plus High Resolution Camera, and it was found to have evolved to the emission line phase and the continuum was gone (Starrfield et al. 2000b). This extremely interesting observation implies that nuclear burning turned off only 9 months after discovery and, in addition, that the nova took less than 6 weeks to decline.

V1494 Aql was discovered in December 1999 and it was observed by CHANDRA at 4 different times. The first three were ACIS-I observations. The first two of these visits showed only emission lines. However, the third observation, obtained about 8 months into outburst, showed that its spectrum had evolved to the Super Soft phase. A Low Energy Grating observation was obtained at the end of September 2000 and this spectrum showed the characteristic hot atmosphere of a Super Soft Source. It qualitatively resembled the XMM spectrum of Cal 83 (Paerels et al. 2001). The analysis of these data with stellar atmospheres is in progress (Hauschildt et al. 2002). At least as interesting, the light curve of these data shows both a burst and an oscillation with a period of about 2500 seconds (Starrfield et al. 2001). Neither of these phenomena is explained by theory. The source of such a burst is completely unknown. The source of the oscillations may be similar to the cause of the oscillations in the pulsating central stars of planetary nebula nuclei since the structure of the two types of systems is similar.

RECURRENT NOVAE

The IUE studies of recurrent novae have shown that there are two types of these interesting stars. One class, which consists of U Sco, V394 CrA, and LMC 1990 #2 is composed of a massive white dwarf and (probably) the stripped core of an evolved star. In order for the outburst to occur as often as it does for U Sco (Kahabka et al 1999), the white dwarf must be massive and luminous and the secondary must be evolved (Starrfield et al 1985, 1988). At least as interesting, studies of the accretion disk in U Sco show no hydrogen which suggests that the secondary is transferring hydrogen poor material (Shore et al. 1991). Nevertheless, hydrogen is ejected in the outburst and its source has yet to be determined. The second class of recurrent nova includes T CrB, RS Oph, V3890 Sgr, and V745 Sco. In these systems the primary is a massive white dwarf but the secondary is a cool giant, and the white dwarf is orbiting within the outermost layers of the giant when the explosion occurs. Observations of the outburst of RS Oph in 1985 by IUE showed that an initial UV burst ionized the giant envelope and the material ejected by the white dwarf formed a blast wave which moved through the envelope over a few months

(Shore et al. 1996). For recurrent novae, both observations and theory suggest that far less mass is ejected during the outburst than is accreted by the white dwarf, and its mass is probably growing toward the Chandrasekhar limit. They could, therefore, be one of the progenitors of SNe Ia (Starrfield et al. 1985, 1988; Shore et al. 1996). One difficulty is that there might be insufficient numbers of recurrent novae to agree with SNe Ia rates. However, in the past two years there have been two new recurrent novae discovered in outburst, CI Aql and IM Nor (Kato et al. 2002). Their outburst characteristics do not resemble either of the above classes of recurrent novae.

SUMMARY AND UNSOLVED PROBLEMS

We have come a long way in our understanding of the cause and evolution of the nova outburst since the beginning of the last century. We have found that a nova emits energy from x-rays to radio and they have been studied at these wavelengths and all wavelengths in between. It has been predicted that novae emit at γ-ray wavelengths and there are preparations underway to observe them with INTEGRAL (Hernanz et al. 2002). Although attempts were made with COMPTON, there were no nearby novae (< 1 kpc) during the time it was in orbit. We have found that a mechanism (as yet unknown) mixes accreted material with core material and the explosion blows the mixed gas into space where it can be studied and the composition determined. As a result of continuing determinations of nova abundances, it is clear that in most cases the nova ejects more material than it has accreted, and the mass of the white dwarf must be shrinking as a result of the nova phenomena. Therefore, novae with extreme amounts of core material in their ejecta cannot be the progenitors of SNe Ia. In contrast, recurrent novae are now thought to be the result of explosions on extremely massive white dwarfs and they accrete more material than is ejected into space. They could be one type of SN Ia progenitor (Starrfield et al. 1985, 1988). Novae participate in the cycle of Galactic chemical evolution in which grains and metal enriched gas in their ejecta, supplementing those of supernovae, AGB stars, and WR stars are a source of heavy elements for the ISM. The large amount of ejecta mass, determined from observations, and their discovery in external galaxies, imply that they are an important contributor to Galactic chemical evolution.

I end this review with a list of outstanding problems that must be solved before we can claim a complete understanding of novae. Some of these have been discussed in this review while all of them have been discussed at one time or another during this meeting.

- Although there have been numerous studies of Cataclysmic Variables at minimum and the masses of the white dwarfs in some systems are known with reasonable precision, at this time there is not a single old nova system with an accepted white dwarf mass.
- The mass accretion rate onto the white dwarf is an important parameter that strongly affects the evolution to runaway. While the mass accretion rate is poorly determined for nova systems, it appears to be sufficiently high that, according to the theoretical calculations, a nova outburst will not occur.

- Observations of nova ejecta show that there is a large amount of core material in the ejecta. How and when is the core material mixed up into the accreted material? If it is mixed up too early in the evolution to the thermonuclear runaway, then the runaway will occur too early with too small an envelope mass.
- Observations imply that a far larger amount of material is ejected than is predicted by the theoretical calculations. What is the cause of this discrepancy? This problem is exacerbated by the observed mass accretion rates that, when used in the simulations, result in predicted nova explosions that do not resemble the observations.
- If novae are ejecting more mass than predicted, as the observations indicate, then how much do the actually contribute to Galactic chemical evolution. Note that most studies of the contribution of novae to Galactic chemical evolution use predicted masses rather than observed ejecta masses.
- Are Classical Novae the source of ^{26}Al in the Galaxy?
- How does a nova maintain Super-Eddington Luminosities for more than two weeks (LMC 1991). Note that in the LMC, with a smaller metallicity than Solar, all fast novae are Super-Eddington at maximum light.
- How does the non-spherical nature of the novae ejecta influence the nebular analyses?
- Why do some novae form dust and others not form dust?
- How does the same nova form different types of dust?
- What was the cause of the burst and the oscillations observed in the CHANDRA x-ray light curve of V1494 Aql? The BEPPOSAX light curve of V382 Vel also showed rapid variations in count rate during the observations (Orio et al. 2002).
- Is there a relationship between recurrent novae and SN Ia?
- How much, if any, of the secondary is entrained by the material blown off the white dwarf during the explosion and does the outburst pollute the secondary ?

I would like acknowledge the contributions of Cecelia Payne-Gaposchkin, Dean McLaughlin, Evry Schatzman, and Bob Kraft to the studies of novae in the past century. I feel very fortunate to have been able to discuss my nova work with Payne-Gaposchkin and present my work to Schatzman at his 70th birthday party in Paris. Although Kraft has not worked actively in this area for a number of years, he has counseled me on numerous occasions. His most recent questions have driven me to study the effects of metallicity on the outburst. I have greatly benefited from 30 years of collaborating with Warren Sparks and Jim Truran. I am also grateful for theoretical collaborations with Giora Shaviv and Peter Hauschildt. Our theoretical work has gone hand in hand with the observations of Bob Gehrz, Joachim Krautter, Steve Shore, Ed Sion, and Mark Wagner with whom I have had the pleasure of collaborating for many years. I would also like to acknowledge partial support from NASA and NSF grants to ASU and part of this review was written while I received support from the Institute for Nuclear Theory at the University of Washington.

REFERENCES

1. Aller, L. H., *ApJ* **432**, 427 (1994).
2. Amari, S. et al., *ApJ* **551**, 1065 (2001).
3. Burwitz, V. et al., these proceedings, 2002.
4. Code, A. D., *Scientific Results from OAO-2*, ed. A. D. Code, NASA Spec. Publ. 310, 1972, p. 535.
5. Crawford, J. A., Kraft, R. P., *ApJ* **123**, 44 (1956).
6. Della Valle, M., Rosino, L., Bianchini, A., Livio, M., *A&A* **287**, 403 (1994).
7. Evans, A. E., et al., these proceedings, 2002.
8. Ferland, G., *PASP* **110**, 761 (1998).
9. Gallagher, J. S., & Code, A. D., *ApJ* **189**, 303 (1974).
10. Gallagher, J. S., Hege, E. K., Kopriva, D. A., Butcher, H. R., Williams, R. E., *ApJ* **237**, 55 (1980).
11. Gallagher, J. S., & Ney, E. P., *ApJ* **204** L35 (1976).
12. Gallagher, J. S., and Starrfield, S., *MNRAS*, **176**, 53 (1976).
13. Gallagher, J. S., and Starrfield, S., *ARAA* **16**, 171 (1978).
14. Gehrz, R. D., *ARAA* **26**, 377 (1988).
15. Gehrz, R. D., Truran, J. W., Williams, R. E., Starrfield, S., *PASP* **110**, 3 (1998).
16. Geisel, S. L., Kleinmann, D. E., Low, F. J., *ApJL* **161**, L101 (1970).
17. Giannone, P., Weigert, A., *Zs. f. Ap.* **67**, 41 (1967).
18. Gómez-Gomar, Jordi; Hernanz, Margarita; José, Jordi; Isern, Jordi, *MNRAS* **296**, 913 (1998).
19. Hauschildt, P. H., Wehrse, R., Starrfield, S., Shaviv, G., *ApJ* **393**, 307 (1992).
20. Hauschildt, P. H., Shore, S. N., Schwarz, G., Baron, E., Starrfield, S., Allard, F., *ApJ* **490**, 803 (1997).
21. Hauschildt, P. H. et al., these proceedings, 2002.
22. Hernanz, Margarita; José, Jordi; Coc, Alain; Isern, Jordi, *ApJ* **465**, L27 (1996).
23. Hernanz, Margarita; José, Jordi; Coc, Alain; Gómez-Gomar, Jordi; Isern, Jordi, *ApJ* **526**, L27 (1999).
24. Hernanz, Margarita, et al., these proceedings, 2002.
25. Hyland, A. R., Neugebauer, G., *ApJL* **160**, L177 (1970).
26. José, J., & Hernanz, M., *ApJ* **494**, 680 (1998).
27. José, Jordi; Hernanz, Margarita; Coc, Alain, *ApJ* **479**, L55 (1997).
28. José, Jordi; Coc, Alain; Hernanz, Margarita, *ApJ* **520**, 347 (1999).
29. Joy, A. H., *PASP* **55**, 283 (1943).
30. Joy, A. H., *Stellar Atmospheres*, ed. J. L. Greenstein, Chicago, U. Chicago Press, 1960, p. 653.
31. Kahabka, P.; Hartmann, H. W.; Parmar, A. N.; Negueruela, I 1999, *A&A* **347**, L43 (1999).
32. Kato, T., et al., astro-ph/0204354 (2002).
33. Kovetz, A., & Prialnik, D., *ApJ* **291**, 812 (1985).
34. Kraft, R. P., *ApJ* **139**, 457 (1964).
35. Krautter, J., & Williams, R. E., *ApJ* **341**, 968 (1989).
36. Krautter, J., Ögelman, H., Starrfield, S., Wichmann, R., & Pfeffermann, E., *ApJ* **456**, 788 (1996).
37. MacDonald, J., Fujimoto, M. Y., Truran, J. W., *ApJ* **294**, 263 (1985).
38. McLaughlin, D. B. 1960, *Stellar Atmospheres*, ed. J. L. Greenstein, Chicago, U. Chicago Press, 1960, p. 585.
39. Mukai, K., & Ishida, M., *ApJ* **551**, 1024 (2001).
40. Ney, E. P. & Hatfield, B. F., *ApJ* **219**, L111 (1978).
41. Ögelman, H. B., Beuermann, K., and Krautter, J., *ApJ* **287**, L31 (1984).
42. Ögelman, H. B., Beuermann, K., and Krautter, J., *A&A* **177**, 110 (1987).
43. Orio, M., *Physics Reports* **311**, 419 (1999).
44. Orio, M., Covington, J., Ögelman, H., *A&A* **373**, 542 (2001).
45. Orio, M., et al., *MNRAS* **326**, L13 (2001).
46. Orio, M., et al., *MNRAS* **333**, L11 (2002).
47. Paerels, F. et al., *A&A* **365**, L298 (2001).
48. Payne-Gaposchkin, C., *The Galactic Novae*, New York, Dover (1957, reprinted 1964).
49. Prialnik, D., *ApJ* **310**, 222 (1986).
50. Prialnik, D., & Kovetz, A., *ApJ* **281**, 367 (1984).
51. Prialnik, D., Livio, M., Shaviv, G., Kovetz, A., *ApJ* **257**, 312 (1982).
52. Prialnik, D., Shara, M. M., Shaviv, G., *A&A* **62**, 339 (1978).
53. Prialnik, D., Shara, M. M., Shaviv, G., *A&A* **72**, 192 (1979).

54. Rawlings, J. M. C., & Evans A. E., these proceedings, 2002.
55. Schatzman, E. 1965, *Stellar Structure*, ed. L. H. Aller & D. B. McLaughlin, Chicago, U. Chicago Press, 1965, p. 327.
56. Shara, M. M., Prialnik, D., Shaviv, G., *A&A* **61**, 363 (1977).
57. Shara, M. M., Prialnik, D., Shaviv, G., *ApJ* **239**, 586 (1980).
58. Schwarz, G. J., et al., *MNRAS* **320**, 103 (2001).
59. Shore, S. N., Kenyon, S. J., Starrfield, S., Sonneborn, G., *ApJ* **456**, 717 (1996).
60. Shore, S. N., et al., *ApJ* **370**, 193 (1991).
61. Shore, S. N., Sonneborn, G., Starrfield, S., Riestra-Gonzalez, R., Ake, T. B., *AJ* **106**, 2408 (1993).
62. Shore, S. N., Starrfield, S., Ake, T. B., Hauschildt, P. H., *ApJ* **490**, 393 (1997).
63. Shore, S. N., et al., *NATURE* **369**, 539 (1994).
64. Sparks, W. M., *ApJ* **156**, 569 (1969).
65. Starrfield, S., *MNRAS* **152**, 307 (1971).
66. Starrfield, S., in *White Dwarfs and Variable Degenerate Stars*, ed. H. Van Horn & V. Weidemann, Rochester, University Press, 1979, p. 377.
67. Starrfield, S., in *Multiwavelength Observations in Astrophysics*, ed. F. A. Cordova, Cambridge, University Press, 1988, p. 159.
68. Starrfield, S., in *Classical Novae*, ed. M. Bode & A. Evans, New York, Wiley, 1989, p. 39.
69. Starrfield, S., Drake, J. et al., *BAAS* **33**, 804 (2001).
70. Starrfield, S., Schwarz, G. J., Truran, J. W., Sparks, W. M., *BAAS* **31**, 977 (1999a).
71. Starrfield, S., Schwarz, G. J., Truran, J. W., Sparks, W. M., Cosmic Explosions, S. Holt and W. Zhang, New York, AIP Proceedings #522, 2000, p. 379.
72. Starrfield, S., Shore, S. N., et al., *BAAS* **32**, 1253 (2000b).
73. Starrfield, S., Snijders, M. A. J., *Exploring the Universe with the IUE Satellite*, Y. Kondo, Dordrecht, Reidel, 1987, p. 377.
74. Starrfield, S., Sparks, W. M., Truran, J., W., *ApJS* **28**, 247 (1974).
75. Starrfield, S., Sparks, W. M., Truran, J. W., *ApJ* **291**, 136 (1985).
76. Starrfield, S., Sparks, W. M., Truran, J. W., *ApJ* **303**, L5 (1986).
77. Starrfield, S., Sparks, W. M., Shaviv, G., *ApJL* **325**, L35 (1988).
78. Starrfield, S., Truran, J., Sparks, W. M., and Kutter, G. S., *ApJ* **176**, 169 (1972).
79. Starrfield, S., Truran, J. W., Sparks, W. M., *ApJ* **226**, 186 (1978).
80. Starrfield, S., Truran, J. W., Sparks, W. M., Krautter, J.,*Extreme Ultraviolet Astronomy*, R. Malina, S. Bowyer, New York, Pergamon, 1991, p. 168.
81. Starrfield, S., et al., in *UV Space Astronomy Beyond HST*, ed. Jon A. Morse, J. Michael Shull, & Anne L. Kinney, 1999b, p.131
82. Struve, O., *Sky and Telescope* **14**, 275 (1995).
83. Tapia, S., *IAU Circular* **#2987** (1976).
84. Vanlandingham, K., et al., *MNRAS* **282**, 563 (1996).
85. Vanlandingham, K., Starrfield, S., Shore, S. N., Sonneborn G.,*MNRAS* **308**, 577 (1999).
86. Vanlandingham, K. et al., these proceedings, 2002
87. Walker, M. F., *PASP* **66**, 230 (1954).
88. Walker, M. F., *ApJ* **123**, 68 (1956).
89. Warner, B., *Cataclysmic Variable Stars*, Cambridge, University Press, 1995.
90. Williams, R. E., Gallagher, J. S., *ApJ* **228**, 482 (1979).
91. Williams, R. E., Ney, E. P., Sparks, W. M., Starrfield, S., Truran, J. W., *MNRAS* **212**, 753 (1985).
92. Williams, R. E., Woolf, N. J., Hege, E. K., Moore, R. L., & Kopriva, D. A., *ApJ* **224**, 171 (1978).

H-Accreting CO WDs: Accretion Regimes and Final Outcomes

Luciano Piersanti*, Santi Cassisi*, Icko Iben Jr† and Amedeo Tornambé*

*INAF-Osservatorio Astronomico di Teramo, via M. Maggini, 64100, Teramo, ITALY;
piersanti,cassisi,tornambe@te.astro.it
†Astronomy Department, University of Illinois, 1002 W. Green Street, Urbana, IL 61801;
icko@astro.uiuc.edu

Abstract. The long term evolution of Carbon-Oxygen White Dwarfs accreting Hydrogen rich matter at a rate typical for a recurrent mild H-flashes regime is reviewed to determine the dependence of the various accretion regimes on both the accretion rate and the chemical composition of the accreted matter. On the basis of these results semi-analytical relations defining the steady state regime are provided for different values of the metallicity in the accreted matter. In addition it has been found that models accreting H-rich matter in a pulsating regime can undergo an explosive dynamical He-ignition in the He-buffer accreted as a by-product of the overlying H-burning shell.

INTRODUCTION

During the last thirty years several works has been devoted to study the thermal properties and the evolution of Carbon-Oxygen (CO) White Dwarfs (WDs) accreting H-rich matter ([4], [15], [19], [20], [21], [22], [12], [17], [5], [18], [8], [6]). The current state of art can be summarized by referring to the work by Cassisi, Iben & Tornambé (hereinafter CIT - [1]) which have computed the long term evolution of H-accreting WDs. According to their results, a WD accreting H-rich matter with solar chemical composition can experience four different accretion regimes which depend mainly on the adopted accretion rate. For high values of the accretion rate, namely larger than $1 - 3 \times 10^{-7} M_\odot yr^{-1}$, the accreting star expands almost immediately and interacts with its companion already expanded, thus producing a Common Envelope (CE) episode; as a consequence a large part, if not all, of the matter transferred to the WD is lost from the system (see [10], [11], [9], [16]) so that the accreting WD can not grow in mass at all. For lower accretion rates ($10^{-7} \leq \dot{M} < 4 \times 10^{-8} M_\odot yr^{-1}$) the accreted Hydrogen is converted into Helium at the same rate at which matter is accreted. As a consequence an He-buffer is piled-up onto the CO WD up to the onset of the He-flash. For all the models computed in [1] the He-flash is non-dynamical, but in any case it is so energetic to induce the expansion of both the He- and H-rich layers, thus producing also in this case a CE episode which, in turn, prevents the CO core to grow in mass efficiently. Lowering the accretion rate still further, the accreting WD experiences recurrent mild H-flashes which pile-up an He-buffer on the CO core. The long term evolution of these models has not been computed by CIT due to the very large computing time required. At the end, for very low values of $\dot{M}$ (lower than $10^{-9} M_\odot yr^{-1}$) the accreting WD experiences strong nova-like

H-flashes; in this case matter is ejected from the system by a combination of dynamical acceleration, wind mass loss and common envelope action.

In the following the CIT's work will be extended, by considering H-accreting models with different metallicity and He content, in order to define the different accretion regimes as a function of the chemical composition of the accreted matter. In addition the final fate of models accreting H-rich matter in a pulsating regime will be discussed.

ACCRETION REGIMES AS A FUNCTION OF METALLICITY

In the left panels of Fig. 1 (*Panels a and b*) the well known evolution in the HR diagram is reported for models with the same total mass accreting H-rich matter at a rate typical for a recurrent mild H-pulses regime but with different chemical composition (see the labels); in the right panels (*Panels c and d*) the evolution of the mass coordinate of the H-burning shell is reported for the same models. As it can be noticed, by adopting different chemical composition for the accreted matter, the morphology of the tracks in the HR diagram does not change; however some differences there exist. First of all, during the low luminosity branch all the tracks are the same since, when H-burning is almost switched off, the models lye along an iso-radius whose location in the HR diagram is largely independent on the chemical composition of the external layers, being fixed only by the total mass of the star. As it can be noticed in Fig. 1c, for a fixed He content in the accreted matter, decreasing the metallicity the mass of the H-buffer increases since more matter has to be accreted for a new H-flash to be ignited. During the flash episode, decreasing the metallicity the tracks become redder: this occurrence is due to the different extensions in mass of the H-rich buffer. For the same reason the bluest point along the loop becomes redder decreasing the metallicity. On the other hand the reddest point becomes bluer since lower the metal content in the accreting matter, less strong the H-flash and smaller the delivered amount of nuclear energy: as a consequence the star has to expand to smaller radii to dissipate the energy excess produced by the flash itself. By an analysis of Fig. 1b it results that, for a fixed value of metallicity in the accreted matter, during the flash episode the tracks becomes redder decreasing the He content: this is due to the fact that lower the He abundance in the surface layers larger the surface opacity. For the same reason the bluest point becomes redder decreasing the He content. At the end, the reddest point becomes bluer decreasing the He content since, higher the He abundance, higher the molecular weight so that for a fixed value of the H-rich buffer, the H-burning shell results hotter and denser and, hence, the H-flash is stronger.

As it is well known ([5], [2], [3], [1]) during the high luminosity plateau phase the accreting model is in steady state: this means that the H-burning shell, whose luminosity is determined only by the mass extension of the underlying core, provides the whole energy radiated from the stellar surface. The bluest and the reddest points along the loop represent a sort of bifurcation points for model accreting H-rich matter (see [5], [14]): the bluest one represents the transition point between models in steady approximation and models which are cooling down; this transition is due to the fact that the H-burning rate ($\dot{M}_{H-sh}$) is higher than $\dot{M}$ so that the H-shell becomes too external and cool and the H-burning efficiency drops down; on the other hand the reddest point represents

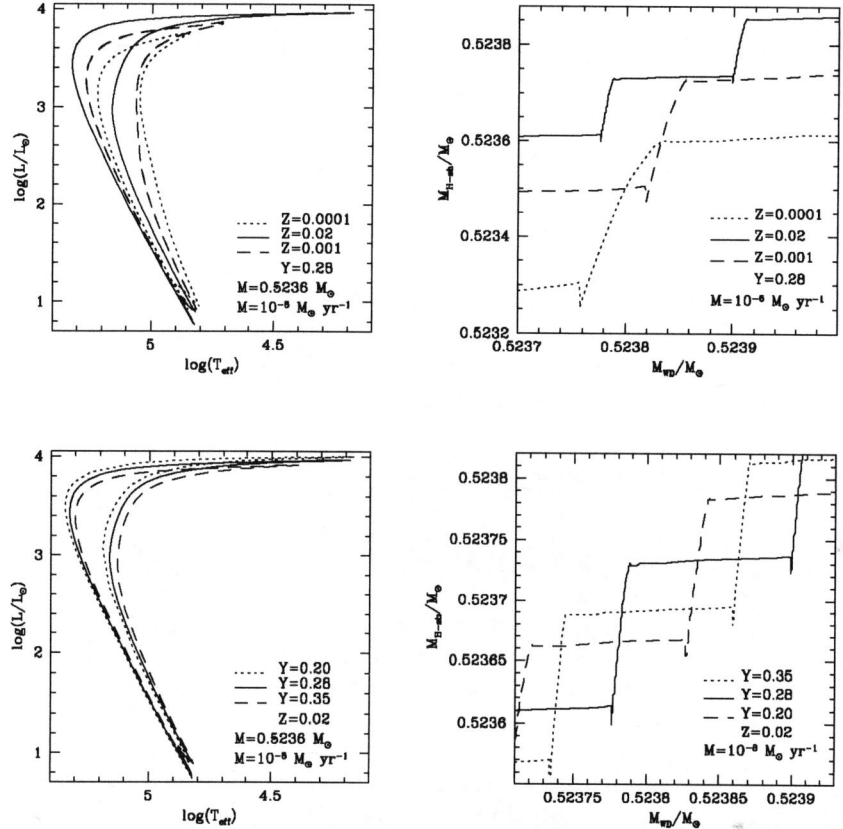

FIGURE 1. *Panel a*: Evolution in the HR diagram of accreting models with the same total mass and $\dot{M}$, but with different metallicity in the accreted matter; *Panel b*: as in *Panel a* but with different He abundance in the accreted matter; *Panel c*: time evolution of the H-burning shell during a complete H-flash for the models in *Panel a*; *Panel d*: as in *Panel c* but for the models in *Panel b*.

the transition point between models in steady state and models in Giant configuration. In fact, if $\dot{M}$ is higher than $\dot{M}_{H-sh}$ at the reddest point, the mass extension of the H-buffer continues to increase so that the external layers does not contract and, as a consequence, the model evolves toward the red. According to these considerations, it results that $\dot{M}_{H-sh}$ at the reddest and the bluest points along the loop represent the upper and the lower limits respectively for the accretion rate suitable for steady state regime. By computing the long term evolution of the models previously discussed we have obtained the maximum and minimum values for the accretion rate suitable for a stationary evolution of accreting WDs. A linear interpolation of these data provides semi-analytical relation which describes the location of the steady state zone in the $M_{WD} - \dot{M}$ plane as shown in Fig. 2 (see [14]): as it can be noticed, decreasing the metallicity the steady state zone reduces its extension and moves to lower $\dot{M}$.

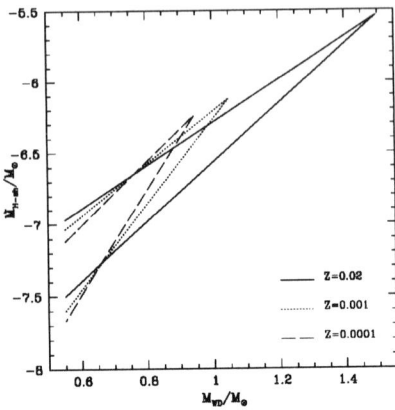

FIGURE 2. The steady state zone in the $M_{WD} - \dot{M}$ plane for models accreting H-rich matter with different metallicity, as labeled inside the figure.

THE LONG TERM EVOLUTION OF FLASHING MODELS

CIT has shown that models accreting H-rich matter in steady state undergo a non-dynamical He-flash ignited at the base of the He-buffer piled-up via H-burning. The comparison of the evolution of these models with those accreting Helium directly at the same accretion rate shows that in H-accreting models the physical conditions suitable for He-ignition are attained sooner since thermal energy is diffused from the H-shell inward, keeping hotter the physical base of the He-shell where the He-flash is ignited. At a first sight, this conclusion could seem to be valid also for models accreting H-rich matter in a pulsating regime since during the high luminosity plateau phase and the following cooling phase thermal energy is diffused from the H-shell inward. We have computed the long term evolution of models accreting H-rich matter at $\dot{M} = 2 \times 10^{-8} M_\odot yr^{-1}$ and we find that at the beginning of the accretion process the He-shell is kept hotter by the overlying H-burning shell. However, when the He-buffer becomes larger than $\sim 0.1 M_\odot$, the He-shell, defined as the point where the energy production via He-burning is at a maximum, jumps inward toward the physical base of the He-shell, defined as the point where the He abundance is 0.5 by mass fraction. This occurrence is due to the fact that the He-rich buffer insulate the two shells so that they thermally decouple. As a consequence the evolution of the He-shell occurs exactly as in the case He-rich matter is accreted directly, so that the final outcome of models accreting H-rich matter through recurrent mild H-flashes is exactly the same as in models accreting He-rich matter at the same rate. A comparison between our models and the extant numerical experiments concerning He-accreting WDs ([7], [6], [23]) suggests that our model will undergo a dynamical He-flash which will produce and explosive event of the He detonation type.

According to our results, we can guess that there exists an area in the $M_{WD} - \dot{M}$ plane suitable for events of the He detonation typology (the shaded region in Fig. 4): the limits in the accretion rate and WD initial mass have been fixed by observing that the accretion

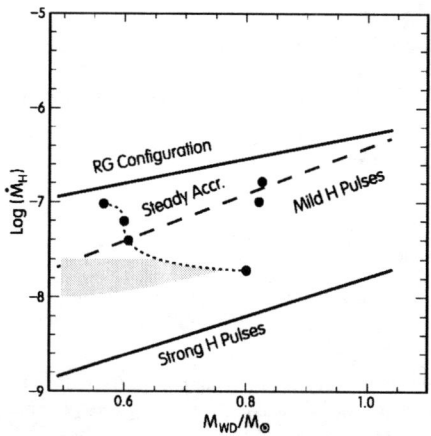

FIGURE 3. $M_{WD} - \dot{M}$ parameter space. The shaded region represents the area suitable for an explosive final outcome triggered by a dynamical He-flash (see text).

rate has to be small enough to guarantee the thermal decoupling of the two burning shells and, in addition, the initial mass can not be too high so that the dynamical He-ignition can occur well before that the accreting model enters in the Strong H-pulses regime. In fact, in the latter case, matter can not be longer added so that the physical conditions suitable for He-ignition can never be attained.

REFERENCES

1. Cassisi, S., Iben, I.Jr. & Tornambé A., 1998, *Ap.J*, **496**, 376
2. Fujimoto, M.Y., 1982, *Ap.J.*, **257**, 752
3. Fujimoto, M.Y., 1982, *Ap.J.*, **257**, 767
4. Giannone, P. & Weigert, A. 1967, *Z.Astrophys.*, **67**, 41
5. Iben, I. Jr., 1980, *Ap.J.*, **259**, 244
6. José, J., Hernanz, M. & Isern, J., 1993, *A.&A.*, **269**, 291
7. Limongi, M. & Tornambé, A., 1991, *Ap. J.*, **371**, 317
8. Livio, M., Prialnik, D. & Regev, O., 1989 *Ap.J.*, **341**, 299
9. Livio, M., Shankar, A., Berkert, A. & Truran, J.W., 1990, *Ap.J.*, **356**, 250
10. MacDonald, J., 1980, *M.N.R.A.S.*, **275**, 828
11. MacDonald, J., 1986, *Ap.J.*, **305**, 251
12. Paczyński, B. & Żytkov, A., 1978, *Ap.J.*, **222**, 604
13. Piersanti, L., Cassisi, S., Iben, I.Jr & Tornambé, A., 1999, *Ap.J.Lett.*, **521**, L59
14. Piersanti, L., Cassisi, S., Iben, I.Jr & Tornambé, A., 2000, *Ap.J.*, **535**, 932
15. Rose, W. K., 1968, *Ap.J.*, **152**, 245
16. Shankar, A., Livio, M. & Truran, J.W., 1991, *Ap.J.*, **374**, 693
17. Sion, E.M., Acierno, M.J. & Tomczyk, S., 1979, *Ap.J. SS.*, **230**, 832
18. Sion, E.M. & Starrfield S., 1986, *Ap.J.*, **303**, 130
19. Starrfield, S., 1971, *M.N.R.A.S.*, **152**, 307
20. Starrfield, S., 1971, *M.N.R.A.S.*, **155**, 129
21. Starrfield S., Sparks, W.M. & Truran, J.W., 1974, *Ap.J. SS.*, **28**, 247
22. Starrfield S., Sparks, W.M. & Truran, J.W., 1974, *Ap.J.*, **192**, 647
23. Woosley, S.E. & Weaver, T.A., 1994, *Ap. J.*, **423**, 371

Nuclear Ashes: Reviewing Thirty Years of Nucleosynthesis in Classical Novae

Jordi José

Dept. Física i Enginyeria Nuclear, Universitat Politècnica de Catalunya, and Institut d'Estudis Espacials de Catalunya (IEEC/UPC), Barcelona, Spain

Abstract. One of the observational evidences in support of the *thermonuclear runaway model* for the classical nova outburst relies on the accompanying nucleosynthesis. In this paper, we stress the relevant role played by nucleosynthesis in our understanding of the nova phenomenon by constraining models through a comparison with both the atomic abundance determinations from the ejecta and the isotopic ratios measured in presolar grains of a likely nova origin. Furthermore, the endpoint of nova nucleosynthesis provides hints for the understanding of the mixing process responsible for the enhanced metallicities found in the ejecta, and reveals also information on the properties of the underlying white dwarf (mass, luminosity...).

We discuss first the interplay between nova outbursts and the Galactic chemical abundances: Classical nova outbursts are expected to be the major source of ^{13}C, ^{15}N and ^{17}O in the Galaxy, and to contribute to the abundances of other species with A < 40, such as ^7Li or ^{26}Al. We describe the main nuclear path during the course of the explosion, with special emphasis on the synthesis of radioactive species, of particular interest for the gamma-ray output predicted from novae (^7Li, ^{18}F, ^{22}Na, ^{26}Al). An overview of the recent discovery of presolar nova candidate grains, as well as a discussion of the role played by nuclear uncertainties associated with key reactions of the NeNa-MgAl and Si-Ca regions, are also given.

GALACTIC ALCHEMY: THE INTERPLAY BETWEEN NOVA OUTBURSTS AND THE GALACTIC ABUNDANCES

The high peak temperatures achieved during nova explosions, $T_{peak} \sim (2-3) \times 10^8$ K, suggest that abundance levels of the intermediate-mass elements in the ejecta must be significantly enhanced, as confirmed by spectroscopic determinations in well-observed nova shells. This raises the issue of the potential contribution of novae to the Galactic abundances, which can be roughly estimated as the product of the Galactic nova rate, the average ejected mass per nova outburst, and the Galaxy's lifetime. This order of magnitude estimate points out that novae scarcely contribute to the Galaxy's overall metallicity (as compared with other major sources, such as supernova explosions), nevertheless they can substantially contribute to the synthesis of some largely overproduced species (see Table 1, for a sample of publications addressing nucleosynthesis in classical novae). Hence, classical novae are likely sites for the synthesis of most of the Galactic ^{13}C, ^{15}N and ^{17}O, whereas they can partially contribute to the Galactic abundances of other species with A < 40, such as ^7Li, ^{19}F, or ^{26}Al [83, 39].

Overproduction factors, relative to solar, corresponding to hydrodynamic calculations

TABLE 1. A sample of publications in refereed journals addressing nucleosynthesis in classical novae

Reference	Model category	Range of nuclei
Arnould & Nørgaard (1975) [4]	Parametric, 1 zone	^{3}He, ^{7}Li, B, C, N, O
Arnould et al. (1980) [5]	Parametric, 1 zone	H-Ar
Boffin et al. (1993) [8]	Parametric, 1 & 2 zones	^{3}He, ^{7}Be, ^{7}Li, ^{8}B, ^{9}C
Coc et al. (1995) [13]	Semianalytic, 1 zone	H-K
Coc et al. (2000) [14]	Hydrodynamic, 1D	O, F
Glasner et al. (1997) [26]	Hydrodynamic, 2D	H-F
Hernanz et al. (1996) [28]	Hydrodynamic, 1D	^{3}He, ^{7}Be, ^{7}Li
Hernanz et al. (1999) [29]	Hydrodynamic, 1D	^{13}N, ^{18}F
Hillebrandt & Thielemann (1982) [32]	Parametric, 1 zone	H-Ar
Iliadis et al. (1999) [33]	Parametric, 1 zone	H-Ca
José et al. (1997) [37]	Hydrodynamic, 1D	Na, Mg, Al
José & Hernanz (1997) [38]	Hydrodynamic, 1D	H-Ca
José & Hernanz (1998) [39]	Hydrodynamic, 1D	H-Ca
José et al. (1999) [40]	Hydrodynamic, 1D	Ne-Na, Mg-Al
José et al. (2001a) [41]	Hydrodynamic, 1D	^{13}N, ^{18}F, ^{7}Be, ^{22}Na, ^{26}Al
José et al. (2001b) [42]	Hydrodynamic, 1D	C,N,O,Al,Mg,Si,Ne
José et al. (2001c) [43]	Hydrodynamic, 1D	Si-Ca
Kercek et al. (1998) [47]	Hydrodynamic, 2D	H-F
Kercek et al. (1999) [48]	Hydrodynamic, 3D	H-F
Kolb & Politano (1997) [49]	Hydrodynamic, 1D	^{22}Na, ^{26}Al
Kovetz & Prialnik (1985) [50]	Hydrodynamic, 1D	H-O
Kovetz & Prialnik (1997) [51]	Hydrodynamic, 1D	H-O
Kudryashov & Tutukov (1995) [52]	Parametric, 1 zone	H-Ar
Kudryashov et al. (2000) [53]	Parametric, 1 zone	H-Ar
Lazareff et al. (1979) [54]	Parametric, 2 zones	C, N, O, F, Ne
Nofar et al. (1991) [56]	Parametric, 1 zone	H-Al
Politano et al. (1995) [59]	Hydrodynamic, 1D	H-Ca
Prialnik et al. (1978) [60]	Hydrodynamic, 1D	H-O
Prialnik et al. (1979) [61]	Hydrodynamic, 1D	H-O
Prialnik (1986) [62]	Hydrodynamic, 1D	H-O
Prialnik & Shara (1986) [63]	Hydrodynamic, 1D	H-Ne
Prialnik & Shara (1995) [64]	Hydrodynamic, 1D	H-P
Shara & Prialnik (1994) [69]	Hydrodynamic, 1D	H-Mg
Sparks et al. (1978) [73]	Hydrodynamic, 1D	H-O
Starrfield et al. (1972) [74]	Hydrodynamic, 1D	H-O
Starrfield et al. (1974a) [75]	Hydrodynamic, 1D	H-O
Starrfield et al. (1974b) [76]	Hydrodynamic, 1D	H-O
Starrfield et al. (1978a) [77]	Hydrodynamic, 1D	^{3}He, ^{7}Li, C, N, O
Starrfield et al. (1978b) [78]	Hydrodynamic, 1D	^{7}Li, C, N, O
Starrfield et al. (1992) [80]	Hydrodynamic, 1D	H-Ar
Starrfield et al. (1993) [81]	Hydrodynamic, 1D	H-Ar
Starrfield et al. (1998) [83]	Hydrodynamic, 1D	H-Ar
Starrfield et al. (2000) [84]	Hydrodynamic, 1D	H-Ar
Starrfield et al. (2001) [85]	Hydrodynamic, 1D	H-S
Vangioni-Flam et al. (1980) [87]	Parametric, 1 zone	Ne, Al
Wallace & Woosley (1981) [89]	Parametric, 1 zone	H-Al
Wanajo et al. (1999) [90]	Semianalytic, 1 zone	H-Ca
Weiss & Truran (1990) [91]	Parametric, 1 zone	H-Ca
Wiescher et al. (1986) [92]	Parametric, 1 zone	H-Ar

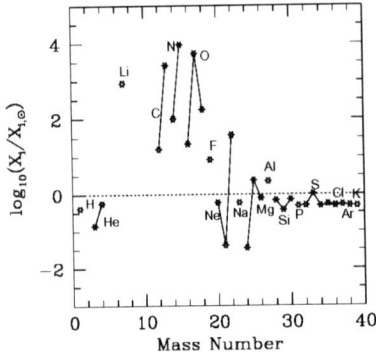 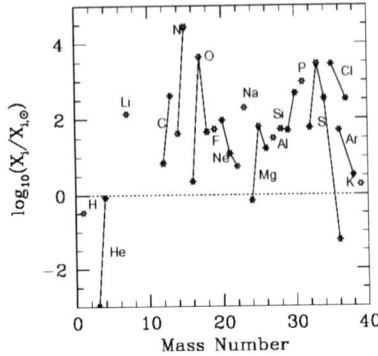

FIGURE 1. (Left) Mean overproduction factors, relative to solar, in the ejecta of a 1.15 $M_\odot$ CO novae. (Right) Same for a 1.35 $M_\odot$ ONe novae. Figures are based on hydrodynamic calculations reported in [39].

of nova outbursts on top of a 1.15 $M_\odot$ CO and a 1.35 $M_\odot$ ONe white dwarf, are shown in Figure 1. Because of the lower peak temperatures achieved in CO models, and also because of the lack of significant amounts of seed nuclei in the NeNa-MgAl region, the main nuclear activity in CO novae does not extend much beyond oxygen, as seen from the overproduction plot. In contrast, ONe models show a much larger nuclear activity, extending up to silicon (1.15 $M_\odot$ ONe) or argon (1.35 $M_\odot$ ONe). Hence, the presence of significantly large amounts of intermediate-mass nuclei in the spectra, such as phosphorus, sulfur, chlorine or argon, may reveal the presence of an underlying massive ONe white dwarf. Another trend derived from the analysis of the nucleosynthesis accompanying nova outbursts is the fact that the O/N and C/N ratios decrease as the mass of the white dwarf (and hence, the peak temperature attained during the explosion) increases.

Abundance Determinations in the Ejecta from Novae

In order to constraint the models, several works have focused on a direct comparison of the atomic abundances inferred from observations of the ejecta with the theoretical nucleosynthetic output (see [39, 83], and references therein). Despite of the problems associated with the modeling of the explosion [86], such as the unknown mechanism responsible for the mixing between the accreted envelope and the outermost shells of the underlying white dwarf [9, 21], or the difficulties to eject as much material as inferred from observations [71], there is an excellent agreement between theory and observations as regards nucleosynthesis (i.e., including atomic abundances -H, He, C, O, Ne, Na-Fe-, and a plausible endpoint for nova nucleosynthesis). In some cases, such as for PW Vul 1984, the agreement between observations and theoretical predictions (see [39], Table 5, for details) is really overwhelming. The reader is referred to [24] for an extended list of abundance determinations in the ejecta from novae, and to [68, 88] for recent efforts to improve the abundance pattern for QU Vul 1984 and V1974 Cyg 1992, respectively.

Since the nuclear path is very sensitive to details of the evolution (chemical composition, extend of convective mixing, thermal history of the envelope...), the agreement between inferred abundances and theoretical yields not only validates the thermonuclear runaway model, but also poses limits on the (yet unknown) mixing mechanism itself: for instance, if mixing occurs very late in the course of the explosion, the accumulation of larger amounts of matter in the envelope will be favored (since the injection of significant amounts of the triggering nucleus ^{12}C will be delayed). Hence, one would expect to end up with a more violent outburst, characterized by a higher T_{peak}, exceeding in some cases 4×10^8 K, and, as a result, a significant enrichment in heavier species, beyond calcium, in the ejecta from novae involving very massive white dwarfs, a pattern never observed so far.

Presolar Grains: Gifts from Heaven

Infrared [22, 24] and ultraviolet observations [70] of the temporal evolution of nova light curves suggest that novae form grains in the expanding nova shells. Both CO and ONe novae behave similarly in the infrared right after the outburst. However, as the ejected envelope expands and becomes optically thin, such behavior dramatically changes: CO novae are typically followed by a phase of dust formation corresponding to a decline in visual light, together with a simultaneous rise in the infrared emission [23, 25]. In contrast, it has been argued that ONe novae (that involve more massive white dwarfs than CO novae) are not so prolific producers of dust as a result of the lower mass, high-velocity ejecta, where the typical densities can be low enough to enable the condensation of appreciable amounts of dust. Hints on the condensation of dust containing silicates, silicon carbide, carbon and hydrocarbons have been reported from a number of novae (see [24] for a recent review).

Up to now, the identification of presolar nova grains, presumably condensed in the shells ejected during the explosion, relied only on low $^{20}Ne/^{22}Ne$ ratios (attributed to ^{22}Na decay), but quite recently five silicon carbide and two graphite grains that exhibit isotopic signatures characteristic of nova nucleosynthesis have been identified [1, 2]. They are characterized by very low $^{12}C/^{13}C$ and $^{14}N/^{15}N$ ratios, ^{30}Si excesses and close-to- or slightly lower-than-solar $^{29}Si/^{28}Si$ ratios, high $^{26}Al/^{27}Al$ ratios (determined only for two grains) and low $^{20}Ne/^{22}Ne$ ratios (only measured in the graphite grain KFB1a-161). Such a promising discovery provides a much valuable source of constraint for nova nucleosynthesis (since contrary to the atomic abundance determinations derived from nova ejecta, measurements provide more accurate isotopic ratios) and opens interesting possibilities for the future.

Theoretical isotopic ratios for a variety of nuclear species, ranging from C to Si, based on hydrodynamic computations of the nova outburst, have been reported by different authors [82, 42, 44, 45]. A more detailed analysis, which focuses on the different chemical pattern expected for CO and ONe novae, will be presented elsewhere [46].

SYNTHESIS OF RADIOACTIVE NUCLEI DURING NOVA OUTBURSTS

Among the isotopes synthesized during classical nova outbursts, several radioactive species deserve a particular attention. Short-lived nuclei, such as ^{14,15}O and ^{17}F (and to some extent ^{13}N) have been identified as the key isotopes that power the expansion and further ejection during a nova outburst through a sudden release of energy, a few minutes after peak temperature [74]. Other isotopes have been extensively investigated in connection with the theoretical gamma-ray output from novae [11, 12, 55]. Hence, ^{13}N and ^{18}F are responsible for the predicted prompt γ-ray emission [29] at and below 511 keV, whereas ^{7}Be and ^{22}Na [27], which decay much later, when the envelope is optically thin, are the sources that power line emission at 478 and 1275 keV, respectively. ^{26}Al is another important radioactive isotope that can be synthesized during nova outbursts, although only its cumulative emission can be observed because of its slow decay.

We will briefly focus on the corresponding nuclear paths leading to the synthesis of the abovementioned gamma-ray emitters, with special emphasis on the nuclear uncertainties associated with the relevant reaction rates. We refer the reader to [30] for a review of the current theoretical predictions of the gamma-ray output from novae and of the chances for a nearby future detection using spacecrafts such as INTEGRAL (see also [31]). As for a comprehensive summary of the main uncertainties affecting nuclear reaction rates for nova temperatures, the reader is referred to [15]. Other recent attempts to fully analyze the impact of nuclear physics uncertainties in nova nucleosynthesis involve parametric one-zone calculations with temperature and density profiles extracted from hydrodynamic models [34, 35], as well as Monte Carlo simulations [72]. We would like to stress, however, that results based on parametric calculations usually tend to overestimate the effect of a given reaction rate uncertainty, as compared with the outcome from hydrodynamic tests. Hence, although these simplified techniques provide a very valuable tool to potentially identify key reactions through an extraordinary large number of tests, final conclusions have to be confirmed through hydrodynamic calculations.

^{7}Be-^{7}Li

^{3}He$(\alpha, \gamma)^7$Be is the main reaction leading to ^{7}Be synthesis (with ^{3}He coming from both the accreted amount plus the contribution in place from ^{1}H(p,e$^+$ $\nu_e)^2$H(p,$\gamma)^3$He), which is transformed into ^{7}Li by means of an electron capture, emitting a γ-ray photon of 478 keV [27].

Its production in classical nova outbursts has been very controversial. Results from the first pioneering calculation, in the framework of a simple parametric model [4], were confirmed by early hydrodynamic simulations [77], assuming however an envelope in-place (the first hydrodynamic nova models that properly included the onset of the accretion phase were not available until the 80s), thus neglecting the impact of the accretion phase on the evolution. These results were refuted later on, in terms of parametric one/two zone models [8], pointing out the critical role played by the ^{9}C(p,γ) reaction, not included in all previous works (i.e., [4, 77]), and claiming therefore for an unlikely synthesis of ^{7}Li in novae. The scenario was recently revisited by [28, 39], who performed new hydrodynamic calculations, taking into account both the accretion and explosion stages, and a full reaction network (including ^{9}C(p,γ)). These studies confirmed that the Be-transport mechanism [10] is able to produce a large overproduction of ^{7}Li in nova explosions.

Among the issues that affect ^{7}Li synthesis in novae, one the most critical ones is the final amount of ^{3}He that survives the early rise in temperature when the thermonuclear runaway ensues. In particular, the different timescales to reach T_{peak} achieved for CO and ONe novae, which deeply depend on the initial ^{12}C content in the envelope, lead to a larger amount of ^{7}Be in CO novae (which survives destruction through ^{7}Be(p,γ)^{8}B because of the very efficient inverse photodisintegration reaction on ^{8}B). No relevant nuclear uncertainties in the domain of nova temperatures affect the corresponding reaction rates.

The puzzling ^{7}Li synthesis in novae, suggested for many years but elusive up to now, has been apparently confirmed for the first time in the spectra of V382 Vel (Nova Velorum 1999) [18], for which an observed feature compatible with the doublet at 6708 Å of LiI has been reported. Although this identification is not yet confirmed (there is no unambiguous interpretation for such a feature), the lithium abundances inferred are fully compatible with the theoretical upper limits given for a fast nova (i.e., [28, 39]).

The potential contribution of classical novae to the Galactic ^{7}Li content turns out to be rather small (i.e., less than 15%), even if the ^{7}Li yield from the most favorable case, a massive CO nova [28], is considered. Nevertheless, a nova contribution is required to match the ^{7}Li content in realistic calculations of Galactic chemical evolution [66, 67].

^{22}Na

The potential role of ^{22}Na for diagnosis of nova outbursts was first suggested by [11]. It decays to a short-lived excited state of ^{22}Ne (with a lifetime of $\tau = 3.75$ yr), which de-excites to its ground state by emitting a γ-ray photon of 1.275 MeV. Through this mechanism, nearby ONe novae within a few kiloparsecs from the Sun may provide detectable γ-ray fluxes [30]. Several experimental verifications of this γ-ray emission at 1.275 MeV from nearby novae have been attempted in the last twenty years, using balloon-borne experiments and detectors on-board satellites such as HEAO-3, SMM, or CGRO, from which upper limits on the ejected ^{22}Na have been derived. In particular, the observations performed with the COMPTEL experiment on-board CGRO of five recent Ne-type novae [36], as well as observations of *standard* CO novae, have led to an upper limit of 3.7×10^{-8} M$_\odot$ for the ^{22}Na mass ejected by any nova in the Galactic disk. A limit that poses some constraints on pre-existing theoretical models of classical nova explosions.

Synthesis of ^{22}Na in novae, extensively investigated in the last two decades [32, 92, 94, 91, 58, 59, 13, 52, 49, 39, 83, 40, 90, 53, 84, 41, 85], proceeds through different reaction paths. In the ^{20}Ne-enriched envelopes of ONe novae [40], it takes place through ^{20}Ne(p,γ)^{21}Na, followed either by another proton capture and a β^+-decay into ^{22}Na (i.e., ^{21}Na(p,γ)^{22}Mg(β^+)^{22}Na), or decaying first into ^{21}Ne before another proton capture ensues (i.e., ^{21}Na(β^+)^{21}Ne(p,γ)^{22}Na). Other potential channels, such as proton captures on the seed nucleus ^{23}Na, play only a marginal role on ^{22}Na synthesis because of the much higher initial ^{20}Ne content in such ONe models. As for the main destruction channel at nova temperatures, ^{22}Na(p,γ)^{23}Mg competes favorably with ^{22}Na(β^+)^{22}Ne. Nuclear uncertainties strongly affect the rates for ^{21}Na(p,γ)^{22}Mg and ^{22}Na(p,γ)^{23}Mg [40], which translate into an uncertainty in the final ^{22}Na yields (and ultimately on the maximum detectability distance of the 1275 keV line expected from nova outbursts). Advances to reduce the uncertainty affecting the ^{21}Na(p,γ) rate have been recently achieved with the DRAGON recoil separator facility at TRIUMF [57, 7].

^{26}Al

Several isotopes should be considered as potential seeds for ^{26}Al synthesis: in particular, ^{24,25}Mg and to some extent ^{23}Na and ^{20,22}Ne [37]. The main nuclear reaction path leading to ^{26}Al synthesis, also investigated in a large number of papers [54, 5, 32, 17, 92, 94, 93, 91, 56, 58, 69, 59, 13, 52, 49, 39, 83, 40, 90, 53, 84, 41, 85], is given by ^{24}Mg(p,γ)^{25}Al(β^+)^{25}Mg(p,γ)^{26}Al^g, whereas destruction is dominated by ^{26}Al^g(p,γ)^{26}Si.

A significant nuclear uncertainty affects the ^{25}Al(p,γ)^{26}Si rate [40], which translates into an uncertainty in the expected contribution of novae to the Galactic ^{26}Al content. A critical issue in order to estimate this contribution is the initial composition of the ONe white dwarf. Whereas calculations by Starrfield et al. [79, 83, 84, 85] assume a core composition based on hydrostatic models of carbon burning nucleosynthesis [3], rather enriched in ^{24}Mg (with a ratio ^{16}O:^{20}Ne:^{24}Mg around 1.5:2.5:1), we adopt more recent values taken from stellar evolution calculations of intermediate-mass stars [65], for which the ^{24}Mg content is much smaller (^{16}O:^{20}Ne:^{24}Mg is 10:6:1). Calculations based on the new ONe white dwarf composition [37, 39, 40, 41], suggest that the contribution of novae to the Galactic ^{26}Al abundance is rather small (i.e., less than 15%), in good agreement with the results derived from the COMPTEL map of the 1809 keV ^{26}Al emission in the Galaxy (see [20]), which points towards young progenitors (type II supernovae and Wolf-Rayet stars).

^{18}F

The predicted gamma-ray emission from novae at and below 511 keV at early epochs after the explosion is basically driven by the amount of ^{18}F present in the envelope [55, 27, 29]. The synthesis of ^{18}F is powered by ^{16}O(p,γ)^{17}F, followed either by ^{17}F(p,γ)^{18}Ne(β^+)^{18}F or by ^{17}F(β^+)^{17}O(p,γ)^{18}F. The dominant destruction channel is ^{18}F(p,α)^{15}O plus a minor contribution from ^{18}F(p,γ)^{19}Ne. The effect of the nuclear uncertainties associated with some of the rates (i.e., ^{18}F(p,α)^{15}O, ^{18}F(p,γ)^{19}Ne, ^{17}O(p,α)^{14}N and ^{17}O(p,γ)^{18}F) [29, 14, 15] is, in this case, quite remarkable: they translate into a large uncertainty in the expected ^{18}F yield, and therefore, in the corresponding gamma-ray flux and maximum detectability distance. Advances to reduce this uncertainty have been reviewed during the Classical Nova Conference [6, 19] and elsewhere [16], and involve several nuclear physics experiments performed in Oak Ridge (USA) and Orsay (France).

The endpoint of nova nucleosynthesis

In agreement with the chemical pattern derived from detailed observations of the ejecta, the theoretical endpoint for nova nucleosynthesis is limited to A < 40 (i.e., calcium), in agreement with current theoretical nucleosynthetic estimates, provided that the temperatures attained in the envelope during the explosion remain limited to $T_{peak} \sim (2-3) \times 10^8$ K.

The nuclear activity in the Si-Ca region has been scarcely analyzed in detail in the context of classical nova outbursts [83, 33, 90, 43]. It is powered by a leakage from the NeNa-MgAl region, where the activity is confined during the early stages of the explosion. The main reaction that drives the nuclear activity towards heavier species (i.e., beyond S) is mainly ^{30}P(p,γ)^{31}S, either followed by ^{31}S(p,γ)^{32}Cl(β^+)^{32}S, or by ^{31}S(β^+)^{31}P(p,γ)^{32}S [43]. The ^{30}P(p,γ) rate is based only on Hauser-Feshbach estimates, which can be rather uncertain at the domain of nova temperatures. To test the effect of this uncertainty on the predicted yields, we have performed a series of hydrodynamic calculations [43], modifying arbitrarily the nominal rate. Hence, for a high ^{30}P(p,γ) rate (i.e., 100 times the nominal one), the final ^{30}Si yields are dramatically reduced by a factor of 30, whereas for a low ^{30}P(p,γ) rate (i.e., 0.01 times the nominal one), the final ^{30}Si yields are slightly increased by a factor of 5, whereas isotopes above silicon are reduced by a factor of ~ 10, with dramatic impact on theoretical estimates ionvolving both the composition of the ejecta and of presolar grains.

The impact of nuclear reaction rate uncertainties on nucleosynthesis calculations points out the need of accurate nuclear physics inputs, and stresses the role played by classical novae as perfect laboratories for nuclear astrophysics.

ACKNOWLEDGMENTS

First, I would like to thank Margarida Hernanz, for a continuous support. I have learned much from her invaluable criticism, encouragement, and advice. I would also like to thank Sachiko Amari, Alain Coc, John D'Auria, Christian Iliadis, Jordi Isern, Jim MacDonald, Steve Shore, Sumner Starrfield, Jim Truran, Michael Wiescher, and Ernst Zinner, among others, for many fruitful and stimulating discussions on several aspects involving classical nova outbursts.

REFERENCES

1. Amari, S., Gao, X., Nittler, L. R., Zinner, E., José, J., Hernanz, M., and Lewis, R. S., *ApJ* **551**, 1065 (2001).
2. Amari, S., *New Astron. Rev.* **46**, 519 (2002).
3. Arnett, W.D., and Truran, J.W., *ApJ* **157**, 339 (1969).
4. Arnould, M., and Nørgaard, H., *A&A* **42**, 55 (1975).
5. Arnould, M., Nørgaard, H., Thielemann, F.-K., and Hillebrandt, W., *ApJ* **237**, 931 (1980).
6. Bardayan, D., et al., these proceedings, 2002.
7. Bishop, S., et al., *NP* A, in press (2002).
8. Boffin, H.M.J., Paulus, G., Arnould, M., and Mowlavi, N., *A&A* **279**, 173 (1993).
9. Calder, A.C., et al., these proceedings, 2002.
10. Cameron, A.G.W., *ApJ* **121**, 144 (1955).
11. Clayton, D.D., and Hoyle, F., *ApJ* **187**, L101 (1974).
12. Clayton D.D., *ApJ* **244**, L97 (1981).
13. Coc, A., Mochkovitch, R., Oberto, Y., Thibaud, J.-P., and Vangioni-Flam, E., *A&A* **299**, 479 (1995).
14. Coc, A., Hernanz, M., José, J., and Thibaud, J.-P., *A&A* **357**, 561 (2000).
15. Coc, A., Smirnova, N., José, J., Hernanz, M., and Thibaud, J.-P., *NP* **A688**, 450 (2001).
16. Coc, A., et al., *NP* A, in press (2002).
17. Delbourgo-Salvador, P., Mochkovitch, R., and Vangioni-Flam, E., in *Recent results on Cataclysmic Variables*, ESA, Bamberg, 1985, p. 229.
18. Della Valle, M., Pasquini, L., Daou, D., and Williams, R.-E., *A&A*, in press (2002).
19. De Séréville, N., et al., these proceedings, 2002.
20. Diehl, R., et al., *A&A* **298**, 445 (1995).
21. Dursi, L.J., et al., these proceedings, 2002.
22. Evans, A., in *The Physics of Classical Novae*, edited by A. Cassatella and R. Viotti, Springer-Verlag, Berlin, 1990, p. 253.
23. Evans, A., and Rawlings, J.M.C., these proceedings, 2002.
24. Gehrz, R.D., Truran, J.W., Williams, R.E., and Starrfield, S., *PASP* **110**, 3 (1998).
25. Gehrz, R.D., these proceedings, 2002.
26. Glasner, S.A., Livne, E., and Truran, J.W., *ApJ* **475**, 754 (1997).
27. Gómez-Gomar, J., Hernanz, M., José, J., and Isern, J., *MNRAS* **296**, 913 (1998).
28. Hernanz, M., José, J., Coc, A., and Isern, J., *ApJ* **465**, L27 (1996).
29. Hernanz, M., José, J., Coc, A., Gómez-Gomar, J., and Isern, J., *ApJ* **526**, L97 (1999).
30. Hernanz, M., these proceedings, 2002.
31. Hernanz, M., Jean, P., José, J., Coc, A., Starrfield, S., Truran, J.W., Isern, J., Sala, G., and Giménez, A., these proceedings, 2002.
32. Hillebrandt, W., and Thielemann, F.-K., *ApJ* **225**, 617 (1982).
33. Iliadis, C., Endt, P., Prantzos, N., and Thompson, W.J., *ApJ* **524**, 434 (1999).
34. Iliadis, C., Champagne, A., José, J., Starrfield, S., and Tupper, P., *ApJS*, in press (2002).
35. Iliadis, C., Champagne, A., José, J., Starrfield, S., and Tupper, P., these proceedings, 2002.
36. Iyudin, A.F., et al., *A&A* **300**, 422 (1995).
37. José, J., Hernanz, M., and Coc, A., *ApJ* **479**, L55 (1997).
38. José, J., and Hernanz, M., *NP* **A621**, 491 (1997).
39. José, J., and Hernanz, M., *ApJ* **494**, 680 (1998).
40. José, J., Coc, A., and Hernanz, M., *ApJ* **520**, 347 (1999).
41. José, J., Hernanz, M., and Coc, A., *NP* **A688**, 118 (2001).
42. José, J., Hernanz, M., Amari, S., and Zinner, E., *NP* **A688**, 439 (2001).
43. José, J., Coc, A., and Hernanz, M., *ApJ* **560**, 897 (2001).
44. José, J., Hernanz, M., Amari, S., and Zinner, E., in *Cosmic Evolution*, edited by E. Vangioni-Flam, R. Ferlet and M. Lemoine, World Scientific, Singapore, 2001, p. 159.
45. José, J., Hernanz, M., Amari, S., and Zinner, E., these proceedings, 2002.
46. José, J., Hernanz, M., Amari, S., and Zinner, E., *ApJ*, in preparation (2002).
47. Kercek, A., Hillebrandt, W., and Truran, J.W., *A&A* **337**, 379 (1998).
48. Kercek, A., Hillebrandt, W., and Truran, J.W., *A&A* **345**, 831 (1999).
49. Kolb, U., and Politano, M., *A&A* **319**, 909 (1997).

50. Kovetz, A., and Prialnik, D., *ApJ* **291**, 812 (1985).
51. Kovetz, A., and Prialnik, D., *ApJ* **477**, 356 (1997).
52. Kudryashov, A.D., and Tutukov, A.V., *Astron. Reps.* **39**, 482 (1995).
53. Kudryashov, A.D., Chugai, N.N., and Tutukov, A.V., *Astron. Reps.* **44**, 170 (2000).
54. Lazareff, B., Audouze, J., Starrfield, S., and Truran, J.W., *ApJ* **228**, 875 (1979).
55. Leising, M.D., and Clayton, D.D., *ApJ* **323**, 159 (1987).
56. Nofar, I., Shaviv, G., and Starrfield, S., *ApJ* **369**, 440 (1991).
57. Olin, A., et al., these proceedings, 2002.
58. Paulus, G., and Forestini, M., in *Gamma-Ray Line Astrophysics*, edited by P. Durouchoux and N. Prantzos, AIP, New York, 1991, p. 183.
59. Politano, M., Starrfield, S, Truran, J.W., Weiss, A., and Sparks, W.M., *ApJ* **448**, 807 (1995).
60. Prialnik, D., Shara, M.M., and Shaviv, G., *A&A* **62**, 339 (1978).
61. Prialnik, D., Shara, M.M., and Shaviv, G., *A&A* **72**, 192 (1979).
62. Prialnik, D., *ApJ* **310**, 222 (1986).
63. Prialnik, D., and Shara, M.M., *ApJ* **311**, 172 (1986).
64. Prialnik, D., and Shara, M.M., *AJ* **109**, 1735 (1995).
65. Ritossa, C., García-Berro, E., and Iben, I. Jr., *ApJ* **460**, 489 (1996).
66. Romano, D., Matteucci, F., Molaro, P., and Bonifacio, P., *A&A* **352**, 117 (1999).
67. Romano, D., and Matteucci, F., these proceedings, 2002.
68. Schwarz, G.J., these proceedings, 2002.
69. Shara, M.M., and Prialnik, D., *AJ* **107**, 1542 (1994).
70. Shore, S.N., Starrfield, S., González-Riestra, R., Hauschildt, P.H., and Sonneborn, G., *Nat.* **369**, 539 (1994).
71. Shore, S.E., these proceedings, 2002.
72. Smith, M.S., et al., these proceedings, 2002.
73. Sparks, W.M., Starrfield, S., and Truran, J.W., *ApJ* **220**, 1063 (1978).
74. Starrfield, S, Truran, J.W., Sparks, W.M., and Kutter, G.S., *ApJ* **176**, 169 (1972).
75. Starrfield, S, Sparks, W.M., and Truran, J.W., *ApJS* **28**, 247 (1974).
76. Starrfield, S, Sparks, W.M., and Truran, J.W., *ApJ* **192**, 647 (1974).
77. Starrfield, S., Truran, J.W., Sparks, W.M., and Arnould, M., *ApJ* **222**, 600 (1978).
78. Starrfield, S, Truran, J.W., and Sparks, W.M., *ApJ* **226**, 186 (1978).
79. Starrfield, S, Sparks, W.M., and Truran, J.W., *ApJ* **303**, L5 (1986).
80. Starrfield, S., Shore, S.N., Sparks, W.M., Sonneborn, G., Truran, J.W., and Politano, M., *ApJ* **391**, L71 (1992).
81. Starrfield, S., Truran, J.W., Politano, M., Sparks, W.M., Nofar, I., and Shaviv, G., *PR* **227**, 223 (1993).
82. Starrfield, S., Gehrz, R., and Truran, J.W., in *Astrophysical implications of the laboratory study of presolar materials*, edited by T. Bernatowicz and E. Zinner, AIP, New York, 1997, p. 203.
83. Starrfield, S., Truran, J.W., Wiescher, M.C., and Sparks, W.M., *MNRAS* **296**, 502 (1998).
84. Starrfield, S., Sparks, W.M., Truran, J.W., and Wiescher, M.C., *ApJS* **127**, 485 (2000).
85. Starrfield, S., Iliadis, C., Truran, J.W., Wiescher, M.C., and Sparks, W.M., *NP* **A688**, 110 (2001).
86. Starrfield, S., these proceedings, 2002.
87. Vangioni-Flam, E., Audouze, J., and Chièze, J.-P., *A&A* **82**, 234 (1980).
88. Vanlandingham, K.M., Starrfield, S., Shore, S.N., and Wagner, R.M., these proceedings, 2002.
89. Wallace, R.K., and Woosley, S.E., *ApJS* **45**, 389 (1981).
90. Wanajo, S., Hashimoto, M., and Nomoto, K., *ApJ* **523**, 409 (1999).
91. Weiss, A., and Truran, J.W., *A&A* **238**, 178 (1990).
92. Wiescher, M., Görres, J., Thielemann, F.-K., and Ritter, H., *A&A* **160**, 56 (1986).
93. Wolf, M.T., and Leising, M.D., in *Gamma-Ray Spectroscopy in Astrophysics*, edited by N. Gehrels and G. Share, AIP, New York, 1988, p. 136.
94. Woosley, S.E., in *Nucleosynthesis and Chemical Evolution*, edited by B. Hauck, A. Maeder, and G. Meynet, Geneva Observatory, Geneva, 1986, p. 1.

The Effects of Thermonuclear Reaction Rate Variations on Nova Nucleosynthesis

Christian Iliadis[*], Art Champagne[*], Jordi José[†], Sumner Starrfield[**] and Paul Tupper[‡]

[*] *Department of Physics and Astronomy, University of North Carolina, Chapel Hill, North Carolina 27599, USA*
[†] *Departament de Física i Enginyeria Nuclear (UPC), Avinguda Victor Balaguer, s/n, E-08800 Vilanova i la Geltrú, Barcelona, Spain*
[**] *Department of Physics and Astronomy, Arizona State University, Tempe, Arizona 85287, USA*
[‡] *Scientific Computing - Computational Mathematics Program, Stanford University, Stanford, California 94305, USA*

Abstract. We investigate the effects of thermonuclear reaction rate uncertainties on nova nucleosynthesis. One–zone nucleosynthesis calculations have been performed by adopting temperature–density–time profiles of the hottest hydrogen–burning zone (i.e., the region in which most of the nucleosynthesis takes place). We obtain our profiles from 7 different, recently published, hydrodynamic nova simulations covering peak temperatures in the range from T_{peak}=0.145–0.418 GK. For each of these profiles, we individually varied the rates of 175 reactions within their associated errors and analyzed the resulting abundance changes of 142 isotopes in the mass range below A=40. In total, we performed ≈7350 nuclear reaction network calculations. We find that present reaction rate estimates are reliable for predictions of Li, Be, C and N abundances in nova nucleosynthesis. However, rate uncertainties of several reactions have to be reduced significantly in order to predict more reliable O, F, Ne, Na, Mg, Al, Si, S, Cl and Ar abundances.

The thermonuclear runaway model reproduces several key features observed in nova outbursts. At present, the most successful calculations involve one–dimensional hydrodynamic codes that are directly coupled to large nuclear reaction networks. However, some outstanding problems remain to be solved. For example, the masses of the underlying white dwarfs are unknown and the rates of mass accretion are poorly constrained. The composition of white dwarfs involved in either CO or ONe novae is far from understood and may vary from outburst to outburst. The mechanism responsible for the mixing of white dwarf core material into the accreted hydrogen envelope is not universally accepted. The amount of mass ejected is controversial. Finally, many nuclear reaction cross sections entering in the hydrodynamic model calculations are uncertain by orders of magnitude.

In the present work, we focus on the effects of reaction rate uncertainties in nova model calculations. In the past, such effects were frequently ignored by stellar modelers who used only one specific set of recommended reaction rates from available libraries. Our calculations are performed with an extended reaction network by using temperature–density–time profiles extracted from recent hydrodynamic nova simulations. The advantage of this procedure is that a single network calculation lasts only a few minutes. This has allowed us to independently vary the rates of 175 reactions by different factors within their uncertainties, and to analyze the resulting abundance variations of 142 isotopes in the mass range below A=40. The procedure is repeated for a number of temperature–density–time profiles obtained from recent hydrodynamic nova simulations involving different white dwarf masses and compositions. We would also like to point out a disadvantage of this procedure. The reaction network is not coupled directly to the hydrodynamics and, consequently, we ignore the important effect of convection on the final nova abundances. As pointed out previously (see, for example, Ref. [8]), convective mixing carries material from the hydrogen burning region to the surface on short time scales. This will cause an increase in ejected abundances of fragile nuclei that would have been destroyed, if they had not been carried to higher and cooler layers. Therefore, our calculations are unsuitable for defining *absolute* isotopic abundances resulting from nova nucleosynthesis. However, we claim that our procedure is adequate for exploring the effects of reaction rate uncertainties on abundance *changes* in the hottest hydrogen burning zone, i.e., the region in which most of the nucleosynthesis takes place.

The nuclear reaction network used in the present work follows the detailed evolution of 142 stable and proton–rich isotopes from hydrogen to calcium. The nuclei are linked by 1265 nuclear processes including weak interactions, reactions of type (p,γ), (p,α), (α,γ) etc., and corresponding reverse reactions.

For the construction of the thermonuclear reaction rate library we have used, with few exceptions, the most recent compiled and evaluated results given in Ref. [1] and Ref. [4] for the mass ranges A=1–20 and A=20–40, respectively. For the reactions $^{13}C(p,\gamma)^{14}N$, $^{14}N(p,\gamma)^{15}O$, $^{16}O(p,\gamma)^{17}F$, $^{18}O(p,\alpha)^{15}N$, $^{19}F(p,\gamma)^{20}Ne$, $^{19}Ne(p,\gamma)^{20}Na$, $^{15}O(\alpha,\gamma)^{19}Ne$ and $^{14}O(\alpha,p)^{17}F$ we still employ the rates from Ref. [2] since changes in recent updates are small (less than 30%). The library used here for nucleosynthesis calculations for the mass range $A \leq 40$ is, in our opinion, the most recent and consistent set of thermonuclear reaction rates available at present.

In addition to the information described above, our one–zone reaction network calculations require assumptions regarding the evolution of temperature and density, and the initial envelope composition. We have used temperature–density–time profiles of the hottest hydrogen burning zone, obtained from recently published hydrodynamic nova simulations. Properties of these evolutionary nova models are described in detail elsewhere (Refs. [6, 9, 10]). Our network calculations, for a specific temperature–density–time profile, have been performed with the same initial isotopic composition as was used in the corresponding hydrodynamic nova simulation.

The investigation of reaction rate sensitivities in nova nucleosynthesis requires the variation of reaction rates within their respective uncertainties. Therefore, quantitative estimates of reaction rate errors are needed. For most reaction rates involving stable or long–lived target nuclei the errors were taken from either Ref. [1] or from Ref. [4]. The reader should realize that it is frequently difficult to assign errors to reaction rates. This situation arises, for example, if Hauser–Feshbach theory is used to calculate a reaction rate, or if a reaction involves a short–lived target nucleus. In the former case, we have generally assumed that reaction rate errors amount to a factor of 100 up and down. The same assumption has been made in the latter case as well, with a few exceptions. In some cases, reaction rate uncertainties are not constant but depend on stellar temperature. If a reaction rate error varied significantly with temperature, for the sake of simplicity we have adopted in our network calculations the maximum reaction rate error in the temperature range of interest to nova nucleosynthesis (T=0.1–0.4 GK). This assumption is conservative since it can overestimate some of our predicted abundance changes.

Among the 1265 nuclear processes included in our network, we varied the rates of 175 selected reactions. Those included all exothermic (p,γ) and (p,α) reactions and the most important (α,γ) and (α,p) reactions, on stable and proton–rich target nuclei with masses A$\leq$40. The rates of those 175 reactions, together with the corresponding reverse reaction rates, have been varied individually by factors of 100, 10, 2, 0.5, 0.1 and 0.01 in successive reaction network calculations. Since we have explored nova nucleosynthesis for seven different temperature–density–time profiles, a total of 175×6×7=7350 network calculations were performed. For each network calculation, the final abundances of 142 isotopes were analyzed. Short–lived isotopes (e.g., ^{13}N, ^{14}O, ^{15}O and ^{17}F) present at the end of a network calculation were assumed to decay to their stable daughter nuclei. It is important to point out that the reaction–rate variations performed in the present work have only a minor influence on the amount of hydrogen consumed, the amount of helium produced, and the total thermonuclear energy released.

In Table 1 we summarize qualitatively some of our results. For complete quantitative results, see Ref. [5]. The table lists isotopes whose abundances change by more than a factor of 2 in at least one of the nova models considered here as a result of varying a particular reaction rate within uncertainties. It is striking that for the vast majority of reactions included in our network calculations, reaction rate variations have an insignificant effect on final isotopic abundances in all nova models. Instead, final abundances are influenced by variations of a restricted number of key reaction rates. Overall we find that present reaction rate estimates are reliable for predictions of Li, Be, C and N abundances in nova nucleosynthesis. However, uncertainties in the rates of several reactions have to be reduced significantly in order to predict more reliable O, F, Ne, Na, Mg, Al, Si, S, Cl and Ar abundances.

It can be seen from Table 1 that reaction rate variations of a few reactions, such as ^{23}Na(p,γ)^{24}Mg, ^{23}Mg(p,γ)^{24}Al, ^{30}P(p,γ)^{31}S and ^{33}S(p,γ)^{34}Cl, influence final abundances of a large number of isotopes. Consequently, new measurements of these reactions could significantly reduce uncertainties of isotopic abundances in nova model calculations. The reader might be surprised by the fact that certain reactions that were previously thought to play a role in nova nucleosynthesis do not appear in Table 1. In agreement with previous work [3], we find insignificant isotopic abundance changes as a result of ^{27}Si(p,γ)^{28}P, ^{31}S(p,γ)^{33}Cl, ^{35}Ar(p,γ)^{36}K and ^{39}Ca(p,γ)^{40}Sc reaction rate variations for all nova models. This result has been confirmed by recent hydrodynamic model calculations [7]. The ^{15}O(α,γ)^{19}Ne and ^{19}Ne(p,γ)^{20}Na reactions, which were thought to cause a breakout of material from the CNO mass region to the region beyond Ne, are also missing in Table 1. Rate variations for both reactions have only small effects on final abundances in all nova models, except in the model which achieves the highest peak temperature (T$_{peak}$=0.418 GK). But even for this rather high peak temperature, no breakout of material from the CNO mass region is observed. This result has been confirmed by recent hydrodynamic model calculations [10]. We also find that (α,γ) and (α,p) reactions in general are not important for nova nucleosynthesis.

TABLE 1. Influence of reaction rate variations on isotopic abundances in nova nucleosynthesis.

Reaction rate variation*	Isotopic abundance change†
CO nova models	
$^{17}O(p,\gamma)^{18}F$	^{18}F
$^{17}O(p,\alpha)^{14}N$	$^{17}O, ^{18}F$
$^{18}F(p,\alpha)^{15}O$	^{18}F
$^{22}Ne(p,\gamma)^{23}Na$	$^{22}Ne, ^{23}Na, ^{24}Mg, ^{25}Mg, ^{26}Al$
$^{23}Na(p,\gamma)^{24}Mg$	^{24}Mg
$^{26}Mg(p,\gamma)^{27}Al$	^{26}Mg
$^{26}Al^g(p,\gamma)^{27}Si$	^{26}Al
ONe nova models	
$^{17}O(p,\gamma)^{18}F$	$^{17}O, ^{18}F$
$^{17}O(p,\alpha)^{14}N$	$^{17}O, ^{18}F$
$^{17}F(p,\gamma)^{18}Ne$	$^{17}O, ^{18}F$
$^{18}F(p,\alpha)^{15}O$	$^{16}O, ^{17}O, ^{18}F$
$^{21}Na(p,\gamma)^{22}Mg$	$^{21}Ne, ^{22}Na, ^{22}Ne$
$^{22}Ne(p,\gamma)^{23}Na$	^{22}Ne
$^{23}Na(p,\gamma)^{24}Mg$	$^{20}Ne, ^{21}Ne, ^{22}Na, ^{23}Na, ^{24}Mg, ^{25}Mg, ^{26}Mg, ^{26}Al, ^{27}Al$
$^{23}Mg(p,\gamma)^{24}Al$	$^{20}Ne, ^{21}Ne, ^{22}Na, ^{23}Na, ^{24}Mg$
$^{26}Mg(p,\gamma)^{27}Al$	^{26}Mg
$^{26}Al^g(p,\gamma)^{27}Si$	^{26}Al
$^{26}Al^m(p,\gamma)^{27}Si$	^{26}Mg
$^{29}Si(p,\gamma)^{30}P$	^{29}Si
$^{30}P(p,\gamma)^{31}S$	$^{30}Si, ^{32}S, ^{33}S, ^{34}S, ^{35}Cl, ^{37}Cl, ^{36}Ar, ^{37}Ar, ^{38}Ar$
$^{33}S(p,\gamma)^{34}Cl$	$^{33}S, ^{34}S, ^{35}Cl, ^{36}Ar$
$^{33}Cl(p,\gamma)^{34}Ar$	^{33}S
$^{34}S(p,\gamma)^{35}Cl$	$^{34}S, ^{35}Cl, ^{36}Ar$
$^{34}Cl(p,\gamma)^{35}Ar$	^{34}S
$^{37}Ar(p,\gamma)^{38}K$	$^{37}Cl, ^{37}Ar, ^{38}Ar$
$^{38}K(p,\gamma)^{39}Ca$	^{38}Ar

* Only those reactions are listed which have a significant influence on isotopic abundances in at least one of the nova models considered in the present work.

† Only those isotopes are listed whose abundances change by more than a factor of 2 as a result of varying the corresponding reaction rates within their adopted errors.

ACKNOWLEDGMENTS

The authors would like to thank A. Coc, M. Hernanz, R. Hix and M. Smith for stimulating discussions. This work was supported in part by the U. S. Department of Energy under Grant No. DE–FG02–97ER41041, by CICYT–PNIE ESP98–1348 and DGES PB98–1183–C03–02, and by Grants from NASA and NSF to ASU.

REFERENCES

1. Angulo, C., et al., *Nucl. Phys. A*, **656**, 3 (1999).

2. Caughlan, G. R., and Fowler, W.A., *At. Data Nucl. Data Tables*, **40**, 283 (1988).
3. Iliadis, C., Endt, P.M., Prantzos, N., and Thompson, W.J., *Astrophys. J.*, **524**, 434 (1999).
4. Iliadis, C., D'Auria, J.M., Starrfield, S., Thompson, W.J., and Wiescher, M., *Astrophys. J. Suppl.*, **134**, 151 (2001).
5. Iliadis, C., Champagne, A.E., José, J., Starrfield, S., and Tupper, P., *Astrophys. J. Suppl., in press* (2002).
6. José, J., Coc, A., and Hernanz, M., *Astrophys. J.*, **520**, 347 (1999).
7. José, J., Coc, A., and Hernanz, M., *Astrophys. J.*, **560**, 897 (2001).
8. Lazareff, B., Audouze, J., Starrfield, S., and Truran, J.W., *Astrophys. J.*, **228**, 875 (1979).
9. Politano, M., Starrfield, S., Truran, J. W., Weiss, A., and Sparks, W. M. , *Astrophys. J.*, **448**, 807 (1995).
10. Starrfield, S., *et al.*, *Astrophys. J., in preparation* (2002).

Nuclear Astrophysics at ISAC with DRAGON: Initial Studies

Art Olin[*][†], Shawn Bishop[**], Lothar Buchmann[*], Mohan L. Chatterjee[‡], Alan Chen[*], John M. D'Auria[**], Sabine Engel[§], Dario Gigliotti[¶], Uwe Greife[‖], Don Hunter[*], Ahmed Hussein[¶], Dave Hutcheon[*,††], Cybele Jewett[‖], Jim King[‡‡], Shigeru Kubono[§§], Michael Lamey[**], Alison M. Laird[*], Rachel Lewis[¶], Wenjie Liu[**], Shin'ichiro Michimasa[§§], Dave Ottewell[*], Peter Parker[¶], Joel Rogers[*], Frank Strieder[§], Michael Wiescher[***] and Chris Wrede[**]

[*]*TRIUMF*
[†]*U. of Victoria*
[**]*Simon Fraser U.*
[‡]*Saha Inst., Calcutta*
[§]*Ruhr-U. Bochum*
[¶]*U. of Northern BC*
[‖]*Colorado School of Mines*
[††]*U. of Alberta*
[‡‡]*U. of Toronto*
[§§]*CNS, U. of Tokyo*
[¶]*Yale U.*
[***]*U. of Notre Dame*

Abstract. The new DRAGON recoil separator facility, designed and built to measure directly the rates of radiative proton and alpha capture reactions important for nuclear astrophysics, is now in operation at the TRIUMF-ISAC radioactive beams facility in Vancouver, Canada. Experiments have been conducted for the first time on the ^{21}Na(p,γ)^{22}Mg reaction. The evolution of nova explosions, and particularly their ^{22}Na abundance, depends sensitively on this reaction rate. The radioactive ^{21}Na beam with an intensity of up to 5 x 10^8 /s was directed onto a windowless hydrogen gas target (3.8 x 10^{18} H atoms/cm^2). Prompt reaction gamma rays were detected using a BGO array and separated reaction products detected using a silicon strip detector at the end of the 20.8 m recoil mass separator. Yield measurements recording simultaneously singles and coincident signals were performed by scanning in energy over the known resonance reported previously in ^{22}Mg at E$_{cm}$ = 212 keV, and in addition, over a strong resonance observed at E$_{cm}$ $\approx$822 keV. Known resonances in the ^{21}Ne(p,γ)^{22}Na, ^{20}Ne(p,γ)^{21}Na, and ^{24}Mg(p,γ)^{25}Al reactions have been used to calibrate the DRAGON. Studies are in progress to further define the performance of the DRAGON facility. Status of the data analysis and results from system performance studies will be presented along with a brief description of the new ISAC and DRAGON facilities.

NUCLEAR ASTROPHYSICS WITH RADIOACTIVE BEAMS

There is increasing interest worldwide in the construction and operation of radioactive beams facilities [1]. One of the stated goals of such facilities is the experimental determination of the rates of nuclear reactions which occur in cataclysmic stellar events. Many of these reactions of astrophysical importance involve radiative capture of either protons or α particles on light or medium mass, radioactive nuclei far below the Coulomb barrier. However, because of the small cross sections of radiative capture reactions in the energy region of interest (compared to particle transfer reactions, for example) it has proved difficult to measure such cross sections. In fact, so far, only one direct measurement [2], $^{13}N(p,\gamma)^{14}O$ has produced positive results in a radiative capture measurement on radioactive isotopes relevant to explosive stellar scenarios. The availability of high intensity radioactive beams makes possible a significant expansion in our knowledge of these rates.

The $^{21}Na(p,\gamma)^{22}Mg$ reaction plays an important role in the production of interstellar ^{22}Na ($T_{1/2}$= 2.6 years)in nova environments [4, 3], a gamma-emitting isotope that could be an observable of nova explosions. The temperature dependence of this reaction rate will determine when the pathway will be dominated by this reaction as compared to beta decay of ^{21}Na ($T_{1/2}$=22.5 s). For nova temperatures the stellar reaction rate is dominated by low lying resonances in ^{22}Mg just above the $^{21}Na + p$ threshold (5.501 MeV), in particular an s-wave resonance at E_{cm}=212 keV[4, 3]. However, for astrophysical scenarios at higher temperatures (e.g. x-ray bursts), higher lying resonances are important.

Here we will report on the initial commissioning studies and our progress in measuring the $^{21}Na(p,\gamma)$ resonances up to E_{cm}=822 keV.

DRAGON AT THE ISAC FACILITY

DRAGON is located in the ISAC hall at TRIUMF on the high energy beam line. ISAC [9, 10] delivers isotopically-pure radioactive heavy-ion beams of energies between 0.15 and 1.5 MeV/u. DRAGON [7, 8] consists of four main components, namely, a windowless, gas target, a gamma-detector array, an electromagnetic recoil separator, and a heavy-ion recoil detection system.

The heavy ion beam enters the target gas cell through a series of differentially pumped tubes. The gas pressure in the cell is regulated to be typically 4.7 Torr, and the gas density is uniform over most of the 11 cm between the innermost apertures. Initial studies using the gas target were performed to determine charge state distributions resulting from the interaction of low energy heavy ions passing through the gas; results of these are reported elsewhere [11]. A measurement of the pressure dependence of the energy loss through the gas target using a ^{21}Ne beam at 275 keV/u was performed both with our standard collimators and with much smaller entrance and exit collimators that confine the gas to the geometric target length. An effective target length of 12.3 ± 0.5 cm is obtained from this measurement.

A gamma detector array comprised of 30 BGO crystals of hexagonal cross section surrounds the target. Monte Carlo simulations predict that the gamma-ray detection efficiency of the array is approximately 40% for 1-10 MeV gamma rays produced in the target.

The DRAGON recoil separator (Figure 1) is designed to accept recoils leaving the target traveling within (±20) mrad of the beam direction and separate them from beam particles. The separator consists of a series of magnetic dipoles(M), magnetic quadrupoles(Q) magnetic sextupoles(S) and electrostatic dipole benders(E). Arranged in the order (QQMSQQQSE)(QQSMQSEQQ), they form two stages of mass separation. After being bent away from the recoils, the separated beam is stopped on slits between the electromagnetic elements. The recoils of a single charge state are transmitted to a double-sided silicon-strip detector (DSSSD) oriented perpendicular to the heavy-ion's trajectory which measures recoil energy and position over a 25 cm^2 area. The average energy resolution at 1.0 MeV/u was 1.4% FWHM and the time resolution was 1.2 ns, both averaged over the 16 front strips. The length of the separator from target to recoil detector is 20.8m. Other detectors of the recoiling reaction product are also being developed for use at the final focus.

Commissioning of the separator using stable beams was done in two distinct modes. In the first, the gas target was replaced by a double-steering magnet so that the beam could be deflected horizontally or vertically through angles comparable to the largest cone angle of recoil particles from capture reactions of interest. By varying the initial angle or the beam energy, it was possible to confirm that the desired focus conditions were obtained at the three slit locations and the achromaticity condition was met at the Mass and Final selection slits with magnet currents very close to the calculated values. The acceptance of the separator remains near unity up to 15 mrad deflection angle.

Separator 'tuning' is done either with the radioactive beam directly or with a stable 'pilot' beam, requiring current of at least 40 pA (electrical). Quadrupoles and sextupoles are set according to the mass, selected charge state, and nominal energy of the beam after passing through the gas target, scaled directly from a developed reference tune. The exact beam energy is then determined by finding the field of our first dipole which puts the beam on-axis at the charge-selection slits. The energy calibration has been determined from measurements of resonances in p($^{21}Ne,\gamma$)^{22}Na with E_{cm}= 259.3 keV, and p($^{20}Ne,\gamma$)^{21}Na with E_{cm}=1112.7 keV. Analysis of additional known reactions is in progress to refine our calibration.

CURRENT STATUS

While the E_{cm}=822 keV ^{21}Na(p,γ) resonance has some relevance to thermonuclear runaway at high temperatures, our primary motivation for investigating this resonance in detail was that its significant strength facilitated commissioning studies using radioactive beam and exploring DRAGON's capabilities and limitations. This broad resonance was recently studied in the elastic scattering channel [12]. The strength enabled us to make a detailed study of the yield curve. As can be seen in Fig 2, the ^{22}Mg recoils can be observed clearly without requirement of a γ in the BGO calorimeter, and a small "leaky beam" signal appears just above it at the full beam energy. There is a well-defined coincidence peak in the γ - heavy ion time signals corresponding to the time-of-flight through the separator. The "leaky beam" signal is suppressed in events satisfying this requirement.

Our preliminary excitation function, shown in Fig 3, corresponds to a resonance with a width of order of our target thickness. From the consistency of the fit to this data we can estimate that uncertainties due to variations in tuning of the accelerator, beam energy

TABLE 1. Present DRAGON Experimental Program

Reaction	E_x (MeV)	Scenario
^{21}Na(p,γ)^{22}Mg	5.714	Novae and Xray Bursters
^{19}Ne(p,γ)^{20}Na	2.646	Hot CNO cycle
^{13}N(p,γ)^{14}O	5.155	Hot CNO cycle
^{25}Al(p,γ)^{26}Si	5.687	Novae
^{15}O(α,γ)^{19}O	4.033	Hot CNO cycle breakout

and width, and tuning of the separator are less than 30%.

The measured yields[6], again preliminary, for the important 212 keV level are shown in Fig 4. This is the first measurement of a sub-mev width resonance successfully observed with radioactive beam. For this measurement the gamma coincidence was required in order to suppress the "leaky beam" signal.

Table 1 shows the present program of experiments at DRAGON. Data taking for the ^{21}Na(p,γ) resonances is complete. We anticipate that the ^{19}Ne(p,γ) and ^{13}N(p,γ) will be accomplished over the next year. The ^{15}O(α,γ) measurement is very challenging due to the very low expected cross section. Efforts are underway to develop the very intense beam required.

ACKNOWLEDGMENTS

We wish to acknowledge the help of TRIUMF staff and operators, especially those at the ISAC facility. We are also thankful to all TRIUMF staff, too numerous to mention here, who helped in the construction of DRAGON. We wish to express our special thanks to M. Dombsky and P. Bricault for providing the radioactive ^{21}Na beam and to B. Laxdal and M. Passini for considerable efforts with beam tuning. Major financial support from NSERC is highly appreciated.

REFERENCES

1. M. Smith and E. Rehm, Annu. Rev. Nucl. Part. Sci. **51**(2001) 91.
2. T.H. Decrock et al., Phys. Rev. C, **48**(1993) 3088.
3. A. A. Chen et al., Phys. Rev. C, **63** (2001) 065807.
4. N.Bateman et al., Phys. Rev. **C63** (2001) 035803.
5. S. Engel et al., Nuclei in the Cosmos conference proceedings, to be published.
6. S. Bishop et al., Nuclei in the Cosmos conference proceedings, to be published.
7. J. M. D'Auria et al. Nucl. Instrum. Meth. **B126** (1997) 262.
8. D. Hutcheon et al., The DRAGON Facility paper, to be published.
9. R.E. Laxdal, Proceedings of the 2001 IEEE Particle Accelerator Conference (PAC2001), Chicago June (2001).
10. R.E. Laxdal et al., ibid.
11. W. Liu, et. al., Nucl. Instrum. Meth. in Phys. Res., submitted for publication, M.Sc. Thesis, Department of Chemistry, Simon Fraser University, 2001.
12. C. Ruiz et al., Phys. Rev. C65, 042801(R).

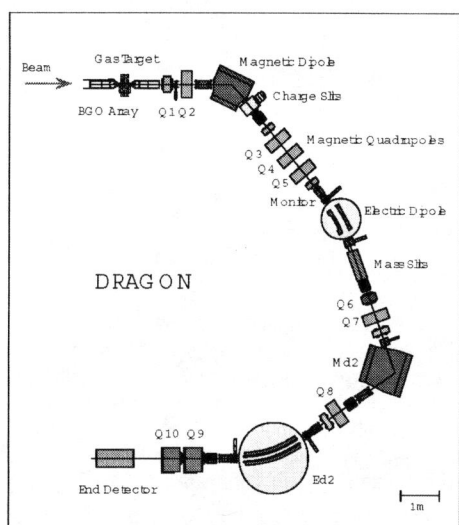

FIGURE 1. The Dragon Recoil Separator

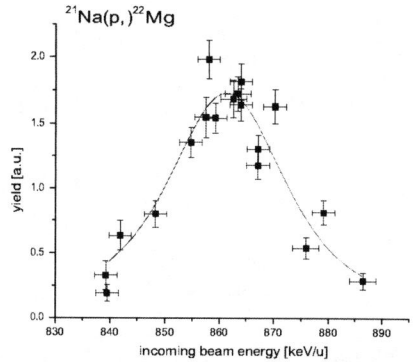

FIGURE 3. Excitation function for the 822 keV/u resonance (preliminary). The curve shows the expected shape for a broad resonance.

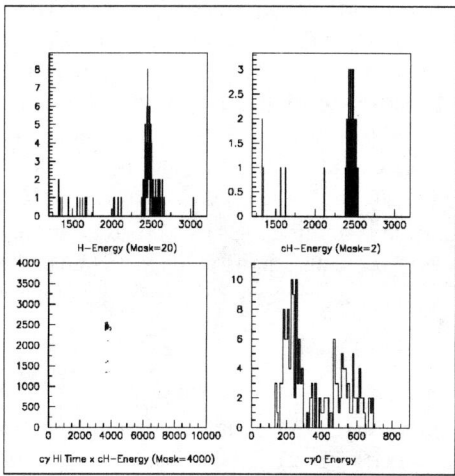

FIGURE 2. The upper panels show the energy deposited in the heavy ion detector with and without requiring a γ in the preceeding 4.5 μs. Below we show the time and γ-energy distributions for coincident events.

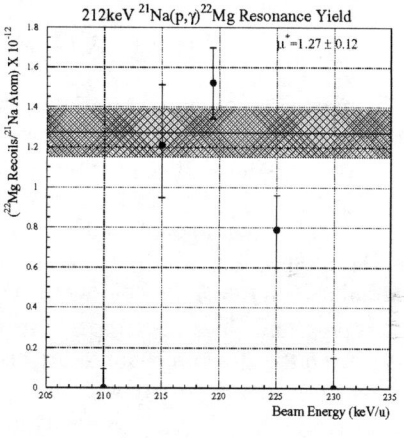

FIGURE 4. Excitation function for the 212 keV resonance (preliminary). The data point at 220 keV/u represents an average of 11 individual runs whose values have the expected Poisson distribution about their cental value. The hatched region shows the 1σ errors for the middle three data points. The extreme points show our limit for off-resonance backgrounds or direct capture yields.

Multidimensional Nova Simulations

Ami Glasner and Eli Livne

Racah Institute of Physics, The Hebrew university, Jerusalem, ISRAEL.

Abstract. We review the history of multidimensional Nova simulations starting with the motivation and first steps of dimensional analysis. A review of the up to date published research is given, summarizing the main conclusions derived from the results of the existing computational tools. We try to analyze the difficulties that such simulations are facing and we suggest that most of the differences between existing calculations are due to specific numerical problems that we can demonstrate. The review ends with a discussion about the analytic and computational improvements needed in order to improve our knowledge of the mixing process.

MOTIVATION AND FIRST STEPS

The most accepted model for Nova outbursts is a thermonuclear runaway (TNR) of accreted, hydrogen rich, envelope on top of a cold white dwarf. This approach is based mostly on research that assumes 1D spherical symmetry. Analyzing the self consistency of this 1D approach we can find multidimensional effects that should be considered. In order to explain the need we examine here a few of the main features of the 1D model that are relevant to the discussion. The accreted hydrogen is accumulated on top of the cold white dwarf core. As matter is accumulated it is heated by compression. Once the temperature at the base is about $2-3 \cdot 10^7$ °K the burning passes from the slow pp chain to the faster CNO cycle and the flow becomes unstable to convection. The runaway will take place once the characteristic time for heating by nuclear reactions is faster than the dynamical time.

The characteristic time for heat diffusion is very long all along (years). The convective velocities according to the 1D mixing length theory (MLT) grows from about 10^6 cm/sec to a few times 10^7 cm/sec and the turnover time decreases accordingly from a few hundred seconds to 10-50 seconds. The burning time decreases rapidly with temperature. For the relevant densities the burning time is about 10^4 seconds at the temperature of $7 \cdot 10^7$ °K and it decreases to about 1 second at temperatures of $10 \cdot 10^7$ °K. The dynamical time scale is about 1 second. Therefore, the runaway is starting at that temperature.

The main issues in the picture we described up to now that call for multidimensional analyzes of the problem are:

    a) Deviation from spherical symmetry by the nature of the convective flow.

    A simple argument concerning the relevant time scales shows that there is a question about the fate of early local temperature fluctuations. Do they lead to local runaways, or are they spread around the whole envelope and quenched by the convective flow.

b) What is the nature of the ignition process?
Is it ignition by local fluctuation that spreads all around the envelope as a deflagration wave. Or, is there a mechanism that leads to a global slow evolution to the runaway stage.

c) The 1D MLT used for describing the convective fluxes can not describe phenomena such as undershoot of the convective flow leading to dredge up of CO (ONeMg) from the cold core to the burning hydrogen rich envelope. It is well known that mixing of CNO elements to the envelope has a great effect on the runaway process and is needed in order to explain the observed abundances of the nova ejected envelope. There is a debate in the published literature about the mechanism that drives this mixing.

The first attempts to study the issue were made in the early eighties. Fryxell&Woosley [1] carried out a dimensional analysis of multidimensional effects for TNRs that occur on thin stellar shells. They claim that for the Nova case there is initiation at a point and a flame that spreads the burning by small scale turbulence with velocity:

$$\mathbf{v} = (\mathbf{h}_p \cdot \mathbf{v}_c)/\tau_b. \tag{1}$$

where, $\mathbf{h}_p$ is the pressure scale height, $\mathbf{v}_c$ is the characteristic convective velocity and τ_b is the burning time. Shara [2] made a study of the time scales for spreading of perturbations as function of the white dwarf mass and the local thermodynamic conditions. Using this timescales he predicts that local TNRs are likely to occure on many degenerate dwarfs. He also predicts that these runaways have a volcanic-like nature. This mechanism, operating at early stages when the whole envelope is still radiative was never studied in details. We intend to examine it in the near future by performing the relevant calculations.

REACTIVE FLOW IN NOVA OUTBURSTS – WHO AND HOW

For solar abundance accretors without mixing prior to runaway the overall evolutionary time of a 1D model from the stage that the envelope becomes convective to the runaway is many hours. All along this period the flow is extremely subsonic whereas the multidimensional codes are mainly explicit. Therefore, the stability restrictions on the time step ($\sim 10^{-4}$ seconds) limit the ability to carry out long calculations. As a first step all of the early studies took the following strategy. A 1D model is evolved up to the dynamical stage (temperature of $\sim 10^8$ °K as we stated before) and then it is mapped to a multidimensional grid.

Simulating Nova In 2D

The first hydrodynamic calculations were carried out by Shankar et. al. [3],[4]. They carry out their study with the code PROMETHEUS Fryxell et. al. [5] which is a purely Eulerian hydro code operating according to the PPM method (Colella & Woodward) [6]. The code is a hydro code with gravity and nuclear burning but it ignores diffusive heat conduction (justified since the relevant time for heat conduction is years). The computational domain is an angular slice of the star. In the radial direction it includes the whole hydrogen rich envelope and a few zones of the cool (CO enriched) core. In the transversal domain it extends to $1°$ near the equator.

The study assumes very big temperature perturbations 100%-600% of a small local region at the base of the envelope. The studied cases differ by the initial profile and the control parameter was the temperature at the base of the envelope. The authors were able to make only very short calculations, up to one second, during this short time (about one dynamical time unit) they found out that the perturbed region always floats up and spreads away.

The main conclusions of this study are:
  (i) Perturbations rise and cool on dynamical time scales.
  (ii) Convection spreads perturbations.

They get the impression that global temperature evolution is the important ignition mechanism.

The next study was made by Glasner et. al. [7],[8]. They carry out their study with the code VULCAN, Livne [9], which is a mixed Eulerian Lagrangian (ALE) hydro code with capability for either explicit or implicit time steps. The code is a hydro code with gravity and nuclear burning. The computational domain is an angular slice of the star. In the radial direction it includes the whole hydrogen rich envelope and a few zones of the cool (CO enriched) core. In the transversal direction it extends to $\sim 20°$ near the equator. The size of a typical grid zone near the base of the envelope is 5X5 Km. In order to make the simulation self consistent the model was relaxed to a state for which the convective flow was fully developed (the 1D model includes only convective heat conduction). They follow the runaway from the ignition stage ($\tau_{burning} = \tau_{dynamic}$) up to a stage just after the maximum in the global energy production rate when expansion quenches the runaway.

The main conclusions of this study are (FIGURE 1):
  (i) Shear instabilities of the transversal component of the convective flow induces mixing of CNO elements from the core.
  (ii) The thermonuclear runaway takes place with very irregular local flames.
  (iii) There is no trace for flame propagation.
  (iv) Convective cells and convective velocities are bigger than those predicted by the 1D MLT.

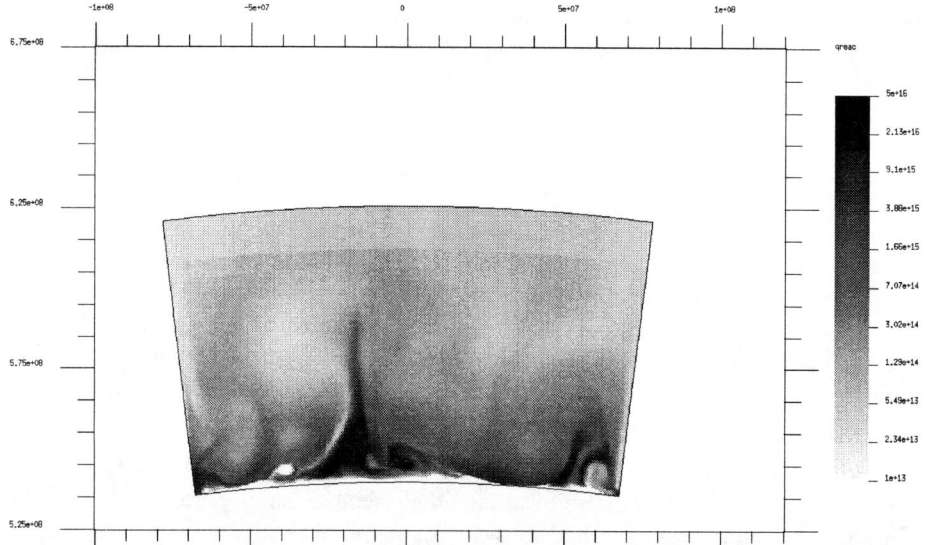

FIGURE 1. Logarithm of the burning rates close to the peak of the runaway.

A few months later Kercek et. al. [10] published their results. They carry out their study with their version of the PROMETHEUS code (Fryxell et. al.) [5]. The initial model and the computational domain are the same as Glasner et. al. [7]. Working with difference equations that differ from those of the VULCAN [9], they had to add a relaxation scheme to assure initially hydrostatic equilibrium. The study also examine effects of resolution by performing models with fine grid 1X1 Km.

The main conclusions of this study are:
(i) Direct numerical 2D simulations of TNRs are feasible.
(ii) There are signs for convergence for higher resolution.
(iii) The general outcome resembles Glasner et. al. [8].
(iv) Convective mixing by dredge-up occurs, but at a later stage.
(v) The runaway is somewhat less violent than the one Glasner et. al. [8] predict for the same initial model.

Simulating Nova In 3D

Up to now there is only one published Nova simulation in 3D. The simulation was made by Kercek et. al. [11], it was performed with the 3D version of the PROMETHEUS code (Fryxell et. al.) [5]. The initial model and the computational domain are the same as their 2D study. The resolution approached the coarser 2D resolution (cells of ~8X8X8 km).

The main results are:
(i) Flow patterns in 3D differ from those in 2D.
(ii) Self enrichment with CO during the outburst is very slow.
(iii) They runaway is not resembling fast Nova.

They therefore come to the conclusion that in order to simulate the observed fast Nova, mixing of CNO elements is essential and it should take place prior to the runaway.

2D SIMULATIONS – THE NEXT STEP

As we already stated, all investigators started their multidimensional study with a 1D model close to the dynamical stage of the runaway. Glasner et. al. [12] made an effort to extend the 2D research in order to examine the runaway from early stages. The aim of the research was to examine the evolution towards the runaway from the stages that convection sets on and to study the fate of early perturbations. In order to achieve this aim we used initial profiles for the 2D code that were extracted from the 1D simulation when the temperature at the base of the envelope is $4 \cdot 10^7, 7 \cdot 10^7$ and $9 \cdot 10^7$ °K. These models were later evolved in 2D. For the case in which the base temperature is $7 \cdot 10^7$ °K we also examined the effects of small perturbations on the evolution. The study was restricted to the most economic simulations we can make, i.e. with the purely Eulerian scheme and with fixed outer boundary conditions. We are aware of the inaccuracy of this assumption and we address and justify this issue and the mixing issue in the next section.

The main results are:
a) We can follow the ignition process almost from the set on of convection up to the runaway itself (FIGURE 2). For the case with the lowest initial temperature, $4 \cdot 10^7$ °K, there is a tendency towards runaway but we stopped the simulation after 700 seconds.
b) Early perturbations are quenched as the pressure profile is restored on a few sound crossing times. Some mixing and a bit enhanced burning is induced by each perturbation (FIGURE 3).
c) Macroscopic overshot mixing exists from the moment the flow becomes unstable to convection. The amount of mixing increases (not linearly) with the approach towards runaway conditions. This tendency is not surprising since we know that the minimal unstable wavelength for the Kelvin-Helmholtz instability for incompressible flow with gravity is growing with the increase in the shear (convective) velocity [13].

$$\lambda_{min} \propto \mathbf{U}^2 /(\mathbf{g} \cdot \Delta\rho) \qquad (2)$$

where: **U** - is the transversal velocity jump at the interface.
g - is the gravitational acceleration.
$\Delta\rho$ - is the density jump at the interface.

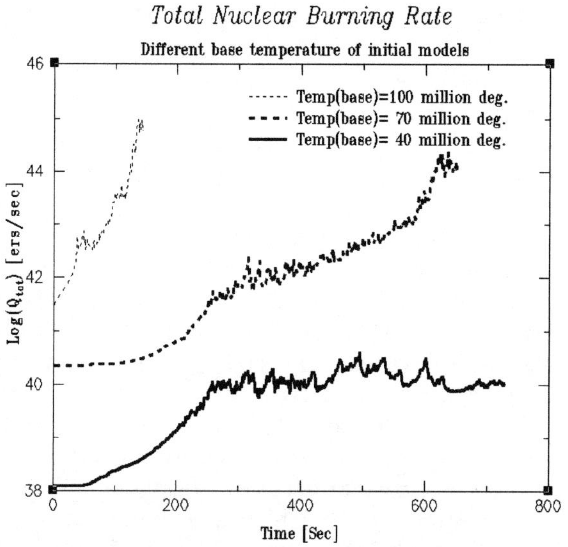

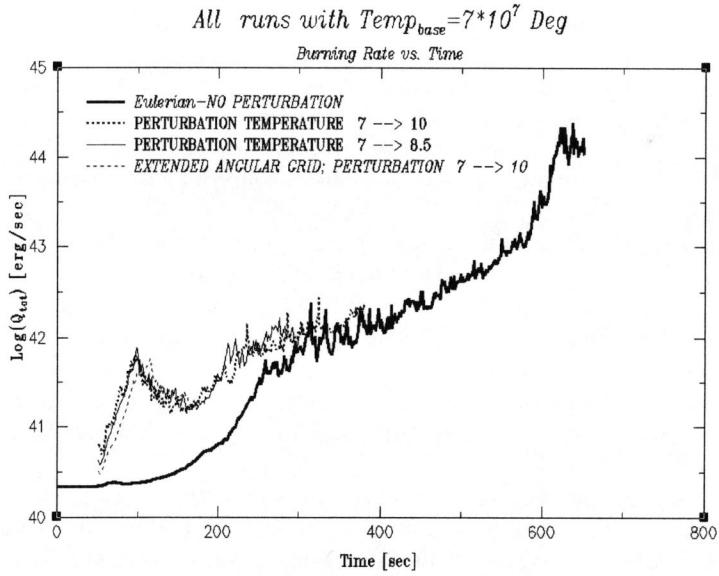

FIGURE 2. Logarithm of the total burning rate [erg/sec] as function of time [sec] for the models with base temperature of 40, 70 and 100 million $°K$.

FIGURE 3. The logarithm of the total energy production rate [erg/sec] for the models with base temperature of 70 million $°K$. Full line-the nominal reference. **Local Perturbations at time=50 sec:** Dotted line- perturbation from 70 to 100 million $°K$, Full light line- perturbation from 70 to 85 million $°K$, Dashed line- perturbation from 70 to 100 million $°K$ of a model with extended angular domain (twice).

THE DIFFICULTIES AND THE DOUBTS

One of the most significant effects found in the 2D and 3D simulations is the dredge-up of CNO rich matter from the core. The buildup of convective cells at the base of the envelope induces shear flow at the core-envelope interface. This shear flow is unstable, the increasing perturbations induce mixing (Glasner et. al. [8], Rosner et. al. [14]). The exact amounts of mixing depend extremely on the details of the numeric prescriptions. The challenge is to predict the amount of mixing at each stage and to be able to differentiate between physical mixing and the mixing induced by numeric effects. We are aware that the models presented here are only the first step since they are marginal concerning the needed resolution and they lack smoothness in all the relevant variables. Yet, we believe that our results are qualitatively correct. More work is needed, as we will show, in order to derive and verify the exact amounts of mixing. We therefore, want at this stage to evaluate the results we reported up to now in this review. In order to do it we try to extract the issues that seem to be delicate. By doing this we can find out where we have to put more effort and improve the models, the research methods and the computational tools. Another outcome of this evaluation is also a hint as to the reason for the differences between the few studies we report here. In the following paragraphs we give a few specific examples:

a) The first obstacle the multidimensional simulations face is the mapping of the 1D initial model to 2D. The 1D model is almost in hydrostatic equilibrium. Small deviations from this equilibrium caused by rezoning or by differences in the numerical details of the 1D and the 2D code can create fictitious velocities that enhance mixing and mass loss. A deviation of less than one per-cent in the derivation of the acceleration term immediately (within seconds) gives rise to a directional flow (inflow or outflow) with amplitudes bigger than the amplitude of the convective velocities. For the VULCAN simulations we got perfect accuracies concerning this issue. The PROMETHEUS code reports on some difficulties concerning hydrostatic equilibrium of the initial model.

b) Discontinuities in the initial model arise from the abundance jump between the cold CNO matter of the core and the hydrogen rich envelope (solar abundances) of the 1D model. In such a profile the pressure passes continuously whereas all other variables are discontinuous. The density can jump by a factor of seven or even more. This discontinuity is an extreme obstacle to the numeric simulation. Any reasonable numerical approach demands continuous changes on scales of a few zones. The existence of such a jump imposes numerical mixing and it doesn't let us make convergence tests.

c) The mixing process has two effects. On one hand the mixing with CNO elements enhances the burning rates, on the other hand the CNO matter is much colder and therefore, once it mixes with the hydrogen, it cools the mixture and tends to quench the burning. There is a great sensitivity of the burning rates to the exact amount of mixing. Too much mixing (more than a few per-cent) can lead to decrease of the burning rates and to a much slower runaway.

d) In order to get confidence in our simulations we checked the sensitivity of the results to the outer boundary conditions. The bottom line is that there is a great

sensitivity, a result that is surprising in a way. There are three options that we compare:
(i) The usual Eulerian scheme with free outer boundary condition.
   For this option some of the matter with positive radial velocity in the outer zone leaves the computational grid in each step. Although the initial pressure at the outer zone is orders of magnitude less than the pressure at the bottom zone it comes out that the small decrease in pressure on a few dynamic times (the flow is extremely sub-sonic) quenches the runaway.
(ii) An ALE scheme.
   For such a scheme after the Lagrangian time step one is free to define the new grid for the Eulerian stage of the step. For the option we choose at each time step the new GRID is such that points on it move only in the radial direction. Each tube of radial zones moves in a way that assures mass conservation within the tube. In this way the outer boundary is expanding with time (FIGURE 4).
(iii) The usual Eulerian scheme with no flux outer boundary condition.
   For this option matter is forced to stay within the grid. This assumption can be partially justified if we notice that most of the mass that leaves the grid is part of a convective cell. If the grid was originally extended farther that amount of mass should flow back into the grid once the convective turnover of this cell is completed.

The energy production rate as function of time is presented in FIGURE 5. The intermediate scheme is the one that is expected to be the most accurate. We see that the Fixed Eulerian scheme fits quite well with the intermediate scheme, both from the point of view of energy production rates (FIGURE 5) and the topology of the convective cells (FIGURE 4). It is quite surprising to see that the free Eulerian scheme gives significantly different results (essentially no runaway).

We have the impression that most of the differences between the results of Glasner et. al. [8] and Kercek et.al. [10] are due to the four items we discussed above. If Kercek et.al. [10],[11] adopted for their simulations the Eulerian free boundary condition scheme it might be that the outcome is the mild runaway they got. The fact, that within time, all of the hydrogen disappears in their simulations (fig. 8 in Kercek et.al. [10]) supports this assumption.

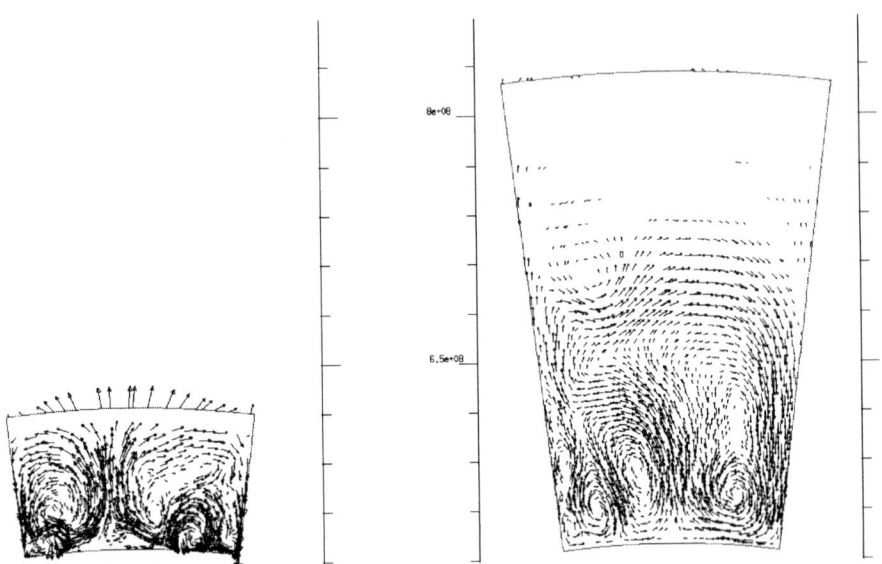

FIGURE 4. The velocity field of the convective flow at a time close to the peak of the runaway. **Left**- Eulerian with fixed outer boundary condition. **Right**- ALE scheme (Eulerian Lagrangian), see text.

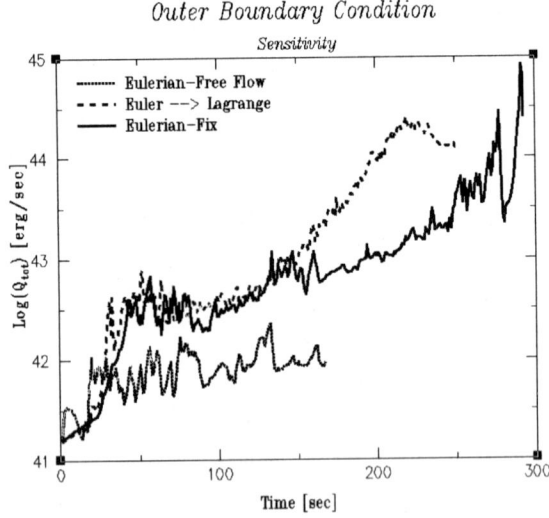

FIGURE 5. Demonstration of the sensitivity to the outer boundary condition. The logarithm of the total energy production rate [erg/sec] as function of time. Dotted line – Eulerian free outer boundary condition – fails to give a runaway. Full line - Eulerian no-flux outer boundary condition. Dashed line- ALE (Eulerian Lagrangian) scheme with mass conservation. Both last cases give a fast runaway and resemble each other.

CONCLUSIONS AND A VIEW TO THE FUTURE

The study of multidimensional reactive flow in Nova outbursts has made a great progress in the last ten years. With the existing numeric tools we were able to demonstrate our ability to attack problems such as the ignition mechanism, the fate of early local perturbations and the topology of the convective flow.

The question of when and how the cold core matter mixes with the hydrogen rich (solar abundant) accreted envelope has a great importance for the Nova research. This problem still faces theoretical and numerical uncertainties. Three mechanisms were suggested: Mixing induced by diffusion prior to the convective stage [15] and references therein, shear mixing induced by rotation [16],[17] with other references therein and shear mixing induced by convection [18],[8],[14]. Our task is to study the third mechanism. As we have pointed out this study is very delicate. Up to now we were able to get qualitative results and to demonstrate the ability of the shear mixing induced by convection to predict amounts of mixing that resemble the abundances observed in Nova ejecta. An effort is made now by us and by the Flash group in Chicago to make a linear analysis of the instabilities in order to find the relevant control parameters and the relevant unstable modes. We intend to use the acquired knowledge in order to perform 'wise' simulations putting efforts to resolve the relevant modes. Our aim is to give quantitative predictions of the amounts of mixing as function of time from the onset of convection to the peak of the runaway.

ACKNOWLEDGMENTS

On this occasion we want to thank here the local organization committee of the Sitges Nova conference for all the efforts they made to make the meeting so successful.

REFERENCES

1. Fryxell, B. A.., and Woosley, S.E., *ApJ*, **261**, 332 (1982).
2. Shara, M.M., *ApJ*, **261**, 649-660 (1982).
3. Shankar, A.,Arnett,W.D., and Fryxell,B.A.,*ApJ*, **394**, L13 (1992).
4. Shankar, A.,Arnett,W.D.,,*ApJ*, **433**, 216 (1994).
5. Fryxell,B.A.,Muller,E.,and Arnett,W.D.,1989,Max-Planck-Institut fur Astrophsik.
6. Colella, P., and Woodward, P., *J.Comput.Phys.*, **54**, 174 (1984).
7. Glasner, S.,A.,Livne,E .,*ApJ*, **445**, L149 (1995).
8. Glasner, S.,A.,Livne,E .,and Truran,J.,W.,*ApJ*, **475**, 754 (1997).
9. Livne,E .,*ApJ*, **412**, 634 (1993).
10. Kercek,A.,Hillebrandt,W .,and Truran,J.W.,*A&A*, **337**, 379 (1998).
11. Kercek,A.,Hillebrandt,W.,and Truran,J.W,*A&A*, **345**, 831 (1999).
12. Glasner, S.,A.,Livne,E .,and Truran,J.W, *in preparation.*
12. Chandrasekhar S., "The Kelvin-Helmholtz Instability," in *Hydrodynamic and Hydromagnetic Stability*, Oxford Univ. Press, Oxford, 1961, pp. 481.
14. Rosner, R.,Alexakis,A.,Young,Y.,N.,Truran,J.W.,and Hillebrandt,W .,*ApJ*, **562**, L177 (2001).
15. Prialnik, D.,and Kovetz,A .,*ApJ*, **281**, 367 (1984).
16. Kippenhahn,R.,and Thomas,H.-C.,*A&A*, **63**, 265 (1978).
17. Livio, M.,and Truran,J.W.,*ApJ*, **318**, 316 (1987).
18. Woosley,S.E.,*Nucleosynthesis and Chemical Evolution*,(1986).

Mixing by Non-linear Gravity Wave Breaking on a White Dwarf Surface

A. C. Calder[*†], A. Alexakis[**], L. J. Dursi[*†], R. Rosner[*†], J. W. Truran[*†], B. Fryxell[*‡], P. Ricker[*†], M. Zingale[*§], K. Olson[¶], F. X. Timmes[*†] and P. MacNeice[¶]

[*]*Center for Astrophysical Thermonuclear Flashes, The University of Chicago, Chicago, IL 60637*
[†]*Department of Astronomy and Astrophysics, The University of Chicago, Chicago, IL 60637*
[**]*Department of Physics, The University of Chicago, Chicago, IL 60637*
[‡]*Enrico Fermi Institute, The University of Chicago, Chicago, IL 60637*
[§]*Department of Astronomy and Astrophysics, The University of California, Santa Cruz, CA 95064*
[¶]*UMBC/GEST Center, NASA/GSFC, Greenbelt, MD 20771*

Abstract. We present the results of a simulation of a wind-driven non-linear gravity wave breaking on the surface of a white dwarf. The "wind" consists of H/He from an accreted envelope, and the simulation demonstrates that this breaking wave mechanism can produce a well-mixed layer of H/He with C/O from the white dwarf above the surface. Material from this mixed layer may then be transported throughout the accreted envelope by convection, which would enrich the C/O abundance of the envelope as is expected from observations of novae.

INTRODUCTION

Classical novae result from the ignition (and subsequent explosive thermonuclear burning) of a ($\sim 10^4$ m) layer of hydrogen-rich material that has accreted from a main sequence companion onto the surface of a white dwarf [1, 2, 3, 4]. Observed abundances and explosion energies estimated from observations indicate that there must be significant mixing of the heavier material of the C/O or O/Ne white dwarf into the lighter accreted material (H/He). This mixing is critical because otherwise hydrogen burning would be too slow to reproduce observed nova characteristics in outburst. Further, without this mixing it is difficult to understand the observed abundances of intermediate-mass nuclei in the ejecta. Accordingly, nova models must incorporate a mechanism that will dredge up the heavier white dwarf material [5, and references therein].

A recently proposed mixing mechanism is the breaking of non-linear resonant gravity waves at the C/O surface [6, 7, 5]. The gravity waves, driven by the "wind" of accreted material, can break, forming a layer of well-mixed material. This mixed layer may then be transported upward by convection, thereby enriching the accreted material. Because the length scale of this mixed layer may be very small (much smaller than the length scale of convection), previous precursor simulations have not captured this effect.

In this manuscript, we present a simulation of a wind-driven non-linear gravity wave breaking on the surface of a white dwarf. The simulation was performed with FLASH, a parallel, adaptive-mesh simulation code for the compressible, reactive flows found in

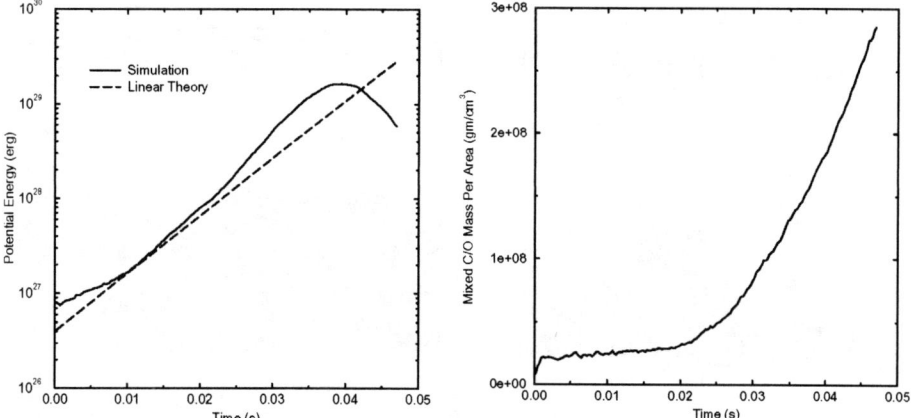

FIGURE 1. Results from the simulation of a wind-driven gravity wave. The left panel shows the potential energy of the wave vs. time. Also shown is the potential energy predicted by the linear theory. The right panel shows the mixed C/O mass per unit area vs. time.

many astrophysical environments [8, 9]. This simulation is part of an ongoing study of this mechanism to assess its efficacy for mixing white dwarf material with envelope material.

THEORY AND SIMULATION DETAILS

This work applies the theory of gravity wave generation (originally developed for the air over water interface) [10, and references therein] to a white dwarf with an accreted envelope. The control parameters for this model are $G = g\delta/U_{max}^2$, which is the ratio of potential to kinetic energy of the wind (g is the constant gravitational acceleration, U_{max} is the maximum wind speed, and δ is the characteristic length scale of the wind profile) and the ratio of densities at the interface, $r = \rho_1/\rho_2$. For this simulation, the domain was 1.0×10^6 cm by 1.0×10^6 cm, $g = 4.5 \times 10^9$ cm/s, and $r = 10$, with a density of the white dwarf material at the interface of 10^4 gm/cm^3. The white dwarf and accreted envelope materials are modeled as simple $\gamma = 5/3$ gases with the white dwarf material composed (by mass) of a 50/50 C/O mix and the accreted material composed of a 75/25 H/He mix.

The density and pressure profiles were obtained by integrating the equation of hydrostatic equilibrium

$$\frac{dp}{dy} = -\rho g \hat{\mathbf{k}}, \tag{1}$$

which for the case of a compressible, gamma-law gas gives

$$\rho = \rho_i \left[1 - (\gamma - 1)\frac{g\rho_i y}{P_0 \gamma}\right]^{\frac{1}{\gamma - 1}} \tag{2}$$

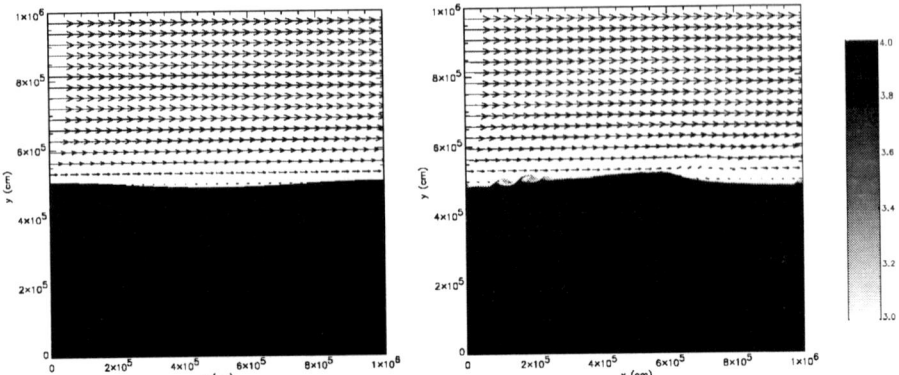

FIGURE 2. Images of the log of density with velocity vectors at earlier times during the simulation. The units of density are g/cm^3 and the length of the velocity arrows is proportional to the magnitude of the velocity, with a maximum of 2×10^8 cm/s. The left panel shows the initial conditions, and the right panel shows the simulation at t = 0.015 s.

and

$$P = P_0 \left[1 - (\gamma - 1) \frac{g \rho_i y}{P_0 \gamma} \right]^{\frac{\gamma}{\gamma - 1}}. \qquad (3)$$

Here P_0 is the pressure at the interface and ρ_i is the density immediately above or below the interface. The wave was created by forcing the interface to be sinusoidal and perturbing the pressure and adding a velocity via a prescription similar to linear theory. The wind profile is given by

$$U(y) = U_{\max} \left(1 - e^{-y/\delta} \right) \qquad (4)$$

with $\delta = 1.0 \times 10^5$ cm and $U_{\max} = 2 \times 10^8$ cm/s. The boundary conditions were isothermal hydrostatic on the upper and lower boundaries and periodic on the sides. The simulation was performed with seven levels of adaptive mesh refinement, for an effective resolution of 512×512 zones.

RESULTS

Figure 1 shows the potential energy of the wave and the amount of mixed material during the course of the simulation. The left panel is a plot of potential energy vs. time showing the result from the simulation and the potential energy given by the linear theory as $4.0 \times 10^{26} \exp(140t)$. The growth rate in this expression (140) comes from the control parameter G, and 4.0×10^{26} (erg) is the initial potential energy of the wave [7]. The two potential energies agree reasonably well until the wave breaks at about 0.04 s. The right panel is a plot of mixed C/O mass per unit area vs. time. The figure demonstrates the dramatic increase in the amount of mixed material that occurs above $t = 0.25$, the point during the course of the simulation at which the wave begins to break.

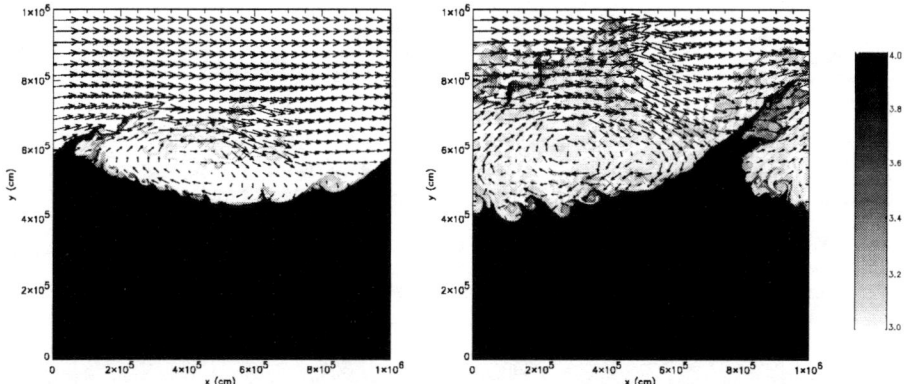

FIGURE 3. Images of the log of density with velocity vectors at later times during the simulation. The units of density are g/cm^3 and the length of the velocity arrows is proportional to the magnitude of the velocity, with a maximum of 2×10^8 cm/s. The left panel shows the simulation at t = 0.030 s, and the right panel shows the simulation at t = 0.045 s.

Figure 2 shows gray scale images of the log of density with velocity vectors at early times in the simulation. The longest velocity arrows, at the top of the images, correspond to 2.0×10^8 cm/s, the maximum wind speed. The left panel shows the initial conditions, and the right panel shows the configuration at $t = 0.015$ s. Visible in the right panel is the development of Kelvin-Helmholtz instability on the upwind (left) side of the wave. Figure 3 shows similar gray scale images of the log of density with velocity vectors at later times in the simulation. By $t = 0.030$ s (left panel), the wave has begun to break, and the images show vorticity downwind from the crest of the wave. At $t = 0.045$ s (right panel), there is substantial mixing and obvious vorticity.

CONCLUSIONS

The simulation presented in this manuscript demonstrates the proposed breaking wind-driven non-linear gravity wave mixing mechanism. The results show the development of well-mixed zone just above the surface of the white dwarf. This simulation will be one part of a study of this mixing mechanism investigating effects of wind profiles and speeds. Complete details of the study will appear in Alexakis, et al. [11].

The expectation is that the breaking of non-linear gravity waves on the surface of the white dwarf will lead to a thin well-mixed layer of material that may then be transported throughout the the envelope by convection. This study investigating the mixing mechanism should provide quantitative information about the mixing rate that will allow for the development of subgrid models that may be applied to multidimensional convection simulations to study the enrichment of the envelope. A preliminary simulation of this kind is also presented in this volume [12].

ACKNOWLEDGMENTS

This work is supported in part by the U.S. Department of Energy (DOE) under Grant No. B341495 to the Center for Astrophysical Thermonuclear Flashes at the University of Chicago. J. W. Truran acknowledges partial support from DOE grant DE-FG02-91ER40606. L. J. Dursi is supported by the Krell Institute CSGF. K. Olson acknowledges partial support from NASA grant NAS5-28524. M. Zingale acknowledges support from the Scientific Discovery through Advanced Computing (SciDAC) program of the DOE, grant number DE-FC02-01ER41176. Additional details about the project and information about requesting a copy of FLASH may be found at http://flash.uchicago.edu.

REFERENCES

1. Truran, J. W., 1982, "Nuclear Theory of Novae," in *Essays in Nuclear Astrophysics,* edited by C. A. Barnes, D. D. Clayton, and D. N. Schramm, Cambridge University Press, Cambridge, 1982, pp. 467-493
2. Shara, M. M., *PASP* **101** 5-31 (1989)
3. Starrfield, S., "Thermonuclear Processes and the Classical Nova Outburst," in *Classical Novae,* edited by N. Evans and M. Bode, Wiley, New York, 1989, pp. 39-60
4. Livio, M., *Mem. Soc. Astron. Ital.*, **65** 49-57 (1994)
5. Rosner, R., Alexakis, A., Young, Y.-N., Truran, J. W., and Hillebrandt, W., *ApJ* **562** L177-L179 (2001)
6. Rosner, R., et al., *BAAS* **32** 1538 (2000)
7. Alexakis, A., Young, Y.-N., and Rosner, R., *Phys. Rev. E* **65** 026313 (2002)
8. Fryxell, B., et al., *ApJS* **131** 273-334 (2000)
9. Calder, A. C., et al., "High Performance Reactive Fluid Flow Simulations Using Adaptive Mesh Refinement on Thousands of Processors," in *Proc. Supercomputing 2000,* 2000, IEEE Computer Soc., http://sc2000.org/proceedings/
10. Miles, J., *J. Fluid Mech.* **3** 185-204 (1957)
11. Alexakis, A., et al., In prep. (2002)
12. Dursi, L. J., et al., This volume (2002)

Onset of Convection on a Pre-Runaway White Dwarf

L.J. Dursi[*†], A.C. Calder[*†], A. Alexakis[*†], J.W. Truran[*†], M. Zingale[**], B. Fryxell[‡], P. Ricker[*†], F.X. Timmes[*†] and K. Olson[§*]

[*]*Center for Astrophysical Thermonuclear Flashes, The University of Chicago, Chicago, IL 60637*
[†]*Department of Astronomy and Astrophysics, The University of Chicago, Chicago, IL 60637*
[**]*Department of Astronomy and Astrophysics, The University of California, Santa Cruz, CA 95064*
[‡]*Enrico Fermi Institute, The University of Chicago, Chicago, IL 60637*
[§]*UMBC/GEST Center, NASA/GSFC, Greenbelt, MD 20771*

Abstract. Observed novae abundances and explosion energies estimated from observations indicate that there must be significant mixing of the heavier material of the white dwarf (C+O) into the lighter accreted material (H+He). Accordingly, nova models must incorporate a mechanism that will dredge up the heavier white dwarf material, and fluid motions from an early convection phase is one proposed mechanism.

We present results from two-dimensional simulations of classical nova precursor models that demonstrate the beginning of a convective phase during the 'simmering' of a Nova precursor. We use a new hydrostatic equilibrium hydrodynamics module recently developed for the adaptive-mesh code FLASH. The two-dimensional models are based on the one-dimensional models of Ami Glasner[1], and were evolved with FLASH from a pre-convective state to the onset of convection.

INTRODUCTION

As a classical nova precursor accretes material from its neighbor, it heats up; by the time its peak temperature becomes roughly 4×10^7 K – well before the final stages of runaway – the accreted atmosphere becomes convectively unstable. The resulting convective motions may be important for the process of dredging up white dwarf material into the accreted atmosphere.

In this paper, we examine the turn-on of convective motions in a white dwarf atmosphere based on one-dimensional early-time models provided to us by Ami Glasner. This initial model is the same used in other multidimensional studies [1, 2], but taken at an earlier time – at the last timestep before the onset of convection in the 1-d model code. We map this model into the multidimensional FLASH code [3] using techniques developed in [4], and perturb the models to investigate the onset of convective motions.

SIMULATIONS

Figure 1 represents early convective motions in the atmosphere. The temperature in a 5 km × 5km region at the hottest (and least convectively unstable) point in the atmosphere is initially increased by 5%. Sound waves are emitted, and a convective roll

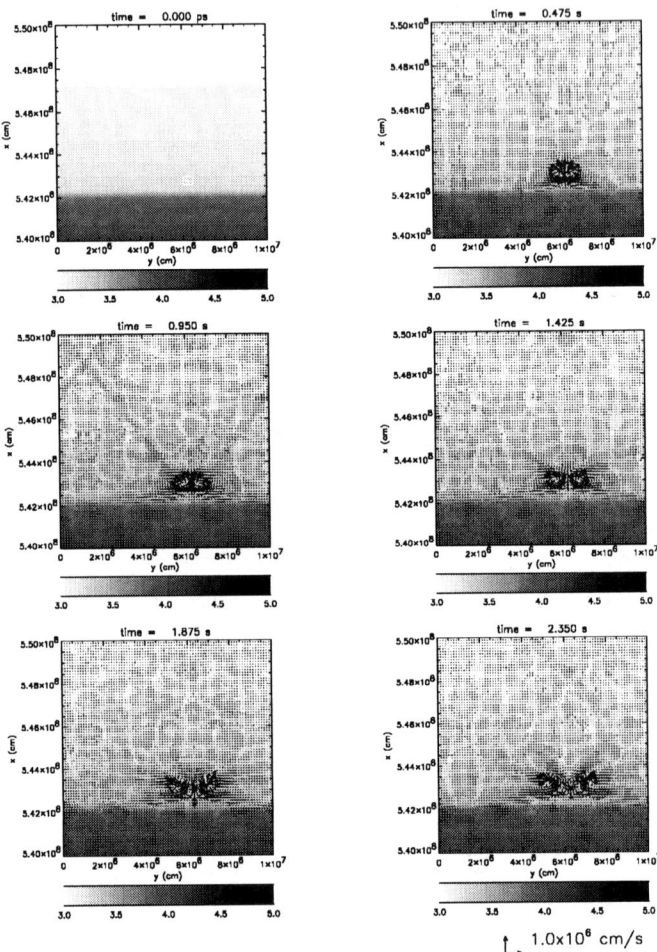

FIGURE 1. The 2D evolution of the first convective roll, seeded by a 5% temperature perturbation at the hottest point in the atmosphere. The white contour surrounds the 'hot spot' – $T > 4.01 \times 10^7$K. Sound-wave transients from the initial perturbation are visible. Typical induced velocities are $v \sim 10^6$ cm s^{-1} in the rolls, $v \sim 10^5$ cm s^{-1} at the interface.

begins. This is shown in Figure 1, through approximately one rollover time, at times $t = 0, .2, .4, .6, .8,$ and 1.0 s after the perturbation is imposed. Note that the velocity field extends to the C+O interface.

The dynamics of the rolls depend on the initial amplitude of the perturbation. Shown

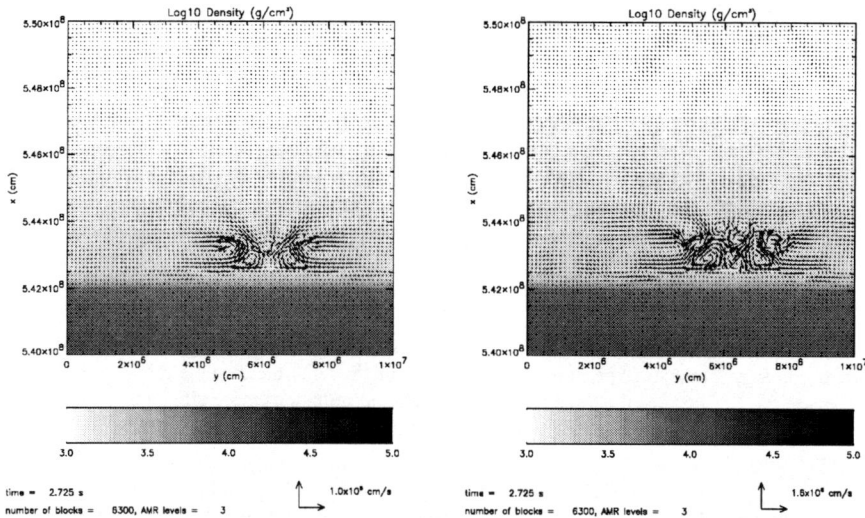

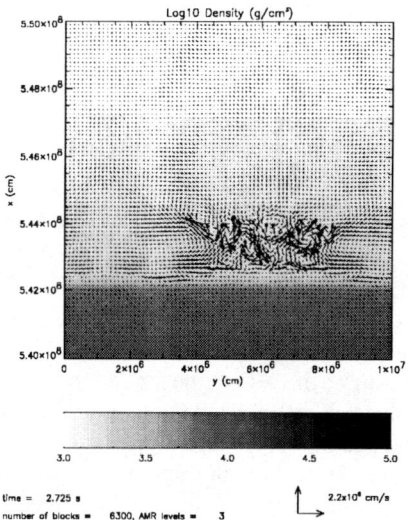

FIGURE 2. Early convective motions as in Figure 1 at time $t = 1.2$s after perturbation, for perturbations in the temperature of 2%, 5%, and 10%. Velocity arrows are scaled to the amount of thermal energy in the perturbation.

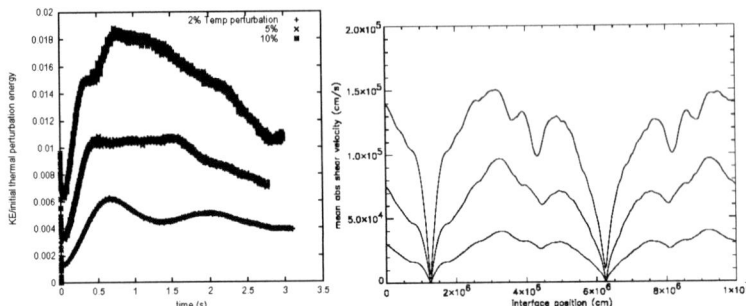

FIGURE 3. On the left, kinetic energy in the atmosphere measured in units of the original thermal perturbation energy for the three perturbation amplitudes; on the right, the absolute values of the interface velocities at time $t = 1.2$s for the three amplitudes. Large perturbations more efficiently cause motions, and the interfacial velocities rise monotonically with the perturbation size.

in Figure 2 are the motions in the atmosphere at time $t = 1.2$ s with temperature perturbations of 2%, 5%, and 10%, respectively. The velocity vectors in the plots are scaled so that the kinetic energy of the motions are scaled to the thermal energy of the perturbation; note that the sound waves have very similar amplitudes in these plots. Even with these scalings, the 10% perturbation generates considerably more motion.

We can quantitatively see the effects of perturbation size on convective motions above. Shown, in Figure 3 are the kinetic energy as a fraction of the initial thermal energy perturbation, and the average of the interfacial shear velocities over the course of the simulation. Larger perturbations produce motions at the interface much more efficiently.

DIRECTION OF FUTURE WORK

Using what we are learning about shear and gravity-wave driven mixing [5], we can hope to model the unresolved mixing due to the interfacial shear generated by these motions. Shown in Figure 4 is a logarithmic plot of metallicity in the atmosphere at time $t = 0.89$ s, where the only source of C+O in the simulation is modelled mixing driven by the velocities at the interface between the white dwarf and the atmosphere. The sub-grid model we have used here is preliminary, based on early results of observed mixing fluxes in small-scale simulations run by A. Alexakis and A. C. Calder described in this volume. As our understanding of the convective motions and the shear-driven mixing improves, we hope to see if this provides a robust dredge-up mechanism. We will incorporate our improved subgrid model into both one-dimensional and multi-dimensinoal simulations, and test our model results with respect to the typical observed levels of enrichment of nova envelopes.

FIGURE 4. A metallicity plot at time $t = 0.89$s of a simulation similar to that shown in Figure 1, but with the C+O white dwarf removed from the simulation domain; all metallicity is either from the accreted material, or from the preliminarily sub-grid model which represents the results of mixing simulations presented elsewhere in this volume.

Acknowledgments

This work is supported in part by the U.S. Department of Energy under Grant No. B341495 to the Center for Astrophysical Thermonuclear Flashes at the University of Chicago. LJD is supported by the Krell Institute CSGF. K. Olson acknowledges partial support from NASA grant NAS5-28524. We greatfully acknowledge Ami Glasner for the initial model on which these calculations were based.

REFERENCES

1. Glasner, A. S., Livne, E., and Truran, J. W., *ApJ*, **475** (1997).
2. Kercek, A., Hillebrandt, A., and Truran, J. W., *AAP*, **337** (1998).
3. Fryxell, B., Olson, K., Ricker, P., Timmes, F. X., Zingale, M., Lamb, D. Q., MacNeice, P., Rosner, R., Truran, J. W., and Tufo, H., *ApJS*, **131**, 273–334 (2000).
4. Zingale, M., Dursi, L. J., ZuHone, J., Caceres, A., Calder, A. C., Fryxell, B., Olson, K., Plewa, T., Ricker, P. M., Riley, K., Rosner, R., Siegel, A., Timmes, F. X., Truran, J. W., and Vladimirova, N., *ApJ* (submitted).
5. Calder, A., Alexakis, A., Dursi, L., Rosner, R., Truran, J., Fryxell, B., Ricker, P., Zingale, M., Olson, K., Timmes, F., and MacNeice, P., "Mixing by Non-linear Gravity Wave Breaking on a White Dwarf Surface," in *Proceedings of the International Conference on Classical Nova Excplosions*, edited by M. Hernanz, AIP, 2002.

The Role of Novae in Galactic Chemical Evolution

Donatella Romano* and Francesca Matteucci*†

*International School for Advanced Studies, SISSA/ISAS, Via Beirut 2-4, I-34014, Trieste, Italy
†Dipartimento di Astronomia, Università di Trieste, Via G.B. Tiepolo 11, I-34131, Trieste, Italy

Abstract. We discuss the role of novae as producers of ^{7}Li and CNO isotopes in the Milky Way. A detailed model for the chemical evolution of the Milky Way including novae, Type Ia supernovae, Type II supernovae as well as single low- and intermediate-mass stars is adopted and the results are compared with the available observational constraints. It is shown that novae are among the most promising candidates in order to explain the steep rise off the Li plateau in the $\log \varepsilon(^7\text{Li}) - [\text{Fe/H}]$ diagram observed for stars in the solar vicinity. We also find that novae are likely to be the main producers of ^{15}N, whereas they should only partly contribute to ^{13}C and ^{17}O.

INTRODUCTION

Both theoretical and observational evidence suggests that classical novae may be important sources of the nuclides ^{7}Li, ^{13}C, ^{15}N and ^{17}O [e.g., 1, 2, 3, 4, 5, 6, 7].

The temporal evolution of ^{7}Li, ^{13}C and ^{15}N in the Galaxy has been studied by [8] and [9] in the framework of a complete, coherent model of Galactic chemical evolution taking into account nova nucleosynthesis. In order to explain the relevant data, [8] and [9] assigned more than 50% of the total ^{7}Li production and 100% of the ^{15}N production to novae. ^{7}Li production during nova outbursts was computed according to the results of [3]; ^{13}C and ^{15}N were assumed to be produced in proportion to ^{7}Li. It is worth emphasizing that in these models ^{13}C is not overproduced only if no ^{13}C from hot-bottom burning in asymptotic giant branch (AGB) stars is included.

Here we use the *two-infall model* for the chemical evolution of the Galaxy [10, 11] in order to reassess the problem of the contribution from novae to the Galactic chemical evolution of ^{7}Li and CNO group nuclei.

THE CHEMICAL EVOLUTION MODEL

The Galaxy is assumed to form out of two main infall episodes. During the first one, the Galactic bulge, inner halo and thick-disk are build up (on a timescale of ~ 1 Gyr); during the second one, the thin-disk is build up in an inside-out way (i.e., on a timescale increasing with increasing Galactocentric distance). The infalling material has a primordial chemical composition. A quite long timescale of ~ 7 Gyr is needed in order to reproduce the metallicity distribution of G-dwarfs at the solar ring. Such a long timescale was first suggested by [10] and then adopted in most Galactic chemical evolution models.

The adopted nucleosynthesis prescriptions are from [12] and [13] for single low- and intermediate-mass stars (LIMS) and massive stars, respectively, and from [14] for Type Ia supernovae (SNeIa). The detailed nucleosynthesis prescriptions for ^{7}Li can be found in [15]; in particular, nucleosynthesis from novae is included according to the results of [5]. Novae only contribute to the Galactic abundances of ^{7}Li and CNO group nuclei.

For a complete description of the basic chemical evolution equations we refer to [10, 11], whereas details about the inclusion of nova system nucleosynthesis in the model can be found in [16, 15] and Romano & Matteucci (in preparation). We only recall here that, in the framework of our model, the nova system formation rate at a given time t is computed as a fraction α of the white dwarf (WD) formation rate at a previous time $t - \tau_m - \Delta t$

$$R_{\text{novae}}(t) = \alpha \int_{0.8}^{8} \psi(t - \tau_m - \Delta t) \varphi(m) \, dm, \qquad (1)$$

τ_m being the lifetime of the star of mass m, $\psi(t)$ the star formation rate and $\varphi(m)$ the initial mass function. Notice that all stars between $m = 0.8$ M$_\odot$ and $m = 8$ M$_\odot$ end up as WDs. Δt is a suitable delay-time which ensures a strong enough nova outburst. α is a parameter, whose value is fixed by the request of reproducing the current rate of nova outbursts in the Galaxy. A value of $\alpha \sim 0.1$ leads to $R_{\text{outbursts}}(t_{\text{Gal}}) \sim 20$ yr^{-1}, to be compared with the value inferred from scalings from extragalactic nova surveys [$R_{\text{outbursts}}(t_{\text{Gal}}) = 15-50$ yr^{-1}; 17, 18][1].

RESULTS

Lithium

In Fig. 1 we compare model results to observations in the $\log \varepsilon(^7\text{Li}) - [\text{Fe/H}]$ diagram for solar neighbourhood dwarfs[2]. Halo stars sharing the same ^{7}Li abundance (Li plateau) are thought to reflect the primordial ^{7}Li abundance [20]. Higher-metallicity stars tracing the upper envelope of the observations can be regarded as records of different enrichment processes occurred during the Galaxy lifetime, under the hypothesis that they did not undergo any destruction/depletion process altering their initial ^{7}Li content.

Chemical evolution models should try to reproduce the upper envelope of the observations. The late, steep rise off the Li plateau occurring at [Fe/H] ~ -0.5 can be explained if some long-lived ^{7}Li sources start to inject important amounts of ^{7}Li into the interstellar medium (ISM) at [Fe/H] ~ -0.5 [16, 15]. Promising candidates are novae and low-mass stars ($m \sim 1-2$ M$_\odot$), as can be seen from Fig. 1, left panel. We take the average ^{7}Li production from nova systems from [5] and find that novae are responsible for less than 20% of the meteoritic ^{7}Li [15]. This average ^{7}Li production from novae is consistent with the upper limit on the ^{7}Li mass ejected from Nova V 382 Vel 1999 recently estimated by [7], provided that the identification of the line is correct.

---

[1] Each nova system is assumed to suffer 10^4 outbursts during its life [19].
[2] $\log \varepsilon(^7\text{Li}) = \log_{10}(N_{^7\text{Li}}/N_\text{H}) + 12$; [Fe/H] = $\log_{10}(\text{Fe/H})_\text{star} - \log_{10}(\text{Fe/H})_\odot$.

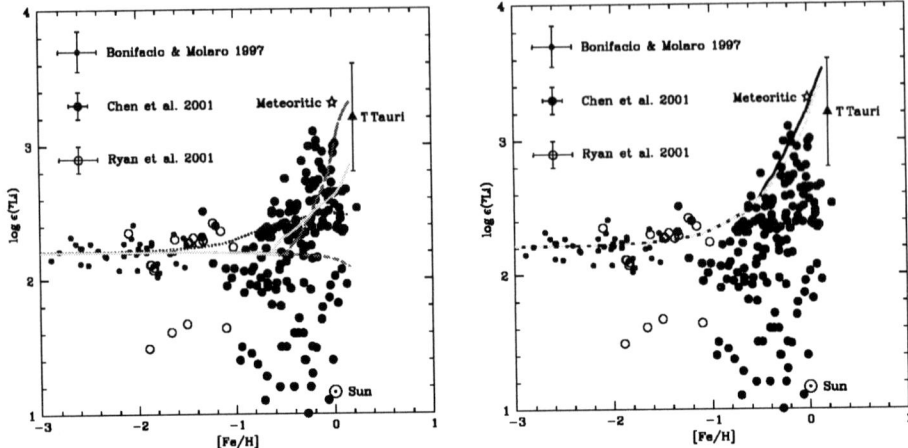

FIGURE 1. Left panel: theoretical $\log\varepsilon(^{7}\mathrm{Li})$ vs. [Fe/H] in the gas in the case of ^{7}Li production from: *i)* low-mass giants *(long-dashed line)*; *ii)* AGB stars *(short-dashed line)*; *iii)* novae *(continuous line)*; *iv)* SNeII *(dotted line)*. Right panel: same as left panel, when all the stellar sources plus cosmic rays *(continuous line)* or all the stellar sources except novae plus cosmic rays *(dashed line)* are added together. Data are non-LTE values from [20, 21, 22].

CNO isotope ratios

The observations clearly show that the ^{12}C/^{13}C and ^{16}O/^{17}O ratios decrease from the time of the Solar System formation up to now. Moreover, for both ratios a positive gradient is observed across the Galactic disk. This is commonly explained as due to the primary nature of ^{12}C and ^{16}O coupled to the primary plus secondary nature of ^{13}C and secondary nature of ^{17}O [e.g., 23][3]. As far as ^{14}N/^{15}N is concerned, the situation is more complicated; previous solar neighbourhood observations suggested that this ratio is increasing with time [24] whereas more recent ones pointed out just the opposite behaviour [25].

In Figs. 2, 3 and 4 we show the results we get if 100% of ^{15}N and ^{17}O comes from novae, while ^{13}C is produced by both novae and single stars (mostly LIMS). ^{13}C, ^{15}N and ^{17}O from novae are treated as secondary elements, i.e., their yields are scaled with the initial abundance of the seed elements. Nevertheless, the yields of ^{15}N and ^{17}O given by [5] for a solar chemical composition of the matter accreted by the WD have to be further reduced in order not to overestimate their solar abundances (Romano & Matteucci, in preparation).

As far as ^{12}C/^{13}C is concerned, both the temporal evolution at the solar radius and

---

[3] We remind the reader that a primary element is always synthesized starting from H and He, while a secondary one is synthesized starting from ^{12}C and/or ^{16}O seeds already present in the composition of the star at the time of its birth. However, a primary element which is restored to the ISM on long timescales behaves almost like a secondary one from the point of view of Galactic chemical evolution.

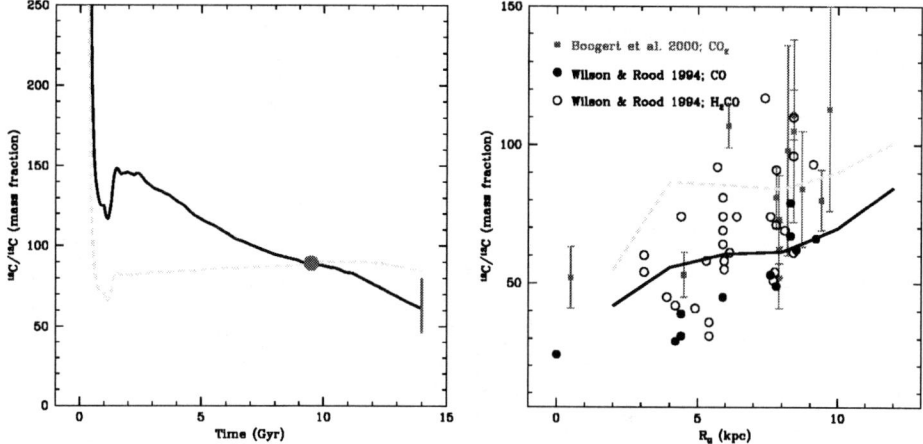

FIGURE 2. Left panel: temporal evolution of the carbon isotope ratio in the solar neighbourhood. The model $^{12}C/^{13}C$ ratio 4.5 Gyr ago is normalised to its solar value. *Continuous line:* ^{13}C production from novae plus single stars; *dashed line:* ^{13}C production from single stars alone. The big dot is the observed meteoritic ratio; the vertical bar is the ratio observed in the local ISM. Right panel: theoretical present-day carbon isotope ratio across the Galactic disk vs. observations.

TABLE 1. Solar abundances by mass of the minor CNO isotopes. The predicted values refer to novae as the only producers of ^{15}N and ^{17}O, whereas ^{13}C is produced by both novae and LIMS (see text).

	^{13}C	^{15}N	^{17}O
Predicted	3.97×10^{-5}	3.87×10^{-6}	3.85×10^{-6}
Observed*	3.65×10^{-5}	4.36×10^{-6}	3.89×10^{-6}

* Anders & Grevesse 1989.

the present-day values across the Galactic disk are well reproduced by a model which accounts for ^{13}C production from both novae and single stars. If single stars are the sole ^{13}C producers, the $^{12}C/^{13}C$ ratio stays nearly constant from the time of the birth of the Sun up to now, at variance with observations (Fig. 2, left panel).

For N and O, a production scenario more complicated than that assumed here seems to emerge: taking novae as the only factories of ^{15}N and ^{17}O allows us to reproduce the solar abundances of these elements (see Table 1), but the local ISM values of the $^{14}N/^{15}N$ and $^{16}O/^{17}O$ ratios turn out to be underestimated (Figs. 3 and 4, left panels). The slope of the $^{14}N/^{15}N$ gradient along the Galactic disk is well reproduced, but we do not match the absolute values of this ratio (Fig. 3, right panel). We conclude that the synthesis of ^{15}N and ^{17}O in stars of different masses deserves further investigation. However, we also notice that the existence of a positive gradient along the disk for both the $^{14}N/^{15}N$ and the $^{16}O/^{17}O$ ratios is consistent with a large fraction of the Galactic ^{15}N

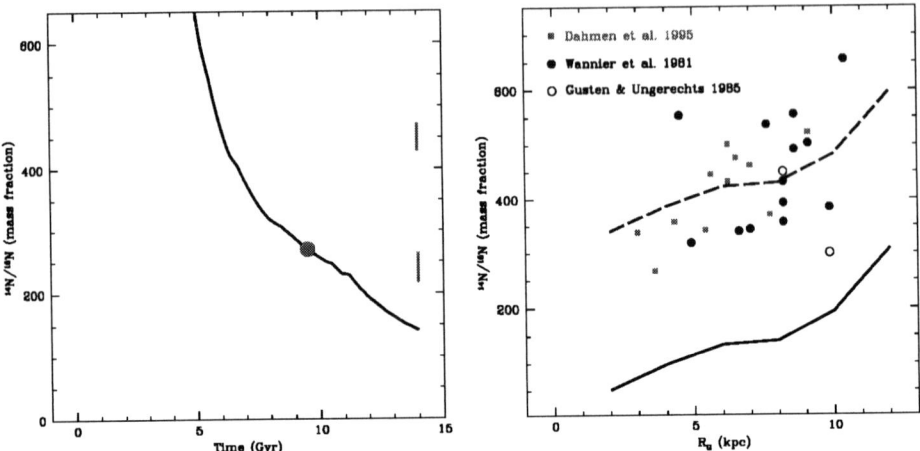

FIGURE 3. Left panel: temporal evolution of the nitrogen isotope ratio in the solar neighbourhood. The model $^{14}N/^{15}N$ ratio 4.5 Gyr ago is normalised to its solar value. ^{15}N *is only produced during nova outbursts*. The big dot is the observed meteoritic ratio; the vertical bars are the ratios observed in local clouds [24, 25]. Right panel: the theoretical present-day gradient *(continuous line)* is arbitrarily offset *(dashed line)* to better compare with the data. The slope of the gradient agrees well with that suggested by the observations.

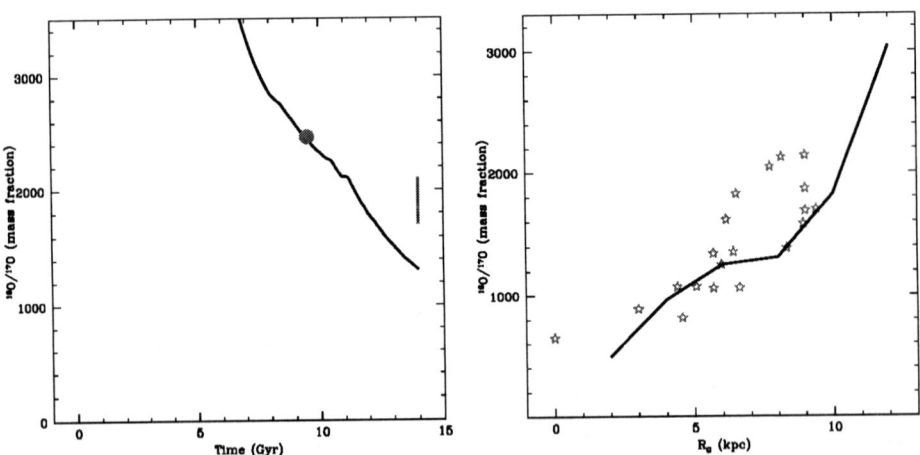

FIGURE 4. Same as Fig. 3 for oxygen. ^{17}O *is only produced during nova outbursts*. Although the model predicts for the local ISM a value of the $^{16}O/^{17}O$ ratio lower than observed, the $^{16}O/^{17}O$ gradient across the disk is satisfactorily reproduced. Data are from [23] and references therein.

and ^{17}O being produced by novae.

CONCLUSIONS

By taking into account nova nucleosynthesis results from [5] in the framework of a complete model for the chemical evolution of the Galaxy, we improve the predictions on the temporal evolution of ^{7}Li and ^{12}C/^{13}C in the solar neighbourhood. Also the gradient of ^{12}C/^{13}C along the Galactic disk is satisfactorily reproduced. In particular, we consider different stellar sources of ^{7}Li and ^{13}C and find that novae are responsible for less than 20% of the meteoritic ^{7}Li and $\sim$ 50% of the meteoritic ^{13}C.

As far as ^{15}N and ^{17}O are concerned, we show that if 100% of ^{15}N and ^{17}O originates in novae the slopes of the gradients of ^{14}N/^{15}N and ^{16}O/^{17}O along the Galactic disk are well reproduced. However, these same ratios are predicted to decline faster than observed from the time of the Sun formation up to now. This suggests that the interplay between novae and some other stellar source(s) should be taken into account.

ACKNOWLEDGMENTS

We thank the organisers for a very fruitful and stimulating meeting.

REFERENCES

1. Starrfield, S., Truran, J. W., Sparks, W. M., and Kutter, G. S., *ApJ*, **176**, 169 (1972).
2. Starrfield, S., Sparks, W. M., and Truran, J. W., *ApJ*, **192**, 647 (1974).
3. Starrfield, S., Truran, J. W., Sparks, W. M., and Arnould, M., *ApJ*, **222**, 600 (1978).
4. Hernanz, M., José, J., Coc, A., and Isern, J., *ApJ*, **465**, L27 (1996).
5. José, J., and Hernanz, M., *ApJ*, **494**, 680 (1998).
6. Gehrz, R. D., Truran, J. W., Williams, R. E., and Starrfield, S., *PASP*, **110**, 3 (1998).
7. della Valle, M., Pasquini, L., Daou, D., and Williams, R. E., *A&A* (2002), in press.
8. D'Antona, F., and Matteucci, F., *A&A*, **248**, 62 (1991).
9. Matteucci, F., and D'Antona, F., *A&A*, **247**, L37 (1991).
10. Chiappini, C., Matteucci, F., and Gratton, R., *ApJ*, **477**, 765 (1997).
11. Chiappini, C., Matteucci, F., and Romano, D., *ApJ*, **554**, 1044 (2001).
12. van den Hoek, L. B., and Groenewegen, M. A. T., *A&AS*, **123**, 305 (1997).
13. Nomoto, K., Hashimoto, M., Tsujimoto, T., Thielemann, F.-K., Kishimoto, N., Kubo, Y., and Nakasato, N., *Nucl. Phys. A*, **616**, 79c (1997).
14. Thielemann, F.-K., Nomoto, K., and Hashimoto, M., "Explosive Nucleosynthesis in Supernovae," in *Origin and Evolution of the Elements*, edited by N. Prantzos, E. Vangioni-Flam, and M. Cassé, Cambridge Univ. Press, Cambridge, 1993, pp. 297–309.
15. Romano, D., Matteucci, F., Ventura, P., and D'Antona, F., *A&A*, **374**, 646 (2001).
16. Romano, D., Matteucci, F., Molaro, P., and Bonifacio, P., *A&A*, **352**, 117 (1999).
17. della Valle, M., and Livio, M., *A&A*, **286**, 786 (1994).
18. Shafter, A. W., *ApJ*, **487**, 226 (1997).
19. Bath, G. T., and Shaviv, G., *MNRAS*, **183**, 515 (1978).
20. Bonifacio, P., and Molaro, P., *MNRAS*, **285**, 847 (1997).
21. Chen, Y. Q., Nissen, P. E., Benoni, T., and Zhao, G., *A&A*, **371**, 943 (2001).
22. Ryan, S. G., Kajino, T., Beers, T. C., Suzuki, T. K., Romano, D., Matteucci, F., and Rosolankova, K., *ApJ*, **549**, 55 (2001).
23. Prantzos, N., Aubert, O., and Audouze, J., *A&A*, **309**, 760 (1996).
24. Wilson, T. L., and Rood, R. T., *ARA&A*, **32**, 191 (1994).
25. Lucas, R., and Liszt, H., *A&A*, **337**, 246 (1998).

The Electrostatic Screening of Nuclear Reactions in Dense Plasma

N. J. Shaviv* and G. Shaviv[†]

*Racah Institute of Physics, The Hebrew University, Jerusalem, Israel
[†]Department of Physics and Asher Space Research Institute, Technion, Haifa, Israel

Abstract. We developed a new method to calculate the electrostatic screening of reacting ions from first principles. We present results indicating that the reactions between light protons and heavy nuclei (like CNO) are supressed while the reactions between protons and themselves are enhanced (as assumed before). We show the derivation and discuss application to thermonuclear runaways.

INTRODUCTION

The classical theory of thermonuclear run aways on While Dwatfs fail to predict correctly the ejected mass as well as the observed super Eddington luminosity. Here we adress to the first problem in an attempt to re rexamine the theory of the electrostatic screening factors which are used in evaluating the nuclear rates in dense plasmas. The screening factors can be quite large in very dense matter and hence the effect may be important.

We first describe the classical screening theory and how important it can be. We then explan why some people claimed that dynamic screening is important and we show why they are wrong. We explain the difficulties with the classical way the screening is defined and hence we present a simple, clear and well understood theory. We show how it follows from first principles.

We next show how the screening is intimately associated with relaxation processes in the gas and why the results of the Molecular Dynamics are a natural consequence of these processes. We show how to calculate the screening analyticaly (in a simplified way) and explain in what way the Molecular Dynamics are accurate.

THE CLASSICAL SCREENING THEORY

The idea of screening was first suggested by Schaztman (1948) and the theory was developped by Salpeter (1954). The basic assumption is that an ion immersed in a plasma feels a mean electrostatic potential. When two ions approach each other, the incoming ion enters into the potential well of the first ion and converts the potential energy into kinetic one and thus enhances the effective relative kinetic energy which enters into the cross section. As the cross section is a sensitive function of the energy, the end result is an enhancement of the rate of nuclear reactions.

The size of the effect can be estimated in the following way: Define the plasma parameter as:

$$\Gamma = \frac{Z_1 Z_2}{<r>kT} \quad \text{where} \quad <r> = n^{-1/3}, \tag{1}$$

where n is the number density. In the weak screening limit $\Gamma \ll 1$, the enhancement of the rate is given by $f = \exp(\Gamma)$. In the case of the pp reaction in the Sun $\Gamma = 0.05$ and the enhancemnet of the pp rate is just $f = 1.06$. On the other hand, in strong screening the effect may dramaticaly bigger.

Let us see how the enhancement is obtained in the classical way; The rate of any reaction is given by:

$$R_{12} = n_1 n_2 \int f_2(1,2) \sigma(E_{kin-rel}) |v_1 - v_2| dv_{12}. \tag{2}$$

The index 2 on f means that $f_2(1,2)$ is the pair distribution function, and the 1,2 means the phase space coordinates of particle 1 and particle 2.

The binary distribution function is then written as (cf. the known BBGKY expansion):

$$f_2(1,2) = f_1(1) f_1(2) \exp(-\psi(\mathbf{x}_1 - \mathbf{x}_2)/kT) \tag{3}$$

where ψ is called the mean potential and is (in the simple electrostatic case) a function of the coordinates only. The substitution in the expression for the rate yields:

$$R_{12} = n_1 n_2 \exp(-\psi(0)/kT) \int f_1(1) f_1(2) \sigma(E_{kin-rel}) |v_1 - v_2| dv_{12}, \tag{4}$$

where we have already substituted the value at the origin as an approximation. This is essentially the Salpeter result in the weak screening limit. The basic problem is that this expansion is valid only when the number of particles in the Debye sphere is large, which is not the case in the solar core ($N_D \approx 3$). As the density increases, the number of particles in the Debye sphere decreases. The relevant densities in the case of Nova thermonuclear run away are about a factor of a hundred higher. Moreover, the mean potential, or the mean field, is not a dynamic quantity but a long time thermodynamic mean. Hence there is a need to approach the problem in a different way. Here we present a direct and clear approach from first principles.

OBTAINING THE SCREENING FROM FIRST PRINCIPLES

The total energy of a pair of particles marked 1 and 2 in the plasma is given by:

$$E^{pair} = E_{kin}(1) + E_{kin}(2) + \frac{Z_1 Z_2 e^2}{r_{12}} + \sum_{i \neq 1} \frac{Z_1 Z_i e^2}{r_{1i}} + \sum_{j \neq 2} \frac{Z_2 Z_j e^2}{r_{1j}} \tag{5}$$

where r_{ij} is the distance between particle i and j. Let now two particles scatter off each other. We define the screening energy as:

$$E_{scr} = E^{pair}(r = r_{min}) - E^{pair}(r = r_{far}), \tag{6}$$

namely, we compare the total energy of the pair when it is at the classical point of return with that at some distance point. Clearly, if the particles were moving in vacuum, $E^{pair} \equiv 0$. The value of r_{far} cannot be too large because the effect of the collision will be lost. Hence we select it to be 2-3 Debye radii.

NUMERICAL METHOD

We adapted the Molecular Dynamic method to the problem at hand. In the calculation, pairs of approaching particles are identified and their relative and abolute state is followed. Once they reach the classical turning point all the dynamic properties are registered. The follow up of the particles continues until the separation distance is few Debye radii. The dynamic properties are registered once more and compared with those at the distance of closest approach to derive the effect of the plasma on the scattering particles.

RESULTS

We show in figure 1 the results for the pp reaction. We note that (a) the screening energy is not constant as a function of the relative kinetic energy. (b) The screening energy is positive for low energies and negative for high energies. Thus, on the average, the reactions between particles with low relative kinetic energy are enhances and vice versa. Let us show that this result is a direct consequence of the thermodynamic equilibrium in the plasma.

THE FOKKER-PLANCK EQUATION FOR THE PLASMA

Let $f(E)$ be the energy distribution function. Let $\Delta E(E)$ be the energy gain/loss of two particles in the plasma during a collision between them. The energy gain is from the plasma and so is the loss. It is important to stress that the energy exchange between the scattering particles and the plasma may be positive or negative. The screening energy is the energy exchange during the approach only. The Fokker-Planck equation is then:

$$\tau \frac{\partial f(E,t)}{\partial t} = -\frac{\partial}{\partial E} f(E,t) \langle \Delta E(E) \rangle + \frac{\partial^2}{\partial E^2} f(E,t) \langle \Delta E^2(E) \rangle \tag{7}$$

where we define the following averages:

$$\langle \Delta E(E) \rangle = \int P(E, \Delta E) \Delta E(E) d\Delta E \tag{8}$$

$$\langle \Delta E^2(E) \rangle = \int P(E, \Delta E) \Delta E^2(E) d\Delta E \tag{9}$$

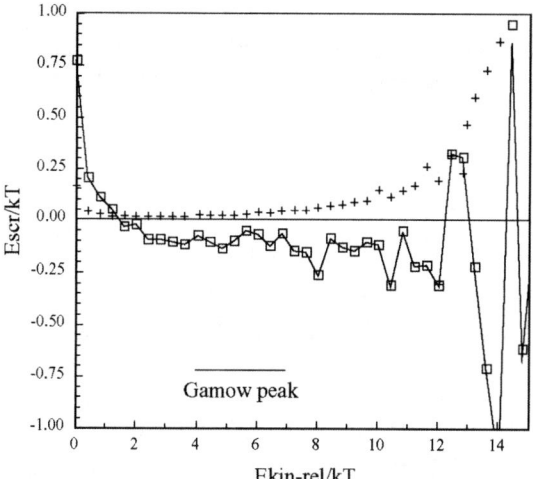

FIGURE 1. The results for the pp reaction in pure Hydrogen plasma at a density of $n = 10^{29}$#/cc and $T = 1.5 \times 10^7$K. Open squares connected with a line are the screening energy in units of kT as a function of the relative kinetic energy in units of kT. The pluses are the corresponding spread in the distribution (σ).

where $P(E, \Delta E)$ is the probability of a particle with energy E to change it energy by ΔE in collision time τ. Assuming a steady state ($\partial/\partial t = 0$) it is clear that in equilibrium we must have:

$$\int f(E) \langle \Delta E(E) \rangle dE = 0 \qquad (10)$$

(Note, $\langle \Delta E(E) \rangle$ is the average energy change for particles with energy E). The latter condition means that the sum of all changes must vanish or else the plasma is not in equilibrium.

The distribution function $f(E)$ in equilibrium is positive definite. Hence $\langle \Delta E(E) \rangle$ must change sign or vanish identically. In the classical Salpeter theory $\langle \Delta E(E) \rangle \equiv 0$ namely, the energy gain in the approach is exactly returned to the plasma as the particles move apart. We see therefore, that the screening energy must change sign at some energy exactly as found in the Molecular Dynamic calculations. We can go one step further. The Fokker-Planck equation is usually written in terms of the velocities rather than the energies. When it is witten as a function of the velocities we find that

$$\langle \Delta \mathbf{v} \rangle = C \int_0^v f(\mathbf{v}_1) \frac{\mathbf{v}_2 - \mathbf{v}_1}{|\mathbf{v}_2 - \mathbf{v}_1|^3} d\mathbf{v}_1 \qquad (11)$$

where C is a constant (cf. Shaviv & Shaviv 2001). The integral is essentially the Rosenbluth potential. v is the absolute value. Expression 11, which is obtained under simplifying assumptions on the collisions and the plasma, described the same physical behavior as that found in the Molecular Dynamics, namely (a) the energy gain/lost by the scattering pair from the plasma, during a collision is not a constant as a function of energy (as would be the result if thermodynamic arguments were used) and (b) the fast particles lose energy on the average (suppresion of screening) while lose particles gain energy.

We remark here that $\langle \Delta E(E) \rangle$ is the energy exchange during the approach *and* separation of the scattering particles while the screening energy is the energy exchange during the approach only. The formalism to treat the screening is obviously similar with similar results.

GETTING THE SCREENING CORRECTION

To obtain the screening correction it is not sufficient to find the average value because the nuclear cross section is very non linear. Thus one has to average over the Gamow peak. Detailed results will be given elsewhere.

ACKNOWLEDGEMENT

GS acknowledges the partial support of the Asher foundation.

REFERENCES

1. Shaviv, N.J. and Shaviv, G. 2001, ApJ **558**,925.
2. Schatzman, E.J. 1948, Phys. Rad. **9**, 46.
3. Salpeter, E.E. 1954, Australian J. Phys. **7**,373.

Movies of Novae Explosions: Restricted Three-Body Dynamics and Geometry of Novae Shells for Purely Gravitational Development

David K. Lynch, S. Mazuk, Eric Campbell, and Catherine C. Venturini

The Aerospace Corporation, El Segundo, CA 90245, USA

Abstract. We present animations of computer simulations of novae shell development. We assume that the shell can be modeled as a large number of massless points that are instantaneously ejected from the surface of the white dwarf (WD) and move ballistically thereafter. Parameters allowed to vary are binary period, ejection velocity and rotation period of the WD. The explosions can be viewed parallel and perpendicular to the orbital plane. At high ejection velocities a nearly spherical shell is produced. At ejection speeds very near the WD's escape velocity, very complicated and ever changing geometries result and the material remains close to the system's barycenter. Each particle is assigned a generic line profile and the integrated shell line profile is obtained by simple addition (which assumes that the shell is optically thin). The ultimate goal of this project is to compare the simulated profiles to those we observe in an attempt to deduce geometrical and dynamical properties of the shell from the spectra.

INTRODUCTION

The shapes of spectral lines in unresolved novae shells are due to many factors including the geometry and kinematics of the shell. Previous studies (see Payne-Gaposhkin 1957 for a review; Solf 1983) have revealed a number of line shapes corresponding to certain geometries: spherical shells, shells with equatorial or polar enhancements, etc. Optically-thin, spherically expanding shells produce rectangular flat-topped profiles. While there is good reason to believe that the thermonuclear runaway (TNR) is, or at least starts out, spherical (See reviews in Bode and Evans 1989), images of resolved novae shells reveal there are significant departures from perfect symmetry in the form of knots, pole/equator asymmetries and internal structure.

In the work described here, we have tried to isolate those aspects of the shell geometry and kinematics that contribute to the line profiles. This is done by modeling discrete (velocity) shells without regard to radiative processes such as temperature, density, excitation and optical depth. The models presented below are based purely on gravitational (Newtonian) expansions of massless particles from a spherical white dwarf (WD) in a circular binary system where the secondary fills its Roche lobe. Hydrodynamics are not included here, nor the time-dependent density and velocity ejection profiles that are known to occur

COMPUTATIONS

The binary star system is modeled in the elliptic restricted three body problem. The two stars, one of which is a white dwarf, are constrained to be in elliptic Keplerian orbits about each other and their motion remains unaffected by the material ejected from the white dwarf. This material results from the uniform explosion on the surface of the white dwarf and is modeled as a set of infinitesimal particles with velocity magnitude Vo in the radial direction. The radial direction is defined relative to the center of the white dwarf. WD rotation is allowed

A representative binary star system $M_S = 0.21 Mo$, $M_P = 0.43 Mo$ (white dwarf) has been studied, where the stars are restricted to lie in circular orbits. Next, it is assumed that the spherical volume of M_S must equal its Roche Lobe limit volume, which can be approximated directly from the non-dimensional mass of the stars. These relations define the semimajor axis of the orbit of the stars and, consequently, the period. In this case, the semimajor axis equals 5.9e+05 km (0.00039 AU) and the period is 2 hrs 41 min. See Figure 1.

The resulting motion of the ejected particles from the surface of the white dwarf is modeled by numerically integrating the elliptic restricted three body problem equations of motion in a non-dimensional form and in the traditional rotating frame using a Runga-Kutta 7/8 integrator. A set of particles is defined that are evenly spaced on the surface of the white dwarf. Each particle begins with a radial velocity of magnitude Vo in the rotating frame and is integrated for a specified time step Δt, assuming only gravitational forces affect its motion. If the ejected particle passes beneath the surface of either star, then integration of the motion of that particle stops. The integration begins anew for the next particle. Using this technique, the gravitational singularity within each star is avoided.

To preserve numerical accuracy, variations in Jacobi Constant are continuously monitored during each integration for abnormally large values (i.e. > 5.0e-09 with the integration tolerance = 1.0e-13.) On average, about one in a thousand integrated trajectories appears to be numerically questionable and these are removed from the computation.

In the simulations below, snap shots from the movies are presented that show the system viewed from three orthogonal directions and the corresponding line profiles below each picture. In all cases shown here, the escape velocity Vesc =3600 km/s.

PRELIMINARY RESULTS

The morphology of the shell and the corresponding line profiles depend strongly on the speed at which material leaves the surface of the WD. This is especially true when the lift-off speed Vo is near Vesc and soon after the explosion. At lift-off speeds well below escape velocity, the material simply falls back onto the WD. At just below Vesc, material leaves the WD and remains in chaotic orbits for some time before falling back and some of it escapes the system altogether as a result of "gravitational boost." When $0.99 <$ Vo/ Vesc <1.01, highly complex geometries develop.

The influence of the secondary's early passage through the expanding shell remains evident at very late stages (and presumably forever), though the effect is less pronounced for higher values of Vo. The "navel" produced as the shell envelopes the secondary is the result of retardation of the velocity. This produces the "tongue" of material reaching inward toward the WD and this material (though still expanding outward) makes the central spike in the line profile. Material from lower velocity lift-offs also produces a spike.

When Vo/ Vesc >1.01, the shell remains roughly spherical even at late times. A spherical expansion produces the expected flat-topped line profiles, though only at the speeds significantly above the WD escape velocity do the rectangular profiles remain. As the lift-off velocities increase beyond Vesc, the line profiles grow flatter and flatter and begin to resemble the classical spherical shell plateaus (Figure 4). Departures from a perfect flat top profiles are due to two effects.

The line profiles depend strongly on the viewing geometry(Figure 2 & 3). This is especially true when Vo is close to the Vesc. Therefore caution must be used in assigning an "expansion velocity" or identifying lines that may have a component that is doppler-shifted relative to the underlying plateau. The polar projection shows a narrow central emission feature with a broad, often flat, underlying plateau. Such profiles are frequently observed (e.g. Lynch et al. 2000). The equatorial projection does not always show the central spike. For an arbitrary viewing angle, the profile will be a linear combination of the two profiles, and therefore the central spike should always be present to some degree.

ACKNOWLEDGMENTS

This work supported by The Aerospace Corporation's Independent Research and Development program and by the US Air Force Space and Missile Systems Center through the Mission Oriented Investigation and Experimentation program, under contract F4701-00-C-0009 with the US Air Force. ATR 2002(8189)-6

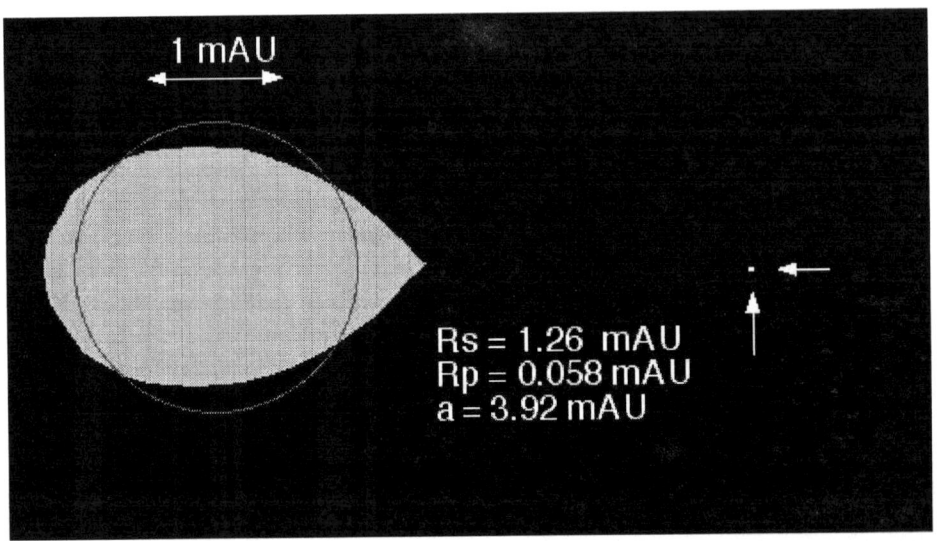

FIGURE 1. The scale of the system is shown above. Although the secondary is distorted, we approximated the potential as spherical, because most of the mass in the interior is nearly spherical.

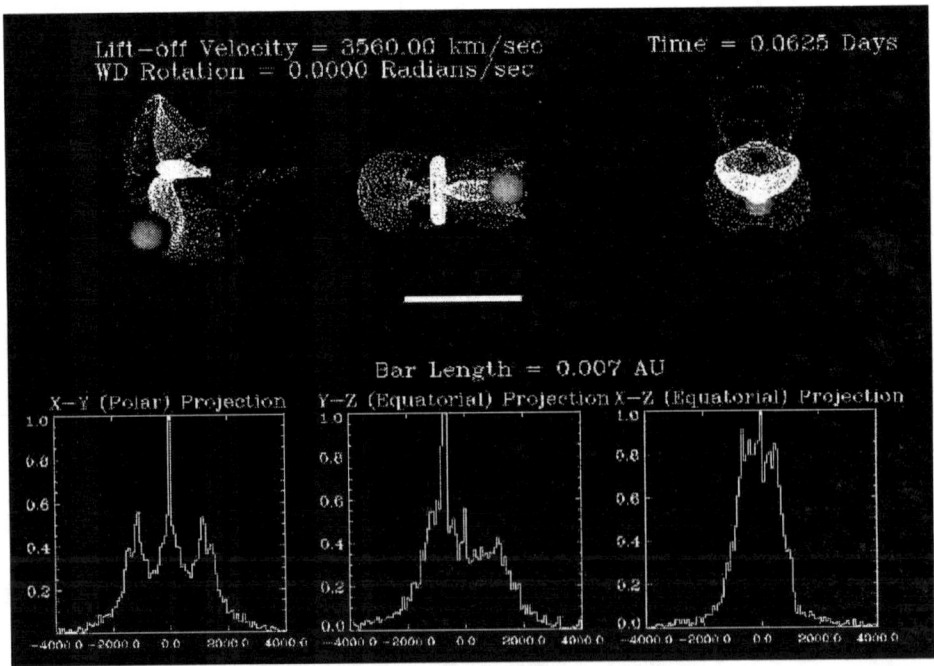

FIGURE 2 Vo = 3560 km/s, Vesc = 3600 km/s, snapshot at T = 0.0625 days Note how viewing geometry affects the observed line profiles for when Vo ≈ Vesc.

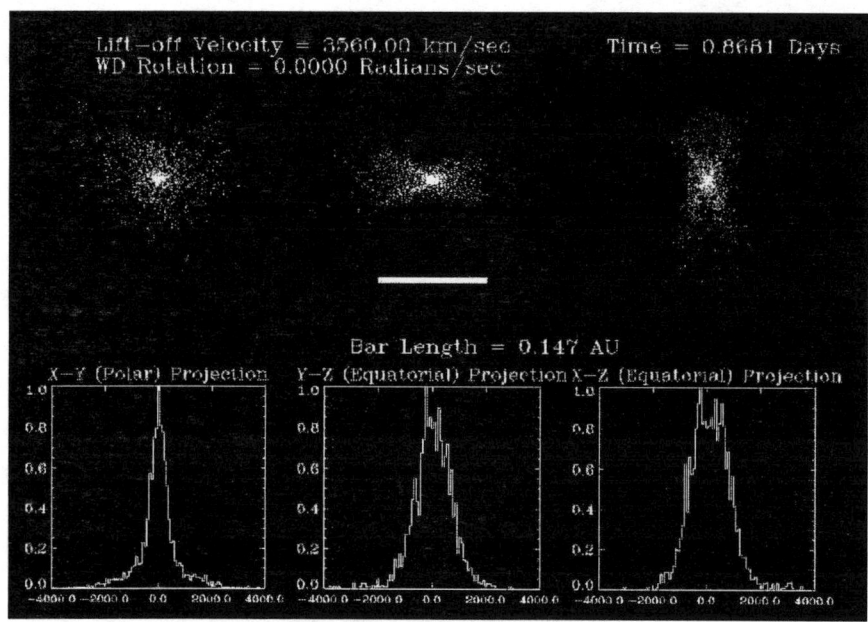

FIGURE 3. Vo = 3560 km/s, Vesc = 3600 km/s, snapshot at T = 0.9681 days. Note that as time goes on and material becomes more homogeneously distributed around the barycenter, the line profiles from the different viewing geometries become more alike.

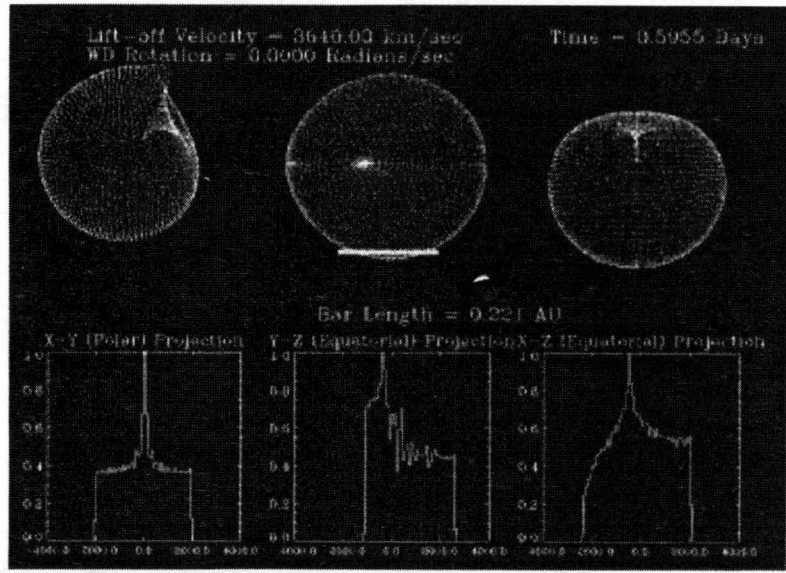

FIGURE 4 Here Vo=3640.. The slight departures from sphericity are due to retardation of material as it passes near the secondary and these are responsible of the central peak in the line profile.

REFERENCES

1. Bode, M. and Evans, A. (eds) 1989 *Classical Novae*, John Wiley and sons, Chichester, England
2. Hutchings, J.B. MNRAS, 158, 177-198 (1972)
3. Lynch, David K., Rudy, R.J., Mazuk, S. and Puetter, R.C. 2000 Astrophys. J. 541, 791 (2000)
4. Payne-Gaposhkin, C. 1957 *The Galactic Novae*, North-Holland Publishing Company: Interscience Publishers, New York,
5. Solf, J. ApJ 273 647-659 (1983)

Nova Nucleosynthesis Calculations: Robust Uncertainties, Sensitivities, and Radioactive Ion Beam Measurements

Michael S. Smith*, W.R. Hix*,†, S. Parete-Koon*,**, L. Dessieux**, M.W. Guidry**, D.W. Bardayan*, S. Starrfield‡, D.L. Smith§ and A. Mezzacappa*

*Physics Division, Oak Ridge National Laboratory, Oak Ridge, TN, USA
†Department of Physics & Astronomy, University of Tennessee, Knoxville, TN, USA, & Joint Institute for Heavy Ion Research, Oak Ridge, TN, USA
**Department of Physics & Astronomy, University of Tennessee, Knoxville, TN, USA
‡Department of Physics & Astronomy, Arizona State University, Tempe, AZ, USA
§Technology Development Division, Argonne National Laboratory, Argonne, IL, USA

Abstract.
We examine the quantitative impact of nuclear physics uncertainties on predictions of nova models via Monte Carlo simulations wherein, for the first time, the uncertainties of all relevant nuclear reactions are considered simultaneously. We determine uncertainties in predictions of isotope synthesis - including radioisotopes which may be observable tracers of novae - resulting from uncertainties in the input nuclear physics. We also detail the reaction rate sensitivity of radioisotope production, and discuss reactions which need further study. Finally, we examine the influence on nova nucleosynthesis of two new reaction rates - $^{17}F(p,\gamma)^{18}Ne$ and $^{14}O(\alpha,2p)^{16}O$ - that were studied in recent ORNL measurements with radioactive ion beams.

MOTIVATION

The very high temperatures and densities of nova outbursts - greater than 10^8 K and 10^4 g/cm^3, respectively - enable unstable nuclei produced by capture [e.g., (p,γ)] reactions to undergo further reactions before they decay. The resulting sequences of reactions (the hot CNO cycles and rapid proton capture process) differ substantially from reaction sequences in non-explosive stellar environments [1], with energy generation rate up to 100 times greater. Furthermore, the resultant pattern of synthesized abundances are also different from solar, with overabundances of ^{13}C, ^{15}N, and ^{17}O, as well as Ne, Na, and heavier elements in some explosions. Long-lived radioactive nuclei such as ^{18}F and ^{22}Na are also synthesized and ejected, and their observation may provide stringent constraints on nova models [2, 3].

Progress in probing the nova phenomena can be made with improved determinations of the rates of the important reactions as a function of temperature. The sensitivity of nova model predictions of energy generation and element synthesis to selected reaction rates has been demonstrated in numerous studies (e.g., [4, 5]). Especially important are reactions involving proton-rich radioactive nuclei with relatively short ($\lesssim$ 1 minute) lifetimes [6, 7]. Experimental determinations of some of these reaction rates are now becoming possible with the availability of beams of radioactive nuclei at a number of facilities worldwide [7]. Since these beams are difficult to produce, it is crucial to guide experimental programs by determining which nuclear reactions have the largest impact on predictions of nova outburst simulations. Ideally, the correlation between all relevant reaction rates and all synthesized isotopes should be determined. Furthermore, it is desirable to determine to what accuracy any given rate should be measured, especially in light of the uncertainty of all the other reaction rates. We have addressed these issues with our Monte Carlo studies of nova nucleosynthesis, where - for the first time - we simultaneously consider the uncertainty of *all* input nuclear reaction rates to determine *all* correlations between rates and abundance synthesis predictions. It is also important to determine the impact of new determinations of reaction rates, and we have addressed this by varying individual rates and examining the changes in model predictions. Finally, our study is the first to quantitatively determine the uncertainties on abundance predictions in novae considering the uncertainties of all nuclear reaction rates. This enables statistically robust comparisons of nova model predictions to observations.

NOVA NUCLEOSYNTHESIS CALCULATIONS

The temporal evolution of the isotopic composition in novae was followed using a nuclear reaction network [8] containing 169 isotopes, from hydrogen to ^{54}Cr with nuclear reaction rates drawn from REACLIB [9]. We have examined the nucleosynthesis in 3 nova models: on a 1.00 $M_\odot$ CO White Dwarf (WD), and on 1.25 $M_\odot$ and 1.35 $M_\odot$ ONeMg WDs. The first two are representative of the most prevalent classes nova, while the third represents a more energetic outburst. The calculations for the 1.25 $M_\odot$ and 1.35 $M_\odot$ ONeMg WDs begin with a set of initial abundances for 169 nuclides, adopted from Politano et al. [10]. They assumed a solar composition mixed with 50% by mass oxygen, neon and magnesium. The initial abundances for the 1.00 $M_\odot$ CO WD nova was 50% solar and 50% products of He burning (an equal mix of ^{12}C and ^{16}O with a trace of ^{22}Ne). The enhancement in each case was representative of the envelope material mixing with the matter from the underlying white dwarf [11].

We utilized a post-processing approach where the nucleosynthesis is decoupled from the hydrodynamics of the burst. Since our reaction variations did not appreciably change the nuclear energy generation nor, therefore, the temperature and density history of the explosion, this approach was deemed valid. We simulated the explosion by extracting hydrodynamic trajectories – time histories of the temperature and density – from one-dimensional hydrodynamic calculations with a limited reaction rate network for outbursts on 1.0, 1.25, and 1.35 $M_\odot$ WDs similar to Ref. [4]. Different mass elements ("zones") of the envelope at different radii generate unique trajectories. In our simulations, the ejecta of each of the nova models consists of between 26-31 zones. For our studies of the sensitivity of individual reactions, separate nuclear reaction network calculations with the full complement of nuclei and nuclear reactions were carried out to study the nucleosynthesis details within each zone; no mixing of the zones was included. To calculate the final total predicted abundances in the ejecta of each explosion, a sum was made of abundances over the zones, weighted by the ratio of the zone mass to the total envelope mass. These calculations are more realistic than previous post-processing nucleosynthesis calculations which used constant temperatures and densities and those which considered only the hottest zones of the explosion (e.g., [5]).

MONTE CARLO SIMULATIONS

To investigate the extent to which nuclear reaction uncertainties translate into abundance variations, we use a Monte Carlo technique which assigns a different, uncorrelated, random enhancement factor to *each* reaction rate in the simulation. The nucleosynthesis is calculated with these modified reaction rates, the results stored, and the process repeated with different enhancement factors. After 10000 iterations, the mean values and 90% confidence limits are determined from the distribution of abundance predictions. In this paper, we present results from the innermost zone of a 1.25 $M_\odot$ WD nova model, though the method can examine the impact on entire outburst models. A representative example for the radionuclide ^{22}Na is shown in Figure 1.A., where the upper and lower 90 % confidence limits differ by a factor of 3.6. Monte Carlo methods have been employed with great success in the analysis of Big Bang nucleosynthesis [12], but have not previously been applied to other thermonuclear burning scenarios. These confidence limits are the first statistically robust uncertainties determined for nova nucleosynthesis. They have important implications for determining the sensitivity of orbital observatories (e.g., INTEGRAL) for detection of gamma rays from novae.

The reaction rate enhancement factors are distributed according to the log-normal distribution, which is the correct uncertainty distribution for quantities like reaction rates which are manifestly positive [13],

$$p_{log-normal}(x) = \frac{1}{\sqrt{2\pi}\beta x} \exp\left(-\frac{(\ln x - \alpha)^2}{2\beta^2}\right), \qquad (1)$$

where α and β are the (logarithmic) mean and standard deviation, respectively, and $p(x)dx$ is the probability of finding p in the range x to $x+dx$. Our use of the log-normal distribution for the reaction rates represents a significant improvement over previous Monte Carlo calculations. For small uncertainties ($< 20\%$), the difference between the log-normal distribution and the normal (gaussian) distribution is small. However, for uncertainties of larger sizes like those encountered in this problem, the difference is important. Figure 1 shows the difference between normal and log-normal distributions for the abundance prediction of one isotope, ^{22}Na. We have assigned uncertainties of $\sim 50\%$ [$\beta = \ln(1.5)$] to rates whose measurement would require radioactive ion beams, and uncertainties of a factor of 2 for rates calculated by Hauser-Feshbach methods. Beta decay rates are given their tabulated values [14]. For all other rates we assign $\beta = \ln(1.2)$. We have deliberately used uncertainties that are somewhat underestimated to ensure that our resulting Monte Carlo uncertainties are not unduly inflated. Even so, many of the abundant metals have 90% confidence limits with a width of a factor of 2 or larger, such as ^{16}O (2.7), ^{17}O (2.7), ^{18}O(3.3), and ^{30}Si (5.6). The predicted abundances of radionuclides also have large uncertainties: for ^{18}F, the 90% confidence level spans a factor of 3.3, and for ^{26}Al, it spans a factor of 3.6.

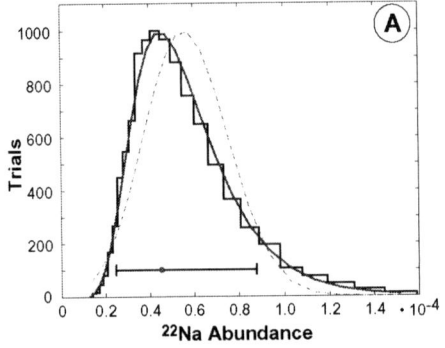 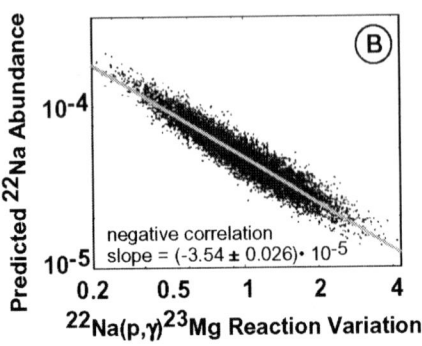

FIGURE 1. A. Histogram of the deviations of the predicted ^{22}Na abundance from the mean in the Monte Carlo simulation. The solid (dashed) curve is the log-normal (normal) distribution with logarithmic (arithmetic) mean and standard deviation from the Monte Carlo. The mean abundance and 90 % confidence levels are shown by the horizontal bar. B. Distribution of the predicted ^{22}Na abundance with the variation in the ^{22}Na(p,γ)^{23}Mg reaction rate. A linear fit and the slope are also indicated.

We also determine the correlation between small variations of *all* relevant reaction rates and *all* synthesized isotopes in the outburst. Figure 1.B. shows a representative result - the distribution of the predicted ^{22}Na abundance with the variation in the ^{22}Na(p,γ)^{23}Mg reaction rate. A linear fit determines that, in this case, there is a negative correlation – as the capture rate increases, the resulting ^{22}Na abundance decreases – and that the correlation is statistically significant – the slope is more than a few standard deviations different from zero. We have used our analysis, for example, to determine a prioritized list of reactions that most influence the production of radioisotopes that may be observable tracers of novae. For ^{18}F, the critical reactions are (in order of importance) ^{17}O(p,γ)^{18}F, ^{17}F(p,γ)^{18}Ne, ^{16}O(p,γ)^{17}F, ^{18}F(p,α)^{15}O, and ^{17}O(p,α)^{14}N. For ^{22}Na, the most important reactions are: ^{22}Na(p,γ)^{23}Mg, ^{20}Ne(p,γ)^{21}Na, ^{23}Na(p,α)^{20}Ne, ^{23}Na(p,γ)^{24}Mg, and ^{21}Na(p,γ)^{22}Mg. For ^{26}Al, the most important reactions are ^{23}Mg(p,γ)^{24}Al, ^{26}Al(p,γ)^{27}Si, ^{23}Na(p,γ)^{24}Mg, ^{23}Na(p,α)^{20}Ne, ^{20}Ne(p,γ)^{21}Na, ^{25}Mg(p,γ)^{26}Al, and ^{25}Al(p,γ)^{26}Si. Radioactive beams are required to study a number of these reactions, and our calculations can be used to set priorities for reaction measurements.

REACTION RATE SENSITIVITY STUDIES

A recent measurement of the excitation function for the ^{1}H(^{17}F,p)^{17}F reaction at ORNL's Holifield Radioactive Ion Beam Facility was used to obtain the first unambiguous evidence for the $J^{\pi} = 3^{+}$ state in ^{18}Ne and precisely determine its energy and total width [15]. This was in turn used to determine a new ^{17}F(p,γ)^{18}Ne reaction rate [16], which was up to a factor of 30 slower than the widely-used rate [17] found in REACLIB. We analyzed these rates in our post-processing nucleosynthesis code to determine the impact of the new HRIBF measurement [18]. The ratios of abundances produced using the

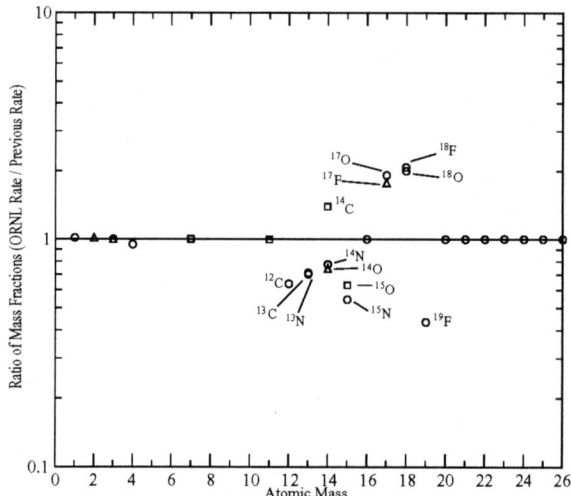

FIGURE 2. The ratio of mass fractions using the ORNL to that from one previous rate estimate [17] plotted against nuclide mass for the entire envelope of a 1.25 $M_\odot$ white dwarf nova. The ORNL rate changes the mass fractions of some nuclei by up to a factor of 2. Changes in the hottest zones are up to a factor of 600. The circle symbols mark species with mass fractions greater than 10^{-8}, the square symbols mass fractions between 10^{-8} and 10^{-16}, and the triangle symbols mass fractions less than 10^{-24}.

two rates is shown in Figure 2. We find that in the 1.25 $M_\odot$ WD nova, for example, the new rate changes the abundances of ^{18}F, ^{18}O, ^{17}F and ^{17}O synthesized in the hottest zones up to a factor of 600 compared to some previous estimates, and produced significant changes in the abundances of ^{18}F, ^{18}O, ^{17}F, ^{17}O, ^{19}F, ^{15}N, ^{15}O, ^{12}C, ^{13}C, ^{13}N, ^{14}N and ^{14}O by up to a factor of 2.1 when averaged over the entire exploding envelope. The changes are even larger (to a factor of 14,000 in the hot zone and 3.7 overall) for the 1.35 $M_\odot$ WD nova, but almost negligible for the cooler 1.0 $M_\odot$ WD nova.

The production of the important, long-lived radionuclide ^{18}F is increased in the hottest zones of the nova by the network using the slower ORNL ^{17}F(p,γ)^{18}Ne rate. This is because a faster ^{17}F(p,γ)^{18}Ne rate creates ^{18}F sooner after the peak of the outburst (from decay of ^{18}Ne) and therefore at higher temperatures – where it is more likely to be destroyed by ^{18}F(p,α)^{15}O. The network with slower ^{17}F(p,γ)^{18}Ne rate delays the production of ^{18}F and therefore creates more of it at a lower temperature – where it is more likely to remain as a mass 18 isotope. This effect does not, however, carry over to the outer zones of the explosion, because the overall lower temperatures of these zones limits the post-peak destruction of freshly-synthesized ^{18}F. If only the hottest zones were considered, an incorrect conclusion would have been drawn regarding the change in the synthesis of ^{18}F – showing the importance of considering the nucleosynthesis throughout the nova model. Mixing of the zones, which was not included in our calculations, could also significantly affect the final calculated abundances.

Another sensitivity study was motivated by an ORNL measurement that found a possible simultaneous two-proton decay out of a resonance in ^{18}Ne via a measurement of the ^{17}F(p,2p)^{16}O reaction using a radioactive ^{17}F beam [19]. This suggests that there is a reaction link $^{14}O(\alpha,2p)^{16}O$, proceeding through this resonance in ^{18}Ne, which is not currently included in the nucleosynthesis network. To determine the possible impact on nova nucleosynthesis that this reaction could have, we varied the rate of this reaction (as multiples of the rate of the competing $^{14}O(\alpha,p)^{17}F$ reaction) and inserted this into our reaction network. The results showed that there is no change in nova nucleosynthesis for strengths of this reaction equal to or weaker than the $^{14}O(\alpha,p)^{17}F$ reaction. Our analysis shows that the sum of the mass flow of nuclei via $^{14}O(\alpha,p)^{17}F$ and via $^{14}O(\alpha,2p)^{16}O(p,\gamma)^{17}F$ is roughly constant. We also carried out detailed sensitivity studies for the $^{14}O(\alpha,p)^{17}F$ reaction and determined that this reaction has little influence on element synthesis in novae [20].

ACKNOWLEDGMENTS

ORNL is managed by UT-Battelle, LLC, for the U.S. Department of Energy under contract DE-AC05-00OR22725. W.R.H. is supported by NASA under contract NAG5-8405, NSF under contract AST-9877130, and by funds from the Joint Institute for Heavy Ion Research.

REFERENCES

1. Wallace, R. K. & Woosley, S. E., *Astrophys. J. Suppl.*, **45**, 389 (1981).
2. Harris, M. J. et al., *Astrophys. J.*, **522**, 424 (1999).
3. Hernanz, M. et al., *Astrophys. J.*, **526**, L97 (1999).
4. Starrfield, S et al., *Mon. Not. Roy. Astron. Soc.*, **296**, 502 (1998).
5. Iliadis C. et al., *Astrophys. J. Suppl.*, in press (2002).
6. Champagne, A. & Wiescher, M., *Ann. Rev. Nucl. Part. Sci.*, **42**, 39 (1992).
7. Smith, M. S. & Rehm, K. E., *Ann. Rev. Nucl. Part. Sci.*, **51**, 91 (2001).
8. Hix, W. R., and Thielemann, F.-K., *J. Comp. Appl. Math.*, **109**, 321 (1999).
9. Thielemann, F.-K., Freiburghaus, C., Rauscher, T., et al. (1995), http://ie.lbl.gov/astro/friedel.html.
10. Politano, M. et al., *Astrophys. J.*, **448**, 807 (1995).
11. Livio M. & Truran, J.W., *Astrophys. J.*, **425**, 797 (1994).
12. Smith, M. S., Kawano, L. H., & Malaney, R. A., *Astrophys. J. Suppl.*, **85**, 219 (1993).
13. D. L. Smith, *Probability, Statistics, and Data Uncertainties in Nuclear Science and Technology* (LaGrange Park: Am. Nuc. Soc., 1991).
14. Tuli, J. K., Nuclear Wallet Cards, http://www.nndc.bnl.gov/wallet/ (2000).
15. Bardayan, D. W., et al., *Phys. Rev. Lett.*, **83**, 45 (1999).
16. Bardayan, D. W., et al., *Phys. Rev. C*, **62**, 055804 (2000).
17. Wiescher, M., Gorres, J., & Thielemann, F., *Astrophys. J.*, **326**, 384 (1988).
18. Parete-Koon, S. et al., *Astrophys. J., in preparation* (2002); M.S. Thesis, Univ. Tennessee (2002).
19. Gomez del Campo, J. et al., *Phys. Rev. Lett.*, **86**, 43 (2001).
20. Dessieux, L. et al., in preparation (2002).

The Imprint of Nova Nucleosynthesis in Presolar Grains

Jordi José*, Margarita Hernanz†, Sachiko Amari** and Ernst Zinner**

*Dept. Física i Enginyeria Nuclear, Universitat Politècnica de Catalunya, and Institut d'Estudis Espacials de Catalunya (IEEC/UPC), Barcelona, Spain
†Institut d'Estudis Espacials de Catalunya (IEEC/CSIC), Barcelona, Spain
**Laboratory for Space Sciences and the Physics Department, Washington University, St. Louis, MO, USA

Abstract. Five silicon carbide and two graphite presolar grains, whose isotopic signatures point toward a nova origin, are described. In addition, theoretical isotopic ratios of several nuclear species, ranging from C to Si, computed with hydrodynamic models of nova outbursts, are shown. Such theoretical information can be useful for future identification of nova candidate grains.

INTRODUCTION

Primitive meteorites contain stardust (presolar dust) that predates the Solar System. Presolar grains have huge isotopic anomalies that can only be explained by nucleosynthesis in stars. Detailed studies of these grains have opened up a new field of astronomy [11]. Silicon carbide (SiC), graphite (C), diamond (C), silicon nitride (Si_3N_4), and oxides have been identified as presolar grains. Ion probe analyses of single presolar grains have revealed AGB stars and supernovae as stellar sources [11]. SiC grains have been most extensively studied and can be classified into different populations based on their C, N, and Si isotopic ratios (Figs. 1 and 2).

THE DISCOVERY OF NOVA CANDIDATE GRAINS

Here we report on 5 SiC and 2 graphite grains that exhibit the isotopic signature of nova nucleosynthesis (Table 1) [1, 2]. The SiC grains have very low $^{12}C/^{13}C$ and $^{14}N/^{15}N$ ratios, while the graphite grains have low $^{12}C/^{13}C$, but normal $^{14}N/^{15}N$ ratios. Because there is evidence that indigenous N in presolar graphite has been isotopically equilibrated [3], the original $^{14}N/^{15}N$ ratios of the two graphite grains could be much lower. $^{26}Al/^{27}Al$ ratios have been determined for only two grains and are $>10^{-2}$. These isotopic features, low $^{12}C/^{13}C$, low $^{14}N/^{15}N$, and high $^{26}Al/^{27}Al$ ratios, are the signature of hot H-burning, and therefore, classical novae are possible stellar sources of these grains.

The $^{20}Ne/^{22}Ne$ ratio of graphite grain KFB1a-161 (<0.01, [8]) is considerably lower than ratios predicted by nova models (see following section). This suggests that ^{22}Ne in the grain is not trapped Ne from the ejecta, but originated from the decay of ^{22}Na

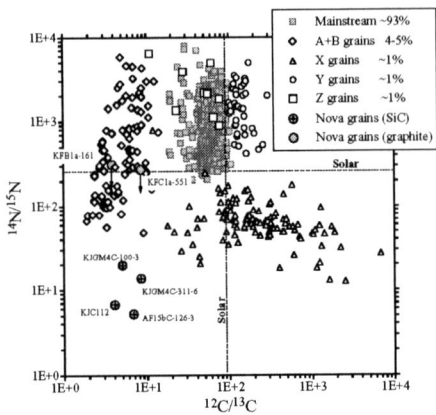

FIGURE 1. Carbon and nitrogen isotopic diagram of nova candidate grains and SiC grains.

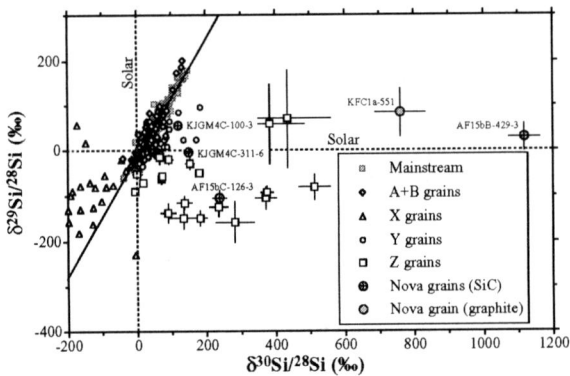

FIGURE 2. Silicon isotopic ratios of five nova candidate grains and SiC grains. Ratios are expressed as delta values, deviations from the solar Si isotopic ratios in permil (see Table 1 for definition.)

($t_{1/2}$=2.6 yr). Silicon isotopic ratios of the 5 SiC grains are characterized by ^{30}Si excesses and close-to- or slightly lower-than-solar ^{29}Si/^{28}Si ratios. CO nova models [5, 7, 9] predict close-to-solar ^{30}Si/^{28}Si and close-to- or lower-than-solar ^{29}Si/^{28}Si, whereas huge enrichments of ^{30}Si and close-to-solar ^{29}Si/^{28}Si ratios are obtained for ONe novae [5, 6, 10]. Comparison between grain data and nova models indicates that the grains formed in ONe novae with a white dwarf (WD) mass of at least 1.25M$_\odot$.

The isotopic signature of the grains qualitatively agrees with predictions of nova models. However, in order to quantitatively explain the grain data, one has to assume that material newly synthesized in the nova outburst was mixed with more than ten times as much unprocessed, isotopically close-to-solar, material before grain formation.

TABLE 1. Presolar grains with an inferred nova origin

Grain*	$^{12}C/^{13}C$	$^{14}N/^{15}N$	$\delta(^{29}Si/^{28}Si)$	$\delta(^{30}Si/^{28}Si)$	$^{26}Al/^{27}Al$
AF15bB-429-3 [1]	9.4 ± 0.2		28 ± 30	1118 ± 44	
AF15bC-126-3 [1]	6.8 ± 0.2	5.22 ± 0.11	−105 ± 17	237 ± 20	
KJGM4C-100-3 [1]	5.1 ± 0.1	19.7 ± 0.3	55 ± 5	119 ± 6	0.0114
KJGM4C-311-6 [1]	8.4 ± 0.1	13.7 ± 0.1	−4 ± 5	149 ± 6	>0.08
KJC112 [4]	4.0 ± 0.2	6.7 ± 0.3			
KFC1a-511 [1]	8.5 ± 0.1	273 ± 8	84 ± 54	761 ± 72	
KFC1a-161 [8] †	3.8 ± 0.1	312 ± 43			
Solar**	89	272			
Nova Models‡	0.2 – 3	0.1 – 120	(−1000) – 2000	(−1000) – 10000	0.02 – 0.9

* The first five grains are SiC and the last two grains are graphite. Errors are 1σ.
$\delta^i Si/^{28}Si \equiv [(^iSi/^{28}Si)_{grain}/(^iSi/^{28}Si)_\odot - 1] \times 1000$
† $^{20}Ne/^{22}Ne = <0.01$; $^{22}Na/C = 9 \times 10^{-6}$
** The N ratio is that of the air. In the text, it is indicated as solar; $(^{20}Ne/^{22}Ne)_\odot = 13.6$
‡ $^{20}Ne/^{22}Ne = 0.01 - 3000$

THEORETICAL ISOTOPIC RATIOS: CO VERSUS ONE NOVAE

Carbon and nitrogen

CO novae have higher $^{14}N/^{15}N$ and lower $^{12}C/^{13}C$ ratios than ONe novae (see Fig. 3). The higher initial ^{12}C content in CO nova models, resulting from mixing between the solar-like accreted matter and material from the C-rich outermost layers of the WD core, favors ^{13}C synthesis through $^{12}C(p,\gamma)^{13}N(\beta^+)^{13}C$, thus leading to low $^{12}C/^{13}C$ ratios. In contrast, the higher peak temperatures achieved in ONe models allow (p,γ) reactions on ^{13}C to produce ^{15}N through the chain $^{13}C(p,\gamma)^{14}N(p,\gamma)^{15}O(\beta^+)^{15}N$. Because peak temperatures increase with the WD mass, the $^{14}N/^{15}N$ ratio decreases with increasing WD mass for both CO and ONe nova models. Moreover, the $^{14}N/^{15}N$ ratio can be used to distinguish between CO and ONe novae: large ratios ($\sim 10 - 100$) are only achieved in models with low-mass CO white dwarfs.

Silicon excesses

Massive ONe novae produce large ^{30}Si excesses, whereas $^{30}Si/^{28}Si$ ratios are close to solar for all CO novae (see Fig. 3). Both CO and ONe novae yield ^{29}Si deficits or close-to-solar $^{29}Si/^{28}Si$ ratios. Hydrodynamic models show that CO novae have a very limited nuclear activity beyond the CNO cycle, thus explaining the close-to-solar Si content in the ejecta. In contrast, the higher peak temperatures achieved in ONe novae, together with the higher initial ^{27}Al content, allow significant ^{28}Si synthesis, first through $^{27}Al(p,\gamma)^{28}Si$, and later through $^{26}Al^{m,g}(p,\gamma)^{27}Si(p,\gamma)^{28}P(\beta^+)^{28}Si$. To overcome the large Coulomb potentials, temperatures above 2×10^8 K are required for any significant synthesis of $^{29,30}Si$ from ^{28}Si.

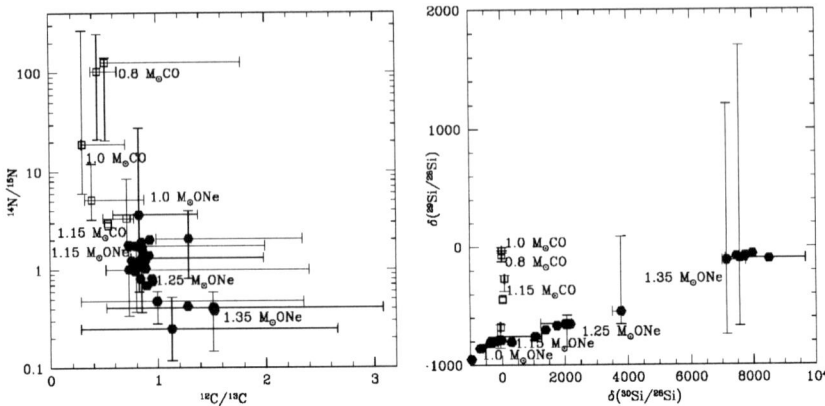

FIGURE 3. (Left) Theoretical C and N isotopic ratios from hydrodynamic calculations of nova outbursts ([5, 6], and unpublished data). Open squares (CO novae) and filled hexagones (ONe novae) represent mean mass-averaged values calculated for the total ejecta. Bars indicate variations found when individual shells are taken into account. (Right) Theoretical $^{29}Si/^{28}Si$ and $^{30}Si/^{28}Si$ ratios plotted as deviations from solar in permil.

Aluminum and magnesium

Similar $^{26}Al/^{27}Al$ ratios are found for both CO and ONe novae (typically, $\sim 0.1-0.6$). Although ^{26}Al is efficiently synthesized in ONe novae (where favorable conditions – high peak temperatures and presence of 'seed' nuclei– are found), the larger initial ^{27}Al abundance (not strongly modified during the outburst) results in $^{26}Al/^{27}Al$ ratios similar to those obtained for CO novae, where the scarcity of 'seed' nuclei for ^{26}Al synthesis is coupled with a much lower initial ^{27}Al content. Therefore, the $^{26}Al/^{27}Al$ ratio does not provide a robust criterion for distinguishing between CO and ONe novae. Both CO and ONe novae yield small $^{26}Mg/^{25}Mg$ and $^{24}Mg/^{25}Mg$ ratios (around $\sim 0.1-0.2$), much below solar values (1.1 and 7.9, respectively). Most models show some synthesis of ^{25}Mg through $^{24}Mg(p,\gamma)^{25}Al(\beta^+)^{25}Mg$, accompanied by a decrease in both $^{24,26}Mg$ due to (p,γ) reactions.

Oxygen and neon

In contrast to the Mg isotopes, $^{16}O/^{17}O$ and $^{16}O/^{18}O$ ratios strongly depend on the nova type as well as on the WD mass. The $^{16}O/^{17}O$ ratios obtained are always much below the solar value (i.e., 2622), especially in ONe novae ($\sim 1-10$). Since $^{17,18}O$ synthesis proceeds through (p,γ) reactions on the initial ^{16}O, the lower $^{16}O/^{17}O$ and $^{16}O/^{18}O$ ratios in ONe novae reflect the higher peak temperatures achieved in these models, which are required to burn ^{16}O.

Much higher $^{20}Ne/^{22}Ne$ ratios are obtained for ONe ($\sim 100-2000$) than for CO novae ($\sim 0.1-1$). In addition, moderate to high $^{20}Ne/^{21}Ne$ ratios are also found for ONe novae ($\sim 100 - 10000$), whereas CO novae show only high ratios ($\sim 2000 - 50000$). The differences in the ^{20}Ne/^{22}Ne ratios are due to the higher initial ^{20}Ne content in ONe novae, since ^{20}Ne is barely modified by the outburst for most nova models, whereas ^{22}Ne is substantially destroyed by proton-capture reactions. ^{21}Ne, a very fragile isotope, is strongly depleted, leading to large ^{20}Ne/^{21}Ne ratios for both nova types.

DISCUSSION

We have compared our theoretical predictions with recently published results from hydrodynamic models (see [7] for CO novae, and [9, 10] for both nova types). There is general agreement with the calculations reported in [7], in particular for the ^{12}C/^{13}C and ^{16}O/^{17}O ratios. However, a much wider range of ^{14}N/^{15}N ratios are predicted in [9]. This is consistent with results from models with low-mass white dwarfs (i.e., ~ 0.6 M$_\odot$), not considered in our calculations, for which ^{14}N(p,γ)^{15}O(β^+)^{15}N is not very efficient, thus leading to high ^{14}N/^{15}N ratios.

In addition, for many isotopic ratios, including C, Al and Si, there is excellent agreement with the results reported in [9, 10]. Some discrepancies can be attributed to differences in parameters adopted in the calculations, in particular the different prescriptions for the initial amount of O, Ne and Mg in the outer shells of the ONe white dwarfs, where mixing with the accreted material takes place.

REFERENCES

1. Amari, S., Gao, X., Nittler, L. R., Zinner, E., José, J., Hernanz, M., and Lewis, R. S., *The Astrophysical Jour.* **551**, 1065 (2001).
2. Amari, S., *New Astronomy Rev.* **46**, 519 (2002).
3. Hoppe, P., Amari, S., Zinner, E. and Lewis, R. S., Geochim. Cosmochim. Acta **59**, 4029 (1995).
4. Hoppe, P., Kocher, T. A., Strebel, R., Eberhardt, P., Amari, S., and Lewis, R. S., Lunar Planet. Sci. **27**, 561 (1995).
5. José, J., and Hernanz, M., *The Astrophysical Jour.* **494**, 680 (1998).
6. José, J., Coc, A., and Hernanz, M., *The Astrophysical Jour.* **520**, 347 (1999).
7. Kovetz, A. & Prialnik, D., *The Astrophysical Jour.* **477**, 356 (1997).
8. Nichols, R. J., Jr., Kehm, K., Hohenberg, C. M., Amari, S., and Lewis, R. S., *Geochim. Cosmochim. Acta*, submitted (2002).
9. Starrfield, S., Gehrz, R., and Truran, J.W., in *Astrophysical implications of the laboratory study of presolar materials*, edited by T. Bernatowicz and E. Zinner, AIP, New York, 1997, p. 203
10. Starrfield, S., Truran, J.W., Wiescher, M.C. & Sparks, W.M., *Monthly Notices Royal Astronomical Soc.* **296**, 502 (1998).
11. Zinner, E., *Annual Rev. Earth and Planet. Sci.* **26**, 147 (1998).

OBSERVATIONS: OPTICAL, UV, IR, AND RADIO SPECTRA

Panchromatic Study of Novae in Outburst: Phenomenology and Physics

Steven N. Shore

Dept. of Physics and Astronomy, Indiana University South Bend, South Bend, IN 46634-7111 USA;
Osservatorio Astrofisico di Arcetri; Dipartimento di Fisica, Università di Pisa

Abstract. This paper reviews general features common to all novae, classical or recurrent, from a multi-wavelength perspective, although concentrating on the 1000 - 8000Å range, for which the greatest wealth of high resolution data exists. Particular information derived from line profiles will be highlighted, e.g. abundances, mixing within the ejecta, fossilized instabilities and symmetries of the expanding gas, radial dependence of the velocity and density, and time dependent ionization. Energetics of the outburst, especially the prolonged constant luminosity phase following initial outburst, will be discussed along with the evidence for super-Eddington phases of the later outburst. Finally, contrasts between Galactic and Magellanic Cloud novae are discussed with a view toward extending studies to extragalactic novae with future ground-based and satellite observatories.[1]

AN OVERVIEW OF THE PHYSICAL ENVIRONMENT: WHERE DOES THE LIGHT COME FROM?

The nova outburst presents one of the most challenging real time analysis problems in astrophysics. They vary on many timescales, from minutes to years, and display complex wavelength-dependent line variations that require a comprehensive examination of the whole spectrum.[2] Radiative transfer in a stellar wind is difficult enough to model. But when the medium is changing on the photon diffusion time it is far more difficult to model. Nonetheless, the physical processes are not that obscure and this section describes some of the basics.

In the (unfortunately) prototypical example of the explosion of a nuclear device, we encounter the range of radiative and dynamical processes that mimic nova spectra. The energy release is very fast, less than 0.1 msec. The initial radiative pulse propagates

---

[1] I thank the organizers of this meeting for their kind invitation, and especially Jordi José and Margarita Hernanz for jam sessions. Much of the work described here has been in collaboration with Sumner Starrfield, Greg Schwarz, Karen Vanlandingham, Peter Hauschildt, Jason Aufdenberg, Joachim Krautter, Bob Gehrz, Howard Bond, Ron Downes, Tom Ake, and George Sonneborn; to all my thanks, and also to Massimo Della Valle, Ami Glassner, G. Anupama, Dave Lynch, Nye Evans, Nir Shaviv, Christian Iliadis, Alan Calder, and Jonathan Dursi and many other participants for wonderful discussions during the meeting. This work was supported, in part, by NASA through *Chandra* and *Hubble Space Telescope* guest observer programs and awards of observing time as *Director's Discretionary Time* from HST.

[2] This paper, which as closely as possible recapitulates the talk, emphasizing work since the 1989 meeting and how our picture has evolved. It is intended to be complementary to the other papers in this volume and less an comprehensive overview of the literature which is well covered elsewhere in these proceedings.

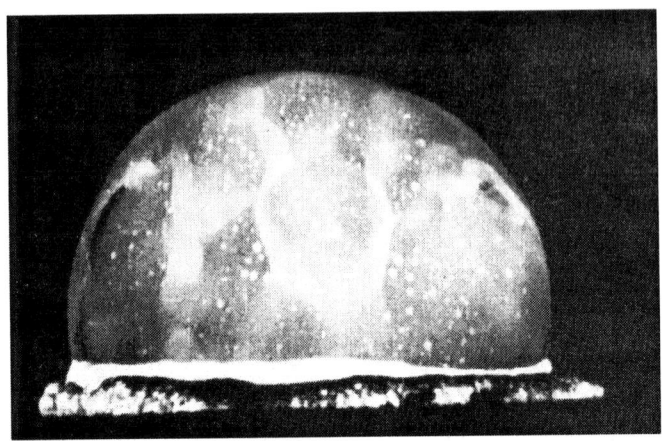

FIGURE 1. Greenhouse George (Enwetak Atoll, 1951) fireball at 20 msec showing the development of structure in the earliest stages of the expansion ($\approx$100 m); both the Rayleigh-Taylor and Richtmyer-Meshkov instabilities produce fragmentation during the expansion. Notice that the pseudo-photosphere is smaller than the shock radius, seen by the Mach collar and dust cloud beyond the fireball.

into the surrounding medium which becomes a time dependent H II region whose radius initially increases linearly with time having a characteristic timescale $t_c^{-1} \sim n_e \alpha$, where n_e is the electron density and α is the recombination coefficient. This stage ends when the medium exhausts the input ionizing photons. The overpressured hot sphere now expands supersonically; this is the *fireball* shown in Fig. 1.[3] On expansion, the combined effects of decreasing temperature and recombination decouple of shock from the radiating surface (breakout): the pseudo-photosphere first stalls and then recedes relative to the compressional front. The first hard radiative (γ, XR, and UV) pulse triggers an ionization-induced vapor cloud that dissipates quickly but may have analogs in some types of novae (see below). Ultimately, the blast turns transparent and the more slowly moving debris of the ejecta become visible within the fireball. This early stage, within the first 0.1 sec, is also always accompanied by structure formation at the fireball surface, regardless of the means of detonation (ground/airburst); it is likely due to acceleration-induced instabilities (Rayleigh-Taylor, Richtmyer-Meshkov).

So how does this relate to novae? The ejecta are optically thick in the earliest stages of the expansion. The principal difference between the bomb and a nova is the survival of the central site for the explosion. In a nuclear explosion, this is destroyed and the evolution of the ejecta depends only on the total energy injected at the instant of the explosion. Not so for a classical nova, and that is why the analogy with a time dependent H II region is so interesting. The white dwarf remains active as a nuclear engine for some

---

[3] This stage is almost iconographic for our generations "thanks" to H bomb tests in the 1950's. Among the best set of fireball and early explosion images are reproduced with length scales in Glasstone, S. and Dolan, P. J. (eds.) 1977, *The Effects of Nuclear Weapons, 3rd Ed.* (Washington, DC: DOD) (http://nuketesting.enviroweb.org/nukeffct/enw77.htm#Bottom) and http://www.saintjoe.edu/~pfaff/nukes.html.

time and continues to irradiate the ejecta even as they freely expand[4]. This radiatively induces a temperature gradient that dominates the spectrum formation. Novae also differ in an important way from supernovae, despite the similarities in the dynamical structure of the ejecta since they have no distributed heating sources (no mixed radionuclides power novae as they do for a SN Ia) nor is there a propagating shock (as in SN II) that heats the gas and drives a precursor. Since nova ejecta act, in essence, as a passive radiative filter for the emission from the underlying star, the ejecta cool initially if the expansion is fast compared to the radiative diffusion timescale. This adiabatic behavior is the origin of the infrared *fireball* stage (see Gehrz, these proceedings).

When $T_{rad} < (few) \times 10^4 K$, recombination actually *increases* the opacity since it's due mainly from line absorption by relatively low ionization species (such as Fe II and other heavy metals) and continuum absorption by CNO. This is the *iron (Fe) curtain* stage. Strong absorption bands redistribute the flux toward less opaque (longer) wavelengths and optical Fe II and [Fe II] emission appears (absorption from the upper levels of optically thick UV transitions also appear in absorption); this stage persists as long as the UV transitions remain thick. With continuing expansion, however, the optical depth τ decreases (roughly as t^{-2} or faster) and the emitted spectrum becomes progressively harder as the pseudo-photosphere moves inward toward higher radiation temperature while the bolometric luminosity of the source, L_{bol}, remains constant. This exacerbates the ionization of the shell, further reducing its opacity, and the Fe curtain finally disappears. Although models predict a comparatively short duration for the constant L_{bol} phase, observations show this can extend for weeks after the initial outburst.[5] Ultimately, since the ejecta have fixed mass in the absence of a wind, the pseudo-photosphere "falls out the bottom": the entire shell becomes transparent and completely ionizes. This is the *nebular* stage and is marked by the appearance of nebular and coronal forbidden lines from very highly ionized species such as Ne VII and Fe VII - XII (see Lynch, these proceedings).

The ejecta velocity field is set by processes that occur during the explosion and it and any density structure provide a (nearly) frozen record of the event. Models indicate that the explosion (and possibly a subsequent fast wind) impose a "Hubble flow" (constant velocity gradient) on the matter, in marked contrast to a typical stellar wind outflow where the gradient in the most rapidly expanding outermost layers of the wind asymptotically approaches zero. I will return to this point below when discussing line profile formation. Since $v = v_{max}(r/r_{max})$, for a radial distance r, time t, and expansion velocity v, and r and t are interchangeable, a constant total mass M_{ej}, has a density law $n(r) \sim r^{-3} \sim t^{-3}$. Dynamically, then, novae are comparatively simple but spectrum modeling is still extremely computationally intensive (see Hauschildt, this volume).

---

[4] More importantly, the WD continues at roughly constant nuclear luminosity. the duration being strongly a function of M_{WD}, and its surface temperature increases as it relaxes to stable burning.

[5] It may be even longer. EUVE didn't detect the brightest sources because of extinction, V1974 Cyg was only weakly detected by Voyager 2, and FUSE observations of V382 Vel reveal only a small additional energy range. Only recurrents, particularly RS Oph, and the most rapid ONeMg novae – such as V838 Her – show constant L_{bol} phases of less than one week.

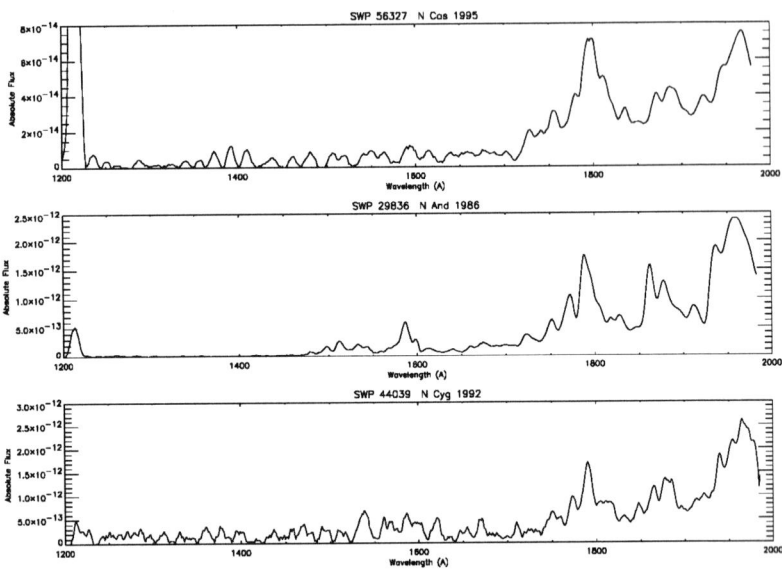

FIGURE 2. Short wavelength UV during the Fe curtain phase for three novae: OS And 1986 (CO), V1974 Cyg 1992 (ONeMg), and Cas 1995 (CO).

STAGES OF THE SPECTRAL DEVELOPMENT AND THE LIGHT CURVE

This constant L_{bol} phase greatly simplifies the picture of how individual spectral regimes develop. The infrared, having the lowest opacity, is the first to experience a rise from both the increase in UV opacity and the color temperature of the optically thick surface. It is here too that emission will inevitably first appear. Remember, the ejecta are quite distended and emission appears as soon as there is sufficient excitation of the appropriate transition and a low enough optical depth; in NLTE comparatively high ionization can be seen in the IR before it is observed in other wavelength regimes.

The ultraviolet is really driving everything in the early stage, once the line opacity kicks in. V1974 Cygni 1992 showed us most clearly what is happening. The fireball stage is much like that seen for SN IIe; the rise at longer wavelength is simply redistributed flux and the timescale is fast because of the sensitivity of the ionization to the local conditions in the cooling ejecta. The two characteristic optical timescales, t_2 and t_3 that define the "speed class"[9] are also linked to the UV. The region longward of 2000Å becomes progressively less opaque, followed by the shorter wavelengths, and t_3 is approximately the time when (using the old IUE terminology) the SWP (1200 - 2000Å) integrated flux becomes about equal to the LWP integrated flux (2000 - 3300Å).[6]

---

[6] It is reasonable to still, even without the invaluable continuing contributions of this satellite, to divide these two regions because of the distribution of absorption systems from iron peak species in the UV.

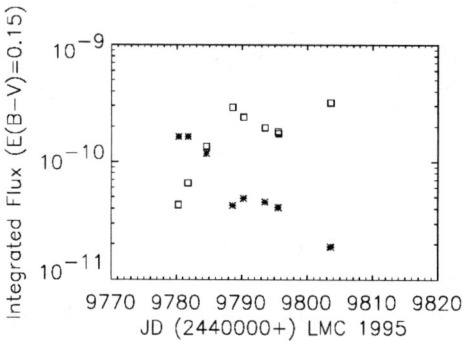

FIGURE 3. Light curves for LMC 1995 (CO Nova) showing correspondence between the UV (1200-3200Å) (squares) and V light curves (stars).

The centimeter radio continuum turns transparent very early in the outburst [7]. The principal opacity source at these long wavelengths, through the near infrared, is due to thermal bremsstrahlung and therefore depends only on the change in the column density so the spectral turn-over frequency (where $\tau_v = 1$) decreases as the expansion progresses.[7] The flux at each wavelength declines as t^{-3} and, because of the time delay there is a characteristic decrease in the peak flux with decreasing frequency. For the optical and UV, this is not so simple since the combined effects of photo-ionization and line transitions make these wavelengths much more sensitive to both the filling factor of the dense gas and the ionizing flux of the central source. Millimeter observations, especially interferometric imaging, is really needed here as it is for stellar winds, and arrays such as ALMA will be very important for rapid responses to bright novae.

Two optical phenomenological spectral classes have been distinguished [22]. The Fe II-type show strong iron emission in the optical and Balmer emission (the Fe II and [Fe II] lines are the more transparent transitions from highly excited states pumped from the UV), and the He-type, which almost immediately display He I/II and lines from more excited or ionized species. The same behavior holds for the NIR 2.2μ Na I lines that are pumped at around 3000Å. These subgroups, which also have (as always in our business) transitional forms, have been used to frame much of the discussion at this meeting but it's important to note the connections with the UV and ejecta opacity. If the shell is strongly fragmented early, or if the ejecta are completely photo-ionized to near nebular conditions very quickly, the Fe curtain stage may not occur or lines may appear from more than one density regime in the ejecta. In the (now near) absence of UV data during the start of the outburst, other wavelength regions must serve as proxy measures of the energetics and structure, guided by the experience of the last decade of multiwavelength observations.

Mass Determinations: With the exception of the recurrents, in the optically thick

---

[7] Recall that for a stationary isothermal stellar wind the frequency dependence of the opacity means the solid angle is a function of frequency which leads to the observed spectral curvature.

stage almost all classical novae look similar. Flux redistribution universally requires the material to reach column densities in excess of 10^{24} cm^{-2} which means the masses (assuming large filling factors) are generally at least 10^{-5} to $10^{-4} M_\odot$.[8] With this column density one achieves an approximate scaling, $M_{ej} \sim 6 \times 10^{-7} f N_{H,24} v_3^2 t_3^2$ $M_\odot$, where f is the filling factor (usually of order 0.1), v_3 is in 10^3km s^{-1}, and t_3 is in days. This estimate agrees with the results from IR measurements (Gehrz, these proceedings) and also almost all mass determinations from more precise atmosphere and photo-ionization analyses. In other words, we have a problem since, as Starrfield has emphasized in these proceedings even with comparatively low filling factors: such large ejections are not obtained from models.

STRUCTURE OF THE EJECTA

Even before the shell is resolved, much can be gleaned from the line profiles alone about the structure of the ejecta, *e.g.* [4]. A universal feature of classical novae at all stages, and on essentially all transitions, is the appearance of knots and broad, non-Gaussian wings as soon as the emission appears. These *may* depend on the stage of the outburst and the specific line since, depending on the excitation and parent ion, the emission comes from a specific density and temperature regime in the shell. The earliest optical and IR profiles are nearly always broader than those well after emerging t_3, and in some cases at a single epoch the profile can depend on the depth of formation of the transition (Della Valle & Williams 2002, *A& A, in press*, see also Della Valle, these proceedings).

A further consequence of this profile structure is fundamental for modeling and interpreting the light curve: the ejecta cannot in general be either spherically symmetric [20], completely uniform, or have large filling factors. This is also well known for novae from observations of resolved structures (the well known HST images of V1974 Cyg [10] and T Pyx [11] and the survey [2]) but also from the images of η Car and LBVs, systems that are at or possibly even above L_{Edd} at outburst.[9] Radio images (*e.g.* Bode, these proceedings) also show temporal evolution of the symmetry of the resolved ejecta. Line profile analyses of several novae for which high resolution UV spectra were obtained in the earliest stages show the emergence of structure as the fastest moving material becomes transparent. The picture that emerges is of a more nearly spherical (or elliptical) fast component of low density, within which is imbedded a more fragmented, axisymmetric (or simply less spherical) slower moving, higher density component. It's also clear, from optical and UV persistence of O I and other neutral and low ionization species emission, that FUV-shielded dense regions coexist along with tenuous hot gas long after the onset of the nebular stage.

---

[8] Only a few novae have been observed early enough in the UV catch the fireball spectrum, as the Fe curtain drops, (Nova Cas 1995, LMC 1991, V1974 Cyg) and the density gradient required for the models seems to be steeper than r^{-3}.

[9] A sample detailed spatially resolved dynamical analysis based on multifiber spectroscopy for HR Del is consistent with bipolar ejection for the bulk of the material (Hillwig, T. 2001, *PhD Thesis - Astronomy*, Indiana Univ. and references therein); see also [18] for details.

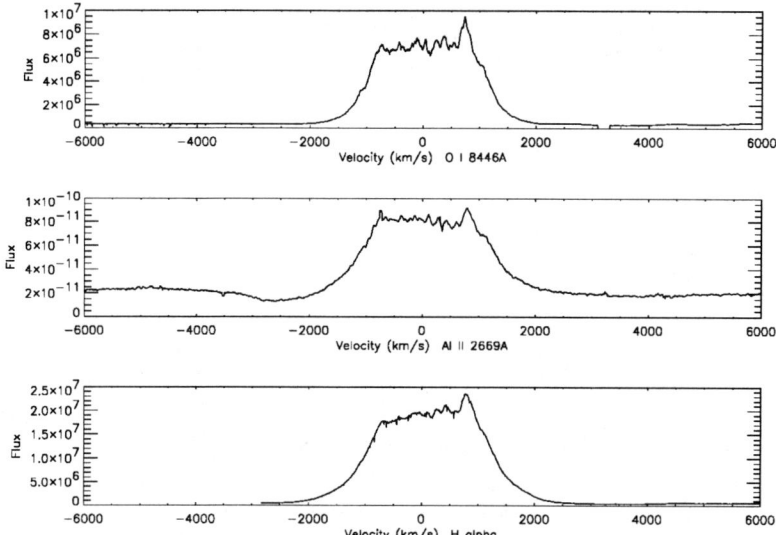

FIGURE 4. V382 Vel, 26/6/99 (ESO; Della Valle), 29/6/99 (STIS; O5JV); Simultaneous line profiles for O I 8446Å, Al II 2669Å and Hα for Hα illustrating similarities and excitation and ionization dependent structure in profiles. Note weak P Cyg absorption on Al II line to ≈ -3500 km s^{-1}, about the same as Hα at maximum P Cyg absorption phase one month earlier.

What about the *optical discrete absorption line systems* described by Payne-Gaposchkin, Maclaughlin, Hutchings, and others? Are these observed in recent novae? For V705 Cas, the Na I D line showed *broad* (100 km s^{-1}) discrete absorption separated from the P Cyg trough at -800 km s^{-1} before the dust forming event.[10] A similar event was seen for DQ Her but with *narrow* components. Archival reports of absorption line *systems* have yet to be confirmed in the optical using flux calibrated data since the myriad emission/absorption features during the Fe curtain phase can easily produce the impression of velocity-variable systems. There is a possibility, yet to be confirmed, that systems of absorption features can be detected in the high resolution UV spectra (*cf.* Casatella, these proceedings).

Spectropolarimetry: This method has been little used in the analysis of structure during the earliest stages of outburst but it is potentially a *very* valuable tool for elucidating structure [21][1] [3]. Differential polarization of emission lines has been observed in the UV relative to the continuum for BY Cir 1995 [8] although UBV observations [3] obtained within the first week after outburst are consistent with an initially spherical ejection.[11]

P Cygni Profiles: The strongest optical P Cygni lines on the Balmer series show v_{max}

---

[10] A spectrum is shown in 1994, *S&T*, **87**(4), 42.
[11] This is also true for V382 Vel (S. Potter and B-G Andersson, private communication).

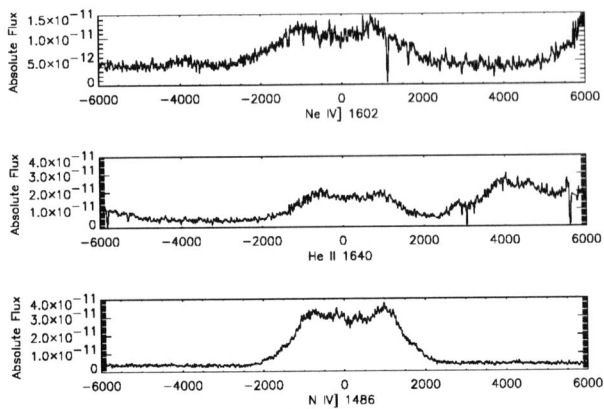

FIGURE 5. Line profiles during the nebular stage of V382 Vel (Aug. 1999)

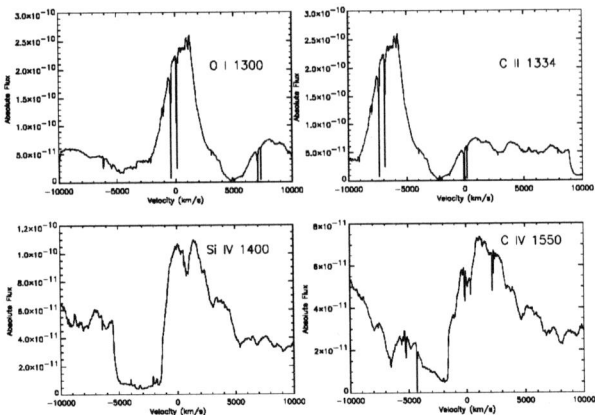

FIGURE 6. P Cyg profiles on UV resonance lines for V382 Vel at about 40 days after outburst (STIS, O5JV02). Note the residual Fe curtain absorption overlying the region of emission line formation.

decreasing with increasing upper level, similar to Be stars. It's important, though, to note the difference in the velocity law in the ejecta compared with a continuous outflow: novae will not show the same saturation effects at large r.[12]

Because of the details of line formation, maximum velocities differ depending the spectral region. For recurrents, the UV and optical velocities agree and are quite high, of order 10^4 km s^{-1}. The highest velocities are determined from the UV resonance

---

[12] For a wind, the optical depth at the terminal velocity is limited by the intrinsic linewidth, Δv, which is usually interpreted as "turbulent" flows while for a nova the constant gradient produces a power law decrease in τ.

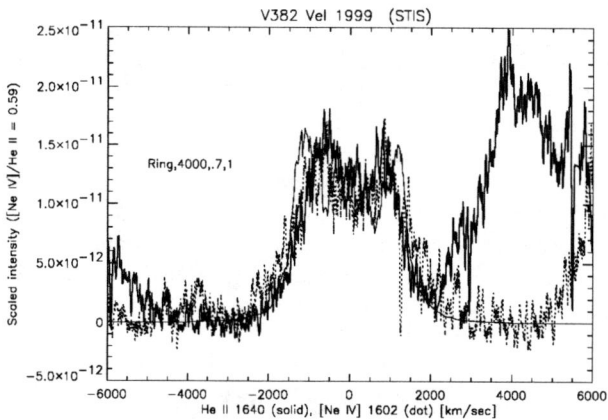

FIGURE 7. Comparison of Ne IV] 1602Å, He II 1640Å and sample ring calculation for V382 Vel, STIS spectrum, 29/8/99 (after entering the nebular stage). Similarities between He II and Ne IV] indicate chemical homogeneity in the ejecta.

lines. For ONeMg novae, the peak UV velocity is about 7000 km s^{-1} (V838 Her) and generally between 4000 km s^{-1} (V1974 Cyg) and 5500 km s^{-1} (V382 Vel, LMC 1990# 1, LMC 2000). CO novae show lower velocities, generally around 2000 - 3000 km s^{-1} (OS And, LMC 1988# 1). These are typically about a factor of 1.5 to 2 higher than seen from Hα or other optical lines at the same stage. The line profiles, as I've mentioned, narrow with time as the outermost regions contribute progressively less to the emission measure.[13] In general, the emission wings for the P Cyg stage extend nearly to the maximum absorption velocity but these too narrow with time.

Evidence for Homogeneity of the Ejecta: This is a hard question to address since it generally requires either completely spatially resolved shells or very high S/N spectra. At least for V1974 Cyg [17], there is strong evidence for variable Ne/C between the knots based on small aperture GHRS spectra of the brightest knot in the resolved ring. Individual knots have, in general, not been analyzed and the usual approach to photoionization modeling (using integrated fluxes) precludes such analyses. Instead, a recommendation is to use the $\tau(v)$ approach now employed for interstellar and Lyα forest absorption lines by taking line ratios for individual profiles to obtain abundance ratios as a function of velocity.

XR and UV Correlations: Recombination Analysis: One trick recently applied to novae is the use of recombination lines to determine when the XR emission has ceased for the white dwarf. This is a proxy measure, based ion the assumption that when the luminosity and temperature of the WD falls below the critical ionization level, the ejecta recombine with the rate asymptotically slowing down due to their free expansion, [19]

---

[13] Unlike a stationary outflow, the relative covering factor of either the ejecta's pseudo-photosphere or the actual white dwarf surface compared with the increasingly distended optically thin volume causes the P Cyg absorption trough to weaken rapidly as well as decrease in width.

(and refs. therein). Two features emerge from this picture that apply throughout the spectrum: the ionization level is fixed, reflecting the radiative conditions at the time of turnoff of the central source, and the ionization structure asymptotically freezes with time as the recombination timescale increases beyond the expansion timescale, which for a Hubble-type flow remains constant. Before direct XR detections of the SSS phase, the UV strongly pointed to the onset of this ionizing continuum and, ultimately, signaled its exhaustion (for V1974 Cyg it was possible to show that individual knots differ in density by examining the disappearance of specific profile features with time. Again, without continuing UV and XR simultaneous observations, optical and IR surrogates will have to suffice.

CLASSES OF NOVAE

Recurrent novae: Anupama (these proceedings) has reviewed the basic characteristics of recurrents so I will make only some remarks about their development. The mass problem encountered with classical outbursts doesn't seem to be seen for these systems. For RS Oph 1985, the ejected mass was about $10^{-7} M_\odot$, while for U Sco (1979, 1999) and LMC 1990 # 2, it was about the same. These were very different binary systems, the first having a long period (about one year) and the others being in short period systems. In general, He/H is at least solar and usually higher, up to a factor of 2 above solar. The recurrents in giant systems, RS Oph, T CrB, V745 Sco, and V3890 Sgr, all show similar spectral development that is initially due to the ejecta but quickly transfers to decaying emission from the environment. In fact, these systems are the closest physical analogs to the previous description of a nuclear blast: the environment ionizes quickly and then recombines as the ejecta move outward. The strongest emission lines are those from the time dependent H II region formed within the enveloping stellar wind; these initially produce a characteristic emission line profile at all wavelengths that has broad wings and sharp cores with no strong blueshifted absorption. For the compact systems (U Sco, LMC 1990 # 2, T Pyx, and the newly identified systems CI Aql and IM Nor), a peculiar feature is the early appearance of P Cyg profiles on the UV resonance lines without the previous extended iron curtain phase, although Anupama discusses the appearance of P Cyg structure on the optical lines of IM Nor (which is identified as having a subgiant or giant companion). This certainly accounts for the swiftness of the optical light curve, it's likely the ejecta are not spherical (this is certainly the case based on radio observations of jetlike structures or blobs from the 1985 outburst of RS Oph).

CO novae: This class is the most diverse in light curve behavior and spectral development (contrast VSNET and AAVSO light curves of, for example, HR Del and Cas 1995 [large brief optical flares], DQ Her and V705 Cas [opaque dust formers], and OS And [boringly uniform]). Dust formation is a marked feature of this class, although it isn't an exclusive property. The most recent strong dust forming episode was observed in V705 Cas 1993, which was the first to be observed throughout the $0.1 - 10\mu$ region. It's worth a moment to explain the analysis of this vent because it is the only time we have captured a DQ Her-like transition at all wavelengths. The UV and optical flux dropped by a factor >10 within one week, while the infrared simultaneously increased. The equality of this

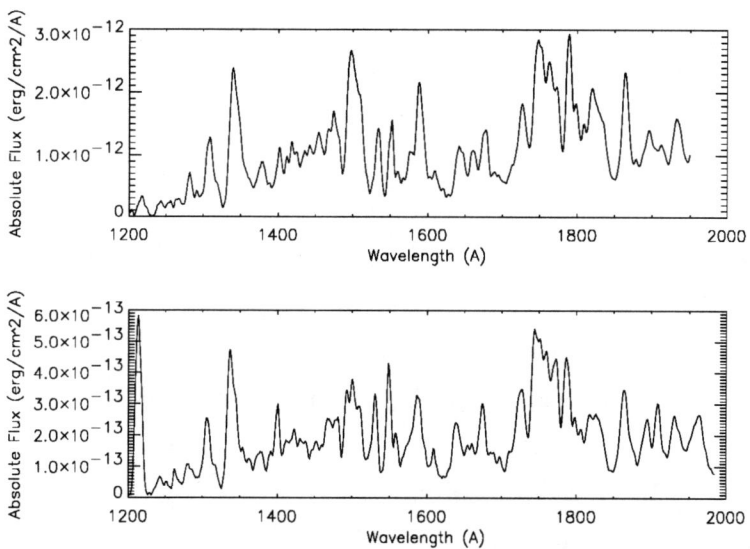

FIGURE 8. Ultraviolet spectrum of V705 Cas 1993 during dust forming event. No change was observed in the spectrum, but the flux level has dropped by a factor of 8.

short wavelength deficit with the IR rise provides one of the few clues we have that the dust is homogeneously distributed with large filling factor in the *outer* ejecta, at temperatures below the Debye temperature for silicate and carbonaceous grains. I emphasize the locale because the UV spectrum was otherwise unchanged *except* for the flux level. Unlike SN II spectra, in particular SN 1987A, there was no change in line symmetries and the extinction law was consistent with silicates (the later observation of the infrared broad 11μ SiC feature is an important check on this assertion) [6].

The CO class also appears to show evidence for some sort of wind phase based on both the UV and optical spectrum (*e.g.* [5], [12]). Two CO novae have been observed during the fireball stage, Cas 1995 and LMC 1991, and this stage requires a different density profile, with a much steeper radial gradient ($d\ln\rho/d\ln r \approx -6$) than a constant mass model to reproduce the spectra.

ONeMg Novae: A surprising number of recent novae have been of this subclass, including two of the brightest novae of the century (V1974 Cyg 1992 and V382 Vel 1999). The class is remarkably uniform in its characteristics, even novae observed in the Large Magellanic Cloud (LMC 1990 #1, LMC 2000) have shown nearly identical spectral evolution to their galactic counterparts.[14] A somewhat puzzling, but general, spectral development is a phase of strong P Cyg profiles on all UV lines (first detected on V693 CrA), following the disappearance of the iron curtain, that persists for days or up to

---

[14] Unlike our experience with UV observations of CO novae, for ONeMg type we have been able to scale the exposure times and spectral development using only t_3 and v_{max}.

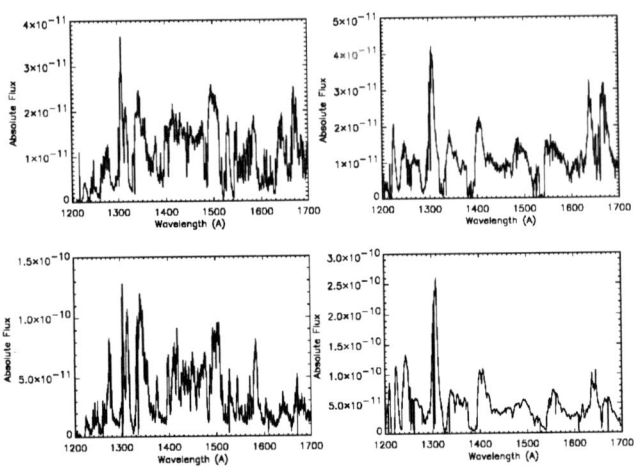

FIGURE 9. Line profiles for two ONeMg novae during the Fe curtain and P Cyg stages: (top) V1974 Cyg (at days 21 and 50), (bottom) V382 Vel (at days 15 and 45).

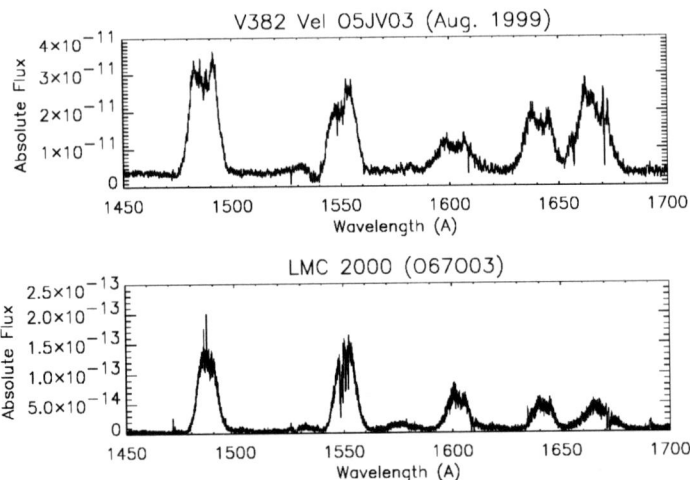

FIGURE 10. Two ONeMg novae in different galaxies at the same spectral stage: (top) V382 Vel 1999, (bottom) LMC 2000. Both are STIS spectra during the early nebular stage.

one week before the transition to the nebular spectrum. The observed terminal velocity is always larger than that obtained from optical absorption troughs (up to twice) and always exceeds the velocity width of the subsequent emission profiles at *all* wavelengths. The UV resonance lines are often saturated and nearly black in the absorption troughs, implying large covering factors as well as large column densities; these narrow with time and v_{max} decreases (seen best for V1974 Cyg).

Superluminous Freaks: Several novae fall into this category, notably LMC 1991. Samples of extragalactic novae (*e.g.* Shara, Shafter, and Della Valle, these proceedings) contain a small number of exceedingly bright sources, but since these surveys are optical the actual peak luminosity depends on shell geometry and filling factors and may be higher than inferred. LMC 1991 had a peak bolometric luminosity more than 5 times L_{Edd} for a 1.4 $M_\odot$ white dwarf [13]. The full examination of UV and visual data for the LMC novae remains to be completed, however, and there are no comparable observations for any other extragalactic novae.

REFERENCES

1. Bjorkman, K. et al., *ApJ*, **425**, 247 (1994).
2. Downes, R. A., and Duerbeck, H. W., *AJ*, **120**, 2007 (2000).
3. Evans, A. et al., *A&A*, **384**, 504 (2002).
4. Gill, C. D. and O'Brian, T. J. *MNRAS*, **307**, 677 (1999)
5. Hauschildt, P. H. et al., *AJ*, **108**, 1008 (1994).
6. Hayward, T. L. et al., *ApJ*, **469**, 854 (1992).
7. Hjellming, R. M. et al., *AJ*, **84**, 1619 (1979).
8. Johnson, J. J., et al., *AJ*, **113**, 2200 (1997).
9. Payne-Gaposchkin, C., *The Galactic Novae*, North-Holland: Amsterdam (1957).
10. Paresce, F. et al., *A& A*, **299**, 823 (1995).
11. Shara, M. M., et al., *AJ*, **114**, 258 (1997).
12. Schwarz, G. J. et al., *MNRAS*, **300**, 931 (1998)
13. Schwarz, G. J. et al., *MNRAS*, **320**, 103 (2001)
14. Shore, S. N., et al., *AJ*, **106**, 2408 (1993).
15. Shore, S. N., et al., *ApJ*, **421**, 344 (1994).
16. Shore, S. N. et al., *ApJL*, **463**, L21 (1996)
17. Shore, S. N. et al., *ApJ*, **490**, 393 (1997)
18. Solf, J., *ApJ*, **273**, 2647 (1983).
19. Vanlandingham, K., et al., *AJ*, **121**, 1126 (2001).
20. Wade, R. A., et al., *PASP*, **112**, 614 (2000).
21. Whitney, B. A., and Clayton, G. C., *AJ*, **98**, 297 (1989).
22. Williams, R. E., *AJ*, **104**, 725 (1992).

Some clues to the understanding of the ultraviolet spectra of novae

Angelo Cassatella* and Rosario González–Riestra[†]

*Istituto Astrofisica Spaziale e Fondamentale, Via del Fosso del Cavaliere 100, Roma, Italia
[†]XMM Observatory, Villafranca Satellite Tracking Station, Madrid, Spain

Abstract. A lot more work is needed to properly understand the spectra of classical novae near outburst due to their complexity, especially in the ultraviolet range. The detailed *IUE* monitoring data of V1974 Cyg 1992 allow, anyhow, to put some constraints on model spectra, and are expected to provide important information about the dynamics of the ejected envelope. The problems here addressed are related with the identification of the absorption systems in this nova and with the interpretation of the "emission" features in the early spectra.

THE ABSORPTION SYSTEMS

In spite of the important progresses made in modeling the spectra of novae in outburst[1], several questions remain still open about the interpretation of the observations, especially in the ultraviolet range, where line blanketing is severe.

The complexity of a nova spectrum is well illustrated in the example of Fig. 1, which shows a rebinned high resolution spectrum of V1974 Cyg 1992 on day 12 after discovery. To understand how the several emission–like peaks and absorption dips are originated, we will consider the 1500–1590 Å region in two spectra of V1974 Cyg taken on days 10.6 and 28 after discovery, which are plotted at full resolution in Fig. 2. The figure indicates that virtually all absorption features can be accounted for by two systems of shortward–shifted absorption lines arising from FeII multiplets uv44–45 and SiII uv2: a low velocity system (the principal system), and higher velocity system (the diffuse–enhanced system). The velocity of the former is ≈ 1.6 times larger than that of the latter, in agreement with optical and ultraviolet observations of other novae. At the same date, these systems do account for most if not all the absorption features in the UV spectrum, as demonstrated in Fig. 3, showing the 1740–1810 Å region on day 10.6, and the region around 2600 Å on day 27.

By comparing the spectra of days 10.6 and 28 (Fig. 2), we deduce that the radial velocity of the principal and diffuse–enhanced systems has become more negative by roughly 300 and 500 km/s in 17.4 days, respectively. More in general, a careful analysis of all the *IUE* high resolution data of V1974 Cyg available, indicates that the radial velocity of the two systems tends to increase with time (in absolute value), until a near–plateau value is reached, following a well–defined law [2]. The principal system endures longer than the diffuse–enhanced system, which suddenly vanishes around day 60. In any case, the simultaneous presence of the two systems until day 60 implies that, before that date, synthetic spectra should take into account the presence of *both* systems.

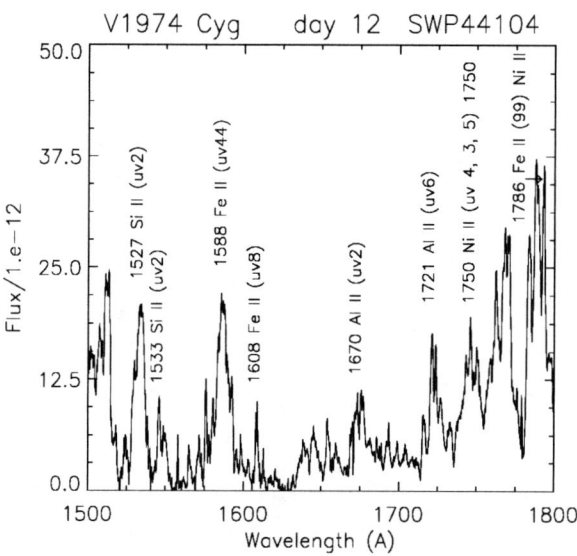

FIGURE 1. Rebinned *IUE* high resolution spectrum of V1974 Cyg obtained on day 12 after discovery. The reality of the "emission" peaks is discussed in the text

Once the presence of two shortward–shifted absorption systems has been established in V1974 Cyg, one might invoke, for example, the occurrence of asymmetric ejection with different velocities, e.g. in the polar and equatorial directions to explain their simultaneous presence in terms of projection effects. Although this possibility cannot be ruled out a priori, its applicability to novae in general seems unlikely for statistical reasons, since the orbital inclination angles of nova systems are expected to be randomly distributed, while the observed ratio of the high to the low velocity components is typically in the range 1.5 to 2[3].

EMISSIONS OR LINE OPACITY MINIMA?

Another interesting question is whether the flux peaks indicated in Fig. 1 are real emission features or are just due to local line opacity minima, as suggested [1]. To clarify this point, let us consider the FeII uv1 multiplet around 2600 Å in Fig. 3. The shortward shifted components from the principal and diffuse–enhanced systems at -1610 and -2560 km/s are all well visible in the spectrum but, of course, no absorption components are seen at zero velocity, except for the pure resonance lines (from the 0 eV level), due to their interstellar/circumstellar origin. A similar behaviour is shown by the MgII doublet and by other resonance lines showing a P Cygni profile. It is then reasonable to think that the emission–absorption feature peaking around 2630 Å is a P Cygni profile similar to that observed in the MgII doublet, rather than the result of an opacity minimum.

Let us now consider how the continuum and the emission lines evolve with time in

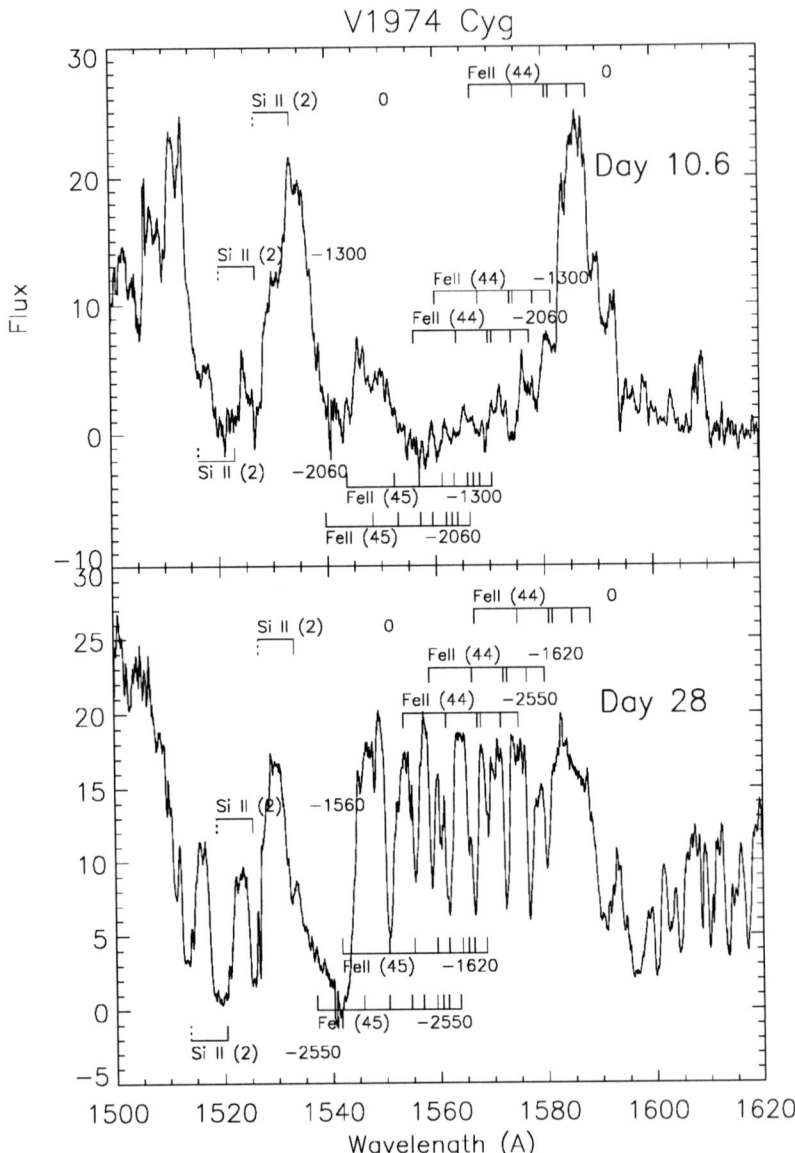

FIGURE 2. Details of *IUE* high resolution spectra of V1974 Cyg on day 10.6 and 28. The overlays mark the positions and indicate the radial velocity in km/s of the principal and diffuse-enhanced systems of FeII uv44–45 and SiII uv2. Vertical dotted bars indicate zero voltage resonance lines

V1974 Cyg. This is shown in Fig. 4, where the visual magnitude, the flux in the 1455 Å continuum and in the emission lines from MgII 2800 Å, OI 1300 Å, CIII 1909 Å, OIII 3132 Å and NV 1240 Å are plotted as a function of time after discovery. It appears from the figure that the time needed for a given emission line to reach a flux maximum

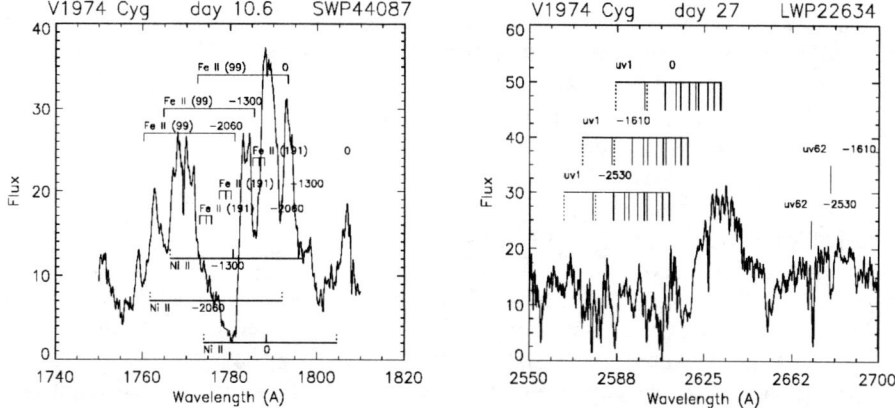

FIGURE 3. The 1740–1820 Å region in the spectrum of V1974 Cyg of day 10.6 (left–hand panel) and the 2550–2700 Å region showing the FeII uv1 and uv62 multiplets on day 27 (right–hand panel)

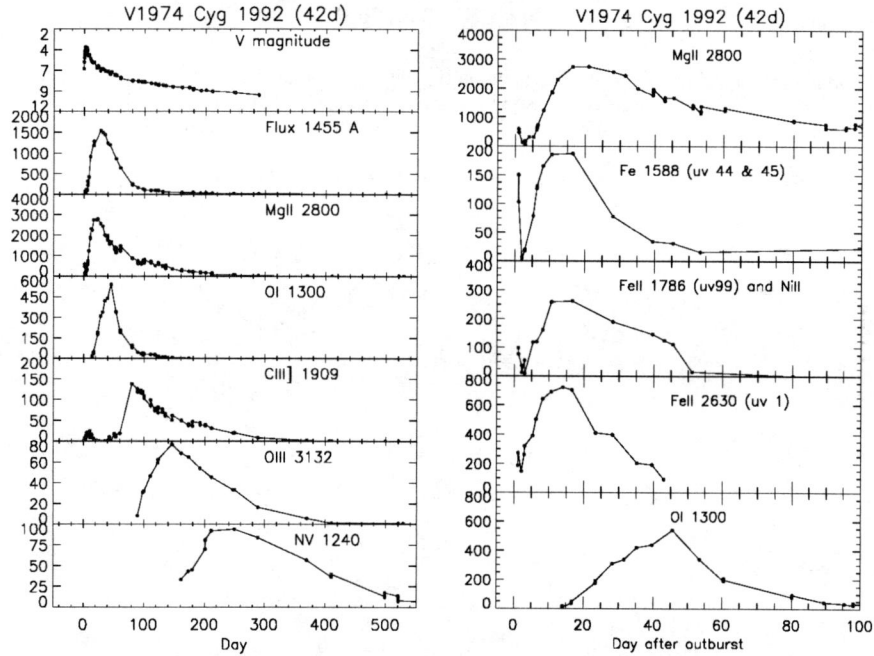

FIGURE 4. Evolution of the emission lines intensity in V1974 Cyg. In the left–hand panel it is shown that the time of maximum increases with increasing the ionization energy of the emission line considered. The right–hand panel shows that the flux in the FeII emissions at 1588 Å and 1786 Å varies in phase with the MgII 2800 Å flux, but it is not correlated with the flux in the OI line (which in turn follows closely the behaviour of the 1455 Å continuum)

increases with increasing its ionization potential. Hence, lines with similar ionization potential as are FeII and MgII, are expected to peak at about the same time. In the right–hand panel of Fig. 4, it is shown that the flux in the FeII 2630 Å emission reaches a maximum at roughly the same time as the flux in the MgII 2800 Å line and that the same happens for the flux in the FeII feature at 1588 Å (uv 44 and 45) and in the 1786 Å blend due to NiII and FeII (uv99). The flux variations of these features are then correlated, while there is no correlation with the flux in the OI line (which in turn follows closely the behaviour of the 1455 Å continuum [4]).

We then conclude, without excluding the efficiency of line blanketing effects, that emission at zero velocity is needed to account for the observed FeII and MgII features.

In any case, should the "emission peaks" just reflect a local opacity minima, their central wavelength would not be stable, due to the rapidly changing level of ionization in the ejecta and to the large velocity variations, contrary to what indicated by the observations. In fact, a test made using 17 *IUE* observations taken until day 109 shows that the 1588 emission blend is quite stable in wavelength: 1588 ± 4 A.

REFERENCES

1. Short C.I., Hauschildt P.H., Starrfield S., Baron E., 2001, ApJ **547**, 2001
2. Cassatella A., Rossi C., González–Riestra R., Altamore A., 2002, in preparation
3. McLaughlin D.B., 1960, in *Stars and Stellar Systems* ed. J.L. Greenstein, Stellar Atmospheres (Chicago Univ. Press)
4. Cassatella A., Altamore A., González–Riestra R., 2002, A&A **384**, 1023

Spectroscopic monitoring of classical and recurrent novae at Asiago Observatory

T. Iijima

*Astronomical Observatory of Padova, Asiago Section,
I-36012 Asiago (Vicenza) ITALY*

Abstract. In the course of the monitoring of the spectral evolutions of classical and recurrent novae on the outbursts, 12 classical novae and 2 recurrent novae (U Sco, CI Aql) were observed at Padova-Asiago Observatory in the years from 1990 to 2002. Here, I will mainly report of Nova Aql 1999-2 (V1494 Aql). This fast nova showed unexpected large variations of the intensity ratio of Hα/Hβ in the early decline phase and variable high velocity jets were observed more than 100 days after light maximum.

OBSERVATIONS AT ASIAGO

The monitoring of spectral evolution of novae is a traditional work of Padova-Asiago Observatory. The observations have been carried out using three spectrographs and two telescopes in the recent years. The high dispersion spectra, $\lambda/\Delta\lambda \cong 8000$, were taken with the Reosc Echelle spectrograph mounted on the 182 cm telescope at the Mount Ekar station of the Astronomical Observatory of Padova. The spectral range of most echelle spectra is from 435 nm to 690 nm. The medium dispersion spectra, $\lambda/\Delta\lambda \cong 800$ with a grating of 600 lines/mm, were taken with a Boller & Chivens spectrograph. This spectrograph was mounted on the 182 cm telescope until the end of 1997, then was translated to the 122 cm telescope of the Asiago Observatory of the University of Padova. The observation with the Boller & Chivens spectrograph on the 122 cm telescope started in August 1998. Until this time, a prismatic spectrograph Camera VI, whose resolutions are 10 nm/mm at Hβ and 30 nm/mm at Hα, was mounted on the 122 cm telescope. Now we use the Reosc Echelle spectrograph mounted on the 182 cm telescope and the Boller & Chivens spectrograph mounted on the 122 cm telescope.

Table 1 gives names of novae, dates of discovery, dates of first and last observations in our observatory, and numbers of spectra taken with respective spectrographs. The term "continue" in the column of last observation means the object is still monitored. The recurrent novae T CrB and RS Oph are also frequently monitored in our observatory, but they are not included in this list, because no explosion was observed in the last decade.

Among these objects, detailed reports of V1974 Cyg were made by Rafanelli et al. [6] and Rosino et al. [7], and the spectral evolution of V723 Cas in the extremely long pre-maximum stage was reported by Iijima et al. [2]. Recently, Iijima [3] analyzed the spectra of U Sco on the outbursts in 1999 and estimated the helium abundance of the ejecta to be N(He)/N(H)= 0.16 ± 0.02, which means this object may not be helium rich.

TABLE 1. Spectroscopic observations of classical and recurrent novae at Asiago

Name	year	discovery	first obs.	last obs.	B&C	Echelle	C. VI
V838 Her	1991	March 24	March 31	1991, July 2	6		
V1974 Cyg	1992	Feb. 19	Feb. 24	1996, August 9	15	2	
V1425 Aql	1995	Feb. 7	Feb. 10	1995, Oct. 23			22
V723 Cas	1995	August 24	August 26	continue	49	36	93
V2487 Oph	1998	June 15	July 9	1998, July 9		1	
V1493 Aql	1999	July 13	July 25	1999, July 25	1		
V1494 Aql	1999	Dec. 1	Dec. 3	2000, Sept. 16	32	17	
V445 Pup	2000	Dec. 30	Jan. 14	2001, April 2	8	2	
V1548 Aql	2001	May 11	May 19	2001, May 19	1		
V2274 Cyg	2001	July 13	July 17	2001, August 28	20		
V2275 Cyg	2001	August 18	August 24	continue	14	1	
V838 Mon	2002	Jan. 7	Jan. 8	continue	18	11	
U Sco	1999	Feb. 25	Feb. 26	1999, March 15	2	5	
CI Aql	2000	April 28	May 13	2000, July 26	4	9	

Here, I will report of some topics of Nova (V1494) Aql 1999-2.

V1494 AQL 1999-2

This nova was discovered on December 1, 1999 [5]. Kiss & Thomson [4] estimated the date of light maximum to be December 3.4 UT (JD 2451515.9 ± 0.1). The first spectra in our observatory were taken on December 5, 1999, then 32 medium dispersion spectra and 17 high dispersion spectra were obtained until September 16, 2000.

The early spectroscopic and photometric evolutions of this nova were reported by Kiss & Thomson [4]. They estimated the absolute magnitude at light maximum to be $M_V = -8.8 \pm 0.2$ mag and the distance to be 3.6 ± 0.3 kpc, but the effect of the interstellar extinction was not taken into account in their work.

Intensity ratio of Hα/Hβ

The intensity ratio of Hα to Hβ was measured to make an estimate of the amount of the interstellar extinction. This nova, however, showed unexpected large variations of the intensity ratio. The results of the spectroscopic observations are summarized in Table 2, where "days" is the number of days after light maximum and "FWHM" is full width at half maximum of the emission component of Hβ. The column "Hα/Hβ" gives the intensity ratio of Hα to Hβ.

As seen in Table 2, the intensity ratios of Hα/Hβ on the first two spectra were not so different from theoretical values [1], but very high ratios were found on the third and successive spectra. The intensity ratio lasted high more than 100 days then slowly decreased.

TABLE 2. Spectroscopic observations of V1494 Aql

Date	JD	days	Instr.	Sp range	FWHM	mV	Hα/Hβ	E(B–V)
	2450000+			nm	km/s			±0.1
1999								
Dec. 5	1518.2	2.3	B&C	350–710	2460	5.0	3.4	0.24
Dec. 6	1519.2	3.3	B&C	400–705	2490	5.3	3.8	0.34
Dec. 17	1530.2	14.3	B&C	400–720	3130	6.7	8.9	1.15
2000								
Feb. 6	1580.7	64.8	B&C	400–515	2790	9.6		
Feb. 21	1595.7	79.8	Ech	435–690	2730	9.4	8.1	0.99
March 7	1610.7	94.8	B&C	400–515	2790			
March 16	1619.7	103.8	Ech	435–690	2740	8.6	9.2	1.11
March 17	1620.7	104.8	Ech	435–690	2710	8.9	9.1	1.10
April 23	1657.6	141.7	Ech	435–690	2720	9.9	8.2	1.00
April 26	1660.6	144.7	Ech	435–690	2710	10.0	7.3	0.89
May 15	1679.6	163.7	Ech	435–690	2700	9.9	6.7	0.82
May 16	1680.6	164.7	Ech	435–690	2680	9.9	6.4	0.77
June 9	1704.6	188.7	Ech	435–690	2720	10.4	6.5	0.78
Sept. 9	1797.3	281.5	B&C	400–738	2820	10.5	5.6	0.64
Sept. 16	1804.4	288.5	Ech	435–690	2690	11.0	6.1	0.73

The intensity of [N II] 658.4 nm emission line of this nova was only 5% of Hα even in the nebular stage (Iijima & Esenoglu in preparation). Therefore, the high intensity ratios of Hα/Hβ were not due to the blending of [N II] lines. Probably, a high velocity mass ejection occurred between the dates of the second and third observations, because the width of the emission component of Hβ significantly increased. A new ejection of matter, however, may not change the intensity ratio so much even when the physical condition of the new ejecta was very different, because the intensity ratios of Hα/Hβ remain roughly in the range 2.6 − 3.0 for various conditions of electron density and temperature of the nebulosity [1].

If the high intensity ratios of Hα/Hβ were due to interstellar and circumstellar extinctions, our results may suggest a large condensation of obscuring matter in the early stage of the explosion. This nova, however, did not show a DQ Her type sudden drop of luminosity when the ratio of Hα/Hβ increased.

The last column of Table 2 gives the amount of the reddening estimated from the deviations of the observed intensity ratios of Hα/Hβ from the theoretical ones [1]. The intensity ratio of Hα/Hβ = 2.62 at $T_e = 10^4$ K, $N_e = 10^9$ cm^{-3} is assumed for the first three spectra when no forbidden line was seen, while Hα/Hβ = 2.81 at $T_e = 10^4$ K, $N_e = 10^6$ cm^{-3} is used for the latter spectra.

Fig. 1 shows V mag of this nova, where dots are the observed V mag collected by VS-Net, while open squares are those after the correction of the extinction. In contrast to the rapid decline of the observed luminosity, the corrected luminosity increased in the earliest stage, and its maximum was found 14.3 days after the observed light maximum. If the corrected luminosities represent the real light variation of this nova, we may have to change our idea for the classical novae explosions. For example, the estimate of the absolute magnitude at light maxima from the decline rates of luminosity will become nonsense. On the other hand, if the large intensity ratios of Hα/Hβ were not due to the

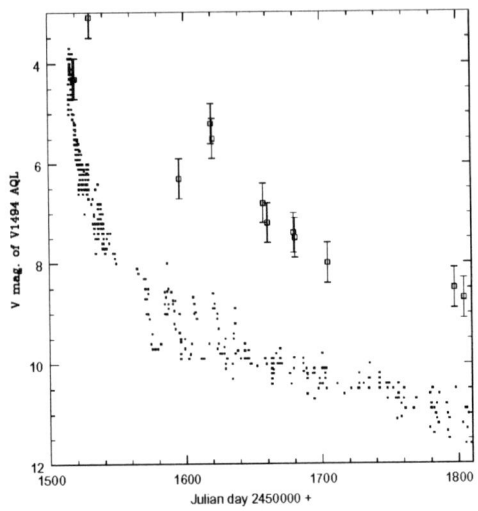

FIGURE 1. Light curve of V1494 Aql

extinction, there should be an unknown process to produce the very high intensity ratios of $H\alpha/H\beta$.

High velocity jets

Another mysterious phenomenon was observed 104.8 days after maximum. Fig. 2 shows traces of the $H\beta$ region of the spectra taken on March 16 and 17, 2000. The spectrum of March 17 showed broad emission components of $H\beta$. The radial velocity of the blue component was $-2860 \, \text{km s}^{-1}$, and that of the red component was $2850 \, \text{km s}^{-1}$. These emission components were very faint one day before. Unfortunately, we were not able to observe this nova more than one month after the observation on March 17. When the next spectrum was taken on April 23, no trace of the broad emission components was seen.

The radial velocities of the broad emission components were much higher than the expansion velocity of the nova shell, $2000 \, \text{km s}^{-1}$, estimated by Kiss & Thomson [4]. At least as the author knows, no any previous nova showed similar high velocity jets in a later stage of explosion. We have no idea to explain this high velocity jets at the present time, but this phenomenon could be a good clue to understand the structure of this system. Further works are waited.

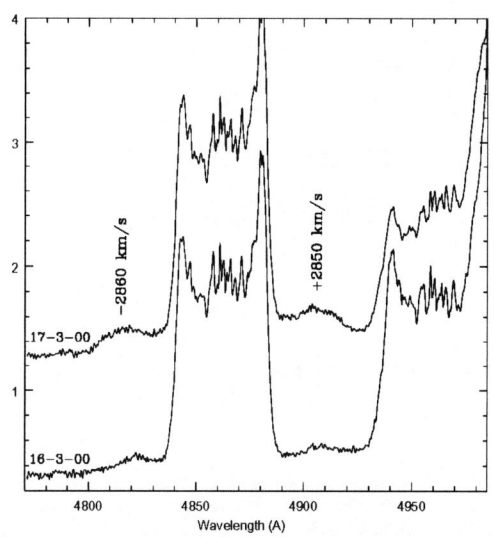

FIGURE 2. Traces of Hβ region on March 16 (down) and 17 (up), 2000

REFERENCES

1. Hummer D.G., Storey P.J., 1987, MNRAS, 224, 801
2. Iijima T., Rosino L., Della Valle M., 1998, A&Ap, 338, 1006
3. Iijima T., 2002, A&Ap, 2002, A&Ap, 387, 1013
4. Kiss L.L., Thomson J.R., 2000, A&Ap, 355, L9
5. Pereira A., 1999, IAU Circ., No. 7323
6. Rafanelli P., Rosino L., Radovich M., 1995, A&Ap, 294, 488
7. Rosino L., Iijima T., Rafanelli P., et al., 1996, A&Ap, 315, 463

Infrared and Radio Observations of Classical Novae: Physical Parameters and Abundances in the Ejecta

Robert D. Gehrz

Astronomy Department, University of Minnesota, 116 Church Street, S.E., Minneapolis, MN 55455

Abstract. Infrared and radio photometric, spectroscopic, and imaging observations of classical nova explosions can be used to determine physical parameters associated with the outburst and elemental abundances in the ejecta. It is shown that some recent CO and ONeMg novae ejected shells with significantly enhanced abundances of CNO, Ne, Na, Mg, Al, Si, and S. The properties of nova dust grains are reviewed and compared to those of the grains in Interplanetary Dust Particles (IDP's) and comet comae. The possibility that classical novae may be a potential source of chemical anomalies and grains in the ISM and the primitive Solar system is considered. Future investigations of remaining unresolved issues are suggested.

INTRODUCTION

There is mounting evidence that Galactic classical nova explosions may play an important role in the chemical evolution of the Galaxy by contributing significantly to the dust and intermediate mass metal content of some ISM clouds in which new stars are forming [1,2,3,4]. In this review we describe how infrared (IR) and radio photometric, spectroscopic, and imaging observations provide quantitative measurements of some of the principal physical parameters that characterize nova explosions, the properties of grains that condense in nova winds, and the abundances of the elements in nova ejecta. We discuss the implications of the results of recent IR/radio studies as they apply to the potential contributions of classical novae to the contents of the Interstellar Medium (ISM) and the primitive Solar system.

PHYSICAL PARAMETERS

A robust characterization of a nova explosion requires an accurate determination of the distance, D (kpc), so that the temporal development of the luminosity of the event, $L_o(t)$ in $L_\odot$, the physical size of the ejected shell, and the mass of gas ejected, M_{gas} in $M_\odot$, can be specified. Other critical parameters are the size and composition of the grains that form in the wind and the abundances of the solid phase and gas phase elements in the ejecta.

Distance

IR/radio photometric and imaging observations can lead to the determination of D in several independent ways. Immediately following the initial thermonuclear runaway (TNR), the hot gas expelled in the explosion is initially observed as an expanding, optically thick photosphere, or "fireball," that emits most of its energy in the visible and IR. The blackbody temperature T_{BB} (K) of the spectral energy distribution (SED) during this phase is 5000-7000 K, the fireball expands at constant velocity V_o (km s^{-1}) that can be obtained from optical/IR spectroscopy, and the flux rises rapidly as the area of the photosphere increases. In cases where the fireball phase is observed early enough, the blackbody radius, θ_r (arcseconds), and the distance, D (kpc), of the fireball can be determined by the expansion parallax method:

$$\theta_r(t) \approx 10^{11} [(\lambda f_\lambda(t))_{max}]^{1/2}, \quad D \approx 5.75 \times 10^{-4} V_o (d\theta_r/dt)^{-1} \qquad (1)$$

where $(\lambda f_\lambda(t))_{max}$ is the time dependent apparent flux from the fireball (W cm^{-2}) and the angular expansion rate $d\theta_r/dt$ (milliarcseconds day^{-1}) can be determined by extrapolating $\theta_r(t)$ back to the origin to give the time, t_o (JD), of onset of the TNR. This method has been applied successfully to the cases of V1500 Cyg [5] and QV Vul [6]. Since the fireball becomes optically thin within a few days to a week after the outburst, the successful application of this technique requires frequent observations at the earliest possible time. A handful of nova explosions that occur on low mass CO white dwarfs (WD's) have developed optically thick dust shells that act as calorimeters of the luminosity of the central engine. The most outstanding examples of this behavior were NQ Vul [7], LW Ser [8], and QV Vul [6]. In cases such as these, equations (1) can be used to yield a blackbody parallax distance to the nova. The blackbody parallax method for both fireballs and dust shells must be applied with caution, for one must have confidence that the velocity being used actually refers to the expansion of the optically thick region. In the case of QV Vul, for example, the fireball and dust forming ejecta seemed to expand at different rates [6].

A more satisfying way of determining distances is to use a combination of IR/radio imaging and spectroscopy. The very high spatial resolution available with the VLA A configuration, MERLIN, the HST IR camera NICMOS, and IR imagers and spectrometers operating on the new generation of large ground-based telescopes can resolve details as small as 0.05 to 0.2 arcseconds across. Thus, a typical nova shell expanding at V_o = 1000 km s^{-1} at a distance of D = 2 kpc will have an angular expansion rate of $d\theta_r/dt \approx 0.1$ arcseconds yr^{-1} and can be well resolved by these facilities after t $\approx$ 1-5 years. Again, the technique uses the concept embodied in equations (1), except that $d\theta_r/dt$ is determined by direct measurement of the angular size of the ejected shell from the images. Expansion parallaxes derived from the radio images are somewhat uncertain because they require the assignment of an expansion velocity from either visual or IR spectroscopy and some assumptions about the velocity gradients in the ejecta. On the other hand, expansion parallaxes derived from IR images hold more promise of giving accurate values of D because IR images can be made in the light of the very hydrogen lines whose profiles at spectral resolutions of $R = \lambda/\Delta\lambda \approx 1000$-$2000$ (150-300 km s^{-1}) can be used to determine V_o at various positions within the ejecta using IR long-slit spectroscopy [3]. Radio expansion parallaxes from VLA and MERLIN images have been determined for QU Vul [9] and V1974 Cyg [10]. IR expansion parallaxes from HST NICMOS images are available for QU Vul, QV Vul, and V1974 Cyg [11].

A complicating factor in comparing distances obtained from radio and IR imaging is that radio imaging has sensitivity that shows both the inner, slow-moving, high-density ejecta and outer, fast-moving, low-density ejecta in the "Hubble flow" of the nova wind. On the other hand, IR images refer primarily to the inner, slow ejecta. Because of the sensitivity of the radio data to faint flux levels and the high spatial resolution of the VLA and MERLIN, radio imaging is proving especially useful for studying the geometry of the ejecta of novae after one to five years [12]. The development of the geometrical non-uniformities that are clues to the physical processes that shaped the remnant are spatially well defined by this time [9,10, 13].

Luminosity

Once D has been established, the luminosity follows immediately from:

$$L(t) = 4\pi D^2 f(t), \quad L_o \approx 3 \times 10^{17} D^2 [\lambda f_\lambda(t)]_{max} \quad \text{in } L_\odot \tag{2}$$

where f(t) is the integrated apparent luminosity and where the equation on the right applies to the luminosity of the central engine for blackbody fireballs and optically thick dust shells. Gehrz and Ney [14] have shown that $f(t) = 1.36(\lambda f_\lambda(t))_{max}$ for the blackbody case. The maximum outburst value of L_o for expanding fireballs, in all instances where it has been well determined, has approached or exceeded the Eddington luminosity $L_{Edd} = 4\pi \, cGM_\odot \kappa_T^{-1} = 6.8 \times 10^4 \, L_\odot$ for a one solar mass white dwarf made of A/Z = 2 material for which the Thompson scattering opacity would be $\kappa_T \approx 0.2$ cm^2 gm^{-1}.

Ejected Mass

An assessment of the contributions of nova explosions to ISM and Solar System abundances is dependent upon an accurate determination of the total mass, M_{gas} ($M_\odot$), ejected in a single event as well as the recurrence time between events. IR/radio observations of the free-free emission from fireballs that have become optically thin can be used in two ways to estimate a lower limit to the ejected mass. IR light curves during the free-free emission phase show a sharp transition between rising and falling flux [4], and the SED begins to deviate from that of a blackbody as the fireball becomes optically thin at the shortest wavelengths. At this time, t_T days after the shell ejection, the ejecta are still hot enough that the opacity is dominated by Thompson scattering, in this case for hydrogen-rich material so that $\kappa_T \approx 0.4$ cm^2 gm^{-1}, and the mass of the ionized gas in the shell is given by:

$$M_{gas} \approx \pi R^2 \kappa_T^{-1} \approx 3.3 \times 10^{-13} (V_o t_T)^2 \quad \text{in } M_\odot. \quad (3)$$

Later, the optically thin ejecta show a free-free energy distribution that becomes self-absorbed at longer wavelengths. The wavelength λ_c (μm), where the free-free and Rayleigh Jean's tails of the SED intersect, can be clearly identified as it moves through the IR and out into the radio region. This wavelength moves through the IR in a few days to a few weeks and can be shown to measure the ion density of a relatively compact, constant density, thin shell of radius R (cm) and thickness ℓ (cm) where $f = \ell/R$ by the relationship $(n_H \lambda_c)^2 \ell = (n_H \lambda_c)^2 f R \approx 1.3 \times 10^{36}$ (cm^{-3}) [4]. For $f \ll 1$, the mass of ionized gas is given by the relationship:

$$M_{gas} \approx 4\pi R^2 \ell n_H H \approx 4\pi R^3 f n_e H \approx 8.2 \times 10^{-14} f^{1/2} (V_o t)^{5/2} \lambda_c^{-1} \quad \text{in } M_\odot, \quad (4)$$

where $H = 1.67 \times 10^{-24}$ gm is the mass of the nucleon. Equation (4) reduces to the case of a constant density sphere for $f = 1/3$, which seems to be a reasonably good approximation for very young ejecta.

Radio light curves can be used to model the outflow to yield a gas mass. However, when the radio intensity peaks, some 100 to several thousand days after the outburst depending upon the frequency being observed, the interpretation of the data is complicated because velocity gradients have begun to dominate the shell thickness and density distribution. Hjellming et al. [10, 14] have shown that the expansion is best modeled by a "Hubble flow" at this point, where the velocity increases outward linearly from the base to the outer edge of the flow and where the shell density assumes a r^{-3} dependence. The technique is similar to the simplified theory given in equation (4) except that one must integrate outward through the shell over the distributions for $V_o(r)$ and $n_H(r)$ from the inner to the outer radii. These radii are well resolved in VLA and MERLIN images after several years for novae closer than 1-2 kpc.

All three of the methods of mass determination (IR Thompson scattering, IR Free-free, and radio free-free) yield comparable estimates for the ejected mass for a given nova. Because radio observations detect lower-density, faster-moving, outer ejecta than do IR measurements, the radio masses are somewhat higher, but by less than a factor of two. All methods show that the ionized ejecta of typical CO novae are in the range 1-10x10^{-5} M$_\odot$ and that those of typical ONeMg novae are in the range 1-4x10^{-4} M$_\odot$. These values are uncomfortably high compared with values predicted by current theories of the TNR. In fact, the problem of reconciling theoretical and observed ejected masses is exacerbated because a large amount of neutral gas may remain undetected by the observations described here. Saizar and Ferland [16] and Ferland [17] have suggested the possibility that these large ejected masses make novae potential competitors with supernovae as processors of ISM gas and contributors to ISM abundances.

ABUNDANCES FROM IR CORONAL EMISSION LINES

The IR spectra of ONeMg novae are dominated by strong forbidden ("coronal") emission lines from highly ionized intermediate mass metals (CNO, Ne, Na, Mg, Al, Si, S, Ar, etc.) for more than a year following the outburst. It is possible to derive abundances of these species using several analytical techniques such as the CLOUDY [18], NEBULAR [19], and NEBU [20] algorithms as well as detailed level balancing [21]. Elementary collisional excitation theory is especially useful for establishing lower limits to the masses of species that have strong IR cooling lines. For example, when the cooling of the shell is dominated by [Ne II] 12.8 µm emission, a lower limit to the neon abundance is given by [3]:

$$n_{Ne\ II}/n_H = 3.5 \times 10^{-11}(I_{12.8}\ n_H)/(\lambda F_\lambda|_{\lambda = 12.8\ \mu m}) \tag{5}$$

Gas phase abundances of recent bright novae determined from IR spectroscopy are summarized in Table 1. It is apparent that both CO novae like LW Ser, V705 Cas, and V1425 Aql, and ONeMg novae like QU Vul and V1974 Cyg can produce copious overabundances of C, N, O, Ne, and Al. Given that hydrogen is typically depleted by the nova TNR [2,4], these large abundances are all the more striking and suggest the possibility that novae process a substantial fraction of the CNO in the ISM.

DUST FORMATION AND GRAIN PROPERTIES

Carbon dust formation in ejecta produced by a TNR on a CO novae has been shown to be responsible for a sudden visual extinction event accompanied by a transfer of the luminosity into the thermal IR in DQ Her class novae [1, 2, 4]. The characteristics of the grain condensation event are summarized by the following relationships:

$$T_c \approx 1000\ K, \qquad R_c = [L_o/16\pi\sigma T_c^4]^{1/2}, \qquad t_c \approx R_c V_o^{-1} \tag{6}$$

where L_o (erg s^{-1}) is the luminosity of the WD, T_c is the condensation temperature, R_c (cm) is the radius of the bottom of the dust condensation zone, t_c (days) is the time at which the ejecta first reach R_c, and $\sigma = 5.7 \times 10^{-5}$ erg s^{-1} cm^{-2} K^{-4}. Typical values of R_c and t_c are 3-6$\times 10^{14}$ cm and 30-100 days respectively. The dust shells for the most extreme examples of dust forming novae become optically thick from visual wavelengths well into the thermal IR. In such cases, the shell measures the luminosity of the central engine and the temporal development of the IR luminosity, L_{IR} (in $L_\odot$), can be used to assess the post-outburst temporal development of the luminosity of the WD. The depth of the visual extinction and L_{IR} can be used to deduce the dust mass and the size of the grains through:

$$M_{dust} \approx 1.6 \times 10^{-11} \rho_{gr} V_o^2 t^2 T_{BB}^{-6} (M_\odot), \quad a_{gr} \approx 2 \times 10^{22} L_o V_o^{-2} (t_{(IR\ max)}^{-2}) T_{BB}^{-6} (\mu m) \tag{7}$$

Carbon grains in CO novae grow to radii as large 0.2 to 0.7 μm. Given the mineral and chemical properties of the grains, equations (3), (4), and (7) can be used to deduce the abundance of the condensable species (see Table 1). Such abundances for the condensables must be regarded as lower limits to the true abundances, since they do not take into account condensables remaining in the gas phase.

At this point, IR observations have confirmed that nova ejecta can produce all of the astrophysical grains that have been observed in circumstellar and ISM environments, including amorphous carbon, silicates, SiC, and hydrocarbons. In several cases, the

TABLE 1. Chemical Abundances in the Ejecta of Classical Novae from IR Observations

Nova	X^a	Y	$\dfrac{[n_X/n_Y]_{nova}}{[n_X/n_Y]_{Solar}}$	References[b]
LW Ser	carbon dust	H	≥ 15	(1)
QU Vul	Ne	H	≥ 1.2	(2)
"	Ne	Si	≥ 6	(3)
"	Mg	Si	≈ 5	(4)
"	Al	Si	≈ 70	(4)
QV Vul	carbon dust	H	≥ 7	(5)
V1974 Cyg	C	H	≈ 12	(6)
"	N	H	≈ 50	(6)
"	O	H	≈ 25	(6)
"	Ne	H	≥ 10	(7)
"	Ne	H	≈ 50	(6)
"	Ne	O	≈ 4	(8)
"	Ne	Si	≈ 35	(7)
"	Mg	H	≈ 5	(6)
"	Mg	Si	≥ 3	(9)
"	Al	H	≈ 5	(6)
"	Al	Si	≈ 5	(9)
"	Si	H	≈ 6	(6)
"	S	H	≈ 5	(6)
"	Ar	H	≈ 5	(6)
"	Fe	H	≈ 4	(6)
V705 Cas	silicate dust	H	≈ 15	(10)
"	carbon dust	H	20	(11)
"	O	H	≥ 25	(12)
"	Ca	H	20	(13)
V1425 Aql	C	He	≈ 9	(14)
"	N	He	≈ 100	(14)
"	O	He	≈ 9	(14)

[a] The abundances cited for "dust" are calculated using the amount of gas condensed into grains (see equation (7)).
[b] References: (1) Gehrz et al. 1980, Ap. J, 237, 855; (2) Gehrz et al. 1985, Ap. J., 298 L47; (3) Greenhouse et al. 1990, Ap. J., 352, 307; (4) Greenhouse et al. 1988, A. J., 95, 172; (5) Gehrz et al. 1992, Ap. J. 400, 671; (6) Hayward et al. 1996, Ap. J., 469, 854; (7) Gehrz et al. 1994, Ap. J. 421, 762; (8) Salama et al. 1996, A&A, 315, L209; (9) Woodward et al. 1995, Ap. J., 438, 921; (10) Gehrz et al. 1995, Ap. J., 448, L119; (11) Mason et al. 1998, Ap. J., 494, 783; (12) Salama et al. 1999, MNRAS, 304, L20; (13) Salama et at. 1998, in The Universe as Seen by ISO, ESA Publication SP-427, V. 1, p 233; (14) Lyke et al. 2001, AJ, 122, 3305.

signatures of three or four different grain components have been observed during the temporal development of the ejecta of a single nova event, suggesting either that there are large chemical gradients in the ejecta and/or that the standard picture of equilibrium kinetic grain growth does not apply to the nova environment [6]. A recent breakthrough in the understanding of the IR SED's of dusty novae was achieved when Evans et al. [22] and Mason et al. [23] independently recognized that the so called unidentified Infrared (UIR) emission lines represented a strong hydrocarbon emission component that had to be accounted for in modeling the shape of the silicate and SiC emission features. It now seems possible that the strange, strong 10 µm emission feature seen in the SED of V1370 Aql [24] might have best been explained by modeling a SiC emission feature with UIR contributions because there was no evidence of any 20 µm emission feature that would have been expected silicates were present in the ejecta.

CONTRIBUTIONS OF CLASSICAL NOVAE TO THE ISM

Gehrz et al. [2] have argued that, for classical novae to compete with other stellar populations as processors of ISM material, their ejecta must contain abundance elevations of several hundred with respect to solar and approximately ten with respect to Type II supernova abundances. If one accepts the premise proposed by Saizar and Ferland [16] and Ferland [17] that novae may be a much more important source of mass than hitherto suspected, then these abundance elevation requirements may be relaxed accordingly. The theories of the TNR events expected to occur on both extreme CO and ONeMg WD's show that such large elevations might be expected for the intermediate mass metals (CNO, Ne, NA, Mg, Al, Si, S, etc.). The results of abundance determinations from recent IR spectroscopic studies of novae, summarized in Table 1, clearly indicate that the ionized gas in nova ejecta is sufficiently enriched in these elements to suggest that novae may affect ISM abundance on local scales. For example, the Galactic orbital trajectories might be expected to take nova-producing binary star systems through or near regions of star formation. This may have been the case for the primitive Solar system [25, 26].

An issue of particular interest is whether novae are associated with the production of certain extinct radio isotopes that are believed to have been present in the primitive solar system. Namely, these are the ^{22}Ne (Neon-E) and ^{26}Mg anomalies seen in meteorite inclusions [2, 25]. The former is produced largely through the beta decay of ^{22}Na through the reaction ^{22}Na $\rightarrow$ ^{22}Ne + e^+ + γ (mean life = 3.75 yr) and the latter by the beta decay of ^{26}Al $\rightarrow$ ^{26}Mg + e^+ + γ (mean life = 10^6 yr). Although the TNR theory suggests that both CO and ONeMg are capable of contributing significantly to these anomalies, the high velocities and dynamical structure in the nova winds make an isotopic analysis through IR spectroscopy untenable. However, a number of IR coronal forbidden lines of Ne, Na, Al, and Mg will be accessible using SIRTF at resolutions of $R = \lambda/\Delta\lambda$ of 60 to 600 (500-5000 km s^{-1}) to estimate overall elemental abundances [27]. Another primary way of assessing the contributions of novae to these chemical anomalies on a Galactic scale is to detect the beta decay gamma rays from a nova in outburst, a possibility that may be explored upon the launch of the International Gamma Ray Observatory (INTEGRAL) in October of 2002 [2, 28].

CLASSICAL NOVAE AND THE PRIMITIVE SOLAR SYSTEM

There is growing evidence to support the notion that one or more nova explosions produced material that was directly incorporated into the primitive Solar system during the early collapse or pre-collapse phases. First, there is now substantial evidence that the grains formed in nova outflows are similar in many ways to the small particles that make up the structure of the Interplanetary Dust Particles (IDP's) that are believed to have been released from comet nuclei by insolation during their perihelion passage [29]. IDP's are composed of hundreds of sub-micron grains assembled in a "cluster of grapes" fractal pattern and these small substructure grains are similar in physical size and mineral composition to the stardust made in nova explosions. Both nova grains and IDP's have carbon, silicate, and hydrocarbon structures. There is both an amorphous component and a crystalline, annealed component in the IDP grains. While nova grains appear to have the spectral structure of amorphous particles early in their growth, there is evidence that they may be subjected to harsh radiation that could anneal them later [4, 30].

Second, there are many remarkable similarities between the grains that are observed to grow in nova winds and the grains that populate the comae and tails of comets in terms of the composition, mineralogy, and size distributions. The grain composition and size in comets can be inferred from the albedo, A, superheat, S, and 10 μm silicate emission feature intensity, Δm_{10}, where these quantities are defined by the relationships:

$$A \approx F_v/F_{IR}, \quad S = T_{gr}/T_{BB} = T_{gr} r^{1/2}/278, \quad \Delta m_{10} = 2.5\log_{10}(F_{10}/F_c), \quad (8)$$

where F_v and F_{IR} (W cm^{-2}) are the integrated reflected and thermal SED's respectively, T_{gr} is the observed coma/tail grain temperature at heliocentric distance r (AU), T_{BB} is the temperature that a black, conducting sphere would assume at r, and F_{10} and F_c (W cm^{-2} μm^{-1}) are the fluxes of the silicate emission feature peak and the continuum at 10 μm [14]. Gehrz and Ney [14] and Williams et al. [31] have shown that there is a correlation between the superheat and silicate feature strength that is consistent with the presence of both carbon and silicate grains, and that the grains have radii ranging from 0.2 to 1 μm. Some comets grains also have a hydrocarbon component as evidenced by a 3.28 μm C-H stretch emission feature [32]. All of these properties are consistent with those of the grains that grow in nova winds. The suspicion that comet grains and IDP's have a common source is strengthened by the fact that the variation of albedo with phase angle, $A(\theta_{ph})$, is consistent with the small grain components being aggregated into larger, fluffy structures similar in size to IDP's [14, 33]. Finally, the theoretical and observational studies cited in the previous section [2] show that nova TNR's create ejecta containing the intermediate mass metals associated with the Neon-E and ^{26}Mg anomalies seen in Solar system meteorite inclusions [25].

The live ^{22}Na and ^{26}Al produced in the nova TNR could easily have been implanted into the grains during their condensation in the nova wind and transported into the Solar system as solids that were incorporated into planetesimals and meteorites. There is a final, compelling connection between nova grains and the grains found in meteorite inclusions. In a seminal paper presented at this conference, José et al. [30] report the identification of five SiC and two graphite presolar grains whose isotopic signatures are consistent with the interpretation that they were formed in a nova wind. This is the first direct evidence that some grains in the primitive Solar system were imprinted with the signature of nova TNR nucleosynthesis.

OUTSTANDING QUESTIONS AND FUTURE WORK

Several unresolved issues remain that can be addressed by IR and radio observations of novae. Three of the foremost are listed below with some comments about how future studies might address them:

1) What is the total amount of mass ejected in a nova event?: An assessment of the total mass of the ejecta depends upon a knowledge of the neutral material as well as the ionized material. IR spectroscopy at high resolutions during the early free-free emission phase when there are many recombination lines of HI and HeI may provide some answers here. Modeling of the shell kinematics and density structure using a combination of IR/radio imaging and IR photometry and spectroscopy to provide a self-consistent temporal development model may also prove useful. The velocity structures observed in ISO spectra of several bright novae [35, 36, 37] remain to be fully modeled. Gehrz and Hjellming [37] have obtained a large body of contemporaneous IR/radio data on recent bright novae that remains to be analyzed.

2) Why and how do grains of different mineral compositions form in the ejecta of some novae?: Rawlings and Evans [38] and Evans et al. [39] have described ongoing theoretical efforts to model the formation and evolution of grains in nova ejecta.

3) What is the quantitative effect of nova ejecta on local and galactic ISM scales?: This can be addressed by the SIRTF IRS spectrometers which can provide data for extensive IR forbidden line abundance analysis for novae in our own and other galaxies. R. Gehrz and his collaborators plan a global study of nova rates and nova ejecta chemical abundances in M33 using SIRTF Guaranteed Time Observations.

ACKNOWLEDGMENTS

The author acknowledges support from NASA, the NASA SIRTF Project Office, the NSF, the US Air Force, and many collaborators over the last 27 years of conducting IR and radio observations of classical novae. J. E. Lyke provided valuable suggestions.

REFERENCES

1. Bode, M. F., and Evans, A. N., eds., *Classical Novae* and the references therein, John Wiley and Sons: London, 1989.

2. Gehrz, R. D., Truran J. W., Williams, R. E., and Starrfield, S. G., *Pub. A.. S. P.*, 110, 3-26 (1998).
3. Starrfield, S. G., Truran, J. W., and Gehrz, R. D., "Dust Formation and Nucleosynthesis in the Nova Outburst", in *Astrophysical Implications of the Laboratory Study of Presolar Materials*, eds. T. J. Bernatowicz and E. Zinner, AIP: New York, 1997, pp 203-234..
4. Gehrz, R. D., *Physics Reports*, 311, 405-418 (1999).
5. Ennis, D. et al., Ap. J., 214, 478-487 (1977).
6. Gehrz, R. D., Jones, T. J., Woodward, C. E. et al.., Ap. J., 400, 671-680 (1992).
7. Ney, E. P. and Hatfield, B. F., Ap. J., 219, L111-L115 (1978).
8. Gehrz. R. D., Grasdalen, G. L., Hackwell, J. A., and Ney, E. P., Ap. J., 237, 855-865 (1980a).
9. Taylor, A. R., Hjellming, R. M., Seaquist, E. R., and Gehrz, R. D., Nature, 335, 235-238 (1988).
10. Hjellming, R. M., "Radio Images and Light Curves for Nova Cygni 1992", in *Cataclysmic Variables*, eds. A Bianchini et al., Kluwer: Dordrecht, 1995, pp. 139-144.
11. Krautter, J., Woodward, C. E., Schuster, M. T., Gehrz, R. D., et al., A. J., in press (2002).
12. Bode, M. F., "Review: The Evolution of Nova Remnants", in *Classical Nova Explosions*, eds, M. Hernanz and J. José, AIP:New York, 2002.
13. Heywood, I, "Radio Emission from Nova Cassiopeia 1995", in *Classical Nova Explosions*, eds, M. Hernanz and J. José, AIP:New York, 2002.
14. Gehrz, R. D., and Ney, E. P., Icarus, 100, 162-186 (1992).
15. Hjellming, R. M., Wade, C. M., Vandenberg, N. R.; and Newell, R. T., A. J., 84, 1619-1631 (1979).
16. Saizar, P., and Ferland, G. J., Ap. J., 425, 755-766 (1994).
17. Ferland, G. J., 1997, in *Proceedings of the 13th North American Cataclysmic Variable Workshop*, eds. S. Howell, E.Kuulkers, and C. E. Woodward, Astronomical Society of the Pacific, in press.
18. Ferland, G. J., 1996, Department of Physics and Astronomy Internal Report (U. Of Kentucky).
19. Shaw, R. A., and Dufour, R. J., Pub. A. S. P., 107, 896-906 (1995).
20. Petitjean, P., Boisson, C., and Pequignot, D, Astron. And Astrophys., 240, 433-452 (1990).
21. Osterbrock, D. E., 1989, *Astrophysics of Gaseous Nebulae and Active Galactic Nuclei* Mill Valley: University Science Books.
22. Evans, A. N. Et al, M. N. R. A. S., 292, 192-204 (1997).
23. Mason, C. G., Gehrz, R. D., Woodward, C. E., Smilowitz, J. B., Hayward, T. L., and Houck, J. R., Ap. J., 494, 783-791 (1998).
24. Gehrz, R. D., Ney, E. P., Grasdalen, G. L., et al., Ap. .J., 281, 303-312 (1984).
25. Clayton, D. D., Quart. J. R. A. S., 23, 174-212 (1982).
26. Prantzos, N., Ap. J., 405, L55-L58 (1993).
27. See: http://www.mpe-garching.mpg.de/iso/linelists/index.html
28. Hernanz, M., "Future INTEGRAL Observations of Classical Novae", in *Classical Nova Explosions*, eds, M. Hernanz and J. José, AIP: New York, 2002.
29. Brownlee, D. E., Royal Soc. Phil. Trans., Series A, 323, 305-312 (1987).
30. José, J., Hernanz, M., Amari, S.., and Zinner, E., "The Imprint of Nova Nucleosynthesis in Pre-Solar Grains", in *Classical Nova Explosions*, eds, M. Hernanz and J. José, AIP: New York, 2002, in press (these proceedings).
31. Williams, D. M., Mason, C. G., Gehrz, R. D., Jones, T. J., Woodward, C. E., Harker, D. E., Hanner, M. S., Wooden, D. H., Witteborn, F. C., and Butner, H. M., Ap. J., 489, L91-L94 (1997)
32. Brooke, T. Y., et al., Bull. A. A. S., 25, 1065 (1993).
33. Wooden, D. H., et al., Icarus, 143, 126-137 (2000).
34. Salama et al., Astron & Astroph., 315, L209-L212 (1996)
35. Salama et al. Mon. Not. R. A. S., 304, L20-L24 (1999).
36. Salama et at. 1998, in *The Universe as Seen by ISO*, ESA Publication SP-427, V. 1, p 233.
37. Gehrz, R. D., and Hjellming, R. M, 2002, unpublished data.
38. Rawlings, J. M. C., and Evans, A. "Formation and Evolution of Dust in Novae", in *Classical Nova Explosions*, eds, M. Hernanz and J. José, AIP: New York, 2002.
39. Evans, A., Smith, O., Tyne, V. H., and Rawlings, J. M. C. "The Properties of the Dust Around Nova V705 Cas", in *Classical Nova Explosions*, eds, M. Hernanz and J. José, AIP: New York, 2002.

0.8-2.5 µm Spectroscopy of Novae

David K. Lynch*, Richard J. Rudy*, Catherine C. Venturini*, S. Mazuk*, William L. Dimpfl*, John C. Wilson[†], Neal A. Miller[¥], Richard Puetter[°]

The Aerospace Corporation, El Segundo, CA 90245 USA
†Cornell University, Ithaca, NY 14853 USA
¥NASA GSFC, Greenbelt, MD 20771 USA
°UCSD/CASS, La Jolla, CA 92093 USA

Abstract. We present new infrared results on a number of recent novae including V2274 Cyg (Nova Cygni 2001), CI Aquilae, and V723 Cas (Nova Cassiopeiae 1995), and review some of the outstanding aspects of their infrared spectra. The discussion includes the identification of coronal lines and their interpretation and structure in the CO emission band at 2.3 µm.

INTRODUCTION

The 0.8 – 2.5 µm region provides a wealth of diagnostic detail about novae and the evolution of their shells because of the number and variety of lines that occur. Low excitation hydrogen Paschen and Brackett lines provide a background against which to assess abundances, electron densities and temperatures. The intrinsic flux ratio of the OI lines at 0.8446 µm and 1.1287 µm is well known and therefore any departure from it is usually a good indicator of interstellar reddening (Rudy et al. 1991). Neutral and low ionization recombination lines of nitrogen, oxygen, carbon, phosphorus, silicon, calcium, magnesium and iron often provide crucial abundance information which in turn can be related to the initial composition, evolutionary state and surface mixing of the white dwarf (WD) before and during the thermonuclear runaway.

In this paper we will review a number of recent novae in view of these considerations. The data reported here were taken using the Aerospace Corporation's Near InfraRed Imaging Spectrograph (NIRIS - Rudy et al. 1991) on the Lick 3 meter telescope and CorMASS (Wilson et al. 2001) on the Palomar 1.5 m telescope.

CORONAL LINES IN V723 CAS (NOVA CAS 1995)

The term "coronal line" originally referred to any line seen in a star that is also seen in the solar corona. Nowadays, the definition has been refined by Greenhouse et al. (1990) who defined coronal lines as *fine structure transitions to the ground term for species whose ionization energies exceed 100 eV.* Coronal lines are also seen in planetary nebulae, AGNs and Seyfert galaxies. The first IR coronal lines were observed in Nova Cyg 1975 by Grasdalen and Joyce (1976) who reported lines of Si, Al, Ca, and Mg. Since that time the list of lines has grown steadily (Ferland, Lambert, & Woodman 1986; Greenhouse et al. 1990, 1993; Benjamin and Dinerstein 1990; Woodward et al. 1995; Mason et al. 1996; Wagner & Depoy 1996; Rudy et al. 2002a.) Based on the energetics, we have assembled an updated list of IR coronal lines (Table I.) As of this writing (2002) approximately fifteen IR coronal lines have been identified in novae between 0.8 and 2.5 µm and these are shown in **bold text** in Table I. Many of the lines in Table I have not been detected in any astronomical source.

Figure 1 shows two NIRIS spectra of V723 Cas (Nova Cas 1995), a bright, long-lasting, narrow-line (FWHM ~ 500 km sec^{-1}) novae that was well-suited for determining wavelengths of unidentified emission lines that appear in the nebular/coronal stages of novae (Iijima, Rosino, & Della Valle, 1998; Rudy et al. 2002a). The spectra were taken in Aug 1999 and July 2000 (1372 days and 1698 days after peak brightness, respectively.) We found 16 previously unobserved or unidentified emission lines. One of these, [Ti VII] 2.2050 µm, is a new coronal line that de-excites directly to the ground state. Three other lines are associated with transitions from collisionally excited, low-lying, metastable levels of Fe^{+5}. Of the remaining lines, five were at wavelengths of 0.8926, 1.1110, 1.1900, 1.5545, and 2.0996 µm and these have all been observed in other novae and remain unidentified. Additionally, V723 Cas showed an as yet unidentified line at 2.218 µm.

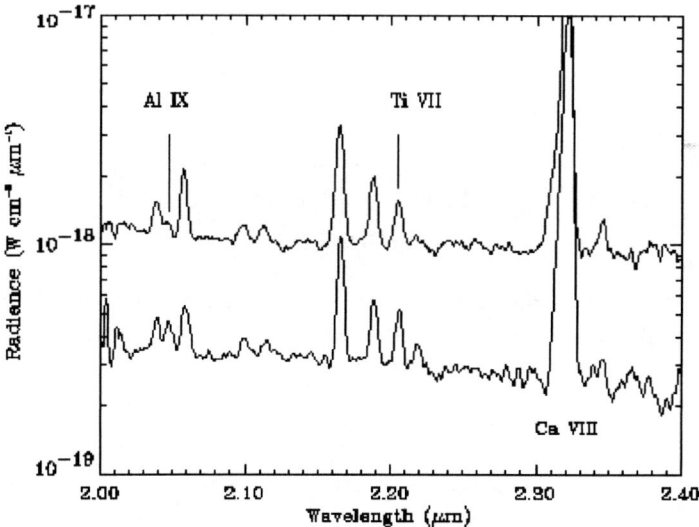

FIGURE 1. V723 Cas (Nova Cas 1995). Note the new coronal line [Ti VII] and the as yet unidentified line at 2.218 µm.

CI AQL

CI Aql is a broad-line, recurrent novae in U Sco subclass. It was observed at Palomar (1.5 meter with CorMASS) and Lick (3 meter with NIRIS) from 3 days to 391 days during its most recent outburst (Lynch et al. 2003). The spectrum (Figure 2) showed a rapid evolution to the nebular stage, the first coronal lines appearing less than a month after peak brightness. As the novae faded during the next six months, the continuum maintained a constant slope that did not fit a Planck function, even after de-reddening.

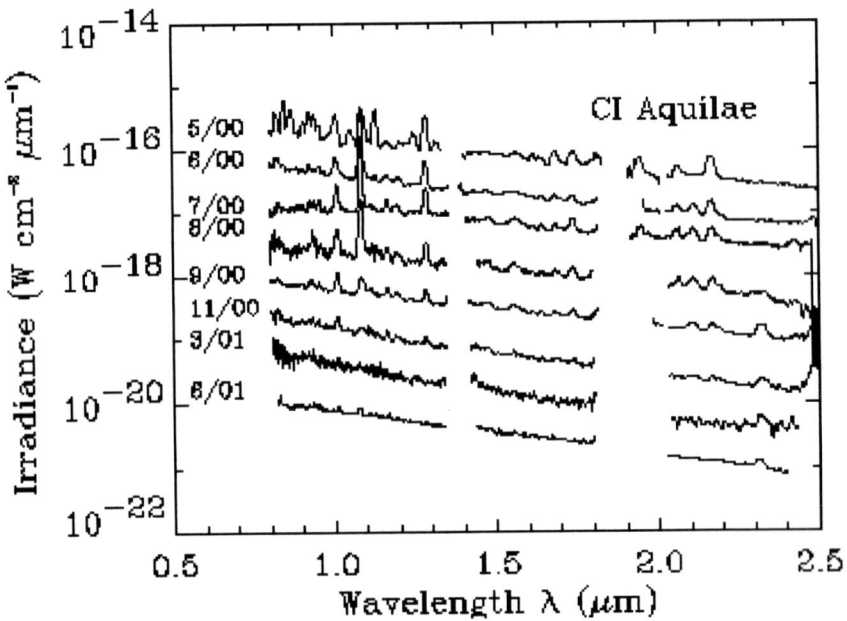

FIGURE 2 CI Aquilae spectra between May 9 2000 and June 2 2001 (days 3 and 391 respectively). Note that at first the HI and neutral metal lines were very strong, but these quickly fade and were replaced by the high ionization and coronal lines.

V2274 CYG (N CYG 2001)

V2274 Cyg (N Cyg 2001) was observed on July 31, 2001 UT, approximately two weeks after peak brightness. In addition to the usual emission lines, the spectrum showed first-overtone emission from carbon monoxide (CO) around 2.3 µm. The feature showed distinct band heads, indicating high molecular rotational energy (Rudy et al. 2002b). A spectrum-synthesizing program, allowing independent temperatures to be assigned to molecular rotations and vibrations, was adapted to generate overtone spectra for CO. As shown in Figure 3, the nova spectrum was fit by including contributions from two isotopes of carbon and using different rotational and vibrational temperatures (non-local thermodynamic equilibrium). The temperatures used were 4000 K for rotational state populations and 1700 K for the vibrational populations. The vibrational temperature fit indicated the population ratios of the 2nd and 3rd vibrational states, characterized by the first two band heads which are clearly resolved and not perturbed by the second isotope. The deviation of the fit beyond 2.35 μm is influenced by the increasing compensation for strong absorption by atmospheric water vapor.

CO in the environment of the nova is excited through collisions with electrons, which are present at a number density that balances the principal ionic species, H$^+$. The estimated electron number density, 10^{12} cm^{-3}, indicates that the time between vibrational state changing collisions is comparable with the radiative lifetime of the vibrational states. This leads to a strong competition between collisions and radiation in establishing the vibrational state distribution in the steady-state environment of the gases ejected by the nova. In contrast, translational and rotational energy of CO is dominated by collisions and the resulting state distributions reflect the local thermal environment. One consequence of the competing processes is that an independent estimate of the number density of the energy exchange collision partner can be gained through modeling the competition, in this case 4×10^{10} cm^{-3}.

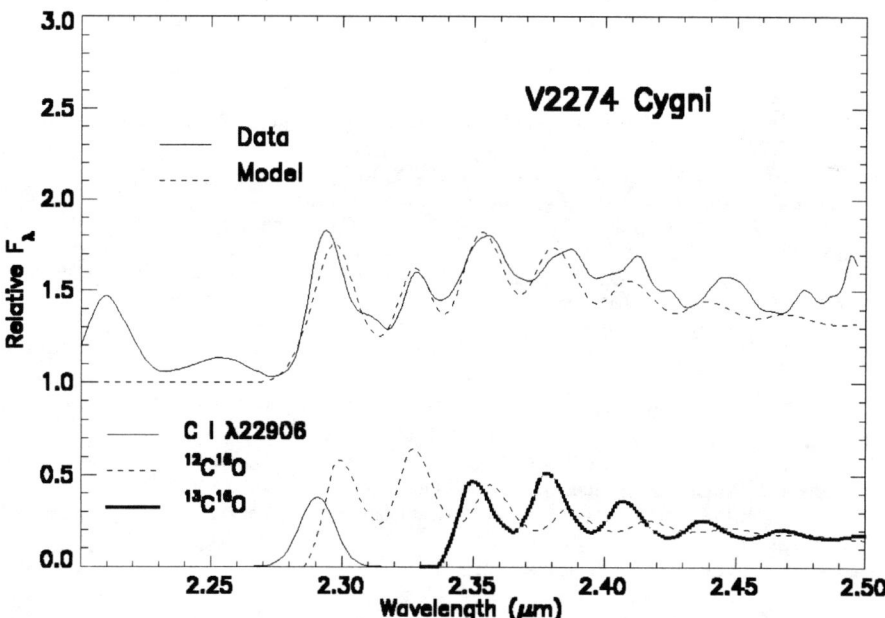

FIGURE 3. Fit to Nova Cyg 2001 spectrum by NLTE CO band model. The best fit results in $^{12}C^{16}O/^{13}C^{16}O \approx 3$.

CONCLUSIONS

Novae are a very inhomogeneous class of objects whose spectral variations are a function of time after outburst, mass of the white dwarf and secondary, orbital period and structure in the accretion disk. Despite a large amount of data, modeling novae is challenging because of spatial variations in the shell and lack of thermodynamic equilibrium. Future studies should include more frequent observations and observations at high spectral resolution. These will be particularly important in assessing the properties of the shell during the early phases when molecules are present, and during the latter phases when coronal lines dominate.

ACKNOWLEDGMENTS

This work supported by NASA, The Aerospace Corporation's Independent Research and Development program and by the US Air Force Space and Missile Systems Center through the Mission Oriented Investigation and Experimentation program, under contract F4701-00-C-0009 with the US Air Force. ATR 2002 (8189)-3

REFERENCES

1. Cox, A.N. (ed) *Astrophysical Quantities*, AIP Press, New York, 2000. 4th edition.
2. Dinerstein, H. L. 2001 *ApJL* **500**, L223
3. Ferland, G. J., Lambert, D. L., & Woodman, J. H. 1986 *ApJS* **60**, 375
4. Greenhouse, M. A., Grasdalen, G. L., Woodward, C. E., Benson, J., Gehrz, R. D., Rosenthal, E., Skrutskie, M. F. 1990 *ApJ* **352**, 307
5. Greenhouse, M. A., Feldman, U., Smith, H. A., Klapisch, M., Bhatia, A. K., & Bar-Shalom, A. 1993, *ApJS* **88**, 23
6. Grasdalen, G. L., & Joyce, R. R. 1976, *Nature*, **259**, 187
7. Iijima, T., Rosino, L., & Della Valle, M. 1998 *A&A* **338**, 1006
8. Lynch, D.K., J. C. Wilson. N. A. Miller. R. J. Rudy, C.C. Venturini, S. Mazuk and R. C. Puetter, 2003 (in preparation)
9. Mason, C. G., Gherz, R. D., Woodward, C. E., Smilowitz, J. B., Greenhouse, M. A., Hayward, T. L., & Houck, J. R. 1996 *ApJ* **470**, 577
10. Rudy, R. J., Puetter, R. C., & Mazuk, S. 1999 *AJ* **118**, 666
11. Rudy, Richard J., Catherine C. Venturini, David K. Lynch, S. Mazuk, and R. C. Puetter 2002a ApJ (in press)
12. Rudy, R.J. W. L Dimpfl, D.K. Lynch and J.C. Wilson, 2002b *ApJ* (in press)
13. Wagner, R. M., & DePoy, D. L. 1996, *ApJ* **467**, 86
14. Wilson, J. C., Skrutskie, M. F., Colonno, M. R., Enos, A. T., Smith, J. D., Henderson, C. P., Gizis, J. E., Monet, D. G., Houck, J. R. 2001 *Pub. A.S.P*, 113, 780, 227-239.
15. Williams, R.E, Hamuy, M., Phillips, MM., Heathcote, S.R., Wells, L.. and Navarette, M. 1991 *ApJ* **376**, 721-737
16. Woodward, C. E. et al. 1995, *ApJ* **438**, 921

Table I

Known and Expected Coronal Lines Between 0.8 and 2.5 Microns

λair (Å)	λair (vac)	ID	Terms	Ji-Jk	Lower cm-1	Upper cm-1	Ionization Potential
8154	8156	[Cr XII]	2P°-2P°	1/2-3/2	0	12261	270.7
8310	8313	[Co XIV]	3P-3P	0-1	0	12030	379
8325	8327	[Ar XIII]	3P-3P	1-2	9859	21893	618
9215	9217	[Cl X]	3P-3P	2-1	0	10849	400.1
9913	**9916**	**[S VIII]**	**2P°-2P°**	**3/2-1/2**	**0**	**10085**	**280.95**
9978	9981	[Mn X]	3P-3P	2-1	0	10019	221.8
10106	10109	[Cr VIII]	2P°-2P°	3/2-1/2	0	9892	161.18
10140	10143	[Ar XIII]	3P-3P	0-1	0	9859	618
10308	10310	[P XI]	2P°-2P°	1/2-3/2	0	9699	424.4
10311	10314	[V XI]	2P°-2P°	1/2-3/2	0	9696	230.5
10669	10672	[Cl XII]	3P-3P	1-2	7240	16629	529.3
10747	10750	[Fe XIII]	3P-3P	0-1	0	9302.5	331
10798	10801	[Fe XIII]	3P-3P	1-2	9302.5	18561	331
12520	**12523**	**[S IX]**	**3P-3P**	**2-1**	**0**	**7985**	**328.75**
12783	12786	[Cr IX]	3P-3P	2-1	0	7821	184.7
12817	12821	[Mn XII]	3P-3P	1-2	7200	15000	286
13038	13041	[V VII]	2P°-2P°	3/2-1/2	0	7668	128.1
13250	**13257**	**[Ti X]**	**2P°-2P°**	**1/2-3/2**	**0**	**7543**	**192.1**
13746	**13749**	**[P VII]**	**2P°-2P°**	**3/2-1/2**	**0**	**7273**	**220.42**
13836	13840	[Cl XII]	3P-3P	0-1	0	7240	529.3
13885	13889	[Mn XII]	3P-3P	0-1	0	7200	286
13924	**13927**	**[S XI]**	**3P-3P**	**1-2**	**5208**	**12388.1**	**447.5**
14221	14225	[Mn XVIII]	3P-3P	0-1	66560	73590	
14301	**14305**	**[Si X]**	**2P°-2P°**	**1/2-3/2**	**0**	**6990.6**	**351.1**
15514	15518	[Cr XI]	3P-3P	1-2	5536	11980	244.4
16641	16645	[V VIII]	3P-3P	2-1	0	6007.8	150.6
17151	**17156**	**[Ti VI]**	**2P°-2P°**	**3/2-1/2**	**0**	**5829**	**99.3**
17353	17358	[Sc IX]	2P°-2P°	1/2-3/2	0	5761.1	158.1
17356	**17361**	**[P VIII]**	**3P-3P**	**2-1**	**0**	**5760**	**263.57**
18059	18064	[Cr XI]	3P-3P	0-1	0	5536	244.4
18676	18681	[P X]	3P-3P	1-2	3692	9045	372.1
19075	19080	[V X]	3P-3P	1-2	4180	9421	205.8
19196	19201	[S XI]	3P-3P	0-1	0	5208	447.5
19641	**19646**	**[Si VI]**	**2P°-2P°**	**3/2-1/2**	**0**	**5090**	**166.77**
19662	19667	[Zn XII]	2D-2D	3/2-5/2	0	5095	274
20445	**20450**	**[Al IX]**	**2P°-2P°**	**1/2-3/2**	**0**	**4890**	**284.66**
22050	**22056**	**[Ti VII]**	**3P-3P**	**2-1**	**0**	**4534**	**119.53**
22578	22584	[Ca XIII]	3P-3P	1-0	24460	28888	657
23205	**23211**	**[Ca VIII]**	**2P°-2P°**	**1/2-3/2**	**0**	**4308.3**	**127.2**
23504	**23560**	**[K XII]**	**3P-3P**	**1-0**	**18954**	**23207**	**564.7**
23917	23923	[V X]	3P-3P	0-1	0	4180	205.8
23935	23941	[Sc XIV]	3P-3P	1-0	31174	35351	757
24015	24021	[Ti IX]	3P-3P	1-2	3119	7282	170.4
24673	24680	[Cu XI]	2D-2D	3/2-5/2	0	4060	232
24807	**24814**	**[Si VII]**	**3P-3P**	**2-1**	**0**	**4030**	**205.27**

Data from NIST (http://physics.nist.gov/cgi-bin/AtData/main_asd). Ionization potentials from Cox (2000). Lines that are **bolded** have been seen in novae.

Spectral evolution of galactic novae

Elena Mason*, Massimo Della Valle[†] and Antonio Bianchini**

*ESO, Alonso de Cordova 3107, Vitacura, Casilla 19001, Santiago, Chile
[†]Osservatorio Astrofisico di Arcetri, L.go E. Fermi 5, 50125 FI, Italy
**Dipartimento di Astronomia, Università di Padova, Vicolo dell'osservatorio 5, 35122 PD, Italy

Abstract. We present low resolution spectroscopy of five galactic novae occurred in the years 1991 and 1992. Spectra spread over several months since outburst and show the spectroscopic evolution of the ejecta. We classify each nova according to the Cerro Tololo scheme. The spectroscopic classification is then considered in relation to the nova photometric evolution aiming to provide further evidence for a two nova populations in our Galaxy.

INTRODUCTION

In the '90s spectroscopic surveys by Williams et al. [1, 2] allowed the identification of the following spectroscopic classes of novae: i) the *Fe* novae, ii) the *He/N* novae, and iii) the *Fe-b* or *Hybrid*.

The observation that fast and slow novae have different spatial distribution within our Galaxy together with the newly identified spectroscopic classes, triggered the search for possible correlation between photometric and spectroscopic classification, thus for intrinsically different nova populations within the Galaxy.

Della Valle and Livio [3] first provide evidence for such a correlation, and defined the two following nova classes: i) *Fast novae*: bright maximum ($M_V < -9$), fast decline ($t_2 < 12$ and $t_3 < 20$ days), typically characterized by He/N spectra at maximum and small height above the galactic plane (≤ 200 pc), ii) *Slow novae*: relatively weak maximum magnitude ($M_V < -7.5$), $t_3 > 20$-25 days, FeII as the most prominent non Balmer lines at maximum, and larger heights above the galactic plane (up to ~ 1000 pc).

Data presented here are unpublished spectra of five galactic novae occurred in 1991 and 1992 (Nova Her 1991, Nova Pup 1991, Nova Sgr 1991, Nova Sgr 1992 N.1, Nova Sgr 1992 N.2), which were analyzed to provide evidence of such two galactic nova populations.

DATA

For each object we present both the spectroscopic and photometric evolution. Spectroscopic data were taken at La Silla telescopes during target of opportunity time. The number of spectra taken vary from object to object depending on their brightness, visibility, and the telescope schedule. A detailed log of the observation for the spectra presented here is in Table 1. Light curves and photometric evolution of each object were derived

TABLE 1. The observed novae and their log of observations

object	obs. date	days since max	telescope + inst.	grism/grating	exptime (sec)
V838 Her*	28/03/91	5	1.5+B&C	grt21	20,120,25
	10/04/91	18	1.5+B&C	grt21	300
	05/07/91	104	3.6+EFOSC	B300,R300	900,600
	30/05/93	798	NTT+EMMI	grs2	3600
V4160 Sgr	01/08/91	3	2.2+EFOSC	grs4,7,8,9	180,300,240,180
V351 Pup	09/01/92	13	2.2+EFOSC	grs3,4	60,15
	04/03/95	1162	NTT+EMMI	grs2,5,6	300,600,600
V4157 Sgr	22/10/92	251	2.2+EFOSC	grs6	300
	02/07/93	504	NTT+EMMI	grs5	1800
	13/08/94	911	1.5+B&C	grt21	2×3600
V4169 Sgr	22/10/92	104	2.2+EFOSC	grs3,5,6	300,180,30
	02/07/93	357	NTT+EMMI	grs5	1800
	13/08/94	764	1.5+B&C	grt21	3600

* For this object only a selection of the whole spectroscopic data set is presented. Thus, the *observed spectral evolution* in Table 2 is more complete than what shown in the Figure 1.

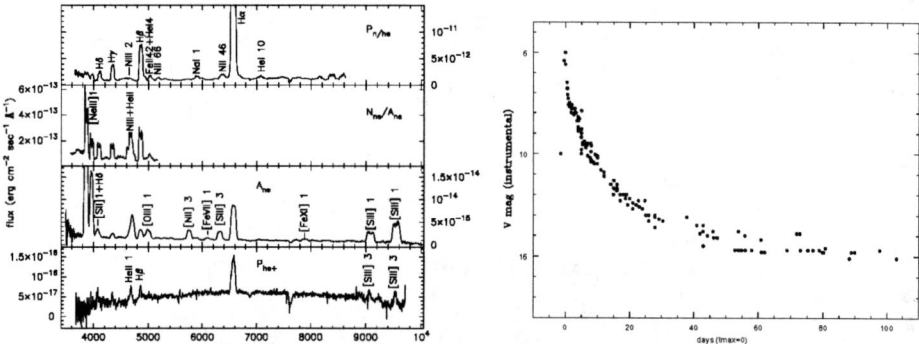

FIGURE 1. Spectroscopic (left) and photometric (right) evolution of nova Her 1991 (V838 Her).

combining photometric data by IAUC, AVVSO, and authors' broad band photometry.

Figures 1, 2, 3, 4, and 5 refer, respectively, to nova Her 1991, nova Sgr 1991, nova Pup 1991, nova Sgr 1992 N.1, and nova Sgr 1992 N.2. Each figure shows, on the left, the observed spectral evolution, and, on the right, the light curve. Spectra taken during the same run but with different set up have been combined in the figures. Light curves have been plotted using the same scale in the abscissa (140 days, with day 0 being the time of the observed maximum), to better show the different speed classes.

DISCUSSION

Line identification of the spectra at different epochs was used to classify each nova according to the Cerro Tololo scheme [1, 2]. The light curves were used to derive the

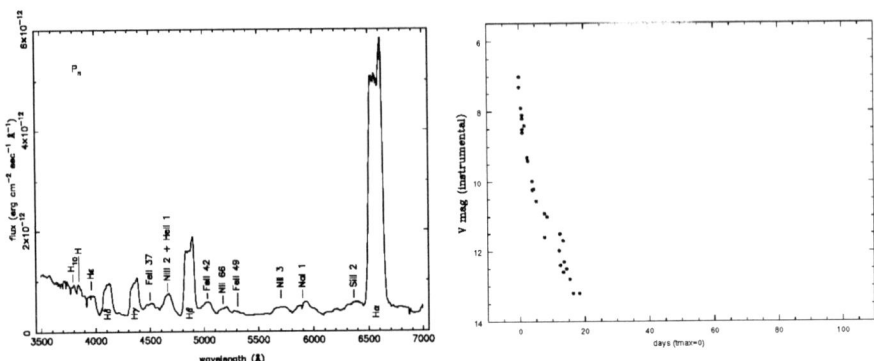

FIGURE 2. Spectroscopic (left) and photometric (right) evolution of nova Sgr 1991 (V4160 Sgr).

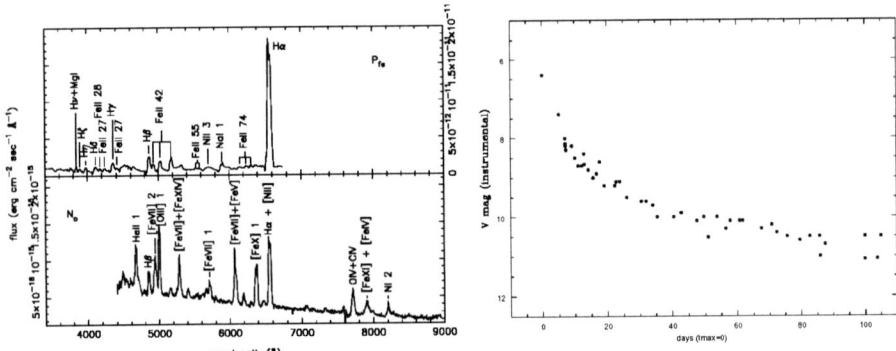

FIGURE 3. Spectroscopic (left) and photometric (right) evolution of nova Pup 1991 (V351 Pup).

times t_2 and t_3, thus the speed class of each nova.

In table 2 we report the information derived by the data analysis of each object. The absolute magnitude, M_V, was computed using the MMRD relation by Della Valle and Livio [4], and then used to derive the distance d and the galactic height z. Distances were computed using an average value for A_V from different determination either in the literature and by us [see 5, for details].

REFERENCES

1. Williams, R.E., Hamuy, M., Phillips, M., Heatcote, S., Wells, L., 1991, *ApJ*, **376**, 721
2. Williams, R.E., Phillips, M., Hamuy, M., 1994, *ApJS*, **90**, 297
3. Della Valle, M., Livio, M., 1998, *ApJ*, **506**, 818
4. Della Valle, M., Livio, M., 1995, *ApJ*, **452**, 704
5. Mason, E., 1996, master thesis, University of Padova

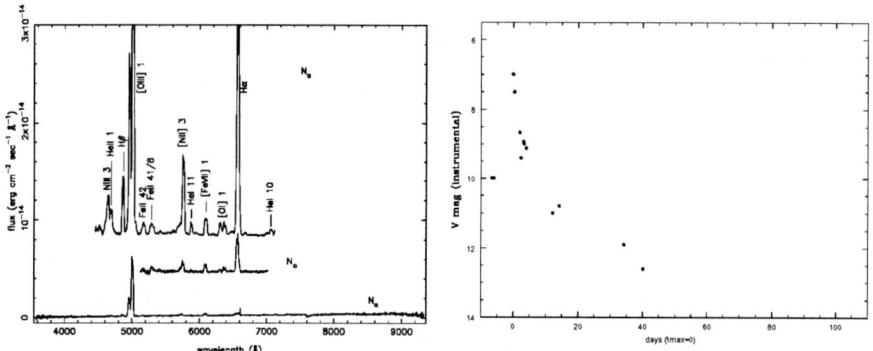

FIGURE 4. Spectroscopic (left) and photometric (right) evolution of nova Sgr 1992 N.1 (V4157 Sgr).

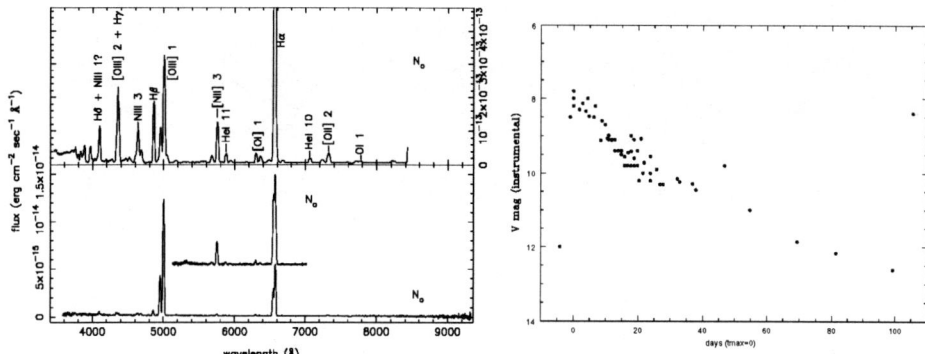

FIGURE 5. Spectroscopic (left) and photometric (right) evolution of nova Sgr 1992 N.2 (V4169 Sgr).

TABLE 2. Summary of the characteristics derived for each nova from the analysis of their spectra and light curves.

	V838 Her	V4160 Sgr	V351 Pup	V4157 Sgr	V4169 Sgr
l	43.32	0.20	252.72	5.32	4.56
b	6.62	-6.97	-0.73	-3.07	-6.96
$t_2(d)$	1.5	2	12.5	3.3	21.5
$t_3(d)$	4	4	25	9	21.5
$m_V(max)$	5.4	7.0	6.4	7.0	7.8
$M_V(max)$	-9.03	-9.01	-8.54	-8.97	-7.78
A_V	1.22±0.41	1.09±0.13*	2.25±0.54	2.66±0.56	1.19±0.34
$d(kpc)$	4.4	9.6	3.5	4.6	7.9
$z(pc)$	507	1164	44	246	957
spectral ev.[†]	$P_{fe,he,(ne)}$ N_{ne} A_{ne} P_{he}	P_n (C_{ne}) (A_{ne})	P_{fe} (P_n^o) N_o (N_{ne}) (A_{ne})	($P_{fe,n}$) N_o (A_n)	N_o
FWHM (km/s)	2500	2500	1500	2500	2500
class	Fast&Hybrid	Fast&He/N	Slow&FeII	Fast&FeII	Slow&FeII

* This unlikely small value for nova Sgr 1991 results from the average of only two determinations. The computation based on the EW of the NaI interstellar absorption line is certainly biased by the nearby HeI emission of the maximum spectrum.

[†] Spectral phases reported within parenthesis refers to observations in the literature and not by us.

A photoionization model analysis of the ONeMg nova QU Vul

Greg J. Schwarz

University of Arizona, Steward Observatory

Abstract. This paper presents a new analysis of the moderately fast ONeMg nova QU Vul 1984 using published ultraviolet, optical, infrared and radio observations. The technique uses an optimization code in combination with Cloudy photoionization models to fit the emission line spectra obtained at different epochs during the nebular phase of the outburst. Previous studies of QU Vul yielded considerable differences in the derived elemental abundances, ejected mass and distance. This analysis, which is consistent with all the available data and over multiple epochs, generally finds the lowest abundance enhancements of the previous studies. The distance most compatible with the all of the available distance determinations is 2.4 kpc. At this distance, the ejected mass is extremely high with a value of $\sim 6 \times 10^{-4}$ $M_\odot$. Coupling these factors with the long lightcurve decay time and nuclear burning timescale implies that the outburst occurred on a low mass ONeMg white dwarf.

INTRODUCTION

QU Vul was discovered on 22.13 December 1984 and reached maximum visual magnitude of ~ 5.6 three days later. The decline from maximum was "fast" with a t_2 of 25 days. QU Vul was widely observed over many wavelengths from X-ray to radio. This makes it an ideal candidate to model since the multiwavelength observations provide important constraints on the model parameters.

The Cloudy 94.00 photoionization code [1] is used to model the observed UV to infrared emission line ratios given in Sazair et al. [2] and Greenhouse et al. [3]. Cloudy simultaneously solves the equations of thermal and statistical equilibrium for a model emission nebula. The predicted emission line fluxes are compared against the observed nebular spectra. The χ^2 fit between the observed and the predicted emission lines is then accessed by an optimization code, MINUIT 94.1 [4], which searchs parameter space for the best agreement. The best fit Cloudy parameters for the modeled dates are given in Table 1.

RESULTS

The average abundances from this analysis and the other published abundances, relative to hydrogen, are given in Figure 1. For a given element, the earlier works generally span a large range in abundance with differences of a factor of two being common. The previous analyses only agree in a significantly greater elemental enhancement than was

TABLE 1. Best fit CLOUDY model parameters

Parameter	Day 470	Day 550	Day 630
T_{BB} ($\times 10^5$ K)	$3.2^{+0.1}_{-0.3}$	$3.2^{+0.1}_{-0.3}$	3.2 ± 0.2
Source luminosity ($\times 10^{37}$ erg s^{-1})	$2.5^{+0.6}_{-0.3}$	$3.7^{+0.5}_{-0.6}$	2.5 ± 0.3
Hydrogen density ($\times 10^6$ cm^{-3})	6.8 ± 0.8	$5.6^{+0.7}_{-0.2}$	$3.4^{+0.6}_{-0.2}$
Inner radius ($\times 10^{15}$ cm)	4.0	4.7	5.4
Outer radius ($\times 10^{16}$ cm)	2.0	2.3	2.7
He/He$_\odot$	$1.25^{+0.75}_{-0.30}$	$1.00^{+0.7}_{-0.25}$	$1.25^{+1.0}_{-0.3}$
C/C$_\odot$	0.2 ± 0.1	0.2 ± 0.15	$0.2^{+0.2}_{-0.1}$
N/N$_\odot$	11^{+3}_{-4}	13 ± 6	12^{+5}_{-6}
O/O$_\odot$	$2.5^{+0.5}_{-1.0}$	$2.5^{+1.5}_{-1.3}$	$2.5^{+1.8}_{-0.7}$
Ne/Ne$_\odot$	20^{+5}_{-8}	$20^{+10.0}_{-6.5}$	$25^{+6.5}_{-10.0}$
Mg/Mg$_\odot$	10 ± 6	10^{+10}_{-4}	10^{+2}_{-3}
Al/Al$_\odot$	68^{+22}_{-48}	24^{+16}_{-19}	68^{+32}_{-44}
Si/Si$_\odot$	$2^{+1.8}_{-1.3}$	$2^{+3.0}_{-1.5}$	2 ± 1.5
Ar/Ar$_\odot$	$0.3^{+0.3}_{-0.25}$	$0.25^{+0.15}_{-0.20}$	$0.3^{+1.0}_{-0.25}$
Fe/Fe$_\odot$	$0.5^{+0.4}_{-0.25}$	$0.5^{+0.25}_{-0.20}$	$0.6^{+0.4}_{-0.35}$
Ejected Mass ($\times 10^{-4}$ M$_\odot$)	7.4 ± 0.8	$9.7^{+0.4}_{-0.3}$	$9.1^{+1.0}_{-0.6}$
Total χ^2	51.9	44.4	46.2
Degrees of freedom	20	20	21

determined in this study.

Sazair et al. (hereafter S92) [2] and Andrea, Drechsel & Starrfield (hereafter ADS94) [5] assumed a homogenous ejecta and used Ionization Correction Factors (ICFs) to obtain their elemental abundances. The different solutions of these similar studies is due to the number of epoches modeled and the difference in the determination of the electron density and temperature. Given the restrictive assumptions employed in the ICF method, a more sophisticated analysis using a photoionization code was tried by Austin et al. (hereafter A96) [6]. A Metropolis algorithm was used to adjust the Cloudy 84.0 parameters to map the multi-dimensional χ^2 space without becoming stuck in local minima. Unfortunately, this analysis generally produced the highest elemental abundance estimates because it used significantly fewer emission lines from only one date to constrain the solution.

The enhancements in the intermediate-mass elements, particularly neon, magnesium and aluminum, are compatible with hydrodynamical models of outbursts on ONeMg white dwarfs (WDs). These models produce significant amounts of these elements from mixing with the WD and from nucleosynthesis during the TNR [7]. The low O/N ratio is consistent with a lower mass WD and the relatively slow decline in the visual lightcurve. The subsolar carbon abundance is surprising since the hydrodynamical models imply it should be created during a TNR. Adding a carbon dust mass of 7×10^{-7}M$_\odot$ [8] only increases the carbon number abundance by about a factor of 2.5. In addition, the oxygen and silicon abundances are also lower than expected when compared to the low mass ONe WD models of Jose & Hernanz [9] but this may reflect more of the particular parameters of the nucleosynthesis model.

The abundance solution from this study is an improvement over the previous analyses

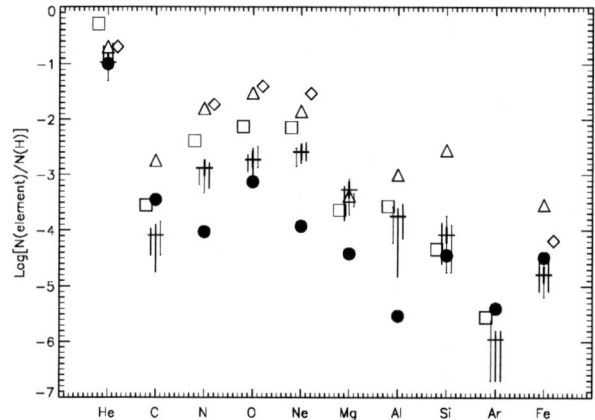

FIGURE 1. Comparison between the elemental abundances (by number) in the literature and this study. Some of the symbols are offset slightly to improve readablilty. The three vertical lines represent the abundance determined in this analysis. From left to right, the lines provide the elemental abundance range for the day 470, 550, and 630 models. The squares, triangles, and diamonds represent the S92, ADS94, and the A96 estimates, repectively, while the filled circle gives the corresponding solar values.

TABLE 2. Distance determinations

Method	Distance (kpc)
Della Valle & Livio (1995) MMRD [10] *	$1.9^{+0.9}_{-0.6}$
Quiescent distance modulus †	2.4
$L_{Eddington}$ **	~ 2
Photoionization model fits ‡	2.9±0.3

* Using t_2 = 25 days, $E(B-V)$ = 0.61, and V_{max} = 5.6.
† Assuming M_V = 5.3 ($i \sim 80°$) and $V \sim 19$.
** Using $L_{Edd} = 6.4 \times 10^4 \, M_{WD}/(1+X) \, L_\odot$, where M_{WD} = 1.4 $M_\odot$ and X = 0.6. Note that at 2.4 kpc and M_{WD} = 1.2 $M_\odot$ the luminosity is about twice the Eddington luminosity.
‡ Assuming $C = 1$ otherwise the distace scales as $C^{0.5}$.

because it uses an improved physical model constrained by data from the UV to the radio. The best models are also consistent over multiple dates obtained and reproduce a wide range of ionizations. This large dataset is very important since it reduces problems introduced by random observational errors and serves as a check on each date's solution.

Distances based on the expansion parallax of the ejecta range from 1.75 kpc [11] to 3.6 kpc [12]. This difference illustrates the problem of associating the correct velocity with the shell images. In addition, Wade, Harlow & Ciardullo [13] showed that these distance determinations can be biased due to projection effects as the assumption of a spheroidal ejection breaks down. In Table 2 four other methods of obtaining the distance are shown. The methods are the Maximum Magnitude vs Rate of Decay relationship of Della Valle

TABLE 3. Ejected masses from the literature and this analysis

Study	Wavelength	Mass (10^{-4} M$_\odot$)
Taylor et al. (1987) [14]	Radio	$8\times(D/3.6\text{ kpc})^{5/2}$
Taylor et al. (1988) [12]	Radio	$(3.6\pm0.5)\times(D/3.6\text{ kpc})^{5/2}$
Shin et al. (1998) [15]	IR	$(5-8)\times(D/3.6\text{ kpc})^{5/2}$
Greenhouse et al. (1988) [3]	IR	$>9\times(D/3.6\text{ kpc})^{5/2}$
Sazair et al. (1992) [2]	UV - Optical	$(0.2-15)\times(D/3.5\text{ kpc})^2$
This study	UV - Radio	$(8.7\pm1.2)\times(D/2.9\text{ kpc})^2$

and Livio [10], the distance modulus from the observed pre-outburst magnitude and the absolute, inclination corrected, quiescent magnitude, assuming that the nova radiated at its Eddington luminosity early in the outburst, and from model fits to the observed observed Hβ flux and radio observations. From these methods a distance of ~ 2.4 kpc is found to be the most consistent with all of the observations.

There are five ejecta masses for QU Vul in the literature with an average of about 1×10^{-3} M$_\odot$ (see Table 3). On the surface these mass estimates imply an extreme ejection event for QU Vul. However, each of the previous mass estimates is dependent on the distance. At 2.4 kpc the average of the published masses declines to $\sim 4\times10^{-4}$ M$_\odot$ and $\sim 6\times10^{-4}$ M$_\odot$ for this study (covering factor = 0.68). These lower mass determinations still present a challenge to nova theory which consistently predicts significantly lower masses for novae of this speed class.

The behavior of the UV emission lines provides an indirect probe of the underlying photoionization conditions and can be used to determine the turnoff time [16]. In Figure 2 the dereddened He II (1640Å) lightcurve for QU Vul using the IUE data (error bars) from Vanlandingham et al. [17] are presented and supplemented with dereddened He II (4686Å) data (filled circles) from the literature. The He II (4686Å) fluxes have been converted to He II (1640Å) using a 1640Å/4686Å case B recombination ratio of ~ 7, typical of nebular conditions. The new data clearly shows the departure in the evolution of the lightcurve expected from a turnoff of the photoionization source. The line in Figure 3 is the helium recombination model assuming a turnoff date of 1540 days and a electron density at turnoff derived from the best fit models. This turnoff time is significantly longer than those determined by Vanlandingham et al. for the other ONeMg novae in their analysis.

CONCLUSIONS

This paper present the results of a new analysis of the fast ONeMg nova QU Vul. The best fit models are consistent with the numerous observations obtained during the nebular phase which were obtained over a wide range of wavelengths and ionizations. Relative to other well studied ONeMg novae, QU Vul had a lower abundance enhancement, a more massive mass ejection event, a longer turnoff time, and one of the slowest optical declines. These results imply that the outburst occurred on a relatively low mass WD. An important test of this hypothesis would be a determination of the WD mass in this

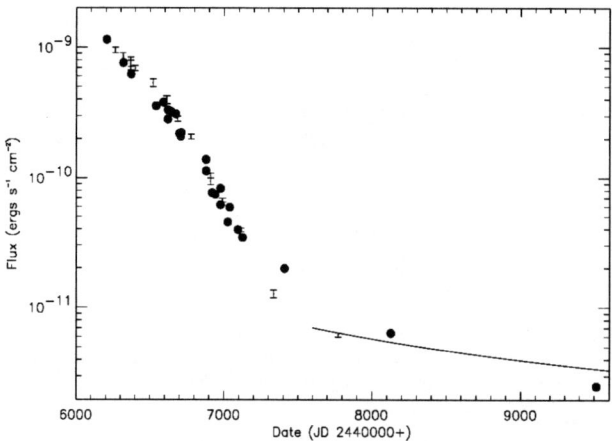

FIGURE 2. The dereddened He II (1640) lightcurve and recombination model. The error bars give the IUE fluxes and the filled circles are the converted optical fluxes.

system. The ejected mass estimate is also over an order magnitude greater than predicted by TNR models of ONeMg novae with gross properties similar to QU Vul. Given that the ejecta are also enriched with ≥ 10 times the solar abundances of nitrogen, neon, magnesium, and aluminum, these findings suggest that QU Vul-like outbursts may be important contributors to the local ISM and to the isotopic anomalies seen in meteorites.

REFERENCES

1. Ferland, G., Korista, K., Verner, D., Ferguson, J., Kingdon, J., and Verner, E., *PASP*, **110**, 761 (1998).
2. Sazair, P., Starrfield, S., Ferland, G., Wagner, J., R.M. Truran, Kenyon, S., Sparks, W., Williams, R., and Stryker, L., *ApJ*, **398**, 651 (1992).
3. Greenhouse, M., Grasdalen, G., Hayward, T., Gehrz, R., and Jones, T., *AJ*, **95**, 172 (1988).
4. James, F., and Roos, M., MINUIT, Tech. rep., CERN Program Library D506 (1993).
5. Andrea, J., Drechsel, H., and Starrfield, S., *A&A*, **291**, 869 (1994).
6. Austin, S., Wagner, R., Starrfield, S., Shore, S., Sonneborn, G., and Bertram, R., *AJ*, **111**, 869 (1996).
7. Starrfield, S., Schwarz, G., Truran, J., and Sparks, W., "The Effects of New Nuclear Reaction Rates and Opacities on Hydrodynamic Simulations of the Nova Outburs," in *Cosmic Explosions*, edited by S. Holdt, AIP Proceedings, 2000, p. 212.
8. Smith, C., Aitken, D., Roche, P., and Wright, C., *MRAS*, **277**, 259 (1995).
9. Jose, J., and Hernanz, M., *ApJ*, **494**, 680 (1998).
10. Della Valle, M., and Livio, M., *ApJ*, **452**, 704 (1995).
11. Downes, R., and Duerbeck, H., *AJ*, **120**, 2007 (2000).
12. Taylor, A., Hjellming, R., Seaquist, E., and Gehrz, R., *Nature*, **335**, 235 (1988).
13. Wade, R., Harlow, J., and Ciardullo, R., *PASP*, **112**, 614 (2000).
14. Taylor, A., Seaquist, E., Hollins, J., and Pottasch, S., *A&A*, **183**, 38 (1987).
15. Shin, J.-Y., Gehrz, R., Jones, T., Krautter, J., and Heidt, J., *AJ*, **116**, 1966 (1998).
16. Shore, S., Starrfield, S., and Sonneborn, G., *ApJL*, **463**, 21 (1996).
17. Vanlandingham, K., Schwarz, G., Shore, S., and Starrfield, S., *AJ*, **121**, 1126 (2001).

Elemental Abundance Analysis of Nova Cygni 1992

Karen M. Vanlandingham*, Sumner Starrfield[†], Steve N. Shore** and R. Mark Wagner[‡]

*Columbia University
[†]Arizona State University
**Indiana University South Bend
[‡]Arizona State University, Steward Observatory

Abstract. We present the initial results of our analysis of V1974 Nova Cygni 1992 (Cyg 92). Cyg 92 was discovered on 1992 February 19.1. It was a fast ONeMg nova, with $t_2 \sim 15$ days and $t_3 \sim 45$ days. Austin et al. [1] determined the reddening for the nova to be $E(B-V) = 0.36 \pm 0.04$ and the distance to be 2.8 ± 0.7 kpc. We have spectroscopic observations of Cyg 92 in both the visible and ultraviolet. Optical observations were taken from 4 to 540 days after outburst. The UV observations, made by IUE, cover 1 to 500 days after outburst. We have combined the photoionization code CLOUDY 94 and the minimization routine MINUIT to determine physical parameters and elemental abundances for the nova ejecta, as we have done for several other ONeMg novae in the past [2] [3] [4] [5]. Our results for Cyg 92 show enhanced abundances, relative to solar material, for helium, nitrogen, oxygen, neon, magnesium, aluminum, silicon, and sulfur. The carbon abundance is approximately solar and iron is slightly below solar. These results are based on independent fits to three different sets of optical+UV observations taken at 300, 400, and 500 days after outburst. Comparing our results with those of previous analyses of Austin et al. [1] and Moro-Martin et al. [6], we find that our abundances are significantly lower than the others. The analysis by Austin et al. [1] suffered from an error in the dereddening of the UV spectra, resulting in inaccurate line flux ratios. Unfortunately, Moro-Martin used Austin's abundances as a starting point for their analysis. While many of the abundances obtained by Moro-Martin are lower than those of Austin's they are still generally much larger than our values, particularly for oxygen and neon. We believe an artificially high helium abundance, resulting from starting with Austin's values, lead to their higher abundances.

METHOD

We have used the photoionization code CLOUDY 94.0 [7] in combination with the minimization routine MINUIT [8] in order to determine the physical parameters of Nova Cyg 92. CLOUDY simultaneously solves the equations of statistical and thermal equilibrium. Given values of the physical conditions of the gas (ionization, density, temperature, and chemical composition), CLOUDY predicts the resulting emission-line spectrum. We initially used individual CLOUDY runs, adjusting the parameters between the runs, to obtain a rough fit to the observations. Once a rough fit was obtained we then used MINUIT to fine-tune the model. MINUIT uses the Davidson-Fletcher-Powell (DFP) variable-metric algorithm, a quasi-Newton's method. MINUIT uses the function values and their derivatives to find the steepest slope in parameter space and thus the quickest descent to the minimum. Given a set of initial parameters, CLOUDY predicts the emission line fluxes. The predicted line fluxes are compared to the observed fluxes and the χ^2 is calculated. The value of the χ^2 is then passed back to MINUIT, which chooses a new set of parameters which best minimize the χ^2. The χ^2 is calculated as

$$\chi^2 = \sum \left(\frac{O_i - M_i}{\sigma_i}\right)^2 \qquad (1)$$

where M_i is the predicted line flux ratio, O_i is the observed line flux ratio, and σ_i is the error in the measurement of the observed line flux for each line. In order to ensure the minimum found is a true, global minimum, as opposed to a local minimum, we use multiple observations on different dates. If all dates converge to the same abundance values then the solution is more reliable.

EJECTED MASS

We can calculate the ejected mass of hydrogen based on the densities and abundances we find from our three modeled dates. These values give an ejected hydrogen mass of 2.2×10^{-4} $M_\odot$. Shore et al. [9] derived an ejected mass of $10^{-4} Y^{-1/2} M_\odot$ where Y is the helium abundance enhancement factor. Using our average helium abundance, this equation gives an ejected mass of $8.4 \times 10^{-5} M_\odot$. Both of these values are larger than what is predicted by current hydrodynamical models for ONeMg novae [10]. This is a long-standing problem with the hydro codes and indicates that the physics of the nova outburst is still not fully understood.

CONCLUSIONS

Our initial analysis of Cyg 92 gives abundance values that are lower than those found by other groups. Much of the discrepancy is due to the error in reddening correction made by Austin et al. [1], whose abundances were used as starting points for subsequent studies. Shore et al. [11] analyzed much later spectra taken by the GHRS on HST and found a carbon abundance that was 10 times lower than the Austin et al. value. The abundances we find, however, are still enhanced relative to solar values and show trends that agree with those predicted by the hydrodynamical codes for ONeMg novae [12]. In addition, the results for Cyg 92 fall in line with those of other ONeMg that we have analyzed using the same techniques.

REFERENCES

1. Austin, S., Wagner, R., Starrfield, S., Shore, S., Sonneborn, G., and Bertram, R., *AJ*, **111**, 869 (1996).
2. Vanlandingham, K., Starrfield, S., Shore, S., Wagner, R., and Sonneborn, G., *MNRAS*, **282**, 563 (1996).
3. Vanlandingham, K., Starrfield, S., and Shore, S., *MNRAS*, **290**, 87 (1997).
4. Vanlandingham, K., Starrfield, S., Shore, S., and Sonneborn, G., *MNRAS*, **308**, 577 (1999).
5. Schwarz, G., *ApJ* (2002).

TABLE 1. Nova Cygni 1992 - new CLOUDY fits

Parameter	Day 300	Day 400	Day 500	Austin	Moro-Martin [*]
$\log(T_{BB})$	5.6036	5.609	5.633		
$\log(L)$	38.006	38.053	38.021		
$\log(H_{den})$	7.2490	6.9834	6.7896		
α	-3.0	-3.0	-3.0		
$\log(R_{in})$	15.317	15.44	15.54		
$\log(R_{out})$	15.715	15.84	15.94		
Fill	0.11	0.11	0.114		
Power	0.0	0.0	0.0		
He	1.4 (3)	1.3 (3)	1.5 (3)	4.4	4.5
C	1.2 (2)	0.9 (2)	1.0 (2)	-	70.6[†]
N	46.0 (6)	56.0 (6)	46.9 (6)	282	50.0
O	19.6 (7)	21.0 (8)	15.0 (8)	110	80.0
Ne	64.5 (9)	57.0 (9)	66.2 (9)	250	250.0
Mg	2.3 (1)	0.5 (2)	9.4 (3)	-	129.4[†]
Al	55.6 (2)	66.2 (0)	60.1 (2)	-	127.5[†]
Si	47.3 (1)	34.4 (1)	49.0 (2)	-	146.6[†]
S	26.3 (1)	0.3 (1)	85.0 (1)	-	1.0
Fe	0.9 (1)	0.6 (2)	2.0 (1)	16	8.0
χ^2	81.9	115.1	103		
Total number of lines	41	38	37		
# of free parameters	15	14	15		
Reduced χ^2	26	24	22		

[*] Started with Austin et al's [1] values
[†] No lines in spectra

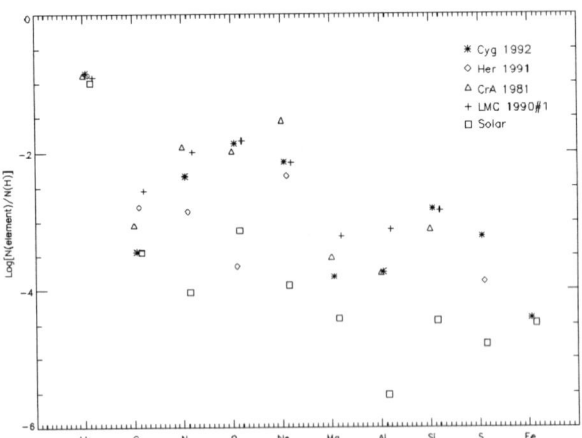

FIGURE 1. Plot of the best values of the elemental abundances as determined from the three dates we analyzed. For comparison the solar values have been plotted, as well as those of three other ONeMg novae which we have previously analyzed using the same methods.

6. Moro-Martin, A., Garnavich, P., and Noriega-Crespe, A., *AJ*, **121**, 1636 (2001).
7. Ferland, G., Korista, K., Verner, D., Ferguson, J., Kingdon, J., and Verner, E., *PASP*, **110**, 761 (1998).
8. James, F., and Roos, M., MINUIT, Tech. rep., CERN Program Library D506 (1993).
9. Shore, S., Sonneborn, G., Starrfield, S., Riestra-Gonzales, R., and Ake, T., *AJ*, **106**, 2408 (1993).
10. Starrfield, S., Wiescher, M., and Sparks, W., *MNRAS*, **296**, 502 (1998).
11. Shore, S., Starrfield, S., Ake, T., and Hauschildt, P., *ApJ*, **490**, 393 (1997).
12. Starrfield, S., Shore, S., Sparks, W., Sonneborn, G., Truran, J., and Politano, M., *ApJ*, **391**, 71 (1992).

The INES Guide for Classical Novae

R. González–Riestra* and A. Cassatella[†]

XMM Science Operations Center, VILSPA, Madrid, Spain
[†]*CNR, Istituto di Astrofisica Spaziale and Dipartimento di Fisica E. Amaldi, Universita Roma Tre, Roma, Italy*

Abstract. IUE ULDA/INES Access Guides have been published by ESA since 1989 with the purpose of facilitating the use of IUE data to scientists interested in a specific class of astronomical objects. We present here the "INES Guide for Classical Novae", in which we shall collect all the available IUE data for a total of 36 novae in outburst and 20 old novae. The Guide will include basic information about each object, observing logs, visual light curves, and representative high and low dispersion spectra.

THE ULDA AND INES ACCESS GUIDES

The value of archival data was recognized from the very beginning of the IUE project. Already in 1989 ESA started to publish the IUE–ULDA Access Guides, a series of of subject–oriented books in which a specialist in the field collected all the IUE low–resolution spectra, and compiled useful related information with the aim of facilitating the use of the data to scientists not familiar with the specific subject.

Eight ULDA guides were published in the period 1989–1996. After the availability of the INES (IUE Newly Extracted Spectra, Wamsteker et al. 2000) System, it was decided to continue with the publication of similar guides. At the time of writing, two INES Access Guides have been published, and three more are in preparation (see References).

THE INES GUIDE FOR CLASSICAL NOVAE

A total of 36 classical novae were observed with IUE during their outbursts. To this dataset we must add twenty old novae (see Tables 1 and 2). There is a large scatter both in the number of pointings and in the time period covered by the observations for the different objects. While some of them were observed only at the time of the outburst, in other cases the observations spanned several years (more than 11 for GQ Mus). The record in the number of spectra is for Nova Cygni 1992: 327 along three years.

The IUE–INES archive contains around 2000 spectra of novae ($\approx$ 80% taken in low dispersion mode). Most of the low resolution spectra have been extensively used, but only a limited fraction of the high resolution spectra have been analyzed. Despite the poor sampling of some of the objects, the large amount of data collected has made possible to derive a general picture of the the evolution of a nova in the UV during outburst (see González-Riestra and Krautter, 1998, for a review).

TABLE 1. Novae in Outburst included in the INES Guide

Name	Year	Name	Year	Name	Year	Name	Year
V1500 Cyg	1975	QU Vul	1984	LMC 1990 a	1990	LMC 1992	1992
LW Ser	1978	OS And	1986	V351 Pup	1991	V992 Sco	1992
V1668 Cyg	1978	V842 Cen	1986	LMC 1991	1991	V4327 Sgr	1993
V693 CrA	1981	LMC 1988 a	1988	V4160 Sgr	1991	V705 Cas	1993
V4077 Sgr	1982	LMC 1988 b	1988	V838 Her	1991	V723 Cas	1995
V1370 Aql	1982	QV Vul	1987	V444 Sct	1991	LMC 1995	1995
GQ Mus	1983	V827 Her	1987	V4169 Sgr	1992	V888 Cen	1995
MU Ser	1983	V443 Sct	1989	V4171 Sgr	1992	BY Cir	1995
PW Vul	1984	V977 Sco	1989	V1974 Cyg	1992	V1425 Aql	1995

TABLE 2. Old Novae included in the INES Guide

Name	Year	Name	Year	Name	Year	Name	Year
CK Vul	1670	X Ser	1903	HR Lyr	1919	CP Pup	1942
V841 Oph	1848	DI Lac	1910	RR Pic	1925	DK Lac	1950
Q Cyg	1876	DN Gem	1912	DQ Her	1934	V446 Her	1960
T Aur	1892	GI Mon	1918	CP Lac	1936	V533 Her	1963
GK Per	1901	V603 Aql	1918	BT Mon	1939	HR Del	1967

The IUE spectra of classical novae form an unique dataset. It constitutes a valuable reference for the study of the multifrequency behaviour of these objects, and for the planning and analysis of the observations of novae in the UV range with future experiments. We therefore considered that the compilation of a "Classical Novae" INES Access Guide would be of great help to researchers in the field.

The "Classical Novae" INES Access Guide will provide the following information for each object (see Figs. 1, 3 and 3):

- Basic parameters (whenever available): coordinates, date of outburst, date of visual maximum, typical decay times t_2 and t_3, and orbital period, including bibliographical references.
- List of the different names of the object, obtained from CDS.
- IUE log of observations, including Image number, dispersion, aperture, exposure time, day and date, day after outburst, FES visual magnitude and Exposure Classification Code.
- List of publications which have made use of IUE data.
- Plots of representative spectra at different phases of the outburst.

Work is currently in progress, and we expect that the Guide will we published shortly. We also plan to produce a WEB-based version of the guide, which will allow an interactive usage, inspection of individual spectra and on–line access to bibliography.

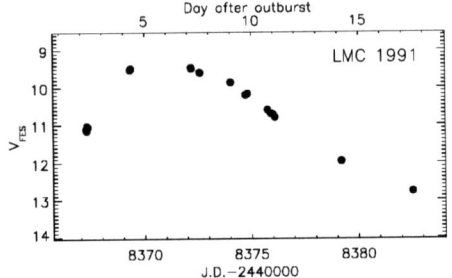

FIGURE 1. Example of the contents of the INES Guide: Basic Parameters

ACKNOWLEDGMENTS

The data included in this Guide have been obtained from the INES Principal Centre (http://ines.laeff.esa.es), at the Laboratory for Astrophysics and Fundamental Physics (LAEFF) of the Spanish National Institute for Aerospace Technology (INTA).

REFERENCES

1. Cappellaro, E. et al.,"IUE-ULDA Access Guide for Supernovae", ESA-SP 1189 (1995)

Image	D	A	Date	U.T.	H.J.D.$_{mid}$ -2440000.	Day after Outburst	V$_{FES}$	T$_{exp}$ (sec)	ECC
LWP20237	L	L	1991-04-26	02:35:16	8372.60943	7.609	9.60	300	562
LWP20244	L	L	1991-04-27	12:38:34	8374.02754	9.028	9.86	150	452
SWP41491	L	L	1991-04-27	12:48:18	8374.04037	9.040	9.87	1200	530
SWP41494	L	L	1991-04-28	05:06:59	8374.71828	9.718	10.20	900	451
LWP20252	H	L	1991-04-28	06:41:14	8374.82227	9.822	10.17	7560	332
SWP41506	H	L	1991-04-29	03:50:09	8375.75764	10.758	10.60	16920	332
LWP20265	L	L	1991-04-29	08:16:15	8375.84591	10.846		240	551
LWP20266	L	L	1991-04-29	09:34:55	8375.90037	10.900	10.70	210	452
LWP20267	H	L	1991-04-29	10:23:58	8376.00961	11.010	10.72	13200	
SWP41513	L	L	1991-04-29	14:13:35	8376.09823	11.098	10.80	960	430
LWP20289	L	L	1991-05-02	16:28:50	8379.18784	14.188	11.96	210	342
SWP41550	L	L	1991-05-02	16:40:21	8379.20018	14.200	11.96	960	330
LWP20317	L	L	1991-05-06	00:27:40	8382.52336	17.523	12.77	720	352
SWP41580	L	L	1991-05-06	00:43:29	8382.54059	17.541		1800	341
LWP20318	L	L	1991-05-06	01:28:34	8382.57190	17.572		1800	562
SWP41581	L	L	1991-05-06	02:07:13	8382.60673	17.607		3180	451
SWP41675	L	L	1991-05-23	00:22:47	8399.58195	34.582		11400	260
LWP20428	L	L	1991-05-23	03:39:16	8399.71666	34.717		11100	251
SWP41829	L	L	1991-06-13	09:52:59	8420.97284	55.973		10500	51
LWP20638	L	L	1991-06-18	22:31:29	8426.54312	61.543		18000	332
SWP41870	L	L	1991-06-19	03:36:28	8426.67678	61.677		4500	240
LWP20866	L	L	1991-07-22	20:13:35	8460.51005	95.510		28800	337
SWP42116	L	L	1991-07-23	04:41:26	8460.82106	95.821		21600	2X4
SWP42123	L	L	1991-07-25	10:27:48	8462.97306	97.973		6300	56
SWP42294	L	L	1991-08-20	10:37:33	8488.97116	123.971		4800	41
SWP42369	L	L	1991-09-01	11:51:31	8501.02943	136.029		6000	44
SWP42581	L	L	1991-09-29	04:45:05	8528.74006	163.740		7200	251
SWP42927	L	L	1991-10-28	10:07:15	8557.97740	192.977		9600	52
LWP21675	L	L	1991-11-10	01:47:31	8570.59723	205.597		3900	03
SWP43310	L	L	1991-12-03	18:22:53	8594.30730	229.307		7200	31
SWP43311	L	L	1991-12-03	20:55:39	8594.45331	229.453		14100	53
SWP43902	L	L	1992-01-31	12:20:18	8653.12805	288.128		19800	345
SWP44230	L	L	1992-03-25	12:01:17	8707.14282	342.143		24600	35

IUE Papers

Nova in the Large Magellanic Cloud 1991 (1991)
González-Riestra, R., Clavel, J., Cassatella, A.
IAU Circular 5253

Nova in the Large Magellanic Cloud 1991 (1991)
Shore, S.N., Starrfield, S.G., Sonneborn, G.
IAU Circular 5257

Nova in the Large Magellanic Cloud 1991 (1991)
Della Valle, M., Leisy, P., McNaught, R.H., Savage, A., Hartley, M., Hughes, S.M., Garradd, G.
IAU Circular 5260

FIGURE 2. Example of the contents of the INES Guide: Observing Log and Bibliography.

2. Cassatella, A. and González–Riestra, R., "INES Access Guide for Classical Novae", in preparation
3. Courvoisier, T. J.-L.and Paltani, S., "IUE-ULDA Access Guide for Active Galactic Nuclei", ESA-SP 1153 (1992)
4. la Dous, C., "IUE-ULDA Access Guide for Dwarf Novae", ESA-SP 1134 (1989)
5. la Dous, C. and Giménez, A., "IUE-ULDA Access Guide for Chromospherically Active Binary Stars", ESA-SP 1181 (1994)
6. Festou, M., "IUE-ULDA Access Guide for Comets", ESA-SP 1114 (1990)
7. Festou, M., "INES Access Guide for Comets", in preparation
8. Formiggini, L. and Brosch, N., "INES Access Guide for Normal Galaxies", ESA-SP 1243 (2000)
9. Franchini, M. et al., "IUE-ULDA Access Guide for K Stars", ESA-SP 1203 (1996)
10. Gómez de Castro, A. I. and Franqueira, M., "IUE-ULDA Access Guide for T Tauri stars", ESA-SP 1205 (1997)
11. Gómez de Castro, A. I. and Robles A., "INES Access Guide for Herbig Haro objects", ESA-SP 1237 (1999)

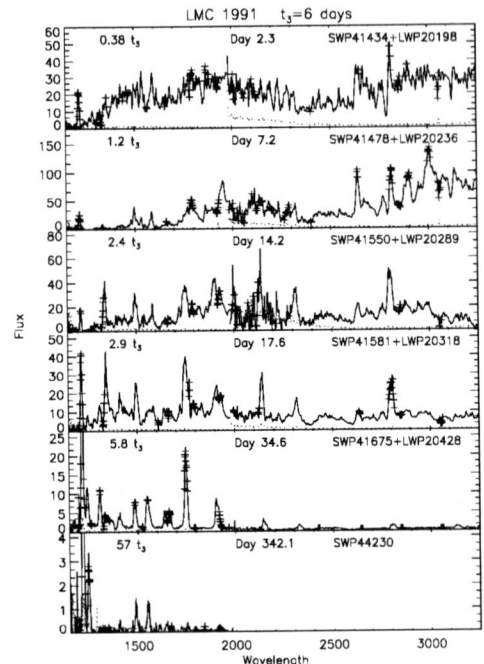

FIGURE 3. Example of the contents of the INES Guide: Representative low–resolution spectra.

12. González–Riestra, R. and Krautter, J., in "Ultraviolet Astrophysics beyond the IUE Final Archive", ESA-SP 413 (1998)
13. Longo, G. and Cappacioli, M., "IUE-ULDA Access Guide for Normal Galaxies",ESA-SP 1152 (1992)
14. Niedzielski, A., "INES Access Guide for Wolf Rayet stars", in preparation
15. Wamsteker, W., Skillen, I., Ponz, J.D., de la Fuente, A., Barylak, M., Yurrita, I., Astrophysics & Space Science, **273**, 155 (2000)

The Spectral Evolution of Nova Velorum 1999 (V382 Vel)[1]

A. Augusto and M. Diaz

Departamento de Astronomia – IAG - Universidade de São Paulo, Caixa Postal: 3386, CEP: 01060 - 970, São Paulo SP – BRASIL, e_mail: anselmo@astro.iag.usp.br

Abstract. Nova Velorum 1999 is the brightest nova that appeared in the last 24 years, being a good candidate for study of its shell. In this work we present the spectral evolution of the Nova Velorum 1999 shell during the first 3 years after the visual maximum. An average shell ejection velocity of 1600 km/s which decreases with time was found. The main results of this work suggest that the shell is inhomogeneous with a small filling factor (<0.1). We also estimate the electron temperatures and densities in the shell as well as the central source temperature. With this information we estimate the He numerical abundance and lower limits to the N, O, Ne, S, Ar and Fe numerical abundances. The results confirm that this nova is a neon nova. It is also found that iron is possibly enhanced in the shell, when compared to the solar value.

INTRODUCTION

Nova Velorum 1999 is the brightest nova since V1500 Cyg reaching V < 3.0 at maximum in 1999 May 22.5 UT [1]. It is also a fast nova with $t_3 \sim 12.3$ d [2]. This object evolved from FeII and FeII-b types according to Tololo Classification [3], to a nebular neon nova. [Ne II] $\lambda 12.81$ μm was found at $\Delta t = 42.5$ d (days after outburst) [4]. Neon nova are believed to be systems harboring an O-Ne-Mg white dwarf primary. They present a short time of decline, a bright maximum and a neon rich shell. This enhancement is probably due to some dredge-up of material from the white dwarf [5].

OBSERVATIONS

Our data has been acquired with Cassegrain spectrographs coupled to the 1.6 m telescope at LNA and to the 1.5 m telescope at ESO. A resolution of approximately 9 Å was achieved with a coverage from 3300 Å to 10200 Å. We also observed the Hα line with medium dispersion (1.8 Å FWHM) at LNA on the 670th day. The observation sequence consisted of three or four narrow-slit exposures (~2") in order to obtain the best spectral resolution. We then used a wide-slit exposure (~9") to establish the flux comparison. The fluxes of the standard stars were taken from [6] and an average

---

[1] Observations reported in this paper were performed at the Laboratório Nacional de Astrofisica-CNPq/LNA.

extinction curve to the site was used. High-dispersion spectroscopy in the infrared ($\lambda_c = 1.19$ μm) was also taken on the 70th day. Data reduction procedures were performed using IRAF[2] package.

SPECTRAL EVOLUTION

We observed this nova at nebular phase. The first spectra showed strong permitted lines but we classified these spectra as Nebular-Ne because at the 86th day the strongest line was [NeIII] λ3869. [OIII] λλ4959,5007 were very strong too. Several FeII multiplets and strong Balmer lines were also found. On the 221th day an increase in ionization was noticed. Several permitted lines virtually disappear while several forbidden lines strengthened. Ionization increased up to [FeVII] λ6087 (X_{ion} = 100 eV). An increasing blue continuum was observed from the 86th day (Fig. 1). In the last spectrum, (Δt = 731 d) the lines weakened and decreased in FWHM, which made possible the measurement of some blended lines such as HeII λ4686.

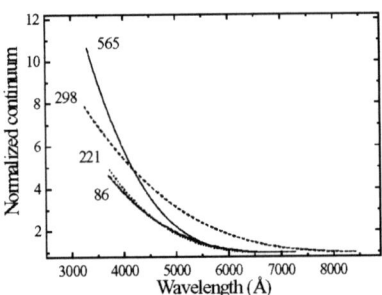

FIGURE 1. Continuum fitting of observed spectra. Numbers are the time after outburst.

ANALYSIS

Dynamical evidences for circumstellar envelope

Neon forbidden lines were used to estimate the shell ejection speed. We considered these the best lines to perform this task because they were optically thin, relatively isolated and very strong. Our measurements are shown in Figure 2. We have obtained two fits, one to the red line side and the other to blue line side. Considering an exponential law, we obtained v_0, that is the average initial speed of the material and a "damping" time scale, τ (the time at which the observed speed drops by a factor 1/e). The results are v_0 = 1740 (70) km/s and τ = 3300 (1000) d to the red side line and v_0 = 1400 (60) km/s and τ = 3000 (1000) d to the blue side line. The decrement in

---

2 IRAF is distributed by the National Optical Astronomy Observatories, wich is operated by the Association of Universities for Research in Astronomy (AURA), Inc., Under cooperative agreement with the National Science Foundation.

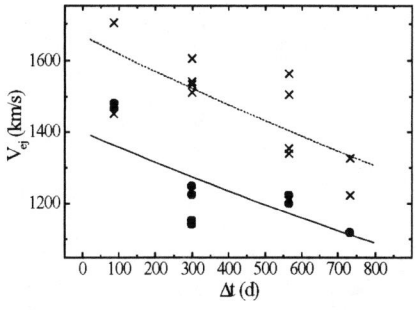
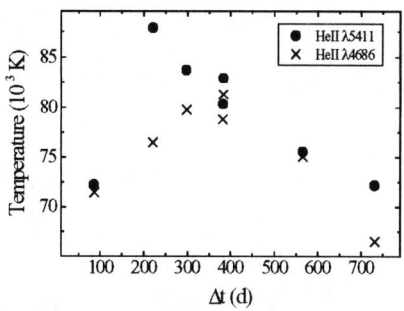

FIGURE 2. FWHM for neon lines. The red side is represented by dashed line and crosses. Blue side is continuous line and circles.

FIGURE 3. Evolution of the central source Zanstra temperature.

velocity might be due to the interaction with circunstellar matter or by the high ionization region, where the neon lines are formed, moving backwards during the shell evolution.

Physical Properties

We have evaluated the temperatures and numerical abundances by using line ratios corrected by interstellar reddening, considering an interstellar extinction E(B-V) = 0.2 [7,8]. The electron density in the shell was constrained by considering a homogeneous spherical shell in isotropic expansion with theoretical ejected masses taken from [9]. With these densities, the electron temperature was constrained using the classical oxygen and argon line ratios. The values found for electron densities ranged between 10^3 and 10^7 cm^{-3} while the electron temperatures varied between 6000 and 26000 K.

We used the HeII lines at 4686 Å and 5411 Å and the classical Zanstra method to estimate the central source temperature. The best values are derived using the 5411 Å line because it is a relatively isolated feature. The 4686 Å line was blended with NIII λ4640 and [NeIV] λ4721. One finds T_* (max) >80 000 K at Δt = 350 d and a significant decay after that epoch (Fig. 3). Comparing Zanstra temperatures achieved from Hβ and HeI with that found from HeII we constrain the covering-filling factor product (ηξ) to less than 0.1.

Numerical Abundances Estimates

We estimated the numerical abundances using the observed intensities line and solving the equations of statistical equilibrium for 5 levels with the electron densities and temperatures discussed above. At the first few points in the evolution of the ejecta, the error bars are larger because of the high densities expected at the beginning of the expansion. This leads to large temperature uncertainties. During the last days the density errors had not influenced the temperature in such a way. The resulting absolute

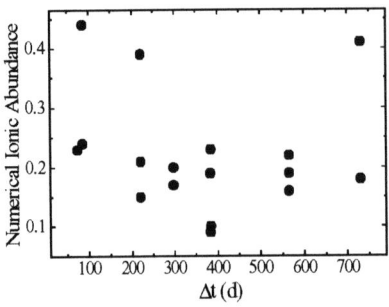
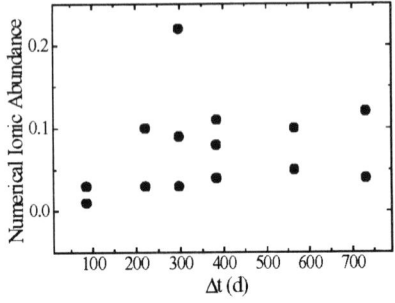

FIGURE 4. He I abundance by number.   **FIGURE 5.** He II abundance by number.

abundance values are shown in Table 1. Helium abundances were obtained for all days using all available lines. With a sum between HeI (Fig. 4) and HeII (Fig. 5) averages for each day we calculated the average between all days which represent the estimated Helium abundance for the ejecta. Oxygen (Fig. 6) and neon (Fig. 7) abundances were obtained with the same procedure used for helium. There are lines available for estimating 3 ionization stages; O^0, O^+ and O^{++} for oxygen and Ne^{++}, Ne^{+3} and Ne^{+4} for neon. Nitrogen, sulphur and argon abundances were obtained with blended and/or weak lines, yielding larger errors. The only line used to estimate the nitrogen abundance was [NII] $\lambda 5755$, which was very weak at the last day. There was only Ar^{++} to estimate the argon abundance. Sulphur was the worst estimate because there were only two points, one in $\Delta t = 86$ d ([SII] $\lambda\lambda 4069, 4076$ Å) and another in $\Delta t = 298$ d ([SIII] $\lambda\lambda 9069, 9532$ Å).

TABLE 1. Numerical abundances.

Element	He	O	Ne	Fe	N	Ar
log (N/N_H^+)	$\approx -0.57 \pm 0.08$	$> -3.83 \pm 0.03$	$> -2.98 \pm 0.02$	$> -3.65 \pm 0.02$	$> -3.98^{+0.18}_{-0.3}$	$> -5.7^{+0.16}_{-0.24}$

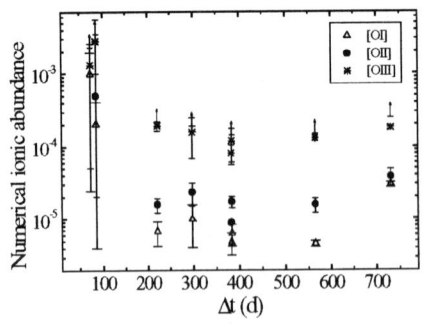
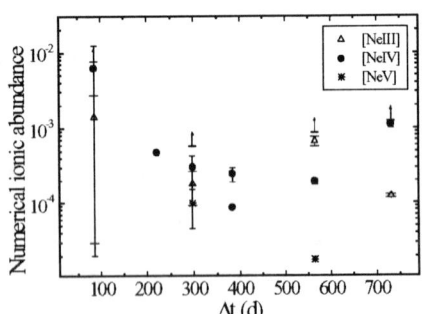

FIGURE 6. Oxygen abundance measurements.   **FIGURE 7.** Neon abundance measurements.

CONCLUSIONS

Our spectral analysis brought up indications that this object is a neon enhanced nova. We found a neon numerical abundance >10.8 times the solar values, greater than the average of the value found for "ordinary" novae (~6.8) [9]. An explanation for the origin of this system would be the enhancement by a O-Ne-Mg white dwarf progenitor. On the other hand, the very bright maximum, the fast decline, the initial spectrum (FeII- b), and the short nuclear turnoff time-scale had corroborated this conclusion. The iron abundance found is about 7 times solar and about 70 times larger than those suggested by the modelling of the X-ray emission between 0.8-10 keV [10,11]. The oxygen and neon abundance limits found are relatively low when compared to other neon nova. The neon abundance is high when compared with other heavy elements in this object. We found an abundance ratio of 0.14 between oxygen and neon. The plot in figure 8, may suggest that there are two distinct novae populations with distinct O/Ne ratios. The estimates for nitrogen, argon and sulphur were statistically poor because of the only one line available and ionization degree, severe blends, or the absence of diagnostic lines in some epochs. Another important uncertainty factor was the assumption that the shell is homogeneous and spherically symmetric. This may lead to some discrepancies in comparision with the abundance determination using more realistic density distribuitions.

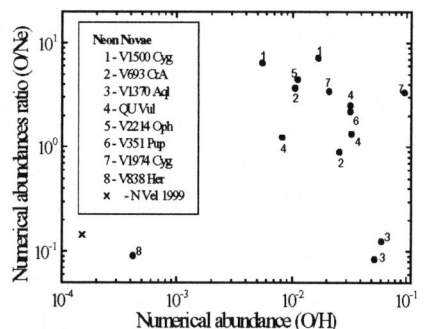

FIGURE 8. O/Ne ratio in neon nova.

ACKNOWLEDGMENTS

Work partially supported by FAPESP 02/00805-4 and 99/06261-1 and CAPES.

REFERENCES

1. Williams, P.; Gilmore, A.C., *IAU Circular 7176* (1999).
2. Liller, W.; Jones, A. F., *IBVS 5004*, 1-3 (2000).
3. Williams, R. E. et al., *ApJ*, **376**, 721-737 (1991).
4. Woodward et al., *IAU Circular 7220* (1999).
5. Starrfield et al., *MNRAS*, **296**, 502-522 (1998).
6. Hamuy, M. et al., *PASP*, **104**, 533-552 (1992).
7. Shore et al., *IAU Circular 7192*, (1999).
8. Amôres & Lèpine, (*private communication*), (2002).
9. Gehrz, R. D. et al., *PASP*, **110**, 3-26 (1998).
10. Mukai K. & Ishida M., *ApJ*, **551**, 1024-1030 (2001).
11. Orio M., et al., *MNRAS*, **326**, L13-L17 (2001).

WSO/UV: World Space Observatory/Ultraviolet

M. Hernanz[a], R. González-Riestra[b], W. Wamsteker[c,d], B. Shustov[c,e], M. Barstow[c,f], N. Brosch[c,g], C. Fu-Zhen[c,h], M. Dennefeld[c,i], M. Dopita[c,j], A.I. Gómez de Castro[c,k], N. Kappelmann[c,l], I. Pagano[c,m], J. Sahade[c,n], H. Haubold[c,o], J.-E. Solheim[c,p], P. Martínez[c,q]

[a]*IEEC/CSIC, Spain,* [b]*XMM SOC, VILSPA, Spain,* [c]*WIC: WSO Implementation Committee,* [d]*ESA-VILSPA, Spain,* [e]*INASAN, Russia,* [f]*University of Leicester, United Kingdom,* [g]*TAU, Israel,* [h]*USTC-CfA, China,* [i]*IAP, France,* [j]*ANU, Australia,* [k]*UCM, Spain,* [l]*IAAT, Germany,* [m]*CNR, Italy,* [n]*CONAE, Argentina,* [o]*UN-OOSA, Vienna, Austria,* [p]*University of Tromso, Norway,* [q]*SAAO, South Africa*

Abstract. We summarize the capabilities of the World Space Observatory (UV) Project (WSO/UV). An example of the importance of this project (with a planned launch date of 2007/8) for the study of Novae is given.

INTRODUCTION

The World Space Observatory/Ultraviolet (WSO/UV) project is a mission development initiated to overcome the severe lack of capabilities for UV observations for Astrophysics [1] in the planning time lines of the major Space Agencies (extending at least beyond the year 2014). Apart from its scientific importance, it represents a new mission implementation model for large space missions which can be applied to the need for large light collection power required to keep *space astrophysics* complementary to the continuously increasing sensitivity of *ground-based telescopes*. One of the assumptions, associated with the WSO concept is to avoid the excessive complexity required for multipurpose missions. Although there may exist purely technological or programmatic policy issues, which would suggest such complex missions to be more attractive, many other aspects, which do not need to be explored here, argue against such mission model. Following this precept, the first implementation model for a World Space Observatory has been done for the ultraviolet domain WSO/UV [2, 3, 4]. An innovative aspect of the WSO implementation model is the truly world-wide participation, already in the early stages of the project, generating fully independent access for astronomers from *all countries* through its open operations model.

THE WSO/UV MISSION

The *primary mission* of the World Space Observatory is a spectroscopic capability extending over the wavelength range: $\Delta\lambda$ **103 - 310 nm.** Three individual spectrographs will be mounted at a distance of 50 mm from the optical axis of the telescope (T-170M) with an image quality of < 0.5 arcsec. Two resolution modes are supported: $R(\lambda/\delta\lambda) \approx 55,000$ for point source (Ø 1 arcsec), and $R(\lambda/\delta\lambda) \approx 1,000$ for Long Slit (1 x 70 arcsec). In the high resolution mode, the wavelength range has to be split in two sections to be able to match the spectrum on the detector (MCP) face. The long slit-low resolution mode is foreseen to cover the full range from $\Delta\lambda$ 103 - 310 nm in a single exposure.

As *secondary instrumentation* there will be a suite of 2 UV imagers with 6 filters (High Sensitivity Imager: **HSI** and High Resolution Imager: **HRI**) with redundancy; an optical imager (**OI**); and a special UV imaging device (**CFC**) placed at the optical axis of the telescope to utilize the high optical quality of the T-170M. In principle the operational mode will be an observing program driven by the spectrographic demand, with continued exposures for the imagers. Of course also specific imaging programs are foreseen, and will be supported. The field-of-view of the imagers, their wavelength coverage and focal ratio are indicated in table 1.

The *orbit* for WSO/UV is planned to be a halo type orbit around the second Earth-Sun Lagrangian point (L2).

The combined WSO system, which is based on current top-of-the-line technology (Technology Readiness Level TRL > 6); high throughput of the telescope/spectrograph; a limited number of primary instruments on board; and the orbit efficiency of the WSO/UV, will be such that an order of magnitude improvement will be obtained with respect to that associated with the Hubble Space Telescope. A very important reason for WSO/UV is supplied by the fact that, at its scheduled start of operations phase, the results of the GALEX UV survey will be available. No other mission is planned which would allow to use its results for further studies of the GALEX Catalogue, which will contain many objects which will be of great importance for the understanding of cosmological evolution of the overall content of the Universe.

AN EXAMPLE OF WHAT WSO/UV WILL BE ABLE TO CONTRIBUTE TO NOVA STUDIES

To illustrate the importance of the observing capabilities supplied by the WSO/UV mission for the study of Novae, we show in Figure 1 the results obtained for Nova V1974 Cygni 1992 where, through the availability of the International Ultraviolet Explorer, a completely new insight in the evolution of the nova phenomenon was obtained, revolutionizing our understanding of the explosive processes driving a nova [5].

The capabilities of WSO will allow to obtain high resolution UV spectra of novae of similar luminosity in M31 with a signal-to-noise ratio of 10 in only 15 minutes. Even for novae in M81 (at 3.3 Mpc) useful spectra can be obtained with reasonable exposure times (of the order of 1-2 hours).

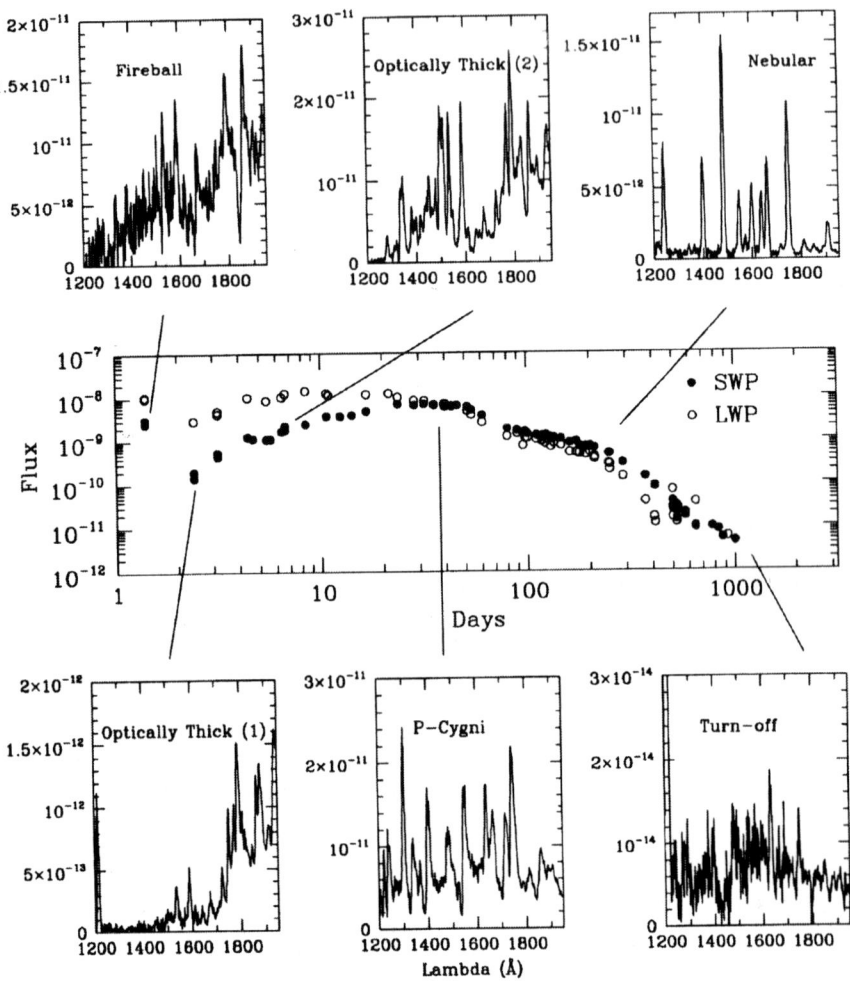

FIGURE 1. This figure shows the photometric and spectroscopic ultraviolet evolution of the bright Nova V1974 Cygni 1992 (from IUE data), the best observed nova so far over all the electromagnetic spectrum. The complete time coverage of the UV observations of this nova allowed to characterize the different phases of its evolution, even identifying some of them never observed before. Particularly well seen in this object is the "Constant Bolometric Luminosity Phase", predicted by the standard Thermonuclear Runaway Theory. The peak bolometric luminosity of this nova during this phase was of the order of 10^{38} erg s^{-1}.

TABLE 1. Imaging with WSO/UV.

Camera	HRI (f/50)	HIS (f/10)	CFC (f/10)	OI (f/10/f/50)
$\Delta\lambda$ (nm)	$\lambda\lambda 115 - \lambda\lambda 310$			$\lambda\lambda 360 - \lambda\lambda 800$
F.O.V. (arcsec)	60	300	300	240

[D80 (λ 630 nm) = 0.35 arcsec]

REFERENCES

1. Wamsteker, W. et al., in *Hubble's Science Legacy: Future Optical-Ultraviolet Astronomy from Space*, eds. K.R. Sembach, J.C. Blades, G.D. Illingworth, & R.C. Kennicutt, ASP Conference Series, 2002, in press.
2. *WSO/UV Assessment Study*, ESA [CDF-05 (A)], 2000.
3. Rodríguez-Pascual, P.M., UE-CEES PX0116000 [CDF-05 (C)], 2000.
4. *APD Team Assessment of ESA WSO proposal*, [APD Report., CL#01-1168], 2001.
5. R. González-Riestra, and J. Krautter, 1998, in *UV Astrophysics beyond the IUE Final Archive*, eds, R. González-Riestra and W.Wamsteker, ESA SP-143, 1998, p. 367.

Radio Emission from V723 Cas

I. Heywood*, T.J. O'Brien*, S.P.S. Eyres[†], M.F. Bode** and R.J. Davis*

*Jodrell Bank Observatory, The University of Manchester, Macclesfield, SK11 9DL, UK
[†]Centre for Astrophysics, The University of Central Lancashire, Preston, PR1 2HE, UK
**Astrophysics Research Institute, Liverpool John Moores University, Twelve Quays House, Birkenhead, CH41 1LD, UK

Abstract. V723 Cas (Nova Cassiopeiae 1995) is an unusually slow nova discovered on 1995 August 24. MERLIN has been used to follow the evolution of the radio emission from this nova providing ten epochs of data and nine radio maps between 1996 and 2001. The radio light curve shows an initial optically thick rise followed by a turn-over into optically thin emission and a corresponding decline in flux as the shell expands. The shell is resolved in the radio by 1998 March and subsequently breaks up into several compact components. Later maps display an unusual effect involving a 90 degree change in the preferential emission axis between successive epochs. A simple radiative transfer model is applied to the light curve in order to determine physical parameters for the shell. These results are presented together with a discussion of the radio maps in terms of an expanding non-uniform shell with varying optical depth.

INTRODUCTION

V723 Cas (Nova Cassiopeiae 1995) was discovered at $V = 9.2$ mag on 1995 August 24 by M. Yamamoto [1] and reached visual maximum $V = 7.1$ mag on 1995 December 17 which is defined as day zero. The optical light curve displays unusually slow evolution with a long pre-maximum halt and has led to V723 Cas being classified alongside HR Del and RR Pic [2]. Oscillations in the light curve are evident post-maximum, a feature also seen in the light curve of HR Del. The typical mechanism for the production of radio emission in novae is a thermal free-free process although non-thermal emission has been detected in a few novae, the most notable example being GK Per [3]. The MERLIN interferometer was used to observe V723 Cas between 1996 and 2001 resulting in nine epochs of data at a wavelength of 6cm and one at 18cm. Details of the observations can be seen in Table 1. Interpretations of the flux measurements and the 6cm maps follow.

RADIO FLUX MEASUREMENTS

As expected from previous observations of novae the radio light curve for V723 Cas undergoes an initial rise followed by a turn over and then a decline, Figure 1. The dashed lines on this figure are the 3-phase model fits as described below. The error on each flux measurement is taken to be the rms noise (σ) measured from each map.

TABLE 1. Summary of the MERLIN observations of V723 Cas.

Observation date	Wavelength (cm)	Days since optical maximum	Flux density (mJy)	Peak brightness (mJy beam^{-1})
1996 Dec 13	6	362	0.91 ($\pm$ 0.14)	0.61
1997 Jan 25	6	405	1.46 ($\pm$ 0.24)	0.94
1998 Mar 04	6	808	6.57 ($\pm$ 0.07)	1.58
1998 Apr 06	18	841	0.66 ($\pm$ 0.06)	
1998 Dec 07	6	1086	6.75 ($\pm$ 0.11)	1.37
2000 Feb 27	6	1533	5.38 ($\pm$ 0.13)	0.94
2000 Apr 16	6	1582	5.21 ($\pm$ 0.1)	2.03
2001 Jan 29	6	1870	4.44 ($\pm$ 0.2)	0.97
2001 Jun 03	6	1995	3.15 ($\pm$ 0.13)	0.55
2001 Oct 26	6	2140	2.94 ($\pm$ 0.15)	0.49

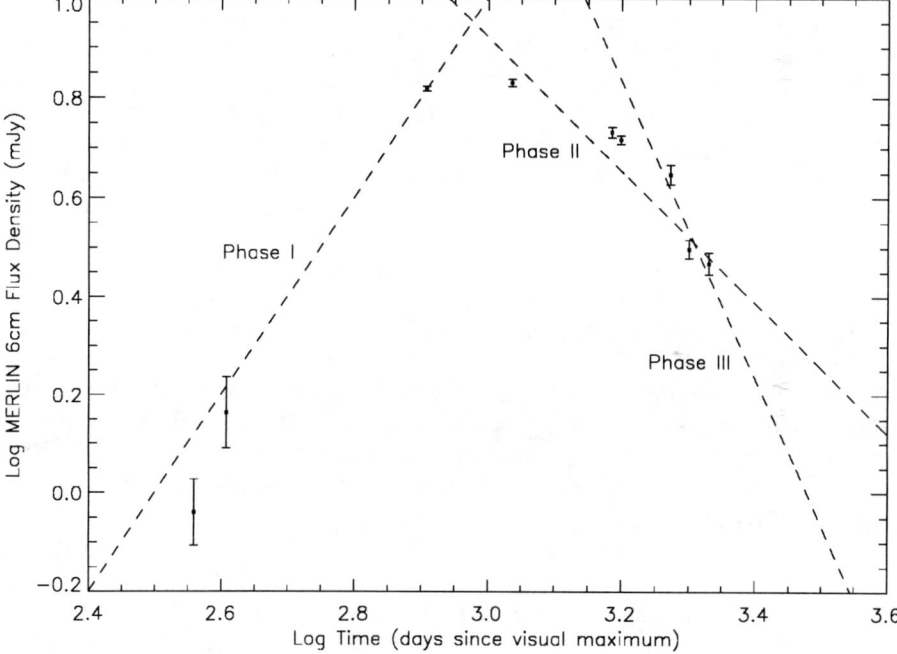

FIGURE 1. The radio light curve of V723 Cas as measured by MERLIN at a wavelength of 6cm from one to six years after outburst. The dashed lines represent predictions of the model described in the text with parameter values given in Table 2.

Light Curve Modelling

The model used to describe the light curve is that of Hjellming and collaborators [4] and referred to by Gehrz and Bode in these proceedings. In this model after outburst the

TABLE 2. Shell parameters for V723 Cas.

Temperature (K)	23000
Outer shell velocity v_2 (kms^{-1})	370
Inner shell velocity v_1 (kms^{-1})	210
Ejected mass ($M_\odot$)	2.5×10^{-4}
Switch-off time t_s (days)	145

outer edge of an isothermal spherical shell of ejecta expands with constant velocity v_2. The density within the shell is described by a "Hubble flow" resulting in a $1/r^3$ dependence. Ejection terminates at time t_s and an inner boundary to the shell is formed which subsequently expands with velocity v_1. Solutions of the radiative transfer equations for this geometry can yield physical parameters for the shell when fitted to the data.

This model gives the "classic" three phase evolution for the shell. During phase I the source is entirely optically thick. The shell expands with the radio photosphere coincident with the outer boundary of the shell giving the flux a $\nu^2 t^2$ dependence. The 6cm flux from 1998 March and the 18cm flux from 1998 April yield a spectral index of 2.08 which is consistent with the assumption that the source is optically thick at this stage and therefore in phase I of its evolution. The onset of phase II is marked by the time at which the photosphere begins to lag behind the outer edge of the expanding shell and the source becomes partially optically thin (although emission is still dominated by the optically thick region inside the photosphere). Solution of the equations for this phase gives a $\nu^{0.6} t^{-4/3}$ dependence but relies on an extreme approximation [5]. During the final phase the shell is optically thin. Fading due to expansion and decreasing density means the flux goes as $\nu^{-0.1} t^{-3}$.

The dashed lines on the light curve of Figure 1 represent best fit lines for the three phases. The model appears to be a good description of phases I and III and the physical parameters presented in Table 2 are estimated from this initial fit. The poorer fit for Phase II may be a result of the extreme assumptions used in its derivation or may represent limitations of the overall model.

RADIO MAPS

The nine radio maps at 6cm are presented in Figure 2. There is no map presented for the 18cm observation as MERLIN is unable to resolve V723 Cas at this wavelength. The contour levels are (-3, 3, 4, 5, 6...)$\times \sigma$ except for the 1998 March and December epochs which have (-3, 3, 6, 9, 12...)$\times \sigma$ levels. The greyscale runs from 2σ to the peak flux value in the map. The first two maps show that the source is essentially unresolved by MERLIN at these stages but becomes resolved by 1998 March. Further expansion of the shell and the transition to the partly optically thin phase II of its evolution leads to structure becoming evident in the fifth epoch, 2000 February. The north-south ridge evident from 1998 March through 2000 February divides into two peaks by 2000 April.

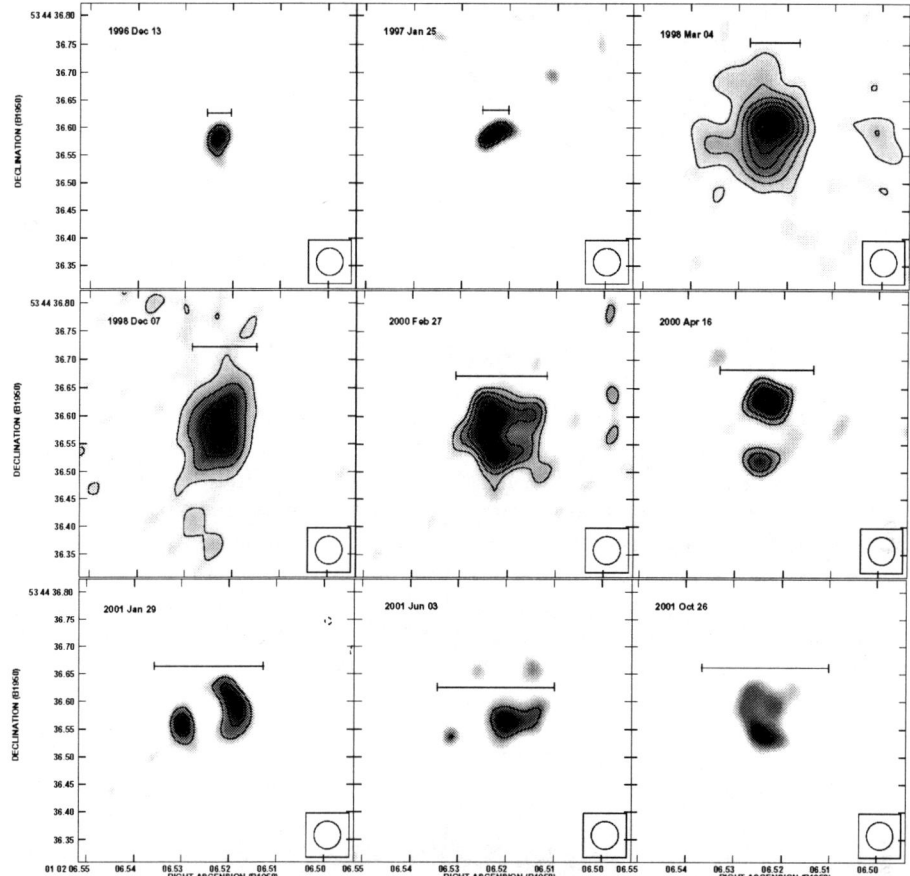

FIGURE 2. The 6cm MERLIN radio maps for the nine epochs of Table 1 with the earliest map at top left and time increasing to the right along each row. In each case a horizontal bar indicates the expected size of the shell for a typical distance and ejection velocity (see text). The circle at the bottom right of each map shows the 50 mas restoring beam.

Some 9 months later the 2001 January map shows the emission dominated by two components aligned on an east-west axis, almost the exact reverse of the previous epoch. The emission then gradually fades over the last two epochs.

Nova shells are known to exhibit density enhancements such as polar caps and equatorial and tropical rings [6] and the favoured explanation for the non-uniform structure seen in the radio is that higher density regions of the shell appear as peaks in the maps as the photosphere recedes. The switching of emission features from north-south to east-west can be seen in the radio maps of other novae such as V1974 Cyg [7]. In this latter case the 18cm maps show this effect occuring later than in the 6cm observations. Since the 18cm emission remains optically thick for longer than the 6cm emission this may support the assumption that this is also an optical depth effect.

The horizontal bars on the maps show the angular size predicted by assuming a distance of 2.4 kpc [2] and estimating an expansion velocity from the separation of the two peaks seen in the [FeVII] (6087Å) line. This forbidden line is entirely optically thin with a well defined double peak structure. The ejection velocity is measured to be $210 kms^{-1}$. This gives a lower estimate for the extent of the shell as higher velocity components seen as wings on this line profile are also present. However as Figure 2 shows the predicted size of the shell is consistent with the observed extent of the radio emission. It is assumed that any acceleration is negligible for this calculation.

CONCLUSIONS

V723 Cas is only the fourth nova to have been resolved in the radio (see Bode's review in these proceedings). The results presented here further demonstrate the potential for simple emission models to describe the radio lightcurves of novae and hence allow the estimation of values for a number of important parameters, most notably the mass of the ejected shell. However they also demonstrate that these models are rather too simplistic in their assumptions of spherical, smooth shells of ejecta. The radio imaging, much like the optical imaging presented elsewhere in these proceedings, shows that the shells are clumpy and that varying optical depth through the ejecta can have a significant effect on its appearance. We intend to develop existing models to incorporate asphericity and clumping for comparison with MERLIN and VLA observations. We will also investigate how such shells will be seen by the next generation of more sensitive radio interferometers such as e-MERLIN.

REFERENCES

1. Yamamoto, M., *IAU Circ.* **6213**, (1995).
2. Chochol, D., and Pribulla, T., *Contrib. Astron. Soc. Skalnate Pleso* **27**, 53-69 (1997).
3. Reynolds, S.P., and Chevalier, R.A., *ApJ* **281**, L33-35 (1984).
4. Hjellming, R.M., Wade, C.M., Vandenberg, N.R., and Newell, R.T., *AJ* **84**, 1619-1631 (1979).
5. Hjellming, R.M., "Images and Lightcurves of the Radio Remnants of Novae," in *Physics of Classical Novae*, edited by A. Cassatella and R. Viotti, IAU Coll. 122, Springer-Verlag, Berlin, 1990, pp. 169-178.
6. Gill, C.D., and O'Brien, T.J., *MNRAS* **314**, 175-182 (2000).
7. Eyres, S.P.S., Davis, R.J., and Bode, M.F., *MNRAS* **279**, 249-256 (1996).

MODEL ATMOSPHERES AND LIGHT CURVES

Nova Model Atmopheres

Peter H. Hauschildt[*], G. Schwarz[†], C. Ian Short[**], E. Baron[‡] and S. Starrfield[§]

[*]*Dept. of Physics and Astronomy & Center for Simulational Physics, The University of Georgia, Athens, GA 30602-2451*
[†]*Steward Observatory, 933 N. Cherry, Tucson, AZ 85721*
[**]*Department of Physics, Florida Atlantic University, Boca Raton, FL 33431-09910*
[‡]*Dept. of Physics and Astronomy, University of Oklahoma, 440 W. Brooks, Rm 131, Norman, OK 73019-0225*
[§]*Department of Physics and Astronomy, Arizona State University, P.O. Box 871504, Tempe, AZ 85287-1504*

Abstract. We review the basics physics of nova atmospheres and discuss the physics that has to be included for detailed models of their early spectra. We also present NLTE calculations for model atmospheres of novae during outburst. This fully self-consistent NLTE treatment for a number of model atoms includes > 5000 NLTE levels and > 57000 NLTE primary transitions. We discuss the implication of departures from LTE for the strengths of the lines in nova spectra. The new results show that our large set of NLTE lines constitute the majority of the total line blanketing opacity in nova atmospheres. Although we include LTE background lines, their effects are small on the model structures and on the synthetic spectra. We demonstrate that the assumption of LTE leads to incorrect synthetic spectra and that NLTE calculations are required for reliably modeling nova spectra. In addition, we show that detailed NLTE treatment for a number of ionization stages of iron changes the results of previous calculations and improves the fit to observed nova spectra. These new models have also been used to fit the lightcurves of novae during the early phases of the nova outburst.

INTRODUCTION

In a series of papers [1, 2, 3, 4, 5, 6, 7, 8, 9] we have developed detailed spherical, expanding NLTE model atmospheres to treat the optically thick, early stages of nova outbursts and have also analyzed a number of observed nova spectra. The models were computed self-consistently, employing the equation of transfer in a special relativistic framework [10, 11], energy conservation in the co-moving frame calculated with a modified Unsöld-Lucy method [12, 13], NLTE effects for a large number of atomic and molecular species using numerical methods that we have developed to treat very large and detailed model atoms [14, 15, 16, 17, 18], and extensive NLTE and LTE line blanketing by several million lines that are dynamically selected from the Kurucz lists [19] and other sources (for molecular lines). These models have been extremely successful in fitting observed nova spectra [3, 6], and, in particular, have been used to identify the UV signature of the "fireball" phase of Nova Cygni 1992 [2, 9] and LMC 1991 [20].

Although we treat nova atmospheres as steady-state expanding shells (winds) [cf. 21], the time development of the nova spectra can be simulated by a series of nova atmospheres at constant luminosity and increasing model temperatures during the phase of constant bolometric luminosity of the nova. Therefore, the analysis of a time sequence of observations can provide insight, for instance, into the development of the velocity field of the shell. Such sequences can also be used to check the results of the analysis for consistency and to obtain an estimate on the statistical errors inherent in the analysis process. This means that grids of nova atmosphere models and synthetic spectra are important to investigate the time evolution of a nova outburst.

METHODS AND MODELS

The basic assumptions of our nova models are the same as used in [7]. The expanding nova shell is assumed to have a power law density of the form $\rho \propto r^{-N}$ with $N = 3$ for the models presented in this paper. The velocity law is derived from the condition of constant mass-loss rate (in radius) with a prescribed maximum velocity v_{max} (we use here $v_{max} = 2000\,\text{km}\,\text{s}^{-1}$), consistent with typical values observed in classical novae. We further parameterize the models with the "model temperature" T_{model} through the relation $L = 4\pi R^2 \sigma T_{model}^4$, where L is the luminosity of the model (here set to $L = 50,000\,L_\odot$ for all models, the absolute value of the luminosity does *not* affect the spectra, see [22, 4]). R is the radius of the shell at $\tau = 1$ in the bound-free (hereafter, b-f) continuum at 5000Å. The model temperature is comparable to, but should not be confused with the effective temperature T_{eff}, which is well defined only for PP atmospheres [4].

We solve the radiative transfer equation consistently for lines and continua (allowing for arbitrary overlaps) with the method discussed in [10] and [4] rather than employing the Sobolev approximation. [23] have shown that this simpler method *cannot* be used in nova atmospheres due to the large number of overlapping lines as well as the strong coupling between lines and continua. Such complications require that the multi-level NLTE rate equations be solved self-consistently and simultaneously with the radiative transfer and energy equations, and the equations must include the effects of both line blanketing and expansion of the nova atmosphere.

We use our general purpose stellar atmosphere code PHOENIX to model nova atmospheres. PHOENIX [version 13, 24, 18] uses a special relativistic spherical radiative transfer for nova models and an equation of state (EOS) which includes more than 300 ions of 40 elements (with up to 26 ionization stages each). The temperature correction is based on a variant of the Unsöld-Lucy method that has been modified to include NLTE and scattering. This algorithm converges very quickly and is highly stable. Both the NLTE and LTE background lines (see below) are treated with a direct opacity sampling method. We do *not* use pre-computed opacity sampling tables, but instead dynamically select the relevant LTE background lines from master line lists at the beginning of each iteration and sum the contribution of every line within a search window to compute the total line opacity at *arbitrary* wavelength points. The latter feature is crucial in NLTE calculations in which the wavelength grid is both irregular and variable from iteration to iteration due to changes in the physical conditions. This approach also allows detailed and depth dependent line profiles to be used during the iterations. To make this method

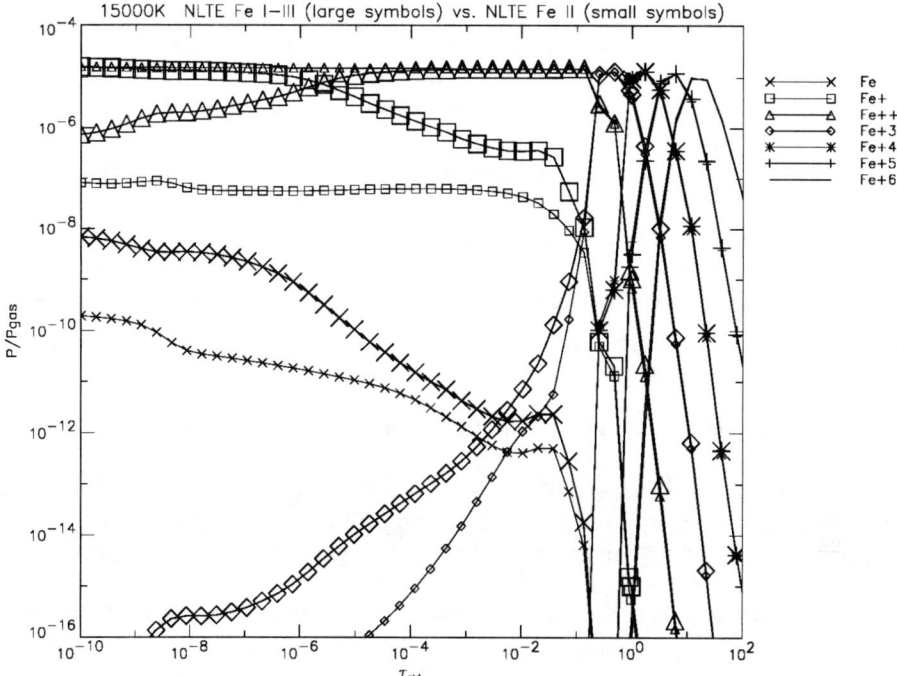

FIGURE 1. Ionization balance for iron for a nova model with $T_{\text{model}} = 15000\,\text{K}$. The plot shows the relative concentration P_i/P_{gas} for some iron ions. The large symbols show a NLTE model where Fe I–III are treated in NLTE, the small symbols show the results for a model where Fe II is the only Fe ion treated in NLTE. All other NLTE species are identical for both models.

computationally efficient, we employ modern numerical techniques, e.g., vectorized and parallel block algorithms with high data locality [18, and references therein], and we use high-end workstations or supercomputers for the model calculations.

In the calculations presented in this paper, we have significantly enlarged the set of NLTE species, and include a total of ~ 5000 NLTE levels and ~ 55000 NLTE primary lines in the calculations presented here. Details of the model atoms and the numerical methods used to directly treat these large number of NLTE lines and levels can be found in [7] and [24], respectively.

RESULTS

We have computed a grid of models to investigate the effects of NLTE on the atmospheric structure and the spectra of novae. All models include the NLTE treatment as discussed above as well as the standard PHOENIX equation of state and additional LTE background lines (about 2 million atomic lines). The NLTE effects are included in both the temperature iterations (so that the structure of the models includes NLTE effects) and all radiative transfer calculations. For primary NLTE lines we add 3 to 5 wavelength points within their profiles to the global wavelength grid (the transfer equation, the rate operators and the approximate rate operators are computed for every wavelength point that falls into the profile of each of each primary line, resulting in substantially more wavelength points per line due to line crowding). This procedure typically leads to about 120,000–190,000 wavelength points for the model and the synthetic spectrum calculations. We iterated the models until radiative energy conservation was fulfilled better than 1% for both the co-moving frame flux and its derivative and the NLTE departure coefficients were converged to better than 1% at the same time. We have found in test calculations that this is necessary and sufficient to obtain a converged solution [see also 14].

NLTE effects on the concentration of atoms and ions

In the nova atmosphere model with $T_{model} = 15000\,K$, hydrogen recombines in the optically thin regions of the atmosphere. Therefore, NLTE effects lead to changes in the location and size of the recombination zone as compared to the LTE models. In particular, they lead to an earlier recombination of H II to H I at an optical depth of $\tau_{std} \approx 3 \times 10^{-3}$ compared to the LTE location of the H recombination zone around $\tau_{std} \approx 10^{-5}$. In the outer atmosphere, however, the NLTE model shows a higher electron density compared to the LTE structure. Here the NLTE effects cause a slight over-ionization of hydrogen. To a lesser degree this is also the case for He I. We remind the reader that any one of these models is actually a snapshot of the atmosphere *in time*, so the differences between the LTE and NLTE models will be important for the comparison between theory and observations in a sequence of spectra for novae in outburst. In general, the effects of NLTE on the ionization balance are relatively small for C, N, and O, see [7] for a more thorough discussion of the ionization effects on these and other species.

For iron, the NLTE effects on the ionization balance are small but significant. NLTE reduces the concentration of Fe^+ ions in the Fe II line forming region ($10^{-4} \le \tau_{std} \le 10^{-2}$), thus reducing the overall strength of the Fe II lines compared to the LTE case. These results differ from those we reported previously [5] because, in these new models, we also include NLTE effects for Fe^0 and Fe^{+2}. In particular, the new treatment of Fe III reduces the NLTE effects of the Fe^+/Fe^{+2} ionization balance. The full treatment of Fe III reduces the NLTE over-ionization for Fe II compared to our previous calculations. We show this in Fig. 1, where we compare the full NLTE model (large symbols) with a model where Fe II is the *only* iron species treated in NLTE (but all other NLTE species are included). This shows that nova atmosphere calculations require a comprehensive NLTE treatment with many ions of the NLTE species. The new models fit the observed spectra of novae better than previous model generations [20].

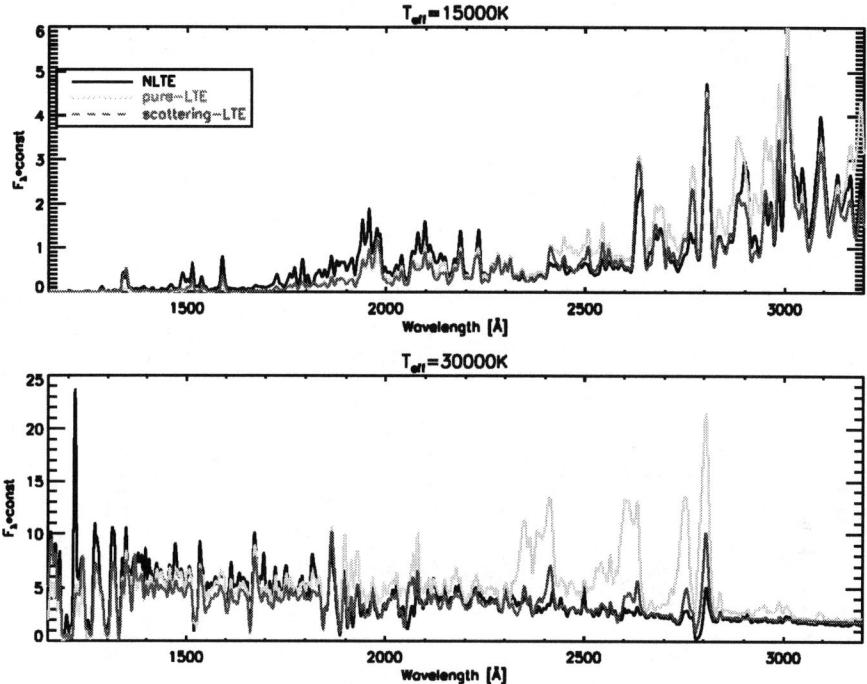

FIGURE 2. Comparison of synthetic spectra for nova model atmospheres with $T_{\mathrm{model}} = 15000\,\mathrm{K}$ and $T_{\mathrm{model}} = 30000\,\mathrm{K}$ in the IUE wavelength range. The spectra have been boxcar smoothed to the low resolution of the IUE satellite ($\simeq 6\,\mathrm{\AA}$). In the "pure-LTE" spectrum all NLTE departure coefficients have been set to unity whereas in the "scattering-LTE" spectrum the lines are assumed to have, in addition, an albedo for single scattering of 0.95. The latter corresponds to the assumptions for the LTE background lines. All spectra use the structure of the full NLTE model.

NLTE effects on the synthetic spectra

Figure 2 shows the progressive changes in the spectra with increasing temperature of the atmosphere. The ultraviolet region is particularly sensitive to such changes. In order to understand which lines are responsible for this sensitivity, we show in Fig. 3 synthetic spectra obtained by omitting LTE background lines so that only NLTE lines provide line blanketing. The spectra are very similar to the full spectra shown in Fig. 3. This indicates that the NLTE lines provide the bulk of the line blanketing and that the LTE background lines, even though numerous, constitute a comparatively small addition to the total line opacity. The fact that our detailed NLTE treatment allows us to handle the majority of the line blanketing in full NLTE is an important feature of the calculations, and the result that they constitute the majority of the line opacity is a basic result of this paper. The LTE background lines are, however, still important in localized spectral regions and must be included for the most accurate model and synthetic spectrum calculations. An additional result is that the NLTE effects on the synthetic spectra are far greater than the effect of the

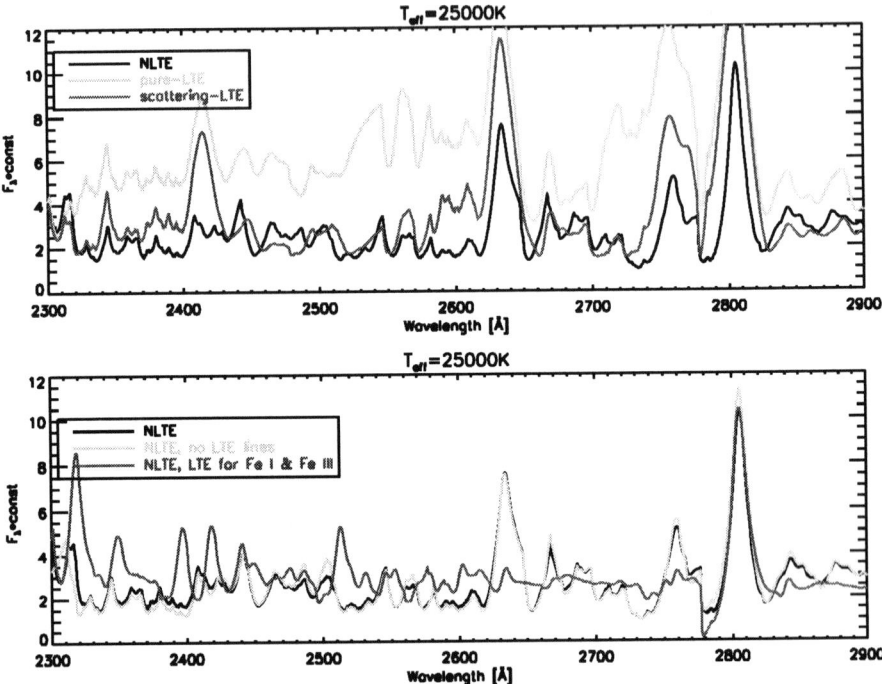

FIGURE 3. Comparison of synthetic spectra for nova model atmospheres with $T_{\text{model}} = 25000\,\text{K}$. In the "pure-LTE" spectrum all NLTE departure coefficients have been set to unity whereas in the "scattering-LTE" spectrum the lines are assumed to have, in addition, an albedo for single scattering of 0.95. The latter corresponds to the assumptions for the LTE background lines. In the "NLTE, no LTE lines" spectrum, all LTE background lines have been neglected. The "NLTE, LTE for Fe I and III" spectrum uses LTE for Fe I and III (with an albedo for single scattering of 0.95) but includes all other NLTE species, in particular Fe II). This model has been fully iterated with its set of NLTE species. All other spectra use the structure of the full NLTE model.

LTE background lines. In Fig. 3 we also show the synthetic spectra obtained by setting all departure coefficients to unity, i.e., the pure LTE assumption, and show synthetic spectra obtained by using a single scattering line albedo of 0.95 for all lines, *i.e.*, the NLTE lines are artificially treated in the same manner as the LTE background lines. An inspection of the plots shows that these synthetic spectra are very different from the full NLTE results shown. In [7] we show that this is also true for the infrared lines, which are much stronger in the pure-LTE spectra than in the NLTE spectra. For the IR lines, the NLTE case falls somewhere in between the pure-LTE and scattering-LTE cases.

The IR lines are mainly isolated, single transitions. The situation is *very different* in the UV, where the lines overlap strongly due to the level distributions in the iron peak ions and form a pseudo-continuum. Individual lines are frequently so blended that is impossible to assign a single identification to a feature in the synthetic spectrum [see 4, for discussion of this effect]. Although one might think that the strong line overlap would reduce NLTE effects, the computations show that the LTE and NLTE UV spectra

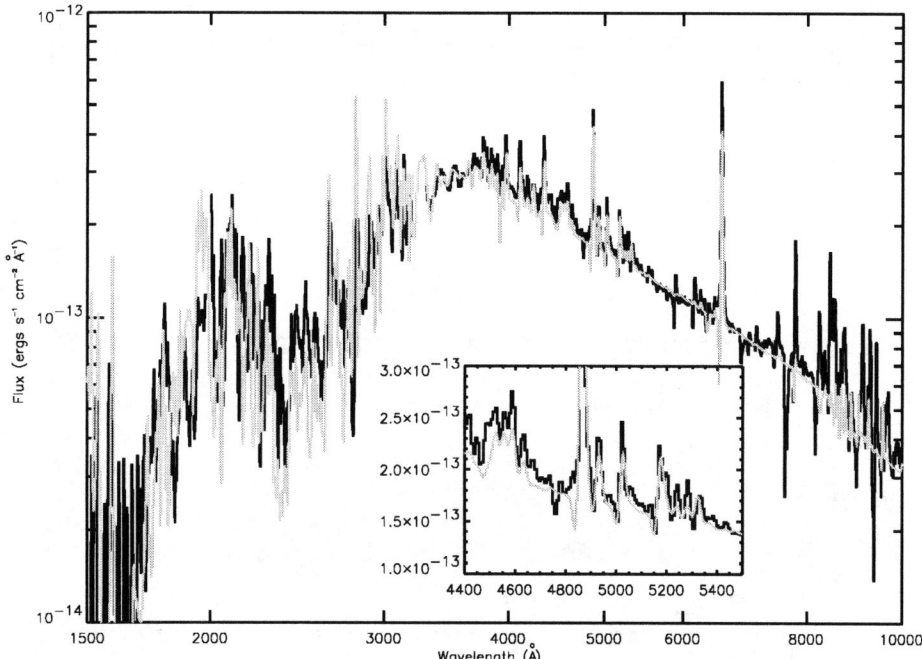

FIGURE 4. Fit to the UV+optical spectrum of nova LMC 1988#1 at optical maximum. The model atmosphere has the parameters $T_{\mathrm{model}} = 14{,}000\,\mathrm{K}$, $v_0 = 2000\,\mathrm{km\,s^{-1}}$, $N = 3$, and 1/3 solar metal abundances.

are very different for all models that we have calculated. In both the pure-LTE and the scattering-LTE spectra, the UV flux blocking is more severe than in the NLTE models. For $T_{\mathrm{model}} \leq 20000\,\mathrm{K}$, the LTE spectra produce less flux by a factor of 10 in the wavelength region blueward of about 1500Å than the NLTE spectra. Whereas for $T_{\mathrm{model}} = 15000\,\mathrm{K}$ the pure-LTE and scattering-LTE spectra are similar to each other, the pure-LTE spectra show enormous emission lines[1] for larger model temperatures ($\approx$ 23,000 to 30,000 K) in the 2000Å to 3000Å wavelength range that are absent in the NLTE spectra. These features are caused by clusters of iron lines (mostly Fe II) and the Mg II h+k doublet. These lines are much weaker in the scattering-LTE spectra, and they are nearly completely suppressed in the NLTE spectrum. This is extremely important for the evaluation of the optical taxonomy of nova spectra [see 25], since the "Fe II" classification is based on the appearance of the optical emission features.

Thus, pure-LTE is an extremely poor assumption in nova atmospheres, although scattering-LTE is a somewhat better description. However, *neither approach can replace a detailed and complete NLTE calculation* — the complexity of the line formation in a

---

[1] Note that these are real emission features, not gaps in the pseudo-continuum.

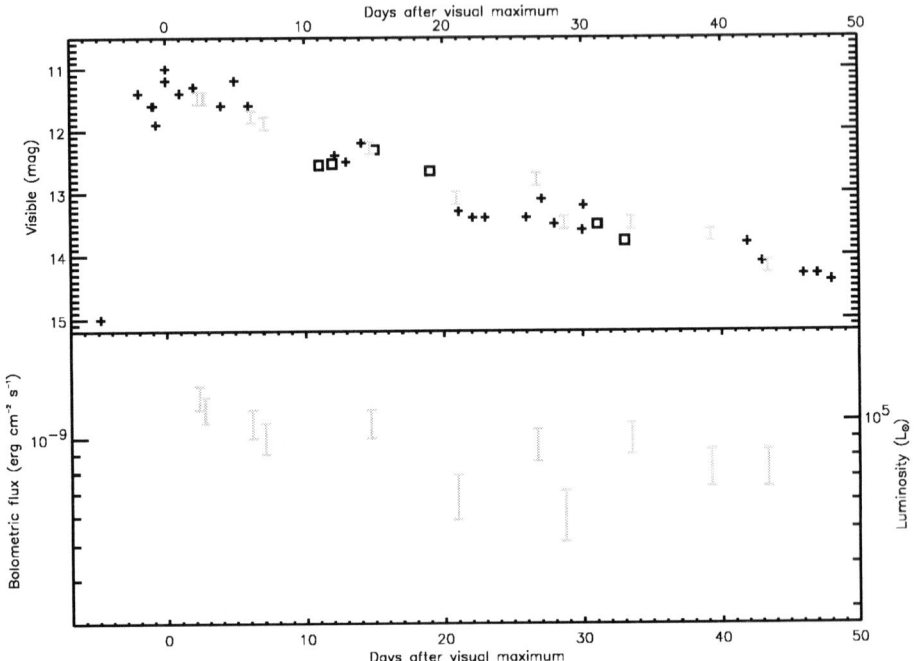

FIGURE 5. Fit to lightcurve of nova LMC 1988#1 (pluses are visual estimates and squares are V band photometry taken from the IAUC) with a sequence of nova model atmospheres. The bolometric lightcurve was constructed by using the result of multi-epoch model atmosphere analyses of the observed IUE and optical spectra. The error bars on the synthetic V band magnitudes were computed from the best fit PHOENIX models.

nova atmosphere can only be adequately described by these NLTE model calculations.

Results for nova LMC 1988#1

In Fig. 4 we show a typical fit of a synthetic spectrum to an observed nova spectrum, in this case for nova LMC 1988#1. The quality of the fit is very good, although only small parameter adjustments have been made in the model temperature T_{model} and the overall metal abundance (here set to 1/3 the solar metal abundances). The optical iron lines are fit very well, this is only possible with these very detailed NLTE model calculations. It is very important to simultaneously fit the multi-wavelength spectrum of novae in order to minimize errors in the determination of parameters [see 22, for details]. Fitting models to a multi-epoch sequence of observations allows us to prove that the bolometric luminosity of the nova is constant, as shown in Fig. 5. This also allows us to better constrain the derived parameters (such as abundances) and to connect the early phases to later nebular phase analyses. We have done analyses for a number of novae, see [20] for results for nova LMC 1991.

SUMMARY AND CONCLUSIONS

In this paper, we have presented a new set of model atmospheres for novae during the early phase of the outburst, with significantly improved input physics. Modern numerical methods that allow the detailed treatment of thousands of NLTE levels make it now possible to include the majority of the line blanketing opacity in detailed NLTE. LTE background lines from species that we do not treat explicitly in NLTE are not important for the structure of the atmosphere. They do, however, contribute some of the opacity and are necessary in order to reproduce some of the details in the synthetic spectra, particularly in the UV.

We have discussed the synthetic spectra for a number of models in detail. The comparison with LTE and simplified NLTE models for a number of specific wavelength ranges shows that (unfortunately) the detailed models cannot easily be approximated by simplified approaches. The addition of a detailed NLTE treatment for Fe I and (more importantly) Fe III has changed the synthetic spectra compared to our previous results [5], which in turn improves the fits to observed nova spectra (Schwarz et al, in preparation). Therefore, it is important to include a number of ionization stages in detailed NLTE to correctly model the line strengths. Our models span a wide range of (model) temperatures. These correspond to different times of the evolution of novae: In the early wind phase we have $T_{model} \approx 10000-15000\,\mathrm{K}$, while in the later pre-nebular stage $T_{model} \approx 30000\,\mathrm{K}$. Thus the analysis of a time sequence of nova spectra can be used to reduce the error bars of, e.g., abundances and to check for the internal consistency of the solution.

ACKNOWLEDGMENTS

We thank our many collaborators who have contributed to the development of PHOENIX, in particular we would like to thank David Branch, Steve Shore, Jason Aufdenberg, Andreas Schweitzer, and France Allard. This work was supported in part by NASA grants NAG5-9396, NAG5-9222, NAG5-8425, and NAG5-3619, NSF grants AST-0086246 and AST-9720704, as well as HST-GO-08671.09-A and GO-0-1013C to the University of Georgia; and by NASA grants NAG5-3505 & NAG5-12127 and an IBM SUR grant to the University of Oklahoma. This work was supported in part by the Pôle Scientifique de Modélisation Numérique at ENS-Lyon. Some of the calculations presented in this paper were performed on the IBM SP2 and SGI Origin 2000 of the UGA UCNS, at the San Diego Supercomputer Center (SDSC), and at the National Center for Supercomputing Applications (NCSA), with support from the National Science Foundation, and at the NERSC with support from the DoE. We thank all these institutions for a generous allocation of computer time.

REFERENCES

1. Hauschildt, P. H., Wehrse, R., Starrfield, S., and Shaviv, G., *ApJ*, **393**, 307 (1992).
2. Hauschildt, P. H., Starrfield, S., Austin, S. J., Wagner, R. M., Shore, S. N., and Sonneborn, G., *ApJ*, **422**, 831 (1994).
3. Hauschildt, P. H., Starrfield, S., Shore, S. N., Gonzales-Riestra, R., Sonneborn, G., and Allard, F., *AJ*, **108**, 1008 (1994).
4. Hauschildt, P. H., Starrfield, S., Shore, S. N., Allard, F., and Baron, E., *ApJ*, **447**, 829 (1995).
5. Hauschildt, P., Baron, E., Starrfield, S., and Allard, F., *ApJ*, **462**, 386 (1996).
6. Schwarz, G. J., Hauschildt, P. H., Starrfield, S., Baron, E., Allard, F., Shore, S. N., and Sonneborn, G., *MNRAS*, **284**, 669 (1997).
7. Hauschildt, P. H., Schwarz, G. J., Baron, E., Starrfield, S., Shore, S., and Allard, F., *ApJ*, **490**, 803 (1997).
8. Short, C. I., Hauschildt, P. H., and Baron, E., *ApJ*, **525**, 375–385 (1999), URL http://adsabs.harvard.edu/cgi-bin/nph-bib_query?bibcode=1999A%pJ...525..375S&db_key=AST.
9. Short, C. I., Hauschildt, P. H., Starrfield, S., and Baron, E., *ApJ*, **547**, 1057–1070 (2001), URL http://adsabs.harvard.edu/cgi-bin/nph-bib_query?bibcode=2001A%pJ...547.1057S&db_key=AST.
10. Hauschildt, P. H., *JQSRT*, **47**, 433 (1992).
11. Hauschildt, P. H., Störzer, H., and Baron, E., *JQSRT*, **51**, 875 (1994).
12. Allard, F., and Hauschildt, P. H., *ApJ*, **445**, 433 (1995).
13. Barman, T. S., *Ph.D. Thesis* (2002).
14. Hauschildt, P. H., *JQSRT*, **50**, 301 (1993).
15. Hauschildt, P. H., and Baron, E., *JQSRT*, **54**, 987 (1995).
16. Hauschildt, P. H., Baron, E., and Allard, F., *ApJ*, **483**, 390 (1997).
17. Baron, E., and Hauschildt, P. H., *ApJ*, **495**, 370 (1998).
18. Hauschildt, P. H., Lowenthal, D. K., and Baron, E., *ApJS*, **134**, 323–329 (2001), URL http://adsabs.harvard.edu/cgi-bin/nph-bib_query?bibcode=2001A%pJS..134..323H&db_key=AST.
19. Kurucz, R. L., Atomic data for opacity calculations, Kurucz CD-ROM No. 1 (1994).
20. Schwarz, G. J., Shore, S. N., Starrfield, S., Hauschildt, P. H., Della Valle, M., and Baron, E., *MNRAS*, **320**, 103–123 (2001), URL http://adsabs.harvard.edu/cgi-bin/nph-bib_query?bibcode=2001M%NRAS.320..103S&db_key=AST.
21. Bath, G. T., and Shaviv, G., *MNRAS*, **197**, 305 (1976).
22. Pistinner, S., Shaviv, G., Hauschildt, P. H., and Starrfield, S., *ApJ*, **451**, 451 (1995).
23. Baron, E., Hauschildt, P. H., and Mezzacapa, A., *MNRAS*, **278**, 763 (1996).
24. Hauschildt, P. H., and Baron, E., *Journal of Computational and Applied Mathematics*, **102**, 41–63 (1999).
25. Williams, R. E., Hamuy, M., Phillips, M. M., Heathcote, S. R., Wells, L., and Navarette, M., *ApJ*, **376**, 721 (1991).

Classical Novae as Super-Eddington Objects

Nir J. Shaviv

Racah Institute of Physics, Hebrew University, Jerusalem 91904, Israel

Abstract. Several of the inconsistencies plaguing the field of novae are resolved once we consider novae to be steady state super-Eddington objects. In particular, we show that the super-Eddington shell burning state is a natural consequence of the equations of stellar structure, and that the predicted mass loss in the super-Eddington state agrees with nova observations. We also find that the transition phase of novae can be naturally explained as "stagnating" winds.

INTRODUCTION

Classical novae exhibit long duration super-Eddington luminosities while in their eruptive state. At least, this is the conclusion that should be reached when combining that the peak luminosity of novae (with $M_{WD} \gtrsim 0.5 M_\odot$, Livio 1992) is always super-Eddington and that in all cases where the bolometric evolution was recovered (using UV observations), it was shown to decay slowly (e.g., Friedjung 1987, Schwarz et al. 2001, Shaviv 2001b). Thus, if classical novae are super-Eddington for durations much longer than any relevant dynamical time scale, two basic questions arise:
• *How can super-Eddington objects exist for durations much longer than their dynamical time scale?*
• *Even if such super-Eddington states exist, why do novae choose this state instead of following the known core-mass luminosity relation (CMLR)?*
  A seemingly unrelated question, which we try to address as well, is
• *Why do the theoretical simulations consistently under predict the amount of ejected mass, as compared with actual observational determinations of ejecta mass?*

The notion that novae are super-Eddington contradicts the common wisdom usually invoked in which objects cannot shine beyond their classical Eddington limit, L_{Edd}, since no hydrostatic solution exists. In other words, if objects do pass L_{Edd}, they are highly dynamic. They have no steady state, and a huge mass loss should occur since their atmospheres are then gravitationally unbound and should therefore be expelled. Thus, classical novae according to this picture, can pass L_{Edd} but only for a short duration corresponding to the time it takes them to dynamically stabilize after the onset of the thermonuclear runaway (TNR). This is indeed seen in detailed 1D numerical simulations of nova TNRs, where novae can be super-Eddington but only for several thousand seconds (e.g., Starrfield, 1989). However, once they do stabilize, they are expected and indeed do reach in the simulations a state given by the CMLR which describes the Hydrogen shell burning state. Namely, we naively expect to find no steady state super-Eddington atmospheres. This, however, is not the case in nature.

THE SUPER-EDDINGTON STATE

To existence of a super-Eddington state becomes natural, once we consider that:

1. Atmospheres become unstable as they approach the Eddington limit. In addition to instabilities that operate under various special conditions (e.g., Photon bubbles in strong magnetic fields or the s-mode instability under special opacity laws), two instabilities were found to operate in Thomson scattering atmospheres (Shaviv 2001a). Moreover, one of these instabilities does not depend on the boundary conditions and is therefore extremely general. It implies that *all atmospheres will become unstable already before reaching the Eddington limit*.
2. The effective opacity relevant for the calculation of the radiative force on an inhomogeneous atmosphere is not necessarily the microscopic opacity (Shaviv 1998). Instead, it is given by $\kappa_V^{eff} \equiv \langle F\kappa_V \rangle_V / \langle F \rangle_V$. The situation is very similar to the Rosseland vs. Force opacity means used in non-gray atmospheres, where the inhomogeneities are in frequency space as opposed to real space. For the special case of Thomson scattering, the effective opacity is always reduced.

To summarize, we find that as atmospheres approach their classical Eddington limit, they will necessarily become inhomogeneous. These inhomogeneities will necessarily reduce their effective opacity such that the effective Eddington limit will not be surpassed even though the luminosity can be super-classical-Eddington. This takes place in the external part of luminous objects, where the radiation diffusion time scale is shorter than the dynamical time scale of the atmosphere. Further inside the object, convection is necessarily excited such that the total energy flux may be super-Eddington, with the radiative part of it necessarily being sub-Eddington and with the convective flux carrying the difference.

Nevertheless, one of the key features of super-Eddington atmospheres is that a wind will necessarily be accelerated. It is not the catastrophic wind naively expected from super-Eddington conditions, however, it is going to be significant.

SUPER-EDDINGTON WINDS

The atmosphere can remain effectively sub-Eddington while being classically super-Eddington, only as long as the inhomogeneities comprising the atmosphere are optically thick. Clearly however, this assumption should break at some point where the density is low enough. From that radius upwards, the radiative force overcomes the gravitational pull and a wind is generated.

The mass loss rate can then be obtained by identifying the sonic point of a steady state wind with the critical point, which is the radius where the radiative and gravitational forces balance each other. We then have $\dot{M} = 4\pi R^2 \rho_{critical} v_{sonic}$. Furthermore, this can generally be reduced to the form of

$$\dot{M} = \frac{W(\Gamma)(L - L_{Edd})}{cv_s}, \tag{1}$$

where W is a dimensionless wind "function". In principle, W can be calculated from first principles only after the nonlinear state of the inhomogeneities is fully understood. This however is still lacking as it requires elaborate 3D numerical simulations of the nonlinear steady state. Nevertheless, it can be done in several phenomenological models which only depend on geometrical parameters such as the average size of the inhomogeneities in units of the scale height ($\beta \equiv d/l_p$), the average ratio between the surface area and volume of the blobs in units of the blob size (Ξ), and the volume filling factor α of the dense blobs. For example, in the limit where the blobs are optically thick, one obtains that $W = 3\Xi/32\sqrt{\nu}\alpha\beta(1-\alpha)^2$ (Shaviv 2001b). Here ν is the ratio between the effective and adiabatic speeds of sound. Thus, W depends only on geometrical factors. It does not depend explicitly on the Eddington parameter Γ. Moreover, typical values of $W \sim 1$-10 are obtained.

The mass loss predicted by the super-Eddington theory (eq. 1) was compared with observations of super-Eddington objects which have good observational data. These were two novae which are not very fast and which have the best determined *absolute* bolometric evolution: FH-Ser and LMC 1988 #1. The theory was also applied to the Luminous Blue Variable star η-Car.

For the two novae, we find that the predicted mass loss rates agree with their observations if $W \approx 10 \pm 5$, which is clearly consistent with the theoretical estimate for W. The agreement is also with the temporal evolution of the velocities, if those are taken to be the primary absorption line component.

Using $W \approx 10 \pm 5$, the mass loss equation can also be applied to η-Car, which is an entirely different object from novae (in mass, mass loss rate and duration, photospheric size etc), yet, the predicted integrated mass loss is in total agreement with the observed 1-$2 M_\odot$ of ejecta, while the terminal velocity is consistently predicted as well. For more information on the models, see Shaviv (2001b)

STEADY STATE SHELL BURNING

The next step is to show *why* novae become super-Eddington to begin with. In particular, the nova shell burnings steady state should be given by the CMLR (Paczynski 1970), counter to observations. Given the contradiction, we look for new solutions to the stellar structure equations for systems with shell burning. This should describe the steady state of novae after they undergo a TNR. Unlike the standard derivation of the CMLR, we allow the atmospheres to be inhomogeneous. Namely, if the Eddington parameter is larger than a threshold (taken to be 0.85) the effective opacity is reduced. We still do not know the exact behavior of κ_{eff} as a function of Γ. This could later be obtained by comparing the steady state obtained to the actual observations of novae, or through detailed 3D radiative hydro simulations. At this point however, we want to show that using a reasonable behavior of $\kappa_{eff}(\Gamma)$, a super-Eddington steady state does exist.

We take $\kappa_{eff}(\Gamma > \Gamma_{crit} = 0.85) = \kappa_0 \Gamma_{crit}/\Gamma$ and $W = 10$ (somewhat different choices do not change the conclusions). We find that the CMLR obtains a super-Eddington branch. The main differences between the structure in this branch and the structure in the sub-Eddington branch are the following: (1) The super-Eddington branch has a SEW at the top of it, with the photosphere located in the wind. The sub-Eddington branch has no wind (though it could have one if a non-Thomson opacity is considered). Since the wind is "heavy" the actual luminosity at the photosphere could be sub-Eddington. (2) The super-Eddington branch has a convective layer that penetrates into the burning shell (most of the energy is actually released in the convection zone). In the sub-Eddington branch, the burning shell is all radiative. (3) The luminosity in the sub-Eddington branch is not a function of the core mass. It is a function of the core mass in the super-Eddington branch.

Except for the very fast novae, the shell burning evolution can then be described as a steady state which slowly evolves due to mass loss (e.g., as is done by Kato & Hachisu 1994, for sub-Eddington evolution and opacity driven winds). The preliminary results show that predictions for $L(t)$, $T(t)$, v_∞ and M_{ejecta} agree with observations.

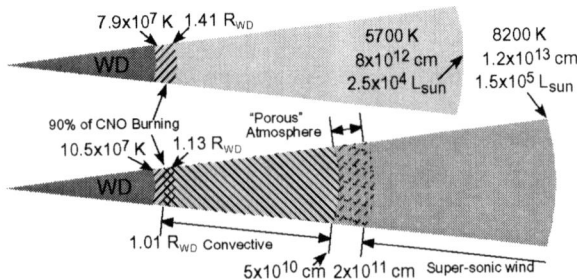

FIGURE 1. The shell structure of a CNO shell burning white dwarf with a $10^{-4} M_\odot$ envelope. Unlike the top panel, the bottom panel describes an atmosphere in which the radiative instability is taken into account, such that it can become porous. This allows the occurrence of the described steady state. The luminosity is super-Eddington. A convection zone arises and its inner boundary is well within the nuclear burning zone. A continuum driven wind is launched such that the photosphere is in a wind.

A TRANSITION PHASE?

An interesting situation can potentially arise if the mass loss predicted requires too large a luminosity to push the ejecta to infinity. This situation is in fact a realization of the prediction by Owocki and Gayley (1997), who first suggested that a hypothetically large mass loss rate would result with a stagnating wind. Since no steady state can obviously be obtained, they suggested that a variable state will exit.

The first analysis performed is a 1D numerical hydro-simulation, which will be elaborated elsewhere (Owocki & Shaviv 2002). A sample result is given in fig. 2. Although there are no indications that the structure should be purely radial, the limited (and preliminary) 1D results appear to qualitatively give a typical transition phase behavior. Thus, we hypothesize that this is the origin of the transition phase (excluding of course the big dips which presumably arise from dust formation).

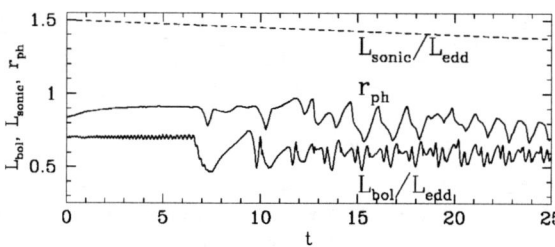

FIGURE 2. Preliminary results of 1D hydro-simulations (Owocki & Shaviv 2002) where the luminosity of a system is slowly reduced, such that at some point the wind stagnates before reaching $r \to \infty$. The radial and temporal units are $10 r_{sonic}$ and r_{sonic}/v_s. The result resembles the behavior in the transition phase.

DISCUSSION

To summarize, several of the open questions pertaining to classical nova eruptions are resolved if we understand them as steady state super-Eddington objects. As such, the observed super-Eddington luminosities, the temperature and velocity evolution, as well as the mass loss rate are all naturally explained. Moreover, we have seen that by allowing atmospheres to become porous, not only is a steady state possible, it also becomes a viable solution of the shell burning stellar structure equations.

One of the interesting results, is that the velocities predicted from the super-Eddington states are those observed as the primary absorption component. These are typically half as those observed for example in the "diffuse enhanced" or "Orion lines". Thus, the prediction is that the bulk of the mass outflow which arises in the super-Eddington steady state, should be with the primary velocity. Indeed, if one looks for example at the ejecta of FH Ser years after the eruption, the bulk velocity is similar to the primary component velocity. According to this picture, the higher speed components are therefore significantly less important in the mass loss budget. This however does not solve the question of their origin.

One possibility is that high velocity components are the result of super-Eddington radiative driving of material that is already present outside the critical radius *after* steady state super-Eddington shell burning is established, but *before* the wind has had time to be established. As a result, this external material is accelerated with the full luminosity, which later falls down with radius as it is "consumed" to accelerate the primary component (which therefore reaches lower velocities). The pre-existing tenuous material could be a result of a shock wind from the early TNR/dynamic phase. It is hard to estimate from first principles the mass that should be contained in it, but an upper limit can be given. It cannot be larger than the amount of material that will exist in the steady state wind over a radius of order the sonic radius.

Nevertheless, one can estimate from Shaviv (2000b) the velocity that the high component should have. Optically thin material sitting above the photosphere before the wind begins to accelerate will itself accelerate with the base flux of $\Gamma_0 L_{edd}$. We can compare this to the acceleration in the primary component in which energy was used to up accelerate the "heavy" wind, thereby leaving only an energy of flux of $\Gamma_{obs} L_{edd}$ at $r \to \infty$. Therefore, at $r \to \infty$ the initially tenuous material will have a larger terminal kinetic energy such that: $\frac{1}{2}\dot{M}v_\infty^2 \approx \frac{1}{2}\dot{M}v_{\infty,0}^2 + (\Gamma_0 - \Gamma_{obs})L_{edd}$, where $v_{\infty,0}$ is the velocity of the steady state wind, that is, of the primary component. Taking the data for FH Ser on day 9 (which is the earliest for which velocity components are present) and *nominal* values for the FH Ser super-Eddington state from Shaviv (2000b), we obtain $v_\infty \approx 1375$ km/s. This is similar to the $1300 - 1800$ km/s seen for this component. Moreover, as a wind begins to buildup above the sonic point, initially lower material will be accelerated with a progressively lower flux. This will give rise to a Hubble flow in the tenuous fast component, in which material at larger radii has larger velocities.

Another noteworthy point is that the integrated ejecta masses observed agree with the mass loss rate predicted by the super-Eddington theory. This implies that if one assumes the envelope masses at TNR to be roughly the observed ejected mass, then consistent nova eruptions are obtained with proper luminosity, temperature, mass loss rate and velocity evolution. The catch is that this envelope mass is typically an order of magnitude larger than the threshold masses required to trigger a TNR. The problem cannot be resolved by "dredging up" ten times as much mass as the envelope from the WD because the hydrogen fraction seen in the ejecta is not extremely small. Thus, we can conclude that the observers properly measure the ejected masses which are consistent with the super-Eddington theory describing the steady state *after* TNR. However, we theoreticians do not fully understand the mechanisms regulating the TNR threshold, which is obviously suppressed enough to allow more accumulation of mass before triggering a TNR.

TABLE 1: The emerging "super-Eddington" picture.

Phase	Schematic Diagram (not to scale!)
TNR. The nova eruption begins with a TNR. Initially, the system is dynamic. Some mass loss can occur from shocks or radiation driving. However, the amount of mass lost this way is insignificant compared to the total mass ejected.	
Formation of the super-Eddington steady state. Even before the wind reaches an equilibrium, a steady state is reached with shell burning, convection and a "porous" envelope. Before the wind builds up, any tenuous material *already* present above the critical point will accelerate to *high* velocities.	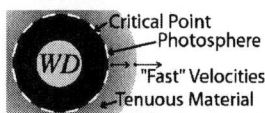
Acceleration of a thick wind. Once the critical point is established in the envelope, a super-Eddington thick wind is accelerated. With time, the steady state extends to include the wind. Since the wind is heavy, the radiation field is "used up" such that the wind reaches slower ("primary") terminal velocities.	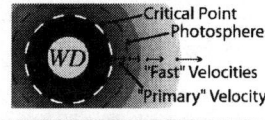
Stagnating wind conditions (Optional!). If the sonic radius shrinks fast enough, but the luminosity does not, conditions may form for a stagnating wind. That is, a wind too heavy to be pushed to $r \to \infty$ is predicted such that there is no steady state. This could explain of the "oscillatory" type transition phase.	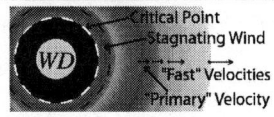
Sub-Eddington state. Once a sub-Eddington state is re-established, the thick wind stops and the bare object emerges. The ejected gas is predominantly expanding with the "primary" velocity. A tenuous part has velocities typically twice as high, with a high dispersion – its outer parts accelerate to higher velocities. Thus, a Hubble flow will exist in the tenuous material.	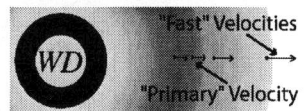

REFERENCES

1. Friedjung, M. 1987, Astron. Astrophys., 179, 164
2. Kato, M., & Hachisu, I. 1994, Astrophys. J., 437, 802
3. Livio, M. 1992, Astrophys. J., 393, 516
4. Owocki, S. P., & Gayley, K. G. 1997, in Nota, A., & Lamers, H., eds, ASP Conf. Ser. Vol. 120: Massive Stars in Transition, p. 121
5. Owocki, S. P., & Shaviv, N. J., in preparation
6. Paczynski, B. 1970, Acta Astronomica, 20, 47
7. Schwarz, G. J., Shore, S. N., Starrfield, S., Hauschildt, P. H., Della Valle, M., & Baron, E. 2001, Mon. Not. Roy. Astron. Soc., 320, 103
8. Shaviv, N. J. 1998, Astrophys. J., 494, L193
9. Shaviv, N. J. 2000, Astrophys. J., 532, L137
10. Shaviv, N. J. 2001a, Astrophys. J., 549, 1093
11. Shaviv, N. J. 2001b, Mon. Not. Roy. Astron. Soc., to appear
12. Starrfield, S. 1989, in Bode M. F., & Evans A., eds, Classical Novae. John Wiley & Sons, Oxford

A few comments on nova models

Michael Friedjung

Institut d'Astrophysique (CNRS), 98 bis Boulevard Arago, F-75014 Paris, France

Abstract. Some basic conditions, to be satisfied by any model of the ejected envelope, are given. Firstly it is recalled that reasons exist, for believing, that in early stages after optical maximum the expansion velocities in the envelope are generally larger at smaller radial distances. Indications, suggesting the opposite conclusion from the examination of emission lines, can be misleading. The highest velocities appear to be those of a wind. However Hubble flows are probably present in other stages for fast novae such as before optical maximum, while in addition the collisions between and overtaking of layers, can lead to a re-establishment of such flows in late stages. On the theoretical side the Kato and Hachisu models are physically consistent, but cannot explain the flows when luminosities are above the Thomson scattering Eddington limit. Finally published models based on detailed non-LTE radiative transfer calculations need to be modified, so as to become dynamically self consistent.

GENERAL INFORMATION FROM THE OBSERVATIONS

I would like to be provocative; the attempt of Sumner Starrfield to do this yesterday, was a flop. In addition, what I am going to say, has been changed to a certain extent, following discussions and what I have heard at this meeting. Before discussing theoretical models, it is still necessary to clear up, what I believe to be, misunderstandings. I am sorry if in places, particularly at the beginning, I am repeating what I have said on other occasions.

The old work of McLaughlin is often neglected by those making theoretical calculations, leading to the proposal of unrealistic models. Firstly he [6],[8] gave good reasons for believing that soon after optical maximum, layers deeper in the envelope at smaller radial distances from the centre of the envelope, had larger outwards expansion velocities. These reasons will not be repeated here, but I must emphasize that they are still valid. Accepting or not accepting them, has a decisive influence on what type of model, one believes in. His arguments are to a considerable extent based on the behaviour of line absorption, produced by the expanding material.

The observed narrowing of emission lines, considered alone, has however been used to support the idea that nova envelopes are like those of supernovae; the motions in supernova envelopes appear to be Hubble flows, where velocity increases with distance from the centre. The explanation of emission line narrowing is then that as the envelope becomes optically thinner, deeper material expanding at lower velocities becomes visible, while the emission of higher velocity material becomes negligible. Such narrowing can be misleading, because of two effects. Firstly observers may not take many spectra near optical maximum of very fast novae; the apparent emission line narrowing can then be just be that due to the progressive decline of a fast wind (not seen immediately after maximum), producing broad wings which fade (see [4]). In addition when ionisation is mainly due to radiation from a photosphere, an inrease of temperature of the photosphere will lead to an increase in the flux of more ionising radiation. In this situation regions with material at a certain stage of ionisation will move outwards with respect to the expanding material, so when the expansion velocity is smaller in layers at greater radial distances, the emisssion lines will narrow. In fact nova photospheres generally emit more and more radiation at shorter and shorter wavelengths, during the develpment of novae after optical maximum. I proposed the existence of such an effect many years ago for RR Tel, now thought of as being a aymbiotic nova [2]; the explanation is probably wrong for that object, but relevant to the understanding of classical novae. We all make mistakes.

I must also mention that important information can be obtained when nova fading is not continuous, that is when the optical brightness oscillates. The properties of the spectrum can then often be well correlated with the optical brightness, so when a nova brightens, the nova returns to an earlier stage of its development. Now the velocities of the different layers vary after optical maximum, usually tending to increase with time; such variatins can be large for the layers with the highest velocity. When there are oscillations, the highest velocities also oscillate, a classical example being the oscillation of the "Orion" absorption components of V603 Aql. In very old work of mine [1], I found a reasonable correlation between Orion velocities and photospheric radii calculated from Zanstra temperatures for three novae, with the slower "diffuse enhanced" velocity corresponding to an Orion velocity at an infinite radius. Obviously modern radiative transfer models, combined with multiwavelength observations, can give much more reliable radii, of for instance the radii where radiation is emitted at 5000 Å. In any case, such results suggest that the Orion velocities are, at least often, those of a continuously ejected wind, present in the deepest layers from which optical radiation can escape, that is near the photosphere (see [7]). Let us remember that such a wind does not need to be uniform; it could contain inhomogeneities.

We may note that the highest velocity "Orion" layers do not always show absorption due to hydrogen Balmer lines [9]. This may be partly due to the ionisation and excitation conditions, but could also indicate an underabundance in hydrogen. In such cases the inneremost last ejected layers, might be largely composed of material belonging to the white dwarf, ejected later than the hydrogen rich material, previously accreted from the companion star.

Let it finally be mentioned in this section, that observations indicate, that the material ejected by fast novae before optical maximum, may possess a Hubble flow. Such ejection could be due to an initial shock. The interaction of the wind with the initially ejected material might lead to the production of other layers, including that containing most of the ejected mass (which has usually been believed to be that reponsible for producing the line emission and absorption of what is called the "principal system"). Collisions could, as a result of instabilities for instance, result in the formation of separate clouds. Faster clouds could then overtake slower ones, leading in later stages to the re-establishment of something similar to a Hubble type flow of the ejected material. However the processes between these two stages are very important and cannot be neglected.

THEORETICAL CONSIDERATIONS

Several theories have been suggested to explain the continuously ejected winds of novae after optical maximum. The optically thick wind theory of [5] has been much criticized, but it is physically consistent. If I understand this theory correctly, the white dwarf component, umdergoing nuclear burning near its surface, has a luminosity slightly below the Thomson scattering Eddington limit. The temperature is large at the base of the wind, while the scale height of the atmosphere at the base is not very much less than its distance from the centre of, what is believed to be, an expanded white dwarf. The acceleration is then initially thermal as a result of gas pressure. Just after the critical point of the wind the OPAL opacity increases, so the luminosity is locally above the effective Eddington limit. The result is a strong acceleration by radiation pressure at large optical depths. The opacity increase is due to the presence of iron absorption lines. In stages of the development of a nova, when such acceleration is important in regions where the wind velocity difference over a mean photon free path is not much less than the thermal velocity, the different behaviour of line and continuous absorption, may lead to the calculations being inaccurate. In any case this type of theory cannot explain what occurs when the luminosity is above the Thomson scattering Eddington limit.

Nir Shaviv [10] (and proceedings of this conference) has proposed a wind theory for objects having super-Eddington luminosities. We shall have to wait for future developments of this theory. However Nir Shaviv's theory, like that of Kato and Hachisu and also like what I proposed in an old paper [3], does not appear to predict the high Orion ejection velocities, which appear to be characteristic of the wind. Certain colleagues would prefer, for theoretical reasons, the wind velocity to be that of the lower velocity principal system. As already explained, such a situation appears to be very unlikely. Something appears to be missing in theories proposed up to now.

In recent years a lot of effort has been put into the develpment of quite rigorous radiative transfer calculations for nova envelopes, so as to explain observed spectra. Non LTE effects have been largely taken into account. This is of course in principle much better than assuming, for instance, a black body energy distribution of a hard to define continuum, as I did in the past. Such calculations have however, up to now, assumed that the expansion velocity of the wind increases with increasing radius from the ejecting object. As we have seen, such a situation is unlikely.

In fact, in a recent paper [11] describing such calculations for Nova V1974 Cyg (Nova Cyg 1992), the calculated spectra agreed fairly well with observed low resolution ultraviolet spectra. The high resolution spectra did not agree so well, suggesting either a wrong velocity law or inhomogeneities or both. There are other problems concerning the wind dynamics in what is, in spite of such problems, a very interesting paper. According to fig. 5b of that paper, the radiative acceleration at the $\tau=1$ at 5000 Å photosphere is near $1\ 10^2$ cm s^{-2} at about 8 days after the maximum radius of that photosphere and less at earlier dates. If the acceleration stayed constant, the wind would, for such an acceleration, reach the assumed terminal velocity of 2000 km s^{-1} at about $2\ 10^{14}$ cm or more than 10^2 of the photospheric radius. Proper calculations of accelerations at different radial distances need to made, but my impression is that radiative acceleration would become less effective at larger radii as, following re-emission at other wavelengths, radiation escaped through optically thin windows. In addition, if we look at the calculated mass loss rates, one finds that the kinetic energy flux acquired by that part of the wind first ejected about 5 days after the maximum photospheric radius, is near $2 \; 10^{38}$ ergs s^{-1}. The assumed luminosity of 50 000 solar luminosities is $1.9 \; 10^{38}$ ergs s^{-1}. The radiative flux would have had difficulties in accelerating the wind, if the acceleration was radiative, while the loss of radiation necessary to produce acceleration, would have a significant effect on the radiative transfer. Though multiple scattereing can lead to a wind having a much larger momentum than the radiation accelerating it, it is much more difficult to violate the conservation of energy! It might be possible to overcome such problems by changing some parameters, but the most reasonable solution is for an optically thick wind to be mainly accelerated at large optical depths, far below the photosphere, as for instance in my very old work [1] or that of Kato and Hachisu.

I must conclude that much work remains to be done, before processes in nova envelopes soon after optical maximum, are properly understood.

REFERENCES

1. Friedjung, M., *MNRAS*, **132**, 317-336 (1966)
2. Friedjung, M., *MNRAS*, **133**, 401-410 (1966)
3. Friedjung, M., *AA*, **31**, 373-381 (1981)
4. Friedjung, M., Mikolajewska, J., Mikolajewsi, M., *A&A*, **348** 475-478 (1999)
5. Kato, M., Hachisu, I., *ApJ*, **437**, 802-826 (1994)
6. McLaughlin, D.B., *PASP*, **59**, 244-249 (1947)
7. McLaughlin, D.B., "The Spectra of Novae", in *Stellar Atmospheres*, edited by Jesse L. Greenstein, (Vol VI of *Stars and Stellar Systems*), The University of Chicago, Press, Chicago, 1960, pp 585-652
8. McLaughlin, D.B., *Ann. d'Ap*, **27**, 496-497 (1964)
9. Payne-Gaposchkin, C., *The Galactic Novae*, North Holland Publishing Company, Amsterdam 1957
10. Shaviv, N., *MNRAS*, **326**, 126-146, 2001
11. Short, C.I., Hauschildt, P.H., Starrfield, S., Baron, E., *ApJ*, **547**, 1057-1070 (2001)

Formation and evolution of dust in novae

J. M. C. Rawlings* and A. Evans[†]

*Department of Physics and Astronomy, University College London, Gower Street, London, WC1E 6BT, UK
[†]Department of Physics, Keele University, Keele, Staffordshire, ST5 5BG, UK

Abstract. We discuss the formation and evolution dust in nova winds. The ability of a nova to produce carbon and/or silicate dust is determined by the propagation of ionization fronts through the ejected material rather than by a single nova parameter (such as speed class). We also describe the changing physical and chemical properties of the dust as the stellar and gaseous remnants evolve.

INTRODUCTION

The formation of dust in nova winds is a well-observed phenomenon, both at infrared (e.g. [3, 7, 9] and references therein) and ultraviolet (e.g. [18]) wavelengths. While carbon is a major grain component in novae, it is clear that some form of silicate, a 'UIR' carrier and silicon carbide are also often present; a summary of the dust production characteristics of recent novae is given by Gehrz et al. [8]. An exciting development is the confirmation that nova grains are a constituent of carbonaceous chondrite meteorites [1, 10].

Any dust in the nova environment must be heavily processed, both by the ultraviolet (UV) radiation field (e.g. [4]) and by the ambient gas [11, 12, 13]. Here we consider the case of a carbon-rich wind (C>O) in which carbon dusts are the main condensate.

CONDITIONS FOR DUST FORMATION

We first note that the assumption that nova ejecta are spherically homogeneous at all times is almost certainly incorrect: observations of nova remnants show the ejecta to be shell-like and clumpy. Therefore ejecta parameters inferred from observations (e.g. density, temperature and elemental abundances), and averaged over the ejecta, are *unlikely to be representative of the environment in which dust forms.*

Rawlings [15] and co-workers [14, 16] have carried out detailed examination of the molecular chemistry that leads to the formation of nucleation sites in nova ejecta. They conclude that, while dust grains are resilient to the intense nova UV radiation field, *the initiating molecular chemistry and the nucleation sites themselves are highly vulnerable to the UV flux and hence, to the state of ionization of the ejecta.*

Rawlings & Williams [16] argued that, for carbon dust formation, the chemically destructive oxygen must be locked up in CO, otherwise atomic oxygen is available to 'burn' the grains. This requires that the chemistry occurs in the 'C I' region, where CO is

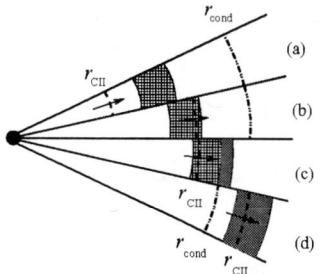

FIGURE 1. The dust formating potential of novae with different ionization strutures: (a) and (b) show the structure within the ejecta of a dustless nova, whilst (c) and (d) show the development of a dust-forming nova. Thick dot-dash lines are location of condensation radius r_c, hatched areas are the density-enhanced shell of ejecta at the head of the outflow in which grain formation occurs, thick dashed lines are the location of the carbon (C II) ionization front r_{CII} which is moving through the ejecta as indicated; the shaded areas repesent the regions of dust formation.

shielded; as the lowest states of ionization are located furthest from the stellar remnant, grain formation and growth must occur *in a thin dense shell at the outer edge of the ejecta*. Such a configuration is naturally provided by a wind that declines with time (e.g. [13]) and is also consistent with the visual light curves of several novae. For example the light curves of V842 Cen and V705 Cas showed deep dust minima; however, the minima lasted for only a few tens of days, indicative of the break-up of a thin shell into clumps.

The kinetic pathways to the formation of silicates are poorly understood. However, the environment in which silicates form must have O>C by number, so that excess carbon is locked up in CO. It is likely that a key stage in the nucleation of silicates involves the formation of simple diatomic molecules (e.g. SiO) whose chemistry (like that of CO) is driven by H_2 in all but the densest environments. SiO is less strongly bound than CO so the need to shield SiO from the radiation field is even greater than it is for CO. In the case of SiO the rôle of the neutral carbon in shielding the SiO region must be played by a species that (i) is sufficiently abundant and (ii) has sufficiently low ionization potential (IP). We suggest that this rôle is played by silicon itself: this would mean that basic molecular chemistry cannot proceed, and SiO cannot survive, in a Si II region.

THE DUST-FORMING POTENTIAL OF NOVAE

The observed timescale for dust formation in novae is short, typically $\sim 1 - 20$ days. In view of the inhomogeneity of the ejecta, simple ionization models severely underestimate the time it takes to ionize the ejecta. The formation of dust will result in a sharp increase in the opacity to ionizing radiation, further retarding the advance of the damaging ionization fronts. This implies that, not only dust nucleation *but also growth* must, to a large extent, take place in the thin dense, neutral shell at the outer edge of the wind.

The ionization/density structure in nova winds must be fairly complex. Fig. 1 shows the devolpment of these structures in both dustless and dust-forming novae. In Fig. 1a the dense shell is expanding outwards towards the condensation radius r_c and is being 'pursued' by the carbon ionization front. If this front traverses the shell before the shell reaches r_c (Fig. 1b) then the ejecta are carbon-ionized and carbon dust formation will not ensue. If, on the other hand, the shell remains at least partially carbon neutral by the time that carbon ionization front reaches r_c (Fig. 1c) then dust formation will occur.

Similar considerations apply to the formation of silicate dust, but the relevant ionization front is that of silicon: complete ionization of silicon before the (oxygen-rich) ejecta reach the appropriate r_c inhibits the condensation of silicate grains. The fact that not all *carbon* dust producing novae also produce silicates may be related to the lower IP of silicon. The abundance of Si in the ejecta would be a crucial factor in determing whether or not a nova produces silicate dust, partly from the point of view of there being sufficient Si to condense into grains but primarily to retard the advance of the ionization front into the grain-forming region.

It is clear from the above that the crucial feature in determining dust formation is whether or not the ejecta are ionized before they arrive at the condensation distance appropriate to the nova, its wind and the condensate. This is likely to be determined by several parameters, such as the ejecta mass, the abundance of elements (like S, which has lower IP than C, or Mg/Fe/Na, which have lower IP than Si) which will tend to retard the advance of the ionization fronts, the bolometric luminosity of the nova etc. As noted by Evans [3], it is very unlikely that the dust-producing potential of a nova may can be assessed on the basis of a single nova parameter. Indeed, because of this it is not possible to draw any general general conclusions about how any nova parameter (for example, the speed class) can *directly* affect the viability of dust formation and its subsequent evolution. There are many free parameters and the effects of their variation must be considered in what is a very large parameter space.

PROCESSING OF CARBON DUST IN NOVAE

As the dust shell disperses its environment becomes increasingly more ionized and carbon grains experience a number of physical and chemical changes (cf. [4]), which relate both to the gas-phase chemistry and to the nature and behaviour of the grains. For carbon grains the most important effects of the change in the ionization on the grains are as follows:

1. The grains 'see' more UV flux as the ionization continua become optically thin. Consequently the equilibrium grain temperature rises.
2. Carbon grains become increasingly annealed. In a hot, dark, (atomic) hydrogen-rich environment, H-atoms will attack and penetrate the carbon lattice leading to a conversion of graphitic (sp^2) to polymeric (sp^3) hydrogenated amorphous carbon (HAC) [2].
3. Continued hydrogen attack will result in chemical erosion with simple species such as CH_4 as products. On the other hand the intense UV field anneals the grains, expelling hydrogen and converting sp^3 HAC to the sp^2 form and thence a hydrogen-poor amorphous carbon.

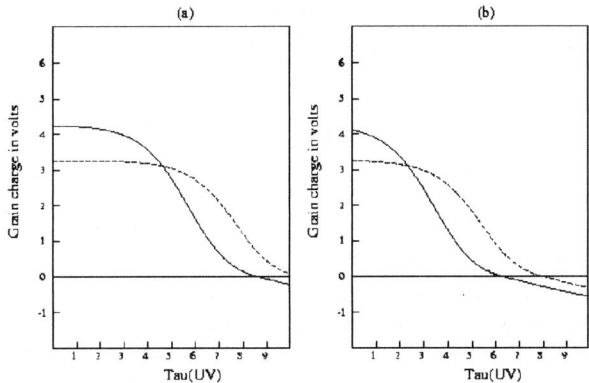

FIGURE 2. Grain charge in a nova wind. (a) Gas density $10^9\,\text{cm}^{-3}$, (b) gas density $10^{10}\,\text{cm}^{-3}$; gas temperature is 2000 K in both cases. Solid line, sp^2 carbon, dashed line, sp^3. See text for details.

As the ionization level rises the sp^2:sp^3 ratio increases. The effects of this are that

1. the band gap falls so that the 'extended red emission' (ERE) [17] associated with sp^3 carbon vanishes;
2. the grains are 'photodarkened'. This causes a change in their optical properties, which also results in a change in the grain temperature;
3. the photoelectric properties of the grains change, with implications for grain charge and survivability.

GRAIN CHARGE

In the nova environment it is likely that grains will have an electrostatic charge which may be large and significant with regards to their stability. Normally grain charge is determined by three factors: (i) accretion of electrons from the gas phase, (ii) accretion of singly charged (positive) ions from the gas phase, (iii) photoelectric emission of electrons from the grains by the UV radiation field. The timescales for these processes are very much shorter than the physical and chemical timescales in the ejecta of novae and we may approximate the grain charge at any point by its equilibrium value.

We have use standard formulae to determine the equilibrium charge on a grain due to electron and ion impact, and the photoelectric effect, taking into account the changing nature of the grains as they evolve and the changing state of ionization of the ejecta; typical results are shown in Fig. 2 for the C I region in the ejecta of a nova with $t_3 = 60$ days. Note the large positive charge on the grains at low optical depths, and the negative charge for $\tau_{uv} \gtrsim 6$. Note also the *change in the sign of charge* for some values of τ_{uv} that occurs as the grains are annealed. This behaviour has a major effect on the accretion of monomers from the gas phase, and hence on the rate at which the grains grow; these effects may be apparent in the growth of the carbon grains around V705 Cas [6].

CONCLUDING REMARKS

The regions in nova ejecta in which the chemical pathways to grain nucleation occur must be carbon and/or silicon neutral. Grains may or may not form in nova winds depending on whether the ejecta are carbon- (carbon grains) or silicon- (silicates) ionized before they reach the relevant condensation radius.

Once formed, the grains are resilient, even when exposed to the radiation field of the nova. However they undergo significant processing, with consequences for the growth of grains and their observational properties (e.g. UIR and ERE emission, β index).

Full details of this work will be published elsewhere.

REFERENCES

1. Amari S., Xia G., Nittlert L. R., Zinner, E., José, J., Hernanz, M., Lewis, R., 2001, ApJ, 551, 1065
2. Duley, W. W., Williams, D. A., 1988, MNRAS, 230, 1P
3. Evans, A., 1990. In *The Physics of Classical Novae*, p. 253, Eds A. Cassatella, R. Viotti, Springer Verlag, Berlin.
4. Evans, A., Rawlings, J. M. C., 1994, MNRAS, 269, 427
5. Evans, A., Geballe, T. R., Rawlings, J. M. C., Eyres, S. P. S., Davies, J. K., 1997, MNRAS, 292, 204
6. Evans, A. Smith, O., Tyne, V. H., Rawlings, J. M. C., Geballe, T. R., 2002, these proceedings
7. Gehrz, R. D. 1990. In *The Physics of Classical Novae*, p. 138, Eds A. Cassatella, R. Viotti, Springer Verlag, Berlin.
8. Gehrz, R. D., Truran, J. W., Williams, R. E., Starrfield, S., 1998, PASP, 110, 3.
9. Gehrz, R. D., 2002, these proceedings
10. José, J., Hernanz, M., Amari, S., Zinner, E., 2002, these proceedings
11. Mitchell, R. M., Evans, A., 1984, MNRAS, 209, 945
12. Mitchell, R. M., Evans, A., Albinson, J. S., 1986, MNRAS, 221, 663
13. Mitchell, R. M., Evans, A., Bode, M. F., 1983, MNRAS, 205, 1141
14. Pontefract, M., Rawlings, J. M. C., 2002. MNRAS submitted
15. Rawlings, J. M. C., 1988, MNRAS, 232, 507
16. Rawlings, J. M. C., Williams, D. A., 1990, MNRAS, 246, 208
17. Scott, A. D., Evans A., Rawlings, J. M. C., 1994, MNRAS,
18. Shore, S. N., Starrfield, S., Gonzalez-Riestra, R., Hauschildt, P. H., Sonneborn, G., 1994, Nature, 369, 539

The properties of the dust around Nova V705 Cas

A. Evans, O. Smith, V. H. Tyne*, J. M. C. Rawlings[†], T. R. Geballe** and S. P. S. Eyres[‡]

Department of Physics, Keele University, Keele, Staffordshire, ST5 5BG, UK
[†]*Department of Physics and Astronomy, University College London, Gower Street, London, WC1E 6BT, UK*
**Gemini Observatory, 670 N. A'ohoku Place, Hilo, HI 96720, USA*
[‡]*Centre for Astrophysics, University of Central Lancashire, Preston, PR1 2HE, UK*

Abstract. Nova V705 Cas (1993) was an archetypical dust-forming nova. It displayed a deep minimum in the visual light curve, and spectroscopic evidence for carbon, hydrocarbon and silicate dust. We report preliminary results of work in progress, in which we use the DUSTY code to determine the properties of the dust. Our results have implications for the rate of grain growth in the ejecta, and for the origin of the UIR features; the latter likely arise in hydrogenated amorphous carbon grains rather than in free-flying PAH molecules.

INTRODUCTION

V705 Cas (1993) was a typical dust-forming nova which showed evidence for carbon, silicate and hydrocarbon dust (Evans et al. 1997, Mason et al. 1998). Early IUE observations (Shore et al. 1994) showed that the grains grew to $\sim 0.2\,\mu$m shortly after dust condensation. Mason et al. (1998) modelled broadband infrared (IR) data using a combination of dust types, and concluded that the circumstellar dust shell consisted of carbon and silicate grains in the ratio $\sim 8.5:1$ by mass.

Evans et al. (1997) fitted a simple function of the form $B(T,\lambda)\lambda^{-\beta}$, where B is the Planck function and β ($\simeq 1$) is a constant (the so-called β-index for the dust), to IR spectra in the range 2–24 μm, and concluded that the grains eventually grew to $\sim 0.7\,\mu$m. However, given the quality of the data and the availability of the DUSTY code (Ivezić & Elitzur 1997, Ivezić, Nenkova & Elitzur 1999), we are now in a position to make a more sophisticated attempt at fitting the data. Here we report ongoing work in which we are using the DUSTY code to model the dust around V705 Cas.

METHODOLOGY

We are modelling the UKIRT data for 1994 August and October/November (Evans et al. 1997), for which we have near-simultaneous data over the wavelength range 2–4 μm, 7.5–13 μm, 16–24 μm, at resolution $\sim 300 - 1000$ (K,L bands) and ~ 60 (N,Q bands).

For a given central source temperature and dust composition, DUSTY calculates the radiative transfer through a spherically symmetric dust shell and determines the observed

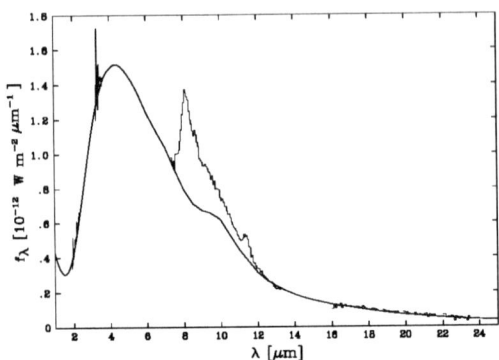

FIGURE 1. Fit of DUSTY model to 1994 August data for nova V705 Cas; thick curve is DUSTY model. Note the strong UIR featres in the 10 μm window; the apparent 'noise' at $\sim 2\,\mu$m and at $\sim 3\,\mu$m are the nebular lines, and the UIR features at 3.28 μm and 3.4 μm, respectively.

spectral energy distribution. We use a downhill simplex routine (Press et al. 1992) to explore the multi-dimensional parameter space and to determine the best fit between the data and the DUSTY output (see Tyne et al. 2002 for details).

In our preliminary investigations we have approximated the star by a blackbody with temperature 70 000 K (it turns out that the model IR spectrum is not sensitive to the temperature of the central object), and an amorphous carbon/warm silicate mix with a r^{-2} dust density distribution; we use optical constants from Hanner (1988; amorphous carbon) and Ossenkopff et al. (1992; warm silicate). We optimize the fit by varying the carbon:silicate ratio, the dust temperature at the inner edge of the dust shell, the grain size distribution, and the geometric and optical thickness of the dust shell. Further, as the relative placement of the spectra in the K, L, N and Q bands is uncertain by $\sim 10-20\%$ (a consequence of the uncertainty in the flux calibration), the placement of individual spectral components are adjusted in an iterative way to optimize the fit. Spectral features, namely the 'unidentified' IR (UIR) features and nebular lines, were removed before fitting the DUSTY output to the data.

The best fit to the 1994 August data is shown in Fig. 1, in which the spectral features (see above) have been restored; a similar fit has been obtained to the 1994 October/November data.

RESULTS AND DISCUSSION

We find that both sets of data are best fitted by a $\sim 0.85:0.15$ ($\pm \sim 0.1$) carbon:silicate mix, dust temperature at the inner boundary ~ 670 K, and a grain size distribution $n(a)\,da \propto a^{-q}\,da$, with $q \simeq 2.5$. Although the maximum and minimum grain sizes are not well constrained by the modelling, we find that

(i) grain growth does indeed occur but appears to be much more rapid than that expected on the basis of a uniform wind (see Evans et al. 1997);
(ii) the smallest grains have dimensions $\sim 0.003\,\mu$m.

Given that the charging of grains will have a significant impact on the rate at which grains grow (Rawlings & Evans 2002), the former conclusion is not unexpected. Furthermore, our determination of the minimum grain size is significant in determining the origin of the UIR features; these may in principle arise either from free-flying polycyclic aromatic hydrocarbon (PAH) molecules, or from the vibration of C−H bonds on the surface of hydrogenated amorphous carbon (HAC) grains (see Evans & Rawlings 1994 for a detailed discussion). The apparent dimensions of the smallest carbon grains in the size distribution suggest particles that are significantly larger than PAH molecules, consistent with the conclusion of Evans & Rawlings (1994) that free-flying PAH molecules would not survive the harsh radiation field of the nova and that the UIR features most likely arise in HAC.

CONCLUDING REMARKS

The co-existence of carbon and silicate dust ('chemical dichotomy') in the IR spectra of evolved stars is usually ascribed to the trapping of silicate dust in a disc arising from mass-loss during an earlier phase of evolution when the star was oxygen-rich, while carbon dust arises from more recent or current (carbon-rich) mass-loss (see e.g. Zijlstra et al. 2001 and references therein). In the case of V705 Cas, UIR emission was present in the 10 μm band when silicate was weak or absent (Evans et al. 1997). Novae may also have two well-separated mass-loss episodes with different chemistries to produce their chemical dichotomies or (more likely) there are steep abundance gradients in the ejecta. Either way there are interesting implications for the thermonuclear runaway and the nature of the nova explosion.

We stress that the work presented here is preliminary, in that the IR flux distribution can not be considered in isolation: the IR (and optical) nebular lines place constraints on the temperature of the stellar remnant and the model is not yet therefore in a 'joined up' state.

Full details of this work will be published elsewhere.

ACKNOWLEDGMENTS

TRG is supported by the Gemini Observatory, which is operated by the Association of Universities for Research in Astronomy, Inc., on behalf of the international Gemini partnership of Argentina, Australia, Brazil, Canada, Chile, the United Kingdom, and the United States of America.

REFERENCES

1. Evans, A., Rawlings, J. M. C., 1994, MNRAS, 269, 427
2. Evans, A., Geballe, T. R., Rawlings, J. M. C., Eyres, S. P. S., Davies, J. K., 1997, MNRAS, 292, 204
3. Hanner, M. S., 1988, NASA Conference Publication 3004, p. 22
4. Ivezić, Ž., Elitzur, M., 1997, MNRAS, 8287, 799
5. Ivezić, Ž., Nenkova, M., Elitzur, M., 1999, User Manual for DUSTY, University of Kentucky Internal Report
6. Mason, C. G., Gehrz, R. D., Woodward, C. E., Smilowitz, J. B., Hayward, T. L., Houck, J. R., 1998, ApJ, 494, 783
7. Ossenkopf, V., Henning, Th., Mathis, J. S., 1992, A&A, 261, 567
8. Press, W. H., Teukolsky, S. A., Vetterling, W. T., Flannery, B. P., 1992, Numerical Recipes in Fortran, Cambridge University Press
9. Rawlings, J. M. C., Evans, A., 2002, these proceedings
10. Shore, S. N., Starrfield, S., Gonzalez-Riestra, R., Hauschildt, P. H., Sonneborn, G., 1994, Nature, 369, 539
11. Tyne, V. H., Evans, A., Geballe, T. R., Eyres, S. P. S., Smalley, B., Dürbeck, H. W., 2002, MNRAS, in press
12. Zijlstra, A. A., Chapman, J. M., te Linkel Hekkert, P, Likkel, L., Comeron, F., Norris, R. P., Molster, F. J., Cohen, R. J., 2001, MNRAS, 322, 280

A solution to the transition phase in classical novae

Alon Retter

School of Physics, University of Sydney, NSW, 2006, Australia

Abstract. One century after the discovery of quasi-periodic oscillations in the optical light curve of Nova GK Per 1901 the cause of the transition phase in a certain part of the nova population is still unknown. Three years ago we suggested a solution for this problem and proposed a possible connection between the transition phase and intermediate polars (IPs). About 10% of the cataclysmic variable population are classified as IPs, which is consistent with the rarity ($\sim$15%) of the transition phase in novae. Recent observations of three novae seem to support our prediction. The connection is explained as follows: The nova outburst disrupts the accretion disc only in IPs. The recovery of the disc, a few weeks-months after the eruption, causes strong winds that block the radiation from the white dwarf, thus dust is not destroyed. If the winds are very strong as is probably the case in DQ Her (perhaps since its spin period is very short) this leads to a dust minimum.

INTRODUCTION

The optical light curve of a classical nova is typically characterized by a smooth decline. Certain novae show, however, a DQ Her-like deep minimum in the light curve while others have slow oscillations, during the so called 'transition phase' (Fig. 1). The minimum is understood by the formation of a dust envelope around the binary system. It is still not known, however, what causes the oscillations during the transition phase (Warner 1995) and why only a small fraction (about 15%) of the nova population has the transition phase.

The models offered so far to the oscillations during the transition phase (Bode & Evans 1989; Leibowitz 1993; Warner 1995) are:

- Oscillations of the common envelope that surrounds the binary system after the nova outburst.
- Dwarf-nova outbursts.
- Formation of dust blobs that move in and out of line of sight to the nova.
- Oscillations in the wind (see also Shaviv, these proceedings).
- Stellar oscillations of the hot white dwarf.

The first model can be rejected very easily as the common envelope phase ('fireball') lasts less than 1-2 days (Hauschildt, personal communication), so it is much shorter than the typical time scale of the transition phase. The second idea is almost certainly wrong as the accretion discs in post-novae are thermally stable (e.g. Retter & Naylor 2000), so they cannot have dwarf-nova outbursts for at least many decades following the nova eruption.

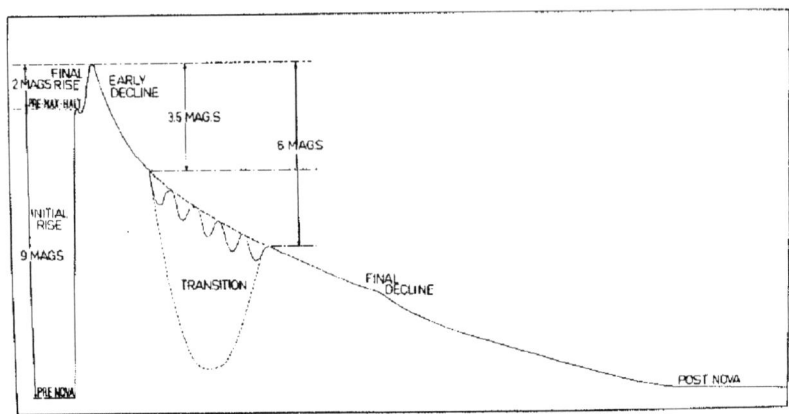

FIGURE 1. A schematic diagram of the optical light curve of classical novae (adapted from Warner 1995)

Retter, Liller & Gerradd (2000a) suggested another solution for this problem and we show here that it can be combined with two of the models listed above. The observations of Nova LZ Mus 1998, which had oscillations during the transition phase revealed a few periodicities in its optical light curve. Retter et al. thus classified LZ Mus as an IP candidate. IPs are cataclysmic variables in which the primary white dwarf has a moderate magnetic field and thus spins around its axis with a period shorter than the orbital period. Hernandz & Sala (this volume) detected X-ray emission from LZ Mus three years after its outburst despite its extremely large distance. This fact is consistent with the IP classification.

Retter et al. further proposed a possible connection between the transition phase and IPs and predicted that novae that have a transition phase should be IPs. About 10% of the cataclysmic variable population are IPs, which is consistent with the rarity ($\sim$15%) of the transition phase in novae. We note that these numbers represent lower limits as not all systems have been well studied. The correct ratios may be as high as $\sim$30%. Recent observations of three young novae seem to support our idea.

THE EARLY PRESENCE OF THE DISC IN YOUNG NOVAE

It was believed that the accretion disc is destroyed by the nova event and that it takes only a few decades for the disc to re-establish. In my Ph.D. thesis I studied this claim. Contrary to the common belief we found very strong evidence in at least two cases for the presence of the accretion disc only a few months after the nova outburst (Retter, Leibowitz & Ofek 1997; Retter, Leibowitz & Kovo-Kariti 1998). We note that the time-scales of the transition phase are similar to the time-scales of the appearance of the accretion disc in post-novae. It is still unknown whether the disc can survive the nova explosion. The various bright sources, which contribute to the optical light curve of postnovae, may, however, overcast the light from the disc and complicate the observations. We may have found now the solution to this question. This work suggests that the accretion disc is destroyed by the nova outburst only in IPs and that discs in other subclasses of CVs survive the nova event.

OBSERVATIONS AND ANALYSIS

Recent observations of three young novae seem to support the suggested connection between the transition phase and IPs. The optical light curve of Nova V1494 Aql 1999#2 showed transition phase oscillations with a quasi-period of $\sim$7 days (Kiss & Thomson 2000). A period of 3.2 h (presumably the orbital period) was discovered in its optical light curve (Retter et al. 2000b). Krautter et al. (2001) and Drake et al. (2002, in preparations) detected a 2523-s periodicity in two X-ray runs using Chandra. The short period can be interpreted as the spin period of the binary system and the nova can thus be classified as an IP. We note, however, the suggestion of an alternative model to the variation – stellar oscillations of the white dwarf (Krautter et al., these proceedings).

On the other hand, Nova V382 Vel 1999 had a smooth decline in the optical (Liller & Jones 2000) and X-ray observations did not reveal any short-term periodicity (Orio et al. 2001). So, it is unlikely that it is an IP.

In addition, Nova V1039 Cen 2001 had oscillations during the transition phase as well (http://vsnet.kusastro.kyoto-u.ac.jp/vsnet/etc/drawobs.cgi?text=CENnova2001). Preliminary analysis of 6 nights of optical photometry using small telescopes + CCDs shows several peaks in its power spectrum (Fig. 2). They can be explained within the IP model, so this nova could also be an IP, consistent with our prediction.

Additional possible support for our idea might come from the detection of oscillations in an X-ray source, which is located near the nucleus of M31. Osborne et al. (2001) proposed that the $\sim$865-sec periodicity represent the spin period of a magnetic white dwarf. The high X-ray luminosity and the transient nature of the object led them to propose that this is a recent novae. King, Osborne & Schenker (2002), however, argued against this suggestion.

DISCUSSION

There seems to be a strong connection between the transition phase in classical novae and IPs. Two famous cases are Nova GK Per 1901 (oscillations) and Nova DQ Her 1934 (minimum), which are well-known IPs. Another possible example is Nova V603 Aql 1918 (oscillations), which was suggested several times as an IP candidate (e.g. Schwarzenberg-Czerny, Udalski & Monier 1992) and whose X-ray light curve show strong variations (Mukai et al., this volume). We note, however, that there has been a long argument whether this nova is indeed an IP (e.g. Mukai et al., these proceedings).

We suggest that this link is connected with the presence of the accretion disc in post-novae. In IPs, the magnetic field truncates the inner part of the disc, making it less

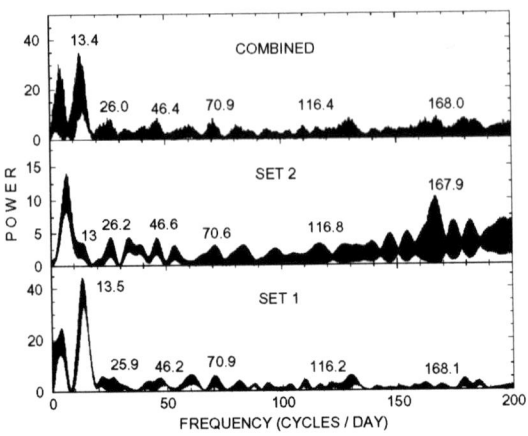

FIGURE 2. Power spectra of six nights of photometry of Nova V1039 Cen 2001 obtained in 2002. The lower two panels show two different sets and the top panel displays the combined data. The several groups of peaks may correspond to the orbital period, the spin period and some combinations of the two. V1039 Cen may thus be grouped with IPs.

massive than in non-magnetic systems (this is, by the way, the reason why IPs do not have dwarf-nova outbursts or only have short and weak outbursts). The nova outburst can, therefore, disrupt the disc in IPs. Its re-establishment and the subsequent interaction with the magnetosphere of the primary white dwarf is a violent process that forms strong winds at the inner part of the disc. The winds block the radiation from the hot white dwarf and the dust is not destroyed. The accretion disc oscillates until finally reaching stability at the end of the transition phase. In non-magnetic systems the disc is barely disturbed and becomes stable much faster and in polars (AM Her systems) there is no accretion disc at all. In both groups there is no transition phase since dust cannot be formed.

The long-term oscillations during the transition phase could be explained if the inner accretion disc lies very close to the co-rotation radius, so the disc and the magnetic field would rotate (almost) together at that point. The relative rotational timescale (the beat period) could be very long – of the order weeks, as observed in the transition phase.

If the winds are very strong, as might be the case in DQ Her (perhaps since its spin period is very short) this leads to the formation of more dust and to a minimum in the light curve. IPs with longer spin periods should show oscillations. Naturally, our model has a strong dependence on the inclination angle.

Further X-ray and optical observations of novae that have a transition phase should confirm or refute our suggestion.

ACKNOWLEDGMENTS

We thank Graham Wynn for a very useful discussion on the cause of the long-term oscillations and Marina Orio for pointing to us Osborne et al.'s work. AR is supported by an ARC grant.

REFERENCES

1. Bode M.F., Evans A., 1989, *Classical Novae*, Alden Press, Oxford
2. King A.R., Osborne J.P., Schenker K., 2002, *MNRAS*, 329, 43
3. Kiss, L.L., & Thomson, J.R. 2000, *A&A*, 355, L9
4. Krautter J. et al., 2001, *AGAb*, 18, 5
5. Leibowitz E.M., 1993, *ApJ*, 411, L29
6. Liller W., Jones A.F., 2000, *IBVS*, 5004, 1
7. Orio M. et al., 2001, *MNRAS*, 326, L13
8. Osborne J.P. et al., 2001, *A&A*, 378, 800
9. Retter A. & Naylor T., 2000, *MNRAS*, 319, 510
10. Retter A., Leibowitz E.M., Ofek E.O., 1997, *MNRAS*, 286, 745
11. Retter A., Leibowitz E.M., Kovo-Kariti O., 1998, *MNRAS*, 293, 145
12. Retter, A., Liller, W., & Gerradd, G. 2000a, in "Cataclysmic Variables: a 60th Birthday Symposium in Honour of Brian Warner", eds. Charles, P., King, A., O'Donoghue, D., Elsevier Science, *New Astronomy Review*, 40/1-2, p65
13. Retter, A., Cook, L., Novák, R., Saxton, J.M., Jensen, L.T., Korčaková, D., & Janík, J. 2000b, *IAUC*, 7537
14. Schwarzenberg-Czerny A., Udalski A., R. Monier, 1992, *ApJ*, 401, L19
15. Warner B., 1995, *Cataclysmic Variable Stars*, Cambridge University Press, Cambridge

Recurrent Novae as a progenitor system of Type Ia Supernovae

Izumi Hachisu* and Mariko Kato[†]

Department of Earth Science and Astronomy, University of Tokyo, Tokyo 153-8902, Japan
[†]*Department of Astronomy, Keio University, Yokohama 223-8521, Japan*

Abstract. Theoretical light curves of 8 recurrent novae in outburst are modeled to obtain various physical parameters. Light curve calculation includes reflection effects of the companion star and the accretion disk by the photosphere of the white dwarf (WD). We also include a radiation-induced warping instability of the accretion disk in order to reproduce the second peak of T CrB outbursts. The early visual light curves are well reproduced with an expanded WD photosphere of a thermonuclear runaway model on a very massive WD close to the Chandrasekhar mass limit, i.e., $M_{\rm WD} = 1.35 - 1.37 \, M_\odot$ except for CI Aql. The white dwarf mass of CI Aql is estimated to be $M_{\rm WD} = 1.2 \pm 0.05 \, M_\odot$. Optically thick winds, which blow from the WDs during the outbursts, play a key role in determining the nova duration and the speed of decline because the wind quickly reduces the envelope mass on the WD. Each envelope mass at the optical maximum is estimated, which indicates an average mass accretion rate on to the WDs during the quiescent phase before the last outburst. Although a large part of the envelope mass is blown in the wind, each WD can retain a substantial part of the envelope mass after hydrogen burning ends. Thus, we have obtained net mass-increasing rates of the WDs. These obtained values strongly indicate that the WDs in the recurrent novae have now grown up to near the Chandrasekhar mass limit and will soon explode as a Type Ia supernova if the WDs consist of carbon and oxygen. We have also clarified the reason why only T CrB shows a secondary maximum.

LIGHT CURVE ANALYSIS OF RECURRENT NOVAE

Recurrent novae are characterized by nova eruptions with its recurrence time scale from a decade to a century. Recurrent novae show some characteristic differences from classical novae: (1) heavy elements such as carbon, oxygen and neon are not enriched in ejecta but similar to the solar values, indicating that the white dwarf is not eroded; (2) very short recurrence periods from a decade to a century theoretically require very massive white dwarfs close to the Chandrasekhar mass limit. These two arguments indicate that the mass of the white dwarfs in recurrent novae increases toward the Chandrasekhar mass limit. We expect that, if the white dwarf is made of carbon and oxygen, the white dwarf ignites carbon at the center and explodes as a Type Ia supernova when it reaches the critical mass of $M_{\rm Ia} = 1.38 M_\odot$ [5].

In this contribution, we summarize our recent results on the light curve fittings of recurrent novae and examine whether or not the recurrent novae are a progenitor system of Type Ia supernovae. Eight among ten known recurrent novae, i.e., U Sco, V394 CrA, LMC RN, CI Aql, T CrB, RS Oph, V745 Sco, and V3890 Sgr are modeled to obtain various physical parameters. The first four systems have a main-sequence (or slightly-evolved main-sequence) companion and are further classified into the U Sco subclass.

The latter four systems have a red giant companion and are subgrouped into the T CrB subclass.

In the thermonuclear runaway model, white dwarf (WD) envelopes quickly expand to $\sim 10-100\ R_\odot$ or more and then the photospheric radius gradually shrinks to the original size of the white dwarfs. Correspondingly, the optical luminosity reaches its maximum at the maximum expansion of the photosphere and then decays toward the level in quiescent phase keeping the bolometric luminosity almost constant. Optically thick winds, blowing from the WD in the decay phase of novae, play a key role to determine the nova duration because a large part of the envelope mass is blown in the wind. When the envelope mass decreases to below the critical mass, the wind stops, and after that, the envelope mass is decreased only by nuclear burning. When the envelope mass decreases further, hydrogen shell-burning disappears and the WD enters a cooling phase.

When the WD photosphere is large enough, the visual light curve is mainly determined by the WD photosphere (early phase). As it shrinks to much smaller than a Roche lobe size (later phase), the irradiations of the accretion disk and the companion star dominates. The mass of the WD is closely related to the decline rate in this early phase as clearly shown by [4]. We have already developed a method to estimate the mass of the white dwarf from fitting between our theoretical light curves and observational ones [1].

U Sco sub-class (U Sco, V394 CrA, LMC RN, CI Aql)

U Scorpii is one of the best observed recurrent novae, characterized by the shortest recurrence period $\sim 8-12$ yr, the fastest decline of its light curve 0.6 mag per day, its extremely helium-rich ejecta He/H~ 2 by number. Historically, the outbursts of U Sco were observed in 1863, 1906, 1936, 1979, 1987, and the latest in 1999. Especially, the 1999 outburst was well observed from the rising phase to the cooling phase by many observers including eclipses, thus providing us a unique opportunity to construct a comprehensive model of U Sco during the outburst.

Fitting of the light curve in the early decay phase ($t \sim 0-10$ days after maximum) indicates that the mass of the WD is as massive as $M_{\rm WD} = 1.37 \pm 0.01\ M_\odot$ as shown in Fig. 1. However, we need the irradiation by the accretion disk to produce the mid-plateau phase ($t \sim 10-30$ days). The size of the accretion disk is also required to be as large as 1.4 times the Roche lobe size. This may indicate that a surface layer of the accretion disk is dragged outward by the strong wind. The envelope mass at the optical maximum is estimated to be as massive as $3 \times 10^{-6} M_\odot$, which is indicating the average mass accretion rate of $2.5 \times 10^{-7} M_\odot$ yr^{-1} between 1987 and 1999. About 60% of the envelope mass is blown in the wind while 40% remains on the WD. The net mass growing rate of the WD is about $1 \times 10^{-7} M_\odot$ yr^{-1}. Therefore, we may conclude that the WD will explode as a Type Ia supernova in a near future. The other results are tabulated in Table 1 [see also 2, 3]. The mass of the WD has recently been estimated to be $M_{\rm WD} > 1.31 M_\odot$ by Thoroughgood et al. [8] from double line features during the 1999 outburst. This strongly supports our methods.

V394 CrA is a twin system to U Sco. The early visual light curve ($t \sim 1-10$

days) is well reproduced by a thermonuclear runaway model on a very massive WD of $1.37 \pm 0.01\ M_\odot$. The ensuing plateau phase ($t \sim 10-30$ days) is also reproduced by the irradiation of a slightly irradiated main-sequence (MS) star and a fully irradiated flaring-up disk with a radius of ~ 1.4 times the Roche lobe size. The other parameters are summarized in Table 1. About 77% of the envelope mass is blown in the wind while 23% is left on the WD. So that the net mass growing rate of the WD is $3.4 \times 10^{-8} M_\odot\ \mathrm{yr}^{-1}$. Therefore, we expect that the WD will explode as a Type Ia supernova.

LMC RN: The observational (optical) data are very sparse so that the mass determination of the WD is not so good compared with the other cases. We have obtained $M_\mathrm{WD} = 1.37 \pm 0.05 M_\odot$. The other estimated values are tabulated in Table 1.

CI Aql: The second recorded nova outburst of CI Aquilae was discovered on 2000 April 28 UT, 83 years after the first recorded outburst in 1917. It has entered the final decline phase a bit before May of 2001, about 300 days after the optical maximum, showing the slowest evolution among recurrent novae. We have estimated, from fitting our theoretical light curves with the 1917 and 2000 outbursts, the WD mass to be $M_\mathrm{WD} = 1.2 \pm 0.05\ M_\odot$, the helium enrichment of the envelope to be He/H~ 0.5 by number, the mass of the hydrogen-rich envelope on the WD at the optical maximum to be $8.6 \times 10^{-6} M_\odot$, and the average mass accretion rate to be $1.0 \times 10^{-7} M_\odot\ \mathrm{yr}^{-1}$ during the quiescent phase between the 1917 and 2000 outbursts. Then, the turn-off time of the CI Aql 2000 outburst is in late March of 2001 after a luminous supersoft X-ray source phase lasts ~ 150 days from November of 2000 until March of 2001. The rather low X-ray fluxes in June and August of 2001 (J. Greiner, in this volume) may be consistent with our analysis. If the donor star is massive enough, CI Aql will explode as a Type Ia supernova in a future.

T CrB subclass (T CrB, RS Oph, V745 Sco, V3890 Sgr)

The very fast decay of the visual light curve during the first 10 days can be well reproduced by a WD photosphere of $M_\mathrm{WD} \sim 1.37\ M_\odot$. The second peak is well reproduced if an irradiated tilting accretion disk around the WD is introduced together with the partly irradiated M-giant companion. Such a radiation-induced, tilting instability of an accretion disk has been suggested by Pringle [6] when a central star is as luminous as the Eddington limit and the condition

$$\frac{\dot{M}_\mathrm{acc}}{10^{-7}\ M_\odot\ \mathrm{yr}^{-1}} < \left(\frac{R_\mathrm{disk}}{10\ R_\odot}\right)^{1/2} \left(\frac{L_\mathrm{bol}}{2 \times 10^{38}\ \mathrm{ergs\ s}^{-1}}\right) \times \left(\frac{R_\mathrm{WD}}{0.003\ R_\odot}\right)^{1/2} \left(\frac{M_\mathrm{WD}}{1.37\ M_\odot}\right)^{-1/2} \quad (1)$$

is satisfied [7], where R_disk is the radius at the edge of the optically thick accretion disk. The accretion rate to the WD in T CrB is as low as $\dot{M}_\mathrm{acc} \sim 0.25 \times 10^{-7} M_\odot\ \mathrm{yr}^{-1}$, which meets this condition. Therefore, it is likely that the radiation induced instability grows during the outburst in T CrB.

RS Oph is also one of the well-observed recurrent novae. It is characterized by a long orbital period of 460 days and a relatively short recurrence period of $\sim$ 10—20 yrs compared with 80 yrs of T CrB. It has been suggested that a companion (M-giant) star is underfilling the Roche lobe and losing its mass by massive cool winds. RS Oph underwent five recorded outbursts (in 1898, 1933, 1958, 1967, and 1985), with the light curves being very similar to each other. The latest (1985) outburst has been observed at all wave lengths from radio to X-rays.

The visual light curve in the first several days around the optical maximum is well reproduced by a massive WD of $M_{WD} = 1.35\ M_\odot$ with the solar metallicity. (Assuming $Z = 0.004$ for RS Oph, we have obtained $M_{WD} = 1.377\ M_\odot$.) The envelope mass at the optical maximum is estimated to be $\Delta M = 2.2 \times 10^{-6} M_\odot$, which indicates a mass accretion rate of $1.2 \times 10^{-7} M_\odot\ \mathrm{yr}^{-1}$ during the quiescent phase between the 1968 and the 1985 outbursts.

The gradual shrinkage of the accretion disk (the size is decreased from 0.1 to 0.008 times the Roche lobe size) is necessary for reproducing the late phase light curve as shown in Fig. 1. It is very likely that the surface of the accretion disk has been dragged by the strong wind and been gradually blown in the strong wind. No matter is supplied from the cool red giant wind because it has already been blown off by the nova ejecta. As a result, the accretion disk has become smaller during the strong wind phase ($\sim$ 70 days after maximum). The observational UV fluxes are also well reproduced by the same model as shown in Fig. 1.

V745 Sco: V745 Scorpii underwent the second recorded outburst in 1989 (the first recorded outburst in 1937). Its light curve is characterized by a very rapid rise (in about half a day) and decline ($t_3 \sim 9$ days, t_3 is the time taken to drop 3 mag from maximum). The very fast decay during the first 20 days can be well reproduced by a WD photosphere of $M_{WD} \sim 1.35\ M_\odot$. The other parameters are summarized in Table 1.

V3890 Sgr underwent the second recorded outburst in 1990 since the 1962 first recorded outburst. Only a bloated WD photosphere cannot reproduce the V light curve even in the early phase of the outbursts. Moreover, our simple picture of the WD photosphere never shows a sharp spike on the light curve as seen in Fig. 1. If the size of the accretion disk is as large as $\sim 20 R_\odot$ at the early phase and then becomes smaller than $0.1 R_\odot$ in about 30 days, we are able to reproduce the light curve for a reasonable set of system parameters as shown in Fig. 1 and Table 1, that is, $M_{WD} = 1.35\ M_\odot$ and $\Delta M = 3 \times 10^{-6}\ M_\odot$.

REFERENCES

1. Hachisu, I., and Kato, M., *ApJ* **558**, 323-350 (2001).
2. Hachisu, I., Kato, M., Kato, T., and Matsumoto, K., *ApJ* **528**, L97-L100 (2000a).
3. Hachisu, I., Kato, M., Kato, T., Matsumoto, K., and Nomoto, K., *ApJ* **534**, L189-L192 (2000b).
4. Kato, M., and Hachisu, I., *ApJ* **437**, 802-826 (1994).
5. Nomoto, K., Thielemann, F., & Yokoi, K., *ApJ* **286**, 644-658 (1984).
6. Pringle, J. E., *MNRAS* **281**, 357 (1996).
7. Southwell, K. A., Livio, M., and Pringle, J. E., *ApJ* **478**, L29-L31 (1997).
8. Thoroughgood, T. D., Dhillon, V. S., Littlefair, S. P., Marsh, T. R., and Smith, D. A., *MNRAS* **327**, 1323-1333 (2001).

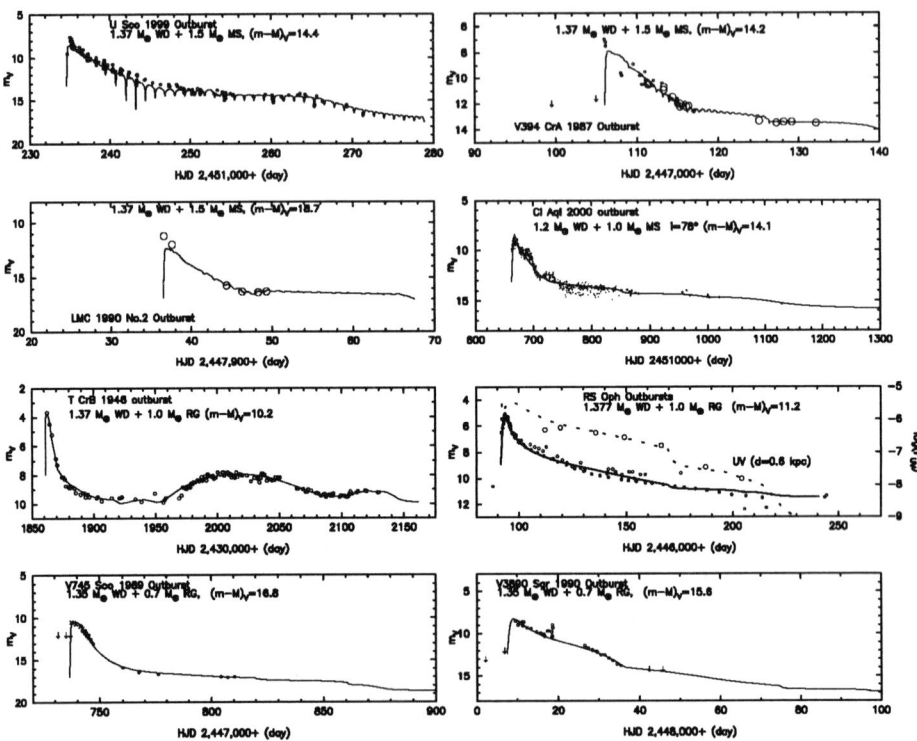

FIGURE 1. Light curves of 8 recurrent novae

TABLE 1. Physical properties of recurrent novae

Star	M_{WD} $(M_\odot)$	X	ΔM_{max} $(M_\odot)$	τ_{rec} (yr)	$\dot{M}_{acc}$ $(M_\odot\,yr^{-1})$	η_H	$\dot{M}_{He}$ $(M_\odot\,yr^{-1})$	Fate
U Sco	1.37	0.05	3×10^{-6}	8	2.5×10^{-7}	0.40	1.0×10^{-7}	SN Ia
V394 CrA	1.37	0.05	6×10^{-6}	39	1.5×10^{-7}	0.23	3.4×10^{-8}	SN Ia
LMC RN	1.37	0.10	5×10^{-6}	22	2.0×10^{-7}	0.19	3.8×10^{-8}	SN Ia
CI Aql	1.20	0.35	9×10^{-6}	83	1.0×10^{-7}	0.17	1.8×10^{-8}	WD(?)
T CrB	1.37	0.70	3×10^{-6}	80	4.1×10^{-8}	0.10	0.4×10^{-8}	SN Ia
RS Oph	1.35	0.70	2×10^{-6}	18	1.2×10^{-7}	0.10	1.2×10^{-8}	SN Ia
V745 Sco	1.35	0.70	5×10^{-6}	52	9.0×10^{-8}	0.05	0.5×10^{-8}	SN Ia
V3890 Sgr	1.35	0.70	3×10^{-6}	28	1.1×10^{-7}	0.10	1.1×10^{-8}	SN Ia

IM Normae and N Sgr 2002: CCD Spectroscopy and Photometry

William Liller

Centro de Estudios de Novae, Instituto Isaac Newton, Casilla 5022 Reñaca, Viña del Mar, Chile

Abstract. CCD photometry and spectrometry were carried out during the outbursts of the recurrent nova IM Normae and Nova Sagittarii 2002 = V4741 Sgr. The former in its first eruption since 1920 is, together with T Pyx, a second example of a slow recurrent nova. IM Nor started as a typical "Fe II" class nova, but after about two months, changed to a "He/N" nova (Williams [1]). N Sgr 2002 was discovered shortly before this conference began, and the results of only three nights are included. All the observations are presented and discussed in detail.

INTRODUCTION

Objective prism spectroscopy of bright novae ($V < 11$) is now being routinely carried out in Viña del Mar using a 100-mm diameter 12° UV-transmitting prism and a 305-mm focal length Schmidt camera with a Texas Instruments TC-237 CCD at the "Newtonian" focus. Covering the wavelength range from just shortward of $\lambda 4000$ to approximately $\lambda 10,000$, the prismatic dispersion causes the spectral resolution to decrease with increasing wavelength amounting to 4.5 A / pixel at $\lambda 4000$, 9.3 A /pixel at $\lambda 5000$, 20.2 A/pixel at $\lambda 6563$, and 45.0 A/pixel at $\lambda 9000$. However, poor seeing as well as imperfect focussing and guiding can degrade these values somewhat. (The angular resolution of the CCD is 5.0"/pixel.).

In addition to these observations, photometry is carried out using the Schmidt in the direct camera mode at f/1.5 with the same CCD and various filters. Details of the photometry have been described elsewhere [2].

It should be noted that this modest observing facility is virtually 100% dedicated to nova studies permitting almost continuous spectral and photometric observations to go on indefinitely. In addition, the nova search program, in operation since late 1982, produces on the average 2.2 discoveries per year.

IM NORMAE

Immediately following the outburst of IM Nor [3], wide-band BV photometry was carried out whenever possible, and eight spectra were obtained from 12 January to 14 April. The resulting light curve is shown in Figure 1; the spectral scans appear in

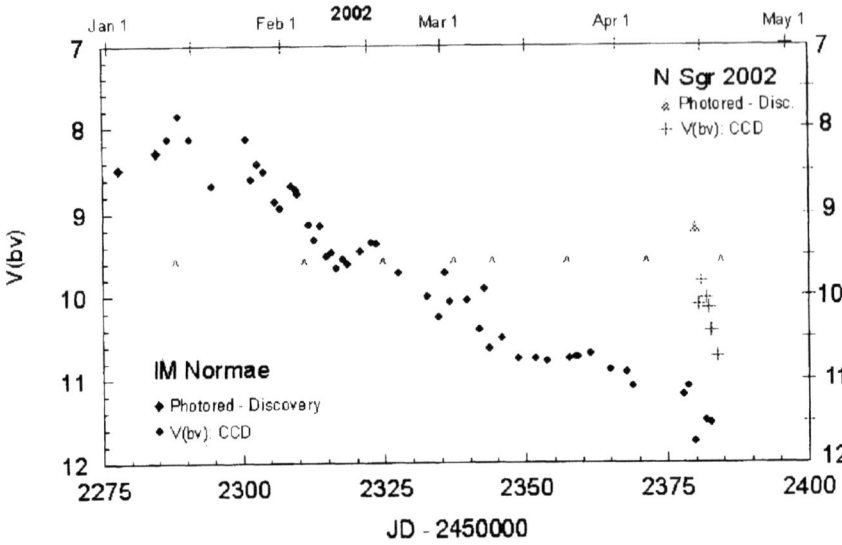

FIGURE 1. The CCD light curves of IM Normae and N Sgr 2002. The discovery photographs were taken with Kodak TP film and an orange filter. The CCD measurements were made through a filter that included the **B** and **V** passbands. Inverted "v's" indicate where the eight IM Nor spectra were obtained.

Figure 2. The nova rose slowly to maximum brightness near 14 January and then faded more or less smoothly becoming 3.0 magnitudes fainter some 75 days later.

Hα, not included in Fig. 2 (it would appear in the space between the two panels), was by a large margin the strongest emission line at all times. At shorter wavelengths Hβ was at first the most conspicuous feature followed closely by the group of Fe II lines to longer wavelengths, thereby marking IM Nor as a type "Fe II" nova [1]. No other prominent lines appeared shortward of Hα in the first spectrum taken two days before the nova reached maximum brightness.

Subsequent spectra show first the Fe II lines increasing in intensity, but after 49 days they quickly fade, and the He I λ5012 line makes it appearance, as do several blends such as the well-known "Bowen blend" of He II λ4686, N III λ4634–42 and C III λ4647. Other blends included the pair of lines N II λ5680 and [N II] λ5755; lines of He I λ5876, Na I λ5890 and λ5896 possibly with O I λ5938; and the [O I] λ6300 and λ6363 lines. By this time IM Normae clearly had changed over from an "Fe" type to an "He/N" becoming a "hydrid" nova [1]. On the last scan [N II] λ5755 remains strong and now, perhaps, [O III] λ5007 is beginning to make an appearance.

Redward of Hα, line identification is more difficult owing to the reduced resolution, but the three bands centered near λ8500 would seem to be blends of Mg II at λ8232 and λ8253; He I λ8486 and Ca II at λ8460; and also the Ca II lines near λ8650 plus N I λ8692. Other lines present are tentatively identified as C II λ7236, He I λ7478, O I λ7772, and He II at λ9162 and λ9383. The final spectrum shows a strong

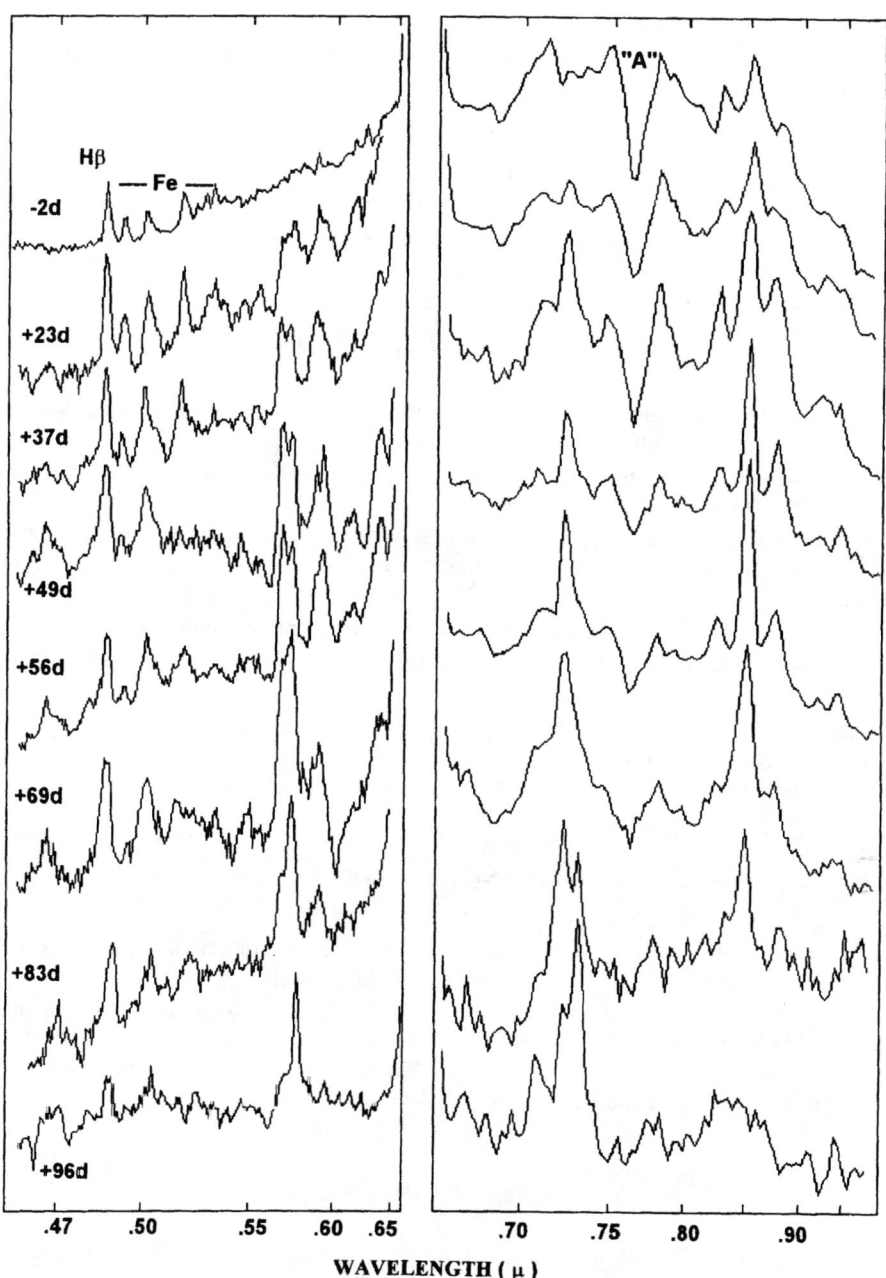

FIGURE 2. Spectra of IM Normae. The vertical scale corresponds to the relative number of CCD counts. At the left are the number of days from peak brightness, in days. Hα, off scale, would appear in the space between panels.

line near λ7250; it is conceivably [Ar IV] with [Ar III] appearing at λ7136. In the first several spectra, the strong atmospheric "A" band of molecular oxygen served as a wavelength standard in the far red, but by the end of the series, it is lost in the noise of the fading continuum. As a result, the determination of wavelength becomes difficult.

A medium-resolution spectrogram obtained 106 days following maximum has been described by Woodward, Miller and Starrfield [4]. Noting that the [O III] lines are clearly present together with the coronal line [Fe III] λ6086, they conclude that IM Nor has arrived at the nebular phase with the development of a nova shell.

NOVA SAGITTARII 2002

Discovered on 15 April [5] at or shortly after maximum brightness, Nova Sgr = V4741 Sgr, faded quickly from its peak brightness with a t_3 of about 8 days. The light curve also appears in Figure 1 (crosses at right) and shows the extremely rapid decline compared to IM Nor.

On the night following discovery, spectral observations commenced, and the scans from three nights appear in Figure 3. Again Hα is the strongest emission line (it would appear in the space between the two panels), and the appearance of the Fe II lines, seen most clearly in the second spectrum, indicates that N Sgr belongs to the Fe II class. Shortward of Hβ there is contamination from the overlapping spectrum of another star, but in the red and infrared, the spectra are less contaminated and closely resemble the early spectra of IM Nor. (The prominent feature that appears near λ9350 is also the result of star contamination.) After 7 days observations were halted owing to the star's faintness (and the departure of the author for this conference).

CONCLUSIONS

Despite the differences between these two novae, IM Nor recurrent and slow, N Sgr classical and fast, the early spectra are remarkably similar with both showing the trademark of the "Fe II" class of novae. IM Nor changed to a "He/N" nova no later than 49 days following peak brightness and evolved to the nebular-shell phase after about 100 days. Subsequently, on 27 May, having become optically thin, IM Nor was clearly detected by the Chandra X-ray observatory [6].

ACKNOWLEDGEMENTS

Special thanks are due to Sumner Starrfield who both encouraged me and advised me in the spectroscopic phase of these observations. Thanks also go to Joachim Krautter, Massimo Della Valle, Elena Mason and Hilmar Duerbeck who patiently assisted me in the interpretations thereof.

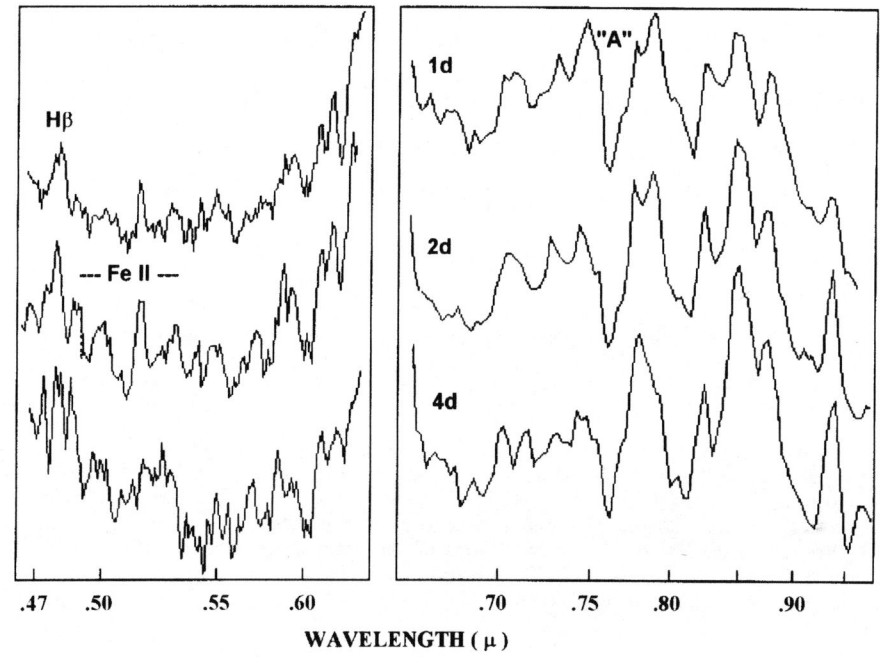

FIGURE 3. Spectra of Nova Sgr = V4741 Sgr. The vertical scale corresponds to the relative number of CCD counts. Hα, off scale, would appear in the space between the two panels. Also shown are the number of days since the nova reached peak brightness.

REFERENCES

1. Williams, R.E., *Astron. J.* **104**, 725-733 (1992).
2. Liller, W., and Jones, A.F., *IBVS*, No. 4403, 1-2 (1996).
3. Liller, W., *IAU Circulars*, No. 7789 (2002).
4. Woodward, C.E., Miller, B., and Starrfield, S.G., *IAU Circulars*, 7896 (2002).
5. Liller, W., *IAU Circulars*, No. 7878 (2002).
6. See Starrfield, S.G., in "Concluding Remarks", this conference.

The Recurrent Nova IM Nor: A Representative Example of the Use of a New Classification of Novae by the Shape of their Light Curves

A.E. Rosenbush

Main Astronomical Observatory of the National Academy of Sciences of Ukraine; Zabolotnogo Str.27, Kyiv-127, 03680 Ukraine.
E-mail: mijush@mao.kiev.ua

Abstract. Identity of light curves of a recurrent nova in repeating outbursts suggests the identity of the physical conditions causing and accompanying the outburst. This concept forms the basis of the classification of recurrent novae proposed in this study. We propose our division of recurrent novae into four groups, which means a similarity of some main properties.
  We discuss the history of the identification of IM Nor in quiescence. The too high outburst amplitude taken from Duerbeck catalog makes this object an exceptional nova, and this casts some doubt upon its identification in the catalog.

The determination of a nova as a recurrent one by an once registered outburst is of interest to the understanding of the outburst nature. At present this is assumed by an outburst amplitude of $7-9^m$ which is a consequence of Kukarkin-Parenago relation.
  There is a morphological classification of recurrent novae into three groups by the secondary star [1]: red giant companion (RS Oph, T CrB, V745 Sco, V3890 Sgr); slightly evolved main-sequence companion (U Sco, V394 CrA, LMC 1990, CI Aql); dwarf companion (T Pyx).
  The classification of classical novae proposed by us relies on the light curve shape on the "outburst amplitude, logarithm of ejected shell radius" axes. In the case of a recurrent nova outburst the shell ejection does not occur, but, as one can see from available data, it is possible that the ejection velocity of nova shell is related linearly both to the star expansion velocity before the outburst maximum and to the star compression velocity after the maximum; therefore we logically come to the logarithmic time scale. The logarithm of velocity is a constant shift of the light curve along the horizontal axis. For the first time the use of the light curve plotted versus log t was proposed by Vorontsov-Vel'yaminov in the 40s of the past century.
  When elaborating our classification principles [2, 3] we were not able to classify the light curve of the old nova IM Nor (1920) because of a very fragmentary light curve presented only by a prolonged light maximum. The early information on the recurrent outburst in 2002 also did not bring any clarity. The field charts of IM Nor gave at least three different identifications of the star [4, 5, 6]. The Duerbeck catalogue [5] gives a brightness of about 21^m in the quiescence, which gives an outburst amplitude of $>13^m$

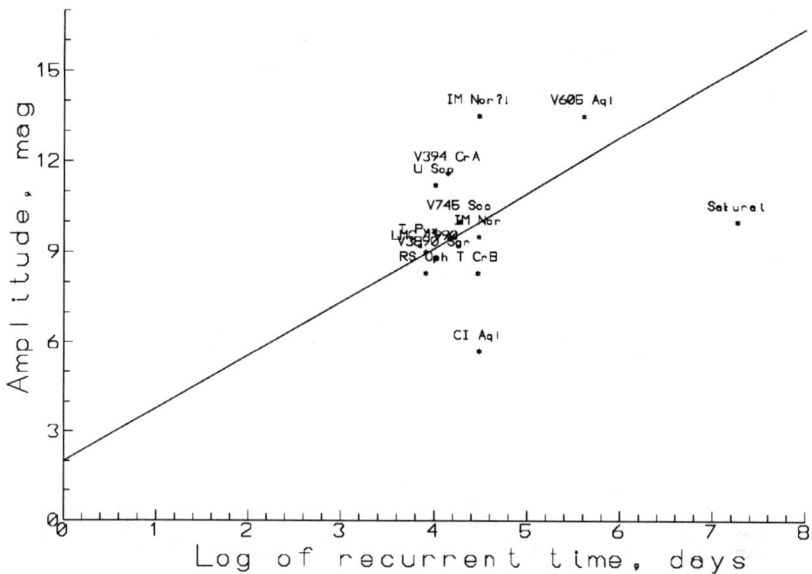

FIGURE 1. Kukarkin-Parenago relation [8] and the position of recurrent novae. V4334 Sgr (object Sakurai) and V605 Aql positions are given according to our study [9]. Two positions are given for IM Nor: the amplitude is according to Duerbeck catalogue (?!) and according to the present study.

and leads to a reasonable expansion of the cone of Kukarkin-Parenago relation (Fig.1). (Added here are 80 years of the recurrent period of CI Aql.) The classification of IM Nor by its high amplitude of outburst according to our scheme was difficult: we failed to establish with confidence that this nova belongs to any group recognized by us. Having the modern light curve, we are inclined to believe that the pre-nova IM Nor is the right star of three ones located to the north-west of the pair discussed in [4], i.e., we agree with the identification of in the catalogue of Downes et al. [6] and with the identification of Kato et al. [7].

When comparing light curves of recurrent novae in the new coordinate axes, we come up against the fact that above-mentioned three groups of recurrent novae can not be distinguished (Table 1). On our opinion the grouping proposed by us represents the commonness of physical processes during an outburst. The recurrent novae show close identity of light curves in subsequent outbursts. And, if we observe the identity of light curves in two recurrent novae, then, probably, the physical conditions in them are also identical. (In classical novae one also can note the identity of light curves.)

The difference between two groups T CrB and U Sco lies only in the presence of a "step" on the light curve in the first group. Two other groups have more pronounced differences (Fig.2). A lesser inclination of the light curve of RS Oph in comparison with other members of the U Sco group should be noted, and this may be a consequence of the well-known ejection of a small shell. The greater inclination of light curves differs the recurrent novae of the U Sco type from the classical novae of the CP Lac type. The latter feature is typical for the light curves in the continuum of classical novae as, for

TABLE 1. Groups of recurrent novae.

T CrB group	U Sco group	IM Nor group	CI Aql group
V606 Aql: (1899) *	RS Oph5	IM Nor	CI Aql
6.7(4.4)-17.3†	4.3-12.5	7.8-17.5	8.8-15.6
V394 CrA	U Sco	T Pyx	V2487 Oph (1998)
7.2-18.8	8.8-19.5	6.5-15.3	9.5-17.7
T CrB	V697 Sco (1941)	V3890 Sgr	
2.0-11.3	10.2-17	8.4-17.2	
V630 Sgr (1936)	V745 Sco	V3964 Sgr (1975)	
4(1.6)-17.6		9.4-<17	
LMC1991	V4739 Sgr (2001)	V4092 Sgr (1984)	
9-?	7.5-?	10.5-?	
		V373 Sct (1975)	
		7.1-18.8	

* The year of documented outburst for assumed recurrent novae is given.
† The brightness of nova at a maximum and at a quiescenceis given as a difference; for V630 Sgr and V606 Aql the estimates of brightness at maximum are given in brackets [5, 8].

example, in the classical nova V1668 Cyg = N 1978 Cyg [10]. It is known that T Pyx has a shell. A shell was also discovered in IM Nor [11], i.e., we also see here a tendency that the members of a group have similar principal properties (the structure of circumstellar environment, the outburst amplitude, the luminosity, etc.).

We should noted that there are a number of classical novae with light curves similar to the light curve of WZ Sge: IV Cep (1971), $7.5^m-17.2^m$ (the brightness at outburst maximum and at the quiescence, respectively); V1330 Cyg (1970), 9p–18.5B; HR Lyr (1919), 6.5–15.6; V723 Sco (1952), 9.8–19. This is a fact yet to be studied.

The classification of low-amplitude, recurrent and others novae requires a caution because the same outburst amplitude or outburst energy are typical of several types of cataclysmic stars, but the light curves do not always differ considerably. The X-ray novae also are recurrent ones. An effective criterion for classification is required, as such, for instance, proposed by us [12]. A scatter of one magnitude in the outburst amplitude is quite natural in view of different orientations of the pre-nova bynary systems relatively to the line of sight and their configuration (accretion disc, etc.). Another reason for the difference may be a realistic spread of the outburst energy. The normalization of outburst amplitude is possible in the case of light curves with characteristic features on them. For instances, this is the step in the T CrB novae group. In this case the outburst amplitude of T CrB itself should be increased by 3 magnitudes because of the contribution of the secondary (M3, 10.2^m) which was not, probably, taken completely into account. The LMC 1991 nova may be a calibrator of absolute magnitudes for this novae group.

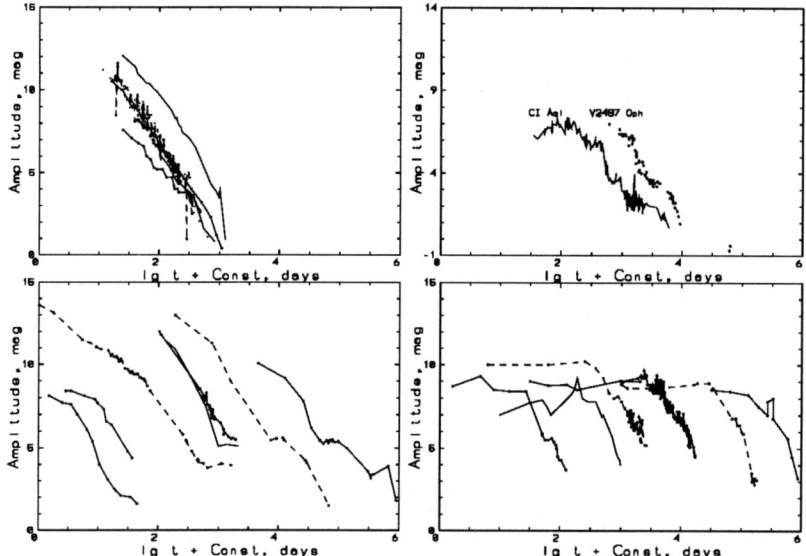

FIGURE 2. Light curves of recurrent novae groups. The T CrB group (left bottom): from left to right: T CrB, V745 Sco, V630 Sgr, V394 CrA (the outburst 1949 – solid line, 1987 – solid line with dots), N LMC 1991, V606 Aql. The U Sco group (left top): the extreme left curve is RS Oph; the other novae demonstrate an equal slope of light curves within the limit of errors of U Sco 1999 visual observations, solid line – 1979 outburst; the right solid line with dots – V697 Sco; the dashed line with dots – V4739 Sgr; the solid line on the right is V1668 Cyg (see the text). The IM Nor group (right bottom): from left to right V4092 Sgr, T Pyx, V373 Sct, IM Nor (outburst of 1920 and 2002), V3964 Sgr, V3890 Sgr. The CI Aql group (right top).

The foundation of light curve building procedure and the classification of novae can be found in [2, 3, 9, 10]. The work in this direction should be continued with invoking of data on physical stellar parameters.

The similarity of light curves opens a possibility for the prediction of the future behavior of a nova, if one finds such similarity since first days of the outburst (in the outburst amplitude, the slope of light curve). Corresponding corrections are brought in the predicted light curve at a later time.

ACKNOWLEDGMENTS

I am very gratefull to numerous collaborators of VSNET for the observations.

REFERENCES

1. Hachisu, I., and Kato, M., *Astrophys.J.* **558**, 323–350 (2001).
2. Rosenbush, A.E., *Astrophysics* **42**, 43–53 (1999).

3. Rosenbush, A.E., *Astrophysics* **42**, 140–148 (1999).
4. Wyckoff, S., and Wehinger, P.A., *Publs Astron. Soc. Pacific* **91**, 173–175 (1979).
5. Duerbeck, H.W., *Space Sci.Rev.* **45**, 1–212 (1987).
6. Downes, R.A., Webbink, R.F., Shara, M.M., Ritter, H., Kolb, U., and Duerbeck, H.W. *Publs. Astron. Soc. Pacific* **113**, 764–768 (2001). URL http://icarus.stsci.edu/~downes/cvcat/.
7. Kato, T., Yamaoka, H., Liller, W., and Monard, B., *Astron. & Astrophys.* (2002) (accepted), (astro-ph/0204354).
8. Payne-Gaposchkin, C., *The Galactic Novae*, North-Holland, Amsterdam, 1957, pp.i-x+1-336.
9. Rosenbush, A.E., *Astrophysics* **42**, 425–430 (1999).
10. Kaler, J.B., *Publs Astron. Soc. Pacific* **98**, 243–245 (1986).
11. Woodward, C.E., Miller, B., and Starrfield, S.G., *IAU Circ.* No 7896, (2002).
12. Rosenbush (Rozenbush), A.E., *Kinematics and Physics Celectial Bodies* **15**, 373–378 (1999).

Spectroscopic and photometric observations of the recurrent nova IM Normae

H.W. Duerbeck and C. Sterken*, R. Baptista[†], M.P. Diaz and C.M. Dutra**,
L. Freyhammer and H. Hensberge,[‡] and A.F. Jones[§]

*WE/OBS, Vrije Universiteit Brussel, Belgium
[†]Departamento de Fisica, Universidade Federal de Sta. Catarina, Brazil
**Departamento de Astronomia, Universidade de Sao Paulo, Brazil
[‡]Royal Observatory of Belgium
[§]VSS/RASNZ, 31 Ranui Road, Stoke, Nelson, New Zealand

Abstract. The outburst of the recurrent nova IM Nor was followed visually and spectroscopically. In the beginning of the outburst, the nova belonged to Williams' Fe II class, but one month later, the spectrum changed to the He/N class. Thus IM Nor is a member of the hybrid class of novae, and has many spectroscopic similarities with CI Aql and T Pyx, the two other slow recurrent novae observed until now. In early February, the radial velocity of the Fe II emission lines was $+220 \pm 35$ km/s, the main absorption had -682 ± 4 km/s, and a faint absorption component -902 ± 1 km/s. The strengths of the interstellar Ca II, Na I, and K I lines as well as those of the diffuse interstellar bands indicate an interstellar extinction $E_{B-V} \geq 0.8$ and a distance $d \geq 2.5$ kpc. The (uncertain) conclusion is that the absolute magnitude of IM Nor is comparable to that of classical novae of the same speed class.

INTRODUCTION

The somewhat elusive group of recurrent novae has grown in number in recent years. Until 2000, most recognized recurrent novae were very fast ones, either with dwarf (U Sco, V394 CrA) or giant (T CrB, RS Oph, V3890 Sgr, V745 Sco) secondary components, while the slow recurrent nova T Pyx appeared as an exotic exception. It came as a surprise that the new members of this class (CI Aql in 2000, and IM Nor in 2002) share some properties with T Pyx.

OBSERVATIONS

The second outburst of IM Nor was detected by Liller (2002) on 2002 Jan 10, 82 years after the first eruption. Visual observations were made by A.F. Jones in New Zealand. The light curve, shown in Fig. 1, gives these data as filled circles, and also contains early V-band CCD observations by W. Liller (open circles). Maximum brightness is poorly determined, and can be as high as $V = 7.7$. Spectroscopic observations were made with the ESO/Brazil 1.52m telescope (with FEROS and B&Ch spectrographs) and the 1.54m Danish telescope (with DFOSC) at La Silla Observatory, Chile. Fig. 2 shows a sequence of spectra obtained between 2002 January 16 and March 30.

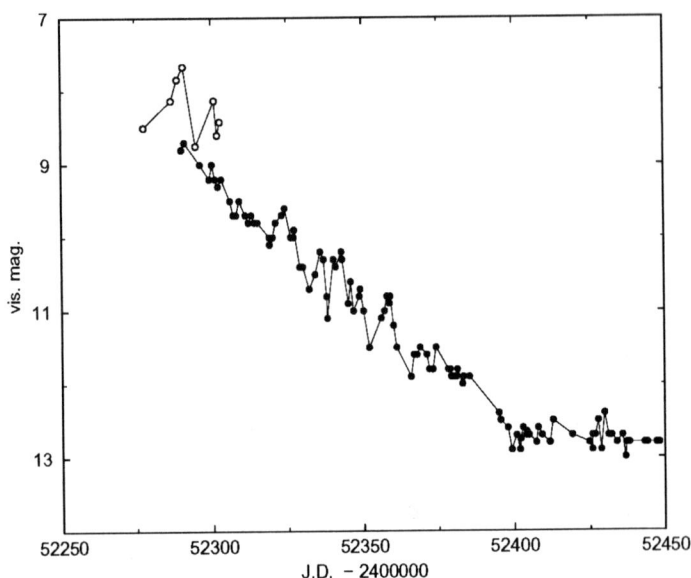

FIGURE 1. Visual observations of the 2002 outburst of IM Nor, made by A.F. Jones (filled circles). Some early CCD V-observations by W. Liller are shown as open circles.

Observations with the B&Ch and DFOSC spectrographs give medium-resolved spectra, which permit a convenient description of features over a wide wavelength range; and they are useful in classifying the stages of the nova in Williams' classification scheme (Williams et al. 1991).

The FEROS spectroscopy of IM Nor, made on 2002 February 9, about 40 days after the begin of the outburst, is the most detailed one. The spectrum is just changing from the Fe II to the He/N stage; both the Fe II lines (multiplet 42) and the C III/N III (Bowen blend) at 4640 is seen, but its wavelength indicates that N III is weak and there may be a contribution of Fe II.

ANALYSIS I: THE LIGHT CURVE

The light curve (Fig. 1) shows a fairly smooth decline from maximum. The t_2 and t_3 times are 60 and 90 days, respectively; thus IM Nor is a moderately fast nova. Generally, such novae show strong brightness fluctuations in their early phases, and/or signs for dust formation. IM Nor, however, has a remarkably smooth light curve; after JD 2452400 (June 5), the nova entered a plateau phase at $V = 12.73 \pm 0.13$.

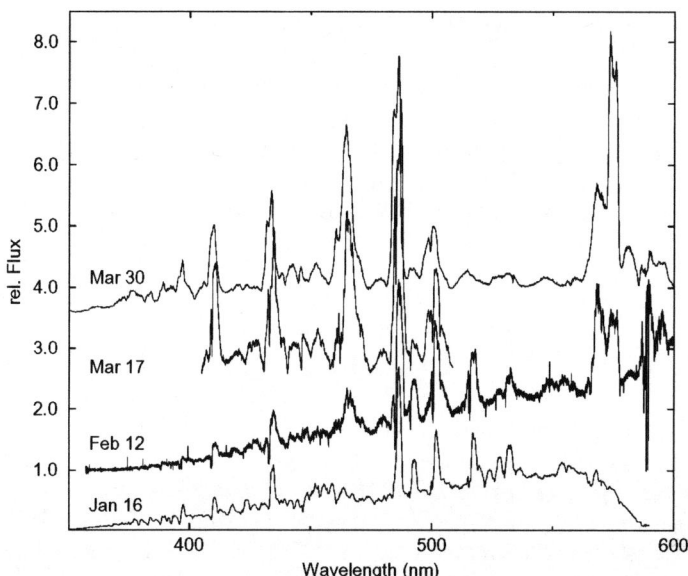

FIGURE 2. Blue spectra of IM Nor taken during the 2002 outburst. The spectra have been scaled and shifted arbitrarily to clearly show the evolution of spectral features. The Jan 16 spectrum shows, besides hydrogen Balmer lines, mainly permitted lines of Fe II. The Feb 12 spectrum still shows strong Fe II, but the Bowen blend (N III/C III) at 4640 Å and a strong emission of N II 5679/[N II] 5755 appear. On Mar 17, the Bowen blend has become stronger than the Fe II emission. This trend continues in the spectrum of Mar 30.

ANALYSIS II: THE INTERSTELLAR LINES

The (uncertain) relations between interstellar line strength and distance by Allen (1973) yield distance estimates of $r = 2.35$ and $r = 1.9$ kpc for Ca II K and Na I $D_{1,2}$, respectively. Using the galactic extinction model by Hakkila et al. (1997), E_{B-V} should be between 0.3 and 0.45 for a distance near 2 kpc. The strength of diffuse interstellar bands (Benvenuti & Porceddu 1989) yields $E_{B-V} = 0.7 \pm 0.1$, and the relations between Na I and K I line strength and extinction (Munari & Zwitter 1997) favour a value $E_{B-V} = 1.1$. Such a high extinction indicates a distance beyond 2.5 kpc, or an absolute visual magnitudes near or above $M_V = -6$. Thus the luminosity of IM Nor may be comparable to classical novae of the same speed class (Downes & Duerbeck 2000). Additional extinction estimates from nova emission line ratios, and a detailed reddening study of the surroundings of IM Nor will be necessary to arrive at a more reliable luminosity value.

ANALYSIS III: THE STELLAR LINES

The FEROS spectrum permitted the best measurement of stellar line profiles. The Balmer lines show a progression of radial velocities that is indicative of the lesser strength of emission at higher series members, converging to the absorption line velocity of the Fe II lines. A fainter, faster absorption component with -901 km/s is only seen in Fe II. Hα is the strongest line, with a FWZI of 5160 km/s.

The temporal development of velocities and line strengths will be discussed in a future paper.

CONCLUSION

Until a few years ago, there were basically only two types of recurrent novae – very fast ones with and without giant companions, and the "exotic" slow recurrent nova T Pyx. *IM Nor is a slow recurrent novae that shows many similarities with other slow, albeit spectroscopically poorly observed recurrent novae* like T Pyx (Catchpole 1969) and CI Aql (Kiss et al. 2001, Matsumoto et al. 2001). Actually, the t_2-times of CI Aql and IM Nor show that these are "moderately fast" novae; but until more slowly evolving recurrent novae are known, they should be counted among the "slow" ones (comprising the speed classes F,MF,S,VS). Taking into account overlooked brightenings, recurrence timescales may be as short as $20 - 25$ years in these systems, as has been observed for T Pyx as well as for CI Aql (Schaefer 2001). The determination of white dwarf masses, accretion rates, and shell masses is important. Such data will put our present understanding of nova explosions on trial.

ACKNOWLEDGMENTS

This research was supported by the Flemish Ministry for Foreign Policy, European Affairs, Science and Technology, and IUAP P4/05 by the Belgian DWTC/SSTC.

REFERENCES

1. Allen, C.W., 1973, Astrophysical Quantities. London, Athlone Press
2. Benvenuti, P., Porceddu, I., 1989, *Astr. Astrophysics*, **223**, 320-335.
3. Catchpole, R.M., 1969, *Mon. Not. R. Astr. Soc.*, **142**, 119-128
4. Downes, R.A., Duerbeck, H.W., 2000, *Astr. J.*, **120**, 2007-2037.
5. Hakkila, J., Myers, J.M., Stidham, B.J., Hartmann, D.H., 1997, *Astr. J.*, **114**, 2043-2053.
6. Kiss, L.L., Thomson, J.R., Ogloza, W., Furész, G., Sziládi, K., 2001, *Astr. Astrophys.*, **366**, 858-864.
7. Liller, W., 2002, *IAU Circ.*, 7789
8. Matsumoto, K., Uemura, M., Kato, T., Kiyota, S., Ayani, K., Kawabata, T., Král, L., Havlík, T., Kolasa, M., Novák, R., Masi, G., 2001, *Astr. Astrophys.*, **378**, 487-494.
9. Munari, U., Zwitter, T., 1997, *Astr. Astrophysics*, **318**, 269-274.
10. Schaefer, B.E., 2001, *IAU Circ.*, 7750.
11. Williams, R.E., Hamuy, M., Phillips, M.M., Heathcote, S.R., Wells, L., Navarrete, M., 1991, *Astrophys. J.*, **367**, 721-737.

Nova Monocerotis 2002 (V838 Mon) in the Early Stages of its Outburst

E.A. Barsukova*, N.V. Borisov*, V.P. Goranskij[†], A.V. Kusakin[†], N.V. Metlova[†] and S.Yu. Shugarov[†]

*Special Astrophysical Observatory RAS, Nizhny Arkhyz, Karachai-Cherkesia, Russia
[†]Sternberg Astronomical Istitute, Moscow University, Russia

Abstract. The results of archival search, photometry and medium resolution spectroscopy of peculiar nova V838 Mon are reported. Prediscovery observations suggest that pre-outburst star was blue with $(B-V)_0 = -0^m.03 \pm 0^m.1$ and $M_V = 2^m.9 \pm 0^m.5$ or $L = 5.6 L_\odot$ (at the distance of 1.2 kpc). Our spectroscopy shows that the star has a normal chemical composition, the metal lines being strengthened by dense atmosphere located above photosphere. We have observed the lithium lines, what suggests that the material of star's envelope is enriched by lithium originated from the nuclear fusion at the outburst and ejected into exterior layers. The light and colour curves are demonstrated, and similarlity with nova V1006/7 in M31 is noticed.

V838 Mon, a peculiar nova in Monoceros 2002 was discovered by Brown [1] on January 6, 2002. Nova had expanding light echo [2, 3]. Before outburst, the star had brightness of $15^m.5V$ and was a member of catalogues by numbers GSC 4822.39 and IRAS 07015-0346 [4]. We performed the investigation of pre-nova in DSS, in the digital archives of USNO FS and in Sternberg Astronomical Institute (SAI) plate collection. Image of V838 Mon was found on 26 photographic plates of SAI archives taken with 40-cm astrograph in the time between 1949 and 1994. Brightness estimates are dispersed in the range of 15.7 and 16.0 B, with the mean value of $15^m.85$ and colour $B - R = 1^m.1$.

A total of 100 multicolour $UBVR$ observations were done during the outburst at SAI Crimean station and Tien-Shan Astronomical Observatory by N.V. Metlova and A.V. Kusakin. Few CCD frames in BVR bands were taken at Special Asrtophysical Observatory. In the R band, we have additional eye estimates by V.P. Goranskij using an image tube with a microchannel plate and a 5-cm lens. Photometry published in VSNET (vsnet-campaign-v838mon) and in IAU Circulars was also used in our analysis[1]. These observations have considerable systematic differences. We reduced all the data to most accurate and homogeneous Crimean set taken by N.V. Metlova. The light and colour curves of V838 Mon are shown in Fig. 1.

The observations represent well the apparent pre-maximum stage lasted for a month till JD 2452307 and the rapid outburst with the light maximum on February 6 (JD 2452312) which we explain by arriving a shock wave to star surface. These features are followed by a gradual decay superimposed by second local maximum which, probably,

---

[1] The photometry by P.Corelli, A.Henden, K.Hornoch, S.Kimeswenger, S.Kiyota, L.Kral, C.Lederle, J.M.Llapasset, F.Lomoz, A.Oksanen, O.Pejcha, L.Smelcer, P.Sobotka, D.West, and F.van Wyk.

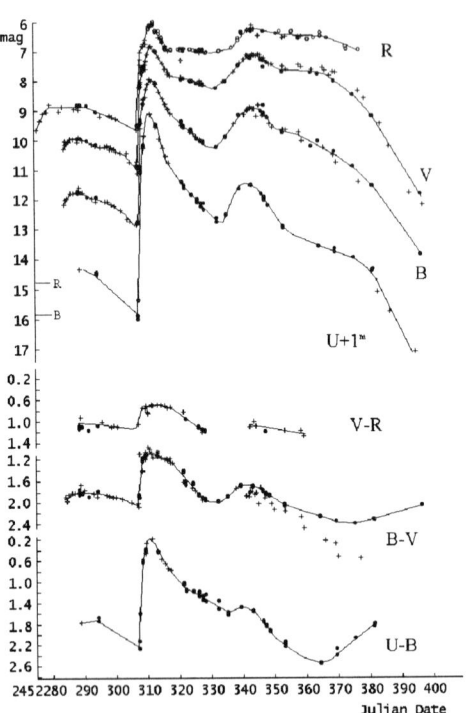

FIGURE 1. Light and colour curves of V838 Mon in $UBVR$ system. Horizontal lines near ordinate axis mark brightness levels in B and R bands before outburst. Our photometry is marked by filled circles. Observations by other cited authors taken from VSNET and IAU Circulars are shown by crosses. Eye estimates with image tube are shown by open circles. Note, that U light curve is displaced downwards by 1 mag. for best representation.

is due to a next weakened shock wave. Note, that the colours in all the stages of outburst were redder, than in the quiet state. The total amplitude was $8^m.07$ in B band (limits of variability are $7^m.93 - 16^m.0$) and $8^m.77$ in R band ($6^m.12 - 14^m.89$). Pre-maximum stage of V838 Mon resembles that of V723 Cas [5], but occured at lower surface temperature.

In the two-colour $(U - B) - (B - V)$ diagram, the star moved along the sequence of red giants displaced by colour excess of $E(B - V) = 0^m.63 \pm 0^m.05$. In the light maximum, the UV excess measured relatively to normal stars is of $0^m.3 - 0^m.5$. Our observations show that in the late decay, the colours became more blue again (Fig. 1), but this tendency does not confirmed by published CCD data, and may be due to the presence of faint nearby stars in the photometer diaphragm.

Spectra with the resolution of 2.4-5.0Å in the wavelength range of λ4100-7300Å were taken with 1-m telescope of Special Astrophysical Observatory by N.V. Borisov on February 21, 23, March 5 and 16, 2002. Primary reduction of the spectra was done in the MIDAS environment by E.A. Barsukova accordingly to standard methods.

One of pre-maximum red spectra on January 23, 2002 is shown in Fig.2a. This is a K type spectrum accordingly to its energy distribution. Weak absorption H_α and Li $\lambda 6707$Å line with P Cyg type profile are seen. In the full spectrum, strong NaI D_1+D_2 doublet is visible along with other strong lines at $\lambda 6500, 6137$, and 5851Å, the identification of which remains debatable [6, 7]. A "forest" of absorption lines is noticeable in the range of $\lambda 4800$- 5600Å, it suppresses continuum radiation at some wavelengths.

We compared the spectrum of V838 Mon with the spectrum of a normal star HD 23524 (K0V). The cross-correlation function was calculated, which showed very high correlation coefficient of 0.68, and low expansion velocity of envelope of -150 km/s for the system of absorption lines. We found that equivalent widths of the absorption lines correlate well, that the same lines predominate in both spectra, but the lines of V838 Mon are 3-4 times more intensive. Comparing the spectra of V838 Mon with barium stars and FG Sge, we have'nt find any similarity between them. On the base of this analysis, we conclude, that V838 Mon has a normal spectrum, but the strengthened lines are forming in the dence atmosphere located above stellar photosphere.

Accepting the low velocity of -150 km/s, we identified strong line at 6500Å as Ba II $\lambda 6496.9$Å with confidence. An identification of this line with Ca I [7] is less probable because there are no traces of more intensive line Ca I $\lambda 6717$Å which is usually observed well in K type spectra. We have detected the following weaker lines of Ba II at the same radial velocity: $\lambda 6769.6$ (with P Cyg profile), doublet 6135.8/6141.7, 5853.7, 5784.2, 4957.2, 4934.1, 4554.0, 4216.0, and 4130.0Å. The other most intensive lines identified in the spectra on January 21 and 22 are of neutral atoms of Mg I, Na I, Fe I, Ti I, Ca I, Cr I and Si I.

In the spectrum taken on February 5, few hours before maximum, the H_α emission line has a narrow component and wide wings. The total width of the wings at continuum level is FWZI = 3100 km/s (Fig. 2a). There is an absorption component in the line profile at the velocity of -300 km/s. Considerable changes in the spectrum reflect an increase of excitation level. Numerous lines of elements which were observed in pre-maximum stage have appeared in that spectrum, but of single-ionized atoms and with pronounced profiles of P Cyg type (Fig. 2b). The velocities of absorptions reached $-180 \div -200$ km/s. He I lines appeared. At last, Ca I $\lambda 6717$Å has evidently appeared in the spectrum, and Li I line disappeared. The strongest line of Ba II 6496.9A weakened. We explane these changes by shock wave arrived to surface. It was accompanied by rise of temperature and increase of ionization degree of atoms in surrounding medium.

One more red spectrum was taken on February 16, in the decay after maximum. H_α emission was narrow and had already no high velocity wings. Deep absorption component at $v_r = -220$ km/s was seen in its profile. The absorptions of Fe II, Si II, Ti II were still visible, but the lines of Ba II and weak absorptions of neutral atoms of Fe I and others appeared again (Fig. 2a). The weak Li I line at 6707Å was seen again, too. Ca I $\lambda 6717$Å remained still visible. So, there were two systems of absorption lines of the same elements both neutral and single-ionized simultaneously. This gives evidence for stratification of the envelope. This dichotomy became more evident in the later spectra of nova published in Internet [2].

---

[2] C.Buil; http://www.astrosurf.com/buil/us/nmon/nmon.html

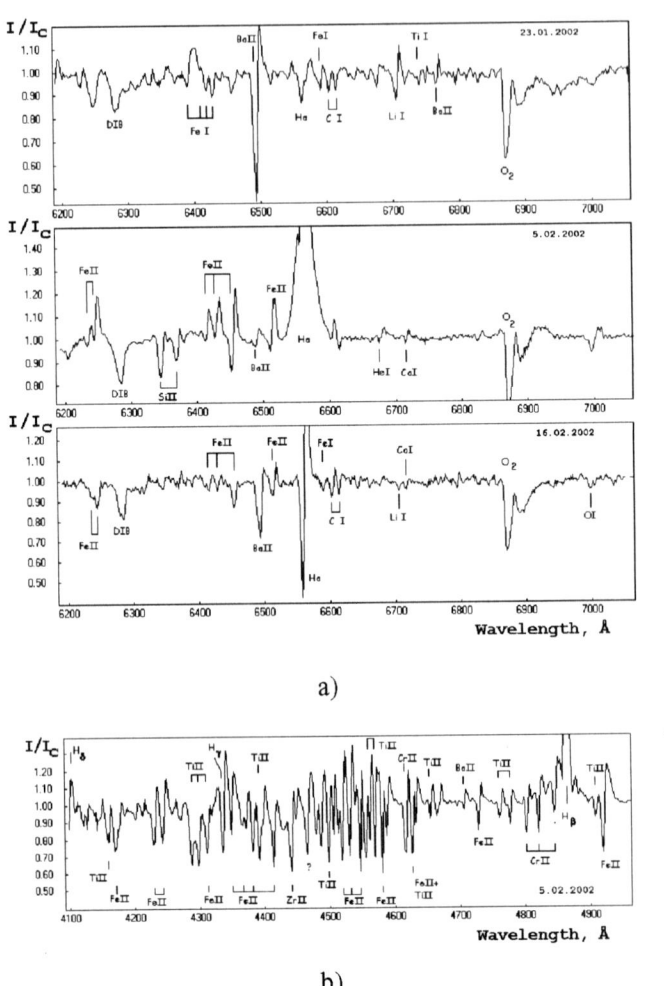

FIGURE 2. a) Comparison of spectra in Hα in pre-maximum stage on January 23, few hours before maximum on February 5 and during decay of outburst on February 16. b) The spectrum taken on February 5 in Hβ line region.

Our photometry and detailed discussion will be published in [8].

Let us consider the conclusions which can be done on the nature of the object and its outburst from our observations. Before the outburst, we may adopt the following photometric parameters derived from B and R magnitudes: $V = 15^m.25 \pm 0^m.1$, $B - V = 0^m.60 \pm 0^m.1$. Assuming the colour excess $E(B-V) = 0^m.63$, we have $(B-V)_0 = -0^m.03 \pm 0^m.1$. $A_V = 1^m.9$, and with the Galactic reddening of $1^m.6$ per 1 kpc[9] we have distance to V838 Mon of 1200 pc. Then the luminosity of pre-nova is $M_V = 2^m.96 \pm 0^m.5$ or $L = 5.6 L_\odot$. Located in the galaxy M31, V838 Mon would have maximum brightness of about $19^m B$.

Nova V1006/7 of M31 (1988), a possible twin of V838 Mon had unusual spectrum M0Ie and reached $18^m.2B$ in maximum. Its spectrum [10] was a duplicate of spectrum V838 Mon, including strong absorptions of Ba II, Na I and Li I $\lambda 6707$Å. It is interesting that one more outburst of that nova in 1968 was discovered [11].

Probably, V838 Mon is a cataclysmic system, although its luminosity is higher than luminosities of the most luminous systems of such type (by 1^m). The interpretation of the star as a white dwarf is excluded, but the star may be a hot subdwarf being on the evolutionary way into white dwarfs. A red giant in AGB-stage is excluded, too.

The appearence of Li I $\lambda 6707$Å line may be the result of fast mixing of slowly expanding envelope, including hydrogen burning leyer. Lithium is an intermediate product of hydrogen burning. The products of nuclear fusion get to the star surface in a mixing episode. Nuclear ignition in the hydrogen rich layer of white dwarf in the cataclysmic system is a widely accepted scenario of novae, but lithium is generated in the case of low mass WD, where the burning temperature is lower. In the case of V838 Mon, Li I line is observed only due to low photospheric temperature. In a standard nova scenario, the transit to optically thin envelope (nebular phase) is forthcoming. Maybe, in the extreme case the expansion velocity is small to loose the envelope. Then the expansion may stop, and the system will return back to binary with common photosphere. After that the system will go to semi-detached configuration. There are no evidences yet in favour of binarity of V838 Mon, but the intense lines of hydrogen confirm this version.

The flash of a shell helium source in the post-AGB stage [6] is another possible nature of V838 Mon phenomenon. In this case we may expect the appearance of spectrum with s-process elements, and later the light weakenings due to formation of large carbon particles in the star envelope.

ACKNOWLEDGMENTS

One of the authors, S.Yu.Sh. thanks the Russian Foundation for Basic Research for support by Grant No.00-15-96553.

REFERENCES

1. Brown, N.J., *IAU Circ.* No. 7785 (2002).
2. Orio, M., Harbeck, D., Gallagher, J., Woodward, C., *IAU Circ.* No. 7792 (2002).
3. Henden, A., Munari, U., Marisse, P., Boshi, F., Corradi, R., *IAU Circ.* No. 7889 (2002).
4. Kato, T., *IAU Circ.* No. 7786 (2002).
5. Munari, U., Goranskij, V.P., Popova, A.A., Shugarov, S.Yu., Tatarnikov, A.M., Yudin, B.F., Karitskaya, E.A., Kusakin, A.V., Zwitter, T., Lepardo, A., Pasuello, R., Sostero, G., Metlova, N.V., Shenavrin, V.I., *Astron. and Astrophys.* **315**, 166-169 (1996).
6. Della Valle, M., Iijima, T., *IAU Circ.* No. 7786 (2002).
7. Zwitter, T., Munari, U., *IAU Circ.* No. 7812 (2002).
8. Goranskij, V.P., Kusakin, A.V., Metlova, N.V., Shugarov, S.Yu., Barsukova, E.A., Borisov, N.V., *Astron. Letters* **28**, No.10 in print (2002).
9. Sharov, A.S., *Astron. Zhurnal (Rus.)* **40**, 900-911 (1963).
10. Rich, R.M., Mould, J., Picard, A., Frogel, J.A., Davies, R., *Astrophys. J.* **341**, L51-L53 (1989).
11. Sharov, A.S., *Soviet Astron. Letters* **16**, 85-86 (1990).

V723 Cas a borderline classical nova

M. Friedjung* and T. Iijima[†]

*Institut d'Astrophysique, 98 bis Boulevard Arago, F-75014 Paris, France
[†]Astronomical Observatory of Padova, Asiago Section, Osservatorio Astrofisico, I-36012 Asiago (Vi), Italy

Abstract. V723 Cas had a light curve similar to that of HR Del before maximum, with a very slow pre-maximum rise, explained according to [2] by the presence of an optically thin wind *before maximum* unlike the optically thick wind generally seen for classical novae *after maximum*. Examination of the Fe II emission lines by the SAC method, is compatible with this also having been the case for V723 Cas.

INTRODUCTION

The novae V723 Cas (1995) and HR Del (1967) had light curves unlike those observed for most classical novae, with a slow rise lasting several months after an initial brightening, a sharp maximum above the light curve plateau, followed by later flares. Like that of HR Del, the spectrum of V723 Cas appears to have been in a pre-maximum stage before the first maximum, the velocity of the main blueshifted absorption decreasing in that stage. According to a previous study (Friedjung 1992), HR Del could, unlike classical novae *after optical maximum*, have then had an optically thin wind, accelerated above an almost stationary photosphere. The white dwarf component of HR Del would appear to have an exceptionally small mass for such a component in a classical nova binary of 0.67 $M_\odot$ [7], this being perhaps connected with its unusual properties.

We attempt here to investigate whether a similar conclusion might be drawn for V723 Cas. We apply the SAC (self-absorption curve) method to the Fe II emission lines, so as to obtain what may be a minimum column density above the photosphere.

APPLICATION OF THE SAC METHOD

In the SAC method graphs of $\log(F\lambda^3/gf)$ are plotted against $\log(gf\lambda)$, where F is the flux of an emission line, λ is its wavelength and f is the oscillator strength. The points for the lines of the same multiplet lie on a curve, as long as the populations of levels inside the same spectroscopic term are proportional to their statistical weights. The curve is horizontal at small values of $\log(gf\lambda)$, then bending downwards because of self-absorption. A fairly early form of the method is described by Friedjung & Muratorio [4]; it was recently applied by [6] to the symbiotic nova RR Tel, using fewer assumptions than previously. When the curves for different multiplets have the same shape, relative vertical shifts give relative upper level populations, while relative horizontal shifts give

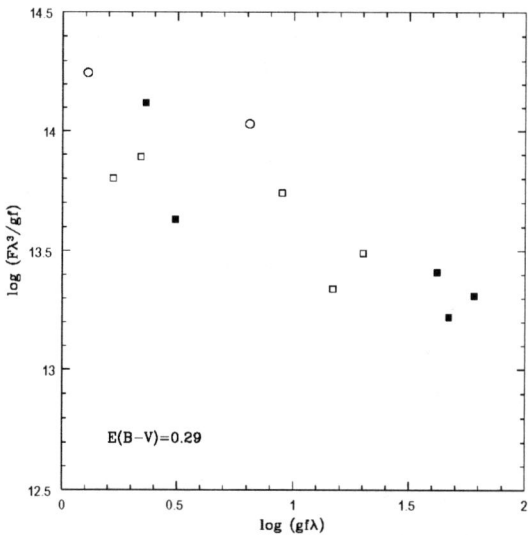

FIGURE 1. Self-Absorption lower E(B-V)

relative lower level populations. Similarity of shape can be tested for empirically when the data are of good quality; fairly rigorous calculations are still needed to test for the range of validity of such an assumption. The lower even metastable levels are probably in LTE [8], this probably is not the case for the odd upper terms. We therefore only consider here horizontal shifts of multiplets having the same upper term.

The data used here are from a high resolution spectrum, measured by Iijima et al. [5]. The relative flux levels in this spectrum of September 12 1995, were calibrated, using the continuum fit of a lower resolution spectrum, taken on September 6 1995 at wavelengths not very far outside the range of the the latter, less resolved, spectrum. There are in addition differences in the estimated amount of reddening; corrections were made to relative fluxes both for E(B-V) = 0.29, corresponding to $A_V = 0.80$ [5] and E(B-V) = 0.78 [1].

The data should be usable in the estimation of the ratio of populations of terms with an excitation potential between 2.73 and 2.88 eV (lower level terms of multiplets 27, 37, 38 and 43) to those of terms with an excitation potential between 3.21 and 3.24 eV (lower level terms of multiplets 48, 49 and 55) from relative horizontal shifts of the self-absorption curves of the different multiplets. We compare populations of multiplets having the same upper term. The attempt to measure the horizontal shift between the curves for the multiplets with an $z^4 D^o$ upper term, 27, 38 and 43 on the one hand and the curve for multiplet 48 on the other, did not lead to a very clear result, as only fluxes for 3 lines of multiplet 48 had been measured, with one weak line of this multiplet being possibly unreliable. The comparison between multiplets with a $z^4 F^o$ upper term, 37 (5 lines) on the one hand amd multiplets 49 and 55 on the other (7 lines), may suggest a shift of the order of 0.2 with both reddening assumptions, as long as the weak

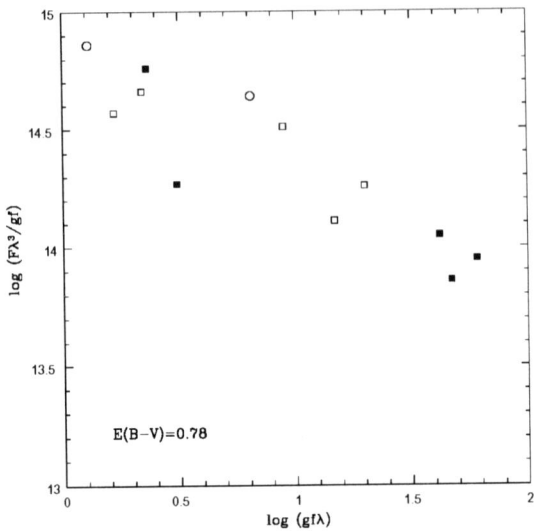

FIGURE 2. Self-Absorption higher E(B-V)

5254.9 Å (multiplet 49) line is neglected. The graphs of the last 3 multiplets, with 2 reddening estimates, are shown in the two figures. In them the open boxes are for the lines of multiplet 37, the closed boxes are for multiplet 49, while the open circles are for multiplet 55. Supposing the lower terms to be in LTE then leads to a θ of the order of 0.5. The optical thickness would appear to be not less than unity at $\log(gf\lambda) = 0.0$ (λ in Å). Assuming solar abundances and noting that all iron need not have been in the form of Fe^+, would for our θ, correspond to a minimum column density of hydrogen of 3 10^{21} cm^{-2}. Such a column density is at least consistent with the presence of an optically thin wind on September 12 1995. More detailed studies of spectra are however needed in the future, for clear conclusions to be drawn.

Another study [3], based on the apparent weakness of Balmer line broadening by electron scattering, indicated a low column density for the ionised part of the wind.

REFERENCES

1. Evans, A., Gehrz, R.D., Geballe, T.R., Woodward, C.E., Salama, A., Barlow, M,, Lyke, J.E., Hayward, T.L., Eyres, S.P.S., Krautter, J., Starrfield, S.G., Gonzales-Riestra, R., Greenhouse, M.A., Hjellming, R.M., Shore, S.N., Wagner, R.M., Williams, R.E., Pequinot, D., 2002, in preparation
2. Friedjung, M., 1992, A&A, 262, 487
3. Friedjung, M., 2002, in the proceedings of the Göttingen meeting on *The Physics of Cataclysmic Variables and Related Objects*, in press
4. Friedjung, M., Muratorio, G., 1987, A&A, 188, 100
5. Iijima, T., Rosino, L., Della Valle, M., 1998, A&A, 338, 1006
6. Kotnik-Karuza, D., Friedjung, M., Selvelli, P.L., 2002, A&A, 381, 507
7. Ritter, H. Kolb, U., 1998, A&AS, 129, 83
8. Verner, E.M., Verner, D.A.,, Korista, K.T., Ferguson, J.W., Hamman, F., Ferland G.j., 1999, ApJS, 120, 101

Radial Pulsation of the Cooling White Dwarf in the Decay of Nova Cassiopeiae 1995 (V723 Cas)

Vitaly P. Goranskij, Natalya V. Metlova, and Sergei Yu. Shugarov

Sternberg Astronomical Institute, Moscow University, 119992, Moscow, Russia

Abstract. The episode of oscillations with the period of 90^m was observed on September 19, 1999 in the light decay stage of V723 Cas. This is a slow nova having highly inclined orbit with the period of 0.693265 days. The phenomenon is explained by pulsations in the expanded envelope of white dwarf. The possibility of pulsations was earlier predicted by K. Schenker.

INTRODUCTION

K. Schenker was the first who discussed the possibility of radial pulsations in the envelopes of classical novae [1]. On the base of linear and nonlinear analysis, he shows that the strong running wave type instabilities occur in the inner static envelope structures, and then rapidly develop into shock waves. Finally, it looks like strongly non-adiabatic radial pulsations. The pulsations may be confirmed in a search of short period variability in UV and soft X-rays during the late decline phase. S.Starrfield and J.Drake [2] observed $41^m.7$ period in soft X-rays in V1494 Aql, and attributed it to pulsation of the nova envelope in non-radial g-modes.

Here we report the optical observations of pulsations in the envelope of slow classical nova V723 Cas.

V723 Cas was discovered by M. Yamamoto on August 24, 1995 [3]. Maximum brightness $V = 7^m.09$ was reached on December 17, 1995 (JD 2450069) in a short-lived flare. Nova has highly inclined orbit with the long orbital period of 0.693265 days. It is one of the longest orbital periods in the classical novae, what suggests relatively large orbital separation of components. Before light maximum, nova passed through three-month-long stable common photosphere phase with a supergiant F-type spectrum [4]. In the light decay, nova underwent few eruptions, which changed radically its photometric colours. In the time interval between October 27, 1996 and May 30, 1997 it entered the nebular phase [5, 6]. The strong forbidden lines of highly ionized elements, such as [Fe VII], [Fe VI], [Ca V], [Ar V], etc., are seen in our spectrum taken on January 2001 with the 6-m telescope. Due to high temperature, He II 4686 Å emission is very bright in this spectrum being brighter than H$_\alpha$. Nova has also a strong, hot and slowly decaying continuum suggesting that the radius of the cooling white dwarf photosphere decreases slowly. In September 1998, the white dwarf photosphere had detached from the secondary star, and the eclipsing light curve became visible [7].

OBSERVATIONS

This study is based on the extensive CCD monitoring which consists of 4100 R band frames, and UBV photometry taken in the time interval of JD 2451020 - 2452261. SBIG CCDs ST-6, ST-7, ST-8, and a single-channel photoelectric UBV photometer were used with the telescopes having the apertures between 30 and 125 cm located in Moscow SAI, Crimean Astrophysical Observatory and SAI Crimean station.

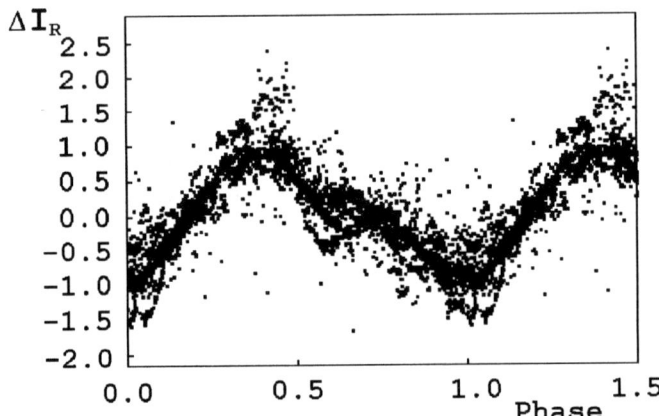

FIGURE 1. Orbital R band light curve of V723 Cas in the intensity scale. The trend of light decay is eliminated. The observation refer to seasons of 1998 - 2001.

Since September 1997 till December 2001, the amplitude of the orbital light variations gradualy increased from $0^m.07$ to $0^m.80$ in R band. Otherwise, the amplitude and shape of the orbital light curve stay approximately constant in the intensity scale. Fig. 1 shows the R band orbital light curve in the intensity scale, the trend of light decay is eliminated. The following elements were used:

$$Min\ I = 2451842.666 + 0^d.693265 E.$$

The orbital light curve is strongly asymmetric, has light maximum at the phase of $0^P.4$ and secondary minimum at the phase of 0^P55. The light curve shape resembles closely that one of recurrent nova CI Aql [8], which is also highly inclined eclipsing binary with large orbital period of 0.618355 day. Both light curves have similar features due to high accretion rate.

PULSATIONS

The episode of highest amplitude pulsation was observed on September 19, 1999 just after an eclipse (Fig. 2). This event occured on 1372 day after light maximum. In one of seasons, the brightest comparison star in the field of view GSC 3668.1121 was suspected in small-amplitude variability in the long time scale of 20 days. To control its constancy in the night of observations, we use two check stars, the light curves of which are also shown in the Figure. We prefer this comparison star because it gives maximum accuracy of nova measurements, of about $0^m.005$. Both check stars do not show variability, and therefore we attribute the oscillations to nova.

Three consequent maxima are detected in that night. The period is equal to $0^d.062$, maximum amplitude is of $0^m.05$ R. But the intrinsic amplitude may be larger because of the big contribution of the surrounding nebular medium.

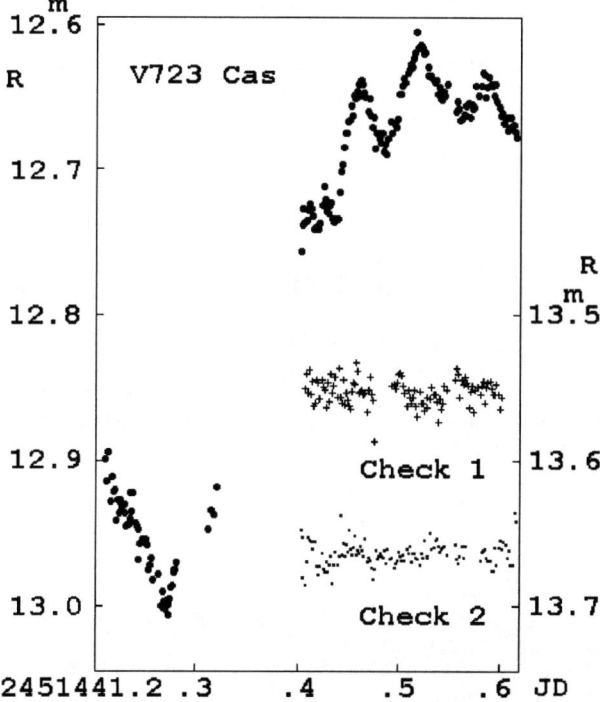

FIGURE 2. Pulsations of V723 Cas observed on September 19, 1999. The light curves of two fainter control stars are also shown.

No more episodes of oscillations were observed in the end of 1999, and in the following seasons of 2000 and 2001. In the nearby nights, however, the lower amplitude pulsations with the identical period are possibly present. The seing conditions of small amplitude oscillations are improving with time due to depression of surrounding nebular component in the same way as those ones of the orbital light variations, but we have not detected the pulsations in the last two seasons. The absence of this phenomenon in the following seasons excludes both magnetic white dwarf rotation with the given period, and disk instabilities. Regular instabilities of the accretion flow or regular brightness changes of the hot spot do not explain these oscillations, too, because they are located on the opposite part of the disk or covered by expanded envelope, and can not be seen in the orbital phases after eclipse egress. The oscillation period is far from the orbital one, so it is not a superhump phenomenon. Otherwise, the excitation of a radial pulsation mode for a short time in the decaying nova envelope (as soon as conditions arise) is more probable. Note, that the pulsation period of V723 Cas is about twice longer than that of V1494 Aql, but this may be explained with larger radius of the envelope of V723 Cas, due to larger separation between its components.

So, our observations confirm Klaus Schenker's analysis.

ACKNOWLEDGMENTS

This study was supported by Grant No.1.4.2.2 of Russian scientific and technical program "Astronomy". S.Yu.Sh. thanks the Russian Foundation for Basic Research for partial support by Grants No. 00-15-96553 and 02-02-17524. V.P.G. and S.Yu.Sh. are thankful to organizers of the Conference for financial support necessary for their participation in this Conference.

REFERENCES

1. Schenker, K., "Insights from Pulsating Nova Envelops" in *Radial and Nonradial Pulsations as Probes of Stellar Physics*, edited by C.Aerts et al., ASP Conference Series **259**, pp 580-583 (2002).
2. Starrfield, S., Drake, J. "The Extraordinary X-ray Light Curve of the Classical Nova V1494 Aql in Outburst: the Discovery of Pulsation and a "Burst" in *Two Years of Science with Chandra*. Abstracts from the Symposium held in Washington, DC, 5-7 September 2001.
3. Hirosawa, K., *IAU Circ*. No.6213 (1995).
4. Munari, U., Goranskij, V.P., Popova, A.A., Shugarov, S.Yu., Tatarnikov, A.M., Yudin, B.F., Karitskaya, E.A., Kusakin, A.V., Zwitter, T., Lepardo, A., Pasuello, R., Sostero, G., Metlova, N.V., Shenavrin, V.I., *Astron. and Astrophys.* **315**, 166-169 (1996).
5. Zhu, Z.-X., Hang, H.-R., *Acta Astron. Sinica* **40**, 247-275 (1999).
6. Iijima, T., Rosino, L., *IAU Circ.* No.6703 (1997).
7. Goranskij, V.P., Shugarov, S.Yu., Katysheva, N.A., Shemmer, O., Retter, A., Chochol, D., Pribulla, T., *Inform. Bull. Var. Stars* No.4852, 1-3 (2000).
8. Matsumoto, K., Uemura, U., Kato, T., Kiyota, S., Ayani, K., Kawabata, T., Kral, L., Havik, T., Kolasa, M., Novak, R., Masi, G., *Astron. and Astrophys.* **378**, 487-494 (2001).

Photometric observations of two Novae in Cygnus

Irina Voloshina*, Helen Rovithis-Livaniou† and Natalia Metlova*

*Sternberg Astronomical Institute, Moscow State university, Universitetskij prospect 13, 119992 Moscow, Russia
†Department of Physics, University of Athens, Panepistimiopolis, GR-157 84 Zografos, Athens, Greece

Abstract. Two Nova came to the outburs in Cygnus during the summer of 2001: V2274 Cyg and V2275 Cyg. Here we present the results of our observations for both of them obtained with 60 cm telescope in Crimea soon after their discovery. For V2274 Cyg UBV and CCD observations in V, R and I bands during two months are available. Unfortunately only the sparse UBV measurements of V2275 Cyg, the second Nova in Cygnus, were carried out because of the bad weather conditions in autumn.

V2274 Cyg, the first Nova in Cygnus in 2001 was discovered by Y.Nakamura on July 13.651 UT [1] near the maximum or soon after it. On July 16.515 UT, its brightness was $V = 11^m.69$.

Our photometric monitoring of this Nova started on July 19 and lasted to the end of August 2001 when the system was on the stage of early decline. At first observations were made with a UBV photometer attached to the 60 cm telescope of Sternberg Astronomical Institute in Crimea, with time resolution 10 sec (EMI 9789 was used as a detector). The star GSC 2683.2526 from the close vicinity of the Nova was used as a local standard with magnitudes: $V = 11^m.88, B = 12^m.69, 13^m.10$.

The results of our UBV observations of V2274 Cyg are given in paper [2]

When V2274 Cyg became too faint we continued our observations with CCD camera ST7 at the same telescope in Crimea. Our observations were carried out in bands V, R, I with exposure 90 s in every band. The results of CCD observations are presented in Table 1.

Using data from our previous paper [2], data given in Table 1 and the measurements of different authors communicated by VSNET, the light curve of V2274 Cyg was constructed. This curve is shown in Figure 1.

Taking into account that the Nova is absent on the DSS images, with limiting magnitude down to 21^m we can suggest that the outburst amplitude of this Nova exceeded 8 magnitudes.

The analysis of UBV observations of Nova carried out during two nights, July 20 and 26, with the goal to find variability on a short time scale clearly shows that there were no periodic variations at that time. We can only note a small brightening on July 26 overlapping the overall decline.

TABLE 1. CCD observations of V2274 Cyg

JD244000	(var-st) V band	JD 2440000	(var-st) R band	JD 2440000	(var-st) I band
52132.3478	1.359	52132.3435	-0.014	52132.3824	-0.730
52132.3490	1.357	52132.3447	-0.016	52133.3836	-0.738
52132.3502	1.335	52132.3459	-0.019	52133.3849	-0.734
52133.3781	1.384	52133.3737	0.006	52134.3691	-0.624
52133.3793	1.376	52133.3749	-0.002	52134.3704	-0.621
52133.3805	1.392	52133.3762	0.00	52134.3716	-0.629
52134.3649	1.564	52134.36037	0.123	52137.3517	-0.510
52134.3661	1.583	52134.3616	0.115	52137.3529	-0.509
52134.3673	1.589	52134.3628	0.116	52137.3542	-0.511
52137.3583	1.772	52137.3469	0.263	52138.3874	-0.355
52137.3599	1.778	52137.3481	0.260	52138.3887	-0.355
52137.3614	1.779	52137.3493	0.262	52138.3899	-0.356
52138.3934	2.032	52138.38327	0.426	52139.3485	-0.264
52138.3950	1.020	52138.3844	0.426	52139.3497	-0.276
52138.3965	2.041	52138.3856	0.430	52139.3510	-0.258
52139.3545	2.121	52139.3427	0.527	52140.3474	-0.223
52139.3561	2.146	52139.3439	0.527	52140.3486	-0.222
52139.3577	2.161	52139.3469	0.519	52140.3498	-0.218
52140.3534	2.165	52140.3431	0.550	52141.3458	-0.243
52140.3550	2.167	52140.3443	0.549	52141.3470	-0.242
52140.3566	2.168	52140.3455	0.546	52141.3483	-0.245
52141.3521	2.082	52141.3415	0.528	52142.3487	-0.356
52141.3537	2.106	52141.3427	0.528	52142.3501	-0.351
52141.3553	2.091	52141.3440	0.528	52142.3515	-0.355
52142.3550	1.838	52142.3431	0.390	52144.2940	-0.179
52142.3566	1.835	52142.3454	0.383	52144.2953	-0.181
52142.3581	1.831	52142.3468	0.380	52144.2965	-0.173
52144.3033	2.093	52144.2897	0.575	52145.3609	-0.161
52144.3053	2.087	52144.2909	0.572	52145.3624	-0.151
52144.3070	2.103	52144.2921	0.575	52145.3639	-0.154
52145.3705	2.202	52145.3536	0.594	52146.3374	-0.148
52145.3723	2.190	52145.3548	0.603	52146.3386	-0.150
52145.3740	2.192	52145.3589	0.611	52146.3399	-0.149
52146.3435	2.142	52146.3327	0.619	52147.3366	0.000
52146.3450	2.140	52146.3340	0.624	52147.3378	-0.001
52146.3466	2.147	52146.3352	0.616	52147.3391	0.001
52147.3433	2.340	52147.3321	0.771		
52147.3448	2.339	52147.3334	0.766		
52147.3464	2.342	52147.3346	0.768		

According to the criteria usually adopted for Novae, we may conclude that V2274 Cyg is a fast Nova.

The second Nova in Cygnus — V2275 Cyg was discovered later by A.Tago on August 18.599 and 18.603 UT [3]. It was brighter than the first one, – about $8^m.8$. K. Ayini [3] performed spectroscopy of this Nova and confirmed its Nova nature in an early phase. P. Schmeer reported that there is UNSO A2.0 star about $18^m.8$ very close to the Nova position[4]. If this star is a protogenitor of this Nova then the amplitude of V2275 Cyg

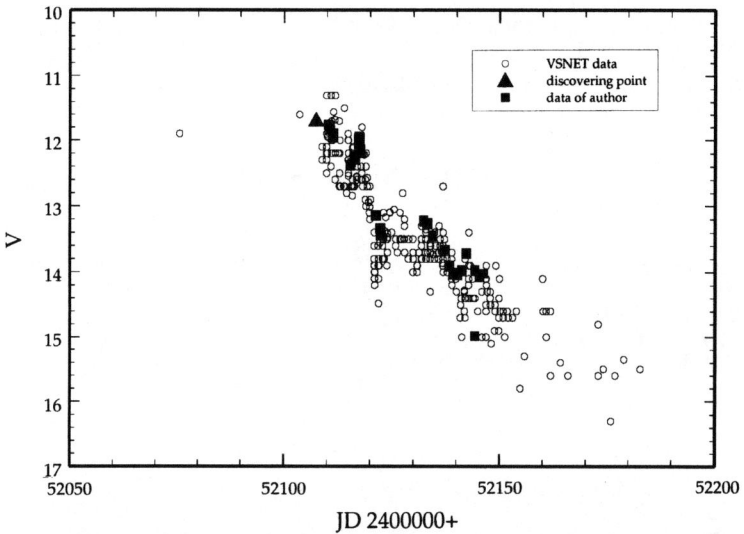

FIGURE 1. The light curve of V2274 Cyg in V-band

TABLE 2. UBV observations of V2275 Cyg

Date	JD	V	B	U
10 October 2001	2452193.2672	12.528	13.080	13.033
25 October 2001	2452208.2786	12.912	13.558	13.553
25 October 2001	2452208.2894	13.011	13.617	13.668
25 October 2001	2452208.3317	13.068	13.655	13.582
26 October 2001	2452209.2244	13.085	13.669	13.513
27 October 2001	2452210.3508	12.954	13.521	13.485
27 October 2001	2452210.3764	13.191	13.578	13.386
28 October 2001	2452211.1872	13.036	13.622	13.459
6 December 2001	2452220.3341	13.201	13.850	13.846
6 December 2001	2452220.3449	13.249	13.853	14.054
9 December 2001	2452223.2705	13.461	14.071	14.098
9 December 2001	2452223.2781	13.279	13.922	13.941
11 December 2002	2452225.2405	13.479	14.099	14.327
11 December 2002	2452225.2475	13.486	14.082	14.169

outburst should be about 12 magnitudes.

Our observations of V2275 Cyg were carried out with UBV photometer on the 60 cm telescope. The accuracy of individual measurements are $0^m.03 - 0^m.04$ in V and B bands and $0^m.04 - 0^m.06$ in U band. Unfortunately only the sparse measurements of V2275 Cyg were carried out because of the bad weather conditions in autumn. The results of our UBV observations for V2275 Cyg are given in Table 2.

We plan to continue our monitoring of these two Nova.

ACKNOWLEDGMENTS

We thank Sergei Antipin for the help with observations.

REFERENCES

1. Nakamura, Y., IAU Circ.7666, 2001
2. Voloshina, I., Metlova N., "UBV photometry of N Cyg 2001 = V2274 Cyg" in *The Physics of Cataclysmic Variables and Related objects*, edited by B.Gansicke, K.Reinsch and K.Beuermann, ASP Conf. Ser., v.261, 2001 (in press)
3. Tago, A., IAU Circ. 7686, 2001
4. Schmeer, P., IAU Circ. 7688, 2001

BVRI photometry of extremely slow nova Aql = V1548 Aql

Nataly V. Primak*, Elena P. Pavlenko† and Sergei Yu. Shugarov, Vitaly P. Goranskij**

*Kiev National University, Ukraine
†Nauchny, 19-17, Crimea 98409, Ukraine
**Sternberg Astronomical Institute, Moscow University, Moscow, 119992, Russia

Abstract. The preliminary result of the detail VRI photometry of Nova Aql 2001 = V1548 Aql during the outburst decline is presented. Nova shows an extremely slow fading with rate of 0.006 mag/day over first 150 days. Quasi-regular brightening with typical time ~ 60 days and amplitude up to 1^m were superposed on the outburst decline. A peculiarities of behaviour of this nova on the color-color diagram are discussed.

OBSERVATIONS

Nova V1548 Aql was discovered by Mike Collins [1]. We observed this star in 2001 – 2002 years in Crimean Astrophysical Observatory, Special Astronomical Observatory and Sternberg Astronomical Institute in the close to Johnson and Morgan VRI fhotometric systems by use SBIG CCD cameras ST-6 and ST-7.

GENERAL FEATURES OF PHOTOMETRIC BEHAVIOR

We combined our $BVRI$ observations with visual observations taken from the VSNET (vsnet.kusastro.kyoto-u.ac.jp/vsnet/lists.html) and presented them in Fig. 1. Over first 150 days since outburst Nova faded with a very slow rate 0.006 mag/day in V. Its t_3 is $> 400\ d$.

V1548 Aql displayed a quasi-periodical brightness variations in all spectral bands superposed on the slow decline. The amplitude of variations was not stable, sometimes it reached 1^m in V. To study the typical time of these variations, we computed a periodogram by Stellingwerf method for our V and available visual data, using the ISDA package [2]. It is presented in Fig. 2. One can see that the strongest peak corresponds to the period of ~ 57-d variations. The data folded on this period show a two-humped shape, so we suggest that the two times shorter period could be more real. All visual data folded on the 28.5-d period are presented in Fig. 3.

During the first 200 days V1548 Aql was redder when fainter, this effect was more prominent in $V - R$ and almost not visible in $V - I$ (see Fig. 4). Later Nova came back very soon to its previous $V - R$ color and during the next 200 d it has been faded without (or almost without) change in $V - R$ and $V - I$.

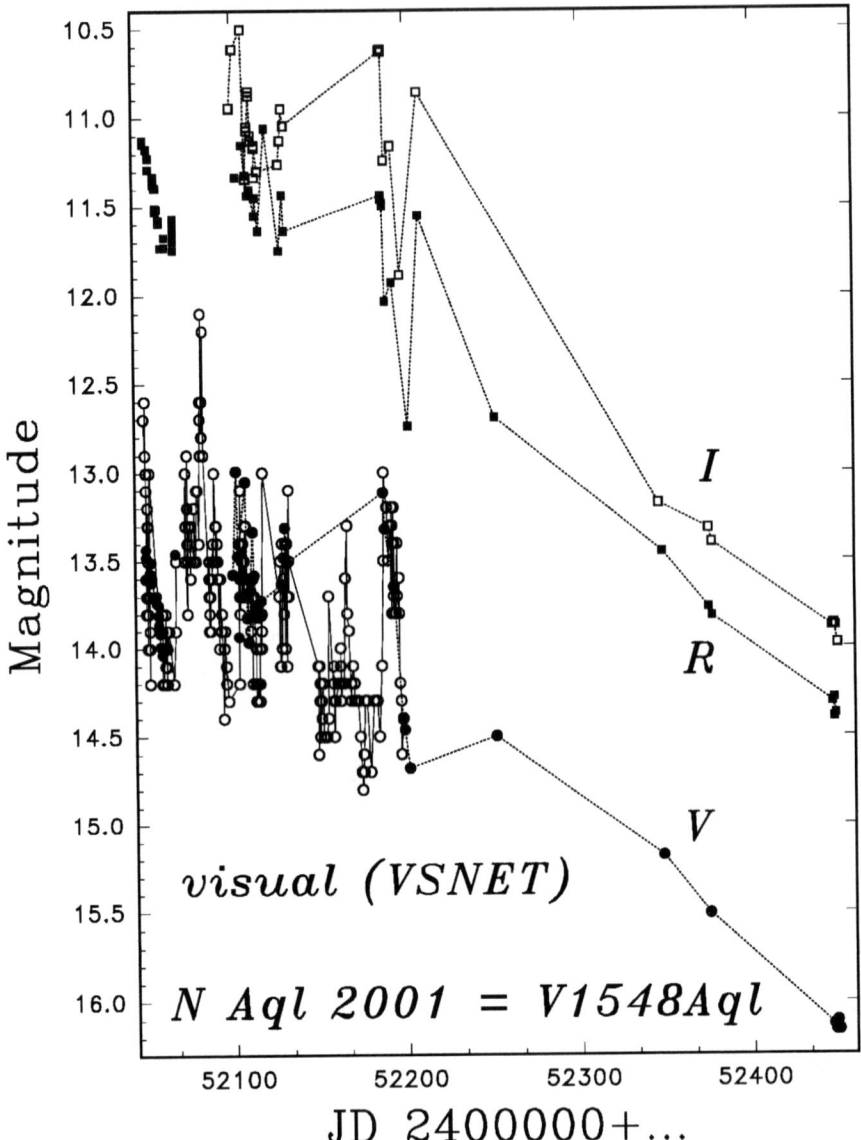

FIGURE 1. Light curve of V1548 Aql in V (filled circles), R (filled squares), I (open squares). Visual data taken from VSNET are shown by open circles.

$V-R$ and $V-I$ colors indicate significant interstellar reddening in direction to the V1548 Aql. Its quasi-periodical behavior during the outburst decline is somewhat similar to those in V723 Cas [3].

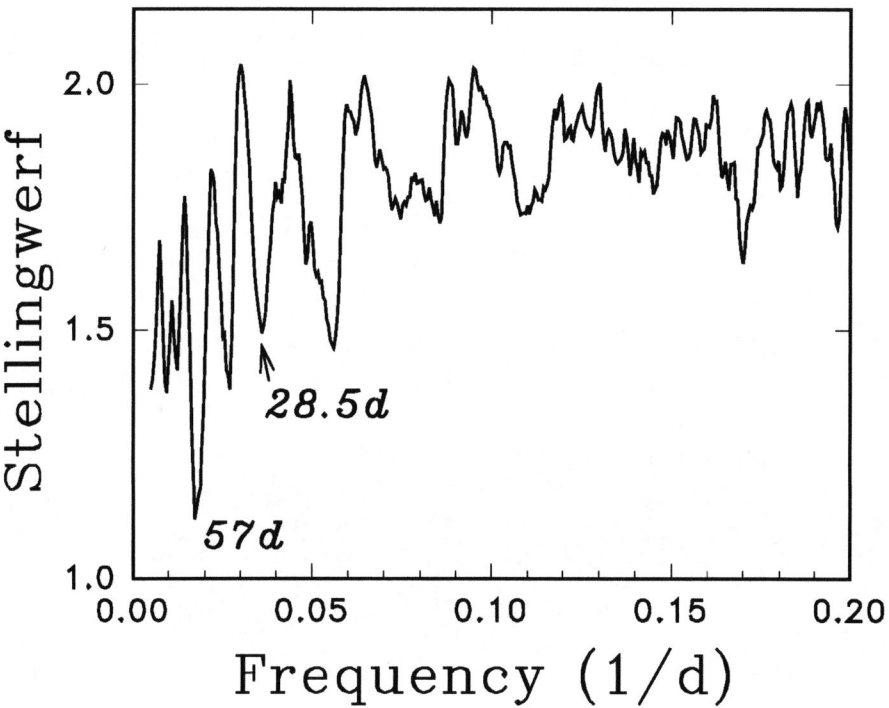

FIGURE 2. Periodogram obtained by Stellingwerf method for the V and visual (VSNET) data.

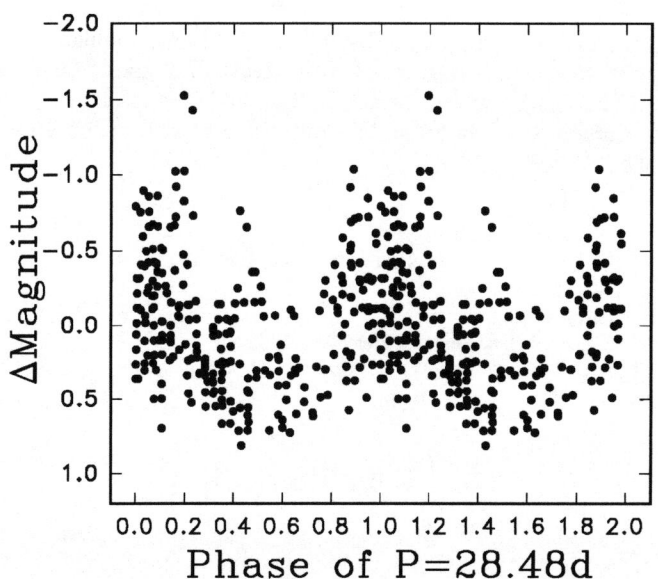

FIGURE 3. V and visual data folded on the period of 28.48 day

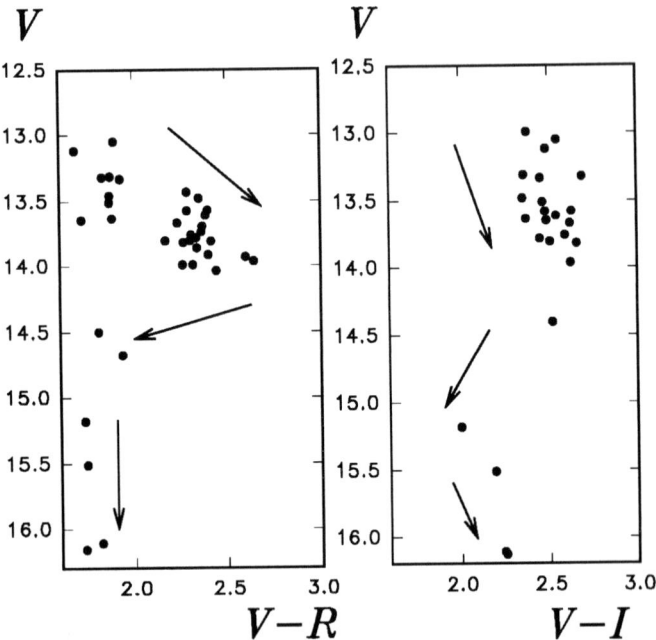

FIGURE 4. Outburst decline in $V, V-R$ and $V, V-I$ diagrams.

ACKNOWLEDGMENTS

This study was partially supported by Ukrainian Fund Fundamental Researches 02/07/00451, by Russian Foundation of Base Research's grants 00-15-96553, 0202-1642, 02-02-26723 and grant No.1.4.2.2 of Russian scientific and technical program "Astronomy". Authors are grateful to the Conference LOC for supporting their participation in the Conference.

REFERENCES

1. VSNET-alert, **5876** (2001).
2. Pelt, J., *Irregular Spaced Data Analysis*, Helsinki (1992).
3. Chochol, D., Pribula, T., *Contribution of the Astronom. Observ. Skalnate Pleso*, **28**, 121-141 (1998).

The Photometry of V1974 Cyg = N Cyg 1992

Sergei Yu. Shugarov*, Vitaly P. Goranskij,* and Elena P. Pavlenko[†]

Sternberg Astronomical Institute, Moscow University, Moscow, 119992, Russia
[†]*Crimean Astrophysical Observatory, Nauchny, Crimean Republic, 98409, Ukraine*

Abstract. The results of photoelectric and CCD *UBVRI* observations and extensive CCD and TV monitoring of V1974 Cas during the decline phase are shown. Strong features in the ultraviolet light and colour curves are caused by the temporal appearance of forbidden neon emissions. The improved orbital period of 0.08125970($\pm$5) day is determined. The "superhump" period of 0.08522 day is still observed, too.

INTRODUCTION

V1974 Cyg is an O-Ne-Mg class nova discovered by P. Collins [1] on February 19, 1992. It reached $4^m.2V$ in maximum. This nova has a faint companion ($B \approx 19^m.3$) located by 2".2 to SE. The observations of light echo and resolved expanded envelope were reported. Nova was bright in the soft X-rays with a continuum temperature of $3.5 \cdot 10^5$K. J.A. DeYoung and R.E. Schmidt [2] have discovered the orbital periodicity with period of about $0^d.08126$. The amplitude of the periodic variations depends on the wavelength being the largest in the near infrared. Later, this period was monitored till 1996 by different observers [3, 4, 5, 6, 7], and we use these data to improve the period.

The secondary "superhump" period of $0^d.085$ was found later with the period relation close to that of other cataclysmic variables having short orbital periods and high accretion rate.

OBSERVATIONS AND LIGHT CURVES

477 multicolour photoelectric and CCD observations of V1974 Cyg were taken with different telescopes and devices at SAI Moscow Observatory, SAI Crimean station, Crimean Astrophysical Observatory and Tien-Shan Astronomical Observatory in *UBV* and *BVR* bands. The observations were carried out in 1992-2001 (JD 2448777-2452254). Observations of the other authors[1] were used for the analysis of the light and colour curves, too. Additionally we have taken 1940 photoelectric, CCD and TV observations in *BVR* bands (17 nights in JD 2450729-52254) to monitor the orbital and

---

[1] The photometry published in IAU Circulars by A. Alonso, A. Ashoka, R. Barrena, D. Bohme, R. Casas, A. Darias, B. Dintinjana, A. Dolsen, J. Gallego, D. Hunzl, F. Hroch, K.P.N. Kutty, H. Mikuz, R. Monella, E. Neureiterova, H. Okura, A.D. Paolantino, A. Piersimoni, J. Wang, as well as data by D. Chochol et al.[8]; E.A. Kolotilov et al.[9]; E.S. Dmitrienko (unpub.).

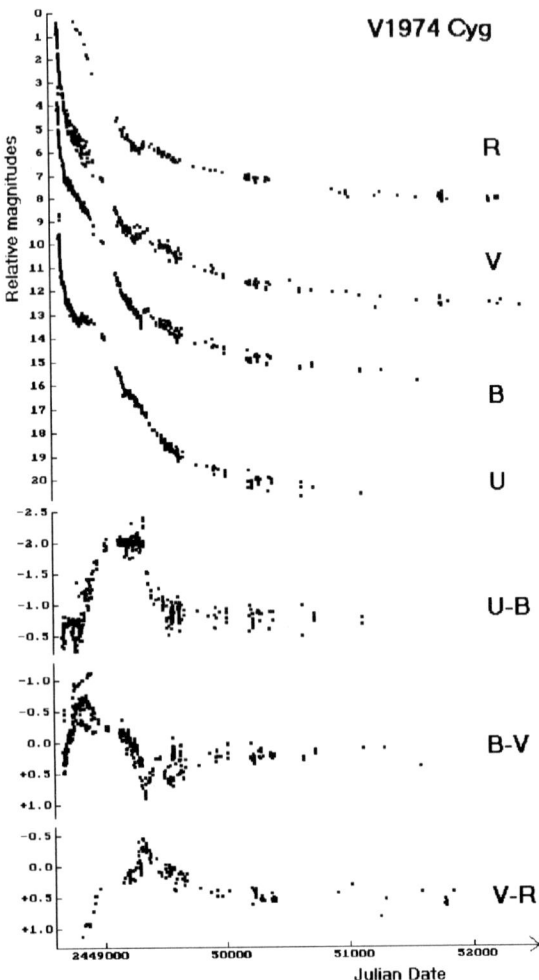

FIGURE 1. The $UBVR$ light and colour curves of V1974 Cyg.

the superhump periodic variability. Due to strong emissions in the spectrum of nova, different observational sets have large systematic errors, which we have tried to take into account with the corrections dependent on time.

The light and colour curves of V1974 Cyg are shown in Fig. 1. The strong hump in the U light curve and $1^m.5$ UV excess in the $U-B$ colour curve are noticeable in the interval of JD 2448900-2449500. We attribute these features to temporal appearance of the strong forbidden lines of [NeV] $\lambda 3346/3426$Å and [NeIII] λ 3869/3968Å in the near UV spectra. The star performed an unusual excursion far to negative UV colours on the two-colour $(U-B)-(B-V)$ diagram, which is the characteristic of only this O-Ne-Mg nova.

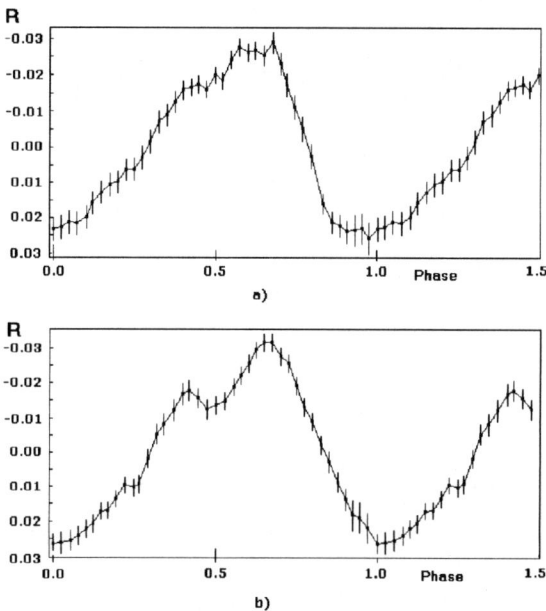

FIGURE 2. (a) The mean orbital light curve folded with the period of 0.0812597 day. (b) The mean superhump light curve folded with the period of 0.08522 day.

Orbital and superhump periods

Our recent monitoring data were subjected to Fourier analysis. The trend of light decay is eliminated. There are many peaks in the amplitude spectrum due to bad window of the set. The early known orbital period of 0.08126 day may be confirmed. But the light curve of this period is strongly dispersed. There are few higher peaks than that of the known period. The averaged light curve with the orbital period is given in Fig. 2a. The mean amplitude of variation is equal to $0^m.05$, but this amplitude is found to be variable, and reaches $0^m.20$ sometimes. In the residuals from that mean orbital light curve, the known secondary period of $0^d.08522$ also may be revealed with Fourier analysis, having the mean light curve shown in the Fig. 2b.

As a result of monitoring, nineteen new times of minima are measured (Table 1).

Using these minima and those ones present in literature, we have revealed the following improved orbital elements:

$$\text{Min Hel.} = 2449234.7346 + 0^d.08125970 \cdot E.$$
$$\pm.0020 \quad \pm 0.00000005$$

The O-C diagram for orbital elements is shown in the Fig. 3. In some of the cited papers, times of light maxima are given, and then we transform them into times of minima, adding the half of period. Our new observations confirm some irregular phase

TABLE 1. Times of Minima of the Orbital Variations.

JD hel. 24...		JD hel. 24...		JD hel. 24...	
50730.320	CCD	50994.490	CCD	52174.304	CCD
50730.408	CCD	51020.425	CCD	52176.256	CCD
50988.339	TV	51112.344	CCD	52176.342	CCD
50988.392	TV	52171.298	CCD	52251.254	CCD
50989.370	TV	52172.270	CCD	52251.338	CCD
50989.440	TV	52173.411	CCD	52254.245	CCD
50994.412	CCD	52174.223	CCD		

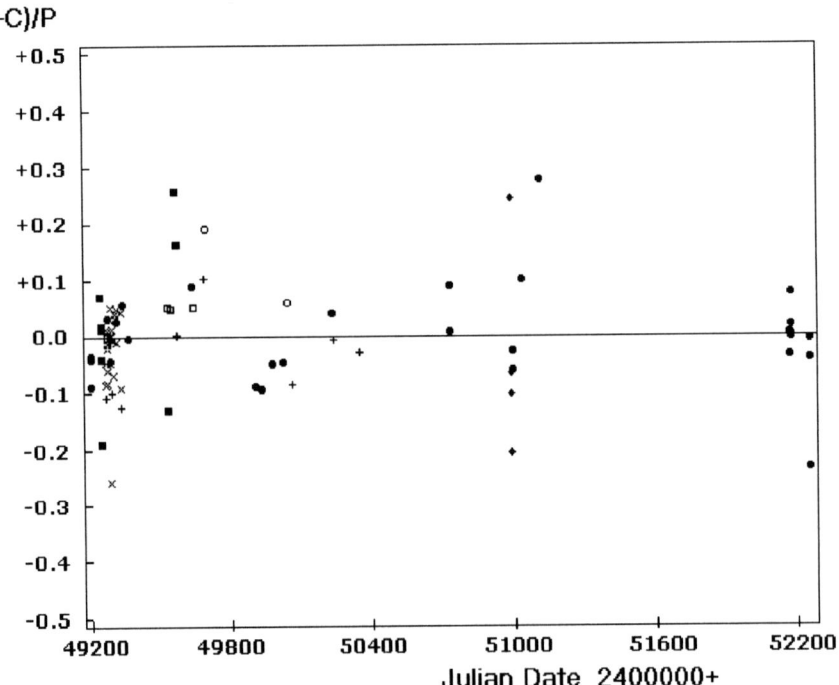

FIGURE 3. The O-C diagram for orbital period. Filled circles are our photoelectric and CCD observations, rhombs are our TV observations. Other symbols are data from cited papers.

deviations of minima relatively to linear formula, which resemble the irregularities seen in the previous seasons. For instance, two minima near JD 2451000 are shifted by phase of $0^p.3$. The shifted minima do not disturb the common tendency what suggests that the orbital period does not change after the nova outburst. Seems that the irregularities are caused by the secondary wave present in the data. The prewhitening of our data set for the orbital period and O-C analysis of the secondary superhump period in the residuals does not give good results that is due to variability of the amplitude of the orbital period.

ACKNOWLEDGMENTS

This study was supported by Grant No.1.4.2.2 of Russian scientific and technical program "Astronomy". One of us, S.Yu.Sh., is thankful for support of Russian Foundation for Basic Research by Grans No. 00-15-96553 and 02-02-16462. Authors are thankful to the Conference LOC for financial support of their participation in the Conference.

REFERENCES

1. Collins, P., *IAU Circ.* No. 5454 (1992)
2. DeYoung, J.A., Schmidt, R.E., *IAU Circ.* No. 5880 (1993).
3. DeYoung, J.A., Schmidt, R.E., *Astrophys. J.* **431**, L37-L49 (1994).
4. Olech, A., Semeniuk, I., Kwast, T., Pych, W., DeYoung, J.A., Schmidt, R.E., Nalezyty, M., *Acta Astron.* **46**, 311-323 (1996).
5. Retter, A., Leibowitz, E.M., Ofek, E.O., *Mon. Not. Roy. Astron. Soc.* **286**, 745-756 (1996).
6. Semeniuk, I., Pych, W., Olech, A., Ruszkowski, M., *Acta Astron.* **44**, 277-289 (1994).
7. Skillman, D.R., Harvey, D., Patterson, J., Vanmunster, T., *Publ. Astr. Soc. Pacific* **109**, 114-124 (1997).
8. Chochol, D., Hric, L., Urban, Z., Komzik, R., Grygar, J., Popousek, J., *Astron. and Astrophys.* **277**, 103-113 (1993).
9. Kolotilov, E.A., Nadzhip, A.E., Shenavrin, V.I., Yudin, B.F., *Astron. Reports* **38**, 548-551 (1994).

The Problem of the Flickering Activity of the Recurrent Nova T CrB

L. Hric[1], K. Petrík[2,3], A. Dobrotka[4] and R. Gális[5]

[1] Astronomical Institute of the Slovak Academy of Sciences, 059 60 Tatranská Lomnica, Slovak Republic, email: hric@ta3.sk
[2] Department of Physics, Faculty of Education, Trnava University, Priemyselná 4, 918 43 Trnava and
[3] Observatory and Planetarium, Sládkovičova 41, 920 01 Hlohovec, Slovak Republic, email: astropet@ta3.sk
[4] Department of Astrophysics, Faculty of Mathematics, Physics and Informatics, Comenius University, Bratislava, Slovak Republic
[5] Faculty of Sciences, University of P.J. Šafárik, Moyzesova 16, 041 54 Košice, Slovak Republic, email: galis@kosice.upjs.sk

Abstract. On the basis of our previous paper (Hric et al. 1998) we can proclaim that the existence of two emission-line regions can remarkably affect the flickering activity in the binary system. We could expect flickering activity dependence on the orbital phase. Deriving the energy of the flickering and comparing the derived value with the energy particularly needed for three possible areas where the source of the flickering could be settled down, we found the most plausible area to be in the unstable accretion onto the white dwarf, i.e. the boundary layer.

INTRODUCTION

The recurrent nova T CrB has become the first cataclysmic variable (CV) studied spectroscopically when it reached the brightness of 2nd magnitude during its outburst in May 1866. The object was studied intensively immediately after its second outburst in February 1946 when T CrB became the brightest recurrent nova in history. Recently, there were published many papers on flickering activity study, improvement of orbital parameters, hot component mass determination and recurrent outbursts interpretation. More detailed summary on the basic characteristics of T CrB with the list of references can be found in our previous papers (Hric et al. 1998, 2000, 2001).

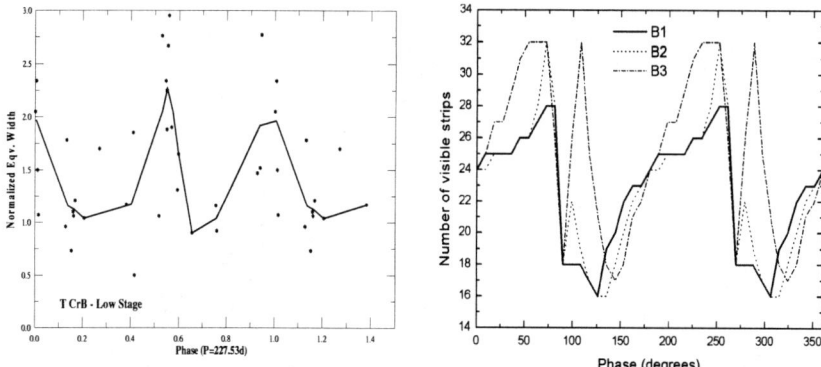

FIGURE 1. (left panel) The equivalent widths of H_α lines dependence on the orbital phase. The discrete points represent discrete values while the solid line represents averaged values (5 points at most).
FIGURE 2. (right panel) Visible surface of ellipsoids' dependence on orbital phase. Edge on view. Partially overlapped regions in contact are depicted by the line B1. Partially overlapped regions in mutual distance of one diameter of the region can be seen as the line B2. B3 line represents very distant regions.

The existence of two emission-line regions proposed by Hric et al. (1998) can remarkably affect the flickering activity in the binary system. We can expect flickering activity dependence on the orbital phase.

In the recent paper we tried to use 3D graphic software to model two overlapped, eccentric, ellipsoidal emission line regions. We used three different configurations of the regions, characterized by two free parameters: overlapped factor and the mutual distance. The best agreement of the configurations with the observations of the equivalent widths of H_α lines appeared to be when the regions are particularly overlapped and are in contact (see observations and 3D model on Figures 1. and 2.).

THE PHOTOMETRIC OBSERVATIONAL MATERIAL

During the long-term photometry (since 1996) we were aiming to obtain longer observational runs (several hours) in order to be able to get some deeper insight into the phenomenology and nature of the rapid light variations (flickering and flares). We have secured 16 runs acceptable for the study of flickering activity. The data discussed in this paper were obtained at the Skalnaté Pleso and Stará Lesná Observatories of the Astronomical Institute of the Slovak Academy of Sciences using the 60-cm reflectors equipped by the single-channel photoelectric photometers (UBV).

RESULTS

Some results after the statistical processing of the data in U filter are depicted in the Fig. 3.

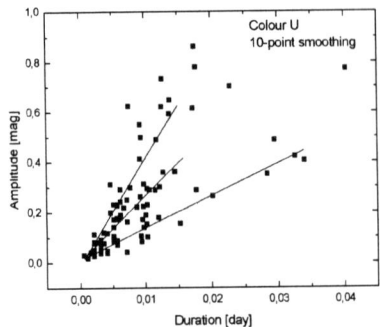 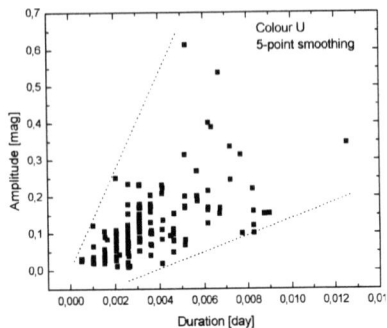

FIGURE 3. Duration vs. amplitude relation in U filter. The fan-shaped distribution of points (dotted line) is to be apparent.

THE ENERGY OF THE FLICKERING

The luminosity of the source of flickering changes while passing from the local minimum to the local maximum and vice versa. The energy of the event could be obtained as the multiplication of the luminosity and the event's duration. The shape of each event is triangle, it is clear that the change of the luminosity L is linear function. We can write:

$$E = \tfrac{1}{2} \Delta L\, \Delta t, \qquad (1)$$

while after the inetgration of the eq.: $dE = L(t)\, dt$ \qquad (2)
We need the difference of the luminosities ΔL between local extremes. Deducing from the Pogson equation we obtain:

$$\Delta L = L_2\, (10^{(\Delta m/2.5)} - 1). \qquad (3)$$

The next step is to deduce the luminosity of the source in the local minimum L_2. Again, we use the Pogson equation and we obtain:

$$L_2 = L_A\, 10^{-0.4(M_{bol} - 4.74)}, \qquad (4)$$

where $M_{bol} = M_V + BC$ is the bolometric magnitude of T CrB in the local minimum, M_V is absolute magnitude of the object in minimum in filter V and BC is bolometric correction. To obtain M_V we have to use Pogson equation again, in a very common form: $m_V - M_V = 5 \log r - 5 + A$, \qquad (5)
where r is the distance of the star in pc and A is interstellar extinction in magnitudes. $m_V = m_2$. Using eqs. 4, 3 and 2 we can write:

$$L_2 = L_A\, 10^{-0.4(m_V - 5 \log r + 5 - A + BC - 4.74)} \qquad (6)$$

$$\Delta L = L_A \, 10^{-0.4(m_V - 5\log r + 5 - A + BC - 4.74)} (10^{(\Delta m/2.5)} - 1) \quad (7)$$
$$E = \tfrac{1}{2} \Delta t \, L_A \, 10^{-0.4(m_V - 5\log r + 5 - A + BC - 4.74)} (10^{(\Delta m/2.5)} - 1) \quad (8)$$

Most of the light in V filter (80 % - 90 %) comes from the red giant of spectral type M3 III (Zamanov & Bruch 1998, Bruch 1992). With respect to this fact we can use BC for the red giant of M3 III. We use the parameters as follows: $A = 0.35 \pm 0,05$ mag (Harrison et al. 1993), $r = 1180$ pc (Bruch 1992), $L_A = 3.845 \times 10^{26}$ W [J.s^{-1}], BC $= -1.91$ mag (Allen 2000), and then from eq. 7 and 8 we obtain:

$$\Delta L = 3.84 \times 10^{0.4(78.36 - m_V)} (10^{(\Delta m/2.5)} - 1) \quad (9)$$
$$E = 1.92 \Delta t \times 10^{0.4(78.36 - m_V)} (10^{(\Delta m/2.5)} - 1) \quad (10)$$

In fact, the real bolometric correction should be a little bit larger because of the contribution of the white dwarf to the overall energy.

From eq. 9 and 10 we can calculate the difference of the luminosities of the events and overall energy loose. But at first we have to take into account the orbital cycle of the binary. We need m_V or m_2 respectively. This value changes with orbital motion. The best value comes from the phase of maximum brightness (the source is not covered).

We found the ratio between maximal magnitude (y_{max}) and the magnitude in given phase (y_i). Corrected m_V in such way was applied to eqs. 9 a 10.

On the basis of our analysis we found the intervals of ΔL (0.2×10^{27} W – 4.2×10^{27} W) and E (0.2×10^{29} J – 34×10^{29} J) of observed flickering events. These values agree well with those of Bruch (1992, Tab. 9).

THE SOURCE OF THE FLICKERING – POSSIBLE SCENARIOS

Nonstable mass transfer from the secondary.

In this scenario the flickering is emitted by the hot spot. It is not confirmed by our results since the energy liberated by this process is too small to explain the amplitudes of the flickering. This result is not in agreement with those of Anupama and Mikolajewska (1999).

Flickering caused by the turbulence in the acretion disk

This mechanism is very probable too explain the flickering activity of T CrB. Calculated energies for dM/dt can be seen in Fig. 5.

Unstable accretion onto the white dwarf

In this case the scenario is fully reliable and could explain the whole range of changes in L or energies, respectively.

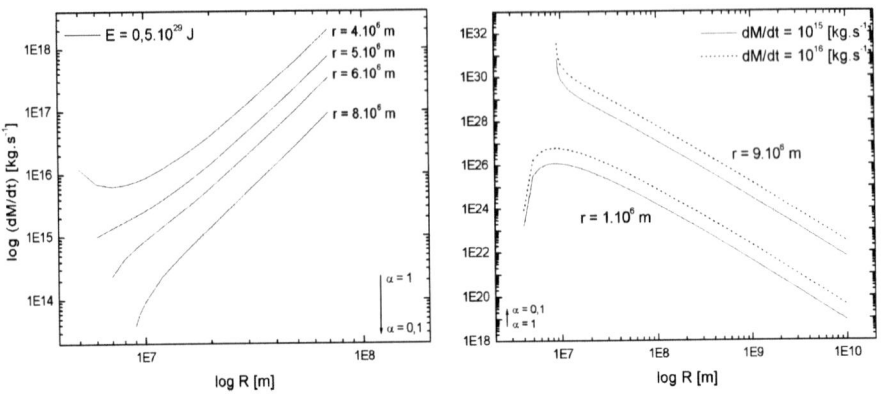

FIGURE 4. (left panel) Mass transfer in the scenario of turbulent regions. The change of the profile starts between diameter of turbulent regions $r = 4.10^6$ m and $r = 5.10^6$ m.
FIGURE 5. (right panel) Energetic losses with respect to the mass accretion and the distance from the center of the disk. Mechanism is reliable for the turbulent elements with larger diameters.

CONCLUSION

On the basis of our results we can locate the source of the flickering to the inner parts of the accretion disk or to the surface of the white dwarf. This is actually the reason why the orbital phase vs. flickering activity dependence probably doesn't exist, since the inclination of the system appears to be very high (i = 68°). It is in contradiction with our previous results (Hric et al. 2001). To solve the problem more observations should be obtained and analyze.

Acknowledgements. This work has been supported through the Slovak Academy of Sciences Grant No. 1008/21 as well as by Faculty of Education, University of Trnava Grant No. 2/02.

REFERENCES

1. Allen, C.W., 2000, Astrophysical Quantities, The Athlone Press, London
2. Anupama, G.C., Mikolajewska, J., 1999, A&A 344, 177
3. Bruch, A., 1992, A&A 266, 237
4. Harrison, T.E., Johnson, J.J., Spyromilio, J., 1993, AJ 105, 320
5. Hric, L., Gális, R., Petrík, K., Niarchos, P., 2001, In: Proc. of the 4th Astron. Conf of HEL.A.S, ed. Seimenis, J., 99
6. Hric, L., Petrík, K., Niarchos, P., Gális, R., 2000, In: Proc. of PhD Conf. "A bridge between generations of variable star observers", eds. Wilson R.E. et al., 145
7. Hric, L., Petrík, K., Urban, Z., Niarchos, P., Anupama, G.C., 1998, A&A 339, 449
8. Zamanov, R.K., Bruch, A., 1998, A&A 338, 988

Activity of the super-soft X-ray source V Sge

Vojtěch Šimon* and Janet A. Mattei[†]

*Astronomical Institute, Academy of Sciences of the Czech Republic, 251 65 Ondřejov, Czech Republic
[†]AAVSO, 25 Birch Street, Cambridge, Massachusetts 02138-1205, USA

Abstract. An analysis of the activity of the super-soft X-ray source (SSXS) V Sge is presented. The long-term (decades) increase of the mean brightness, revealed by [15], stopped and a decay became apparent. A large change of the character of the high/low state behaviour occurred in the last several years. The light curve displays a more complicated profile than in the previous segment of high/low state activity (1991–1995). Nowadays, episodes of long low states occur in a cycle of about 400 days, with superposed brief and less deep fadings. Similarities between the behaviour of V Sge and another SSXS RX J0513.9–6951 are pointed out. We argue that real mass transfer variations occur in V Sge, in variance with the model for the contracting and expanding H-burning white dwarf in J0513, proposed by [11].

INTRODUCTION

Most super-soft X-ray sources (SSXS) are close binary systems in which the mass transfer onto a white dwarf (WD) occurs at a high rate ($\dot{m} \approx 10^{-7}$ $M_\odot$ yr^{-1}). This allows a steady-state hydrogen burning on the surface of the WD [16]. This hydrogen burning is the source of the soft X-ray radiation. The optical/UV radiation originates mostly from the reprocessing of the X-rays in the disk and the donor star [10].

V Sge is a peculiar eclipsing binary ($P_{orb} = 0.514$ d) [5]. There have been accumulated several lines of evidence which strongly support the model of the mass accreting white dwarf (WD) primary from a massive companion [8], [4], [15]. V Sge is therefore a promising candidate for SSXS [13]. V Sge displays strong long-term photometric activity (e.g. [5], [15]).

DATA ACCUMULATION

The data for the years 1961–1995 were obtained from the AAVSO database [7], AFOEV database (1934–1944), and AFOEV and VSNET (1995–2001). The orbital modulation of V Sge was suppressed by rejecting the data inside phases 0.9–1.1 (the primary eclipse; see also [15]). The ephemeris of [12] was used. The data were then binned into one-day means.

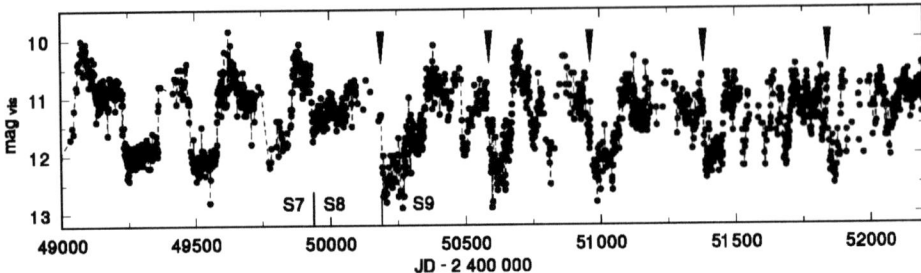

FIGURE 1. The recent light curve of V Sge (one-day means). The respective segments, discussed in text, are marked. The arrows denote the transitions to the long low states in segment S9.

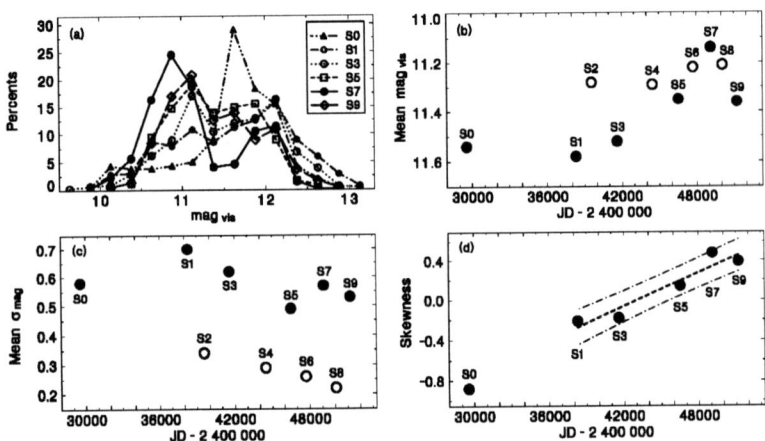

FIGURE 2. (a) The statistical distributions of brightness in the individual active segments, constructed from the one-day means of observations. The binning of 0.25 mag was used. (bcd) Net parameters of each segment (skewness, mean brightness, σ_{mag}) plotted versus JD. Each point is centered on the middle of the appropriate segment. The active and flat segments are marked by the filled and empty circles, respectively.

DATA ANALYSIS

The recent light curve of V Sge is displayed in Figure 1 while the light curve between 1934–1995 can be seen in [14] and [15]. The amplitude significantly varies on the time scale of years. There are definitely intervals of large variations with an amplitude of about 1.5–2.0 mag_{vis} (active segments; see [15] for more). Extended intervals (years) of a flat light curve can be seen in other years (flat segments). The respective segments, mostly according to [15], are marked. Two new segments, S8 and S9, were added. The levels of the high (HS) and low (LS) state are particularly prominent in S7.

Statistical properties of the segments The histograms for the respective active segments are shown in Figure 2a. Notice that generally the maximum gradually shifts toward higher brightness for the later active segments. The net skewness, mean brightness,

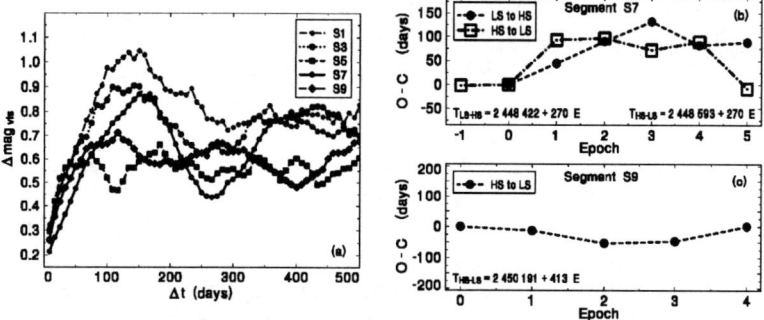

FIGURE 3. (a) The autocorrelation diagram for the active segments S1, S3, S5, S7, S9. Notice the deep minimum at 270 days for segment S7. (b) The $O-C$ diagram for the transitions between the states in segment S7. (c) The same for segment S9, but just the dominant long lows states, marked in Figure 1 by the arrows, are considered. The trends in the $O-C$ variations are apparent in both segments.

and the standard deviation of brightness σ_{mag}, calculated for the individual segments, are shown as a function of time in Figure 2bcd. The mean brightness of the active segments increased in the sequence S1-S3-S5-S7 but the mean level of S9 was lower than in S7. The brightness of the flat segments changed just a little over the covered interval.

Periodicities in the HS/LS transitions A search for periodicities of the high/low state (HS/LS) transitions was undertaken using two methods. The autocorrelation method [9] allows a search for the characteristic time scales or quasi-periods which extend just for several cycles. The autocorrelation diagrams for V Sge are shown in Figure 3a. Notice that quite divergent preferred cycle-lengths can occur in the individual active segments.

The method of the $O-C$ residuals from a reference period was applied to segment S7 and S9 (Figure 3bc). The trends in the variation of the cycle-length are apparent in S7, especially the cycle-length of the HS to LS transitions tends to decrease.

RESULTS

This analysis has shown the evolution of the optical activity of the super-soft X-ray source V Sge, including the changes in the last several years. Nowadays, the light curve displays a more complicated profile than in the previous active segment between 1991–1995. The long-term increase of the mean brightness [15] stopped and a decay became apparent. However, the trend in the net skewness and excess of the active segments continues.

The character of the variations in the active segments of V Sge evolves and depends on the mean level of brightness in the given segment. The low level gives rise to the relatively narrow outbursts [15] while HS/LS transitions occur in the segments with a higher mean brightness. V Sge tends to spend more time in HS than in LS in the segments with a high mean brightness. This suggests that these seemingly different brightness variations in V Sge are a product of a single mechanism.

The activity of another SSXS RX J0513.9–6951 [1] bears some similarity to V Sge. The alternating HSs and LSs resemble those in V Sge (especially in S7, scaled about 1.7 times down). The activity in RX J0513 was interpreted in terms of the limit-cycle model [11] in which the mass outflow rate from the donor star, $\dot{m}$, remains constant, only the radius of the hydrogen burning WD and the inner disk cyclically vary. The radius of the WD increases during the optical HS and the WD strongly irradiates the disk – this drains the disk empty on its viscous time scale. The mass inflow onto the WD therefore decreases, the radius of the WD contracts and the irradiation diminishes – the optical LS sets in. A hole in the inner disk which remains after the contraction of the blown-up WD has to refill during LS on its viscous time scale. The mass inflow onto the WD then resumes and its radius increases again – this leads to the return to HS.

On the contrary, we argue that real variations of $\dot{m}$ play a role in V Sge. Although the light curve of V Sge, especially in S7, is similar to RX J0513, the profiles of HS and LS in V Sge are often more complicated than in RX J0513 (especially in S5 and S9). The recurrence time of HS/LS transitions in S5 and of the superposed brief low states (fadings) in S9 is much shorter than what is explicable by the model by [11]. The short recurrence time of these fadings in V Sge is significantly shorter than the viscous time scale of the disk (even shorter than the time scale of the hole in the inner disk which has to refill during LS in the model by [11], unless the disk viscosity in V Sge is significantly higher than in RX J0513 ($\alpha \approx 1$ versus $\alpha \approx 0.1$). The fluctuations and irregularities in HS/LS in V Sge, including the fluctuations during HS itself [14] do suggest a highly unstable mass flow through the disk. Stream-disk overflow and/or a spray at the impact of the infalling stream can play a role.

The model for the irradiation-driven instability (IDI) of the outer layer of the donor [17] offers a promising mechanism for V Sge (see [15] for more details). IDI allows to obtain double-valued $\dot{m}$ for a distinct range of filling of the donor's lobe, f_{donor}, and the temperature of the WD, T_{WD}. The two levels (HS and LS) and the rapid transitions between them can be interpreted as a direct product of the two-level $\dot{m}$. $\dot{m}$ can vary even if the radius of the donor stays constant because the growing dimensions of the disk in the epoch of the high $\dot{m}$ shield the donor from further irradiation. The striking differences in the orbital modulation between HS and LS [5], [8] really imply large changes of the distribution of the circumstellar matter in V Sge. Also the X-ray spectrum of V Sge in HS and LS can be interpreted by a variable amount of circumstellar matter rather than by a variable radius of the WD [4].

Typical cycle-lengths, but not strict periods, can be traced in the active segments of V Sge. This suggests that the individual episodes of LS are not quite independent of each other. The cycle-length can depend on the rate of growing the disk dimensions, variable disk viscosity and/or the radius or temperature of the WD. The facts that quite divergent cycle-lengths occur in the individual active segments but the given segment tends to keep its characteristic cycle suggest that these parameters keep their typical values during the given segment.

The active segments in V Sge can be regarded as a perturbation of a relatively stable $\dot{m}$, occurring in the flat segments. The IDI model predicts that both double-valued and single-valued $\dot{m}$ can co-exist and that the system can alter between stable and unstable mass transfer, for example if T_{WD} varies. We can offer an explanation for the evolution of activity of V Sge in the framework of IDI if we include the long-term increase of f_{donor}.

The value of $\dot{m}$ depends on the donor's temperature (determined by T_{WD} and f_{donor}). Only variations of T_{WD} suffice for the modulation of $\dot{m}$ inside a given segment. The value of f_{donor} then would determine the mean properties of activity inside the segment.

Generally, the HS/LS transitions appear to be common for the SSX binaries with relatively long orbital periods ($P_{orb} > 10$ hr): V Sge, RX J0513.9–6951 [1], RX J0019.9+2156 [2], [3], and CAL 83 [6].

ACKNOWLEDGMENTS

This research has made use of NASA's Astrophysics Data System Abstract Service and the observations primarily from the AAVSO International database (Massachusetts, USA), and also from the AFOEV database (CDS, France) and VSNET (Japan). We thank the variable star observers worldwide whose decades of observations made this analysis possible. The research of V.Š. is supported by the post-doctoral grant 205/00/P013 of the Grant Agency of the Czech Republic. The support by the project ESA PRODEX INTEGRAL 14527 is also acknowledged.

REFERENCES

1. Alcock, C., et al., 1996, MNRAS, 280, L49
2. Bartolini, C., et al., 1996, 158th IAU Coll., Kluwer, p.427
3. Greiner, J., Wenzel, W., 1995, A&A, 294, L5
4. Greiner, J., van Teeseling, A., 1998, A&A, 339, L21
5. Herbig, G.H., et al., 1965, ApJ, 141, 617 (HPSP)
6. Kahabka, P., 1998, A&A, 331, 328
7. Mattei, J., 1996, AAVSO International database, private com.
8. Patterson, J., et al., 1998, PASP, 110, 380
9. Percy, J.R., et al., 1981, AJ, 86, 53
10. Popham, R., Di Stefano, R., 1996, LNP 472, p.99
11. Reinsch, K., et al., 2000, A&A, 354, L37
12. Smak, J., 1995, Acta Astron., 45, 361
13. Steiner, J.E., Diaz, M.P., 1998, PASP, 110, 276
14. Šimon, V., 1996, A&AS, 118, 421
15. Šimon, V., Mattei, J.A., 1999, A&AS, 139, 75
16. van den Heuvel, E.P.J., et al., 1992, A&A, 262, 97
17. Wu, K., et al., 1995, Publ. Astron. Soc. Aust., 12, 60

The colors and luminosities of the super-soft X-ray sources and classical novae

Vojtěch Šimon

Astronomical Institute, Academy of Sciences of the Czech Republic, 251 65 Ondřejov, Czech Republic

Abstract. An analysis of the color indices and absolute magnitudes M_{V0} of the super-soft X-ray (SSX) binary sources (including the V Sge-type stars) and classical novae in the SSX phase is presented. It can help comparing the properties and configuration of the reprocessing medium in the individual systems. The $(U-B)_0$ and $(B-V)_0$ colors of the novae in SSX phase are similar to those of the classical SSX binaries, but display a larger scatter for the individual objects. The range of M_{V0} of the individual objects with SSX emission is very large, from $M_{V0} = 8.5$ to $M_{V0} = -4$. The SSX phase of the novae V 1974 Cyg and V 382 Vel occurred at significantly brighter M_{V0} than of most SSX binaries while M_{V0} of U Sco was comparable to the SSX binaries. The implications for the reprocessing medium are given.

INTRODUCTION

The super-soft X-ray (SSX) binaries (e.g. CAL 83) are usually close systems in which the mass transfer onto a white dwarf (WD) occurs at a high rate ($\dot{m} \approx 10^{-7}\ M_\odot\ \mathrm{yr}^{-1}$). This allows a steady-state hydrogen burning on the surface of the WD [5]. This hydrogen burning is the source of the soft X-ray radiation. The optical/UV radiation originates mostly from the reprocessing of the X-rays in the disk and the donor star [2]. The V Sge-type stars are close binaries which display very similar properties in the optical as the "classical" SSX binaries. They were defined by [4].

Some symbiotic stars can be SSX sources, too (e.g. AG Dra).

Some classical novae were detected as luminous SSX sources during the late decline phase of their outbursts (e.g. V 1974 Cyg/Nova Cyg 1992, GQ Mus/Nova Mus 1983). The radius of the H-burning WD already shrinks during this phase and its temperature is sufficient to emit SSX and to irradiate the already resumed accretion disk.

It may be interesting to compare the optical properties of the individual categories of SSX sources, particularly their luminosities and color indices. This may help comparing the properties and configuration of the reprocessing medium in the individual systems.

DATA ACCUMULATION

This analysis considers the "classical" SSX binaries, the symbiotic stars with the detected SSX emission, the V Sge-type stars, and the classical novae in the SSX phase. One cataclysmic variable (CV), V 751 Cyg, which displayed SSX emission during its

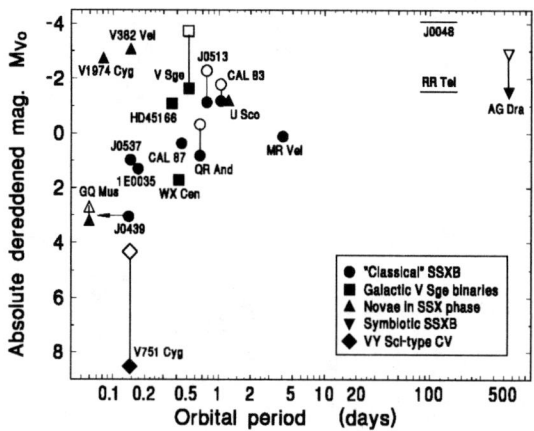

FIGURE 1. The absolute dereddened magnitudes M_{V0} plotted versus the orbital periods P_{orb}. The individual categories of objects are resolved. The lengths of P_{orb} of the symbiotic novae RR Tel and J0048 are unknown.

TABLE 1. The list of the SSX binaries and classical novae with SSX phase

Galaxy: QR And (SSX binary); MR Vel (SSX binary); V Sge (V Sge-type); V 617 Sgr (V Sge-type); WX Cen (V Sge-type); HD 45166 (V Sge-type); V 1974 Cyg/Nova Cyg 1992; GQ Mus/Nova Mus 1983; V 382 Vel (nova); U Sco (recurrent nova); RR Tel (symbiotic nova); AG Dra (symbiotic); V 751 Cyg (VY Scl-type CV)
LMC: RXJ0513-69 (SSX binary); RXJ0439-68 (SSX binary); RXJ0537-70 (SSX binary); CAL 83 (SSX binary); CAL 87 (SSX binary)
SMC: 1E0035-72 (SSX binary); RXJ0048-73 (symbiotic)

optical low state, is also included. The full list of the objects is given in Table 1. Due to the space limitation, the full list of literature used will be published elsewhere.

DATA ANALYSIS

At least one color index could be determined for each object, listed in Table 1. The index was determined from the closely spaced observations, obtained in a single night. In some cases, the mean indices were determined from the folded orbital curve, composed from observations obtained during several nights. The indices were dereddened using $E(B-V)$.

The absolute dereddened visual magnitude M_{V0} was calculated from the distance d and $E(B-V)$. In most cases, $E(B-V)$ was determined from UV spectra. The distances corresponded to the distances of LMC and SMC for the members of these galaxies. The distances of the objects in our Galaxy were usually determined from $E(B-V)$.

Absolute magnitudes In principle, M_{V0} of the SSX binaries can depend on the orbital period P_{orb} because the radius of the Roche lobe of the WD depends on P_{orb}. The larger

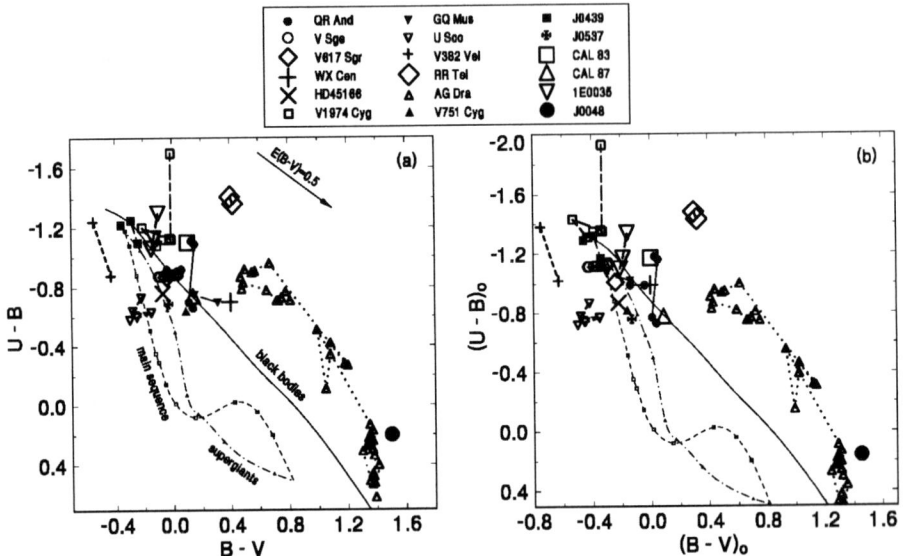

FIGURE 2. (a) $U-B$ vs. $B-V$ diagram. The individual objects are resolved. Notice that some objects lie above the curve of black bodies. This fact is not affected by the possible uncertainties in $E(B-V)$, as can be seen from the comparison with the dereddened colors in (b).

lobe allows a larger disk and therefore more matter for reprocessing of the X-rays. The absolute magnitudes M_{V0} were plotted versus P_{orb} in Figure 1. The individual categories of objects, mentioned in Section 1 and 2 and in Table 1, are resolved. Just M_{V0} which corresponds to the SSX phase is shown for the classical novae and RR Tel. In other cases, the full amplitude of the brightness variations is marked (if available). P_{orb}'s of RR Tel and J0048 are unknown but are supposed to be long, of the order of hundreds of days.

Color indices

$U-B$ vs. $B-V$ diagram (Figure 2a): Most displayed objects form a closed group, except the SSX symbiotic stars, whose colors are significantly different from the remaining objects. Some objects even lie above the curve of black bodies. This fact is not affected by the possible uncertainties in $E(B-V)$ (Figure 2b). The colors of V 751 Cyg correspond to the SSX phase.

$(B-V)_0$ vs. $(V-R)_0$ diagram (Figure 3a): The SSX binaries form a more closed group than the novae in SSX phase and the symbiotics, when their colors are corrected by $E(B-V)$.

$(U-B)_0$ vs. $(B-V)_0$ diagram (classical SSX binaries and the V Sge stars only) (Figure 3b): The mean colors are plotted for each object. The individual objects form a slightly elongated group along the diagonal. There is some tendency for the V Sge stars to possess more negative $(B-V)_0$'s at more positive $(U-B)_0$'s.

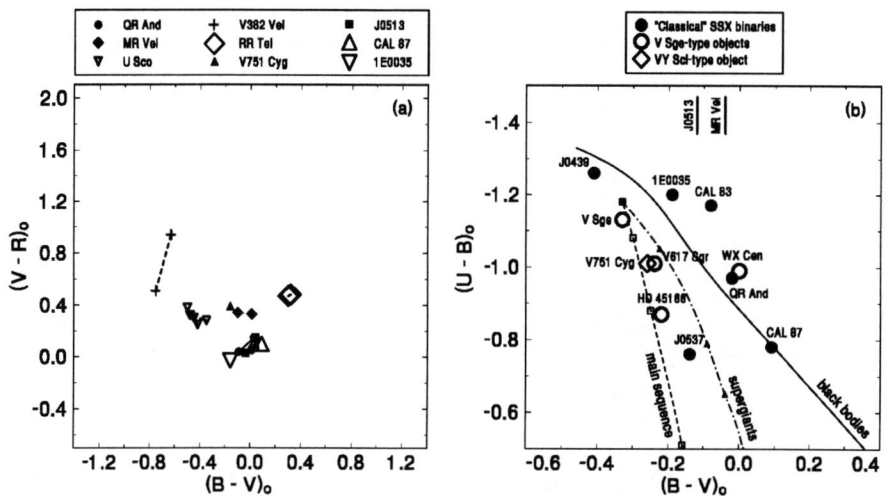

FIGURE 3. (a) $(B-V)_0$ vs. $(V-R)_0$ diagram. (b) $(U-B)_0$ vs. $(B-V)_0$ diagram just for the classical SSX binaries and the V Sge stars. The mean colors are plotted for each object. Just $(B-V)_0$ is available for J0513 and MR Vel.

RESULTS

The range of M_{V0}'s of the individual objects, emitting the SSX-rays, is shown to be very large, from $M_{V0} = 8.5$ (V 751 Cyg) to $M_{V0} = -4$ (J0048). The classical SSX binaries and the V Sge stars tend to display brighter M_{V0} with the increasing P_{orb}. This is in accordance with the hypothesis, outlined above – the longer P_{orb} allows a larger lobe of the WD and hence a larger disk, on which the soft X-rays can be reprocessed into UV/optical radiation.

The high/low state behaviour or outbursts appear to occur for a relatively narrow group of classical SSX binaries and the V Sge stars, within $P_{orb} = 0.5-1.1$ days. The amplitude of the high/low states of such an object is smaller than the total range of M_{V0}'s of the objects plotted in Figure 1.

The novae V 1974 Cyg and V 382 Vel had significantly brighter M_{V0} during their SSX phase than most SSX binaries – this difference is particularly large when objects with similar P_{orb} as these novae are assumed. On the other hand, M_{V0} of the recurrent nova U Sco was comparable to the SSX binaries with similar P_{orb}. This can be interpreted if the ejecta contributed to M_{V0} of V 1974 Cyg and V 382 Vel while a much smaller mass loss occurred during the explosion of U Sco. M_{V0} of U Sco then represents just the contribution of the WD and the resumed disk.

The short-period objects ($P_{orb} \leq 4$ days) form a closed group in $(U-B)_0$ vs. $(B-V)_0$ diagram. The color variations of the individual classical SSXBs and the V Sge stars are significantly smaller than the total range of the color indices in Figure 2. This suggests that the properties of the reprocessing medium are characteristic for a given binary and undergo just relatively small changes (e.g. during various states of activity).

The position of some of the short-period objects above the curve of radiating black bodies suggests a substantial line contribution (e.g. Balmer jump affected by emission).

The indices $(U-B)_0$ and $(B-V)_0$ of J0439 are the most negative of the objects plotted in Figure 3b. At the same time, J0439 has the faintest M_{V0} of the SSX and the V Sge binaries (Figure 1). This speaks in favour of a hot, but spatially small reprocessing medium (disk). This may support the suggestion by [6] that J0439 is a double-degenerate SSX source, with the real P_{orb} shorter than 0.1404 days, determined by [3].

The differences in the positions in the color-color diagrams for the systems with the short P_{orb} (≤ 4 days) may be attributed to the different contributions from the flaring disk and the donor star.

The classical novae in the SSX phase display the $(U-B)_0$ and $(B-V)_0$ colors similar to those of the classical SSX binaries and the V Sge stars, but with a larger scatter for the individual novae. The very negative $(U-B)_0$ index of the last point of V 1974 Cyg can be interpreted as the appearance of very strong [Ne V] 3346 and 3426 A lines which were detected by [1].

The colors of the SSX symbiotic stars are significantly different from the remaining objects. The colors of the symbiotic J0048 are similar to those of AG Dra at quiescence.

M_{V0}'s of the SSX symbiotic stars are comparable to or just slightly brighter than those of the brightest SSXBs and the V Sge stars, which have P_{orb} just $0.5 - 1.1$ days. This may suggest that in spite of a much longer P_{orb} of the symbiotics the dimension of the medium (disk?) in which the reprocessing of the SSX-rays occurs does not increase much with P_{orb}.

ACKNOWLEDGMENTS

This research has made use of NASA's Astrophysics Data System Abstract Service. This research has made use of the electronic catalog of supersoft X-ray sources available at URL http://www.aip.de/~jcg/sss/ssscat.html and maintained by J. Greiner. The support by the post-doctoral grant 205/00/P013 of the Grant Agency of the Czech Republic and the project ESA PRODEX INTEGRAL 14527 is acknowledged.

REFERENCES

1. Barger, A.J., et al., 1993, ApJ, 419, L85
2. Popham, R., Di Stefano, R., 1996, Supersoft X-Ray Sources, LNP, p.99
3. Schmidtke, P.C., Cowley, A.P., 1996, AJ, 112, 167
4. Steiner, J.E., Diaz, M.P., 1998, PASP, 110, 276
5. van den Heuvel, E.P.J., et al., 1992, A&A, 262, 97
6. van Teeseling, A., et al., 1997, A&A, 323, L41

X-RAYS FROM NOVAE

X-ray Observations of Novae

Joachim Krautter

Landessternwarte, Königstuhl, D-69117 Heidelberg, Germany
email: J.Krautter@lsw.uni-heidelberg.de

Abstract. X-ray studies of classical novae which were performed by EXOSAT, ROSAT and most recently by BEPPOSAX, CHANDRA and XMM have turned out to be a very powerful tool to study the hot phases in the outburst of a classical nova. Only X-ray data provide unambiguous information on the evolution of the thermonuclear runaway. X-ray studies so far have identified both a soft component originating on the hydrogen burning white dwarf and a hard component from hot circumstellar material due to shocks in the nova outflow. In addition, emission lines from highly ionized species found in spectra with higher resolution have been observed allowing plasma diagnostics for the hot regions. Here the X-ray observations carried out so far and their implications to our understanding of the nova outburst will be reviewed.

INTRODUCTION

X-ray observations have only recently turned out to be a tool in the study of novae in outburst, since the first X-ray detection of a nova in outburst took place less than twenty years ago. The status of our knowledge in the X-ray regime is, however, quite different from that in other spectral ranges. Only a few objectes were detected and most of them without systematic observations. So the picture which presently emerged from X-ray observations is far less systematic than, e.g. that for the infrared spectral range as described in the review paper by Bob Gehrz (2002), or that for the optical spectral range where systematic observations of a large number of novae have been carried out for nearly a century.

The first nova for which X-rays were discovered in outburst was GQ Mus (N Mus 1983). Ögelman, Beuermann, & Krautter (1984) had observed this object some 460 days after outburst with the CMA on board EXOSAT. (A very lively description of the story of this discovery is given by Ögelman 2002). The interpretation of the data was made difficult by the missing spectral resolution of the CMA as well as by the low count rates. Ögelman, Beuermann, & Krautter found that the count rate was compatible with either soft X-ray radiation from a white dwarf remnant with a surface temperature of $\sim$ 350000 K or shocked circumstellar gas under conditions including a temperature of approximately 10^7 K and a luminosity of approximately 10^{35} ergs s^{-1}. Subsequent observations of GQ Mus, PW Vul (N Vul 1984 #1), and QU Vul (N Vul 1984 #2) gave similar results (Ögelman, Krautter, & Beuermann 1987). The data for PW Vul and QU Vul which were both observed during early phases indicated that the count rate increases during the early phases of the outburst. An extensive discussion of the early observations can be found in Ögelman (1989).

A breakthrough came with the launch of ROSAT (Trümper 1983) and its Position

Sensitive Proportional Counter (PSPC) which had an energy range from 0.1 to 2.4 keV. This combination had a much higher sensitivity than any comparable X-ray telescope/detector before and an energy resolution which, even if very poor, gave some spectral information. With ROSAT most basic X-ray properties could be detected.

A few very valuable observations were obtained with ASCA, RXTE, and Beppo Sax. Another big step forward came recently with Chandra and XMM which both have a higher sensitivity than ROSAT and an energy range which extends, at the hard end, far beyond ROSAT's 2.4 keV. Of particular importance, are the grating spectrometers which allow for X-ray observations with a resolution of up to a thousand $R/\Delta R$ along with a high sensitivity.

In the following I shall review the basic X-ray properties of novae in outburst, the progress made recently as well as open questions using observations of individual objects which were studied more systematically. Earlier reviews on X-ray observations in outburst were given by Ögelman (1989) at the last nova conference in Madrid, Ögelman & Orio (1995), and Orio (1999).

SOURCES OF X-RAYS

A discussion of sources of X-rays from novae in outburst can be found in Ögelman and Orio (1995). Only a short summary and a few additional remarks will be given here.

There are several mechanisms which cause a nova in outburst to emit X-ray radiation. The first occasion when one expects X-ray emission is during the very early phases of the outburst. Once the thermonuclear runaway has started at the bottom of the accreted envelope on top of the white dwarf, the energy created is transported via convection to the surface of the white dwarf and the surface temperature increases. With a luminosity of the order of L_{Edd} and the radius of a white dwarf we expect temperatures of several hundred thousand degrees. During this early 'fireball phase' soft X-ray radiation with the spectral energy distribution (SED) of a hot stellar atmosphere will be emitted. However, after the ejection of the nova shell, the temperature of the expanding pseudophotosphere decreases with increasing radius and the shell will become optically thick for any X-ray radiation created in the underlying hydrogen burning shell. The duration of the X-ray emission during the fireball phase is very short, of the order of hours, and no X-rays have ever been observed. Also in future it will be extremely difficult to detect a nova in X-rays during the fireball phase.

A second phase for X-ray radiation from the white dwarf occurs during later phases of the outburst. As calculations of the TNR model of the nova outburst show (see e.g. Starrfield 1989 and references therein), only part of the ejected envelope has velocities above the escape velocity. The remaining material returns soon to quasistatic equilibrium and forms an envelope around the white dwarf with the initial dimensions of a giant star. Hydrostatic hydrogen burning via the CNO cycle is going on in a thin shell atop the white dwarf core. As the evolution proceeds, the reduction of the shell mass (via nuclear reactions, radiation driven wind and dynamical friction) will yield a shrinking of the radius and an increase of the temperature (Starrfield et al. 1991). In this so called 'phase of constant bolometric luminosity' eventually temperatures of several hundred thousand degrees are reached and the nova becomes again an emitter of soft X-rays which have, like in the early fireball phase, an SED of a hot stellar atmosphere with a luminosity close to L_{Edd}.

A second general mechanism is X-ray emission from hot circumstellar material due to shocks in the nova outflow. In this case one expects harder X-rays with a bremsstrahlung SED. Balman, Krautter & Ögelman (1998) have summarized different ways to produce hard X-ray emission from the nova ejecta: (i) shocks from the interaction of the expanding envelope with circumstellar material like an old nova shell, interstellar matter or a red giant wind; (ii) shocks from density inhomogeneities within the expanding envelope which should resemble X-ray emission from OB star winds; and (iii) shocks from the collision of a fast wind with preexisting slow wind material from the white dwarf remnant. The third general mechanism is coronal line emission from a hot plasma which allows to carry out plasma diagnostics for the line emitting regions.

BASIC X-RAY PROPERTIES OF V1974 CYG

The best paradigm to demonstrate the basic X-ray properties of novae in outburst is V1974 Cyg (N Cyg 1992). V1974 Cyg, a moderately fast ($t_3 \sim 35^d$) ONeMg nova, was discovered in outburst on February 20, 1992 (Collins 1992). With a maximum brightness $V_{max} \sim 4.4$ mag it was the brightest nova since V1500 Cyg (N Cyg 1975). Due to its brightness, V1974 Cyg was observed over all wavelength ranges, from radio to γ-rays. With the PSPC on board ROSAT it was first observed on April 20, 1992, as soon as it had entered the ROSAT observation window (Krautter et al. 1993). Subsequently, V1974 Cyg was observed by ROSAT over nearly two years on a total of 18 occasions (Krautter et al 1996). All observations were carried out with the PSPC.

The lightcurve of the total X-ray countrate of V1974 Cyg in the 0.1-2.4 keV band is shown in Figure 1 by Krautter et al. (1996). Several phases can be distinguished: The intial rise phase from days 63 to 147 is characterized by a strong increase of the count rate from 0.3 to 11.8 counts s^{-1}. From days 255 to 434 the count rate further increased to about 76 counts s^{-1} and remained essentially the same on the next observation on day 511. The last phase, the X-ray decline from sometime after day 511 to day 653 is chracterized by a strong decrease of the X-ray flux down to 0.2 counts s^{-1}. V1974 Cyg has so far the most completely covered X-ray lightcurve, i.e., with reasonably spaced observing epochs from early rise to decline.

The first observation on day 63 yielded a low, but significant detection of a highly absorbed hard spectrum with essentially no photons below 0.7 keV. In the following observations the spectrum got constantly softer, the first indications for a soft component showed up on day 147. A dramatic change occured between days 147 and 255, on which a strong soft component had appeared whose SED exhibited the general characteristics of a supersoft X-ray source (SSS) (e.g. Hasinger 1994; Kahabka & van den Heuvel 1997). With a maximum countrate of 76.5 counts s^{-1} V1974 Cyg was the by far strongest SSS ever observed by ROSAT! In addition to the soft component, the harder component was present on all occasions. While during the decline phase the countrate decreased strongly, the general character of the observed SED with a soft and a hard component did not change.

As Krautter et al. (1996) describe, fits with black body energy distributions did not give any reasonable results. With column density N_H, temperature T_{eff}, and luminosity L as free parameters, totally unrealistic bolometric luminosities of several thousand L_{Edd} for a 1 $M_\odot$ white dwarf were obtained. On the other hand, if one assumes $L=L_{Edd}$, no reasonable fit could be obtained. This result can be generalized for any other SSS (Kahabka & van den Heuvel 1997).

In order to overcome this problem, Balman, Krautter & Ögelman (1998) used, in a reanalysis of the V1974 Cyg data, LTE white dwarf model atmospheres with O-Ne overabundances by MacDonald & Vennes (1991). They could show that the soft component follows an evolution as predicted for the phase of constant bolometric luminosity: The effective temperature increases with decreasing radius at constant luminosity and a strong decrease of T_{eff} and L during the decline phase. From day 255 to day 511, T_{eff} increased from 35.0 eV to a peak temperature of 51 eV ($4.0-5.9 \cdot 10^5$ K) while the white dwarf radius decreased from 2.2 to $0.9(D[kpc]/2) \cdot 10^9$ cm.

For an analysis of the hard component, Balman, Krautter, & Ögelman used Raymond-Smith thermal plasma emission models (Raymond & Smith 1977). The plasma temperatures decreased from 10 keV and 5 keV on days 63 and 91, respectively, to a more or less constant level around $\sim$ 1 keV. However, the values for the first two epochs were not very well defined, since the peak temperatures were outside the ROSAT energy range of 01.-2.4 keV. The shock speed calculated from these temperatures is between 1500 and 3000 km s^{-1}. The hard X-ray flux increased towards a maximum on day 147 with a subsequent decline. The peak unabsorbed flux corresponded to a luminosity of $(0.8-2.0) \cdot 10^{34}$ ergs s^{-1} at a distance of 2-3 kpc. While the temporal evolution of the hard X-ray flux and the plasma temperatures suggest a shock origin of the hard X-ray radiation, the data did not allow to distinguish between the three possibilities described in section II.

The hydrogen column density N_H decreased by about a factor of 10 until it reached a constant level around day 255. This temporal dependence is clear evidence for decreasing circumstellar absorption in the expanding shell. The high circumstellar hydrogen column density during the early phases is the reason for the absence of soft X-ray emission until day 97. A similar result has been found for the fast nova V838 Her (N Her 1991) which was observed with ROSAT five days after optical maximum (Lloyd et al. 1992). Only hard X-ray radiation with no counts below 0.7 keV was found. An analysis with a Raymond-Smith spectrum gave a temperature of 10 keV.

Between days 97 and 147 the expanding envelope started to become transparent for the soft radiation and subsequently the soft flux increased. Once the envelope had become optically thin for the soft X-ray radiation, the X-ray flux remained constant.

The decline of the X-ray flux which started between days 511 and 612 is due to switch-off of the hydrogen burning. Krautter et al. (1996) used the cooling timescale of about six months to estimate a mass of $\sim 10^{-5}$ $M_\odot$ for the hydrogen-exhausted, remnant envelope on the white dwarf.

During its bright phase the point source image of V1974 Cyg was surrounded by an extensive halo caused by scattering from interstellar grains. Mathis et al. (1995) used this halo to study the nature of the interstellar grains.

THE LONG LIFE OF GQ MUS

A so far unique case is GQ Mus. GQ Mus was discovered on January 18, 1983 by William Liller (Liller 1983). As already mentioned above, it was the first novae in outburst discovered in X-rays in 1984. Six years after its last EXOSAT observation GQ Mus was detected at a count rate of 0.143 ± 0.035 counts s^{-1} in the ROSAT All-sky Survey (RASS) on day 3118, i.e., 8.5 years after optical maximum. During the following years several pointed observations were carried out (Ögelman et al. 1993; Shanley et al. 1995) during which the count rate decreased from 0.127 ± 0.006 counts s^{-1} on day 3322 over 0.007 ± 0.002 on day to <0.003 (3σ) counts s^{-1} on day 3871.

As in the case of V1974 Cyg, blackbody fits carried out by Ögelman et al. (1993) did not give satisfying results. Balman & Krautter (2001) carried out an analysis with C-O enhanced LTE atmosphere models by MacDonald & Vennes (1991). Unfortunately, due to the low number of counts in the RASS observation (exposure time of only 150 sec!) only an upper limit $kT_e < 54$ eV ($<6.2\cdot10^5$ K) could be obtained. For day 3322 an effective temperature in the range 38.3-43.3 eV ($4.4-5.1\cdot10^5$ K) was derived. On day 3641 T_{eff} and the unabsorbed flux had further decreased. Balman & Krautter conclude that only the RASS data of day 3118 show evidence for ongoing hydrogen burning. The results imply that on day 3322 hydrogen burning had already turned off. With a duration of 8.5-9.1 years GQ Mus exhibits an exceptionally long phase of hydrostatic hydrogen burning, the longest so far known. The much longer turn-off timescale of GQ Mus, as compared with V1974 Cyg, indicates that the mass of the white dwarf in GQ Mus is lower than that one in V1974 Cyg.

Also in the optical range, GQ Mus had some particular properties (Krautter & Williams 1989), since it exhibited a particularly high ionization with the coronal lines [FeX] $\lambda 6374$ and [FeXI] $\lambda 7892$ being very strong. As Krautter & Williams showed, the high ionization lines were due to photoionization of a hot (several 10^5 K) source. In a spectrum taken on May 14, 1991 (Williams, priv. communication) even [FeXIV] $\lambda 5303$ might be present, [FeXI] $\lambda 7892$ is as strong and [FeX] $\lambda 6374$ about 2.5 as strong as Hα, respectively, which has not been observed - at least to the knowledge of the author - so far in any other astronomical object. In a spectrum taken on March 10, 1993, after the turn-off in the X-ray regime, the coronal lines have disappeared and the ionization is much lower. This shows that the ionization stage of the optical spectrum could be used in some cases as a qualitative indicator for the turn-off of the hydrogen burning.

DURATION OF SOFT PHASE

In order to get more information on the duration of the phase of soft X-ray emission Orio, Covington & Ögelman (2001a) did a search of the ROSAT archive for X-ray emission of postnovae. A total of 350 observations, both in the ROSAT All-sky Survey and serendipitous fields, of 108 different classical and recurrent novae were analysed. Only 3 novae with a super-soft spectrum (V1974 Cyg, GQ Mus, N LMC 1995) were found. For a time of up to 10 years after explosion, 30 galactic and 9 LMC novae were in the ROSAT samples.

After ROSAT a few novae with soft X-ray emission were discovered by recent X-ray satellites (see below). However, it is clear that the vast majority of all postnovae were not observed as soft X-ray sources.

There could be in principle two reasons for the missing soft X-ray radiation: Either novae switch off after a relatively short time or the hydrogen column density for most novae is so high that the soft X-ray radiation gets absorbed. In order to get an estimate of the latter, Nickel (1995) did a literature search of data known for the novae observed in the RASS within 10 years after the outburst. He showed that 21 galactic novae, for which extinction data were available, have an average interstellar extinction $E_{B-V} \sim 0.5$ mag. With this value, only a few novae have interstellar column densities high enough to absorb the whole soft X-rays. This is even more striking for the LMC novae which have much lower interstellar column densities. Circumstellar extinction should not play any role, since most novae were observed in the RASS months or years after maximum when the expanding envelopes should have become optically thin for soft X-rays. On the basis of these data one can conclude that the majority of the novae studied switch off after a relatively short time. Of course, no precise number can be given, but it is probably safe to assume that most novae switch off hydrogen burning after less than two years.

Starrfield et al. (1991) calculated the turn-off time scale as a function of the white dwarf mass assuming that Castor, Abbot, and Klein (1975) radiation pressure driven mass loss is ejecting the accreted envelope. They find that for a 1.25 $M_\odot$ white dwarf a turn-off time of about two years, whereas a nova with a 1.10 $M_\odot$ white dwarf should turn-off after about 100 years. The short turn-off time found from the X-ray observations is clear evidence that most novae have white dwarfs with relatively high masses above 1.2 $M_\odot$. That result can be expected from the TNR model, since according to this model the critical mass in order to start the runaway decreases strongly with increasing white dwarf mass (cf. Starrfield et al. 1989). Statistically, novae with high mass white dwarfs should be seen much more frequently than those with white dwarfs with lower masses.

HIGH RESOLUTION - NEW POSSIBILITIES FOR V382 VEL

V382 Vel (N Vel 1999), a fast ($t_3 \sim 10^d$) ONeMg nova (Shore et al. 1999a,b) was discovered on May 22, 1999 by Williams & Gilmore (1999). With a maximum brightness V_{max}=2.6 mag V382 Vel was one of the brightest novae of the last century and it was extensively observed in many different spectral regimes. In the X-ray range it was observed by RXTE, ASCA, BeppoSAX and Chandra.

The first X-ray observations with RXTE on day 5.7 (Mukai & Swank 1999) did not detect any significant X-ray flux. The first detection came from BeppoSAX observations on day 15 caried out by Orio et al. (2001b) in a very broad band from 0.1-300 keV. They found a hard spectrum between 2 and 10 KeV which they attribute to emission from shocked nebular ejecta at a plasma temperature kT$\sim$6 keV. No soft component was present in the spectrum.

On day 20 Mukai & Ishida (2001) found from ASCA observations a highly absorbed ($N_H \sim 10^{23}$ cm^{-2}) bremsstrahlungsspectrum with a temperature kT$\sim$10 keV. On subsequent observations with RXTE (days 31, 35, 50, and 59), the spectrum became softer, because of decreasing plasma temperatures (kT~4.0 keV to kT~2.4 keV on day 59) and diminishing column density N_H (7.7·10^{22} cm^{-2} on day 31 to 1.7·10^{22} cm^{-2} on day 59). They argue that the X-ray emission must be due to shocks internal to the nova ejecta. Like Orio et al. (2001b) they did not find any soft component.

Six months later Orio et al. (2001b, 2002) obtained a second BeppoSax observation on November 23, 1999, and found both the hard component and a super-soft component. The unabsorbed flux of the hard component had decreased by about a factor of 40. A similar strong decrease of the hard component was reported for V838 Her by O'Brien, Lloyd, and Bode (1994) on a second observation about one year after the first one. Orio et al. (2002) found that fits to the soft component with NLTE models by Hartmann & Heise (1997) and Hartmann et al. (1999) could be significantly improved if at least one emission feature was superimposed. They also found that the flux in the 0.1-0.7 keV range was strongly variable, since it dropped by about a factor of two within less than 1.5 hour. It remained faint for 15 min; no significant spectral changes were found. This behaviour rules out the ejection of an absorbing clump, also orbital variability can be excluded. So far no convincing explanation for this variability could be found.

Chandra ToO observations were reported by Starrfield et al. (2000) and Burwitz (2002). On December 30, 1999, ACIS observations showed the strong soft component. The first high resolution observations of any nova in outburst were carried out by the same group on February 14, 2000, with the Low Energy Transmission Grating (LETG) on board Chandra. Its spectral resolution R~600 surpasses the spectral resolution of all the other X-ray detectors used so far by factors of several ten to more than two orders of magnitude. The spectrum showed a wealth of coronal emission lines of different elements and different ionization stages which can be used for plasma diagnostics for the line emitting regions (Starrfield et al., in preparation). The most prominent emission lines are O VIII, O VII, Ne X, Ne IX, Mg X, Mg XI, N VI, and N VII. The resolution of the LETG spectra is just high enough, to fit to the strongest lines like O VIII λ 19.0 and the O VII doublet $\lambda\lambda$ 21.6, 22.1, several components.

The strong soft component had totally disappeared. So hydrogen burning on top of the white dwarf in V382 Vel must have turned off between December 30, 1999, and February 14, 2000, resulting in a total duration of 7.5-8 months. This indicates that the white dwarf in V382 Vel has a high mass which is, of course, consistent with its ONeMg nature and the short decay time t_3. It is highly probable that the hydrogen burning actually turned off shortly after the Dec. 30 observations, since even in this longest possible case the cooling time of about 6 weeks is extremely short.

NEW SURPRISES OR V1494 AQL

The so far last chapter in the story of X-ray observations has been written by V1494 Aql whose outburst occurred on Dec 2, 1999. With V_{max} ~ 4 mag, V1494 Aql was the brightest nova in the northern sky since V1500 Cyg (1975). After outburst the CHANDRA ToO programme became activated (Starrfield et al. 2002). The first two observations on April 15 and June 7, 2002, obtained with the low resolution ACIS-1 detector yielded a hard spectrum with emission lines, no soft component was seen. After a strong soft component had appeared on August 6, two LETG+HRC observations were carried out on Sep. 28 and Oct. 1 with exposure times of 8 and 17 ksec, respectively.

The high resolution spectra are dominated by a strong soft continuum component on which emission features are superimposed. Since none of these features could so far be identified with any known emission lines, it cannot be excluded that the spectrum is in reality an absorption spectrum where the emission features are only those part with less local absorption. Spectra with the same character are observed during the fireball phase (e.g. Hauschild et al 1992, 1997). On the other hand, the situations during the fireball phases and ten months after outburst, when the high resolution spectra of V1494 Aql were taken, are very different. In the early phases of the nova outburst, an opaque expanding envelope is present which is no longer the case some ten months after outburst. In order to clarify the question whether one sees an emission or an absorption spectrum suitable NLTE models will have to be applied to the observed SEDs. In any case, the spectra clearly demonstrate that nuclear burning near the surface of the white dwarf was still going on.

Drake et al. (2002) analysed the X-ray lightcurve of their LETG observations and found very surprising results, a short time burst and oscillations. During the 17 ksec observations, the X-ray count rate showed for a period of about 1000 second a complex rise and fall with several maxima and minima. Two peaks and other structure could be observed and the peak countrate had increased by about a factor of six as compared with the mean levels before and after the outburst. The spectrum was slightly harder than during the rest of the observations. So far the nature of this outburst remains a puzzle, no reasonable explanation could be found yet.

A periodogram of the X-ray lightcurve revealed periodic variations. A strong signal was found with a best fit period of ~ 2500 s. In addition, additional periods were found. In order to check whether the periodicity found might be related to the spacecraft motion or dithering, Starrfield et al. (2002) performed identical periodogram analyses on light curves of the hot white dwarfs HZ 43 and Sirius B. These periodograms turned out to be flat. In addition, the periodicity was found both in the zeroth order and in the dispersed data. On the basis of these results Starrfield et al. conclude that an instrumental effect can be excluded and that the periodicity is higly probably real.

Starrfield et al. interpret this result as the discovery of non-radial g^+-mode pulsations in the hot, rekindled white dwarf driven by κ/γ effects in the partial ionization zones of C and O near the surface of the white dwarf. The structure of the white dwarf in a nova is very similar to white dwarfs in planetary nebulae. The power spectrum of V1494 Aql and the light curve are very similar to the hot central star of the planetary nebula NGC 1501 (Bond et al. 1996).

CONCLUSIONS

In this review it was shown that X-ray observations of novae offer great opportunities to study novae in outburst. They allow to study the hot white dwarf, shocks in the expanding envelope, derive physical paramerters like T_{eff} and L, and grating spectroscopy will allow us to carry out plasma diagnostics for the line emitting regions. Certainly problems exist, during the early phases only very limited information can be obtained due to the high opacity in the expanding envelope. Second, the interpretation of the results is often not very straightforward. In particular, the variations of the soft X-ray flux are not understood at the present time. From an observational point of view, it is often very difficult to get enough Target of Opportunity time on X-ray satellites. However, the positive assets which were described in the preceding sections are very strong arguments for X-ray observations of novae. The strongest argument is the fact, that X-ray observations are the only possibility to study directly the regions where the energy of the nova outburst is produced.

ACKNOWLEDGMENTS

Among many others there are two colleagues and friends whom I would like to thank particularly, Hakki Ögelman without whom I probably never would have started with X-ray observations of novae, and Sumner Starrfield, from whom I learned within the last 16 years infinitely many things on novae and who critically read this manuscript. I am very grateful to Solen Balman, Vadim Burwitz, Nye Evans, Bob Gehrz, Charo Gonzales-Riestra, Peter Hauschildt, Bill Liller, Jan-Uwe Ness, Marina Orio, Giora Shaviv, Steve Shore, Jim Truran, Mark Wagner, Rainer Wichmann, Bob Williams, and Chick Woodward who were in part collaborators at my work on X-rays and with whom I had many discussions on X-rays from novae and on many other facets of these exiting objects.

REFERENCES

1. Balman, S. & Krautter, J., *MNRAS* **326**, 1441 (2001).
2. Balman, S., Krautter, J., Ögelman, H., *ApJ* **499**, 395 (1998).
3. Bond, H. E., et al., *AJ* **112**, 2699 (1996).
4. Burwitz, V., et al., these proceedings, 2002.
5. Collins, P., *IAU Circ.* no. **5454** (1992).
6. Drake, J., Wagner, R. M., et al., *ApJ*, submitted (2002).
7. Gehrz, R. D., these proceedings, 2002.
8. Hartman, H. W. & Heise, J., *A&A* **322**, 591 (1997).
9. Hartmann, H. W., Heise, J., Kahabka, P., Motch, C., Parmar, A. N., *A&A* **346**, 125 (1999).
10. Hasinger, G., *Rev.Mod.Astron.* **7**, 129 (1994).
11. Hauschildt, P. H., Wehrse, R., Starrfield, S., Shaviv, G., *ApJ* **393**, 307 (1992).
12. Hauschildt, P. H., Shore, S. N., Schwarz, G., Baron, E., Starrfield, S., Allard, F., *ApJ* **490**, 803 (1997).
13. Kahabka, P. & van den Heuvel, E. P. J., *ARAA* **35**, 69 (1997).
14. Krautter, J., & Williams, R. E., *ApJ* **341**, 968 (1989).
15. Krautter, J., Ögelman, H., Starrfield, S., Trümper, J., Wichmann, R., *Ann.Isr.Phys.Soc.* **10**, 28 (1993).
16. Krautter, J., Ögelman, H., Starrfield, S., Wichmann, R., Pfeffermann, E., *ApJ* **456**, 788 (1996).
17. Liller, W. *IAU Circ.* no. **3764** (1983).
18. Lloyd, H. M., et al., *Nature* **356**, 222 (1992).
19. MacDonald, J. & Vennes, S., *ApJ* **373**, L51 (1991).
20. Mathis, J. S., Cohen, D., Finley, J., Krautter, J., *ApJ* **449**, 320 (1995).
21. Mukai, K. & Swank, J., *IAU Circ.* no. **7206** (1999).
22. Mukai, K., & Ishida, M., *ApJ* **551**, 1024 (2001).
23. Nickel, U., Examensarbeit Ruprecht-Karls-Universität Heidelberg, 1995.
24. O'Brien, T. J., Lloyd, H. M., Bode, M. F., *MNRAS* **271**, 155 (1994).

25. Ögelman, H., in *Physics of Classical Novae*, IAU Coll. No.122, eds. A. Cassatella & R. Viotti, Springer-Verlag, Berlin, Germany; New York, 1990, p.148.
26. Ögelman, H., in *From X-ray Binaries to Gamma-Ray Bursts*, Jan van Paradijs Memorial Symposium, eds. E. P.,J. van den Heuvel, L. Kaper & E. Rol, ASP Conference Series, in press, 2002.
27. Ögelman, H. & Orio, M., in *Cataclysmic Variables*, eds. A. Bianchini, M. della Valle, & M. Orio, Astrophys. Space Sci. Lib. 205, 1995, p.11.
28. Ögelman, H., Beuermann, K., Krautter, J., *ApJ* **287** L31 (1984).
29. Ögelman, H., Krautter, J., Beuermann, K., *A&A* **177**, 110 (1987).
30. Ögelman, H. B., Orio, M., Krautter, J., Starrfield, S. 1993, *Nature* **361**, 331
31. Orio, M., *Physics Reports* **311**, 419 (1999).
32. Orio, M., Covington, J., Ögelman, H., *A&A* **373**, 542 (2001a)
33. Orio, M., et al., *MNRAS* **326**, L13 (2001b)
34. Orio, M., et al., *MNRAS* **333**, L11 (2002).
35. Raymond, J. C. & Smith, B. W., *ApJS* **35**, 419 (1977).
36. Shanley, L., Ögelman, H. Gallagher, J., Orio, M., Krautter, J., *ApJ* **438**, L95 (1995).
37. Shore, S. N., et al., *IAU Circ.* no. **7192** (1999).
38. Shore, S. N., et al., *IAU Circ.* no. **7261** (1999).
39. Starrfield, S., in *Classical Novae*, eds. M. Bode & A. Evans, New York, Wiley, 1989, p.39.
40. Starrfield, S., Shore, S. N., et al., *BAAS* **32**, 1253 (2000).
41. Starrfield, S., Truran, J. W., Sparks, W. M., Krautter, J., in *Extreme Ultraviolet Astronomy*, eds. R. Malina, S. Bowyer, New York, Pergamon, 1991, p.168.
42. Trümper, J., *Adv. Space Res.* **2**, 241 (1983).
43. Williams, P. & Gilmore, A. C., *IAU Circ.* no. **7176** (1999).

A XMM-Newton Observation of Nova LMC 1995

Marina Orio*, Jochen Greiner[†], Wouter Hartmann** and Martin Still[‡]

*INAF - Turin Astronomical Observatory, Italy, and
Astronomy Department, U Wisconsin at Madison, USA
[†]Max Planck Institute for Extraterrestrial Physics, Garching bei München, FRG
**SRON Laboratory for Space Research, Utrecht, The Netherlands
[‡]NASA Goddard Space Flight Center, Greenbelt, MD, USA and Universities Space Research Association

Abstract. Nova LMC 1995, detected as a luminous supersoft X-ray source in 1995-1998 with ROSAT, was observed again in December of 2000 with XMM–Newton. It was still a supersoft X-ray source at almost-Eddington luminosity, and the effective temperature had increased to about 450,000 K. We discuss atmospheric models studied for the hot hydrogen burning white dwarf, and whether this nova may be a type Ia supernova progenitor.

INTRODUCTION

NOVA LMC 1995 was the only LMC nova detected as a luminous supersoft X-ray source in the Magellanic Clouds in repeated pointings and a survey of the two galaxies performed by ROSAT (see Orio & Greiner, 1999). It was observed repeatedly since before the eruption, but it was only detected with the ROSAT HRI for the first time 9 months after the outburst. It was detected again in successive HRI serendipitous observations, and in a PSPC observation 3 years after the outburst. At this point, it appeared as a luminous supersoft X-ray source. Orio & Greiner (1995) showed that the data could be fitted with an atmospheric model of a $M \simeq 1.2$ $M_\odot$ white dwarf with an effective temperature, $T \simeq 345,000$ K. The X-ray flux increased in the first 3 post–outburst years, and a likely conclusion was that the atmosphere was still shrinking, as its temperature increased. For this reason, we proposed new X-ray observations of this nova, suggesting that it is a rare "twin" of GQ Mus. GQ Mus, we remind you, was observed to be a luminous supersoft X-ray source for almost 10 years after the outburst, the longest time in this state measured up to now for a post–outburst nova.

In other wavelengths, we have only sparse information about N LMC 1995. It was discovered in outburst at the end of February 1995, it reached V≤10.7 at maximum (above average for a LMC nova), and described as having an expansion velocity of 800–1500 km s^{-1} (Della Valle et al. 1995). The rise to maximum took at least a few days, while the decay by one magnitude in almost 3 days indicated a moderately fast or fast nova (see Liller 1995, Gilmore 1995, Christie 1995).

Novae are detected only rarely as supersoft X-ray sources years after the outburst (Orio et al. 2001, Nedialkov et al. 2002), suggesting that this phase is not very long lived. In Orio & Greiner (1998), we suggested that N LMC 1995 may be the prototype of a rare class of novae that are bound to reach the Chandrasekhar mass, becoming type

Ia SN or even undergoing accretion induced collapse. Consequently it is worthwhile to follow the subsequent evolution of N LMC 1995. Moreover, nova white dwarfs that turn into supersoft X-ray sources offer a unique chance to determine the physical white dwarf parameters using model atmospheres.

NEW ROSAT PSPC OBSERVATIONS

In December 1998, before the PSPC was turned off, N LMC 1995 was observed one last time. The high-voltage drop-out at this stage caused calibration problems. We measured a background corrected count rate 0.030 ± 0.004 cts s^{-1}, lower by a factor of 2 compared to the one of February 1998, however the usable exposure was short (only 2408 s). The source was still supersoft and very luminous. This encouraged us to propose a new X-ray observation with XMM-Newton.

XMM-NEWTON OBSERVATIONS IN DECEMBER OF 2000

N LMC 1995 was observed with XMM-Newton two years later, on December 19 2001, almost 6 years after the outburst. The purpose was to measure the length of the supersoft X-ray phase and to obtain the physical parameters of the white dwarf from the EPIC and RGS spectra. The luminous supersoft X-ray source was still there, and the flux did not vary during about 50 ksec of observation. Fig. 1 shows the count rate spectrum observed with the EPIC PN and MOS (prime mode, full window, thin filter). The solid line is the best blackbody fit, which not surprisingly is not adequate with data of this quality. Fig. 2 shows again the EPIC spectrum and the best atmospheric fit (see below) and Fig. 3 shows the RGS gratings spectrum. The average EPIC PN background corrected count rate, 0.575 ± 0.004 cts s^{-1} in the 0.2-1.5 keV range, is consistent with the February 1998 ROSAT observation. The other measured count rates are given in Table 1. The spectrum of this nova emits most of the flux at energies below 0.5 keV, and is the softest supersoft X-ray sources observed in the Magellanic Clouds. For "softness", it is comparable only with the BeppoSAX spectrum of V382 Vel (Orio et al. 2002).

We note that the RGS spectrum (Fig. 3) has sufficient S/N to rule out isolated and prominent narrow features, in absorption or in emission, due to a surrounding nebula or to a wind. A "forest of atmospheric lines" like observed for Cal 83, (Paerels et al. 2001), on the other hand, would have not been detected. We fitted Non Local Thermodynamical Equilibrium (NLTE) models developed by Hartmann (1997), and other metal enhanced atmospheric models developed along the same lines for this project. No atmospheric models with enhanced carbon can fit the data, probably due to the lack of a significant absorption edge of C VI at 0.49 keV. The best fit to the EPIC spectra, shown in Fig. 2, is obtained with a model atmosphere in NLTE, with log(g)=0.9, at $T \simeq 450,000$ K, and $L_{bol} = 8.9 \times 10^{37}$ erg s^{-1} (assuming a distance 55 kpc to the LMC), $N(H) = 1.5 \times 10^{21}$ cm^{-2} (see Fig. 3) and abundances of Ne, O and Mg that are 10 times enhanced with respect to the solar value. We obtain a reduced χ^2 of 1.2 for the fit with all the EPIC instruments, and $\chi^2 \simeq 1$ with the RGS–1 and RGS–2. A fit with solar abundances

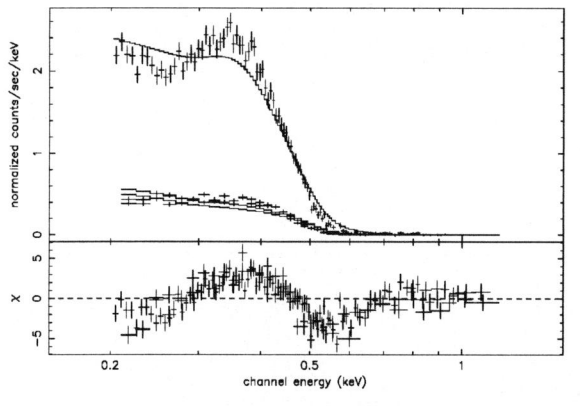

FIGURE 1. The background corrected count rate spectra of N LMC 1995 observed with XMM EPIC-PN and EPIC-MOS on December 19 2001. The solid lines indicate the best-fitting black-body continuum, which is not adequate. The lower panel shows the value of χ^2 as a function of energy.

TABLE 1. Count rates measured with different Newton-XMM instruments.

Instrument	Range (keV)	Count rate (cts s^{-1})
EPIC-PN	0.2-10	0.5754±0.0043
EPIC-MOS 1	0.2-10	0.1025±0.0017
EPIC-MOS 2	0.2-10	0.1144±0.0017
RGS 1	0.3-3.5	0.0196±0.0010
RGS 2	0.3-3.5	0.0221±0.0010

is of slightly inferior quality. The high value of N(H) is consistent with the Galactic absorption + the intrinsic absorption in the LMC evaluated with Points' (2002) ATCA-Parkes H I maps, which show that the N(H) column all the way to the back of the LMC in the direction of the nova is 1.1×10^{21} cm^{-2}.

All the atmospheric models indicate $T_{\text{eff}} \geq 4 \times 10^5$ K, so the WD atmosphere is still shrunking and heating over the past few years. log(g)$\simeq$9 indicates a white dwarf mass $M_{\text{WD}} \geq 1.2$ m$_\odot$. The quality of the fit and the choice of the model abundances is still limited by the uncertainties in the calibration of the EPIC instruments in the 0.2-0.3 keV range where the bulk of the flux is detected. The modest S/N spectra we obtain from the two RGS at $kT \geq 0.3$ keV are important, because the they are calibrated within only a 5–10% uncertainty at the moment. While the abundances are not very well determined yet, we note that the effective temperature is constrained within a narrow range, 4–4.7 $\times 10^5$ K.

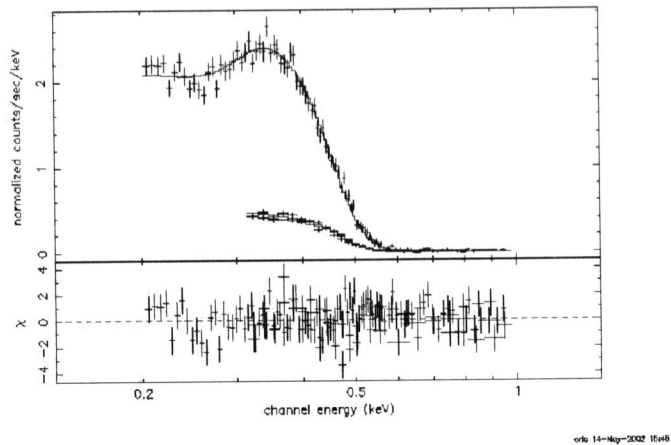

FIGURE 2. The fit to the EPIC spectra (excluding the MOS range 0.-0.3 keV) with a NLTE atmospheric model with T_{eff}=450000 K, L_{bol}=8.9 $\times 10^{37}$ erg s^{-1} enhanced O, Ne and Mg abundances, and log(g)=9 (see text).

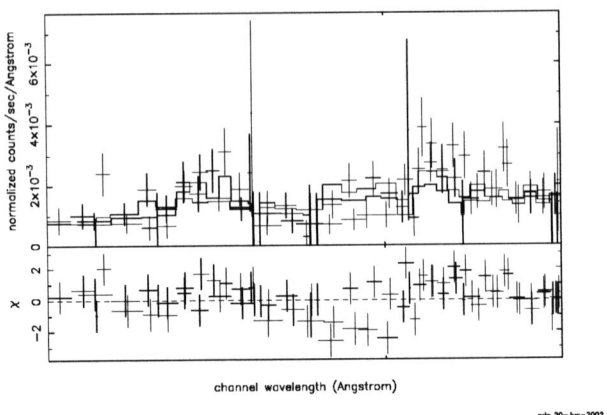

FIGURE 3. The background corrected count rate spectra of N LMC 1995, in the 25-33 Å range where most of the flux is detected, observed with XMM RGS-1 and RGS-2 on December 19 2001. No "obvious" narrow features are detected. The fit with the same NLTE atmospheric model with log(g)=9.0 and enhanced ONeMg abundances used for the EPIC instruments in Fig. 2 is shown.

FINALLY, A "NAKED" WHITE DWARF?

The spectrum of the post-nova supersoft X-ray sources in the first year or two after the outburst can be very complex. The X-ray flux has been observed to vary on short time scales, unlike in the case of N LMC 1995, and this is not quite well understood yet (Orio et al. 2002, Starrfield et al. 2001). The central source is shielded by high ionization lines, which are also produced in the supersoft X-ray range, but do not seem to be due to the photoionizing central white dwarf. Such lines are most likely due to shocks in the shell (see V382 Vel, Orio et al. 2002). Since the BeppoSAX LECS and the ROSAT PSPC could not resolve super-imposed narrow lines from the central source continuum, temperature and effective gravity determined from the observations of V1974 Cyg (Krautter et al. 1996, Balman et al. 1998) and of V382 Vel (Orio et al. 2002) should be considered very uncertain.

Grating observations of another supersoft X-ray source have also shown narrow ionization lines super-imposed on the white dwarf continuum, probably because of ongoing stellar winds (Bearda et al. 2002). Nova LMC 1995 seems to offer a rare opportunity to observe the atmosphere of a supersoft X-ray nova remnant. The nebular emission in X-rays has ceased, and we are finally looking at the "naked" white dwarf. $\log(g)=0.9$ indicates that the white dwarf is quite massive (≥ 1.2 $m_\odot$), indicating that N LMC 1995 may be a type Ia SN progenitor. To be able to answer this question, we may have to wait until a better calibration is obtained for the range 0.2-0.3 keV of the EPIC instruments. Fine tuning the model abundances, and ruling out models with certain sets of abundances, we will be able to assess whether the atmospheric material is enhanced with elements that are typical of the white dwarf core (indicating that convection dredges these elements up), or may instead be closer to a solar composition. The latter possibility would imply that the burning material is still that accreted from the companion, and has been retained after the outburst.

REFERENCES

1. Bearda H., Hartmann H. W., et al., 2002, A&A, 385, 511
2. Christie G.Q., 1995, IAU Circ. 6146
3. Della Valle M., Masetti N., Benetti S. 1995, IAU Circ. 6144
4. Gilmore A.C., 1995, IAU Circ. 6146
5. Hartmann H.W., Heise J., 1997, A&A, 322, 591
6. Krautter J., et al. 1996, ApJ 456, 788
7. Liller W., 1995, IAU Circ. 6143
8. Nedialkov P., Orio M., Birkle K., Conselice C., Della Valle M., Greiner J., Magnier E., Tikhonov N.A., 2002, A&A, in press
9. Ögelman H., Orio M., Krautter J., Starrfield S., 1993, Nature, 361, 331
10. Orio M., Covington J., & Ögelman H. 2001, 373, 542
11. Orio M., Greiner J., 1999, A&A, 344, L13
12. Orio M., Parmar A., Greiner J., Ögelman H., Starrfield S., Trussoni E., 2002 MNRAS, 333, L11
13. Paerels F., et al., 2001, A&A, 365, L308
14. Points S.D., 2002, private communication
15. Shanley L., Ögelman H., Gallagher J., Orio M., Krautter J., 1995, ApJ, 438, L95
16. Starrfield, S., et al., 2001, AAS, 198, 11.09

The Search for Extended Ionization and Reflection Nebulae

Marina Orio* and Gaghik Tovmassian[†]

*INAF - Turin Astronomical Observatory, Italy, and
Astronomy Department, U Wisconsin at Madison, USA
[†]Observatorio Astronomico Nacional and UNAM, Mexico

Abstract. We report on a search for extended ionization and reflection nebulae around nova remnants and their shells. We use the excellent imaging capacities of the WIYN telescope for this project. We have detected at this stage only one ionization nebula around a classical nova, and one associated with a nova-like variable. We discuss the peculiar characteristics of these nebulae and possible reasons for which they are not detected more often. Extended reflection nebulae are also very interesting as probes of the white dwarf temperature and characteristics. We show the example of V838 Mon.

INTRODUCTION

Rappaport et al. (1994) predicted that a new type of ionization nebula should be detected in the regions sourrounding supersoft X-ray sources – including those novae that do turn into supersoft X-ray sources after the outburst for several years. These nebulae should show remarkably distinct and different properties from other ionization nebulae. The main peculiarities should be: gradual transitions from the ionized to the non-ionized state, and exceptionally prominent emission of [OIII] $\lambda 5008$ and He II $\lambda 4686$. In a plot of the flux ratio of [OIII] $\lambda 5008$ and Hβ $\lambda 4862$ versus the flux ratio of He II $\lambda 4686$ and Hβ $\lambda 4862$, the new type of nebulae occupy the upper right corner. The total flux in the [OIII] $\lambda 5008$ should be larger than the Hα flux, even if less extended spatially.

The supersoft X-ray source CAL 83 is sourrounded indeed by one such nebula. It extends asymmetrically to up to 11.6 pc with a bright arc, and up to 16 pc with much fainter emission. The Hα luminosity is 2.7×10^{35} erg s^{-1}. The [O III] luminosity and the Hβ luminosity are in good agreement with the theoretical prediction (Remillard et al. 1995; see Table 1). In sharp contrast, no nebulae were detected with the [O III] filter around 9 other supersoft X-ray sources in the SMC and LMC, neither with the Hα filter around three of them. Upper limits to the flux were a factor of 10 below the Hα flux of CAL 83.

The prediction of Rappaport et al. (1994) prompted Di Stefano et al. (1995) to even speculate that some of these nebulae may have been mistaken for planetary nebulae, while the white dwarf emitting supersoft X-rays remained hidden.

Similarly, we speculate that novae that turned into bright supersoft X-ray sources for more than a few months may be undetected, and that the extended ionization nebula may be the only indication of the white dwarf temperature evolution. We remind that all imaging studies focused instead on resolving the ejected nova shells (which have very limited spatial extensions), and that a project like the one of Remillard et al. (1995) was not possible for the Galactic novae before the use of large field imaging instruments.

The 3.5m WIYN telescope, of whose consortium the University of Wisconsin is a member, is particularly suited for this type of project. At the moment it is equipped with a mini-MOSAIC with a 11 arcmin field of view and an excellent spatial resolution, 0.141 arcsec/picsel. The previous imager, used until 1999, had a 6.7 arcmin field of view and a 0.2 arcsec/picsel spatial resolution. We also hope to detect *reflection* nebulae, like the one discovered expanding at apparently super-luminal velocity around GK Per at the beginning of the century, in observations done in the "clear" skies of Wisconsin (Yerkes telescope, see Richtey 1901). Even if reflection nebulae do not indicate the white dwarf evolution, they are also very interesting because they witness the energy budget and display the "frozen" spectrum at the time of the outburst. We are developing this imaging program through a number of projects and collaborations; several members of the audience have been on our team.

THE UNIQUE NEBULA ASSOCIATED WITH V1974 CYG

Balman et al. (1996) and Rosino et al. (1996) detected this peculiar extended nebula, which was apparently expanding with approximately the speed of light c in 1995 and at $0.35c$ between 1996 and 1999 (Casalegno et al. 2000). The properties of the nebula are described and analysed in detail in Casalegno et al. (2000). It extended from 7.6 to 11.6 parsec away from the nebula, and it was loosing regularity and spherical symmetry as it expanded. Despite the presence of a supersoft X-ray source for ≈ 2 years, its properties are totally different from CAL 83 and from Rappaport et al's (1994) predictions. This nebula is clearly due to ionization and not reflection, as the sharp difference between its spectrum and the one of the nova shell demonstrate. The main surprise is the total absence of detectable flux in [O III], despite an Hα luminosity about half the one of CAL 83 (see Table 1). The upper limits on the flux are at least orders of magnitudes below this level. The only reasonable conclusion that can be drawn is that the white dwarf atmospheric temperature was $T_{eff} \leq 22$ eV for most of the time. It is possible that the higher estimates obtained by Balman et al. (1998) during several post-outburst months were effected by the limited spectral resolution and passband of the *ROSAT PSPC*, which prevented "disentangling" nebular emission, and specially possible nebular lines, from the central source.

Remarkable is also the estimate we obtain for the ionized mass in the nebula: $0.1 m_{\odot}$. This is at least ≈ 3 orders of magnitude larger than the interstellar medium (ISM) mass that can be contained in a few parsecs. It probably indicates mass ejection in a previous common envelope phase, unless the nova system has already undergone at least 1000 outbursts.

Unexpectedly, we have not discovered another nebula at this type associated with any other among 14 Galactic classical novae and two recurrent novae, RS Oph and T CrBor. The list of classical novae includes AB Boo, V603 Aql, Q Cyg, DI Lac, DK Lac, HR Del, QU Vul, V1500 Cyg, CT Ser, U Leo, V4327 Sgr, V705 Cas, V723 Cas, V1494

FIGURE 1. An image of the ionization nebula around V1974 Cyg taken with the the H α filter in September of 2001 at the WIYN telescope. Comparison with the images published in Casalegno et al. (2000) shows that the nebula is still expanding southwards at apparently $\approx$1/3rd the speed of light. The scale of the picture is approximately 7x5.5 arcmin, and we assume a distance of 1.8 kpc to V1974 Cyg.

Aql. 11 of these objects were observed with the Hα filter with the WIYN telescope, and the upper limits are even 5-6 orders of magnitude below the flux measured for V1974 Cyg. We remind that V1494 Aql is indeed known to have been a supersoft X-ray source (see Krautter, this conference). Cohen & Rosenthal (1983) found that after about up to 90 years the deceleration of nova shells' expansion is quite minimal, of order 10% or less, but they estimate that significant deceleration occurs after 500 years. With a typical expansion velocity of 2000 km s^{-1}, an ejected mass 10^{-4} m$_\odot$ is dispersed within 1 parsec after 500 years. If the expansion is later significantly slowed down, it is possible that all the mass ejected in many successive outbursts is still concentrated within $\approx$10 parsecs. For this reason, even if a ionized mass of 0.1 m$_\odot$ like for V1974 Cyg is probably exceptional, 0.01 m$_\odot$ mass (in addition to the ISM mass, orders of magnitude lower) would be present if the system has had 100 previous outbursts, which is not unreasonable. This would be sufficient mass to ionize producing detectable flux, unless the hot white dwarf becomes "cool" in a very short time (and we know this was the case for V1494 Aql). Moreover, for CAL 83 the ionized mass is even consistent with being only ISM, at density of n_e=1-10 cm^{-3}.

INTERMITTENT SUPERSOFT X-RAY SOURCES: A VERY DIFFERENT KIND OF NEBULA

The ionization nebula associated with BZ Cam was also imaged by us with the WIYN telescope (the results are in Greiner et al. 2001). BZ Cam is a CV-like system classified as VY Scl system or "anti-CV". It undergoes optically "low" states instead of outbursts of light. Even if it has never been observed (yet) in X-rays during the optically low

TABLE 1. Characteristics of the three ionization nebulae discussed in the text

Nebula	Hα luminosity (erg s^{-1})	[O III] luminosity (erg s^{-1})	Spatial extension
CAL 83	$\approx 2.7 \times 10^{35}$	1.19×10^{36}	7.6-11 pc
V1974 Cyg	$1.3 - 2.2 \times 10^{35}$	$\leq 1.3 \times 10^{32}$	1-5 pc (in 2001)
BZ Cam	$\approx 10^{31}$	3.94×10^{31}	0.6-1.2 pc

states, we have reason to believe that, like other variables of this type, the white dwarf atmosphere contracts, becoming hotter and turning into a supersoft X-ray source. There is a faint emission nebula with a "bow-shock-like" structure associated with this source, which was attributed mainly to shock ionization (where the ISM is shocked by a stellar wind) by Krautter et al. (1987) and by Hollis et al. (1992). Greiner et al. (2001) obtained optical data during the optically low state and found that the nebula is consistent with mainly photoionization of the central source, even if the bow-shock structure is explained with a shock, due to the high proper motion of the system in the ISM. The flux in Hα (see Table 1) is higher than in the [O III] flux, the latter being more than 4 orders of magnitude lower than CAL 83. This is not surprising: if BZ Cam resembles the related systems, it may be a supersoft X-ray source only for 4-5% of the time (Garnavich & Szkody 1988) and with effective temperature $T_{eff} \leq 25$ eV, while the CAL 83 white dwarf is at $T_{eff} \geq 40$ eV for at least 50% of the time (Kahabka 1998).

Since half or more of the supersoft X-ray sources seem to be intermittent or just transient, there is no "typical" ionization nebula associated with these sources. More common may be intermediate properties between BZ Cam and CAL 83, rather than the characteristics predicted by Rappaport et al. (1994), which were studied for *persistent* sources. We note that the BZ Cam nebula would have NOT been detected by Remillard et al. at LMC distance, but it would have certainly have been detected by us at a typical classical nova distance of $\simeq 2$ Kpc and using the WIYN telescope.

The case of classical and recurrent novae is very special: a strong supersoft X-ray source exists for up to 10 years, but the temperature of this source increases and then decays, and there is relatively high circumstellar density due to previous outbursts and possibly even to a common envelope history of the systems. This situation may yield different and not easily predictable ionization states in the circumstellar medium. However, we still do not know how to explain why we detected only one ionization nebula in 16 systems.

REFLECTION NEBULAE

To conclude, a few words about reflection nebulae. We mentioned in the Introduction that a light echo like the one of SN 1987a was detected for GK Per in 1901, with arc like structures extending up to 3 parsecs away from the nova. Such reflection nebulae may not be uncommon, but they are more difficult to detect at the right time after the outburst. They would not be long lived, and have the spectrum of the nova shell

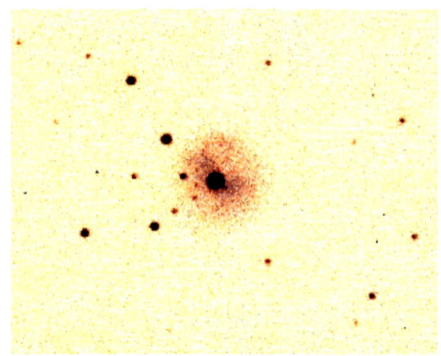
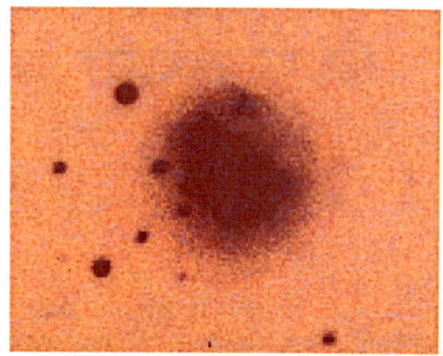

FIGURE 2. The nebula sourrounding V838 Mon, imaged at the WIYN telescope with the narrow band [O III] filter (left) and broad band the V filter (right). The nebula extends to approximately 35 arcsec.

shortly after the outburst, "frozen" in the delayed reflection. Detecting a reflection nebula around a classical nova, we hope to obtain detailed information on the energy budget and maximum luminosity of the nova. Even if we have not been lucky enough to discover one so far, we have recently had a chance to imagine the reflection nebula around V838 Mon (Orio et al. 2002). The jury is still out as to the nature of this intriguing object (see Munari, this conference), but it may be indeed an outburst in a symbiotic nova system, and thus strictly related to the nova phenomenon. On April 22 2002, we observed a filled, approximately circular shell, not limb brightened, with a diameter of 35 arcsec on both nights (Orio et al. 2002, see Fig. 2). The contour of the shell appears irregular, with a broad structure in the shape of an arc and clumps but no small- scale structure. There is more structure in V than in R. Lack of conspicuous $H\alpha$ and [O III] line emission confirms that it is a reflection, rather than an ionization nebula. An R image obtained on April 30 2002 at the WIYN 0.90-m telescope (scale 0.60 arcsec/picsel) still showed a 35 arcsec diameter, so the outer edge of the reflecting material was reached before April 22.

REFERENCES

1. Balman S., Krautter H., Ögelman H., 1998, ApJ, 499, 395
2. Casalegno R., Orio M., et al., 2000, A&A, 361, 725
3. Di Stefano R., Paerels F., Rappaport S., 1995, ApJ, 450, 711
4. Garnavich P., Szkody P., 1988, PASP, 96, 283
5. Greiner J., Tovmassian G., Orio M., et al., 2001, A&A, 376, 1031
6. Kahabka P., 1998, A&A, 331, 328
7. Krautter J., Raadons G., Klaas U., 1987, A&A, 181, 373
8. Hollis J.M., Oliversen R.J., Wagner R.M., Feibelman W.A., 1992, ApJ, 393, 217
9. Orio M., Harbeck D. Gallagher J., Woodward C., 2002, IAU Circ. 7892
10. Rappaport S., Chiang E., Kallman T., Malina R., 1994, ApJ 431, 237
11. Remillard R.A., Rappaport S., Macri L.M., 1995, ApJ 439, 646
12. Richtey G.W., 1901, ApJ 14, 167
13. Rosino L., Iijima T., Rafanelli P., et al., 1996, A&A 315, 463

On the Possibility of Detecting Remnants of Novae in the X-rays and Recovering the Remains of an Explosion After a Century

Şölen Balman

METU Physics Dept., 06531 Ankara, Turkey

Abstract. The nova shell of GK Per is the first classical nova shell resolved and detected in the X-ray wavelengths. I will present the spectral and spatial characteristics of the X-ray nebula derived from a CHANDRA ACIS-S observation. The nebula is brightest on the SW quadrant and towards West with a lumpy morphology. The X-ray shell has a symmetric bow-shaped structure detected in a [NeIX] emission line which shows the shock front most likely expanding into a less dense medium in the NW to SE direction. The X-ray nebula has a low temperature component which is a plasma possibly dominated by emission lines originating from the ejecta with kT=0.11-0.18 keV and an X-ray luminosity of 1.6×10^{31} ergs/s. The higher energy component above 2 keV has kT > 30 keV and an X-ray luminosity of 3.0×10^{31} erg/s, which could be attributed to the emission from accelerated particles behind the shock. This component is found to be heavily absorbed with N_H = 1.7×10^{22} cm^{-2} suggesting that there is a colder shell of material between the two X-ray components in question. The low X-ray luminosity of both components indicates that most of the kinetic energy is still going into expansion of the shell and that the shock is adiabatic. This places the remnant in a late Pre-Sedov/transition to Sedov phase. It is evident from calculations that an adiabatic nova remnant in X-rays would have luminosities in the range 10^{32}-10^{26} ergs/s. Thus, recovering such low luminosity objects with the sensitivity of the present X-ray instruments will be largely limited with distance and surface brightness and thus, should be an intriguing challenge.

INTRODUCTION

An outburst on the surface of a white dwarf in a Cataclysmic binary system (CV) as a result of a thermonuclear runaway in the accreted material causes the ejection of 10^{-3} to 10^{-7} M$_\odot$ of material at velocities up to several thousand kilometers per second which constitutes a shell around the binary system [25], [18]. The studies of resolved old nova shells provide an opportunity of investigating several facts of the evolution of novae long after the initial eruption has subsided. Spectroscopy and imaging of the nebular remnants of old novae have been a diagnostic of physical conditions in (mode of excitation) and the morphology of the nova shells. Optical spectroscopy/imaging of nova shells has been widely used for this purpose. By direct imaging method, the ejecta of over 20 novae have been spatially resolved with evidence/detection of polar blobs and equatorial rings in many of them [24], [15], [16]. The expansion velocities of material in the shell are suggested to vary smoothly as a function of position angle from a maximum in the polar regions to a minimum at the equator, which results in a prolate asymmetry [19]. In addition to this, abundance gradients are expected as a result of TNRs in rotating oblate white dwarfs as detected from observations (eg. RR Pic, DQ Her, V1370 Aql, V838 Her, ect.; [22]) In general, optical and uv wavelength measurements of electron temperatures in the nova shells indicate two groups of old remnants; hot ($T_e > 10^4$ K) and cold ($T_e < 10^4$ K) nova shells [27], [13]. Some of the nova shells have been found to decelerate in time, such as DQ Her, GK Per, V603 Aql, T Pyx, and VG476 Cyg, suggesting the existence of circumstellar interaction [12]. Though there has been no previous detection of old classical nova remnants in the X-ray wavelengths (as in SNRs), classical nova remnants have been detected in the hard X-rays (above 1 keV) as a result of wind-wind interactions it in the outburst stage [3], [20]. Despite these facts there have not been any X-ray or radio wavelength detections of old classical nova remnants (i.e., not in the outburst stage) except for GK Per [6], [1]. At this point, there is also no detailed theoretical framework on the shock evolution in classical nova remnants; however, the circumstellar interaction of nova shells have been modeled for recurrent novae (RS Oph: [7]; T pyx: [9]). An important question is, then, whether the conditions in the nova shells are determined right after the outburst or whether the role played by the circumstellar medium around the novae is significant. The latter seems to be the more prominent factor owing to the lack of X-rays and radio detections of old classical nova remnants.

Nova Persei 1901 (GK Per) is one of the most extensively observed and studied classical nova shell over the entire electromagnetic spectrum. It is known to be a fast ONeMg nova ($V_{eject} \simeq 1200$ km s^{-1}, $M_{eject} \simeq 7 \times 10^{-5}$ $M_\odot$ and distance $\simeq 470$ pc : [23]). Nova Persei (1901) was the first recorded nova to show a light echo due to the reflection of light off the nearby interstellar material [17], [10] as a result of large-scale circumstellar sheets of dust clouds. The optical remnant is 103×90 arcsec^2 and has evolved into series of knots and filaments. The bulk of emission arises in the southwestern quadrant indicating interaction between the nova ejecta and the ambient gas [24], [23]. The detection of the remanant with the VLA at 1.5 & 4.9 GHz as a non-thermal polarized radio source shows the existence of shocked circumstellar or interstellar material [21], [23]. IRAS (60μm, 100μm) observations reveal a symmetric far-IR emission region extending around the nova out to 6 pc suggested to be an ancient Planetary Nebulae associated with the binary [5]. Recent optical and IR wavelength observations of the vicinity of GK Per shows that some of this symmetric nebulosity is produced during the quiescent mass-loss phase of the central binary because of the evolved nature of the secondary [26], [11].

THE CHANDRA DATA OF THE NOVA SHELL OF GK PERSEI

The X-ray nebula extends to 52″ South, 41″ North, 45″ West, and 38″ East of the point source with an irregular shape most counts coming from a region centered at the SW with a count rate of 0.093±0.01 (see Figure 1a). The count rate ratio of the hemisphere centered around NE to the one centered on the SW is 2:3. The X-ray shell has a symmetric bow-shaped structure detected in a [NeIX] emission line which shows the shock front most likely expanding into a less dense medium in the NW to SE direction (see Figure 1b).

The X-ray spectrum consists at least of two thermal components of emission (see Figure 2). The lower temperature component (below 2 keV) has kT=0.11-0.18 keV

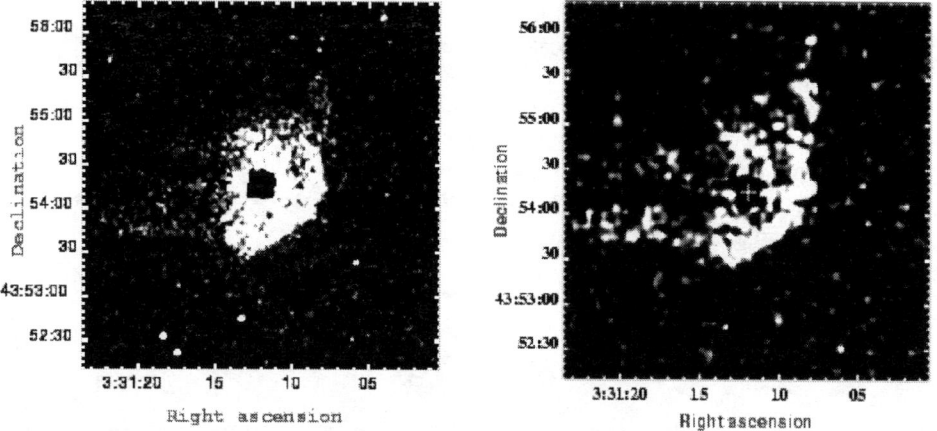

FIGURE 1. (a) The image of the classical nova shell of nova Persei (1901) obtained with the CHANDRA ACIS-S (S3) between 0.3 and 10 keV. The resolution in the image is 1″/pixel and shows emission 2 σ above the background. North is up and West is to the left. The image is smoothed using a Gaussian of σ=2″. (b) The continuum subtracted image of the nebula between energy channels 50-75 within the [NeIX] emission line at the continuum level. Continuum is selected from energy channels 76-100. North is up and West is to the left. The image is smoothed using a Gaussian of σ=2″ after it has been rebinned by 2×2 pixels.

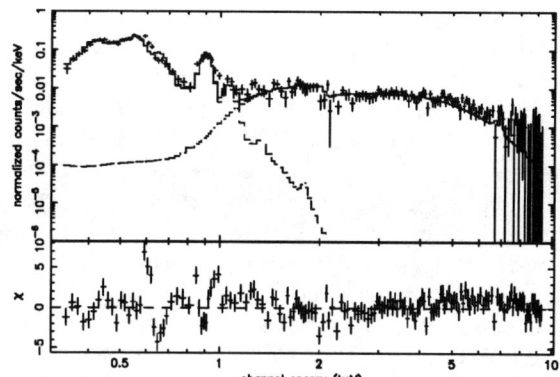

FIGURE 2. The X-ray data fitted with the VMEKAL+NEI emission models including neutral hydrogen absorption. The lower figure shows the residuals between the data and the model in standard deviations. In general, solar elemental abundances are assumed except for Neon and Nitrogen.

with an X-ray flux of 6.5×10^{-13} erg cm^{-2} s^{-1}. There is also a high temperature, kT > 30 keV, embedded, $N_H = 1.7 \pm 0.3 \times 10^{22}$ cm^{-2}, component prominent above 2 keV. The unabsorbed X-ray flux from this component is 1.1×10^{-12} erg cm^{-2} s^{-1}. A distinct emission line of Neon [NeIX] is detected with a flux of 1×10^{-14} erg cm^{-2} s^{-1}. The X-ray emitting plasma is of solar composition except for enhancement in the elemental abundances of Ne and N compared to their solar mass fractions detected as 11±5 and 6±3 respectively. The emission measures yield an average electron density of 0.6-11 cm^{-3} for both of the components (filling factor=1). If the filling factor is decreased to about 5×10^{-5}, the electron density increases to about 500 cm^{-3}. The ionization timescale ($\tau=n_0 t$) translate to ambient densities $n_0 \leq 10$ cm^{-3}. It is most likely that the second component is a hot tenuous gas of circumstellar origin with high filling factor. The shocked mass is calculated in a range $(0.08-5)\times10^{-4}$ M$_\odot$. This is derived assuming a filling factor in a range 5×10^{-5} to 1.

DISCUSSION AND COMPARISONS WITH OTHER WAVELENGTHS

In general, the CHANDRA observation of the shell of nova Persei (1901) have shown that the remnant evolves similar to the young supernova remnants (like in Type II remnants; [8]) as was predicted using the previous observations in different wavelengths. In order to investigate the conditions within the shell and determine the evolutionary stage of the remnant, one needs to look for spectral variations within the shell and compare images and spectra in different wavelengths. Several spectra are constructed to look for spectral variations within the shell [2], However no spectral variation is detected above 2 σ confidence level. This is also constrained with the statistical quality of the data. On the other hand, there are brightness changes within the shell most likely as a results of line intensity variations.

The X-ray image of the shell of nova Persei 1901 is *aligned* with the recent images in the optical and radio wavelengths [4]. The X-ray image in the band 0.3-10 keV is overlayed with the contours derived from the HST image obtained using the [NII] filter (brightest emission line in the optical) (HST image: Shara 2001, private communication). The combined image reveals that the [NII] knots and filaments are in general coincident with the circular region of the X-ray nebula. Both images do not show a well developed shell and are brighter on the Western hemisphere rather than the East. The optical shell, mostly composed of knots and filaments, suggests large-scale density gradients in the vicinity of the classical nova. Though the contours look circular in geometry, the HST data indicates an elliptical velocity gradient in the shell with the fastest material lying in the NW to SE direction (Shara 2001, private communication) which is the direction, the elongation in the X-ray nebula lies as detected in the [NeIX] emission line. The brightest part of the [NII] nebula is slightly offset and farther out than the brightest part of the X-ray nebula which suggest cooling in the forward shock and that the remnant is older than a pre-Sedov remnant.

The X-ray nebula has been aligned also with a recent VLA observation obtained at 1.425 GHz in 1997 (VLA image: Seaquist 2001, private communication). An intriguing result of this combination image is that the peak of non-thermal radio emission falls on, but slightly offset towards inside from the brightest part of the X-ray shell which is the rim [4]. This indicates that the particle acceleration could be occurring mainly at the reverse shock zone. This is also detected in the recent CHANDRA observation of the young supernova remnant 1E 0102.2-7219 [14]. The radio image shows no trace of the wings detected in the X-ray nebula (in the [NeIX] emission line) which is consistent with the lower circumstellar density in the NE to SW direction.

The temperature of the second component implies shock velocities of about 5000 km s^{-1} calculated using the general relation $kT_s = (3/16)\mu m_p (v_s)^2$ assuming Rankine-Huguniot jump conditions in the absence of cosmic-ray acceleration. The calculated expansion speed/shock speed for nova Persei 1901 yields to a total kinetic energy of 1×10^{45}-3×10^{46} ergs for the shocked mass. The size of the nova shell indicates velocities of about 1000-2800 km/s calculated using a linear expansion law. The inconsistency indicates that this component could originate from the accelerated particles and thus the temperatures do not necessarily indicate the conditions in the shocked material. Assuming E/[dE/dt] = t_{max} and 99 years after the outburst, maximum luminosity is [dE/dt]$_{max}$=10^{37-36} ergs s^{-1}. The kinetic energy dumped in the ejecta is usually about 1% of the explosion energy which translates to about 10^{35} ergs s^{-1} for a radiative shell. Thus, given the low luminosity of the shell a few times 10^{31} ergs s^{-1}, the remnant is either in a late pre-Sedov phase or in a transition state to the Sedov phase. This is also supported by the calculated maximum shocked mass which is less than 10 times the ejecta mass using the emission measures and a range of filling factors [2].

In order to investigate the X-ray evolution of nova remnants in detail, it is necessary to construct dynamical (2D/3D geometry) and hydrodynamical models which would predict the spectrum, the geometry and abundances of the nova shells in time. Such models are only being developed (see Bode 2002, O'Brien 2002; this proceeding) in the framework of classical nova studies at this time. The X-ray emission expected from a simple shocked-shell model is of thermal origin. The total power from the shocked-shell, as an X-ray emitting nebula, can be modeled assuming a 0.25 filling factor for the volume of the shell (assumed to be spherical) and $n_e \sim 4n_o$ (from strong shock conditions).

$$L_T \simeq 3.1 \times 10^{33} T_7^{0.5} n_o^2 R_{18.5}^3$$

The temperature T is in units of 10^7 K and the radius of the emitting shell R is in 3.1×10^{18} cm. The fast nova ejecta will start to cool as it sweeps up an equal amount of circumstellar material at R$_{cool}$ and t$_{cool}$. As it sweeps up about 10 times its own mass, it will proceed into the Sedov phase.

$$R_{cool} \simeq 0.03 (M_{-5}/n_o)^{1/3} pc; \quad t_{cool} \simeq 30 (M_{-5}/v_{1000}^3 n_o)^{1/3} yrs$$

$$t_{Sedov} \simeq 64 (M_{-5}/v_{1000}^3 n_o)^{1/3} yrs$$

Finally, If one assumes $n_o \simeq 0.1$-10 cm^{-3}, kT $\simeq 0.1$-10 keV and R $\sim$ R$_{cool}$, the range of X-ray luminosity is afew$\times 10^{34}$ to afew$\times 10^{28}$ ergs/s. Since the remnants are going to be adiabatic, about only 1% of this radiation will be visable. Thus, the maximum limit of X-ray emission is then about 10^{32} ergs/s which is in good agreement with the X-ray luminosity of the remnant of GK Persei (i.e., brightest remnant). As a result, recovering the remains of classical novae/novae remnants is quite a challange with the given sensitivity of the present X-ray instruments. However, CHANDRA ACIS/HRC detectors should be able to recover the remnants that are close by and have high surface brightness in the X-rays.

On the account of looking for other candidates, classical nova Pictoris (1925) is observed by CHANDRA ACIS-S detector. The source and the vicinity of the nova shell is detected with a count rate of 0.06 c/s. The fitted spectra obtained in sectors of 120° centered at the NE and SW of the point source are shown in figures 3a and 3b, respectively. There is significant excess emission towards the SW of the central source which is coincident with the bright SW Polar blob detected in the Hα images [16]. The work is in progress.

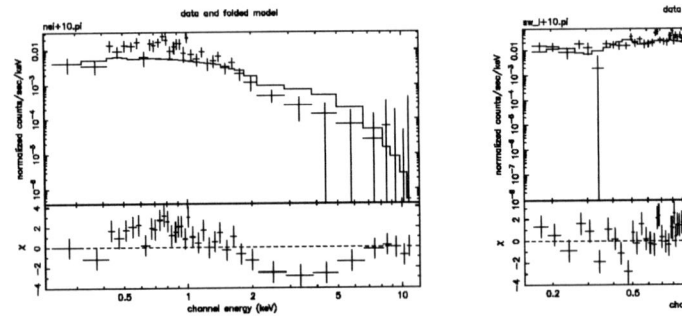

FIGURE 3. (a) The Figure shows the spectrum of a 120° sector centered around the NE direction of the source. The spectrum is fitted using a neutral Hydrogen absorption model (PCFABS) with covering fraction and Bremsstrahlung model of emission. The spectrum includes the central source partially. (b) The Figure shows the spectrum of a 120° sector centered around the SW direction of the source. It is fitted with the same model as in part (a) and includes the same amount of contemination by the central source.

CONCLUSIONS

The nova shell of GK Per is the first classical nova shell resolved and detected in the X-ray wavelengths. The X-ray nebula is brightest on the SW quadrant and towards West with a lumpy morphology. It resembles to the radio shell except for the strong emission line of neon and the region that emits the line. The X-ray shell has a symmetric bow-shaped structure detected in a [NeIX] emission line which shows the shock front most likely expanding into a less dense medium in the NW to SE direction. No significant spectral variation is detected within the X-ray nebula. The X-ray shell is a results of the interaction with the circumstellar medium and the geometrical effects.

The X-ray nebula has a low temperature component that is a thermal plasma possibly dominated by emission lines originating from the ejecta and a higher energy component above 2 keV most likely comming from the accelerated particles in the reverse shock zone or the heated interior of the remnant by shocks. The X-ray shell is predicted to be in a transition to the Sedov Phase.

The standard nova theory predicts enrichment of metal abundances produced prior to the outburst via mixing processes. This observation reveals for the first time enhancement of Neon 8-13 and Nitrogen 3-9 times its solar mass fraction from a remnant in

X-rays that is 100 yrs old. It is also the first calculation of the Neon & Nitrogen abundance of the shell of GK Per.

The ambient density is calculated as ~ 10 cm^{-3}. This indicates that the circumbinary medium of the nova is filled and dense which in turn explains the easy-detection of the X-ray shell in contrast with the other classical nova remnants that are not so quite bright in the X-rays. The predicted luminosity of the CNRs (Classical Nova Remnants) are in a range of 10^{32} - 10^{26} ergs/s. Given the sensitivity of the present X-ray instruments, it should be a challange to recover some of them that are close by and have high surface brightness in the X-rays. The Chandra observation of the X-ray nebula of nova Persei 1901 opens a new door to a new era and a possible new class of X-ray emitters that resemble SNRs, but evolve much faster than SNRs do which in turn should provide unprecedented information on hydrodynamics and shock evolution.

REFERENCES

1. Balman, S. & Ögelman, H.B. 1999, ApJ, 518, L111
2. Balman, S., 2001, in X-rays at Sharp Focus: Chandra Science Symposium, ASP Conf. Ser., S. Vrtilek, E. Schlegel, L. Kuhi (eds), p.34
3. Balman, S., Krautter, J. & Ögelman, H. 1998, ApJ, 499, 395
4. Balman, S. 2002, in preperation
5. Bode, M. F., Seaquist, E. R., Frail, D. A., Roberts, J. A., Whittet, D. C. B., Evans, A. & Albinson, J. S. 1987 Nature, 329, 519
6. Bode, M.F., Seaquist, E. R & Evans, 1987, MNRAS, 227, 217
7. Bode, M. F., Kahn, F. D. 1985, MNRAS, 217, 205
8. Chevalier, R. A., & Fransson, C. 1994, ApJ, 420, 268
9. Contini, M., & Prialnik, D. 1997, ApJ, 475, 803
10. Couderc, P. 1939, Ann. d'Astrophy., 2, 271
11. Dougherty, S. M., Waters, L., Bode, M., LLoyd, H. M., Kester, D. J. & Bontekoe, T. 1996, MNRAS, 306, 547
12. Duerbeck, H. W. 1987, Astr. Space Sci., 131, 461
13. Ferland, G. J., Williams, R. E., Lambert, D. L., Shields, G. A., Slovak M., Gondhalekar, P. M. & Truran, J. W. 1984, ApJ, 281, 194
14. Gaetz, T.J., Butt, Y., Richard, E., Kristoffer E., Plucinsky, P. Schlegel E. & Smith R. 2000, ApJ, L47
15. Gill, C.D.; O'Brien, T.J. 2000, MNRAS, 314, 175
16. Gill, C.D.; O'Brien, T.J. 1998, MNRAS, 300, 221
17. Kapteyn, J. C., 1901, Astr. Nachr., 3756, 201
18. Livio, M., in Interacting Binaries, Saas-Fee Advanced Course 22, ed. H. Nussbaumer & A. Orr (Berlin: Springer), 135
19. Lloyd, H.M., O'Brien T.J., Bode, M. 1997, MNRAS, 284, 137
20. Mukai, M. and Ishida, M. 2001, ApJ, 551, 1024
21. Reynolds, S. P. & Chevalier, R. A. 1984, ApJ, 281, L33
22. Scott, A.D. 2000, MNRAS, 313, 775
23. Seaquist, E. R., Bode, F. M., Frail, D. A., Roberts, J. A., Evans, A. & Albinson, J. S. 1989, ApJ, 344, 805
24. Slavin, A. J., O'Brien, T. J. & Dunlop, J. S. 1995, MNRAS, 276, 353
25. Shara, M. 1989, PASP, 101, 5
26. Tweedy, R. W. 1995, ApJ, 438, 917
27. Williams, R. E. 1982, ApJ, 261, 170

Chandra Observations of Old Novae

Koji Mukai*, Marina Orio[†], Fred Ringwald** and Martin Still*

*Code 662, NASA Goddard Space Flight Center, Greenbelt, MD 20771, USA (also Universities Space Research Association)
[†]Astronomy Dept., Univ. Wisconsin, 475 N. Charter Str., Madison WI 53706, USA (also Osservatorio Astronomico di Torino)
**Department of Physics, California State University, Fresno, 2345 E. San Ramon Ave., M/S MH37, Fresno, CA 93740-8031, USA

Abstract. We present highlights of *Chandra* observations of two old novae, DQ Her and V603 Aql, with the main aim of improving our understanding of the underlying binaries decades after their respective nova eruptions. In DQ Her, we find a partial X-ray eclipse; it is likely that we observe photons scattered in an accretion disk wind. The X-ray spectrum of V603 Aql suggests an origin in multi-temperature plasma; while the low energy lines suggest modest density, the 6.4 keV Fe Kα line suggests that the hard continuum arises in a compact emission region. We also report on our searches for nebular X-ray emission around these old novae.

INTRODUCTION

X-ray observations of old novae are useful in selecting magnetic systems, in constraining the total accretion rate, and potentially in inferring the abundances of the accreting and/or ejected materials. With the increasing capability of X-ray satellites, we can now observe fainter systems or obtain detailed light curves and spectra of brighter systems. Here we present our *Chandra* observations of DQ Her and V603 Aql. The former is X-ray faint, while the latter is among the X-ray brightest old novae known, although both were among the brightest novae of the 20th century.

DQ HER

DQ Her (Nova Herculis 1934, at an estimated distance of $\sim$560 pc; [1]) is the prototype of a subclass of magnetic CVs, "Intermediate Polars" (IPs) or "DQ Her type systems" [2]. Unlike most members, however, DQ Her is not a strong X-ray source as seen from Earth. It was undetected with *Einstein* [3], with an upper limit of 3×10^{30} ergs s^{-1}. Since DQ Her is a deeply eclipsing system in the optical, its white dwarf (the presumed primary X-ray emission site) is likely hidden from our view at all times. DQ Her, however, was detected with *ROSAT* at $\sim 4.0 \times 10^{30}$ ergs s^{-1} [4]. A deep eclipse was not observed, hence these detected X-rays are not from the immediate vicinity of the white dwarf. So what is the origin of the observed X-rays?

We observed DQ Her with *Chandra* ACIS-S in imaging mode (i.e., without a grating) from 2001 July 26 13:00 UT – July 27 02:31 UT and again from 2001 July 29 17:09 UT

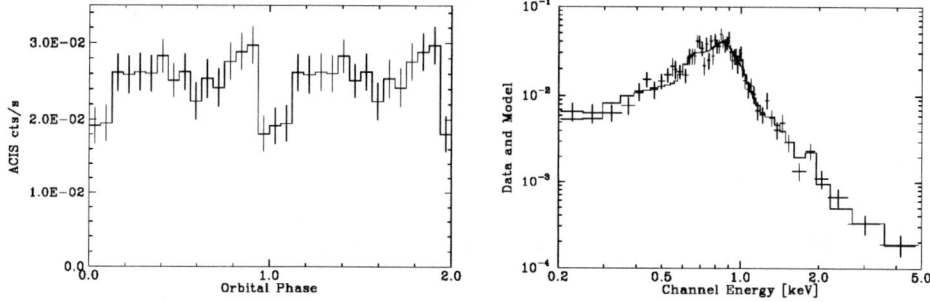

FIGURE 1. (Left) The X-ray light curve of DQ Her, folded on the orbital period (two cycles are shown for clarity). (Right) The X-ray spectrum of DQ Her, as observed with *Chandra* ACIS-S, plotted with a 2-component model (power law plus a kT~0.6 keV thermal plasma model) folded with the ACIS-S response.

– July 30 02:27 UT, for a total integration time of 69 ksec. We detect a moderately strong (0.025 cts s^{-1}, or $\sim 3 \times 10^{30}$ ergs s^{-1}) point source at the optical position of DQ Her in these observations. The light curve, folded on the orbital ephemeris of [5], clearly shows a partial eclipse, lasting just over 0.1 in orbital phase and about 30% deep. The spectrum shows a bump around 0.8–1.0 keV, characteristic of plasma emission with lines of O, Ne, and Fe at these energies (Figure 1). The partial eclipse is a signature of an extended emission region. Material that is ~11 white dwarf radii above the orbital plane should remain uneclipsed, given the known system geometry, suggesting a vertical extent at least this much. These observations are consistent with the idea that the observed X-rays have been scattered in an accretion disk wind, similarly to OY Car in superoutburst [6]. The presence of an accretion disk wind in DQ Her has been inferred from UV observations [7], even though DQ Her lacks a boundary layer. The fact that there is a significant partial X-ray eclipse in DQ Her makes it different from other wind-scattered X-ray sources such as OY Car and UX UMa [8], and may be an indication of differences in their wind structures.

V603 AQL

V603 Aql (Nova Aquilae 1918) is among the X-ray brightest old novae both in intrinsic luminosity (in excess of 10^{32} ergs s^{-1}) and in terms of flux at Earth [9]. We therefore observed V603 Aql with *Chandra* HETG with ACIS-S from 2001 April 19 17:36 UT – April 20 11:53 UT for a total of 64 ksec, to obtain the highest quality X-ray spectrum of an old nova to date.

There have been claims that this is a magnetic system of the IP type, based on polarimetry [10] and on X-ray photometry [11]. However, the former was based on a method which is, for variable sources, susceptible to false detections; on the other hand, [12] established a strict upper limit using a variable-star-safe instrument. Here we examine the claim of X-ray periodicity (Figure 2), using 1st order (dispersed) photons in the *Chandra* HETG data which were taken without interruption (0th order, undispersed,

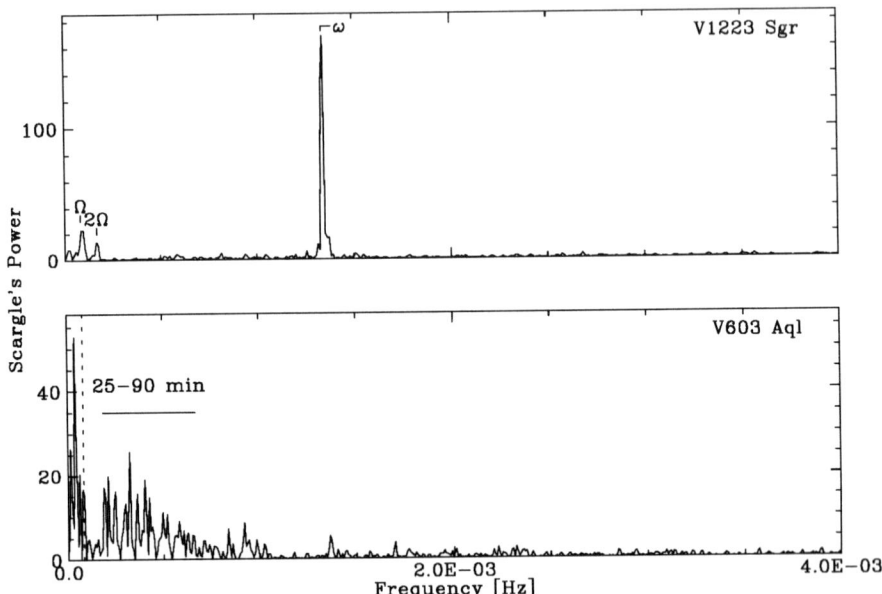

FIGURE 2. Power spectra of the IP, V1223 Sgr (top) and the old nova, V603 Aql (bottom), both from continuous *Chandra* grating observations. The former shows strong peaks at the orbital and spin periods, whereas the latter shows a broad envelope of excess power: the observed X-ray variability is not periodic.

image suffers from a significant pile-up which makes it far less useful for timing analysis). As a comparison, the *Chandra* HETG power spectrum of V1223 Sgr [13], a typical IP, shows prominent peaks at the orbital and spin periods, and in some harmonics, with little power at other frequencies. In contrast, the power spectrum of V603 Aql contains a number of peaks, roughly in the 20 min – 1 hr range. It is highly unlikely that any combinations of harmonics and sidebands can explain all the peaks, given a single underlying clock. We therefore conclude that V603 Aql shows a strong variability, but not a periodicity, and that there is no evidence to date that it is an IP.

The average spectrum of V603 Aql is rich in emission lines; the simultaneous presence of emission lines of Fe, Si, Mg, Ne, and O is a direct evidence for the multitemperature nature of the X-ray emitting plasma. In Figure 3, we show details of selected regions of the spectrum. With the superior spectral resolution of *Chandra* HETG, we have been able to resolve the He-like triplets of Ne IX and Mg XI. The presence of the forbidden (f) component in the latter (and perhaps also the former) sets a limit of the density of the line emitting region at $n_e \leq 10^{13}$ cm^{-3}. At the Fe K region of the spectrum, the resolution of HETG is insufficient to resolve the He-like triplets; what we see are the 6.4 keV fluorescent, 6.7 keV He-like, and 6.97 keV H-like components. The H-like component, from kT$\sim$10 keV plasma, is present but significantly weaker than the He-like component (kT$\sim$5 keV). The fluorescent component has an equivalent width of $\sim$150 eV, consistent with reflection from the white dwarf surface that subtends 2π steradians. This implies that the hard X-ray continuum originates in a compact emission

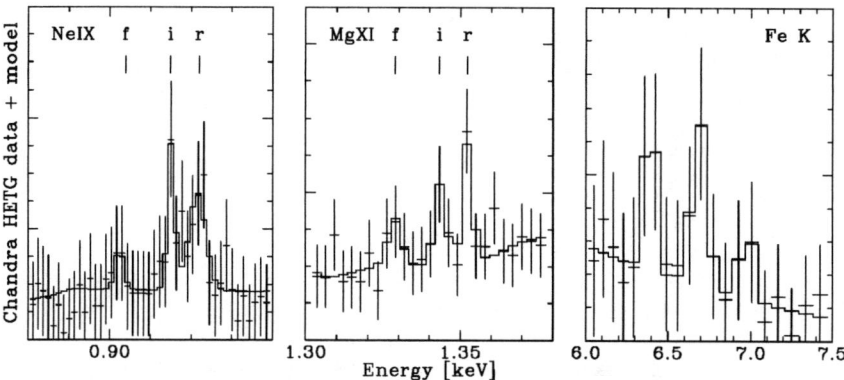

FIGURE 3. Selected X-ray emission lines of V603 Aql. The left two panels show He-like triplets of Ne IX and Mg XI while the right panel shows the Fe K line (~neutral, He-like and H-like).

region, whereas the lower energy lines may originate in a more extended structure.

A SEARCH FOR NEBULAR X-RAY EMISSIONS

Finally, the superior imaging resolution of *Chandra* allows a sensitive search for extended emission, such as that seen around GK Per [14]. In Figure 4, we plot the observed radial profile of the *Chandra* image of DQ Her, and the best-fit model based on the known point-spread function (the fit is strongly constrained by the first few points, which are not shown). We detect an excess of counts 5–10 arcsec away, most prominently at the lowest energies. This excess is most prominent NE of DQ Her itself, and is not consistent with a point source. A comparison with ground-based images of the nova shell [15] suggests a coincidence with an [NII] knot.

Our search for a similar feature around V603 Aql has been inconclusive. This is because we used a grating, which complicates the background and the point-spread-function calibration; and because the central source is much brighter than in DQ Her.

CONCLUSIONS

The majority of X-rays from DQ Her appear to be scattered in the accretion disk wind. Despite its well-credentialed magnetic nature, DQ Her resembles high accretion rate, non-magnetic CVs seen at high inclination. X-ray data may be useful more as a probe of the wind than of the accretion flow immediately around the white dwarf in this system. A small number of X-ray photons (~ 30, a few times 10^{28} ergs s^{-1}) appear to be from the shell around DQ Her.

The X-ray spectrum of V603 Aql is rich in lines and may contain enough clues to advance our understanding of accretion onto the white dwarf. However, we have not yet

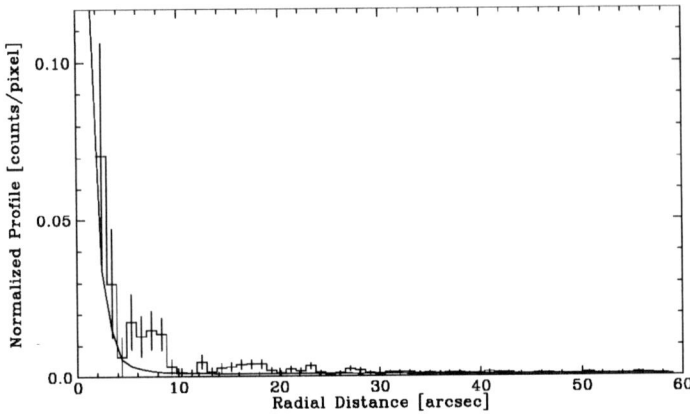

FIGURE 4. The radial profile of the *Chandra* ACIS-S X-ray image of DQ Her in the 0.2–0.5 keV band. The data are plotted as a histogram with errors, while the best-fit model (for 0.3 keV monochromatic X-rays observed on-axis) are plotted as a solid line.

learned how to interpret these clues. It may serve as a template for the studies of non-magnetic CVs in general (previous claims that V603 Aql is magnetic were premature, and quite possibly erroneous).

We have presented a flavor of what X-ray observations of quiescent old novae can do. Observations of additional novae with *Chandra* and *XMM-Newton* should be encouraged.

REFERENCES

1. Herbig, G.H., and Smak, J.I, *Acta Astron.* **42**, 17–28 (1992).
2. Patterson, J., *PASP* **106**, 209–238 (1994).
3. Córdova, F.A., Mason, K.O., and Nelson, J.E., *Ap.J.* **245**, 609–617 (1981).
4. Silber, A.D., Anderson, S.F., Margon, B. & Downes, R.A., *Ap.J.* **462**, 428–438 (1996).
5. Zhang, E., Robinson, E.L., Stiening, R.F., and Horne, K., *Ap.J.* **454**, 447–462 (1995).
6. Mauche, C.W., and Raymond, J.C., *Ap.J.* **541**, 924–936 (2001).
7. Córdova, F.A., and Mason, K.O, *Ap.J.* **290**, 671–682 (1985).
8. Wood, J.H., Naylor, T., and Marsh, T.R., *Mon. Not. R. astr. Soc.* **274**, 31–36 (1995).
9. Becker, R.H., and Marshall, F.E., *Ap.J.Lett.* **244**, L93–L95 (1981).
10. Haefner, R., and Metz, K., *Astron. Ap.* **145**, 311-320 (1985).
11. Udalski, A., and Schwarzenberg-Czerny, A., *Acta Astron.* **39**, 125-138 (1989).
12. Naylor, T., Koch-Miramond, L., Ringwald, F.A., and Evans, A., *Mon. Not. R. astr. Soc.* **282**, 873–876 (1996).
13. Mukai, K., Kallman, T., Schlegel, E., Bruch, A., Handler, G., and Kemp, J., "Chandra HETG Observation of the Magnetic Cataclysmic Variable V1223 Sagittarii," in *New Century of X-ray Astronomy*, edited by H. Inoue and H. Kunieda, ASP Conf. Ser. 251, San Francisco, 2001, pp. 90-93.
14. Balman, S., this volume.
15. Slavin, A.J., O'Brien, T.J., and Dunlop, J.S., *Mon. Not. R. astr. Soc.* **276**, 353–371 (1995).

Chandra ACIS-I and LETGS X-ray observations of Nova 1999 Velorum (V382 Vel)

Vadim Burwitz*, Sumner Starrfield†, Joachim Krautter** and Jan-Uwe Ness‡

*Max-Planck-Institut für extraterrestrische Physik, P.O. Box 1312, D-85741 Garching, Germany [1]
†Department of Physics and Astronomy, P.O. Box 87150, Arizona State University, Tempe AZ85287-1504, USA
**Landessternwarte Königstuhl, 69117 Heidelberg, Germany
‡Universität Hamburg, Gojenbergsweg 112, 21029 Hamburg, Germany

Abstract. The preliminary analysis of the *Chandra* ACIS-I and LETGS observations carried out over a period of 8 months starting 7 months after the outburst of Nova Velorum 1999 (=V382 Vel) is presented here. We find that within a period of less than 6 weeks the flux of the soft X-ray component (0.4-0.8 keV) decreases nearly 200 fold. In addition, we find that the soft component seems to become fainter following a steeper power law than the hard component. Most lines detected with the *Chandra* LETGS are broadened with a corresponding FWHM $\sim$ 2000 km/s that is compatible with the velocity of the expanding shell. The He-like triplets of OVII and NVI allow us to put some constraints on the plasma temperature of 4.5-5.0 10^5 K.

INTRODUCTION

The outburst of Nova Velorum 1999 (=V382 Vel) was first detected on May 22, 1999 independently by Williams [1] and Gilmore [2]. The evolution of this ONeMg nova has been extensively covered with many optical and X-ray observations. In the optical, e.g., Della Valle et al. [3] studied both the photometric and spectroscopic properties. The hard X-ray observations with *RXTE* + *ASCA* and *BeppoSAX* were presented in Mukai and Ishida [4] and Orio et al. [5], respectively. Orio et al. [6] report on a *BeppoSAX* soft X-ray observation $\approx$ 6 weeks prior to our first *Chandra* observation (Fig. 1: 1st mark).

DATA

Four observations of Nova Velorum 1999 were carried out between December 1999 and August 2000 with the first observation 7 months after the outburst in May 1999. In Table 1 an overview of the individual observations is given. The times of our *Chandra* ACIS-I and LETGS observations are marked in Fig. 1 above the optical light curve. Both the *Chandra* LETGS data as well as the ACIS-I data were reprocessed following the

---

[1] burwitz@mpe.mpg.de

TABLE 1. Summary of Observations

#	Obs. ID	Instr.	Exp. [ksec]	Start/Stop Date yyyy/mm/dd	Start Time [UT]	Stop Time [UT]	T-T$_{outburst}$ [days]
1	652	ACIS-I	20.8*	1999/12/30-31	19:25:10	01:16:53	222
2	958	LETGS	24.5	2000/02/14	06:24:13	13:49:08	268
3	73	ACIS-I	5.2	2000/04/21	09:27:30	11:23:29	335
4	1804	ACIS-I	9.6	2000/08/14	10:18:43	13:32:51	450

* the effectively usable exposure here is 144.27 seconds hence the low signal to noise at the high flux level that can be seen in Fig. 2 top.

CIAO threads described on the *Chandra* X-ray Center (CXC) webpages.[2] The spectrum of the first ACIS-I observation #1 had to be treated specially as Nova Velorum 1999 was extremely bright at the time. It exhibited extreme source pile-up in the X-ray CCD image even though it had been placed ∼3 arcmin off-axis. Therefore, in order to extract the source spectrum, the out-of-time events were used yielding an effective exposure time of 144.27 seconds (see Fig. 2 top). The grating observation #2 was extracted using the extraction regions described in the CXC proposers guide (POG) (see Fig. 2 center). The last two observations #3 and #4 were carried out with the source placed ∼7 arcmin off-axis, very close to the X-ray CCD readout nodes, so that one can directly obtain unpiled-up spectra with the highest resolution possible for the CCDs on-board *Chandra*, see (Figs. 2 bottom left and right).

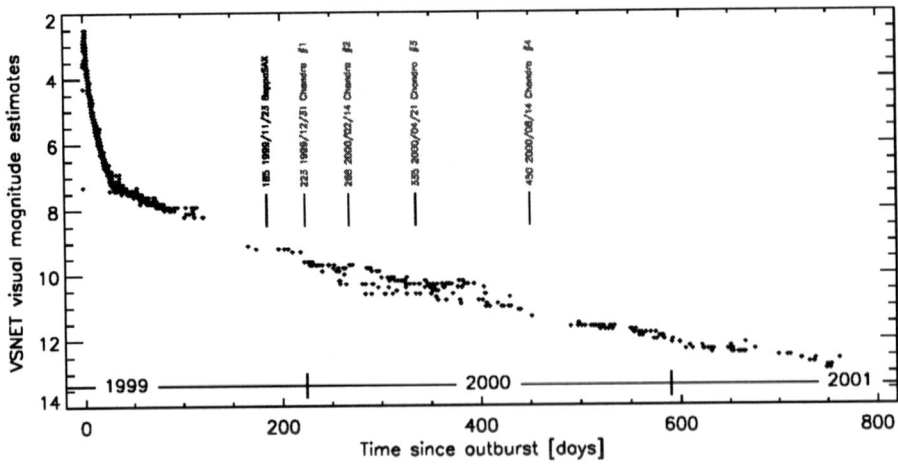

FIGURE 1. X-ray observations relative to the long-term VSNET optical light curve.

---

[2] http://cxc.harvard.edu/ciao

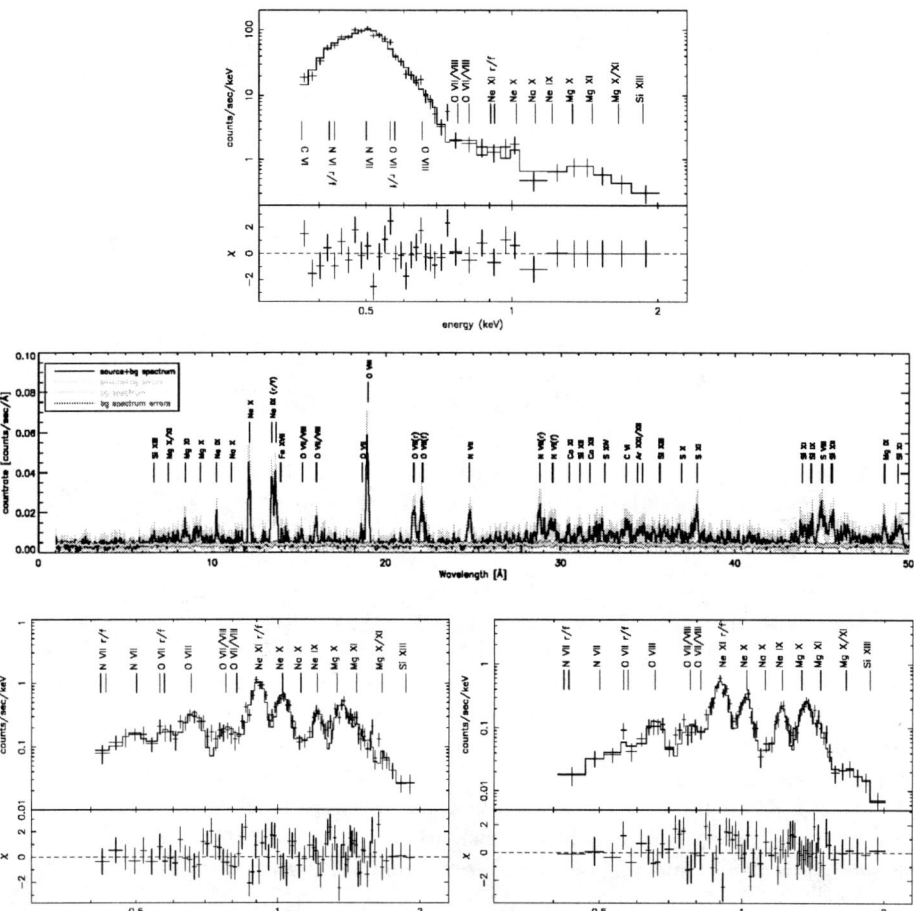

FIGURE 2. Top: The 20.8 ks *Chandra* ACIS-I spectrum from December 30/31, 1999, 222 days after outburst. Center: The 24.5 ks *Chandra* LETGS spectrum February 14, 2000, 268 days after outburst. Bottom Left: The 5.2 ks *Chandra* ACIS-I spectrum from April 21, 2000, 335 days after outburst. Bottom Right: The 9.6 ks *Chandra* ACIS-I spectrum August 14, 2000, 450 days after outburst.

ANALYSIS AND CONCLUSIONS

All spectra show a wealth of emission lines, which mainly comes from the Hydrogen and Helium-like transitions of Mg, Ne, O, N, and C as well as some Si and Na lines. The broad (soft component) centered on 0.5 keV in spectrum #1 is the hard end of the soft component seen in the BeppoSAX observation Orio et al. [6]. If seen at high spectral resolution it would most likely look like the spectrum of Nova 1999 Aquilae (V1494 Aql) (Starrfield et al. [7]) in its X-ray soft state which shows a numerous lines in the region. We find that the intense soft component virtually disappears between

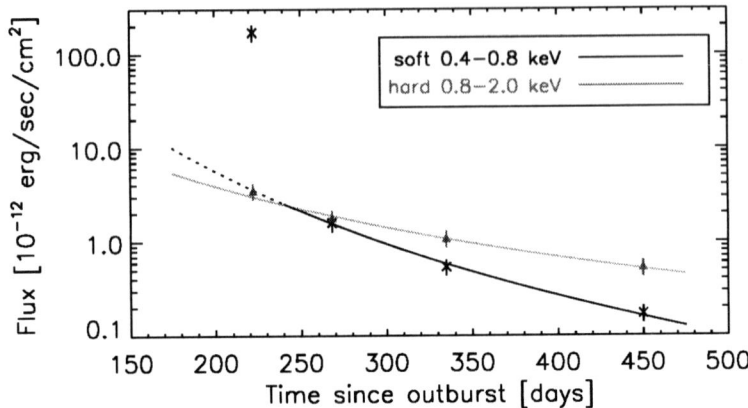

FIGURE 3. Evolution of the X-ray flux with time. From the four *Chandra* observations we obtain for the declining fluxes power-law indices of $n_{hard} \simeq -2.5$ and a steeper $n_{soft} \simeq -4.4$. The former is comparable to that obtained for V1974 Cygni 1992 (Balman et al. [10]).

observations #1 and #2 and that the hard (0.8-2.0 keV) emission gradually decreases (Fig. 3). This hard component is dominated by the emission lines of O, Ne, Mg which are clearly detected in observations #2 to #4. A detailed analysis of the LETGS data shows that most lines are broadened with a corresponding FWHM $\sim$ 2000 km/s that is compatible with the velocity of the expanding shell. The analysis (Porquet and Dubau [8]) of the He-like triplets of OVII and NVI yields some constraints on the temperature (4.5-$5.0\ 10^5$ K) of the plasma. We are working on more detailed modelling of these spectra. Thus we hope to obtain new insights into the evolution of these fascinating objects. More details and analysis will be presented in Burwitz et al. [9]

REFERENCES

1. Williams, P., *IAU Circulars*, **7176** (1999).
2. Gilmore, A., *IAU Circulars*, **7176** (1999).
3. Della Valle, M., Pasquini, L., Daou, D., and Williams, R. E., *A&A*, **390**, 155–166 (2002).
4. Mukai, K., and Ishida, M., *ApJ*, **551**, 1024–1030 (2001).
5. Orio, M., Parmar, A., Benjamin, R., Amati, L., Frontera, F., Greiner, J., Ögelman, H., Mineo, T., Starrfield, S., and Trussoni1, E., *MNRAS*, **326**, L13–L17 (2001).
6. Orio, M., Parmar, A. N., Greiner, J., Ögelman, H., Starrfield, S., and Truss, E., *MNRAS*, **333**, L11–L15 (2002).
7. Starrfield, S., Drake, J., Wagner, R. M., Butt, Y., Hauschildt, P. H., Krautter, J., Gehrz, R. D., Woodward, C. E., M. Della Valle, M. O., Mukai, K., Hernanz, M., Truran, J. W., and Evans, A. E., *AAS*, **198**, 11.09 (2001).
8. Porquet, D., and Dubau, J., *A&AS*, **143**, 495–414 (2000).
9. Burwitz, V., Starrfield, S., Krautter, J., and Ness, J.-U., *A&A*, **x**, in preparation (2002).
10. Balman, S., Krautter, J., and Ögelman, H., *ApJ*, **499**, 395–406 (1998).

XMM-Newton Observations of Classical Novae

Margarita Hernanz and Glòria Sala

Institut d'Estudis Espacials de Catalunya, IEEC-CSIC, C/Gran Capità 2-4, 08034 Barcelona, Spain

Abstract. Observations of five recent Galactic classical novae with XMM-Newton have been performed. The main goal was to monitor (at equal time intervals of around 6 months) if the sources were emitting X-rays and to get their spectral properties. The unprecedented sensitivity of the XMM-Newton instruments allows for detection of recent novae with relatively short observation times, provided that they are still emitting. The broad spectral range of the EPIC cameras, as compared with ROSAT/PSPC, allows for a better coverage of the emission. Four out of five observed sources are still emitting 2 to 4 years after outburst (in the whole range 0.3-8.0 keV), which is a much larger fraction than the one from ROSAT observations. On the other hand, the lack of information below 0.3 keV (as compared with the excellent sensitivity of ROSAT/PSPC) together with the small observation times do not permit to get an unambiguous interpretation of the data.

INTRODUCTION

Detection of X-ray emission from classical novae during their post-outburst stages provides crucial information about the nova phenomenon, as extensively discussed in the review by Krautter[6]. The analysis of the soft X-rays gives a unique insight into the remaining nuclear burning on top of the white dwarf, whereas the hard X-rays reflect the internal and/or external shocks in the ejecta.

It is theoretically predicted, and confirmed through observations, that novae return to hydrostatic equilibrium after a fraction of the accreted envelope is ejected as a consequence of the explosion. It is expected that during this phase, the hot surface of the white dwarf is seen in (soft) X-rays, provided that the envelope has already become transparent to this radiation. The duration of this phase should be, therefore, related to the nuclear burning timescale of the remaining H-rich envelope. This time has been estimated to be as long as hundreds of years, but, of course, its exact value depends on the mass of the remaining envelope which depends, between other factors, on the mass of the white dwarf (see, for instance, table 1 in Gehrz et al.[3]).

Orio et al. (see [11] and references therein) have done a systematic search of X-ray emission from classical novae in their post-outburst stages, in the ROSAT archival data. Contrary to what was expected, only very few novae were discovered as X-ray emitters shortly after their outburst. In fact only 3 novae have shown a soft X-ray spectrum: GQ Mus (Nova Mus 1983, already discovered by EXOSAT in 1983), V1974 Cyg (Nova Cyg 1992), and Nova LMC 1995. GQ Mus was the one with the longest duration of the soft X-ray emission phase (around 9 years, see [8, 14, 1]). More recent observations are reviewed in [6] (see also [12, 2] in this same volume).

One of the conclusions extracted from ROSAT observations was that the duration of the constant bolometric luminosity phase (powered by the soft X-rays) was too short to be compatible with the evolution of the envelope on a purely nuclear timescale [7], indicating some extra mechanism of mass loss, related to the interaction with the companion star and/or to an optically thick wind [5]. In addition, the spectral energy distribution was not well known in general, due to low resolution and/or the limited spectral range of the detectors; therefore, flux determinations and model parameters were not very accurate, leaving the origin of the emission unclear.

Therefore, the monitoring of a sample of novae in X-rays, with unprecedented sensitivity and spectral resolution, is necessary to better understand the post-outburst stages of these explosive phenomena. That is the goal of our current observations of recent Galactic classical novae with XMM-Newton during AO1 GO time. In this paper we present a summary of our preliminary results.

NEW NOVAE DETECTED WITH XMM-NEWTON

In table 1, we show the list of targets we have observed with XMM-Newton, and the epoch of observation. The novae chosen were the most recent ones when we presented the AO1 proposal (April 1999), except for V1141 Sco (N Sco 1997), which we chose because it had been detected by SAX ([9]), but this detection was not confirmed later ([10]) by the same authors. All the novae (except V1141 Sco) have been detected by XMM-Newton at the three epochs of observation with few exceptions: we can not conclude yet if N Sco 1998 and N Mus 1998 were emitting when they were first observed with XMM (i.e., 2 years after outburst). If the answer is no, then reestablished accretion would be the most plausible origin of their emission; if the answer is yes, they automatically would belong to the group of the other detected sources shown in table 1. A new pipeline processing of the data is under way (Hernanz and Sala, in preparation) to help clarify this point.

The sources clearly detected at the three epochs when they were observed are V4633 Sgr (N Sgr 1998) and V2487 Oph (N Oph 1998). Our results concerning the first two observations of V2487 Oph are under referee process now; we will comment here only our observations of V4633 Sgr (N Sgr 1998).

The spectral analysis of N Sgr 1998 we present has made use of the SAS sofware package for XMM-Newton data reduction (version 5.2.0) and of XSPEC from the HEASOFT package (version 11.1.0). By the moment, we have taken the filtered event list from the XMM-Newton Science Survey Centre (SSC) pipeline processing as starting points. We are now reanalyzing all the data from the beginning (i.e., from the observation data files -ODF- delivered to us) with the most recent version of the SAS (v 5.3.2); the analysis, still in progress, seems to give similar results as those presented here (Hernanz and Sala in preparation). However, there are still some differencs between data from EPIC MOS and EPIC PN cameras at low energies, which complicate the analysis of the soft X-ray emission.

The first observation of N Sgr 1998, 934 days after outburst, yielded observed count rates of 2.4×10^{-2}, 2.2×10^{-2} and 7.9×10^{-2} cts/s, for the EPIC MOS1, MOS2 and PN cameras, respectively. In figure 1 we show the EPIC PN observed spectrum. The solid line shows the best-fit model and the corresponding parameters are displayed in table 2. The global luminosity in the 0.3-8.0 keV range is 2×10^{33} erg/s, for a distance of 9 kpc. The data are compatible with blackbody emission with temperature of 47 eV (6×10^5 K), although the flux of this component is not well constrained. An additional two temperature thermal plasma model (Raymond-Smith from XSPEC) is needed (with temperatures $\sim$0.2-0.3 and $\sim$10 keV) in order to fit the intermediate (0.5 to 1 keV) and the harder energy bands. Figure 2 displays the EPIC MOS1, MOS2 and PN observed spectra. The simultaneous fit of the data from the three cameras gives a best fit model fully compatible with that of the EPIC PN alone. Therefore, although we have small statistics, we are confident that the best-fit model is representative of the data.

The second observation of N Sgr 1998, 1083 days after the outburst, had some problems (i.e., no EPIC PN data, bad data for EPIC MOS2). The detection of the nova was clear but a reasonable spectral analysis is not feasible by now. On the contrary, the data of the third observation, 1265 days after outburst, are quite good, taking into account the small observation time. The details of the best-fit model for this third observation are given in table 2. All the temperatures of the best-fit model for the third observation of N Sgr 1998 are smaller than those of the first observation (performed around one year before). This most probably indicates that the burning shell on top of the white dwarf is starting to turn off and that the shocked ejected shell is also cooling. However, we cannot exclude yet that the hard part of the emission is also related with reestablished accretion in the cataclysmic variable where the nova explosion occured, but in that case the high thermal plasma temperature at the third epoch should be similar and not smaller than the one fitting the first observation. Concerning the soft X-ray emission, it would be crucial to have good spectral response below 0.3 keV, in order to better constrain the blackbody model and to better disentangle it from the contribution from the low temperature thermal plasma component. We also are ready to analyze the grism OM data when possible (no SAS procedure allows to do it now); these data will provide hopefully the contemporaneous UV spectra of the sources, thus providing a valuable extra information.

CONCLUSIONS

The first results of our observations of recent Galactic novae with XMM-Newton show that 4 out of 5 objects are still emitting X-rays in the whole range 0.3-8.0 keV as late as 3-4 years after outburst; this is a much larger fraction that the one obtained with ROSAT[11]. It seems quite possible that the sources have emitted continuoulsy after outburst (at least from around 2 years after the explosion). It is still not clear if the novae were still emitting because of residual nuclear burning, because of reestablished accretion or both. In the cases of N Sgr 1998 and N Mus 1998, optical observations suggest that accretion is restablished quite early and that the underlying cataclysmic variable may be an intermediate polar ([4, 13]). Our observations of the X-ray emission from recent classical novae can help to understand the magnetic character of the nova, although larger observation times are needed to confirm or reject that possibility.

TABLE 1. Galactic classical novae observed during XMM-Newton Cycle 1

Target	Discovery date	Date of observation	Time after outburst	XMM-EPIC detection
N Sco 1997	June 5	Oct. 11, 2000	1224 d, 3.4 yr.	NO
(V1141 Sco)		Mar. 24, 2001	1388 d, 3.8 yr.	NO
		Sep. 7, 2001	1555 d, 4.3 yr.	NO
N Sgr 1998	March 22	Oct. 11, 2000	934 d, 2.6 yr.	YES
(V4633 Sgr)		Mar. 9, 2001	1083 d, 3.0 yr.	YES
		Sep. 7, 2001	1265 d, 3.5 yr.	YES
N Oph 1998	June 15	Feb. 25, 2001	986 d, 2.7 yr.	YES
(V2487 Oph)		Sep. 5, 2001	1178 d, 3.2 yr.	YES
		Feb. 26, 2002	1352 d, 3.7 yr.	YES
N Sco 1998	October 21	Oct. 11, 2000	721 d, 2.0 yr.	NO?
(V1142 Sco)		Mar. 24, 2001	885 d, 2.4 yr.	NO?
		Sep. 7, 2001	1052 d, 2.9 yr.	YES
N Mus 1998	December 29	Dec. 28, 2000	730 d, 2.0 yr.	YES?
(LZ Mus)		Jun. 26, 2001	910 d, 2.5 yr.	YES
		Dec. 26, 2001	1093 d, 3.0 yr.	YES

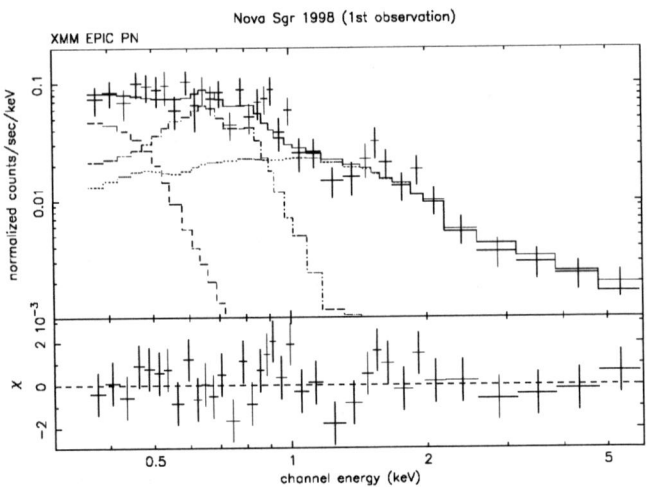

FIGURE 1. The EPIC PN spectrum of N Sgr 1998 (first observation, done 934 days after outburst). The solid line shows the best-fit model, using a blackbody plus a two temperature Raymond-Smith thermal plasma model. The lower panel shows the residuals between the data and the model in units of standard deviations.

ACKNOWLEDGMENTS

We thank the MCYT-PNAYA for funding. This work is based on observations obtained with XMM-Newton, an ESA science mission with instruments and contributions directly funded by ESA Member States and the USA (NASA). We have made use of the SIMBAD database, operated at CDS, Strasbourg, France.

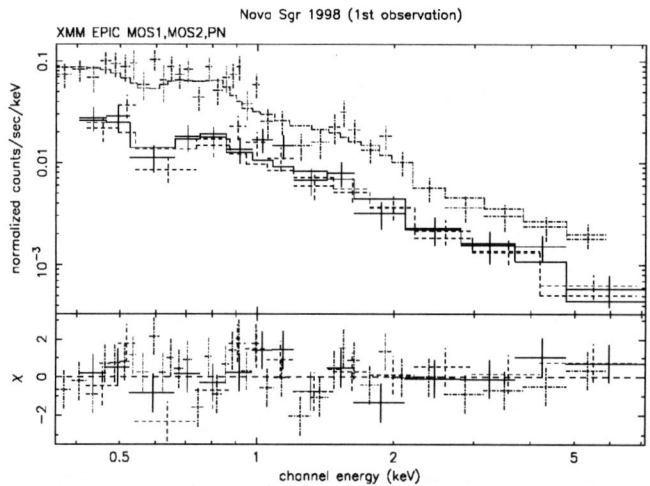

FIGURE 2. Same as figure 1, but for the three EPIC cameras. Solid: MOS1, dashed: MOS2, dot-dashed: PN.

TABLE 2. Parameters of the best-fit models to the XMM-Newton observations of N Sgr 1998. Blackbody and Raymond-Smith plasma model temperatures, as well as model flux are given. The absorption N_H is 0.1×10^{22}cm^{-2}.

Observation	kT_{BB} (eV)	kT_1, RS (keV)	kT_2, RS (keV)	$F_{(0.3-8.0)}$ keV	reduced χ^2 (dof)
1st observation (PN)	47±10	0.23±0.02	10±6	2.5	1.08 (30)
1st observation (MOS1+MOS2+PN)	53±8	0.30±0.04	11±4	2.6	1.11 (58)
3rd observation (PN)	15±0	0.16±0.02	3.7±0.9	1.9	1.29 (21)
3rd observation (MOS1+PN)	15±0	0.16±0.02	3.6±0.8	1.9	1.29 (29)

REFERENCES

1. Balman, S., Krautter, J., *Mon. Not. R.A.S.* **326**, 1441 (2001).
2. Burwitz, V., Starrfield, S., Krautter, J., Ness, J.-U., these proceedings, 2002.
3. Gehrz, R.D., Truran, J.W., Williams, R.E., Starrfield, S., *PASP* **110**, 3 (1998).
4. Lipkin, Y., Leibowitz, E.M., Retter, A., Shemmer, O., *Mon. Not. R.A.S.*, (2001).
5. Kato, M., Hachisu, I., *ApJ* **437**, 802 (1994).
6. Krautter, J., these proceedings, 2002.
7. MacDonald, J., Fujimoto, M.Y., Truran, J.W., *ApJ* **294**, 263 (1985).
8. Ögelman, H., Orio, M., Krautter, J., Starrfield, S., *Nature* **361**, 331 (1993).
9. Orio, M., Trussoni, E., Balman, S., Ögelman, H., Gallagher, J., de Martino, D., della Valle, M., González-Riestra, C., Selvelli, P., *IAU Circ.* **No. 6778** (1997).
10. Orio, M. 1999, *Physics Reports* **311**, 419 (1999).
11. Orio, M., Covington, J., Ögelman, H., *A&A* **373**, 542 (2001).
12. Orio, M., Greiner, J., Hartmann, W., Still, M., these proceedings, 2002.
13. Retter, A., these proceedings, 2002.
14. Shanley, L., Ögelman, H., Gallagher, J., Orio, M., Krautter, J., *ApJ* **438**, L95 (1995).

Photoionization as a Source of X-ray Emission from Classical Novae

Glòria Sala and Margarita Hernanz

*Institut d'Estudis Espacials de Catalunya (IEEC-CSIC),
Edifici Nexus-201, c/Gran Capità 2-4, E-08034 Barcelona, Spain*

Abstract. ROSAT observed 39 classical novae less than 10 years after outburst, but only three of them (Nova Mus 1983 (GQ Mus), Nova Cyg 1992 (V1974 Cyg) and Nova LMC 1995) were found to be soft X-ray sources (0.1-0.4 keV), as a result of steady hydrogen burning on the white dwarf's surface. V1974 Cyg also showed a harder component, and Nova Her 1991 (V838 Her) and Nova Pup 1991 (V351 Pup) were detected only in the ROSAT hard band (0.4-2.4 keV). This hard component was fitted with a thermal plasma model, and the required temperature was understood to be caused by shocks in the ejecta.

In this work, we study the possible contribution to the X-ray spectrum of the photoionization of the ejected gas shell by the central source. We use the XSTAR photoionization code to simulate the X-ray spectrum of a classical nova during the post-outburst phase. We compare these results with ROSAT/PSPC spectra of V1974 Cyg.

INTRODUCTION

In the outburst of a classical nova, a fraction of the accreted material is ejected, while the remaining envelope returns to hydrostatic equilibrium and burns hydrogen steadily. During this H-burning phase, the bolometric luminosity is constant and close to the Eddington value [1]. As the envelope expands, the photospheric radius decreases and the effective temperature increases, shifting the spectrum from optical through UV and X-rays ([2], [3]). This expected soft X-ray emission was detected only in three of all the classical novae observed with ROSAT ([4],[5]), and its spectrum was well fitted with white dwarf atmosphere models ([6], [7], [8]).

The ejected gas shell is believed to be the site of emission of the hard X-rays detected in some novae. In the case of V1974 Cyg, monitored by ROSAT during 2 years after outburst, both the soft and the hard component were present ([6],[9]). The hard component of the emission, fitted with thermal plasma models (a thermal bremsstrahlung in [9] and a Raymond-Smith model, which includes also emission lines of several elements, in [6]), was believed to have its origin in the ejecta, heated by internal or external shocks. The same possible origin was suggested for the X-ray emission of V838 Her, detected in X-rays by ROSAT only 5 days after outburst (see [10]), and for V351 Pup, observed with ROSAT/PSPC 16 months after outburst, [11]. In [12] it was shown that the interaction of different components within the ejecta could account for the heating of the material to the temperatures needed to generate the observed thermal plasma spectrum of V838 Her.

XSTAR PHOTOIONIZATION MODELS FOR CLASSICAL NOVAE

An additional source of X-ray emission can be provided by the photoionization of the ejected nebula by the soft X-ray emission originated at the white dwarf's surface. We have used XSTAR code to simulate this spectrum and to generate table models to fit observational data.

XSTAR ([14], [15]) is a numerical code that simulates the resulting spectrum of a spherical gas shell photoionized by a central X-ray source. The program determines simultaneously the state of the gas and the radiation field as a function of the distance to the central source, keeping a constant density along the gas shell. As shown in [16], when the gas is optically thin, its state, determined by its temperature and by the ionic level populations, depends only on the ionization parameter, $\xi = L/nR^2$, where R is the distance from the source, L is the central source luminosity and n is the hydrogen number density.

In previous works, where we used the XSTAR versions available at that moment (XSTAR 2.0 in [13] and XSTAR 2.1d), we showed that the photoionization was able to explain the whole ROSAT/PSPC X-ray spectra of classical novae. In those models, some recombination emission features appeared at energies higher than 1keV, which could account for the hard X-ray emission detected by ROSAT in the spectra of V1974 Cyg, V838 Her and V351 Pup. But some of those recombination features were due to an error on the XSTAR code (Tim Kallman, private communication). This has been corrected in the newest version of the program, XSTAR 2.1h, released in June 2002.

In this work, we have used the XSTAR 2.1h version to simulate the photoionization of a gas shell with the typical conditions for a classical nova. With the present version of the code, some emission features still contribute to the hard end of the main soft X-ray component of the spectrum (which has been assumed to be a blackbody in this work), but the contribution to the hard band is not so significant as shown in previous works based on oldest XSTAR versions. Since with this new release of the code the spectrum is dominated by the radiation from the central source transmitted through the gas shell and the photoionization adds only some emission features, it will be interesting in the future to simulate the spectrum resulting from the photoionization of the shell with a white dwarf atmosphere model as a central ionizing source.

To compare ROSAT/PSPC spectra of classical novae with photoionization models, a grid of 135 XSTAR spectral models has been generated. In all cases, the spectrum of the central source has been assumed to be a blackbody. Four parameters are allowed to vary within the grid: the effective temperature (T_{eff}), the hydrogen number density of the gas (n), the column density of the gas shell ($column = n\Delta R$, where ΔR is the thickness of the gas shell) and the ionization parameter ξ at the inner radius of the shell. The chemical composition of the gas in the shell was fixed to best-fit values obtained in [17], who used a photoionization model to fit optical and UV spectral data of V1974 Cyg during the nebular phase (a later study in [18] found values similar to the ones obtained in [17], although a more recent study [19] has found different abundances).

The XSPEC package ([20]) has been used to fit ROSAT/PSPC spectra with the XSTAR photoionization table models. We have compared the photoionization spectra simulated with the present XSTAR version to ROSAT/PSPC spectra of V1974 Cyg,

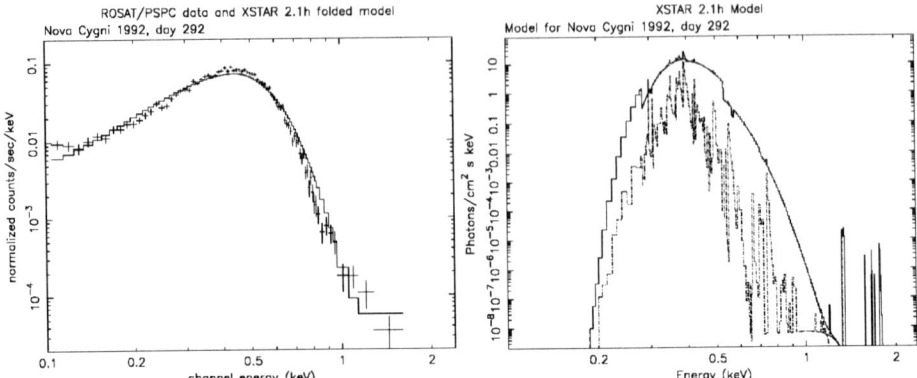

FIGURE 1. Left: ROSAT/PSPC spectrum of V1974 Cyg on day 292 after outburst; the solid line shows the XSTAR model fitted to the data, with galactic $N_H = 3.1 \times 10^{21} cm^{-2}$, $kT = 28 eV$, $n = 10^8 cm^{-3}$, column density $= 10^{18} cm^{-2}$, $\log \xi = 2.6$, $\chi^2_\nu = 6.3$. Right: XSTAR model for this particular observation; the solid line plots the total spectrum, which results from the addition of the radiation transmitted through the shell (dashed line) plus the emission from the photoionized gas in the shell (dash-dotted line).

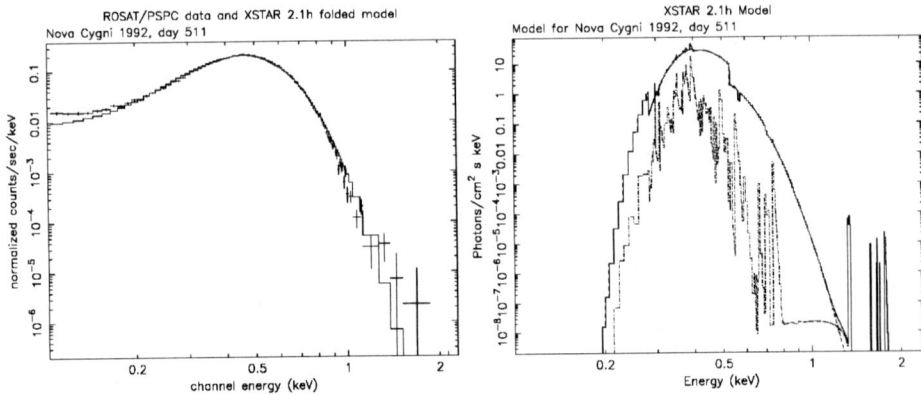

FIGURE 2. Left: ROSAT/PSPC spectrum of V1974 Cyg on day 511 after outburst; the solid line shows the XSTAR model fitted to the data, with galactic $N_H = 3.6 \times 10^{21} cm^{-2}$, $kT = 29 eV$, $n = 10^8 cm^{-3}$, column density $= 10^{18} cm^{-2}$, $\log \xi = 3.0$, $\chi^2_\nu = 7.7$. Right: XSTAR model for this particular observation; the solid line plots the total spectrum, which results from the addition of the radiation transmitted through the shell (dashed line) plus the emission from the photoionized gas in the shell (dash-dotted line).

V838 Her and V351 Pup. The new photoionization models can fit the hard X-ray excess observed in the case of V1974 Cyg (see Figures 1, 2), but cannot explain the hard X-ray emission from V838 Her and V351 Pup.

CONCLUSIONS

The spectral models generated for the X-ray emission of classical novae with the new version of the XSTAR code (XSTAR 2.1h) show that the photoionization of the ejected gas shell modulates and changes the spectrum of the soft X-ray emission from the central white dwarf, but does not contribute significantly to the hard X-ray band. Although this emission can not account for the hard X-ray emission observed by ROSAT in some cases, it adds some emission features at the hard end of the soft X-ray band that can explain the hard X-ray excess detected from V1974 Cyg. With the spectral resolution of ROSAT/PSPC it is not possible to distinguish the contribution of the photoionized shell from other models, but further studies with photoionization codes and using white dwarf atmosphere models as central ionizing source spectrum may be crucial to understand the high-resolution XMM and Chandra data.

ACKNOWLEDGMENTS

We thank Jochen Greiner for his help on the ROSAT data reduction of V1974 Cyg and we thank also the Astrophysikalishes Institut Potsdam for the hospitality offered to one of us (G. Sala). We acknowledge also Tim Kallman for his comments on the XSTAR code. This work has been supported by the Spanish MCYT.

REFERENCES

1. Starrfield, S., "Thermonuclear Processes and the Classical Nova Outburst," in *Classical Novae*, edited by M.F. Bode and A. Evans, John Wiley & Sons Ltd., New York, 1989, pp. 39-60
2. MacDonald, J., and Vennes, S., *ApJ* **373**, L51-L54 (1991).
3. MacDonald, J., "Classical Nova Evolution: Clues from Soft X-ray Emission", in *Cataclysmic Variables and Related Objects*, edited by A. Evans and J.H. Wood, Kluwer Academic Publishers, Dordrecht, 1996, pp. 281-287
4. Orio, M., *Physics Reports* **311**, 419-428 (1999).
5. Orio, M., Covington, J., and Ögelman, H., *A&A* **373**, 542-554 (2001).
6. Balman, S., Krautter, J., and Ögelman, H., *ApJ* **499**, 395-406 (1998).
7. Orio, M., and Greiner, J., *A&A* **344**, L13-L16 (1999).
8. Balman, S., and Krautter, J., *MNRAS* **326**, 1441-1447(2001).
9. Krautter, J., Ögelman, H., Starrfield, S., Wichmann, R., and Pfeffermann, E., *ApJ* **456**, 788-797 (1996).
10. Lloyd, H.M., O'Brien, T.J., Bode, M.F., Predehl, P., Schmitt, J.H.M.M., Trümper, J., Watson, M.G., and Pounds, K.A., *Nature* **356**, 222-224 (1992).
11. Orio, M., Balman, S., Della Valle, M., Gallagher, J., and Ögelman, H., *ApJ* **466**, 410-414 (1996).
12. O'Brien, T., Lloyd, H.M., and Bode, M.F., *MNRAS* **271**, 155-160 (1994).
13. Sala, G., Hernanz, M., and Greiner, J., "Photoionization Models for the X-ray Emission from Classical Novae," in *New Visions of the X-ray Universe in the XMM-Newton and Chandra Era*, edited by F. Jansen and TBD, ESA SP-488, August 2002.
14. Kallman, T.R., *XSTAR User's Guide Version 2.1h*, NASA/GSFC, June 2002.
15. Kallman, T.R., and Bautista, M., *ApJS* **133**, 221-253 (2001).
16. Trater, C.B., Tucker, W.H., and Salpeter, E.E., *ApJ* **156**, 943-951 (1969).
17. Austin, S.J., Wagner, R.M., Starrfield, S., Shore, S.N., Sonneborn, G., and Bertram, R., *AJ* **111**, 869-898 (1996).

18. Moro-Martin, A., Garnavich, P.M., and Noriega-Crespo, A., *AJ* **121**, 1636-1647 (2001).
19. Vanlandingham et al., in this volume, 2002.
20. Arnaud, K., and Dorman, B., *XSPEC User's Guide for version 11.1.x*, HEASARC, NASA/GSFC, Greenbelt, MD 20771, July 2001.

Novae in M31: How Many of them Turn into Supersoft X-Ray Sources?

Marina Orio*, Antonello Dalmazzo† and Petko Nedialkov**

*INAF - Turin Astronomical Observatory, Italy, and
Astronomy Department, U Wisconsin at Madison, USA
†INAF - Turin Astronomical Observatory, Italy
**Sophia University, Bulgaria

Abstract. A large population of supersoft X-ray sources has been detected in M31 with *ROSAT*, *XMM* and *Chandra*. More than half of these sources to are intermittent or transient. How many of them are post-outburst classical or recurrent novae? With this question in mind we compared our recent images of M31 with images taken in past years. We also cross-correlated the observed supersoft X-ray sources with a recent nova survey in M31. So far, only one supersoft X-ray source turns out to have been a classical nova. This confirms that only a small fraction, probably not exceeding 20%, of classical novae become supersoft X-ray sources for more than ≈ 2 of years. Most transient X-ray sources of this type belong to a different class and are not novae.

THE WHITE DWARF DESTINY AFTER THE NOVA OUTBURST

After a classical nova outburst, the remnant turns into an X-ray source. Initially, this is due to thermal bremsstrahlung continuum at temperatures in the range 0.1-10 keV, and emission lines in the same energy range, with luminosity $L_x = 10^{33-34}$ erg s^{-1}. This emission is probably due to shocks in interacting winds, or in the interaction between the ejecta an the circumstellar medium. Apparently rarer, but very interesting to understand the final fate of nova systems, is the luminous "SUPERSOFT" X-ray emission, predicted to be due to residual hydrogen burning in a shell. This emission is observable when the ejecta become optically thin to supersoft X-rays. We expect $L_{bol} = 10^{36-38}$ erg s^{-1}, an atmospheric continuum (in first approximation, blackbody-like) and absorption edges of the white dwarf (or emission edges for high effective temperatures).

The importance of detecting the central source in supersoft X-rays is twofold. First of all, atmospheric models can be fitted to derive the effective temperature, effective gravity and chemical composition of the underlying white dwarf. Second, the length of the supersoft X-ray phase indicates the amount of hydrogen fuel left over after each outburst, hence the likelihood that the white dwarf mass grows towards the Chandrasekhar limit. Ultimately, classical or recurrent novae that turn into supersoft X-ray sources for a significant number of years may be the progenitors of type Ia SN or neutron stars generated by accretion induced collapse. The nature of type Ia SN progenitors is an important open question.

The incidence of classical novae among observed supersoft X-ray sources is also an open question. How long does the residual hydrogen burning last and how often is the post-nova remnant observed as a supersoft X-ray source? Observations in the Galaxy indicate that only up to 20% of all classical and recurrent novae are observed as supersoft X-ray sources for more than few months, and that the supersoft X-ray phase lasts mostly for only up to 2 years. Periods of ≈ 10 years are the exception. Unlike the nebular X-ray emission, the supersoft X-ray emission of the central source is more easily detected in external galaxies of the local group, due to the high intrinsic luminosity of the WD remnant and to the low interstellar absorption away from the Galactic plane. Osborne et al. (2001) suggested that a transient source in the core of M31 may have been a classical nova that was missed by optical observers. Among 21 novae in the LMC in outburst 2-74 years before (16 of which observed only within 22 years), only 1 was observed as a supersoft X-ray source with *ROSAT* (Orio & Greiner 1999, Orio et al. 2002, this conference).

ONE "LUCKY" FINDING: A NOVA OUTSIDE THE BULGE

The supersoft X-ray source RX J0044.0+4118 was observed with *ROSAT* in July of 1991 (Supper et al. 1997). We found that in 1993 the flux had dropped by at least a factor of 4. Members of this team (see Orio et al. 2002 for preliminary results) conduct a project using the WIYN 3.5m telescope located at Kitt Peak, Arizona, to image the fields of the supersoft X-ray sources in M31. We obtain magnitudes of objects in the spatial error boxes of *ROSAT*, in different colours and usually with completeness limit $\simeq$23-24 in different filters. We observe the candidate SuSo in different bands, and as first and most important step we identify the objects with U and B excess and those that are variable. In the course of this project we found that the position of RX J0044.0+4118, $\alpha_{(2000)}$=0h 44m 4.76s and $\delta_{(2000)}$=+41° 18' 20,2", is only 1.6 arcsec distant from an object detected around 18th magnitude during observations done in September 1990, but found to be at V<23.5 in the year 2000 (Nedialkov et al. 2000). Putting together the data obtained by an international network of observers in different sites at different times, we found that the object was certainly a nova, although the data are too sparse to decide whether it a may have been slow or fast nova. The proposed optical counterpart was measured at R$\simeq$17.7 in September of 1990, and it had faded to R>19.2 when it was observed again after 70 days.

This is the first association of an optical counterpart with a luminous supersoft X-ray source in M31. The evidence shows that the associated supersoft X-ray source turned off in the third year after the outburst. We note that no other variable objects with V<23.5 were found in the *ROSAT* spatial error box. We evaluate that the probability that a classical or recurrent nova was in outburst in the *ROSAT* error box in the few years preceding the observation is very small, so the proposed identification is meaningful. Also Hα images independently obtained by Shafter and Irby (2001) confirm the nova nature of this object.

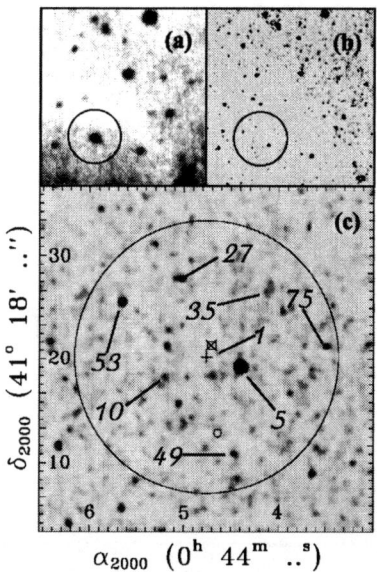

FIGURE 1. A 15" error circle around the *ROSAT* position of the supersoft X-ray source RX J0044.0+4118, observed in July 1991 and NOT detected in 1993, is plotted here on different images. The small image on the left (*a*) was obtained in B band on September 21 1990 and image *c* which is a part of *b* was obtained in the V band with the WIYN 3.5m telescope in 2000 September. The position of the nova, estimated in two different images, is indicated by the open square and by the "X".

SUPERSOFT X-RAY SOURCES IN THE SHAFTER AND IRBY SAMPLE

Shafter and Irby (2001) observed 38 novae between 1990 and 1992. All these nova fields were observed with *ROSAT* few months to three years after the outburst. Most of them were observed with *XMM* 8–10 years after the outburst (Shirey et al., 2002). RX J0044.0+4118 was the only supersoft X-ray source observed with *ROSAT* identified with a classical nova. We did not find supersoft X-ray counterparts for 15 novae (out of 16) in outburst in 1993, and for 24 novae (out of 29) observed in outburst from 1995 to 1997, whose field was observed in the year 2000 with *XMM*. We only notice that the position of Nova 1997-12 marginally overlaps with the 5" spatial errorbox of the *XMM* source XMMU J004304.4+411600, but the latter was a relatively hard and X-ray source, and too luminous to be ascribed to the nova shell. Nedialkov, Veltchev & Orio (2002) found that "Nova 21" of M31 (Sharov et al. 1998) falls in the spatial error box of the *ROSAT* source RX J0044.9+4123 (which was not a supersoft X-ray source). Since "Nova 21" most likely was not a classical nova, but either a luminous blue variable of a symbiotic nova, the coincidence is definitely interesting and could point out at copious X-ray emission in a shocked wind.

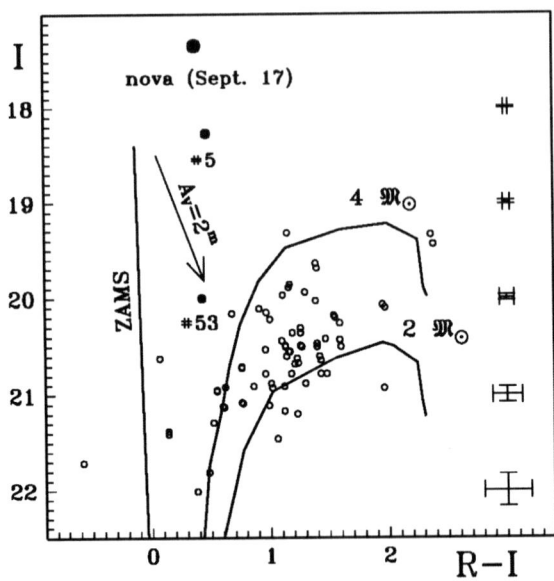

FIGURE 2. Color-magnitude diagram for 77 stars with RI photometry, observed by Nedialkov et al. (2002). The approximate location of ZAMS at the distance modulus $(m-M)_0=24.47$ of M31 is shown. The evolutionary tracks for m=2 and 4 $M_\odot$ are also plotted (see Nedialkov et al. 2002 for all the details). The arrow indicates the reddening vector corresponding to the classical extinction law of Mathis (1990), and $R_V=3.1$. The filled circles represent the magnitudes and colours of the candidate nova on Sept. 17 1990, and of the two brightest stars within the 15" error circle from the nova. On the right the average errors for different values of V are shown. Despite the spread in the data, the candidate nova clearly stands out in the plot.

CONCLUSIONS

Even if we did find one classical nova, we conclude that there is no large overlap between the population of supersoft X-ray sources in M31 and recent novae observed in this galaxy. This is well understood if the fraction of post-outburst nova white dwarf does not exceed ≈20%, as Orio et al. (2002) find for the Galaxy and the LMC. It also means that we have to understand the nature of transient supersoft X-ray sources in a different contest: most of them are not recurrent or classical novae, but a different type of objects.

REFERENCES

1. Mathis J., 1990, ARA&A, 28, 37
2. Nedialkov P.L., Veltchev T.V., Orio M., 2002, in: Meetings in Physics at the University of Sofia, ed. A. Prokyova, Heron Press, Sofia, vol.3 (in press).
3. Nedialkov P., Orio M., Birkle K., Conselice C., Della Valle M., Greiner J., Magnier E., Tikhonov N.A., 2002, A&A, in press.
4. Osborne J.P., et al., 2001, A&A, 378, 800.
5. Orio M., Covington J., & Ögelman H. 2001, 373, 542.

6. Orio M., Greiner J., 1999, A&A, 344, L13.
7. Orio M., Casalegno R., Conselice R., Greiner J., Nedialkov P.L., Tikhonov P.L., 2002, in 'X-Ray Astronomy 2000", ASP Conference Series no. 234, R. Giacconi, S. Serio and L. Stella, eds., in press.
8. Shafter A.W., Irby B.K., 2001, ApJ, 563, 749.
9. Sharov A.S., et al., 1998, AL, 24, 445
10. Shirey R., et al., 2001, A&A, 365, L195.

GAMMA RAYS FROM NOVAE

Gamma-Ray Emission from Classical Novae

Margarita Hernanz

Institut d'Estudis Espacials de Catalunya, IEEC-CSIC, C/Gran Capità 2-4, 08034 Barcelona, Spain

Abstract. The main mechanisms through which classical novae can emit gamma-rays are reviewed. Gamma-ray emission has its origin on the disintegration of some short- and medium-lived radioactive nuclei, like ^{13}N, ^{18}F, ^{7}Be and ^{22}Na. As a result of the disintegration of these nuclei, the emission has two components: lines (511, 478 and 1275 keV) and continuum (between 20-30 keV and 511 keV). The detection of gamma-rays from novae would provide important diagnostics of the explosion mechanism and, specially, a direct insight of the nucleosynthesis processes: the ^{7}Be and ^{22}Na lines at 478 and 1275 keV directly reflect the amount of these nuclei in the ejecta, whereas the 511 keV line and the continuum reflect both the amount of ^{18}F and ^{13}N) and the dynamical properties of the expanding nova envelope. The unsuccessful, to date, attempts to detect gamma-rays from novae are also reviewed, together with the prospects for future detectability.

INTRODUCTION

The potential role of novae as gamma-ray emitters was pointed out long ago by Clayton et al. ([1, 2, 28]). Gamma-ray emission from classical novae has its origin on the disintegration of some short- and medium-lived radioactive nuclei, synthesized during the explosion (see Table 1). Some isotopes, like ^{13}N (τ=862 s) and ^{18}F (τ=158 min), are short-lived. Others, like ^{7}Be (τ=77 days) and ^{22}Na (τ=3.75 yr) are medium-lived. As a result of the disintegration of these nuclei, the emission of gamma-rays from novae has two components, from the spectral point of view: lines (511, 478 and 1275 keV) and continuum (between 20-30 keV and 511 keV). There is another important radioactive nucleus synthesized during nova explosions, ^{26}Al, which is long-lived ($\tau = 10^6$ yr); this isotope emits a photon of 1809 keV when it disintegrates, but its very long lifetime makes it undetectable in individual objects. Although the origin of ^{26}Al, as deduced from the analysis of the of the 1809 keV line emission map (made with the COMPTEL instrument onboard the COMPTON Gamma-Ray Observatory, CGRO), seems more correlated with massive stars than with classical novae, the partial contribution of novae has not been ruled out completely yet (see, for instance, the reviews [32, 4] and references therein; see also the paper by Politano in these proceedings).

From the point of view of temporal behaviour, the emission can be considered as prompt (very early appearance and short duration) or long-lasting. The prompt gamma-ray emission from novae is related to $e^- - e^+$ annihilation, with the positrons coming from ^{13}N and ^{18}F decays. The released positrons annihilate and produce a line at 511 keV and a continuum below it. The continuum is produced both by positronium emission (when it is formed in triplet state) and by Comptonization of the photons emitted in the line; it has a cut-off at around 20-30 keV, related to photoelectric absorption. This

TABLE 1. Radioactive isotopes ejected by novae relevant for gamma-ray emission

Isotope	Main disintegration process	Type of γ-ray emission	Lifetime	Nova type
^{13}N	β^+–decay	511 keV line & continuum	862 s	CO and ONe
^{18}F	β^+–decay	511 keV line & continuum	158 min	CO and ONe
^{7}Be	e^-–capture	478 keV line	77 days	CO
^{22}Na	β^+–decay	1275 keV & 511 keV lines	3.75 years	ONe
^{26}Al	β^+–decay	1809 keV & 511 keV lines	10^6 years	ONe

emission has short duration and is tightly related to the conditions in the expanding envelope (opacity to gamma-rays), in addition to its content of ^{13}N and ^{18}F. The long lasting emission originates in the decay of ^{7}Be and ^{22}Na, which emit lines at 478 and 1275 keV, respectively; their fluxes and their duration directly reflect the content of these isotopes in the expanding envelope and their lifetimes.

There have been many unsuccessful attempts to detect gamma-ray emission from novae. Efforts have been made mainly to detect the ^{22}Na line, at 1275 keV, with CGRO/COMPTEL [21, 22]. Previous attempts to detect the ^{7}Be line, at 478 keV, and the 1275 keV line were made with the GRS instrument onboard the Solar Maximum Mission, SMM, satellite [29, 7].

Other attempts have concentrated on the annihilation emission (511 keV line plus continuum below it), with large field of view instruments, like the Transient Gamma-Ray Spectrometer, TGRS, onboard the WIND satellite [9, 10] and CGRO/BATSE [15], without success.

In addition to search for gamma-ray emission in particular objects, there have been attempts to look for the Galactic accumulated emission at 478 and 1275 keV, with HEAO3, CGRO/OSSE and SMM/GRS [30, 29, 7, 8]. In this case, more flux is accumulated since more sources are contributing, because the typical period between two succesive nova explosions in the Galaxy is shorter than the lifetimes of ^{7}Be and ^{22}Na.

In this paper I will review both the observations and the theoretical predictions of the gamma-ray emission from novae, starting with the 1275 keV ^{22}Na line, continuing with the 478 keV ^{7}Be line and ending with the e^--e^+ annihilation emission.

THE 1275 KEV LINE FROM ^{22}NA DECAY

In the early 90's, Starrfield et al. [34] suggested that Nova Herculis 1991 could be a good target for the CGRO satellite, because of its ^{22}Na line emission at 1275 keV. The theoretical predictions from the nova models available then were more optimistic than the present ones, mainly because of the more favourable old nuclear reaction rates, and also because of the different initial chemical composition of the envelope (see, for instance, [31, 35, 26, 27] for some recent studies of ^{22}Na synthesis in novae). So now we understand the negative results from the observations and we are in a better position to make more accurate predictions of observability with future instruments, like those onboard INTEGRAL (International Gamma-Ray Laboratory), to be launched in October 2002 (see [17, 19]).

Observations

Data accumulated from 1980 to 1987 by the gamma-ray spectrometer onboard the SMM satellite were analyzed to search for 1275 keV emission, as a result of ^{22}Na decay in the Galactic center region and in the ejecta of recent individual novae, by Leising and collaborators [29]. They did not found any evidence of 1275 keV emission from celestial original and obtained an upper limit of 3.10^{-7} $M_\odot$ on the mass of ^{22}Na ejected by the closest recent Ne-rich nova.

More recent observations were made with the CGRO/COMPTEL instrument by Iyudin et al. [21, 22]. Many recent novae were observed during the period from August 1991 to August 1993. Both CO and Ne-type novae were observed. No positive detection was obtained. The 2σ upper limit obtained for Nova Cyg 1992 was $2.3 10^{-5}$ phot/cm^2/s, which translated into a 2σ upper limit of the ^{22}Na ejected mass of $3.0 10^{-8}$ $M_\odot$. This value was in contradiction with the models of the epoch, but is fully compatible with the most recent ones, as commented above and discussed in the section on theoretical models below.

The emission from the ^{22}Na accumulated in the galactic center, as a consequence of many succesive nova explosions, has also been searched with HEAO3 [30], SMM [29], CGRO/OSSE [8] and CGRO/COMPTEL [22, 23], without positive results. Predictions for the future SPI (Spectrometer for INTEGRAL) have been also made [24, 25], updating those made in [20] to explain the upper limits obtained with HEAO3 by [30]. The observation of the cumulative 1275 keV emission from recent novae would not only be an important confirmation of the synthesis of ^{22}Na by novae, but also provide unique information about the Galactic nova distribution, since gamma-rays are not affected by interstellar extinction as are the less energetic optical photons [33].

Theoretical models

The synthesis of ^{22}Na in Ne novae has been predicted theoretically since long ago (see the reviews by Starrfield and José in this volume for details and, in particular, for the different nucleosynthesis in CO and Ne novae). The underlying massive white dwarf was first assumed to be an ONeMg white dwarf, whereas modern stellar evolutionary codes predict massive white dwarfs to be of the ONe type (see paper by García-Berro et al. in this volume and references therein).

In order to simulate the evolution of the gamma-ray spectrum of any nova model, some years ago a gamma-ray transfer code based on the Monte-Carlo technique, which allows to treat the comptonization of high energy photons without approximations, was developed (see [6] for details). The radioactive decays of ^{7}Be, ^{22}Na, ^{13}N and ^{18}F are included to generate the initial photons. Once photons are generated according to their relative isotopic abundances and rates of disintegration, their trip across the expanding ejecta is simulated by taking into account the three different interactions which affect their propagation, i.e., Compton scattering, photoelectric absorption and production of $e^- - e^+$ pairs. The cross sections for all these interactions is computed taking into account the precise composition of the ejecta. Some examples of hydrodynamical models of nova

TABLE 2. Radioactivities in novae ejecta (^{13}N and ^{18}F at 1h after T_{peak})

Nova	$M_{wd}(M_\odot)$	KE (erg/g)	^{13}N ($M_\odot$)	^{18}F ($M_\odot$)	^{7}Be ($M_\odot$)	^{22}Na ($M_\odot$)	$M_{ejec}(M_\odot)$
CO	0.8	8x10^{15}	1.5x10^{-7}	1.8x10^{-9}	6.0x10^{-11}	7.4x10^{-11}	6.2x10^{-5}
CO	1.15	4x10^{16}	2.3x10^{-8}	2.6x10^{-9}	1.1x10^{-10}	1.1x10^{-11}	1.3x10^{-5}
ONe	1.15	3x10^{16}	2.9x10^{-8}	5.9x10^{-9}	1.6x10^{-11}	6.4x10^{-9}	2.6x10^{-5}
ONe	1.25	4x10^{16}	3.8x10^{-8}	4.5x10^{-9}	1.2x10^{-11}	5.9x10^{-9}	1.8x10^{-5}

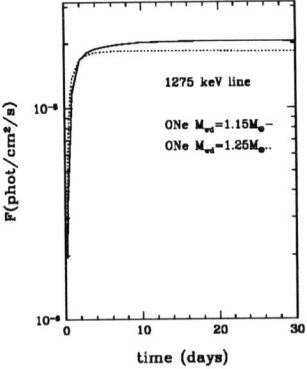 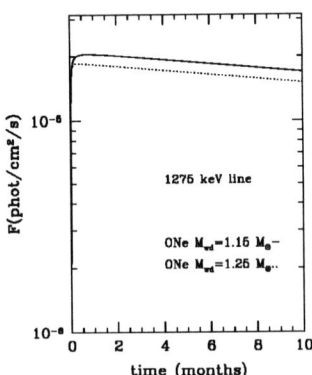

FIGURE 1. (Left). Rise phase of the light curves of the 1275 keV line, produced by ^{22}Na decay, for two ONe nova models of masses 1.15 and 1.25 M$_\odot$. (Right). Decay phase of the same light curves.

explosions (taken mainly from [26]), which have been adopted to compute the gamma-ray spectra are shown in Table 2, just to illustrate the most relevant properties of the ejecta relevant for the gamma-ray emission.

The light curves of the 1275 keV emission for the two ONe novae models of Table 2, are shown in Figure 1. The time origin in this and all the subsequent figures corresponds to the epoch of peak temperature (at the base of the envelope), well before the maximum in visual luminosity. The rise phase corresponds to stages were the expanding envelope is not completely transparent to gamma-rays yet. It lasts 20 days in the 1.15 M$_\odot$ and 12 days in the 1.25 M$_\odot$, because the more massive nova reaches transparency earlier (it is more energetic, as shown in Table 2). Typical fluxes are 2 10^{-5} phot/cm^2/s, for novae at distances of 1 kpc, and the duration is some months; the line width is around 20 keV, FWHM (see [12, 13, 6]). These values of the flux and the corresponding ^{22}Na yields are fully in agreement with the upper limits obtained by Iyudin et al. [21] (see above).

THE 478 KEV LINE FROM ^{7}BE DECAY

Observations

In the 80's, Harris et al. [7] searched for the ^{7}Be emission at 478 keV with SMM, both in individual objects and globally in the central radian of the Galaxy. This emission had been predicted theoretically by Clayton [2]. They obtained 3σ upper limits around 10^{-3} phot/cm^2/s for individual novae and 10^{-4} phot/cm^2/s for the integrated Galactic center emission.

More recently, the same authors have obtained new upper limits to the 478 keV line fluxes, with the WIND/TGRS [11]. More constraining upper limits have been obtained (i.e., 6.310^{-5} phot/cm^2/s for an individual nova, implying an upper limit to the ^{7}Be ejected mass of 6.410^{-8} M$_\odot$), but these are still fully compatible with theoretical models (see Table 2).

Theoretical predictions

The light curves of the 478 keV emission for the two CO novae models of Table 2, are shown in Figure 2. The rise phase lasts 13 and 5 days for the 0.8 and 1.15 M$_\odot$ CO novae, respectively. The difference stems from the larger opacity of the less massive white dwarf, which has a more massive envelope, with smaller expansion velocities (see Table 2). The maximum fluxes are around $1-2\ 10^{-6}$ phot/cm^2/s and the duration of this emission is some weeks; the line width is around 8 keV [6]. The huge maximum at early epochs is related with the Comptonization of the 511 keV line, which has nothing to do with ^{7}Be decay.

THE 511 KEV LINE AND THE CONTINUUM FROM ^{18}F DECAY

Theoretical predictions

In a pioneering work, Leising and Clayton [28] predicted that novae could emit gamma rays as the consequence of positron annihilation, with ^{13}N and ^{18}F being the main contributors to positrons.

Figure 3 shows the 511 keV line light curves for the four nova models in Table 2. The maxima of emission occur at days 5-6 after T_{peak}, with fluxes around 10^{-3} phot/cm^2/s and very short duration (1-2 days). The line width is around 7 keV. In the case of ONe novae, the duration is a bit longer, because of the contribution of the positrons from ^{22}Na decay. These positrons provide a much lower flux which lasts around 1 week, because later on the envelope becomes completely transparent and positrons escape freely without annihilating. There is an early maximum at around 1 hour, which is related to ^{13}N decay. This maximum has very short duration and is very dependent on the distribution of ^{13}N in the outer layers of the envelope (which are the only ones seen at these early epochs).

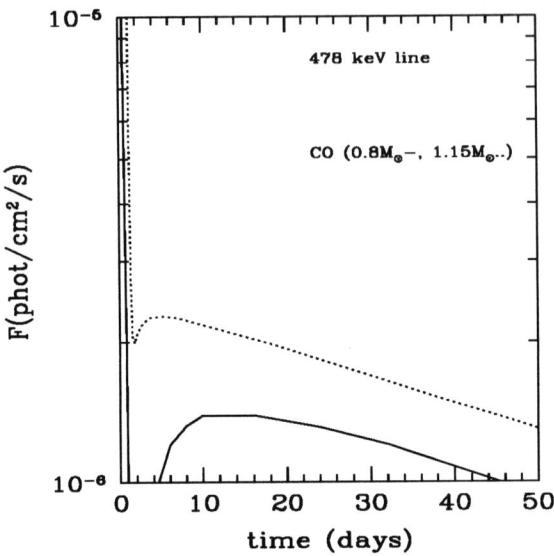

FIGURE 2. Light curve of the 478 keV line, produced by ^{7}Be decay, for two CO nova models of masses 0.8 and 1.15 $M_\odot$.

The continuum emission produced by the positronium continuum plus the Comptonization of the 511 keV line photons, has similar trends than the 511 keV line (see Figure 4). The cutoff at low energies, caused by photoelecric absorption, happens at an energy which depends on the chemical composition of the material (around 20 keV for CO and 30 keV for ONe novae). A complete analysis of the prompt gamma-ray emission from novae, with account for the effect of variable ejected masses and velocities, can be found in [18].

The e^--e^+ annihilation emission from novae strongly reflects the dynamical properties of the ejecta and, of course, its content of ^{18}F and ^{13}N (see [14]); however, the synthesis of ^{18}F during nova explosions depends on the still uncertain rate of ^{18}F+p reactions (see [3] and papers by de Séréville et al. and Bardayan et al. in this volume).

Observations

The prompt emission of gamma-rays related to e^--e^+ annihilation has very short duration and appears well before the maximum in visual luminosity, i.e., before the discovery of the nova (see Figure 5). Therefore, pointed observations of novae performed

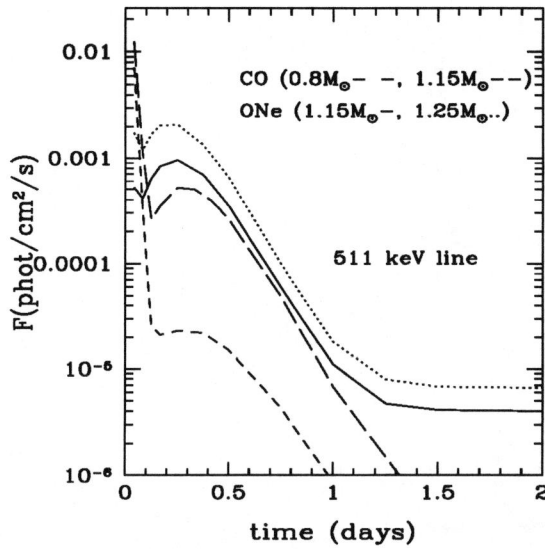

FIGURE 3. Light curve of the 511 keV line for all the models in Table 2: two CO novae of 0.8 and 1.15 $M_\odot$, and two ONe novae of 1.15 and 1.25 $M_\odot$.

after the nova discovery are unable to detect this emission. Wide field of view instruments monitoring the sky (like CGRO/BATSE and WIND/TGRS) are the only ones that have possibilities to detect the prompt gamma-ray emission from novae.

The potential of the BATSE instrument of CGRO to detect this type of emission was pointed out by Fishman et al. [5] prior to CGRO launch. The results from our systematic search in BATSE background data are presented in [15]; the search includes nearby novae that have exploded since CGRO launch. Although no positive detection has been obtained up to now, upper limits to the emitted flux have been derived for some novae; all the limits are compatible with theoretical predictions, except for Nova Sco 1992, for which either the assumed distance (0.8 kpc) is too small or the assumed mass (1.15 $M_\odot$, CO) is too large. From these analyses, still under way, it seems clear that continuum and 511 keV line observations could provide interesting constraints on nova models.

Previous searches of the 511 keV line emission from novae have been performed by [9], with WIND/TGRS. They observed five Galactic novae known to be in the broad TGRS field of view in the period between January 1995 and June 1997. The mean 3σ upper limits they obtained, for an integration time of 6 h, were around 210^{-3} phot/cm^2/s. They also used the non detection of the Galactic center during the whole period mentioned to establish an upper limit to the Galactic nova rate: 123 yr^{-1} for the

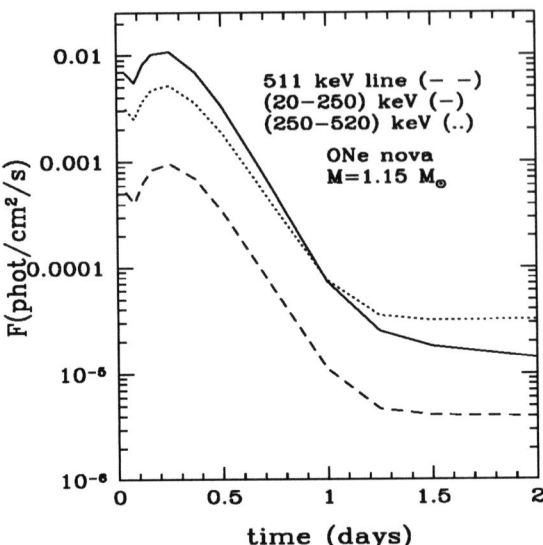

FIGURE 4. Light curve for two bands in the continuum compared with the 511 keV line light curve, for an ONe nova of 1.15 $M_\odot$.

CO novae and 238 yr^{-1} for the ONe novae. Because of the relatively poor sensitivity of TGRS, as compared with theoretical predictions, these upper limits are not constraining the Galactic nova rates, but the method is quite promising for future more sensitive instruments. It is worth mentioning that the continuum emission is more intense than the 511 keV line (see Figures 3 and 4); therefore, instruments sensitive in the hard X (soft gamma) energy range (around 20 to 300 keV), like EXIST, would be more useful than those aimed to detect the 511 keV line alone.

SUMMARY

Classical novae emit gamma-rays of different types, according to the type of nova. CO novae emit 478 and 511 keV lines, plus a continuum between 20 and 511 keV. ONe novae produce lines at 511 keV and 1275 keV, plus a continuum between 30 and 511 keV. The continuum and the 511 keV line are the most intense emissions, but their duration is very short and they appear before visual discovery. Therefore, their detection requires "a posteriori" analyses of observations made with wide FOV intruments (BATSE, TGRS). Future surveys in the soft gamma-ray domain (EXIST?)

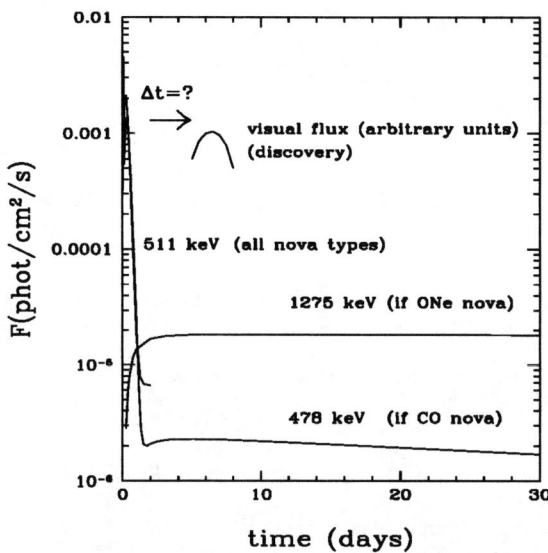

FIGURE 5. Comparison of visual and gamma light curves. Δt represents the unknown delay between visual maximum and maximum in gamma-ray prompt annihilation emission.

could provide a unique information about the Galactic nova distribution.

Detectability distances with INTEGRAL/SPI are 0.5 kpc during weeks for the 478 keV line, 1 kpc during months for the 1275 keV line, and 3 kpc during hours for the 511 keV line plus continuum (see [17, 19] for details).

The future 1275 keV emission map could also provide a direct and unique information about the distribution of ONe novae in the Galaxy.

ACKNOWLEDGMENTS

We thank the MCYT-PNAYA for funding.

REFERENCES

1. Clayton, D.D. & Hoyle, F., *ApJ* **187**, L101 (1974).
2. Clayton, D.D., *ApJ* **244**, L97 (1981).
3. Coc, A., Hernanz M., José J., Thibaud J.P., *A&A* **357**, 561 (2000).
4. Diehl R., Timmes F.X., *PASP* **110**, 637 (1998).

5. Fishman, G.J. et al., in *Gamma-Ray Line Astrophysics*, edited by P. Durouchoux and N. Prantzos, AIP Conference Proceedings 232, New York, 1991, p. 190.
6. Gómez-Gomar, J., Hernanz, M., José, J. & Isern, J., *Mon. Not. R.A.S.* **296**, 913 (1998).
7. Harris, M.J., Leising, M.D., Share, G.H., *ApJ* **375**, 216 (1991).
8. Harris, M.J., et al., *A&A Suppl.* **120**, 343 (1996).
9. Harris, M.J., Naya, J.E., Teegarden, B.J., Cline, T.L., Gehrels, N., Palmer, D.M., Ramaty, R. & Seifert, H., *ApJ* **522**, 424 (1999).
10. Harris, M.J., Teegarden, B.J., Cline, T.L., Gehrels, N., Palmer, D.M., Ramaty, R. & Seifert, H., *ApJ* **542**, 1057 (2000).
11. Harris, M.J., Teegarden, B.J., Weidenspointner, G., Palmer, D.M., Cline, T.L., Gehrels, N., Ramaty, R., *ApJ* **563**, 950 (2001).
12. Hernanz, M., Gómez-Gomar, J., José, J. & Isern, J., in 2nd *INTEGRAL Workshop "The Transparent Universe"*, edited by C. Winkler, T.J.L. Courvoisier and P. Durouchoux, ESA SP-382, Noordwijk, 1997, p. 47.
13. Hernanz, M., Gómez-Gomar, J., José, J. & Isern, J., in *Proc. Fourth COMPTON Symposium*, AIP, New York, 1997, p. 1125.
14. Hernanz, M., J.,José, J., Coc, A., Gómez-Gomar, J., Isern, J., *ApJ* **526**, L97 (1999).
15. Hernanz, M., Smith, D.M., Fishman, G.J., Harmon, A., Gómez-Gomar, J.,José, J., Isern, J., Jean, P., in *The 5th COMPTON Symposium*, edited by M.L. McConnell, J.M. Ryan, AIP Conference Proceedings 510, New York, 2000, p. 82.
16. Hernanz, M., José, J., Coc, A., Gómez-Gomar, J., Isern, J. in *The 5th COMPTON Symposium*, edited by M.L. McConnell, J.M. Ryan, AIP Conference Proceedings 510, New York, 2000, p. 97.
17. Hernanz, M., Gómez-Gomar, J., José, J., Coc, A., *Proceedings of the 4th INTEGRAL Workshop "Exploring the Gamma-Ray Universe"*, ESA SP-459, Noordwijk, 2001, p.65.
18. Hernanz, M., Gómez-Gomar, J., José, J., *New Astron. Rev.* **46**, 559, (2002).
19. Hernanz, M. et al., these proceedings, 2002.
20. Higdon, J.C., Fowler, W.A., *ApJ* **317**, 710 (1987).
21. Iyudin, A.F. et al., *A&A* **300**, 422 (1995).
22. Iyudin, A.F. et al., *Astrophys. Lett. & Comm.* **38**, 371 (1999).
23. Iyudin, A.F. et al., these proceedings, 2002.
24. Jean, P., Hernanz, M., Gómez-Gomar, J & José, J. 2000, *Mon. Not. R.A.S.* **319**, 350 (2000).
25. Jean, P., Hernanz, M., José, J., these proceedings, 2002.
26. José, J. & Hernanz, M., *ApJ* **494**, 680 (1998).
27. José, J., Coc, A. & Hernanz, M., *ApJ* **520**, 347 (1999).
28. Leising, M.D. & Clayton, D.D., *ApJ* **323**, 157 (1987).
29. Leising, M.D., Share, G.H., Chupp, E.L., Kanbach, G., *ApJ* **328**, 755 (1988).
30. Mahoney, W.A., Ling, J.C., Jacobson, A.S., Lingenfelter, R.E., *ApJ* **262**, 742 (1982).
31. Politano, M., Starrfield, S., Truran, J.W., Weiss, A. & Sparks W.M., *ApJ* **448**, 807 (1995).
32. Prantzos N., Diehl R., *Physics Reports* **267**, 1 (1996).
33. Shafter, A.W., *ApJ* **487**, 226 (1997).
34. Starrfield, S., Shore, S.N., Sparks, W.M., Sonneborn, G., Truran, J.W., Politano, M., *ApJ* **391**, L71 (1992).
35. Starrfield, S., Truran, J.W., Wiescher, M.C., Sparks, W.M., *Mon. Not. R.A.S.* **296**, 502 (1998).

ON THE FORMATION OF CVS WITH ONEMG WHITE DWARFS AND THEIR CONTRIBUTION TO THE ^{26}AL PRODUCTION IN THE GALAXY

Michael Politano

Department of Physics and Astronomy, Arizona State University, Tempe, AZ 85287-1504, U.S.A.

Abstract. In this paper, we describe a project to better estimate the amount of ^{26}Al in the Galaxy from nova outbursts on ONeMg white dwarfs. A key component of this project is the development of a quantitative model for the formation of CVs with ONeMg white dwarfs. We review existing scenarios for the formation of CVs with ONeMg white dwarfs and suggest a new evolutionary scenario for the formation of CVs with *low-mass* ONeMg-rich white dwarfs. We also discuss the observational constraints relevant to an estimate of ^{26}Al production from ONeMg novae. We suggest that one of these constraints, the diffuse Galactic 1.8 MeV gamma-ray emission mapped by COMPTEL, may be used to place upper limits on the uncertainties in the nuclear reaction rates relevant to ^{26}Al production.

INTRODUCTION

Classical novae provide the most compelling evidence for the very existence of white dwarfs (WDs) composed of oxygen, neon, and magnesium (ONeMg). This evidence comes from the unusually high enrichments of intermediate mass elements, particularly neon, in the ejecta of a number of classical novae observed over the last 15 years [1]. It is difficult to understand the strength of these enrichments in several cases unless the outburst occurred on a WD composed of ONeMg. The radioactive nucleus, ^{26}Al, is particularly of interest because of the 1.809 MeV gamma-ray it emits when it decays. This 1.809 MeV emission has been mapped by the COMPTEL instrument aboard the Compton Gamma Ray Observatory [2], [3]. This map revealed an extended diffuse emission along the Galactic plane, with a peculiar large-scale asymmetry about the Galactic center and a clumpy structure with several noticeable hot spots. The ~3 solar masses of ^{26}Al inferred from this map is believed to be produced largely in astrophysical environments associated with massive stars (e.g., type II supernovae, Wolf-Rayet stars, and massive AGB stars; cf. [3] for a nice review). Knodlseder confirmed that massive stars are the main contributors to the Galactic ^{26}Al by showing that the COMPTEL map is correlated with the COBE/DMR maps tracing free-free emission [4]. However, none of the above analyses are able to rule out a contribution to the diffuse 1.809 MeV gamma-ray emission by novae or low-mass AGB stars.

PROJECT OVERVIEW

Kolb & Politano estimated that ONeMg novae most likely contribute a total of ~ 0.15 solar mass of ^{26}Al to the Galaxy [5]. We are currently working on a project to better estimate this contribution. This project will proceed in four stages:

1) Develop a quantitative, theoretical model for the formation of CVs with ONeMg WDs that may be used in population synthesis calculations.
2) Measure experimentally improved reaction cross sections for the most significant nuclear reactions related to the production of ^{26}Al.
3) Compute a grid of (~18) ONeMg nova models using improved reaction rates.
4) Combine the production of ^{26}Al per nova calculated from model grid with a model for the classical nova population in the Galaxy, and estimate the Galactic ^{26}Al production from ONeMg novae.

Since phase 1 constitutes a key and novel aspect of this project, we focus our attention on scenarios for the formation of CVs with ONeMg WDs in the remainder of this paper.

EXISTING SCENARIOS FOR THE FORMATION OF CVS WITH ONEMG WDS

(1) Single CE scenario. In the standard model of CV formation, primaries contact their Roche lobe on their ascent of either the giant or asymptotic giant branch (e.g., figure 3 in [7]). Studies of the late stages of evolution of intermediate mass stars (8-12 solar masses) indicate that carbon is ignited non-degenerately once the core mass has reached ~1.1 solar mass. To the best of our knowledge, all previous studies of the formation of CVs using the standard model do not distinguish between CVs with ONeMg WDs and CVs with CO WDs (e.g., [6] – [10]). We propose to extract from stellar models a line demarking carbon ignition as a function of radius and mass, which effectively separates those primaries that, upon contact, have CO cores from those that have ONeMg cores. This line could then be incorporated into population synthesis calculations (for example, by including it in figure 3 in [7]), allowing the formation of CVs with ONeMg WDs to be calculated according to the standard model.

(2) Double CE scenario. A second scenario for the formation of ONeMg white dwarfs in CVs was discussed by Law & Ritter [11]. In this scenario, a primary star in the range ~8-12 solar masses fills its Roche lobe as a subgiant, before the ignition of helium in its core. The companion is a low-mass MS star. A common envelope phase ensues in which the envelope of the subgiant is lost and what remains is the He core of the primary (which becomes a He star) and the original secondary. If the mass of the helium star is between ~0.85 solar mass and ~2 - 4 solar masses, the radius of the He star can increase substantially during the shell helium burning phase

(i.e., after the exhaustion of helium in the core) [11]. This allows the exhausted He star to contact its Roche lobe, initiating a second common envelope phase. The outcome of this second CE phase is a short-period binary consisting of the core of the He star paired with a low-mass MS secondary. The He star degenerate core will be an ONeMg WD if carbon burning had begun before the star contacted its Roche lobe.

(3) Multiple CE scenario. This scenario involves an initial binary system that evolves through multiple common envelope phases, ending as a CV with an ONeMg WD of mass greater than 1.1 solar mass. The details of this scenario are described in the article by Gil-Pons & Garcia-Berro in these proceedings.

(4) The formation of low-mass ONeMg-rich WDs. Observationally-determined ejecta masses for well-studied ONeMg novae are 1 – 2 orders of magnitude *larger* than those predicted by theoretical models (cf. [12]). The cause of this discrepancy is still unknown, but the discrepancy could be reduced if the WD masses in these systems are smaller than 1.1 solar mass.

A mechanism for producing a layer of ONeMg-rich material on a carbon-oxygen (CO) WD of mass as low as 0.75 solar mass was proposed by Shara & Prialnik [13], [14]. They investigated the accretion of hydrogen onto 0.75 - 1.25 solar mass CO WDs at high accretion rates (10^{-6} solar mass/yr). Assuming conservative mass transfer, they find that a Ne- and Mg-rich outer shell with a mass up to ~ 0.1 solar mass can be built up on underlying CO WD [13].

THERMAL TIMESCALE MASS TRANSFER MODEL FOR THE FORMATION OF CVS

In the standard model of the formation of CVs (e.g., [7]), only systems in which the mass transfer will occur on a timescale longer than the thermal timescale of the secondary are included in the population. This condition is generally satisfied by demanding that the mass ratio in the system is *less than* some critical mass ratio, $q < q_{crit}$, where $q = M_{sec}/M_{WD}$ and q_{crit} is a function of M_{sec} (cf. [7]).

However, recently Schenker & King have proposed a thermal timescale mass transfer (TTMT) model for the formation of CVs [15]. Briefly, in their model, systems with mass ratios, $q > q_{crit}$, can become stable after a phase of thermal timescale mass transfer *reduces* the mass ratio *below* the critical value. Subsequent evolution will then occur as in the standard model.

To illustrate their model more clearly, evolutionary tracks for two cases in the TTMT model are shown in Figure 1 (reprinted from [15] with permission). The dashed track is for a low-mass X-ray binary system with initial masses, $M_{WD} = 1.4$ solar mass and $M_{sec} = 1.6$ solar mass. More appropriate to CVs, and to the discussion at hand, is the solid track. Here, the initial component masses are $M_{WD} = 0.7$ solar mass and $M_{sec} = 1.6$ solar mass. The mass ratio in this system is well above the mass ratio for stability against thermal timescale mass transfer (typically, $q_{crit} = 1 - 1.25$). An extended phase of mass transfer at high rates (in excess of 10^{-7} solar

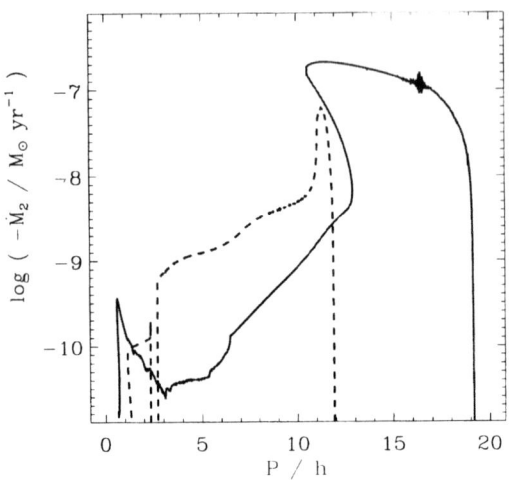

FIGURE 1. Evolution of the mass transfer rate over orbital period for two different cases of thermal timescale mass transfer (reprinted with permission from [15]).

mass/yr) occurs, the orbital period of the system "bounces" at ~10 hrs, the mass transfer rate drops, and the track eventually approaches a "standard" CV secular evolution track as the mass ratio falls below q_{crit}.

AN EVOLUTIONARY SCENARIO FOR THE FORMATION OF CVS WITH LOW-MASS ONEMG-RICH WDS

The mechanism for the formation of low-mass ONeMg-rich WDs proposed by Shara & Prialnik [13] is not possible within the standard model for the formation of CVs, since the initial mass ratio in their scheme must be greater than the critical mass ratio for stability against thermal timescale mass transfer. We propose then that the TTMT model for the formation of CVs provides an evolutionary framework in which the Shara & Prialnik scenario fits quite naturally. Indeed, the mass transfer rate history shown by the solid track in Figure 1 not only provides the necessary high mass transfer rate phase for the formation of an ONeMg-rich layer on the WD, but suggests a natural evolutionary link between supersoft X-ray sources, recurrent novae and ONeMg novae. This is depicted schematically in Figure 2. We note that this scenario also provides a natural explanation for recurrent novae with low-mass WDs (see Naylor, this volume).

The evolutionary scenario we propose for the formation of CVs with low-mass ONeMg-rich WDs depends strongly on the assumed extent of mass loss from the system during the TTMT phase. The TTMT model calculations shown in Figure 1 assume that *none* of the mass accreted by the WD remains on the WD (i.e., the accreted material is presumably ejected periodically in some sort of outburst) [15].

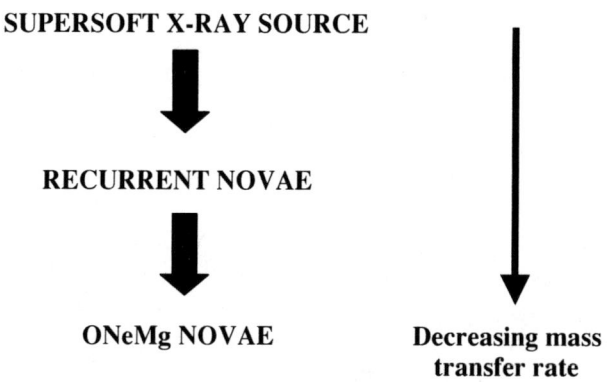

FIGURE 2. Schematic representation of the evolutionary scenario suggested by the combination of the Shara & Prialnik scenario and the TTMT model.

On the other hand, Shara & Prialnik assumed that *all* of the mass accreted by the WD remains on the WD [13]. Physically, the difference in assumptions boils down to whether the accretion of hydrogen onto the WD at rates of 10^{-6} - 10^{-7} solar mass/yr results in strong or weak flashes in the burning shell. Consequently, our proposed scenario remains speculative until these conflicting assumptions are reconciled. We point out that, as a beginning to that end, Schenker, Prialnik and Shara have each noted that the extreme assumptions in their respective models could be relaxed [16].

COMPARISON WITH OBSERVATIONS

Any calculation of the contribution of ONeMg novae to the ^{26}Al production in the Galaxy needs to be consistent with the following constraints (all of which are fairly secure, especially 3):

1) The observed fraction of classical novae that are ONeMg novae is ~1/3. This fraction has remained fairly constant over the past twenty years.

2) The ejecta masses in well-observed ONeMg novae are high (of order 10^{-3} - 10^{-4} solar mass).

3) The contribution of ONeMg novae to the ^{26}Al in the Galaxy must be *less than* the upper limit of 1 solar mass imposed by the COMPTEL map of the diffuse 1.8 MeV gamma-ray emission in the Galaxy.

We note that the third constraint is particularly relevant to the discussion of nuclear reaction rates by Iliadis in these proceedings. As noted there, reaction cross sections for reactions relevant to ^{26}Al production can be uncertain by 3 - 4 orders of magnitude. It is expected that the upper limit of 1 solar mass imposed by the COMPTEL data will provide rather stringent upper limits on the nuclear reaction rates relevant to ^{26}Al production by ONeMg novae. That is, we can clearly eliminate any range of reaction rates that leads to a total production of ^{26}Al well in excess of 1 solar mass.

SUMMARY

1) Forming CVs with high-mass ONeMg WDs (> 1.1 solar masses) according to single or multiple CE scenarios implies that a resolution of the discrepancy between theoretical and observed ejecta masses in ONeMg novae lies <u>with the nova models</u>.

2) The proposed combination of the Shara & Prialnik model for the formation of low-mass ONeMg-rich WDs and the TTMT model may provide a natural evolutionary resolution, if conflicting assumptions about the amount of matter retained by the WD can be reconciled.

3) A <u>crucially-needed</u> observation is the determination of a WD mass in an ONeMg nova.

4) The secure upper limit of 1 solar mass for the amount of ^{26}Al produced by novae from the COMPTEL gamma-ray observations may be used to constrain and possibly reduce significantly the uncertainties in nuclear reaction rates relevant to ^{26}Al production.

ACKNOWLEDGMENTS

This work was funded in part by NASA grant NAG 5-8481 and NSF grant AST-0098742 to Arizona State University.

REFERENCES

1. Starrfield, S., Truran, J.W., Wiescher, M.C., & Sparks, W.M., 1998, *MNRAS*, **296**, 502.
2. Diehl, R. et al. 1995, *A&A*, **298**, 445.
3. Prantzos, N., & Diehl, R. 1996, *Phys. Rep.*, **267**, 1.
4. Knodlseder, J. 1997, Ph.D. thesis, Paul Sabatier Univ. (Toulouse).
5. Kolb, U. & Politano, M. 1997, *A&A*, **319**, 909.
6. Politano, M. 1988, Ph.D. thesis, University of Illinois.
7. Politano, M. 1996, *ApJ*, **465**, 338.
8. deKool, M. 1992, *A&A*, **261**, 188.
9. Han, Z., Podsiadlowski, Ph., & Eggleton, P.P 1995, *MNRAS*, **272**, 900.
10. Howell, S.B., Nelson, L.A., & Rappaport, S. 2001, *ApJ*, **550**, 897.
11. Law, W.Y., & Ritter, H. 1983, *A&A*, **123**, 33.
12. Starrfield, S., these proceedings.
13. Shara, M.M., & Prialnik, D. 1994, *AJ*, **107**, 1542.
14. Shara, M.M. 1994, *AJ*, **107**, 1546.
15. Schenker, K., & King, R.E. 2002, in *The Physics of Cataclysmic Variables and Related Objects*, ASP Conf. Ser., Vol. 261, ed. B.T. Gaensicke, K. Beuermann, & K. Reinsch (ASP: San Francisco), p. 242.
16. Schenker, K., Prialnik, D. & Shara, M.M. 2002 (private communication).

Global Galactic Distribution of Classical Novae

A. F. Iyudin*, V. Schönfelder*, K. Bennett[†], R. Diehl*, W. Hermsen**, G. G. Lichti* and J. Ryan[‡]

Max-Planck-Institut für extraterrestrische Physik, Postfach 1312, 85741 Garching, Germany
[†]*Astrophysics Division, ESTEC, 2200 AG Noordwijk, The Netherlands*
**SRON-Utrecht, Sorbonnelaan 2, 3584 CA Utrecht, The Netherlands*
[‡]*University of New Hampshire, Institute for Studies of Earth, Oceans and Space, Durham NH 03284, USA*

Abstract. Classical novae are considered to be the major source of the ISM enrichment with the isotopes of ^{13}C, ^{15}N, and ^{22}Na. The latter, radioactive ^{22}Na, that decays producing penetrating 1.275 MeV γ-ray photon, could be very useful as a probe of the galactic global distribution of novae. It is expected that the accumulation of ^{22}Na from the frequent novae in the bulge will lead to an observable extended emission. Additionally, a ^{22}Na detection from the single nova can be used to verify predictions of the modern thermonuclear runaway theory (TNR) applied to classical novae. To gain a better insight into this problem we have used a two-way approach. Namely: (1) – we have studied the global galactic distribution of the 1.275 MeV γ-ray line emission assuming that it is mostly originates from decay of the novae-produced ^{22}Na; and (2) – we pursued the ^{22}Na line emission detection from recent individual novae. The combination of both approaches makes possible to tap rather uncertain galactic novae rate by comparing observations of the individual Galactic novae with the integrated ^{22}Na line emission from the disk and/or bulge population.

The COMPTEL telescope on board the Compton Gamma-Ray Observatory (CGRO), due to its combination of imaging and spectroscopic capabilities, is suitable to address the above ideas.

INTRODUCTION

Classical novae are believed to be the source of the ISM enrichment with the isotopes of ^{13}C, ^{15}N, ^{22}Na and ^{26}Al. The latter two, especially radioactive ^{22}Na, that decays producing penetrating 1.275 MeV γ-ray photon, could be used as a check probe of the contemporaneous thermonuclear runaway theory (TNR) of classical novae. To understand the spatial distribution of the ISM enrichment by the novae TNR products one has to know also the global spatial distribution of the galactic novae. Contemporary and early studies of the global galactic distribution of novae suggest the existence of at least two galactic novae populations: a disk and a bulge (or spheroid) population. The properties of these two novae populations are not yet clearly distinguished by the white dwarf type (CO or ONe), or nova speed class (fast, slow), or by some other parameter of binary systems corresponding to the classical novae (CNe) sub-population of cataclysmic variables (CVs). The full model of the global Galactic distribution of 1.275 MeV line emission includes, apart from the volume distribution of the classical novae, also the distribution of different types of novae, e.g., CO-, and ONe- type novae, or, equivalently, slow novae and fast novae. We note, that the disk and bulge novae components might be represented by different types, and/or speed classes of novae! Global Galactic distributions of novae were proposed in papers [4, 13, 15, 22, 28].

To better understand novae properties we have studied them by using a two-way approach, namely: (1) – by deriving CNe global distribution from the distribution of their accumulated ^{22}Na line emission; and (2) – by studying the ^{22}Na line emission from the old and/or more recent individual novae. Contrary to other wavelengths, in the γ-ray band the Galaxy is almost transparent, so that otherwise obscured bulge novae may be detectable up to $A_v \sim 10^3$ in the γ-ray line emission. This makes possible also to investigate the rather uncertain galactic novae rate by comparing observations of the individual Galactic novae with the integrated ^{22}Na line emission from the disk and/or bulge population. The accumulation of ^{22}Na ejected by the frequent novae in the bulge will lead to an observable extended emission. As discussed below, preliminary COMPTEL results point toward a very strong bulge component of the Galactic 1.275 MeV line emission.

The COMPTEL telescope on board the Compton Gamma-Ray Observatory (CGRO), due to its combination of imaging and spectroscopic capabilities, is suitable to verify above ideas. Our latest results on the global galactic distribution of the 1.275 MeV line emissivity clearly points to the assymmetric shape of the galactic bulge, as well, as to the enhanced rate of novae in the bulge.

INSTRUMENT AND ANALYSIS METHODS

COMPTEL, due to its combination of imaging and spectroscopic capabilities [27], provides a unique opportunity to measure line emission from point-like sources or from extended regions (e.g. the Galactic bulge).

Generally, different viewing periods covering the position of the relevant nova were combined to achieve the best possible sensitivity. Imaging and flux evaluation were done in a ±2 σ energy window around the 1.275 MeV line, where σ is the instrumental energy resolution for this line. The maximum-likelihood method was extensively used to derive individual novae fluxes (upper limits), while the maximum-entropy maps were mostly used for the ^{22}Na global Galactic distribution study. For the detection of weak sources (small fluxes) it is essential to optimise the signal-to-noise (S/N) ratio of COMPTEL, which is at the one-percent level only. A powerful tool to optimize the S/N ratio is event selection. The used selections are described by a time-of-flight (ToF) window of 115-130, the minimum Earth-horizon event angle $\zeta \geq 10^o$, and avoid the use of the "faulty" D2-modules, e.g. three D2-modules which have one out of seven photomultipliers (PMs) switched off (see [27] for a detailed instrument description).

The other important requirement for the safe detection of a source is the correct handling of the background underlying the source signal. For detection of the ^{22}Na line-emitting sources one possibility is to model the background in the line interval from the data at adjacent energy intervals (method B1). The background model derived in this way still contains systematic uncertainties due to the underlying continuum emission and small differences in the event distributions in the (χ, ψ) space. A second method derives the background model (B2) from the line energy interval itself [1, 2]. Both methods have been extensively used in the studies of the ^{22}Na line emission from novae.

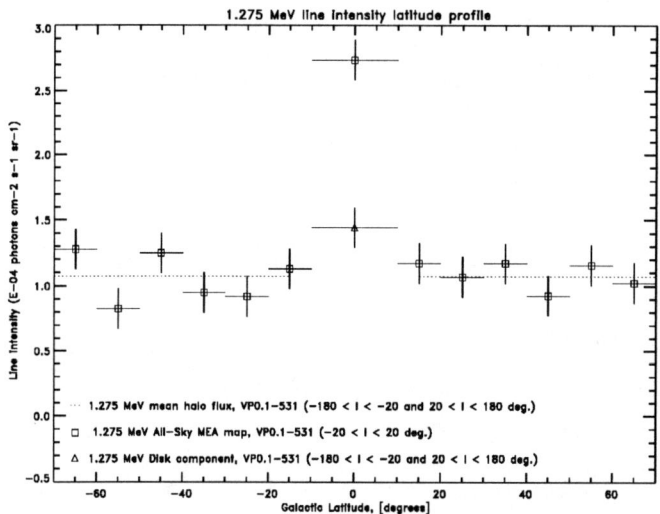

FIGURE 1. Latitude profile of the galactic bulge region in the light of the 1.275 MeV emission is shown by squares for the combination of all observations. The disk component outside the bulge is shown also by a triangle. A dotted line shows a mean halo intensity in the 1.275 MeV line.

GLOBAL GALACTIC DISTRIBUTION OF THE ^{22}NA

COMPTEL results on the global galactic distribution of the ^{22}Na were derived by constructing longitude and/or latitude profiles of the 1.275 MeV line emissivity extracted from the maximum-entropy map. The map itself was made for the combination of all COMPTEL observations till the 2nd CGRO reboost. The latitude profile is shown in Fig. 1.

The bulge component clearly sticks out from the distributions shown. The error bars plotted in the latitude profile were derived from the scatter of the intensity values outside the bulge region.

The main results on the global galactic distribution of the ^{22}Na line emission can be summarized as follows:

• COMPTEL's 1.275 MeV line intensity profiles point toward a bulge shape with the ratio of the major-to-minor bulge axis of ~ 2. This bulge shape is consistent with the bulge model [8, 9] derived from the COBE measurements of the Galactic IR-emissivity;

• Assuming a yield of the ^{22}Na in nova is between 3×10^{-9} $M_\odot$ and 1.2×10^{-8} $M_\odot$ as was modelled in [14, 21] we derive the range for the nova rate in the Galactic bulge as, 20.5 yr$^{-1} \leq R_{CN} \leq 82$ yr^{-1},

where the lower bound value is consistent with the normally quoted [6, 13].

• Further assuming that Galactic bulge novae similar to the M31 case comprises 75 % of all Galactic novae we evaluate the Galactic disk nova rate at ~ 7 yr^{-1}, compared with the value of 5 ± 2.5 yr^{-1}, which was evaluated in [5].

• If one assumes that all bulge 1.275 MeV emission is due to the ONe-type novae, then the space density of active classical novae systems in the bulge derived under this

assumption is very high and therefore unlikely, namely
1.4×10^{-5} pc$^{-3} \leq D_{CN} \leq 5.8 \times 10^{-5}$ pc^{-3},
i.e. more than order of magnitude higher than the value favoured in [26].

• From the above it follows that the bulge novae have to be represented by mostly slow, CO-type novae. This conclusion is consistent with that of [7] and supported by our ^{22}Na detection from very slow nova (see below).

STUDY OF INDIVIDUAL NOVAE

The classical nova outburst has been modelled as a thermonuclear runaway in the accreted hydrogen-rich envelope of the white dwarf companion of a close binary system, e.g. [20, 21, 29]. In general, observations of novae support such models [10, 11].

It is currently believed that novae may be an important source of Galactic ^{22}Na [3]. ^{22}Na decays with a 3.75 yrs life-time to a short lived excited state of ^{22}Ne at 1.275 MeV. In addition to the ^{22}Na line novae are prolific producers of 511 keV γ-ray line emission, which accompanies the decay of the β^+-unstable products of nucleosynthesis in novae, as well as of 478 keV emission originating from the ^{7}Be decay to ^{7}Li. Until now, there are no positive detections of the lines, at 511 keV and 478 keV [12, 24, 25], but the situation with the 1.275 MeV line detection seems to be improving [17, 18, 19].

It is still a common belief that detectable amounts of ^{22}Na are synthesized almost exclusively in the high-mass, fast ONe novae, which have enough Ne seed nuclei. Unfortunately, attempts to measure the fluxes $F_{1.275}$ from a number of recent ONe novae seems to counter this common belief. These results can be summarized as follows: (i) only upper limits were derived with COMPTEL at the level of ($\sim 2 \times 10^{-5}$ photons/(cm^{2}s)) for the ^{22}Na γ-ray line from the studied ONe novae (V693 CrA, V1370 Aql, QU Vul, V838 Her, NSgr 1991, NSct 1991, NPup 1991, V1974 Cyg, NCir 1995, NAql 1995, NCen 1995, NCru 1996 and V382 Vel) [16, 17]; (ii) upper limits from other γ-ray telescopes are usually less constraining [12, 16, 23, 24]; (iii) COMPTEL derived upper limits on the 1.275 MeV fluxes appear to be in general agreement with theoretical predictions [20, 21], provided the adopted estimates of distances to these novae are correct. Thus, all known attempts to detect the 1.275 MeV γ-ray line emission from ^{22}Na produced in the *fast* ONe novae have failed. The only likely positive detection of such emission was found to be related to the *very slow* CO-type Nova Cas 1995 (V723 Cas).

To distinguish between a possible transient source with a 1.275 MeV line emission and a nova with a line due to the decaying ^{22}Na, we have used the method described in more detail in [18]. Namely, we have measured the decay curve of the nova-produced ^{22}Na in the 1.275 MeV line emission, and compared it with a model calculation. This method of following the nova light-curve in the 1.275 MeV line emission for more than 4 years (Fig. 1 in [19]) was crucial in establishing what we claim to be a tentative detection of the ^{22}Na 1.275 MeV line emission from the slow nova NCas 1995. The relative distribution of the $F_{1.275}$ data points (Fig. 1 in [19]) is interpreted as an evidence of the shell ejected by the NCas 1995 first becoming transparent to γ-rays (an initial increase of $F_{1.275}$ at early dates) and, after that how the ^{22}Na synthesized during the nova explosion decayed at later times. An imaging analysis performed in the 1.275 MeV line using all the data yielded a total significance of the NCas 1995 detection of $\sim 4\ \sigma$ (statistical uncertainties only). Based on the fitting results one may state that the most probable evolution of the NCas 1995 flux in the 1.275 MeV line was consistent indeed with that of the slowly expanding shell with a total mass of ^{22}Na of

$$X_{22} \sim 5.4 \times 10^{-8} \left(\frac{D}{2.4 kpc}\right)^2.$$

More discussion of this result is given in our papers [17-19].

ACKNOWLEDGMENTS

The COMPTEL project is supported by the Bundesministerium für Bildung und Forschung / Deutsches Zentrum für Luft- und Raumfahrt (BMBF / DLR), through DLR grant 50 QV 9096 8. AFI acknowledges financial support from the German BMBF through the DLR grant 50 OR 0002.

REFERENCES

1. Bloemen, H. et al., *ApJSS* **92**, 419 (1994).
2. Bloemen, H. et al., *Proc. 5th Compton Symposium*, (1999).
3. Clayton, D.D., & Hoyle, F., *ApJ* **187**, L101 (1974).
4. Dawson, P.C., & Johnson, R.G., *J.R.Astron.Soc.Can.* **88**, 369 (1994).
5. Della Valle, M., & Duerbeck, M., *A&A* **271**, 175 (1993).
6. Della Valle, M., & Livio, M., *A&A* **286**, 786 (1994).
7. Della Valle, M., & Livio, M., *ApJ* **506**, 818 (1998).
8. Dwek, E., et al., *ApJ* **445**, 716 (1995).
9. Freudenreich, H.T., *ApJ* **492**, 495 (1998).
10. Gallagher, J.S. & Starrfield, S., *ARAA* **16**, 171 (1978).
11. Gehrz, R.D., Truran, J.W., Williams, R.E. and Starrfield, S., *PASP* **100**, 3 (1998).
12. Harris, M.D., et al., *ApJ*, **542**, 1057 (2000).
13. Hatano, K., Branch, D., Fisher, A., Starrfield, S., *MNRAS* **290**, 113 (1997).
14. Hernanz, M., Gomez-Gomar, J., Jose, J., Coc, A., Isern, J., *Astrophysical Letters and Communications* **38**, 407 (1999).
15. Higdon, J.C., Fowler, W.A., *ApJ* **317**, 710 (1987).
16. Iyudin, A.F., Bennett, K., Bloemen, H., et al., *A&A* **300**, 422 (1995).
17. Iyudin, A.F., Proceedings of the 10th Workshop on "Nuclear Astrophysics", Ringberg Castle, Tegernsee, Germany, March 20-25, i 2000, *MPA/P12* 118 (2000).
18. Iyudin, A.F., Diehl, R., Lichti, G.G., et al., in Proc. of the 4th INTEGRAL Workshop, Alicante, Spain, September 04-08, 2000, ESA SP-459, 41 (2001a).
19. Iyudin, A.F., Schönfelder, V., Strong, A.W., et al., in Proc. of the Workshop "Gamma-Ray Astronomy 2001", AIP CP-587, 508 (2001b).
20. Jose, J., Hernanz, M., *ApJ* **494**, 680 (1998).
21. Jose, J., Coc, A., Hernanz, M., *ApJ* **520**, 347 (1999).
22. Kent, S.M., Dame, T.M., Fazio, G., *ApJS* **127**, 131 (1991).
23. Leising, M.D., Share, G.H., Chupp, E.L. & Kanbach, G., *ApJ* **328**, 755 (1988).
24. Leising M.D., Clayton D.D., The L.-S., et al., in AIP Conf. Proc., 280, 137 (1993).
25. Mahoney, W.A., et al., *ApJ* **262**, 742 (1982).
26. Patterson, J., *ApJS* **54**, 443 (1984).
27. Schönfelder, V., Aarts, H., Bennett, K., et al., *ApJS* **86**, 657 (1993).
28. Shafter, A.W., *ApJ* **487**, 226 (1997).
29. Starrfield, S., Sparks, W.M. & Truran, J.W., *ApJS* **28**, 247 (1974).

A new experiment for the determination of the ^{18}F(p,α) reaction rate at nova temperatures

N. de Séréville*, A. Coc*, C. Angulo†, M. Assunção*, D. Beaumel**,
B. Bouzid‡, S. Cherubini†, M. Couder†, P. Demaret†,
F. de Oliveira Santos§, P. Figuera¶, S. Fortier**, M. Gaelens†,
F. Hammache‖, J. Kiener*, D. Labar††, A. Lefebvre*, P. Leleux†,
A. Ninane†, M. Loiselet†, S. Ouichaoui‡, G. Ryckewaert†, N. Smirnova‡‡,
V. Tatischeff* and J.-P. Thibaud*

*CSNSM, CNRS/IN2P3/UPS, Bât. 104, 91405 Orsay Campus, France
†CRC and FYNU, UCL, Chemin du Cyclotron 2, B-1248 Louvain La Neuve, Belgium
**Institut de Physique Nucléaire, CNRS/IN2P3/UPS, 91406 Orsay Campus, France
‡USTHB, B.P. 32, El-Alia, Bab Ezzouar, Algiers, Algeria
§GANIL, B.P. 5027, 14021 Caen Cedex, France
¶INFN–Laboratori Nazionali del Sud, Via S. Sofia, 44 - 95123 Catania, Italy
‖GSI mbH, Planckstr. 1, D-64291 Darmstadt, Germany
††Unite de Tomographie Positron, UCL, Chemin du Cyclotron 2, B-1248 Louvain La Neuve, Belgium
‡‡Instituut voor Kern en Stralingsfysika, Celestijnenlaan 200D, B-3001, Leuven, Belgium

Abstract. The ^{18}F(p,α)^{15}O reaction was recognized as one of the most important for gamma ray astronomy in novae as it governs the early 511 keV emission. However, its rate remains largely uncertain at nova temperatures. A direct measurement of the cross section over the full range of nova energies is impossible because of its vanishing value at low energy and of the short ^{18}F lifetime. Therefore, in order to better constrain this reaction rate, we have performed an indirect experiment taking advantage of the availability of a high purity and intense radioactive ^{18}F beam at the Louvain La Neuve RIB facility. We present here the first results of the data analysis and discuss the consequences.

INTRODUCTION

Gamma–ray emission from classical novae is dominated, during the first hours, by positron annihilation resulting from the beta decay of radioactive nuclei. The main contribution comes from the decay of ^{18}F (half–life of 110 mn) and hence is directly related to ^{18}F formation during the outburst. (See the astrophysical discussions in references [1, 2, 3] and by Hernanz in these proceedings.) A good knowledge of the nuclear reaction rates of production and destruction of ^{18}F is required to calculate the amount of ^{18}F synthesized in novae and the resulting gamma–ray emission. The rate (see ref. [4]) relevant for the main mode of ^{18}F destruction (i.e, through ^{18}F(p,α)^{15}O) has been the object of many recent experiments[5, 6] (see also Bardayan in these proceedings and refs. in [3]). However, this rate remains poorly known at nova temperatures (lower than 3.5×10^8 K) due to the scarcity of spectroscopic information for levels near the proton threshold in the compound nucleus ^{19}Ne. This uncertainty is directly related to the unknown proton widths (Γ_p) of the first three levels (E_x, J^π = 6.419 MeV, 3/2$^+$; 6.437 MeV, 1/2$^-$ and 6.449 MeV, 3/2$^+$). The tails of the corresponding resonances (at respectively E_R = 8 keV, 26 keV and 38 keV) can dominate the astrophysical factor in the relevant energy range[3]. As a consequence of these nuclear uncertainties, the ^{18}F production in nova and the early gamma–ray emission is uncertain by a factor of 300[3]. This supports the need of new experimental studies to improve the reliability of the predicted annihilation gamma–ray fluxes from novae.

EXPERIMENT

A direct measurement of the relevant resonance strengths is impossible because they are at least ten orders of magnitude smaller than the weakest directly measured one (at E_R = 330 keV;[7] and Bardayan, these proceedings) due to Coulomb barrier penetrability. Hence, we used an indirect method aiming at the determination of the one nucleon spectroscopic factors (S) in the analog levels of the mirror nucleus (^{19}F) by a neutron transfer reaction: D(^{18}F,p)^{19}F. (Analog, levels expected to have similar nuclear properties have been identified in ^{19}F and ^{19}Ne spectra[8].) From the spectroscopic factors it is possible to calculate the proton widths through the relation $\Gamma_p = S \times \Gamma_{s.p.}$ where $\Gamma_{s.p.}$ is the single particle width readily obtained from a model. The main reason for the choice of a transfer reaction is the much higher reaction cross-section as compared to the direct proton capture. The spectroscopic factors, S, are extracted from the angular distribution of the escaping nucleon via the relation:

$$\left(\frac{d\sigma}{d\Omega}\right)_{exp} = C^2 S \left(\frac{d\sigma}{d\Omega}\right)_{DWBA} \quad (1)$$

Where the $(d\sigma/d\Omega)_{exp}$ is the experimental angular distribution of the protons from the D(^{18}F,p)^{19}F reaction while $(d\sigma/d\Omega)_{DWBA}$ is the theoretical one (Distorted Wave Born Approximation) and C^2 is a known coefficient.

Since ^{18}F is a short lived (110 mn) radioactive isotope, it cannot be used as a target. It must be first produced, then accelerated and directed to the deuterium target (inverse kinematics). We performed the experiment at the *Centre de Recherche du Cyclotron* in Louvain–La–Neuve (Belgium) where such a beam has been developed. The ^{18}F is produced through the ^{18}O(p,n) reaction, chemically extracted to form CH$_3^{18}$F molecules, transferred to the cyclotron source[9] and accelerated to 14 MeV. The targets are made of deuteriated polypropylene (CD$_2$) of $\approx$100 μg/cm^2 thickness. For the energy considered here (1.4 MeV in the center of mass), the deuteron and the outgoing proton are both below the Coulomb barrier. The major advantages is a reduction of the contribution of compound-nucleus reactions leading to a better extraction of spectroscopic factors. The experimental setup is depicted in Figure 1. It consists of two silicon multistrip detectors composed of sectors with 16 concentric strips (of 5 mm width) built by the Louvain–La–Neuve and Edinburgh collaboration[10]. They measure the angle (strip number), energy and time of flight (for particle identification) of the particles. One, LAMP, is positioned 9 cm upstream from the target; it consists of 6 sectors forming a conical

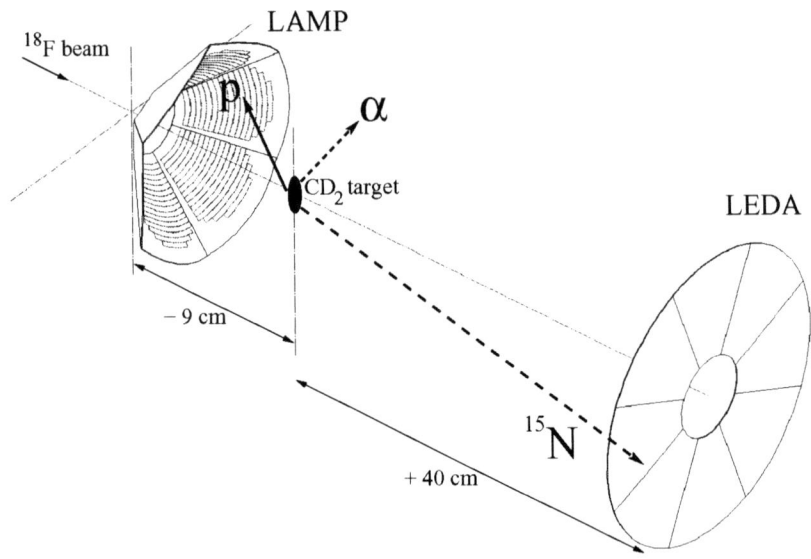

FIGURE 1. Experimental setup.

shape to optimize angular coverage. With such a geometry, it covers laboratory angles between 115° and 160° i.e. forward center of mass angles between 12° and 40° providing a good acceptance for protons in the domain of interest for the differential cross section. Indeed, the proton angular distribution as measured in LAMP is the $(d\sigma/d\Omega)_{exp}$ term in eq. 1. The other detector, LEDA, is made up of 8 sectors forming a disk positioned 40 cm downstream from the target and is used for background reduction and normalization. The levels of interests are situated high above the alpha emission threshold (at 4.013 MeV) and their almost exclusive decay mode is through $^{19}F^* \rightarrow ^{15}N+\alpha$. Hence, to reduce background, we required coincidences between a proton in LAMP and a ^{15}N (or α discriminated by time of flight) in LEDA. Following Monte Carlo simulations, the exact positions of the two detectors have been chosen to optimize resolution and acceptance. The proton detection efficiency is found to be 24% and is only slightly reduced to 19% when the coincidence condition is applied. Rutherford elastic scattering of ^{18}F on Carbon from the target, detected in LEDA, provide the (target thickness) × (beam intensity) normalization.

RESULTS

During the 7 days experiment, 15 bunches of $\lesssim 1$ Ci of ^{18}F were produced providing each a mean beam intensity of 5×10^6 particles per second over a period of ≈ 2 hours. The beam contamination (by ^{18}O) was found to be smaller than 10^{-3}. Thanks to the kinematics, at this low energy, only light particles (p and α from D(^{18}F,p)^{19}F and D($^{18}F,\alpha$)^{16}O)

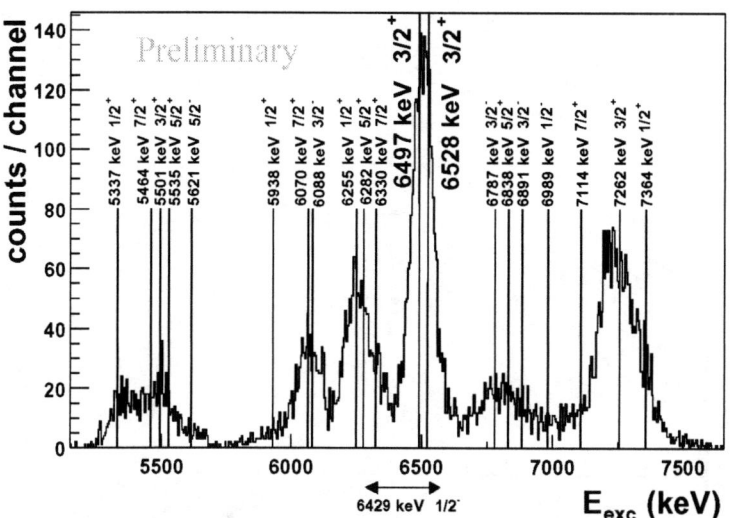

FIGURE 2. Reconstructed ^{19}F spectrum (corresponding to 65% of the total statistics) showing the two $3/2^+$ levels of astrophysical interest around 6.5 MeV of excitation energy.

can reach LAMP while the coincidences with LEDA provide a further selection. The excitation energy of the decaying ^{19}F levels can be kinematically reconstructed from the energies and angles of the detected protons and the known beam energy. The corresponding spectrum is represented in Figure 2 where vertical lines represent the known position of the ^{19}F levels. The resolution is not sufficient to separate the various levels but the two $3/2^+$ levels of interest at 6.497 and 6.528 MeV (the analogs of the $3/2^+$ levels in ^{19}Ne) are well separated from the other groups of levels. There is no peak corresponding to the $1/2^-$ level because it is so broad ($\Gamma_T = 280$ keV) that it cannot be disentangled from the background. The angular distribution, $(d\sigma/d\Omega)_{exp}$, obtained from the data corresponding to the 6.5 MeV peak, i.e. the $3/2^+$ levels, is in good agreement[11] with the theoretical one $(d\sigma/d\Omega)_{DWBA}$ (using nuclear potentials from ref. [12]) providing evidence that the analysis is reliable (e.g. negligible compound nucleus contribution and $\ell = 0$ transferred angular momentum). Since the two $3/2^+$ levels are not resolved, eq. 1 gives the *sum* of the two spectroscopic factors: $S_1 + S_2 \approx 0.2$. The important consequence of this preliminary value is that the contribution of these resonances to the rate *cannot* be neglected but that the nominal rate ($S_1 = S_2 \approx 0.1$) used in gamma–ray flux calculations is not ruled out. However, the extreme case where $S_1 \approx 0.2$, $S_2 = 0$ and $S_1 = 0$, $S_2 \approx 0.2$ have also to be considered to obtain upper and lower rate limits. Figure 3 shows the present reduction on ^{18}F$(p,\alpha)^{15}$O rate uncertainty brought by this experiment. Hopefully, progress in the data analysis (energy calibration and normalization) will further reduce this uncertainty but new experiments are required to obtain a reliable reaction rate for nova gamma–ray flux calculations.

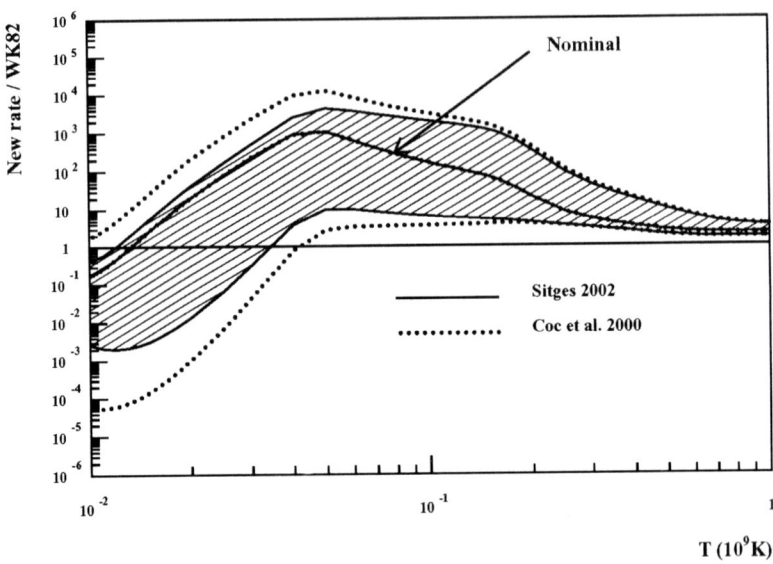

FIGURE 3. Present reduction on rate uncertainties (hatched area) brought by the experiment compared with previous limits[3]. (Ratios are with respect to the Wiescher and Kettner rate[4].) Note that part of the remaining uncertainty is due to the $1/2^-$ resonance.

ACKNOWLEDGMENTS

We thank Alan Shotter and his team for allowing us to use the LEDA and LAMP detectors from the Louvain-La-Neuve and Edinburgh collaboration.

REFERENCES

1. Gómez–Gomar, J., Hernanz, M., José, J. and Isern, J., *MNRAS* **296**, 913 (1998).
2. Hernanz, M., José, J., Coc, A., Gómez–Gomar, J. and Isern, J., *Astrophys. J.* **526**, L97 (1999).
3. Coc, A., Hernanz, M., José, J. and Thibaud, J.P., *Astron. Astrophys.* **357**, 561 (2000).
4. Wiescher M. and Kettner, K.U., *Astrophys. J.* **263**, 891 (1982).
5. Graulich, J.-S., Cherubini, S., Coszach R., et al., *Phys. Rev.* **C63**, 011302(R) (2000).
6. Bardayan, D.W., Blackmon, J.C., Bradfield–Smith, W. et al., *Phys. Rev.* **C63**, 065802 (2001).
7. Graulich, J.-S., Binon, F., Bradfield–Smith, W. et al., *Nucl. Phys.* **A626**, 751 (1997).
8. Utku, S., Ross, J.G., Bateman, N.P.T. et al., *Phys. Rev.* **C57**, 2731 and **C58**, 1354 (1998).
9. Cogneau, M., Decrock, P., Gaelens, M., Labar, D. et al., *Nucl. Inst. and Meth.* **A420**, 489 (1999).
10. Davinson, T., Bradfield–Smith, W. Cherubini, S. et al., *Nucl. Inst. and Meth.* **A454**, 350 (2000).
11. de Séréville, N. et al., in preparation and in "7th Internal Symposium on Nuclei in the Cosmos", proceedings to appear in *Nucl. Phys. A*.
12. de López, M.E.O., Rickards, J. and Mazari, M., *Nucl. Phys.* **51**, 321 (1964).

Study of the ^{18}F(p,α)^{15}O Reaction at Energies Relevant for ^{18}F Nucleosynthesis in Novae

D. W. Bardayan*, J. C. Batchelder[†], J. C. Blackmon*, A. E. Champagne**,
T. Davinson[‡], R. Fitzgerald**, W. R. Hix*,[§], C. Iliadis**, R. L. Kozub[¶],
Z. Ma[§], S. Parete-Koon*,[§], P. D. Parker[||], N. Shu*,[††], M. S. Smith* and
P. J. Woods[‡]

*Physics Division, Oak Ridge National Laboratory, Oak Ridge, Tennessee 37831
[†]UNIRIB, Oak Ridge Associated Universities, Oak Ridge, Tennessee 37831
**Dept. of Physics and Astronomy, Univ. North Carolina, Chapel Hill, North Carolina 27599
[‡]Dept. of Physics and Astronomy, Univ. Edinburgh, Edinburgh EH9 3JZ, United Kingdom
[§]Department of Physics and Astronomy, University of Tennessee, Knoxville, Tennessee 37996
[¶]Physics Department, Tennessee Technological University, Cookeville, Tennessee 38505
[||]A. W. Wright Nuclear Structure Laboratory, Yale University, New Haven, Connecticut 06520
[††]Chinese Institute for Atomic Energy, Beijing, 102413, People's Republic of China

Abstract. Production of the radioisotope ^{18}F in novae is severely constrained by the rate of the ^{18}F(p,α)^{15}O reaction. A resonance at $E_{c.m.} = 330$ keV may strongly enhance the ^{18}F(p,α)^{15}O reaction rate, but its strength has been very uncertain. We have determined the strength of this important resonance by measuring the ^{18}F(p,α)^{15}O cross section on- and off- resonance using a radioactive ^{18}F beam at the ORNL Holifield Radioactive Ion Beam Facility. We find that its resonance strength is 1.48 ± 0.46 eV, and that it dominates the ^{18}F(p,α)^{15}O reaction rate over a wide range of temperatures characteristic of novae.

Nova explosions are some of the most violent events in the universe, exceeded in energy release only by supernovae and gamma-ray bursts [1]. Despite intensive efforts to understand the nova mechanism, significant discrepancies exist between the results of nova models and observations for many global properties such as the ejected envelope mass [1, 2]. A further constraint on nova models could come from observations of the gamma rays emitted from nova ejecta [3, 4]. Missions, such as the soon to be launched INTEGRAL observatory and the planned Advanced Compton Telescope, promise to provide us with the most detailed pictures of the gamma-ray emission from novae ever available. To interpret these observations, however, we must know the relevant thermonuclear reactions rates that affect radioisotope production.

Novae emit gamma rays during the first several hours after the explosion predominantly at energies of 511 keV and below [5]. This emission is produced by electron-positron annihilation in the expanding envelope and the subsequent Comptonization of the resulting gamma-ray photons. Of the possible positron sources, the decay of ^{18}F is the most important because of the relatively large ^{18}F abundance, and because the relatively long length of the ^{18}F half-life ($t_{1/2} = 109.8$ min.) enables positrons to be emitted after the expanding envelope becomes transparent to gamma-ray radiation [5]. The amount of ^{18}F produced (and thus the flux of emitted gamma rays) is severely constrained by its destruction rate via the $^{18}\text{F}(p,\alpha)^{15}\text{O}$ reaction in the burning shells. Recent studies have found that the uncertainties in the $^{18}\text{F}(p,\alpha)^{15}\text{O}$ reaction rate result in a factor of ~ 300 variation in the amount of ^{18}F produced in models [6]. It is difficult to say whether gamma-ray observations of ^{18}F are feasible without a more precise value of the $^{18}\text{F}(p,\alpha)^{15}\text{O}$ reaction rate.

The $^{18}\text{F}(p,\alpha)^{15}\text{O}$ reaction rate may be dominated at nova temperatures (0.1 - 0.4 GK) by a resonance at $E_{c.m.} = 330$ keV [7, 8] that arises from a $J^\pi = \frac{3}{2}^-$ level in ^{19}Ne at $E_x = 6.741$ MeV [7]. Previous estimates of the contribution of this resonance, however, relied on assumptions for the single-particle spectroscopic factor (θ_p^2), upon which the rate depends linearly, and which may be incorrect by an order of magnitude or more [6]. The only experimental constraint on the strength of this resonance comes from a measurement of the energy spectrum of alpha particles emitted when a thick (275 μg/cm^2) polyethylene (CH$_2$) foil was bombarded with a 10 MeV ^{18}F beam [9]. That study, however, was hampered by a significant background which limited their ability to identify the alpha particles associated with the 330-keV resonance. A resonance strength of 3.5 ± 1.6 eV was reported. We have considerabily improved upon the measurement in Ref. [9] by using a coincidence technique along with kinematic reconstruction to produce an essentially background-free measurement of the strength of this important resonance.

We measured the $^{18}\text{F}(p,\alpha)^{15}\text{O}$ cross section at $E_{c.m.} = 330$ keV using a radioactive ^{18}F beam at the ORNL Holifield Radioactive Ion Beam Facility (HRIBF). The average beam current on target was 2×10^5 ^{18}F ions per second. The beam was contaminated with the stable isobar ^{18}O ($^{18}\text{F}/^{18}\text{O} \sim 0.2$), and our experiment had to be designed to overcome this difficulty. The ^{18}F beam bombarded a thin (57 μg/cm^2) polypropylene (CH$_2$) target, and recoil alpha particles and ^{15}O ions were detected in coincidence in the Silicon Detector Array (SIDAR) [10]. The experimental configuration is the same as the one described in Ref. [8] with the exception that the SIDAR covered laboratory angles $18° - 48°$ ($101° < \theta_{c.m.} < 150°$) in this measurement. The beam purity was monitored downstream of the target location by an isobutane-filled gas ionization counter, which provided energy loss information that enabled the proton number of the detected ion to be determined.

The $^1\text{H}(^{18}\text{F},\alpha)^{15}\text{O}$ events were identified by reconstructing the total energy of the reaction products detected in coincidence, as described in Ref. [8]. As a result of the positive Q-value of the $^{18}\text{F}(p,\alpha)^{15}\text{O}$ reaction, the events of interest were readily distinguished from elastic scattering which was the major source of background coincident events. The $^1\text{H}(^{18}\text{F},\alpha)^{15}\text{O}$ events were then further distinguished from $^1\text{H}(^{18}\text{O},\alpha)^{15}\text{N}$ events by plotting (Fig. 1) the lab angles of the detected α particles versus their energies. Through this procedure, the yield of $^1\text{H}(^{18}\text{F},\alpha)^{15}\text{O}$ reaction was measured on resonance [$E(^{18}\text{F}) = 6.6$ MeV] and off resonance [$E(^{18}\text{F}) = 7.5$ MeV]. The data collected during these measurements are shown in Fig. 1, where the off-resonance plot was compiled with $\sim 60\%$ of the incident beam flux used to produce the on-resonance spectrum.

The number of ^{18}F ions incident on target was determined from the measured amount of beam that was elastically scattered into the SIDAR from carbon in the target and using the ratio of ^{18}F to ^{18}O in the beam which was continuously monitored downstream of the target by the ion counter. The total $^{18}\text{F}(p,\alpha)^{15}\text{O}$ cross section was calculated from the

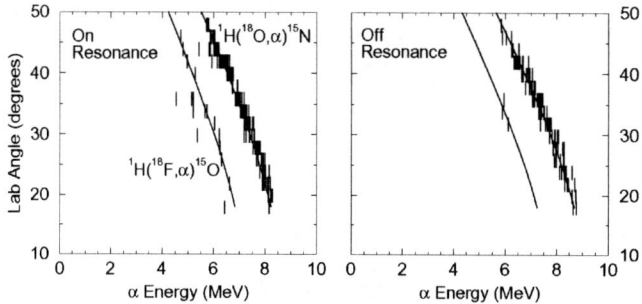

FIGURE 1. The angle of detection of the emitted α particles is plotted as a function of their energies. Owing to the different Q-values of the reactions, the ^{1}H(^{18}F,α)^{15}O events are cleanly distinguished from ^{1}H(^{18}O,α)^{15}N events. Curves have been drawn at the expected energies.

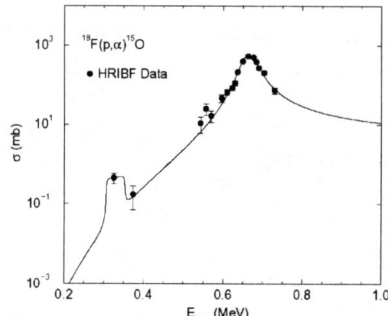

FIGURE 2. The measured ^{1}H(^{18}F,α)^{15}O cross section is shown along with a fit to the data. The 330-keV data are from this work while the 665-keV data are from Ref. [8]. The plotted curve is the calculated cross section which has been averaged over the energy loss in the target for direct comparison with the data. Since the width of the 330-keV resonance is much less than the target energy loss, the curve appears "flat-topped" at these energies.

observed yield assuming an angular distribution characteristic of populating a $J^\pi = \frac{3}{2}^-$ resonance.

The cross sections measured on and off resonance are plotted in Fig. 2 along with the previously measured data for the 665-keV resonance [8]. We also show in Fig. 2 a fit to our data assuming two resonances: a $J^\pi = \frac{3}{2}^-$ resonance at 330 keV and a $\frac{3}{2}^+$ resonance at 665 keV. The parameters of the 665-keV resonance were fixed to those reported in Ref. [8]. The calculated cross section was averaged over the energy loss in the target for direct comparison with the data. The best fit was obtained for a proton width of the 330-keV resonance of $\Gamma_p = 2.22 \pm 0.69$ eV. This value of the proton width is smaller by a factor of ~ 2 than the estimates of Refs. [7, 9], but agrees well with the calculated proton width in Ref. [11] which assumed $\theta_p^2 = 0.01$ for negative parity states. Using our new value of the proton width, we calculate the strength of this important ^{18}F(p,α)^{15}O resonance to be 1.48 ± 0.46 eV. Additionally, we extracted a resonance energy from the

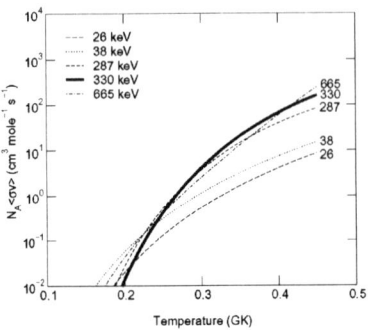

FIGURE 3. The astrophysical $^{18}F(p,\alpha)^{15}O$ reaction rate at nova temperatures labeled with the energies in keV of the contributing resonances.

α-particle angle-energy relationship shown in Fig. 1 of 332 ± 17 keV, in good agreement with 330 ± 6 keV reported previously by Utku et al. [7].

Using our measured resonance parameters, we calculate an improved value for the 330-keV contribution to the $^{18}F(p,\alpha)^{15}O$ reaction rate at nova temperatures. A total reaction rate requires the addition of the contributions from the other known resonances at $E_{c.m.} = 26, 38, 287$, and 665 keV. The astrophysical S-factor was numerically integrated using the resonance parameters given in Table 1 to produce the rates shown in Fig. 3. We find (as shown in Fig. 3) that at the nova temperature range of 0.25-0.4 GK, the 330-keV resonance provides the largest contribution with significant contributions also coming from the 38- and 287-keV resonances at lower temperatures and the 665-keV resonance at higher temperatures. As a result of our improved value of the strength of the 330-keV resonance, we find its contribution to be a factor of ~ 2 lower than reported in Ref. [9]. We find, furthermore, that the total $^{18}F(p,\alpha)^{15}O$ reaction rate (i.e., the sum of the contributions in Fig. 3) is reduced a factor of 2-3 from the rate in Ref. [6] at nova temperatures. We have investigated the effects of our improved $^{18}F(p,\alpha)^{15}O$ rate on the calculated nova nucleosynthesis of ^{18}F by running multi-zone post-processing calculations [12, 13, 14] with hydrodynamic trajectories (the temperature and density as a function of time) similar to those in Ref. [15]. The largest effect observed was in the hottest zone of a 1.35 $M_\odot$ ONeMg white dwarf model ($T_{peak} \simeq 0.43$ GK) where 3 times more ^{18}F was produced using our new rate than when using the Coc et al. rate [6]. Significant changes were also observed for the synthesized abundances of ^{18}O and ^{19}F. When the entire 1.35 $M_\odot$ white dwarf nova model is considered without mixing, the effect is a 34% enhancement in $^{18}F + ^{18}O$.

In conclusion, the astrophysical rate of the $^{18}F(p,\alpha)^{15}O$ reaction at nova temperatures is critical to understanding production of the radioisotope ^{18}F, which may be used to constrain nova models via observations with satellite-based telescopes. The rate of this reaction, however, has been extremely uncertain at nova temperatures because the strength of an important resonance at $E_{c.m.} = 330$ keV was not previously known. We have made the first significant measurement of the strength of this resonance using a radioactive ^{18}F beam at the HRIBF. The results of our measurement in combination

TABLE 1. Resonance parameters used to calculate the $^{18}\text{F}(p,\alpha)^{15}\text{O}$ reaction rate

E_r (keV)	J^π	Γ_p (keV)	Γ_α (keV)	Ref.
26	1/2-	2.8×10^{-20}	220	[6, 7]
38	3/2+	1.1×10^{-14}	4.0	[7, 16]
287	5/2+	2.5×10^{-4}	1.2	[7, 16]
330	3/2-	2.2×10^{-3}	2.7	this work
665	3/2+	15.2	23.8	[8]

with nucleosynthesis calculations indicate that more ^{18}F will be produced than previously thought, where the amount of the enhancement depends on the particular nova model used. While we have measured the strength of the most important resonance, significant contributions could also come at lower temperatures from missing and known resonances at lower energies. Further experiments at the HRIBF will be performed to investigate these effects.

ACKNOWLEDGMENTS

ORNL is managed by UT-Battelle, LLC, for the U.S. Department of Energy (DOE) under contract DE-AC05-00OR22725. UNIRIB is a consortium of universities, the state of Tennessee, ORAU, and ORNL and is partially supported by them. This work was also supported in part by the U.S. DOE under Contract Nos. DE-FG02-97ER41041 with the Univ. of North Carolina at Chapel Hill, DE-FG02-96ER40955 with Tenn. Tech. Univ., and DE-FG02-91ER40609 with Yale Univ. N.S. was partially supported by the NNSF of China (19935030) and by the Major State Public Research Development Program (G20000774). W.R.H. is supported by NASA under contract NAG5-8405, NSF under contract AST-9877130 and by funds from the Joint Institute for Heavy Ion Research.

REFERENCES

1. Starrfield, S., *Phys. Rept.*, **311**, 371 (1999).
2. Wanajo, S., Hashimoto, M., and Nomoto, K., *Astrophys. J.*, **523**, 409 (1999).
3. Leising, M. D., and Clayton, D. D., *Astrophys. J.*, **323**, 159 (1987).
4. Harris, M. J. *et al.*, *Astrophys. J.*, **522**, 424 (1999).
5. Hernanz, M. *et al.*, *Astrophys. J*, **526**, L97 (1999).
6. Coc, A. *et al.*, *Astron. Astrophys.*, **357**, 561 (2000).
7. Utku, S. *et al.*, *Phys. Rev. C*, **57**, 2731 (1998); **58**, 1354(E) (1998).
8. Bardayan, D. W. *et al.*, *Phys. Rev. C*, **63**, 065802 (2001).
9. Graulich, J. S. *et al.*, *Nucl. Phys. A*, **626**, 751 (1997).
10. Bardayan, D. W. *et al.*, *Phys. Rev. C*, **62**, 055804 (2000).
11. Wiescher, M., and Kettner, K.-U., *Astrophys. J.*, **263**, 891 (1982).
12. Hix, W. R., and Thielemann, F.-K., *J. Comp. Appl. Math.*, **109**, 321 (1999).
13. Parete-Koon, S. *et al.*, *Astrophys. J., in preparation* (2002) ;M.S. Thesis, Univ. Tennessee (2002).
14. Smith, M. S. *et al.*, http://www.phy.ornl.gov/hribf/isol01/proceedings (2002).
15. Politano, M. *et al.*, *Astrophys. J.*, **448**, 807 (1995).
16. Shu, N., and et al., *Phys. Rev. C, in preparation* (2002).

The diffuse 1.275 MeV emission from Galactic ONe novae

P. Jean*, M. Hernanz† and J. José†

*Centre d'Etude Spatiale des Rayonnements, CNRS/UPS, 9 avenue colonel Roche, 31028 Toulouse, France
†Institut d'Estudis Espacials de Catalunya, IEEC/CSIC/UPC, Edifici Nexus, C/Gran Capita 2-4, 08034 Barcelona, Spain

Abstract. The radioactive isotope ^{22}Na is one of the primary tracer isotopes that may potentially be detectable by γ-ray spectroscopy. This isotope is predicted to be produced in ONe nova explosions, yet no detection of its 1.275 MeV γ-ray signature has been reported so far. In this paper we will present the implications of the flux upper limits obtained by several instruments on the permissible ^{22}Na yield for ONe novae. We base our analysis on extensive Monte-Carlo simulations of galactic 1.275 MeV emission that were constrained by the most recent results of galactic nova rates and distributions. We demonstrate that the non-detection of the 1.275 MeV line implies a solid upper ^{22}Na yield which is in agreement with current theoretical nucleosynthesis calculations. Sensitivity of the future spectrometer of the INTEGRAL observatory for the observation of this emission is also presented.

INTRODUCTION

Classical ONe nova outbursts are expected to synthesize ^{22}Na and provide a γ-ray signal at 1.275 MeV ([1], [4]). Observations with several γ-ray spectrometers have reported only upper-limits on the 1.275 MeV flux from the Galaxy (HEAO3, SMM, OSSE, COMPTEL) or from individual novae (COMPTEL) and consequently allow to derive an upper-limit of the ^{22}Na mass ejected per ONe nova. Observations with the HEAO3 γ-ray spectrometer provided an upper-limit of the ejected ^{22}Na mass of $3\ 10^{-7}$ $M_\odot$ ([7]). [8], using COMPTEL observations of single novae, estimated an upper-limit of the ejected ^{22}Na mass of $3.7\ 10^{-8}$ $M_\odot$. Using observation of the GC with OSSE, [5] estimated an upper-limit of the ^{22}Na mass of $2\ 10^{-7}$ $M_\odot$. Recently, [9] derived a 2σ upper-limit of the ejected ^{22}Na mass of $3.6\ 10^{-9}$ $M_\odot$ and $2.1\ 10^{-8}$ $M_\odot$ from the COMPTEL observations of the bulge and Nova Cygni 1992, respectively. However, most of the ^{22}Na mass upper-limits derived from the observation of the Galactic diffuse emission have been calculated using overestimated ONe nova frequencies (up to 40 ONe novae yr^{-1} only in the bulge). Current estimations of the ONe nova frequency range from 3 yr^{-1} to 15 yr^{-1} in the whole Galaxy (see the next section). Moreover, the flux at 1.275 MeV in the GC region depends not only on the amount of ^{22}Na ejected per outburst and the rate of Galactic ONe novae but also on their spatial distribution.

In the presented work, we re-estimate the upper-limit of the ejected ^{22}Na mass derived from COMPTEL and HEAO3 observations of the cumulative 1.275 MeV emission taking into account recent results of rates and spatial distributions of novae. Sensitivity

TABLE 1. Characteristics of the models adopted for the simulations.

Models	bulge half radius	disk radial scalelength	disk scaleheight	fraction in the bulge
HF87	2.7 kpc	3.5 kpc	0.106 kpc	0.348
VdK90	2.7 kpc	5.0 kpc	0.300 kpc	0.105
KDF91	1.4 kpc	3.0 kpc	0.170 kpc	0.179
DJ94	1.4 kpc	5.0 kpc	0.350 kpc	0.111

of the future spectrometer SPI of INTEGRAL (INTErnational Gamma-Ray Laboratory) for the observation of this emission is also presented.

DIFFUSE GALACTIC 1.275 MEV EMISSION MODELS

The distribution of the 1.275 MeV emission from galactic ONe novae is calculated with a Monte-Carlo simulation (see [10] for detailed description of the method). The position in galacto-centric coordinates and the age of ONe novae are chosen randomly according to the appropriate distributions and rate. Assuming an ejected ^{22}Na mass per outburst, the 1.275 MeV flux of each nova is computed using its distance and its age to account for the decay of the emission.

For the purpose of this work, a Galaxy model with a disk and a spheroid is convenient to simulate this distribution. Several laws for the spatial distribution of novae in the disk and in the 'spheroid' (representative of the bulge) have been proposed. We have selected four models that differ significantly from each other. The first and older of them is described in [7] - hereafter HF87. The disk and spheroid models are derived from the starlight surface brightness distribution. By scaling the nova rate in the bulge of M31 to the bulge of our Galaxy, HF87 estimate the proportion of novae in the spheroid to be 0.348. The second model has been used by [6] to estimate the occurrence rate of Galactic classical novae and the fraction of novae in the bulge. It is based on a model of the distribution of type Ia supernovae by [3] - hereafter DJ94. The proportion of novae that occur in the bulge is set to 0.111 on the basis of an estimate of the bulge to total galaxy mass ratio. The third model is derived by [13] - hereafter KDF91 - from the Galactic survey of the Spacelab InfraRed Telescope (IRT) that provides a reliable tracer of the distribution of G and K giant stars. Using the total infrared luminosity of the bulge and the disk, the derived proportion of novae occurring in the bulge is 0.179. The last model is taken from [16] - hereafter VdK90. It has been used by [15] to estimate the galactic nova rate. This author assumes that the nova distribution follows the brightness profile of our Galaxy. Under this assumption, the proportion of bulge novae is 0.105. Table 1 summarizes the characteristics of the adopted models.

The Galactic nova rate is poorly known, independently of the underlying white dwarf (WD) composition, because the interstellar extinction prevents us from directly observing more than a small fraction of the novae that explode each year. The most recent estimation has been performed by [15]. He extrapolated the global nova rate from the observed one, accounting for surface brightnesses of the bulge and the disk components and correction factors taking care of any observational incompleteness. With this method, [15] estimated the nova rate to be 35 ± 11 yr^{-1}. [6] found a similar value (41 ± 20 yr^{-1}) using a Monte-Carlo technique with a simple model for the distribution of dust and novae.

The ONe nova rate is obtained by multiplying the total nova rate by the fraction of novae that results from thermonuclear runaways in accreted hydrogen-rich envelopes on an ONe WD. Several authors deduced from observations of abundances in nova ejecta a proportion of ONe novae from 20 to 57 % of observed nova outbursts. [14] reestimated the frequency of occurrence of ONe novae, in light of observations of abundances in nova ejecta. They concluded that, of the 18 classical novae for which detailed abundance analyses were available, only 2 or 3 had a large amount of neon and were ONe novae, whereas 3 other novae showed a modest enrichment in Ne, casting doubt on the type of the underlying WD. Under these considerations, they estimated a fraction of ONe novae between 11 and 33 %.

Therefore, for the calculations presented in the next section, we have adopted frequencies of Galactic novae ranging from 24 yr^{-1} to 46 yr^{-1} and a proportion of ONe novae from 11 % to 33 % corresponding to a lower and upper-limit of the ONe nova rate of 3 yr^{-1} and 15 yr^{-1}, respectively.

OBSERVATION OF THE GALACTIC 1.275 MEV EMISSION

The COMPTEL and HEAO3 measurements provide only upper-limits on the cumulative 1.275 MeV flux which are $3.0 \, 10^{-5}$ photons cm^{-2} s^{-1} in a circle of 10° radius around the GC ([9]) and $4 \, 10^{-4}$ photons cm^{-2} s^{-1} between $l=-30°$ and $l=30°$ (HF87) respectively. In order to derive the ^{22}Na mass upper limit from these observations, we performed several Monte-Carlo simulations of a given ONe-novae frequency-spatial distribution and a fixed value of the ^{22}Na yield per nova. For a large number of Galaxy-tests, the probability for a detection of the galactic emission has been estimated by calculating the fraction of time the simulated flux in the observed region is above the instrument upper limits on the 1.275 MeV flux. The ^{22}Na mass upper limit is defined when the probability of detection by an instrument reaches 90%. It has been calculated for several ONe nova rates and for the proposed spatial distribution models. The results, presented in the figures 1 show that the ^{22}Na mass upper-limit depends strongly on the adopted galactic nova rate and distribution.

Due to the high uncertainty in the nova rate, the most reliable upper-limit should be estimated using the lowest ONe nova rate value and the less favourable spatial-distribution. Indeed, if the yield of ^{22}Na was larger than this upper-limit, spectrometers should have detected the 1.275 MeV line even if the ONe nova rate was larger and even if the distribution was more favorable for its detection. With a rate of 3 yr^{-1} and the VdK90 model we obtain mass upper-limits $M_{22} < 6 \, 10^{-7} \, M_\odot$ with COMPTEL and $M_{22} < 2.5 \, 10^{-6} \, M_\odot$ with HEAO3.

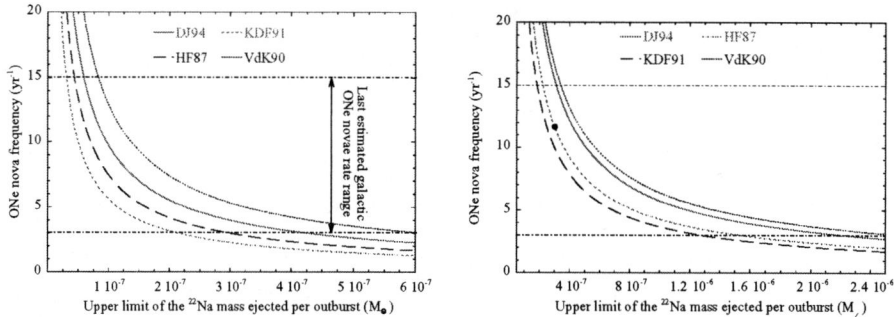

FIGURE 1. Relation between the ^{22}Na mass upper-limit derived from the COMPTEL (left) and HEAO3 (right) 1.275 MeV flux upper-limit, and the Galactic ONe nova rate, for the 4 adopted Galaxy models. The dot shows the estimation of HF87

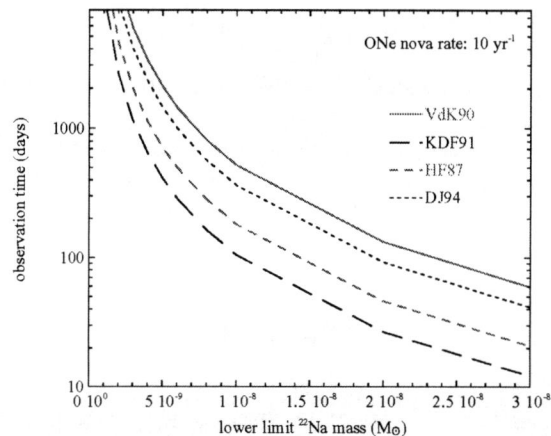

FIGURE 2. SPI observation time necessary to detect the cumulative 1.275 MeV emission from the GC region as a function of the mean ^{22}Na mass ejected per nova.

FUTURE OBSERVATIONS WITH INTEGRAL

[10] evaluated the probability to detect the galactic diffuse 1.275 MeV emission with the spectrometer SPI as a function of the observation time. The method used is similar to the calculations presented in the previous section but the instrumental flux upper-limit is replaced by the estimated γ-ray line sensitivity of SPI. The results presented in figure 2 allow to derive that 80 days of observation of the GC region with SPI could already provide constraints to the mean ^{22}Na yield in ONe novae.

CONCLUSIONS

Several theoretical nucleosynthesis studies have been performed to predict the amount of ^{22}Na ejected in nova outbursts. They have shown that only ONe novae are likely sites for the synthesis of ^{22}Na due to the higher peak temperatures attained during the explosion but also due to the presence of seed nuclei mainly ^{20}Ne, extremely relevant to ^{22}Na synthesis since CNO breakout scarcely takes place at typical nova temperatures. Recent fully hydrodynamical calculations of nova outbursts with updated nuclear reaction rates indicate that novae release about 7 10^{-9} (1.15 $M_\odot$ WD) to 4.4 10^{-9} $M_\odot$ (1.35 $M_\odot$ WD) of ^{22}Na during the explosion ([11], [12]). Such values are below the limits posed by several attempts to find the 1.275 MeV γ-ray signature from novae. Although nuclear uncertainties associated to key reactions relevant to ^{22}Na synthesis exist (i.e., ^{21}Na(p,γ) and ^{22}Na(p,γ)) their incidence in the expected yields is small, namely a factor of 3. A recent reanalysis of the uncertainties associated to ^{21}Na(p,γ) ([2]) suggest even a smaller impact on the ^{22}Na yields.

The most stringent ^{22}Na mass upper-limit obtained so far is presented by [9] who derived a 2σ upper-limit of 2.1 10^{-8} $M_\odot$ from COMPTEL observations of Nova Cygni 1992. This upper-limit still remains above the most optimistic theoretical estimates. Observation with future γ-ray spectrometers (such as SPI) should provide more stringent upper-limits. Gamma-ray observation of novae would provide information not only on their eruption mechanisms and the nucleosynthesis processes involved in their explosion but also on their distribution and their rate in the Galaxy since the problem of the interstellar extinction does not appear at this energy range.

REFERENCES

1. Clayton, D.D & Hoyle, F., 1974, *ApJ*, **187**, L101
2. Coc, A., Smirnova, N., José, J., Hernanz, M. & Thibaud, J.P., 2001, Nuclei in the Cosmos, *Nucl. Phys. A*, **668**, 450
3. Dawson, P. C. & Johnson, R. G., 1994, *JRASC*, **88**, 369
4. Gómez-Gomar, J., Hernanz, M., José, J. & Isern, J., 1998, *MNRAS*, **296**, 913
5. Harris, J. M., 1997, *Proc. of the Fourth Compton Symposium*, AIP conference Proceedings, 410, 1094
6. Hatano, K., Branch, D., Fisher, A. & Starrfield, S., 1997, *MNRAS*, **290**, 113
7. Higdon, J. C., & Fowler, W. A., 1987, *ApJ*, **317**, 710
8. Iyudin, A. F., et al., 1995, *A&A*, **300**, 422
9. Iyudin, A. F., et al., 1999, Proc. of the 3rd INTEGRAL Workshop, *Astro. Lett. & Communications*, **38**, 371
10. Jean, P., Hernanz, M., Gómez-Gomar, J. & José, J., 2000, *MNRAS*, **319**, 350
11. José, J. & Hernanz, M., 1998, *ApJ*, **494**, 680
12. José J., Coc A. & Hernanz M., 1999, *ApJ*, **520**, 347
13. Kent, S. M., Dame, T. M. & Fazio, G., 1991, *ApJ*, **378**, 131
14. Livio, M. & Truran, J. W., 1994, *ApJ*, **425**, 797
15. Shafter, A. W., 1997, *ApJ*, **487**, 226
16. Van der Kruit, P. 1990, in *The Milky Way as a Galaxy*, ed. R. Buser & I.R. King (Mill Valley: University Science Books), p. 331

Future INTEGRAL Observations of Classical Novae

Margarita Hernanz*, Pierre Jean†, Jordi José**, Alain Coc‡, Sumner Starrfield§, Jim Truran¶, Jordi Isern*, Glòria Sala* and Alvaro Giménez‖

*Institut d'Estudis Espacials de Catalunya, IEEC-CSIC, C/Gran Capità 2-4, 08034 Barcelona, Spain
†Centre d'Etude Spatiale des Rayonnements, CNRS/UPS, 9 avenue colonel Roche, 31028 Toulouse, France
**Dept. Física i Enginyeria Nuclear, Universitat Politècnica de Catalunya, and Institut d'Estudis Espacials de Catalunya (IEEC/UPC), Barcelona, Spain
‡CSNSM, CNRS/IN2P3/UPS, Bât. 104, 91405 Orsay Campus, France
§Department of Physics and Astronomy, Arizona State University, P.O. Box 871504 Tempe, AZ 85287-1504, USA
¶Department of Astronomy and Astrophysics, University of Chicago, 5640 South Ellis Ave., Chicago, IL 60637, USA
‖ESA-ESTEC, Noordwijk, The Netherlands

Abstract. The detection of gamma-rays from classical novae would provide a unique proof of the thermonuclear runaway model for these explosions. There has been no positive detection of any nova in gamma-rays up to date. The gamma rays emitted as a consequence of nova outbursts directly reflect the amount of radioactive nuclei synthesized during the explosion; in addition, they provide a direct insight of the dynamic properties of the expanding nova envelope, specially those emitted promptly, when the envelope is not yet completely transparent. The ESA INTEGRAL satellite (launch date October 17th, 2002), will offer an excellent opportunity to observe nearby classical novae, with unprecedented sensitivity and energetic resolution. Both our approved observation in the Guest Observer's Programme, and the foreseen Core Programme observations, will hopefully be able to detect the gamma ray emission from classical nova explosions and to extract the maximum possible scientific results from them.

INTRODUCTION

Classical novae can produce gamma-ray emission of two types, according to its duration: prompt emission, related to electron-positron annihilation (with positrons coming from the short-lived isotopes ^{13}N and ^{18}F) and long-lasting emission, related to ^{7}Be and ^{22}Na decays. The prompt emission has very short duration (2 days), appears very early (before optical maximum) and consists of a line at 511 keV and a continuum below this energy (as pointed out in [19, 3]); this continuum is produced both by positronium emission (when in triplet state) and by Comptonization of the photons emitted in the line. The continuum has a cutoff at 20-30 keV, originated by photoelectric absorption. The long-lasting emission consists of lines (478 keV or 1275 keV, depending on the nova type), which should last around 2 months and 3 years, respectively.

Before INTEGRAL, there have been unsuccessful attempts to detect gamma-ray emission from novae. Efforts have been made mainly to detect the ^{22}Na line, at 1275 keV, with the COMPTEL instrument onboard the Compton Gamma-Ray Observatory, CGRO [13, 14]. In the early 90's, Starrfield et al.[22] already suggested that Nova Herculis 1991 could be a good target for the CGRO satellite, because of its potential ^{22}Na production. Previous attempts to detect the ^{7}Be line, at 478 keV, and the 1275 keV line were made with the GRS instrument onboard the Solar Maximum Mission, SMM, satellite[20, 4]. All these efforts have only provided upper limits, fully compatible with our theoretical predictions (i.e., [17, 23, 18]).

Other attempts have concentrated on the annihilation emission (511 keV line plus continuum below it), with large field of view instruments, like WIND/TGRS[6, 7] and CGRO/BATSE[9], without success and, again, with upper limits compatible with theoretical predictions ([8, 1]). The possible detection of this type of emission from novae with the CGRO/BATSE instrument had been pointed out by Fishman et al.[2] prior to CGRO launch. This emission is more intense than the 478 and 1275 keV lines but has much shorter duration. There have been also attempts to look for the Galactic accumulated emission at 478 and 1275 keV, both with CGRO/OSSE and SMM/GRS[20, 4, 5]. In this case, more flux is accumulated since more sources are contributing, because the typical period between two succesive nova explosions in the Galaxy is shorter than the lifetimes of ^{7}Be and ^{22}Na. But again not enough sensitivity was available. We have recently made predictions about the detectability of this accumulated emission by SPI[15, 16]. More details about the past gamma-ray observations of novae are given in [11].

INTEGRAL will make it much more likely to detect nova explosions. Its sensitivity will be better than that of CGRO/COMPTEL and, furthermore, SPI's excellent spectral resolution will allow for the determination of line shapes (crucial for studying expanding envelopes like those of classical novae). The detection of classical novae will still be quite challenging, since the expected fluxes are small, but much better perspectives for line emission detection with respect to previous missions will be open. This makes it very important to be ready for any possibility (a close enough nova): the fact that there are few chances makes it crucial not to miss any of them. The importance of classical novae as targets for INTEGRAL has led the Integral Science Working Team (ISWT) to include them as sources for the Core Programme. In addition, our Guest Observer's Programme proposal for Cycle 1 has been accepted. We describe it in this paper.

TARGETS

The visual luminosity (i.e., the absolute visual magnitude) of a classical nova at maximum is not directly correlated with its amount of the radioactive nucleus ^{22}Na, or any other radioactive nucleus (in contrast with SNIa, where ^{56}Ni is responsible for both the visual and the gamma-ray luminosities). In order to establish the trigger criteria and to estimate the observation times, we adopt as a representative model a 1.15 $M_\odot$ ONe nova. This nova would emit 1275 keV fluxes up to $\sim 2 \times 10^{-5}$ phot/cm^2/s (see Gómez-Gomar et al.[3] and the recent update in Hernanz et al.[10]), when placed at a distance of 1 kpc. This is the flux taking into account the 20 keV FWHM of the 1275 keV line.

Distance determination

The main factor affecting detectability and, therefore, the main trigger criterion is distance (see table 1), but the distances of novae are not easy to determine accurately. Therefore, some other characteristics, such as apparent visual magnitude, should be used. But, as for any cosmic object, novae which are apparently bright visually can be farther away than novae which are dim, if the visual extinction (intrinsic plus interstellar) of the apparently bright object is much smaller than that of the apparently dim object. Once the preliminary visual light curve and visual extinction are obtained, a distance determination is possible through indirect methods, which suffer from large uncertainties. They depend on various not well known nova properties: *empirical relationship between absolute magnitude at maximum, M_V^{max}, and speed class of the nova (MMRD relation)*: the speed class is measured by the time of decline of the visual magnitude by 2 or 3 magnitudes (t_2 or t_3); *visual extinction of the nova, A_V*, which has intrinsic plus interstellar contributions; the latter varies a lot depending on the location of the nova in the Galaxy. Once M_V^{max} and A_V are known, the derivation of the distance from the apparent magnitude at maximum, m_V^{max}, is straightforward. Therefore, the main uncertainties affecting distance determinations are: general validity of the empirical M_V^{max}-t_2 (or t_3) relationship, determination of A_V, in addition to the determination of t_2 (or t_3) and of m_V^{max} (often it is not known if the nova has been caught at the maximum or after it) from the observations. In figure 1 (left panel), we show a m_V^{max}-distance diagram, for novae discovered in the last century (up to 1995). The data shown are taken from the samples of Shafter[21]. We have superimposed two curves indicating the apparent magnitudes at maximum, m_V^{max}, one could expect, provided that novae are standard candles, with absolute magnitude at maximum M_V^{max}=-7.5, and that visual extinction, A_V, ranges from 0 to 3 magnitudes. For distances up to 1 kpc, m_V^{max} should be smaller (brighter) than 5.5 (for 3 kpc, m_V^{max} ranges from 8 to 5, or brighter if M_V^{max} is < -7.5). If we included novae after 1995, two outstanding points at m_V^{max}=2.8 and 4, and d$\sim$ 2 and 4 kpc (Nova Vel 1999 and Nova Aql 1999b, respectively) would appear (with $M_V^{max} < -7.5$; Nova Vel 1999 probably had $M_V^{max} \sim -8.7$ ([12]), in addition to more "normal" points with distances larger than 5 kpc and m_V^{max} larger than 8. The number of novae discovered during the period 1991-1995 versus m_V^{max} is shown in figure 1 (right panel).

In order to estimate the probability of having a nova at a particular distance, it is instructive to look at figure 2, which shows an histogram of the novae distances for the same nova set mentioned above [21], as well as for the subset of novae in the 1991-1995 period. The sample of years 1991-1995 suffers from small number statistics, but it is more representative of recent more accurate observations. Although the distances have a large uncertainty, some general trends can be extracted: the observed nova rate for novae at distances shorter than **1 kpc** is **1/5=0.20 yr**$^{-1}$ (1991-1995 set), or 16/95=0.17 yr^{-1} (complete set 1901-1995), which is not very large. If we relax the distances of detectability of novae by INTEGRAL by a factor of 3 (i.e., we adopt **3 kpc** instead of 1 kpc, invoking the effect of the uncertain ejected masses), the observed nova rate increases to **6/5=1.20 yr**$^{-1}$ (1991-1995 nova set), or 50/95=0.53 yr^{-1} (complete set 1901-1995). Therefore, there is some chance to have a close nova during INTEGRAL's lifetime (2 to 5 years).

TABLE 1. SPI 3σ detectability distances (in kpc) for lines and continuum

Nova type	$M_{wd}(M_\odot)$	511 keV line	478 keV line	1275 keV line	(170-470) keV
CO	0.8	0.7	0.4	-	0.4
CO	1.15	2.4	0.5	-	2.0
ONe	1.15	3.7	-	1.1	3.0
ONe	1.25	4.3	-	1.1	3.0

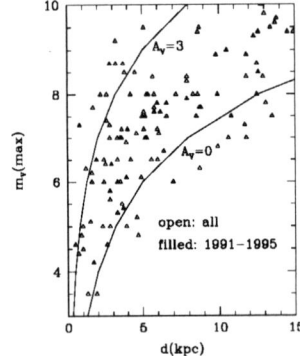 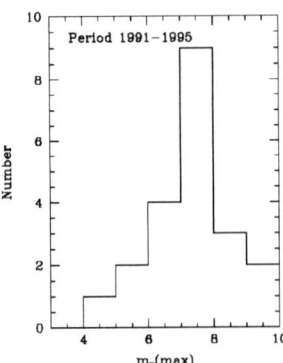

FIGURE 1. (Left) Apparent visual magnitudes at maximum, m_V^{max}, versus distances (see text for details). (Right) Histogram of apparent magnitudes at maximum, for the novae in the period 1991-1995.

Trigger criteria for novae discovered after INTEGRAL launch

With all these data in mind, we suggest the following trigger criteria for classical novae Targets of Opportunity (TOOs) with INTEGRAL: $\mathbf{m_V^{max} < 7}$ is the first required condition; an additional constraint will come from the preliminary determination of distance (from MMRD relation and visual extinction A_V), which would allow us to **discard those novae with distances larger than ∼ 2 kpc**. These criteria are relevant for line emission at 1275 keV (the simultaneous observation of the 478 keV line would allow us to detect the subsample of CO novae closer than 0.5 kpc). The epoch of observation suggested is 1 month after discovery, in order to maximize the probability of detecting any of the 1275 or 478 keV lines (depending on the nova type - ONe or CO - which cannot be known "a priori"). The observation time, estimated with the INTEGRAL Observation Time Estimator, OTE, adopting SPI as prime instrument is **1000 ks**.

The detection of classical novae with INTEGRAL will be quite challenging, but we are sure that, at least, better upper limits for the 1275 KeV emission, which will be more contsraining for the theoretical models, will be obtained.

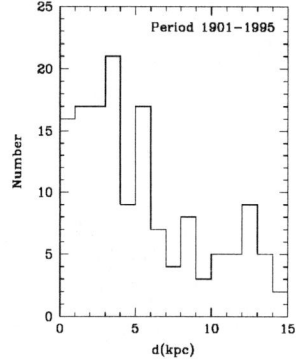
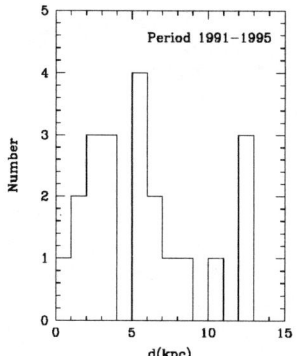

FIGURE 2. (Left) Histogram of distances for the novae discovered in the last century (until 1995). (Right) Same for the subset of the novae in the period 1991-1995

REFERENCES

1. Coc A., Hernanz M., José J., Thibaud J.P., *A&A* **357**, 561 (2000).
2. Fishman, G.J. et al., in *Gamma-Ray Line Astrophysics*, edited by P. Durouchoux and N. Prantzos, AIP Conference Proceedings 232, New York, 1991, p. 190.
3. Gómez–Gomar, J., Hernanz, M., José, J. & Isern, J., *Mon. Not. R.A.S.* **296**, 913 (1998).
4. Harris, M.J., Leising, M.D., Share, G.H., *ApJ* **375**, 216 (1991).
5. Harris, M.J., et al., *A&A Suppl.* **120**, 343 (1996).
6. Harris, M.J., Naya, J.E., Teegarden, B.J., Cline, T.L., Gehrels, N., Palmer, D.M., Ramaty, R. & Seifert, H., *ApJ* **522**, 424 (1999).
7. Harris, M.J., Teegarden, B.J., Cline, T.L., Gehrels, N., Palmer, D.M., Ramaty, R. & Seifert, H., *ApJ* **542**, 1057 (2000).
8. Hernanz, M., J.,José, J., Coc, A., Gómez-Gomar, J., Isern, J., *ApJ* **526**, L97 (1999).
9. Hernanz, M., Smith, D.M., Fishman, G.J., Harmon, A., Gómez-Gomar, J.,José, J., Isern, J., Jean, P., in *The 5th COMPTON Symposium*, edited by M.L. McConnell, J.M. Ryan, AIP Conference Proceedings 510, New York, 2000, p. 82.
10. Hernanz, M., Gómez-Gomar, J., José, J., Coc, A., *Proceedings of the 4th INTEGRAL Workshop "Exploring the Gamma-Ray Universe"*, ESA SP-459, Noordwijk, 2001, p. 65.
11. Hernanz, M., these proceedings, 2002.
12. Della Valle, M., Pasquini, L., Williams, R., Napoleao, T.A., De Souza, W.C., Pearce, A., De S. Aguiar, J.G., *IAUC* **No. 7193** (1999).
13. Iyudin, A.F. et al., *A&A* **300**, 422 (1995).
14. Iyudin, A.F. et al., *Astrophys. Lett. & Comm.* **38**, 371 (1999).
15. Jean, P., Hernanz, M., Gómez–Gomar, J & José, J., *Mon. Not. R.A.S.* **319**, 350 (2000).
16. Jean, P., Hernanz, M., José, J., these proceedings, 2002.
17. José, J. & Hernanz, M., *ApJ* **494**, 680 (1998).
18. José, J., Coc, A. & Hernanz, M., *ApJ* **520**, 347 (1999).
19. Leising, M.D. & Clayton, D.D., *ApJ* **323**, 157 (1987).
20. Leising, M.D., Share, G.H., Chupp, E.L., Kanbach, G., *ApJ* **328**, 755 (1988).
21. Shafter, A.W., *ApJ* **487**, 226 (1997).
22. Starrfield, S., Shore, S.N., Sparks, W.M., Sonneborn, G., Truran, J.W., Politano, M., *ApJ* **391**, L71 (1992).
23. Starrfield, S., Truran, J.W., Wiescher, M.C., Sparks, W.M., *Mon. Not. R.A.S.* **296**, 502 (1998).

NOVA POPULATIONS / NOVAE IN EXTERNAL GALAXIES

Nova Populations

Massimo Della Valle

Osservatorio Astrofisico di Arcetri, Largo E. Fermi 5, 50125 FI, Italy

Abstract. In this article we review the current status of the stellar population assignment for novae. Observations in the Milky Way and in external galaxies point out the existence of two nova populations: fast and bright novae, mainly originated from massive white dwarfs and associated with the *thin disk/spiral arm* stellar population, and slow and faint novae, originated from lighter white dwarfs and associated with *thick-disk/bulge* population.

INTRODUCTION

Baade [1], [2] introduced the concept that different kinds of stellar populations have different spatial distribution within galaxies (see also Oort [3]). We can take advantage of this notion to find out useful hints about the population assignments of the progenitors of novae. Due to their luminosity, $M_V \gtrsim -9$ novae are particularly suited for this purpose because they can be easily identified both in the Milky Way and in external galaxies.

NOVA POPULATION IN THE MILKY WAY

Historical data on galactic novae (e.g. McLaughlin [4],[5],[6] and Payne-Gaposchkin [7]) have received discrepant interpretations. Kukarkin [8], Kopylov [9] and Plaut [10] pointed out the existence of a concentration of novae towards the galactic plane and the galactic center and classified them as belonging to the 'disk population'. Minkowski[11], [12] and [7] showed that the galactic longitudes of novae and planetary nebulae (PNe) have similar distributions and therefore novae, like PNe, belong to Pop II stellar population. Baade [13] assigned novae to Pop II stellar population because of the occurrence of a few ones (e.g. T Sco 1860) in very old stellar population systems, such as the Globular Clusters. Iwanowska and Burnicki [14] suggested that novae are a mixture of Pop I and Pop II objects, and Patterson [15] proposed that novae belong to an 'old disk' population. Tomaney and Shafter [16] found that novae belonging to the bulge of M31 are spectroscopically different from novae observed in the neighborhood of the Sun and deduced that galactic novae are mainly 'disk' objects. Different conclusions were drawn by Della Valle and Duerbeck [17] who compared the cumulative distributions of the rates of decline for M31, LMC and Milky Way nova populations (see Fig. 1) and found that galactic and M31 distributions are indistinguishable, whereas M31 and LMC distributions are different at $\gtrsim 99\%$ significance level. Since the speed class of a nova depends on the mass of the underlying white dwarf (e.g. [18]), systematic differences

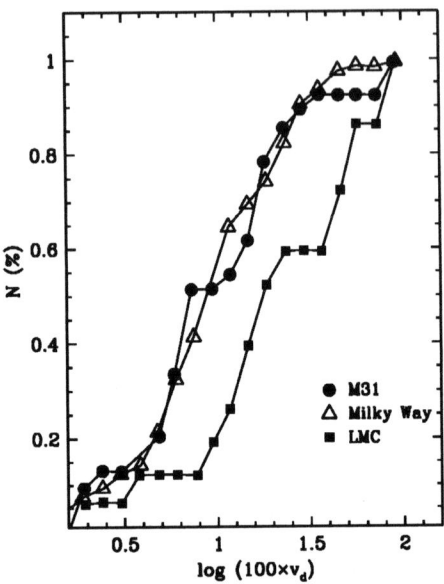

FIGURE 1. Cumulative distributions of the rates of decline for M31, LMC and Galaxy (adapted from[17])

in the distributions of the rates of decline indicate the existence of physical differences between LMC and M31 nova populations. As most novae in M31 are produced in the bulge (Ciardullo et al. [19], Capaccioli et al.[20], Shafter and Irby[21]), one would argue that novae in the Milky Way are also mostly bulge objects.

Disk and Bulge Novae

The quantitative characterization of the concept of nova populations into two classes of objects, i.e. fast and bright 'disk novae' and slow and faint 'bulge novae', has been elaborated in the early 90's by Duerbeck [22] and Della Valle et al.[23],[24], [25],[26]. The former demonstrated that nova counts in the Milky Way do not follow an unique distribution (Fig. 2), the latter authors showed that the rate of decline (which traces the mass of the WD associated with the nova system) correlates with the spatial distributions of the novae inside the Milky Way (Fig. 3 and 4). Fig. 2 shows that nova counts follow two different trends. Dashed and dotted lines are the predictions from simple disk, $\rho(z) = \alpha \times \rho_\circ \exp(-|z|/z_\circ)$, and bulge ($\rho \sim 10^{0.6}$) nova population models [with

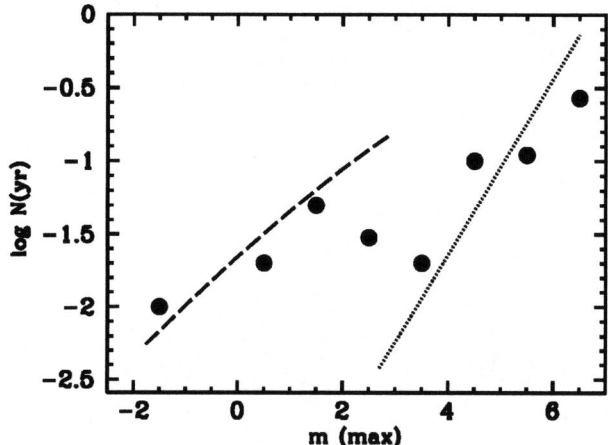

FIGURE 2. Theoretical and observed number of novae. Filled circles: observed rate; dashed line: calculated number counts for a disk population; dotted line: expected slope for a bulge population. Adapted from [22].

$\rho_\circ = 125\text{pc}$ $A_V = 1\text{mag/kpc}^{-1}$, $M_V(\text{max}) = -9$, $\rho_\circ = 10^{-10}$ pc^{-3} yr^{-1} and $\alpha \lesssim 0.1$, see [15], [17], [27] [28]].

A simple consequence of two population distributions is that the rate of decline is expected to correlate with the galactic longitude. Fig. 3 compares the distributions of t3 for galactic novae in the direction of the galactic anti-center ($+90° < l < 270°$, dotted region) and center ($-90° < l < +90°$). The former distribution mainly formed by 'disk' novae, peaks at t3 $\lesssim 20^d$, while the latter one peaks at larger t3 because of the contribution of *slow* 'bulge' novae which are viewed in the direction of the galactic center.

The relationship between the rate of decline and height above the galactic plane for classical novae

If galactic novae originate from both the bulge and thin disk of the Milky Way, then the masses of their WD progenitors are expected to have heights above the galactic plane systematically different because of the initial-mass/final-mass relationship for WDs (e.g.[29]). This fact can be verified in the following way. Theoretical calculations ([30], [31], [32]) have established that the strength of the nova outburst is a strong function of the mass of the underlying WD. On the other hand, the luminosity of the nova at maximum (which is representative of the strength of the outburst) correlates with the rate of decline ([33], [34], [35]). Thus, the distribution of the rates of decline

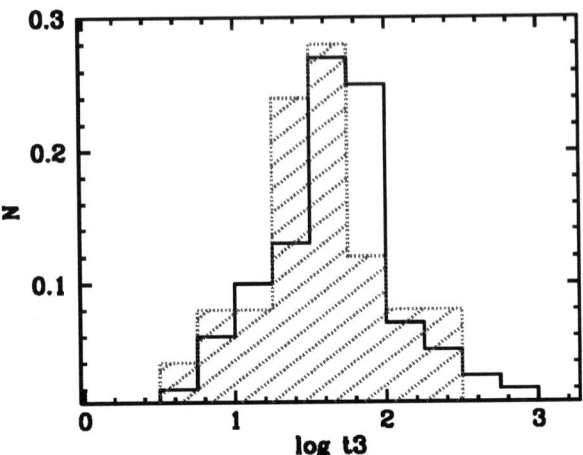

FIGURE 3. The distribution of t3 (the time that the nova takes to fall 3 magnitudes below maximum light) in the direction of the galactic anticenter ($+90° < l < 270°$, shaded region) and center ($-90° < l < +90°$, black). Adapted from [23]

traces the distribution of the masses of the WDs associated with the nova systems. We note that the distance moduli introduced into $z = 10^{[0.2(m-M+5-A)]} \times \sin b$ (where b is the galactic latitude) are derived via expansion parallaxes rather than with the maximum magnitude vs. rate of decline (=MMRD) relationship to avoid circular arguments. For this reason Della Valle et al. [23] restricted their analysis to a fiducial sample of only 19 objects coming from Cohen & Rosenthal [36] and Cohen [37] out of about 100 novae for which the rates of decline have been measured with reasonable accuracy. In this presentation we were able to augment the original sample by 11 new objects from [38], [39] and [40]. Fig. 4 shows the existence (at a confidence level of $\lesssim 3\sigma$) of a relationship between the rate of decline and the height of the novae above the galactic plane, $\log z = -0.58(\pm 0.18) \times \log 100 \times v_d + 2.7(\pm 0.2)$, which can be expressed as $\log z = -1.9(\pm 0.5) \times M_{WD}/M_\odot + 3.7(\pm 0.6)$, after recalling that $\log M_{WD} \propto M_B$ (from [41]) and $M_B \propto \log t2$ (from [35]). The trends illustrated in Fig. 4 and 5 indicate that the fastest nova systems, which contain the most massive WDs, are concentrated close to the galactic disk. It is difficult to understand how such distributions could be due to selection effects, since there is no obvious mechanism to prevent the discovery of fast/bright novae at high z (although we cannot exclude that some slow/faint nova at small z can be heavily absorbed and overlooked). Fig. 6 reports the frequency distribution of the rates of decline (in terms of log t2) for the fiducial sample of novae. The distribution is bimodal and shows that 'disk' novae, i.e. objects characterized by $z \lesssim 150$pc (shaded region), are mostly 'fast' whereas 'bulge' novae are mostly 'slow'. They peak at $M_V = -8.7^{-0.3}_{+0.6}$ and $M_V = -7.2^{-0.9}_{+1.2}$ respectively. The existence of such a bimodality was pointed out by

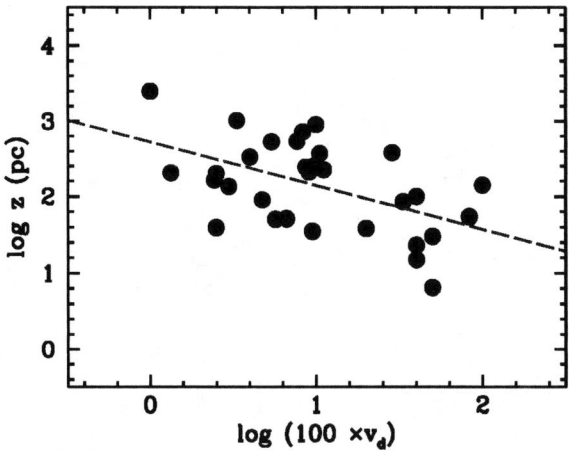

FIGURE 4. The relationship between height above the galactic plane vs. rate of decline, $v_d = 2/t2$. $t2$ is the time that the nova takes to decrease its brightness by 2 magnitudes from maximum.

Arp [42] also for M31 novae (see his Fig. 36), but this fact was neglected afterwards.

The spectroscopic differences between disk and thick-disk/bulge Novae

Williams [43] after studying about two dozen of galactic novae (and a few objects belonging to the LMC) concluded that novae can be broadly divided into two spectroscopic classes of objects: the Fe II and He/N novae. The former are characterized by slow spectroscopic evolution with expansion velocities $\lesssim 2500$ km/s (FWZI) and the Fe II lines as the strongest non-Balmer lines in the early emission spectrum. The latter are fast spectroscopically evolving novae, characterized by high expansion velocity ejecta $\gtrsim 2500$ km/s (FWZI) with He and N lines being the strongest non Balmer lines in the emission spectrum near maximum. Hybrid objects (e.g. V1500 Cyg) that evolve from Fe II to He/N are classified as FeII-b (b=broad) and are physically related to the He/N rather than to FeII class. In Fig. 7 we have plotted the frequency distribution of the heights above the galactic plane of the novae of the fiducial sample after being classified according the Williams' criteria. The top and bottom panels give the distributions of novae which have been classified as He/N (+FeII-b) and Fe II. The histograms show that novae belonging to the He/N class tend to concentrate close to the Galactic plane with a typical scale height $\lesssim 150$ pc, whereas FeII novae are distributed more homogeneously up to $z \sim 1000$ pc and beyond. A K-S test on the data shows that the two distributions are different at $\gtrsim 95\%$ level. Fig. 7 indicates that the objects previously classified as 'disk'

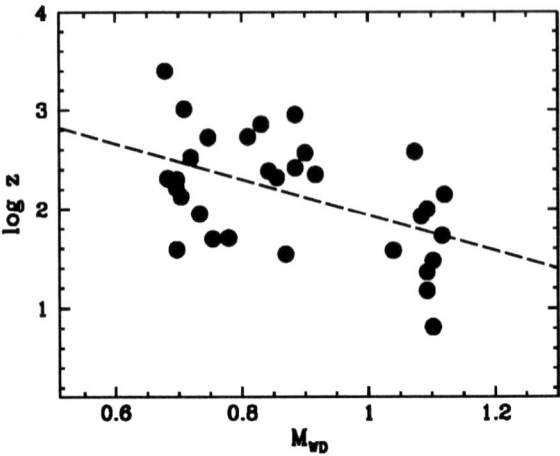

FIGURE 5. The relationship between height above the galactic plane vs. mass of the underlying WD (in solar masses).

and 'bulge/thick-disk' novae tend to correspond to the spectroscopic classes introduced by Williams, indeed about 70% of fast and bright novae belong to the He/N class, while the slow and faint ones form the main bulk of Fe II class. To explain this behavior one should consider the following. The more massive the WD (for a given $\dot{M}$ and T_{WD}) the smaller is the mass of the accreted envelope (in view of $\Delta M_{acc} \propto R_{WD}^4/M_{WD}$) the more violent is the outburst (i.e. shorter t3 and higher expansion velocities) and the larger the fraction $\Delta m_{shock}/\Delta m_{wind}$ where Δm_{shock} is the shell mass ejected at the maximum and Δm_{wind} is the fraction of the shell mass ejected in the subsequent continuous optically-thick wind phase. Since He/N spectra are formed in the shell ejected at the outburst maximum (as one can infer from the top-flatted profiles of the emission lines, for example) it is very likely that He/N novae are generally associated with massive WDs. If this interpretation is correct the distributions reported in Fig. 7 are simple consequences of the trends reported in Fig. 4 and 5.

NOVA POPULATIONS IN EXTERNAL GALAXIES

The use of galactic data to establish the stellar population where novae originate has been often questioned because of observational bias, mainly due to both interstellar absorption in the galactic disk and our position within the Galaxy. These effects can be largely minimized by studying the nova populations in external galaxies, particularly: 1) their spatial distribution; b) the maximum magnitude vs. rate of decline relationship; c) differences (if any) in the nova rates exhibited by galaxies of different Hubble type. The available data are summarized in Table. 1. Col. 1 gives the galaxy identification;

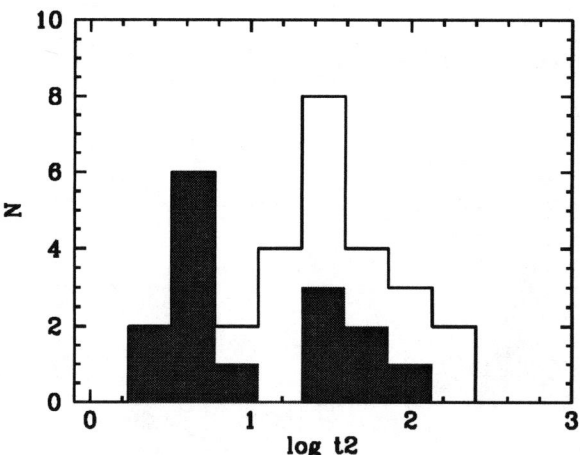

FIGURE 6. Frequency distribution of the rates of decline for the novae of the fiducial sample. The shaded region represents novae with $\lesssim 150$pc.

TABLE 1. Absolute and normalized nova rates

Galaxy	Novae/yr	B_{tot}	(B–K)	(m–M)	ν_K	T
LMC	2.5 ± 0.5	0.57	2.74	18.58 ± 0.1	5.1 ± 1	9
SMC	0.7 ± 0.2	2.28	2.71	19.00 ± 0.1	4.8 ± 1.5	9
M33	4.6 ± 0.9	5.75	2.87	24.64 ± 0.2	3.7 ± 0.9	6
M101	12 ± 4	8.26	3.24	29.35 ± 0.2	0.9 ± 0.3	6
M51	18 ± 7	8.41	3.43	29.60 ± 0.2	1 ± 0.4	4
M31	29 ± 4	3.51	3.85	24.42 ± 0.2	1.5 ± 0.2	3
M81	24 ± 8	7.39	3.99	27.80 ± 0.2	1.7 ± 0.6	2
N5128	28 ± 7	6.32	3.38	27.80 ± 0.2	3.3 ± 0.8	–2
N1316	130 ± 40	9.20	4.15	31.50 ± 0.3	1.7 ± 0.5	–2
M87	91 ± 34	9.49	4.17	30.90 ± 0.2	2.1 ± 0.8	–4
VirgoEs	160 ± 57	9.46	4.26	31.35 ± 0.2	2.2 ± 0.8	–4

col. 2 the nova rate; col. 3 the total B mag of the galaxy corrected for background and foreground absorption; col. 4 the color (B–K); col. 5 the adopted distance modulus; col. 6 the normalized nova rate, i.e. the nova rate per year/$10^{10} L_{K\odot}$; col. 7 the Hubble type. Data come from [24] and [44]. Note that to be consistent with the results of MACHO[45] and EROS[46] teams, who found 4 novae in the SMC after one year of monitoring to search for microlensing events, the nova rate in this system has been estimated, on the basis of simple statistical arguments, to be $\sim$ twice as large as quoted in [44].

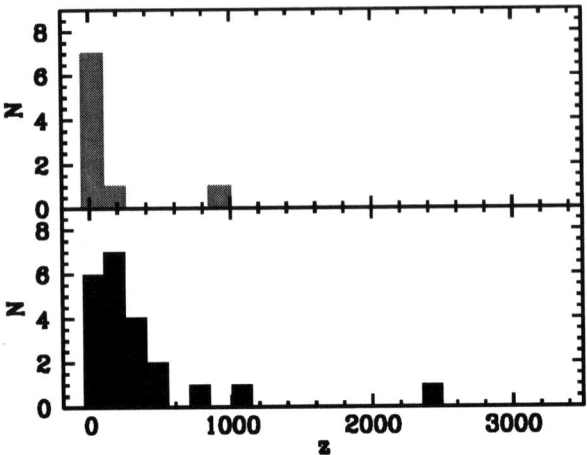

FIGURE 7. Frequency distribution of the height above the galactic plane for He/N (and FeII-b) novae (top panel) and FeII novae (bottom panel).

The Spatial Distribution

Due to its closeness, the M31 nova population has been the only one extensively studied. However a simple skimming through past literature reveals the existence of different ideas on its nova population assignment: disk populations (Arp [42], Rosino [47]); halo population (Wenzel and Meinunger [48]); mainly bulge population (Ciardullo et al. [19], Capaccioli et al. [20] see their Fig. 7); mostly disk population (Hatano et al. [49]).

It is very likely that a fraction of M31 novae considered by [19] and [20] as belonging to the bulge are in fact (according to [49] see their Fig. 1 and 4) physically related to the disk and their allotment to the bulge was a simple consequence of neglecting the geometrical projection effect. However Shafter & Irby [21] have reappraised the Hatano et al's conclusions, after comparing the spatial distribution of novae with the background M31 light (see their Fig. 4). They confirm most M31 novae to be originated in the bulge (about 70% up to a minimum of 50%). Concerning this last point a word of caution seems in order as long as the effects of extinction on nova detections in the M31 disk will be fully quantified.

M33 is an almost bulgeless galaxy [50] and therefore its nova production necessarily originates in the disk. Particularly Fig. 2 of [24] shows that most novae appear superimposed on the arms of the parent galaxy. Similar arguments hold for the LMC.

The analysis of 78 plates obtained in the 1950-55 Palomar campaign for the discovery of novae in M81 has been recently published by Shara, Sandage and Zurek [51]. These authors find that the spatial distribution of novae in M81 is fully consistent with the two

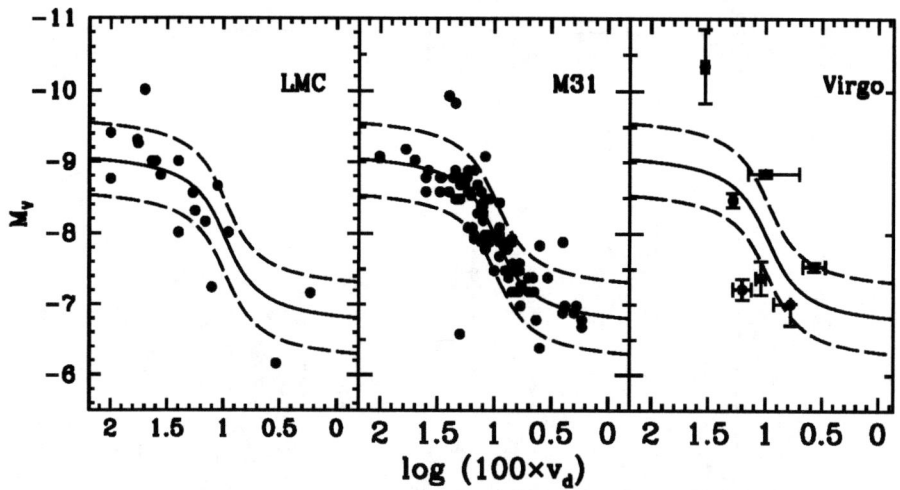

FIGURE 8. The MMRD relationship for LMC, M31 and Virgo nova populations. The solid line indicates the best fit (from [35]). Upper and lower dashed curves are located at $\pm 3\sigma$.

nova populations hypothesis (see their Fig. 1). Particularly they estimate that the fraction of novae belonging to the bulge is not larger than $\sim 60\%$.

The Maximum Magnitude vs. Rate of Decline Relationship

Fig.8 reports the maximum magnitude vs. rate of decline relationship for LMC, M31 and Virgo novae. A simple glance indicates that the nova production in the LMC is clearly biased towards fast and bright novae, whereas the M31 nova population exhibits a prominent 'slow' component. By the same token, admittedly on the basis of a scanty statistic, the trend exhibited by novae discovered in Virgo (Pritchet & van den Bergh [52]). It is not obvious to explain this behavior in terms of an observational bias: indeed the brightest novae are detected in the nearest galaxy (LMC) and would be missed in the more distant ones (Virgo). On the other hand, the differences in the MMRDs find a simple explanation in the framework of the two nova populations scenario: nova systems in disk dominated galaxies are associated with more massive WDs, then resulting in faster and intrinsically brighter nova events.

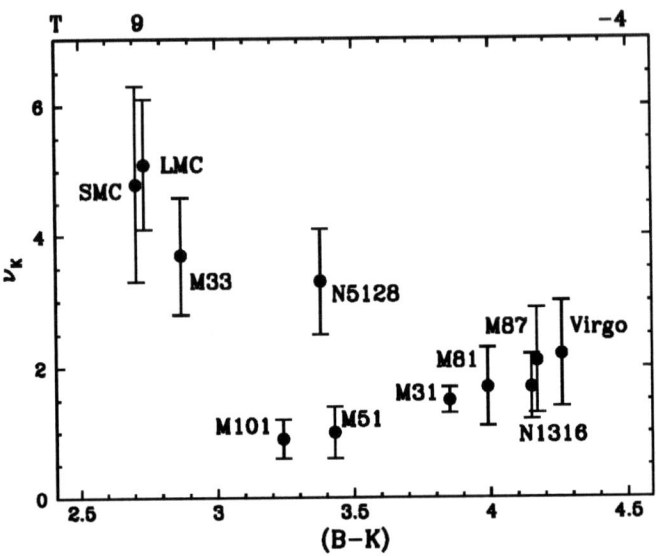

FIGURE 9. The nova rate per unit of K luminosity as a function of the Hubble type of the parent galaxy. The error bars reflect only the poissonian errors affecting the single nova rates.

Is the nova rate depending on the Hubble type of the parent galaxy?

Della Valle et al. [24] suggested that there may exist a systematic difference in the nova rate per unit of K (and H) luminosity in galaxies of different Hubble types (see their Fig. 3 and 4). These authors found that galaxies of late Hubble types are more prolific nova producers than early-type ones by a factor ~ 3. This overproduction of novae could be the consequence of "selection effects" on the nova frequency (Truran & Livio [53], Ritter et al. [54]). For example, if we use the results of [54], we obtain for the ratio $\rho(disk)_{outburst}/\rho(bulge)_{outburst} \sim 4$ in good agreement with the observations. A possible interpretation is that nova systems in disk dominated galaxies result in more frequent nova events due to the shorter nova recurrence time associated with massive WDs (see [55]). However this finding has been put into question by [44] (see also [56]) on the basis of nova rates on M51 and M101. Fig. 9 shows the trend of the normalized nova rate ν_K vs. (B–K) as derived from Tab. 1. With the exception of M101 and M51 (for which the nova rates reported by [44] may be lower limits) the normalized nova rate increases from $\nu_K \lesssim 2$ novae yr$^{-1}/10^{10}L_{K\odot}$ for early types to about $\nu_K \sim 5$ for late types.

CONCLUSIONS

Both observations of galactic and extragalactic novae converge toward the idea that novae are drawn from two different stellar populations. In particular:

1) analysis of the nova counts and rates of decline inside the Milky Way has allowed to derive the notion of disk and bulge/thick disk nova populations. The typical *disk nova* is a fast evolving object whose lightcurve exhibits a bright peak at maximum $M_V \gtrsim -9$ (t2 $\lesssim 13^d$ or t3 $\lesssim 20^d$) a smooth early decline and belongs to the He/N (or Fe IIb) spectroscopic class. The progenitor is preferentially located at small heights above the galactic plane ($\lesssim$ 150pc) and since it is related to (a relatively old) Pop I stellar population, the associate WD is rather massive, $M_{WD} \gtrsim 1\ M_\odot$. The typical bulge/thick disk nova is a slow evolving object whose lightcurve exhibits a fainter peak at maximum, $M_V \sim -7.2$ (t2 $\gtrsim 13^d$ or t3 $\gtrsim 20^d$), often double maxima, dust formation, maximum standstill and belongs to the Fe II spectroscopic class. The progenitors extend up to 1000 pc from the galactic plane and are likely to be related to a Pop II stellar population of the galactic thick-disk/bulge and are therefore associated (on average) with less massive WDs, $M_{WD} \lesssim 1 M_\odot$.

2) analysis of the MMRD relationship for LMC, M31 and Virgo novae confirm the existence of systematic differences in the distributions of the rates of decline of the respective nova populations. 80% of novae in the LMC are bright and fast and therefore associated (on average) with massive WDs while fast novae in M31 are $\lesssim$ 25%. The two distributions are significantly different (K-S gives $\gtrsim$ 99%). The analysis of the spatial distributions suggests that nova populations in M31 and M81 are a mixture of disk and bulge novae ($\sim$ 30% – 70% and 40% – 60% respectively). Novae in the LMC and M33 originate from disk population whereas novae in M87 and NGC 1316 are related 'a fortiori' to bulge population.

3) the previous results imply the existence of differences in the age of the progenitors of disk and bulge novae. This has been recently proven by Subramanian & Anupama [57]. These authors have studied the star formation history of nearby regions around LMC novae and show that most of them occur in the bar or close to it. They conclude that the parent stellar population of the fast and slow novae are likely to be in the range $\lesssim 1-3$ and $\gtrsim 3-10$ Gyr.

4) some of the differences between disk and bulge novae, as described above, are expected on the basis of theoretical arguments (see Kolb [58], Starrfield et al. [59], Kato [60]).

5) Analysis of the distributions of the rates of decline as a function of the absolute magnitude at maximum of LMC, M31 and Virgo, reveals the existence of a scant group ($\lesssim$ 5%) of super-bright novae which deviate systematically from the MMRD relationship by more than one magnitude [61]. One possible explanation is that a *"supernova"* explosion might occur at the end of the life of a CV ([62], see also[63]). The recent study on Nova LMC 1991 ([64]) suggests that the metallicity may be the driving parameter to account for this deviating behavior.

Future studies

Whether or not the nova rate depends on the Hubble type of parent galaxy is still an open question. If low nova rates for M51 and M101 will be confirmed a possible solution of the dilemma has been pointed out by Yungelson, Livio and Tutukov [65] [see also [66]]. These authors noted that small v_k are associated with high-mass spirals such as M51 and M101, while high values of v_k are typical of low mass galaxies such as LMC and M33. If the presently observed rates of disk novae in late type galaxies are mainly determined by the current SFRs, one expects to observe differences in the normalized rates of high- and low-mass spirals. Indeed for the former the SFR in the past was several ($\lesssim 5$) times higher than the present one (Gallagher et al. [67]). For the latter the SFR was nearly constant over the galaxy lifetime (e.g. Gavazzi & Scodeggio [68]). Therefore high-mass spirals contain a higher fraction of old red stars than low-mass ones and consequently also their IR-luminosity is higher (this is actually observed, see col. 4 in Tab.1). As a consequence, since the K luminosity is proportional to the mass in old stars, one may expect v_k in systems such as M33 or LMC to be higher than that in M51 or M101. It is apparent that this issue will be solved as soon as new nova rates in external galaxies will be made available. For example, preliminary results, based on HST observations (e.g. [69]), may suggest high values of nova rates also in elliptical galaxies. We note that nova survey in extragalactic systems, despite their scientific interest, have not been popular among astronomers (although remarkable exceptions, as illustrated above, do exist). Probably the main reason for this is the unpredictable nature of nova events, which made nova surveys considerably (telescope) time consuming. However, the coming into operations of 8-10m class telescopes should change this bias. Della Valle and Gilmozzi [70] have carried out with VLT a pilot programme to discover novae in NGC 1316. They found 4 novae with 3h of observing time (about $\sim 0.8^h$ per nova). A similar programme carried out by [52] with a 4m class telescope on Virgo galaxies (which have comparable nova rates and distance than NGC 1316) discovered 9 novae in 56h ($\sim 6.5^h$ per nova). This experiment teaches us that the 8-10m class telescopes equipped with larger and more efficient detectors are able to improve the payoff per night, in terms of nova detections in galaxies outside the Local Group, by on order of magnitude with respect to the previous generation of telescopes. I'm hopeful that some of the numerous TACs(=Time Allocation Committees) currently circulating through the astronomical community will be sensitive to this last point.

ACKNOWLEDGMENTS

I'm grateful to Nino Panagia and Bob Williams for useful comments.

REFERENCES

1. Baade, W. 1944, ApJ, **100**, 137
2. Baade, W. 1957, Obs., **77**, 164
3. Oort, J.H. 1926, Groningen Pub. **n.40**
4. McLaughlin, D.B. 1942, Popular Astr. **50**, 233
5. McLaughlin, D.B. 1945, PASP **57**, 69
6. McLaughlin, D.B. 1946, AJ, **51**, 136
7. Payne-Gaposchkin, C. 1957, in The Galactic Novae, North-Holland publishing Company, Amsterdam, p. 44

8. Kukarkin, B.V. 1949, in The Investigation of Structure and Evolution of Stellar Systems on the basis of Variable Stars Study
9. Kopylov, I.M. 1955, Izv. Krymsk. Astrofiz. Obs. **13**, 23
10. Plaut, L. 1965, in Stars and Stellar Systems V, A. Blaauw & M. Schmidt (eds.), University of Chicago Press, p. 311
11. Minkowski, R. 1948, ApJ, **107**, 106
12. Minkowski, R. 1950, Publ. Mich. Obs., **10**, 25
13. Baade, W. 1958, Ric. Astr. Specola Vaticana, **5**, 165
14. Iwanowska, W., Burnicki, A. 1962, Bull. Acad. Pol. Sci. Math. Astron. Pys. **10**, 537
15. Patterson, J. 1984, ApJS, **54**, 443
16. Tomaney, A.B., Shafter, A.W. 1992, ApJS, **81**, 683
17. Della Valle, M., Duerbeck, H. 1993, A&A, **271**, 175
18. Shara, M.M. 1981, ApJ, **243**, 926
19. Ciardullo, R., Ford, H., Neill, J.D., Jacoby, G.H., Shafter, A. 1987, ApJ, **318**, 520
20. Capaccioli, M., Della Valle, M., D'Onofrio, M., Rosino, L. 1989, AJ, **97**, 1622
21. Shafter, A.W., Irby, B.K. 2001, ApJ, **563**, 749
22. Duerbeck, H. 1990, in IAU Coll. 122, Physics of Classical Novae, eds. A. Cassatella & R. Viotti (Springer Berlin), p. 34
23. Della Valle, M., Bianchini, A., Livio, M., Orio, M. 1992, A&A, **266**, 232
24. Della Valle, M., Rosino, L., Bianchini, A., Livio, M. 1994, A&A, **287**, 403
25. Della Valle, M. 1995, in Cataclysmic Variables, A. Bianchini, M. Della Valle, M. Orio (eds.), Kluwer Academic Publishers, p. 503
26. Della Valle, M., Livio, M. 1998, ApJ, **506**, 818
27. Duerbeck, H. 1984, ApSS, **99**, 363
28. Naylor, T., Charles, P.A., Mukai, K., Evans, A. 1992, MN, **258**, 449
29. Weidemann, V. 1990, ARA&A, **28**, 103
30. Starrfield, S., Sparks, W.M., Truran, J.W. 1985, ApJ, **291**, 136
31. Kovetz, A., Prialnik, D. 1985, ApJ, **291**, 812
32. Kato, M., Hachisu, I. 1989, ApJ, **340**, 509
33. Zwicky, F. 1936, PASP, **48**, 191
34. McLaughlin, D.B. 1945, PASP **57**, 69
35. Della Valle, M., Livio, M. 1995, ApJ, **452**, 704
36. Cohen, J., Rosenthal, A.J. 1983, ApJ, **268**, 689
37. Cohen, J. 1985, ApJ, **292**, 90
38. Slavin, A. J.; O'Brien, T. J.; Dunlop, J.S. 1995, MNRAS, **276**, 353
39. Downes, R.A., Duerbeck, H. 2000, AJ, **120**, 2007
40. Ringwald, F.A. 2002, this Conference
41. Livio, M. 1992, ApJ, **393**, 516
42. Arp, H.C. 1956, AJ, **61**, 15
43. 43. Williams, R.E. 1992, AJ, **104**, 725
44. Shafter, A.W., Ciardullo, R., Pritchet, C.J. 2000, ApJ, **530**, 193
45. Welch, D.E. et al. 1999, IAUC, 7121, 7308
46. Glicenstein, J.F. et al. 1999, IAUC, 7239, 7286
47. Rosino, L. 1964, Ann. Astrophys., **27**, 498
48. Wenzel, W.., Meinunger, I. 1978, AN, **299**, 237
49. Hatano, K., Branch, D., Fisher, A., Starrfield, S. 1997, ApJ, **487**, L45
50. Bothun, G.D. 1992, AJ, **103**, 104
51. Shara, M.M., Sandage, A., Zurek, D. 1999, PASP, **111**, 1367
52. Pritchet, C.J., van den Bergh, S. 1987, ApJ, **318**, 507
53. Truran, J.W., Livio, M. 1986, ApJ, **308**, 721
54. 54. Ritter, H., Politano, M., Livio, M., Webbink, R. 1991, ApJ, **376**, 177
55. Truran, J.W. 1990, in IAU Coll. 122, Physics of Classical Novae, eds. A. Cassatella & R. Viotti (Springer Berlin), p. 373
56. Sharov, A.S. 1993, AstL., **19**, 147
57. Subramanian, A., Anupama, G.C. 2002, astro-ph/0203098

58. Kolb, U. 1995, in Cataclysmic Variables, A. Bianchini, M. Della Valle, M. Orio (eds.), Kluwer Academic Publishers, p. 511
59. Starrfield, S., Truran, J.W., Wiescher, M.C.,Sparks, W.M. 1998, MNRAS, **296**, 502
60. Kato,M. 1997, ApJS, **113**, 121
61. Della Valle, M. 1991, A&A, 252, L9
62. 62. Iben, I.,Jr. & Tutukov, A.V. 1992, ApJ, **389**, 369
63. Iben, I.Jr., Livio, M. 1993, PASP, **105**, 1373
64. Schwartz, Greg J.; Shore, S. N.; Starrfield, S.; Hauschildt, Peter H.; Della Valle, M.; Baron, E. 2001, MNRAS, **320**, 103
65. Yungelson, L.R., Livio, M., Tutukov, A.V. 1997, ApJ, **481**, 127
66. Tutukov, A.V., Yungelson, L.R. 1995, in Cataclysmic Variables, A. Bianchini, M. Della Valle, M. Orio (eds.), Kluwer Academic Publishers, p. 495
67. Gallagher, J., Hunter, D., Tutukov, A.V.1984, ApJ, **284**, 54
68. Gavazzi, G., Scodeggio, M. 1996, A&A, **312**, L29
69. Shara et al. 2002, this Conference
70. Della Valle, M., Gilmozzi, R. 2002, Science, **296**, 1275

400 Novae in M87

Michael M. Shara* and David R. Zurek*

American Museum of Natural History, Central Park West at 79th St., New York, NY, 10024

Abstract. We report the detection of over 400 classical novae in eruption in the giant elliptical galaxy M87. Ten epochs of observation were assembled from archival Hubble Space Telescope WFPC2 observations. This dataset yielded the spatial distribution of the M87 novae: the novae closely follow the light of the galaxy. The instantaneous luminosity function of the M87 novae was also obtained: it is similar to that of M31 and M81, but with a significant and surprising number of decidedly super-Eddington objects. A 30 day HST campaign on M87 detected a number of equally surprising novae that are always SUB-Eddington.

MOTIVATION

Novae in external galaxies have much to tell us that is very difficult to determine locally. Very large samples of objects, all at the same distance, can be detected with HST imagery at least as far as the Coma Cluster. Among the questions we eventually hope to see answered are:

How are novae spatially distributed in galaxies? What is the speed class distribution of novae? Does it vary systematically with position in a given galaxy, or with Hubble type? What are the global rates of novae in galaxies of different Hubble type?

Can we synthesize nova populations in galaxies from binary stellar evolution theory?

We present here a brief report on recent advances in detecting and characterizing novae in M87 using Hubble Space Telescope data. Complete accounts of our research will be published elsewhere.

OBSERVATIONS

M87 has been observed intensively with the Hubble Space Telescope for a variety of science programs. Two of these efforts have yielded archival datasets that are as valuable for nova searches as for the original projects for which they were proposed. Table 1 lists the epochs of observations and image filters used in our analyses.

At least 40 erupting novae are visible in M87 every time one looks with HST.

TABLE 1. Observations taken by HST/WFPC2

Date	Filters	Total Exposure Times (sec)
1994/02/26	F547M, F658N	800.0, 2700.0
1995/01/23	F702W	280.0
1995/02/03	F555W, F814W	2400.0, 2400.0
1995/05/27	F555W	6700.0
1995/11/23	F814W	3500.0
1996/05/09	F658N	13900.0
1998/02/25	F300W, F450W, F606W, F814W	800.0, 520.0, 520.0, 1400.0
1998/12/17	F300W, F450W, F606W, F814W	800.0, 520.0, 520.0, 1400.0
1998/12/22	F555W, F814W	2000.0, 1800.0
1998/12/25	F555W, F814W	2000.0, 1800.0
1999/05/11	F300W, F450W, F606W, F814W	460.0, 320.0, 320.0, 480.0
2001/05/28	F606W, F814W	400.0, 1040.0 (per day)
2001/06/26	F606W, F814W	400.0, 1040.0 (per day)

RESULTS

Spatial Distribution of the Novae

The spatial distribution of 400 novae detected during 1994-2001 is shown in Figure 1 and plotted in Figure 2, along with the light distribution of the galaxy. The match is remarkably good to within 5 arcseconds of the center of M87 where the detection incompleteness becomes significant. The lesson to be learned from Figure 2 is that the binaries in M87 are distributed in a matter very similar to the single stars. M87 is highly cannibalistic, and it's possible that many (most?) of its stars have been accreted from swallowed galaxies. If so, then the dynamical stirring which must have occurred as each accreted galaxy mixed with M87 was sufficient to avoid obvious clumping or spatial segregation.

Luminosity Function

The instantaneous luminosity function of 400 M87 novae, detected at 10 epochs, is shown in Figure 3, along with the combined distributions of M31 and M81. No dramatic differences are evident, save one: M87 displays a remarkable group of extremely luminous novae, with luminosities up to 10 times the Eddington luminosity for a solar mass white dwarf.

The existence of super-Eddington novae in the Galaxy and in M31 has long been known, and is discussed extensively by Nir Shaviv in these proceedings. The existence of a plethora of such objects in a giant elliptical galaxy is a new result, perhaps somewhat unexpected on simple evolutionary grounds. Yungelson, Livio and their collaborators have suggested that spiral populations should be rich in fast, bright novae, powered by massive white dwarfs from massive progenitors, with a deficit of these objects in old, bulge populations. This is clearly not seen in M87. Possible explanations are that

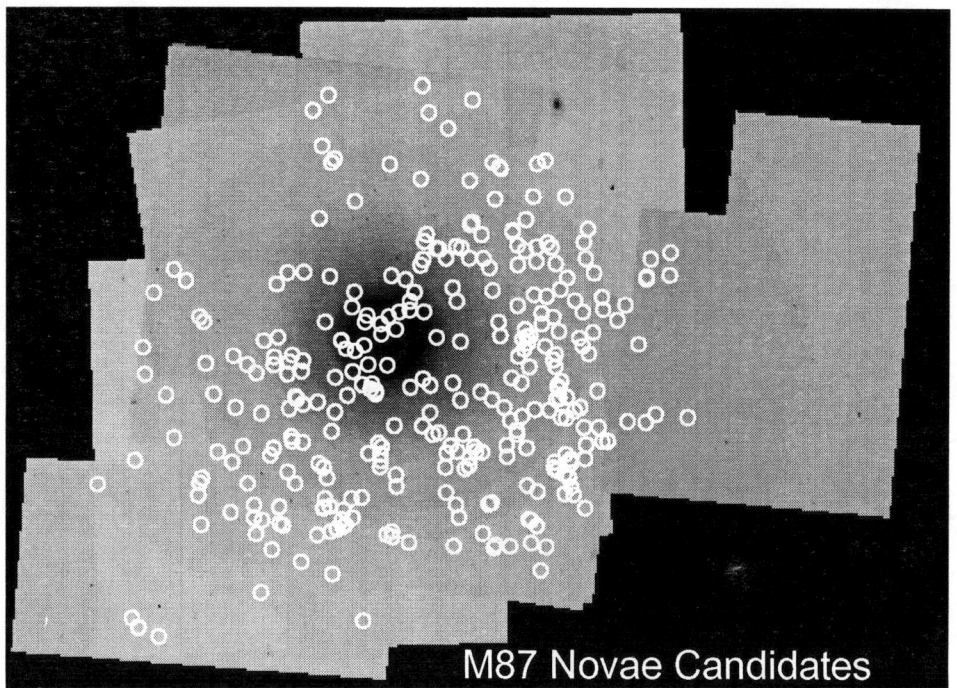

FIGURE 1. Mosaiced image of the various WFPC2 pointings with the positions of the novae candidates indicated.

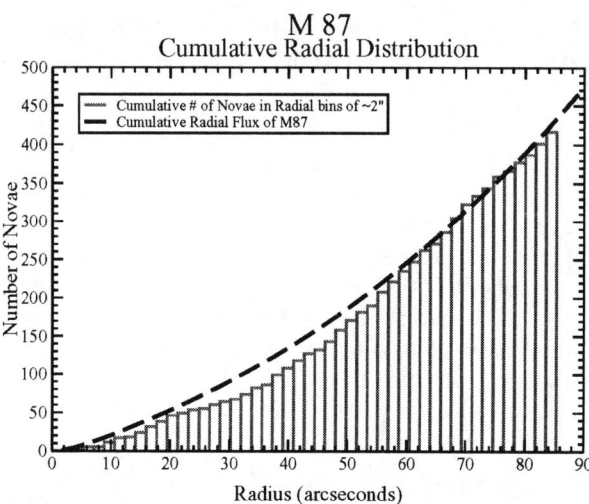

FIGURE 2. Radial distribution of the novae candidates and the radial light profile of M87 normalized to the number of novae candidates.

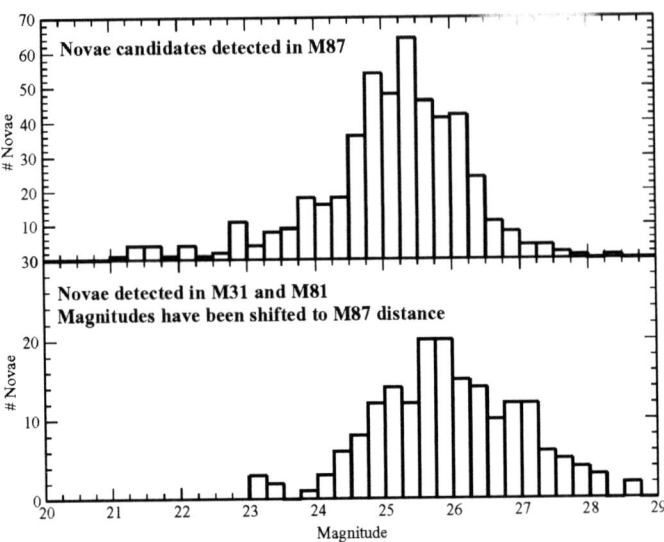

FIGURE 3. Top panel: Instantaneous luminosity function of the ~ 400 novae candidates in M87. Bottom panel: The combined luminosity function of the novae detected in M31 and M81 adjusted to the distance of M87.

most M87 novae seen today have been accreted (perhaps recently) from merging spiral galaxies, or that M87 formed stars (including binaries) until recently, or that novae can be "stored" for up to 10 Gyr before becoming active, or all of the above.

In Figure 4 we display the luminosity function of the M87 novae separated into bins corresponding to angular distance from the nucleus. There is no obvious spatial segregation of brighter and fainter novae seen in M87.

Nova Rate

We have not yet completed detailed completeness simulations, but it's clear that the nova rate in M87 is 300 events per year or more. Comparison with other galaxies' published values suggests that M87 is the most prolific nova producer known, both in absolute numbers AND per unit galaxy luminosity. However, we caution against accepting this result at face value. Our HST nova rate for M87 is considerably higher than groundbased estimates for this galaxy. Frankly, we regard ALL groundbased rates, even those for M31, as rough estimates only. There is not a single published CCD nova survey for any galaxy, including M31, extending nightly over a full observing season, and covering an entire galaxy, to derive rates free of disconcertingly large biasses and correction factors.

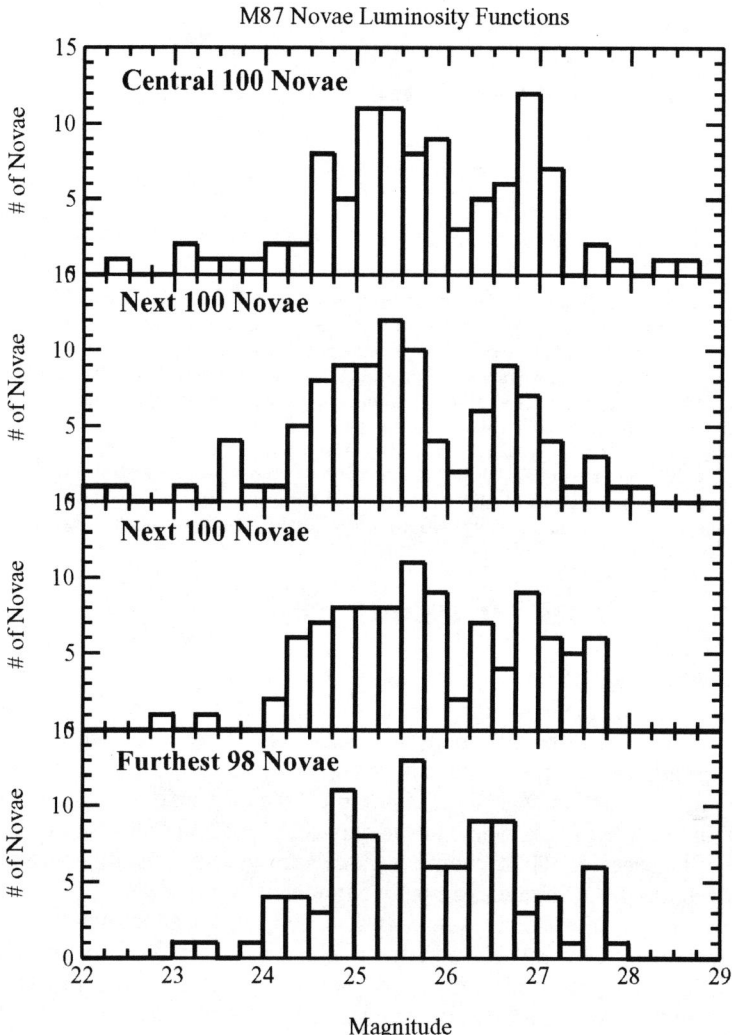

FIGURE 4. Luminosity function of the candidate novae seperated into 4 radial bins containing the same number of candidates.

SUMMARY

Novae are easy to detect in M87 in large numbers. Both highly super-Eddington AND sub-Eddington novae are found. The novae follow the light of M87, and do not segregate according to brightness.

The Galactic Nova Rate

Allen W. Shafter

Department of Astronomy, San Diego State University, San Diego, CA 92182 USA

Abstract. Despite its fundamental importance and a considerable expenditure of observational effort, a reliable estimate of the Galactic nova rate has remained elusive. In this review recent estimates of the Galactic nova rate are critically examined and the problems associated with the determination of the Galactic nova rate are discussed. After considering how the observed properties of novae may depend on the underlying stellar population, a new estimate of the Galactic nova rate is presented based both on Galactic data and on a comparison with the nova rate in M31. Agreement between these two different approaches is found to be satisfactory, with the Galactic data yielding a rate of ~ 30 yr^{-1} and the extragalactic data suggesting a somewhat smaller value of ~ 25 yr^{-1}. After considering likely sources of uncertainty in both the Galactic and extragalactic analyses, a value of 30 ± 10 yr^{-1} is suggested as the current best estimate of the Galactic nova rate.

INTRODUCTION

The Galactic nova rate is relevant to a variety of astrophysical problems. For example, classical nova explosions are thought to play a significant role in the chemical evolution of the Galaxy (e.g., see the contributions in this volume by José and by Romano & Matteucci). In addition to producing a fraction of the ^{7}Li and the short-lived isotopes ^{22}Na and ^{26}Al, novae are believed to be important in the production of the CNO isotopes, particularly in the case of ^{15}N, where novae may account for virtually all of the Galactic abundance of this species. Thus, complete models for Galactic chemical evolution will necessarily require the nova rate as an input parameter.

The rate of novae in a stellar system is also tied to its star formation history [1]. In particular, population synthesis models show that the nova rate is dependent on the parameters governing binary star formation and evolution, such as the initial mass ratio distribution and the loss of angular momentum, both in the common envelope phase and during the secular evolution of the nova system. Thus, in a general sense the present Galactic nova rate can not only help constrain the star formation history of the Milky Way, but the physics of close binary star evolution. If the bulge and disk nova rates can be measured independently, then the formation rate and evolution of close binaries from differing stellar populations can also be studied.

Despite its importance, the Galactic nova rate is not well established. Estimates in the literature have varied widely, ranging from as few as 11 to as many as 260 yr^{-1} [2, 3]. In this paper, I begin by reviewing recent estimates of the Galactic nova rate, including both estimates based on Galactic nova observations and on extragalactic nova surveys. I then present new estimates of the Galactic nova rate based on analyses of data from the Milky Way and M31, taking into account the possible variation of nova properties with stellar population.

TABLE 1. Published Estimates of the Galactic Nova Rate

Author(s)	Year	Method*	Rate (yr^{-1})	reference
Allen	1954	Galactic	100	[4]
Kopylov	1955	Galactic	50†	[5]
Sharov	1972	Galactic	260	[3]
Liller & Mayer	1987	Galactic	73 ± 24	[6]
Ciardullo et al.	1990	Extragalactic	11 – 46	[2]
van den Bergh	1991	Extragalactic	16	[7]
Della Valle	1992	Extragalactic	19 ± 6	[8]
Della Valle & Livio	1994	Extragalactic	15 – 24	[9]
Shafter	1997	Galactic	35 ± 11	[10]
Hatano et al.	1997	Galactic	41 ± 20	[11]
Shafter et al.	2000	Extragalactic	27^{+10}_{-8}	[12]

* See the text for a description.
† Sharov [3] reanalyzed Kopylov's data and found 85 yr^{-1}.

DETERMINING THE GALACTIC NOVA RATE

There are two principal routes to estimating the Galactic nova rate: A "Galactic" method where the global Galactic nova rate is estimated from an extrapolation of the local observed nova rate, and an "Extragalactic" method where the Galactic nova rate is estimated based on a comparison with the nova rates in external galaxies. Previously published estimates of the Galactic nova rate are given in Table 1. Early estimates of the Galactic nova rate, which were based solely on extrapolation of Galactic data, tended to be quite high – on the order of ~ 100 yr^{-1}. More recent estimates, particularly those based on scalings from extragalactic nova surveys, tended to be significantly lower, on the order of 15 – 30 per year. The discrepancy between nova rates based on Galactic data and those based on extragalactic data has largely disappeared in recent years with the addition of more extensive Galactic data [10, 11].

Estimate from Galactic Data

Despite their high optical luminosities ($-10 \lesssim M_V \lesssim -6$), our position within the Galaxy, coupled with the presence of patchy, and often extensive, interstellar extinction results in the discovery of only a small fraction of the novae that erupt each year. Thus, an estimate of the global Galactic nova rate based on Galactic data requires an extrapolation to the entire Galaxy of the observed nova rate in a restricted volume of the Galaxy centered on the Sun. In a spiral galaxy such as the Milky Way, such an extrapolation presents numerous difficulties. It requires an accurate estimate of the volume of space covered in a magnitude-limited nova sample. A knowledge of nova absolute magnitudes and reliable distance estimates to individual novae are therefore needed. It requires a consideration of how nova properties may differ between the bulge and disk populations of the Galaxy. For example, if novae are predominately associated with the Galactic bulge, extrapolation of the nova frequency in the solar neighborhood

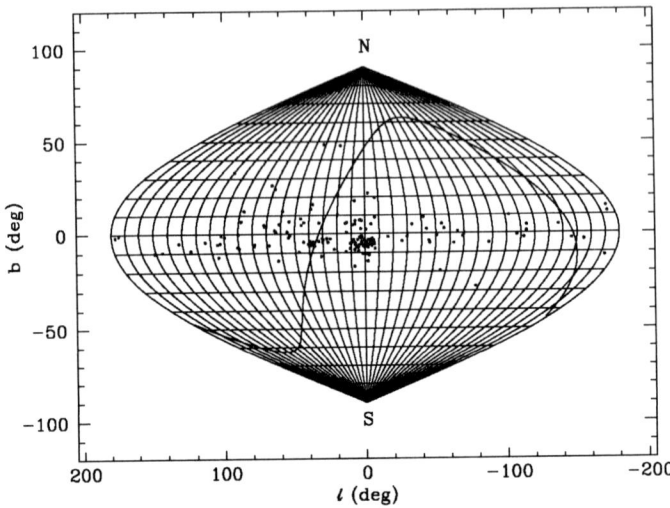

FIGURE 1. The distribution of novae discovered from 1900 to 2000 having $m_V \leq 10$ are shown in Galactic coordinates on an equal-area projection of the sky. The solid line is the Celestial equator.

to the entire Galaxy will yield a different result than if we assume novae arise primarily from a disk population. Furthermore, any differences in the luminosities of bulge and disk novae must be taken into account. Finally, any extrapolation must take into account the physical structure of the Milky Way. Specifically, departures from axisymmetry, such as that caused by the presence of a Galactic Bar, must be taken into account.

Some insight into the population of novae can be gained from a consideration of the distribution of novae on the sky. Figure 1 shows the apparent distribution of novae ($V \leq 10$; 1900–2000) on an equal-area projection of the sky. As has been appreciated for some time, novae appear strongly concentrated toward the Galactic plane, and in the direction of the Galactic center. The longitudinal asymmetry (more novae between $0°$ and $180°$ than between $0°$ and $-180°$) is almost certainly a result of the greater effort spent surveying the Northern sky throughout most of the 20th century. This asymmetry notwithstanding, it appears that novae are associated with both a disk and a bulge population. However, as stressed by Hatano et al. [11] in their Monte Carlo simulations of the Galactic nova distribution, it is possible that many novae in the direction of the Galactic bulge may in fact be foreground disk novae. Thus, in the absence of distance information, the association of novae with either a disk or bulge population based solely on their sky distribution is problematic.

The extent of the incompleteness in Galactic nova observations can be studied by considering how the number of novae discovered increases with apparent magnitude. Figure 2 shows the average annual discovery rate of novae over the past century as a function of limiting magnitude. In a volume of space populated by a uniform density of novae, the observed nova rate should increase with the well-known dependence $N(m) \propto 10^{0.6m}$. Of course, we do not live in a uniform density distribution of stars, but

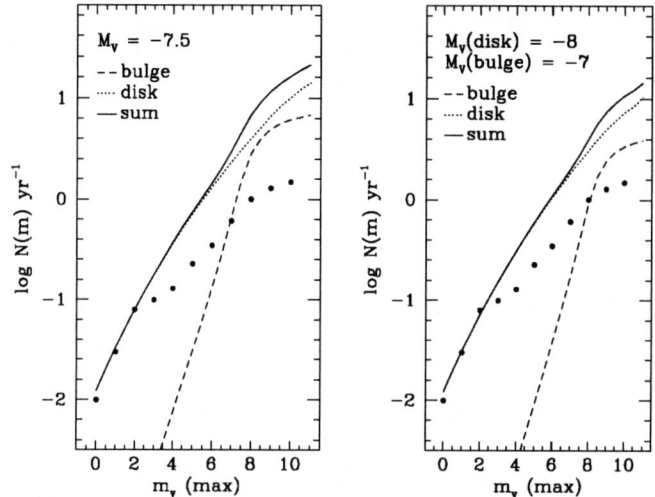

FIGURE 2. Observed and Model Number Counts of Novae. Left panel: the observed number counts (filled circles) for the 101 yr period from 1900–2000 are plotted as a function of limiting magnitude. The solid line shows the expected increase in the number counts for the model described in the text (the bulge and disk contributions to the model are also shown separately). Right Panel: Same as for left panel except that the model assumes differing absolute magnitudes for the bulge and disk novae.

in a more complicated Galactic density distribution that can be characterized by separate disk and spheroidal bulge components.

In order to model the nova rate distribution in the Galaxy, I have adopted the two component model of Bahcall & Soneira [13]. The model assumes that the stellar density in the disk component has an exponential dependence on distance from the Galactic center and on the distance from the Galactic plane, and that the bulge component follows a standard de Vaucouleurs [14] luminosity profile. Following Bahcall & Soneira, we have for the disk component:

$$\rho_d \propto exp[-(z/h_z) - (x - r_0)/h_r], \qquad (1)$$

where z is the distance perpendicular to the Galactic plane, x is the distance from the Galactic center in the plane of the Galaxy, r_0 is the distance from the Sun to the Galactic center (assumed to be 8 kpc), and h_z and h_r are the scale lengths for the exponential distributions of novae perpendicular and parallel to the Galactic plane, respectively. Following Bahcall & Soneira we take $h_r = 3.5$ kpc for the radial scale length and adopt Duerbeck's [15] value of $h_z = 125$ pc for the scale height of classical novae in the Galactic disk. And for the bulge component, we have:

$$\rho_b \propto \frac{exp[-7.669(r/re)^{1/4}}{(r/r_e)^{7/8}}, \qquad (2)$$

where r is the radial distance from the Galactic center and $r_e = 2.7$ kpc is a scale parameter for the Galactic bulge. The luminosity-specific disk-to-bulge nova density ratio, which we define as θ ($= n_{disk}/n_{bulge}$) [16, 17], is allowed to vary in our model. For $\theta = 1$, the ratio is normalized by following Bahcall & Soneira, who noted that in the solar neighborhood the Galactic bulge contributes $\sim 1/800$ of the star density contributed by the disk.

In order to compare the observed nova rate distribution with the model distribution we must specify the average nova absolute magnitude, disk scale height, and model for extinction in the Galactic disk. The absolute magnitudes of novae typically range from -6 to -9, and here we will adopt $M_V = -7.5$ as a reasonable estimate of the average absolute magnitude at maximum. Initially we will assume that there is no difference in absolute magnitude between bulge and disk novae. A nova disk scale height of $z = 125$ pc is assumed [15]. An extinction of 1 mag per kiloparsec is assumed in the Galactic plane, dropping off exponentially with a scale height of 100 pc perpendicular to the Galactic plane.

The solid line in Figure 2 shows the expected increase in the nova rate with limiting apparent magnitude. The disk and bulge contributions to the overall nova rate are shown as dotted and dashed curves, respectively. The curves have been normalized assuming that the discovery of novae brighter than second magnitude is complete. The normalization yields a nova surface density in the solar neighborhood, $\Sigma = 6 \times 10^{-8}$ pc^{-2} yr^{-1} for the uniform absolute magnitude model. Since the bulge contributes so little to the stellar density in the vicinity of the Sun, these are essentially all disk novae. The agreement between the observed data points and the model is excellent for magnitudes brighter than $m_V = 2$, giving us confidence in our normalization, but at fainter magnitudes, incompleteness in the observed nova rate becomes increasingly apparent. The incompleteness levels off at magnitudes fainter than $M_V \simeq 5$, possibly indicating the contribution from the bulge starting to kick in.

The overall nova rate in the Galaxy can be computed from our model by extrapolating the curve to magnitudes sufficiently faint so that the entire Galaxy is sampled. In effect we are taking the nova rate for $m_V = 2$, which is assumed to be complete, and extrapolating it to the entire Galaxy. For the model shown in Figure 2, we arrive at a global Galactic nova rate of 36 ± 13 per year, where the quoted error is based on the Poisson statistics of the number of novae brighter than $m_V = 2$. This value is in excellent agreement with those found by Shafter [10] and Hatano et al.[11].

The Effect of Differing Nova Populations

There is mounting evidence that there may be two populations of novae: A bulge population characterized by relatively slow and dim novae ($< M_V(max) \simeq -7 >$) and a disk population characterized by faster and brighter novae ($< M_V(max) > \simeq -8$). The possibility of two distinct populations of novae was first explored by Della Valle et al. [18], who presented evidence suggesting that fast novae were concentrated closer to the Galactic plane than were slower novae. The evidence has grown stronger recently with additional data and the realization that there appears to be two spectroscopic classes of

TABLE 2. Model Galactic Nova Rate Estimates

θ (n_{disk}/n_{bulge})	M_V (disk)	M_V (bulge)	Galactic Nova Rate (yr^{-1})
One Population			
0.4...	−7.5	−7.5	43 ± 15
1...	−7.5	−7.5	36 ± 13
∞...	−7.5	−7.5	31 ± 11
Two Population			
0.4...	−8.0	−7.0	27 ± 9
1...	−8.0	−7.0	22 ± 8
∞...	−8.0	−7.0	19 ± 7

novae: the Fe II and the He/N novae [19]. Della Valle and Livio [20] showed that the He/N novae cluster close to the Galactic plane and tend to be fast and bright relative to the Fe II class (see the contribution by Della Valle in this volume).

Given the likely existence of two classes of novae, it should be possible to improve on our model nova distribution by considering differences in the average peak luminosities of novae, or in the specific rates of novae, in the Galactic bulge and disk. Figure 2 also shows the result for a two-component model with disk novae characterized by $M_V = -8$ and bulge novae by $M_V = -7$. The principal change from our earlier model is that the brighter disk novae results in a larger volume of space being sampled out to an apparent magnitude $m_V = 2$. Thus, both the surface density of novae in the solar vicinity and the extrapolated global Galactic nova rate are reduced. For this model we find $\Sigma = 4 \times 10^{-8}$ pc^{-2} yr^{-1} in the solar neighborhood and a global Galactic nova rate of 22 ± 8 per year. Table 2 gives the predicted nova rates for additional models where the ratio of the nova density in the disk and bulge, θ, is allowed to vary. As examples, I show a bulge enhanced model ($\theta = 0.4$) representing the value found by Shafter & Irby [17] for novae in M31, and, for comparison, a pure disk model ($\theta = \infty$) is also included.

Estimate from Extragalactic Data

The study of novae in extragalactic systems can shed light on many questions that are difficult to answer from Galactic observations. For example, a determination of the Galactic nova rate based on a comparison with the nova rates in nearby galaxies would appear to avoid, or at least minimize most of the problems encountered by Galactic nova observations. In particular, extragalactic surveys afford the opportunity to study a large sample of equidistant novae with extinction (hopefully) being less of a problem than with Galactic observations. Thus, the luminosities and light curve properties can be studied with relative ease. In addition, by studying novae in a sample of galaxies with differing Hubble types it should be possible to determine if differences exist between novae from differing stellar populations.

Novae have been observed in the nearby spiral galaxy M31 since the pioneering work of Hubble almost a century ago [21]. More recent surveys by Arp [22], Rosino

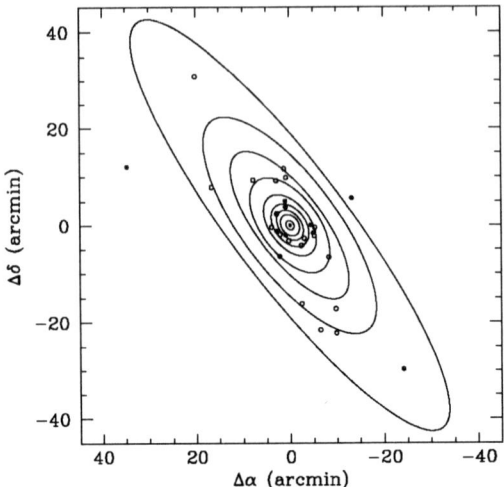

FIGURE 3. The observed spatial distribution of novae in M31. Filled circles indicate "fast" novae with $t_2 \leq 12$ days. There does not appear to be any obvious variation of speed class with position in the galaxy.

[23, 24], Ciardullo et al. [16], and Shafter & Irby [17] have resulted in the discovery of ~ 400 novae. A surprising result of these studies is that the novae appear to be primarily associated with M31's bulge population. This result is unexpected on both observational and theoretical grounds. Observationally, we know from Galactic data that a significant population of novae are associated with the Galactic disk. Moreover, population synthesis models suggest that stellar systems with relatively recent episodes of star formation should produce nova systems with more massive white dwarfs. Such novae are expected to not only be more luminous than their less massive counterparts, but they should erupt more frequently as well [1].

In addition to studying possible differences in the nova rate in M31's different stellar populations, the wealth of data can be exploited to study properties of the nova light curves (peak luminosity and fade rates) as a function of position in the galaxy. The homogeneous sample of novae from Arp's B-band survey is ideal for this purpose. Figure 3 shows the positions of M31 novae with "fast" ($t_2 < 12$ days), and "slow" ($t_2 > 12$ days) shown separately. There is no obvious segregation of nova speed class with position in the galaxy. Specifically, as shown in Figure 4, there is no compelling evidence that the faster novae are more common at large galactocentric radii where the novae are unambiguously associated with the disk.

In recent years novae have been studied in several additional galaxies spanning the Hubble sequence [2, 25, 12]. Despite a wealth of observational data, no general consensus has emerged regarding the stellar population of novae. Ciardullo et al. [2] and Shafter, Ciardullo, & Pritchet [12] have argued that given the uncertainties inherent in the extragalactic nova studies, there is no compelling evidence that the luminosity-specific nova rate varies across the Hubble sequence. Della Valle et al. [25], on the other

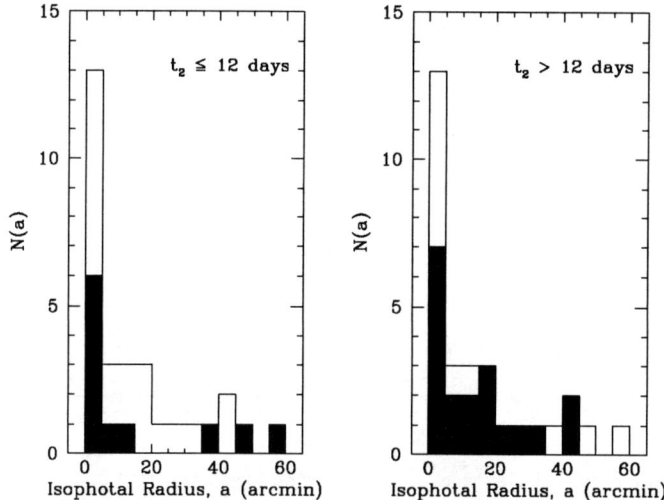

FIGURE 4. The radial distribution of novae in M31. The distribution of the "fast" and "slow" novae are shown as shaded regions. There does not appear to be any significant segregation of fast novae to larger isophotal radii as might be expected if they are predominately associated with M31's disk. If the two distributions are drawn from the same parent distribution, a χ^2 test reveals a probability of 83% that the distributions would differ by more than that observed.

hand, argue that late-type, population I systems such as the LMC and M33 are more prolific nova producers than are early-type galaxies such as M87. A problem with this conclusion is that it is inconsistent with the M31 data, which suggest that the novae in that galaxy are associated with the bulge population. In addition, recent results from HST observations of the giant elliptical galaxy, M87 (see the contribution by Shara in this volume), suggest that the nova rate is at least a factor of two higher than the rate estimated by Shafter, Ciardullo, and Pritchet from ground-based observations. If so, the luminosity-specific nova rate in M87 would be comparable to the LMC and M33.

Despite uncertainties regarding whether novae are primarily associated with population I or population II, to first order the extragalactic surveys have revealed that the nova rates appear to be well correlated with the infrared luminosity of the parent galaxy. Ciardullo et al., Della Valle & Livio [9], and Shafter, Ciardullo, & Pritchet have exploited this correlation, together with an estimate of the Galaxy's infrared luminosity, to estimate the Galactic nova rate. Ciardullo et al. concluded that the nova rate lies between 11 and 46 yr^{-1}, while Della Valle et al. find a similar range of 15–50 yr^{-1}, with lower values between 15 and 24 yr^{-1} strongly favored. Shafter, Ciardullo, and Pritchet deduce a Galactic nova rate of 27^{+10}_{-8} yr^{-1}.

If the nova density is strongly dependent on the underlying stellar population, then this effect should be taken into account when the Galactic nova rate is estimated from comparisons with external galaxies of differing populations. For example, the most heavily studied external galaxy is M31, whose Hubble type (Sb) is somewhat earlier than that of the Milky Way (Sbc). The disk-to-bulge luminosity ratio of M31 is estimated to

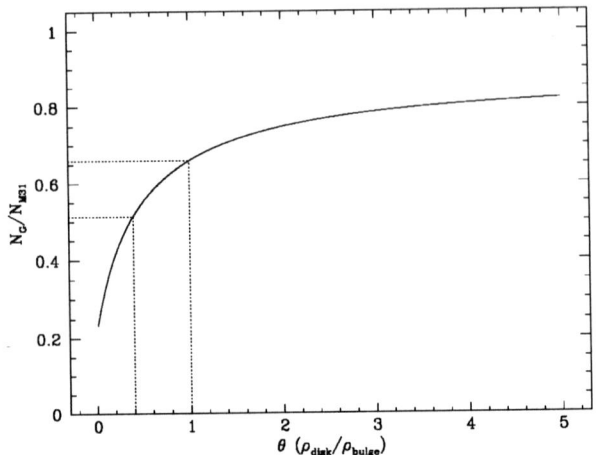

FIGURE 5. The ratio of the Galactic nova rate to that in M31 as a function of the assumed disk-to-bulge nova density, θ. The solid line represents the expected relationship assuming that $(L_{disk}/L_{bulge})_{M31} = 2$ and $(L_{disk}/L_{bulge})_{MW} = 8$. The dashed lines show the expected value of N_G/N_{M31} for $\theta = 1$, and for $\theta = 0.4$ as found by Shafter & Irby [17] in their study of novae in M31.

TABLE 3. Galactic Nova Rate Estimates from M31 Data

M31 Nova Rate (yr^{-1})	θ (n_{disk}/n_{bulge})	Galactic Nova Rate (yr^{-1})
29...	0.4	15 ± 4
	1	20 ± 5
	∞	24 ± 7
37...	0.4	19 ± 5
	1	25 ± 7
	∞	31 ± 9
50...	0.4	26 ± 7
	1	34 ± 9
	∞	42 ± 12

be ~ 2, significantly smaller than the value of ~ 8 for the disk-dominated Milky Way. Thus, if the nova density depends strongly on stellar population, the Galactic nova rate cannot be determined from M31's nova rate by a straightforward scaling of the relative luminosities of the two galaxies. Figure 5 shows how the Galactic nova rate depends on the rate in M31 as a function of the parameter θ, which represents the ratio of the nova density in the disk to that in the bulge. Only for $\theta = 1$ is the ratio of nova rates in the Milky Way and M31 given simply by the ratio of the infrared luminosities of these galaxies, which is taken to be: $(L_{MW}/L_{M31})_K = 2/3$ (see the discussion in Shafter, Ciardullo, & Pritchet [12]). Table 3 shows examples of how the inferred Galactic nova rate depends on differing estimates of the nova rate in M31 and on variations in the disk-to-bulge nova density parameter, θ.

CONCLUSIONS

The Galactic nova rate has been estimated both by extrapolating the observed nova frequency in the solar neighborhood, and by comparing the Milky Way with M31, where the nova rate can be more easily estimated. In both cases, possible variations of nova properties between the disk and bulge populations of the galaxies have been considered. Good agreement is generally found between the two approaches, with the Galactic data yielding a nova rate of ~ 30 yr^{-1}, and the M31 data suggesting a somewhat smaller value of ~ 25 yr^{-1}. The Galactic nova rate estimate is arguably best considered a lower limit since our model assumes that *all* novae reaching magnitudes brighter than $m_V = 2$ during the past century have been discovered. Similarly, M31 nova surveys have assumed that the discovery of novae is relatively unaffected by extinction in the disk. If not, the nova rate based on the comparison with M31 is also likely a lower limit. In view of these caveats, I prefer the higher of the two estimates for the Galactic nova rate, and propose a value of 30 ± 10 yr^{-1} as the best estimate of the nova frequency in the Milky Way.

ACKNOWLEDGMENTS

I thank T. Daub and D. Pollard for comments, and the NSF for grant AST97-31641.

REFERENCES

1. Yungleson, L., Livio, M., and Tutukov, A., *ApJ*, **481**, 127 (1987).
2. Ciardullo, R., Ford, H. C., Williams, R. E., Tamblyn, P., and Jacoby, G. H., *AJ*, **99**, 1079 (1990).
3. Sharov, A. S., *Soviet Astron.*, **16**, 41 (1972).
4. Allen, C. W., *MNRAS*, **114**, 387 (1954).
5. Kopylov, I. M., *Izv. Krymskoi Astrofiz. Obs.*, **13**, 23 (1955).
6. Liller, W., and Mayer, B., *PASP*, **99**, 606 (1987).
7. van den Bergh, S., *PASP*, **103**, 609 (1991).
8. Della Valle, M., "Nova Rate in M33 and the Galaxy," in *Viña Del Mar Workshop on Cataclysmic Variable Stars*, edited by N. Vogt, ASP Conference Series 29, ASP, San Francisio, 1992, p. 292.
9. Della Valle, M., and Livio, M., *A&A*, **286**, 786 (1994).
10. Shafter, A. W., *ApJ*, **487**, 226 (1997).
11. Hatano, K., Branch, D., Fisher, A., and Starrfield, S., *MNRAS*, **290**, 113 (1997).
12. Shafter, A. W., Ciardullo, R., and Pritchet, C. J., *ApJ*, **530**, 193 (2000).
13. Bahcall, J. N., and Soneira, R. M., *ApJS*, **44**, 73 (1980).
14. de Vaucouleurs, G., *Handbuch der Physik*, **53**, 311 (1959).
15. Duerbeck, H. W., *Ap&SS*, **99**, 363 (1984).
16. Ciardullo, R., Ford, H. C., Neill, J. D., Jacoby, G. H., and Shafter, A. W., *ApJ*, **318**, 520 (1987).
17. Shafter, A. W., and Irby, B. K., *ApJ*, **563**, 749 (2001).
18. Della Valle, M., Bianchini, A., Livio, M., and Orio, M., *A&A*, **266**, 232 (1992).
19. Williams, R. E., *AJ*, **104**, 725 (1992).
20. Della Valle, M., and Livio, M., *ApJ*, **506**, 818 (1998).
21. Hubble, E., *ApJ*, **69**, 103 (1929).
22. Arp, H. C., *AJ*, **61**, 15–34 (1956).
23. Rosino, L., *Ann. d'Astrophys*, **27**, 498 (1964).
24. Rosino, L., *A&AS*, **9**, 347 (1973).
25. Della Valle, M., Rosino, L., Bianchini, A., and Livio, M., *A&A*, **287**, 403 (1994).

The Mysterious Eruption of V2434 - LMC

William Liller* and Mati Morel[†]

*Novahound Observatory, Instituto Isaac Newton, Casilla 5022 Reñaca, Viña del Mar, Chile

[†]Morel Astrographics, 6 Blakewell Road, Thornton NSW 2322 Australia

Abstract. A nova search of the LMC revealed that on 6 June, 2001, the 13th magnitude star V2434 or a star in close proximity to it erupted to magnitude 9.7. Twenty-four hours later it was near "normal". Possible explanations for this event include an unusual symbiotic star, a giant flare and a super-fast nova. For reasons to be given, we prefer the last explanations.

INTRODUCTION

The nova search program in Viña del Mar, Chile, in operation since late 1982, has resulted in the independent discovery of 41 novae or nova-like objects to mid-2002. The Southern Milky Way is routinely patrolled with an 85mm Nikon camera, Kodak TP film and an orange filter (to reduce sky brightness and to provide high throughput for the $H\alpha$ emission line); for the Magellanic Clouds, a 20-cm f/1.5 Schmidt camera is used with the same type of film and a yellow or no filter. Exposure times are of the order of a minute or two, made short by the light pollution from the neighboring metropolitan area of well over a million people. Depending on the weather, photographs are taken once every 3 or 4 days of the Magellanic Clouds and Sagittarius, and weekly elsewhere.

A search of the LMC on U.T. May 31.96, 2001, revealed nothing unusual, but on U.T. June 6.96, the 13th magnitude variable star V2434 = GSC 9166 775, located 2.3° SW of NGC 2070, the Tarantula Nebula, appeared at magnitudes 9.7 and 10.0 ($\pm$ 0.2 mags) on two exposures taken 4 minutes apart. The location of the star is well outside the prominent bar in the LMC but in a region of moderately high star density and where other novae have occurred. For example, N LMC 2002 was located 63' due west of V2434. Note that the angular resolution of these photographs is approximately 3" for stars of equal brightness (the focal scale of the Schmidt camera is 677"/mm), and of course, it is entirely possible that it was not V2434 that erupted but the outburst came from a star close to it.

On the following night at U.T. June 7.927, V2434 was once again near "normal"; its broadband V magnitude, Vbb) = 12.73 $\pm$ 0.04. A weakly exposed objective prism spectrum showed $H\alpha$ in emission and nothing else.

DISCUSSION

Henize [1] first noted V2434 as an emission line star and commented in his catalogue, "continuum appears reddish / perhaps a late-type star?" Later investigators also found Hα in emission, but opinions differed as to the spectral type with at least one group calling it a Be star [2]. Humphreys [3] remarked that it was probably a foreground star and possibly a long period variable with hydrogen and calcium H & K in emission. She classified the star as spectral type M2e III and measured V = 12.90, B-V = 1.47. Yamaoka reported [4] that on DSS images V2434 is slightly elongated to the southwest suggesting that it is a close (2-3") double star. He found no significant proper motion of the two stars on four DSS plates taken over a period of 16.9 years, nor was there any apparent change of separation or position angle. V2434 is a well-known X-ray source; its super-soft X-ray designation is 2E IPC5843-5. It has also been detected by ROSAT on numerous occasions. Other names are given for V2434 in the GCVS [5].

Symbiotic stars, which are typically composed of a cool giant such as V2434 and a white dwarf or a low-mass main sequence star [6], occasionally show rapid-rise outbursts, but none to our knowledge recover in a day or less. Joanna Mikolajewska, arguably the leading expert on symbiotics attending the conference, concurs [7].

Were V2434 a low-luminosity Me star, the immediate conclusion would be that a not-unusual 3-magnitude flare caused the outburst. However, unless further spectroscopy reveals that the star is a red dwarf, this explanation must be ruled out since the amount of energy that would have been generated in a flare on a red giant would have been two or three orders of magnitude larger than that reported for the largest flare yet observed [8].

However, it is entirely possible that there is a nearby red dwarf considerably fainter than the more distant red giant but almost in line with it. At quiesence, it would not have been recorded in Humphreys' spectrum. If this eruption did come from a nearby flare star, one would expect to see similar outbursts over time since, of course, flare stars frequently repeat their outbursts; the record holder, UV Cet, averages about 1 flare/day (8). But a survey of some 335 archival photographs from Liller's nova-search program revealed no flares with an upper limit of a few tenths of a magnitudes. See Figure 1. (The total length of the exposures was approximately 8 hours.) A search for a Mira-like periodicity likewise turned up nothing.

Additionally, V2434 was observed by WL with a CCD and a BV filter on other occasions: on 3 nights in June, 2001 following the outburst; on 14 nights between 12 Nov. and 8 Dec., 2001. And finally, during the night of 19-20 Jan., 2002, it was followed continuously for nearly 4 hours. No other outburst has been seen, and the rms fluctuations amounted to only $\pm$ 0.057 mag, close to the uncertainty usually attainable in the the less-than-photometric skies of Viña del Mar. The average brightness of V2434 was V(bv) = 12.70.

If the outburst were caused by a flare, it is perhaps a little unusual that no other flare has been seen before or since despite considerable photometric coverage. Clearly, more

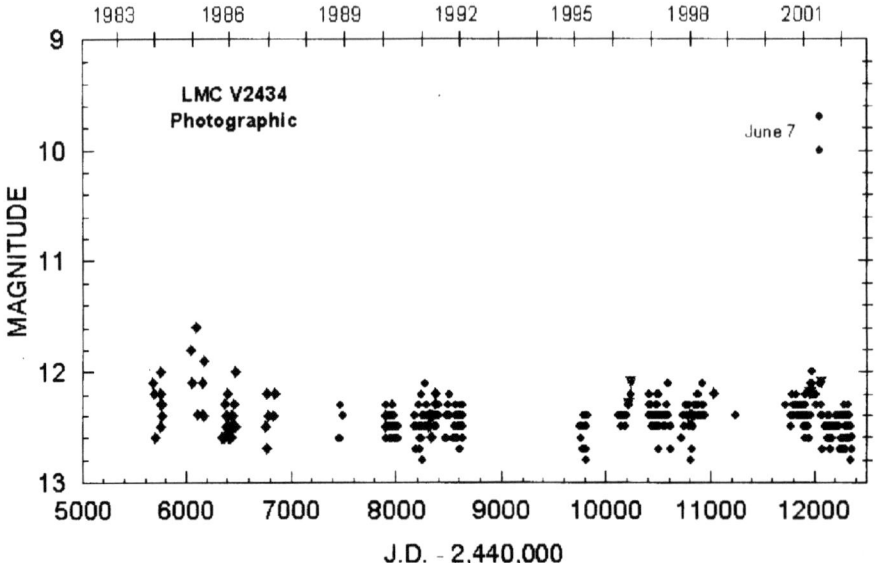

FIGURE 1. 19 years of photographic red magnitudes of V2434 – LMC. Points before 'JD 7000 from Nikon camera images are of lower quality. Later points are from Schmidt photographs. Not shown are several dozen fainter-than estimates mainly from before 'JD 7000.

observations will be needed if this hypothesis is to be confirmed.

Another and much more interesting scenario is that a fast nova occurred in the LMC, again nearly in line with V2434 but this time, of course, much more distant than the giant. With a distance modulus for the LMC of 18.6 mag, the *observed* peak brightness of the outburst would correspond to $M_V = -8.9$ which would, in fact, suggest a fast nova [9]. Furthermore, the value of t_3 would have had to be one day or less to drop from a magnitude of 9.7 to 12.7 or fainter. From an extrapolation of the relationship between absolute magnitude and t_3 published by Cohen [9], a nova with $t_3 = 1.0$ days should have an absolute magnitude of -11.4 meaning that at peak, the observed magnitude of the nova should have been 7.2. And it might have been since it is very likely that the nova was discovered shortly after maximum. If the magnitude at peak was, in fact, 7.2, and $t_3 = 1.0$ days, the peak of the outburst would have occurred no less than 20 hours before discovery.

Shara [10] discussed the possibility that there might be classical novae that reached a maximum luminosity in excess of $M = -10$ which, he noted, would imply an Eddington luminosity greater than 10. Perhaps that is what occurred on 6 June, 2001 in the LMC.

SUMMARY

The outburst that occurred in the direction of V2434 LMC in June 2001 was caused, we argue, either by a flaring red dwarf much nearer to us than V2434, or by a superfast nova in the LMC. The flare hypothesis could be confirmed by further monitoring of the star; the nova hypothesis, while attractive, would be hard to confirm unless, of course, the nova is recurrent.

ACKNOWLEGEMENTS

We should like to thank Taichi Kato, Brian Skiff and Hitoshi Yamaoka who kindly communicated good ideas and suggestions via *vsnet*. In addition, Michael Shara, Hilmar Duerbeck and Joachim Krautter provided valuable thoughts and information.

REFERENCES

1. Henize, K.G., *Ap.J. Suppl.* **2**, 345 (1956).
2. Brunet, J.P., and Imbert, M. *A&A Suppl.* **21**, 109-136 (1975).
3. Humphreys, R.M., *Ap.J.Suppl.* **39**, 389-403 (1979).
4. Yamaoka, H., *vsnet-alert*, no. 5965 (2001).
5. Samus, N.N., *General Catalogue of Variable Stars*, Kosmosinform, Moscow, 1995, pp.1-259.
6. Whitelock, P.A., in *Light curves of variable stars*, edited by C. Sterken and C. Jascheck, Cambridge University Press, 1996, pp.162-167.
7. Mikolajewska, J., private communication at this conference
8. Hall, D.S., in *Allen's Astrophysical Quantities, 4th ed.*, edited by A.N. Cox, Springer-Verlag, New York, 2000, pp.409-410.
9. Cohen, J.G., *Ap.J.* **292**, 90, (1985).
10. Shara, M., a paper presented at the *Int'l Con. on Classical Nova Explosions* (2002).

The Parent Population of Novae in the Large Magellanic Cloud

Annapurni Subramaniam* and G.C. Anupama*

Indian Institute of Astrophysics, II Block Koramangala, Bangalore 560 034, India

Abstract. The parent population of novae in the Large Magellanic Cloud (LMC) are identified by a study of the local, projected, stellar population. The age, density and luminosity function of the stellar population are estimated around 15 novae. The upper limit of the age of the intermediate stellar population is found to be 4 Gyr in all the regions, excepting the region around the slow nova LMC 1948. The star formation history of the underlying population of both the fast and moderately fast novae indicate their parent population to be similar and likely to be in the age range 3.2 – 1.0 Gyr. This is in good agreement with the theoretical age estimates for Galactic cataclysmic variables. The region around the slow nova shows a stellar population in the age range 1 – 10 Gyr, with a good fraction older than 4 Gyr. This indicates that the progenitor might belong to an older population, consistent with the idea that the progenitors of slow novae belong to older population.

INTRODUCTION

Attempts to study the progenitors of novae have, in the past, been based on their spatial distribution (eg. [1], [2], [3], [4]). Population synthesis models of the statistics and properties of Galactic CVs and extragalactic novae indicate that the rate of formation of CVs, the nova rate and the distribution of novae over speed classes depend on the star formation history (SFH) ([5] and references therein). Based on an analysis of the speed classes of Galactic and extragalactic novae and their spatial distribution, the presence of two nova populations: fast, bright disk novae and slow, faint bulge novae have been suggested ([6], [7]). They also suggested the disk novae originate from more massive white dwarfs. Galaxies such as M33 and the Large Magellanic Cloud (LMC) are disk dominated galaxies. Hence most of the nova population in these galaxies are expected from the disk population.

The LMC is one of the very few external galaxies where a large number of novae have been detected and studied in detail over the years. Further, the LMC is believed to have undergone a burst of star formation 3-5 Gyr ago, which probably continued to the present day ([8]). This event of star formation resulted in the majority of the intermediate population seen in the disk of the LMC. Thus, a study of the star formation history may be able to point to the age of the parent population of novae in the LMC.

TABLE 1. List of novae in the Large Magellanic Cloud: coordinates, speed class and type.

No.	Nova	RA (2000)			Dec(2000)			t_3 Days	Remarks
		h	m	s	°	′	″		
1.	LMC 1926	05	14	54.54	−66	48	44.06	200	slow
2.	LMC 1935	03	59	15.90	−67	46	35.71	25.1	fast
3.	LMC 1936	05	07	26.75	−66	39	12.08	31.6	moderately fast
4.	LMC 1937	05	57	04.44	−68	54	47.92	19.9	fast
5.	LMC 1948	05	38	15.38	−70	20	26.23	101.1	slow
6.	LMC 1951	05	12	51.93	−69	58	36.25	6.26	v. fast
7.	LMC 1968	05	09	58.28	−71	39	51.49	5.26	v. fast; USco type RN
8.	LMC 1970#1	05	33	13.25	−70	35	04.41		poor data
9.	LMC 1970#2	05	35	28.90	−70	47	14.32	15.3	fast
10.	LMC 1971#1	04	58	23.23	−68	05	34.02	28.3	moderately fast
11.	LMC 1971#2	05	40	35.22	−66	40	35.23		poor data
12.	LMC 1972	05	28	24.66	−68	49	42.92		poor data
13.	LMC 1973	05	15	18.58	−69	39	46.00		poor data
14.	LMC 1977#1	06	05	45.50	−68	38	12.74		poor data
15.	LMC 1977#2	05	05	10.87	−70	09	01.51	20.7	fast
16.	LMC 1978#1	05	05	52.26	−65	53	02.67	7.8	v. fast
17	LMC 1978#2	05	00	59.65	−67	12	44.81		poor data
18.	LMC 1981	05	32	09.27	−70	22	11.70		v. fast; ONeMg nova
19.	LMC 1987	05	23	50.12	−70	00	23.50	5.26	v. fast
20.	LMC 1988#1	05	35	29.33	−70	21	29.39	39.2	moderately fast
21.	LMC 1988#2	05	08	01.10	−68	37	37.67	9.7	v. fast; ONeMg nova
22.	LMC 1990#1	05	23	21.82	−69	29	48.48	7.69	v. fast; ONeMg nova
23.	LMC 1990#2	05	09	58.28	−71	39	51.49	5.26	Recurrent nova 1968
24.	LMC 1991	05	03	44.98	−70	18	13.64	6±1	v. fast; super bright
25.	LMC 1992	05	19	19.84	−68	54	35.09	11.23	fast
26.	LMC 1995	05	26	50.33	−70	01	23.08		SS X-ray at late phases.
27.	LMC 1997	05	04	26.07	−67	38	38.00		
28.	LMC 1999	05	35	32.77	−69	29	52.01		probably not a nova
29.	LMC 2000	05	25	01.60	−70	14	17.03		
30.	LMC 2002	05	36	46.64	−71	35	34.4	23	fast nova

NOVAE IN THE LMC AND THEIR DISTRIBUTION

Nova outbursts in the LMC have been recorded since 1926 ([9]). Thirty novae have so far been discovered in the LMC, of which one is a recurrent nova. Only 19 novae have photometric/spectroscopic data from which the speed class and other outburst properties are available. The location, speed class (based on [6]) and other known properties of the LMC novae are tabulated in Table 1.

Since the previous study of the distribution of novae in the LMC in 1988 ([10]), ten new novae and one recurrent nova have been discovered. Figure 1 shows the location of all novae in the LMC, identified by their serial number in Table 1. From the figure, it is seen that novae are widely distributed over the face of the LMC, with a lack of novae in the 30 Dor region, and an apparent clumping of novae towards the south east of the bar. It is also seen that most novae located in the bar region belong to the fast category. The

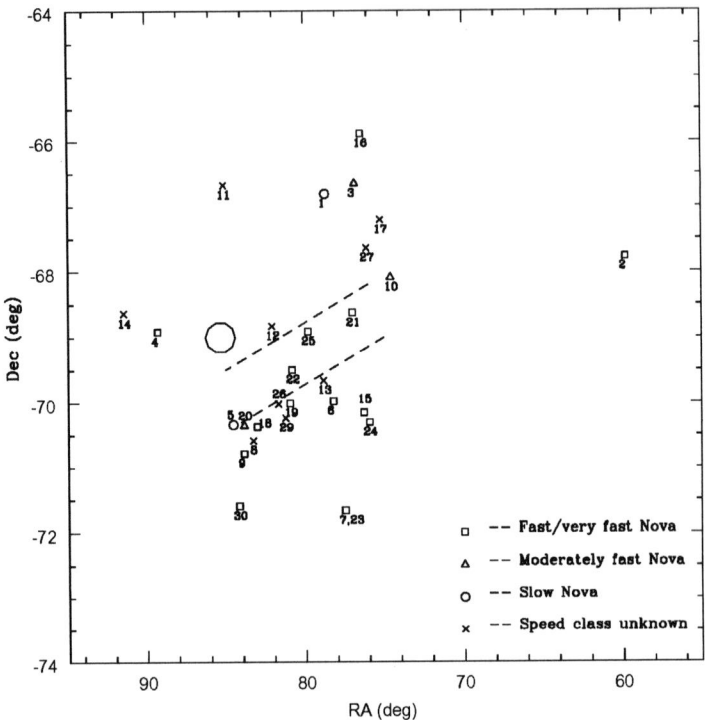

FIGURE 1. Distribution of novae across the face of the Large Magellanic Cloud. The Bar and 30 Dor regions are also shown.

LMC seems to have a high population of fast novae, 72.2% of the novae with known speed classes are fast novae. The slow novae are only 11.1%.

NOVAE NEIGHBOURHOOD STELLAR POPULATION

The ages, density and metallicity of the projected stellar population in the neighbourhood of novae are studied based on the star clusters and the field star population in that region. The data on field stars are obtained from the OGLE II survey ([11]). The data from only one survey are used since it is essential to have a homogeneity of data for an inter-comparison of the different regions. All the available catalogues of the LMC star clusters are used to locate clusters in the neighbourhood of novae.

Field stars within a radius of a few arcmin in the vicinity of novae were identified from the OGLE II survey. Photometric data were found to be available for regions around 15 novae. As more observations were found to be available in the I passband, the V vs $(V-I)$ colour-magnitude diagrams (CMDs) were used in the analyses. The details of

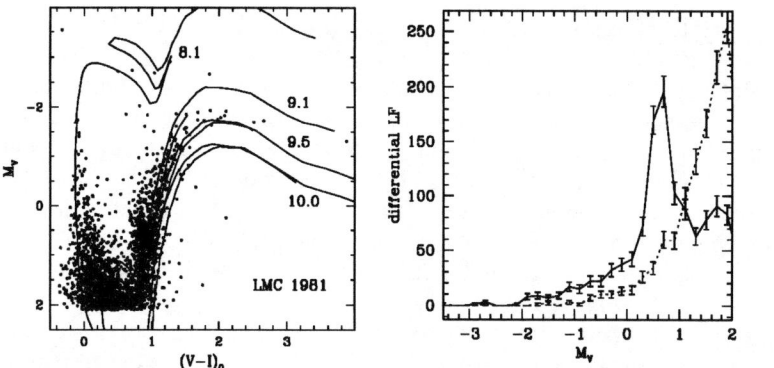

FIGURE 2. Left panel: the CMD of 2472 stars within 3 arcmin from the fast nova, LMC 1981. The isochrones fitted to the CMD, with the corresponding value of log(age) indicated, are also plotted. Right panel: LF of the MS (dotted line) and the red giants (solid line). The error bars indicate the statistical error in the data.

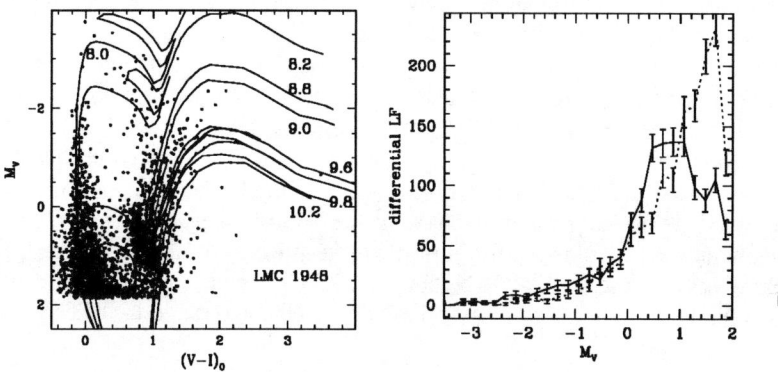

FIGURE 3. Left panel: the CMD of 2762 stars within 2 arcmin from the slow nova, LMC 1948. The isochrones fitted to the CMD, with the corresponding value of log(age) indicated, are also plotted. Right panel: LF of the MS (dotted line) and the red giants (solid line). The error bars indicate the statistical error in the data.

the analyses are presented in [12]. Figures 2 and 3 present examples of the CMDs and the LFs for field stars around a fast novae and a slow nova, respectively.

PROBABLE PARENT POPULATION OF NOVAE

The parent population of the novae is believed to belong to the intermediate age population. In the nova regions studied here, we find that the star formation began a few Gyr ago, though there are traces of old stellar population. The population which is the outcome of this star formation event is likely to be the parent population of the novae.

The SFH of the regions around the 6 fast novae studied here indicate the age range of the parent population of the fast novae is likely to be 3.2 – 1.6 Gyr.

The regions around the two moderately fast novae studied here indicate a parent population in the age range 2.5 – 1.0 Gyr.

The region around only one slow novae has been studied. The striking differences between this region and those around the fast and moderately fast novae are: (a) continuous star formation between 10 Gyr – 630 Myr, and (b) a significant stellar population belonging the age range 4 – 10 Gyr. This implies that the progenitors of slow novae probably belong to an older population, consistent with the idea of slower, fainter, bulge novae in the Galaxy and M31.

The age estimates for the novae for which the speed class is not known, show that the common age range is 1.3 – 2.5 Gyr. If one includes all the ranges, then the age of the intermediate age population is in the range 4 Gyr – 800 Myr.

Combining the results obtained for 14 regions around novae, (excluding the region around slow nova), 28.6% of the regions have 4 Gyr, 42.9% have 3.2 Gyr, 21.4% have 2.5 Gyr and 7% have 2.0 Gyr as the upper limit for age of the intermediate age population. The lower limit for the age of the intermediate age population is found to be 1.6 Gyr for 7% of the regions, 1.3 Gyr for 28.6% of the regions, 1.0 Gyr for 42.9% of the regions and 800 Myr for 21.4% of the regions. The important limit derived in this study is the upper age limit of the parent population of novae in the LMC, which is 4 Gyr. The lower age limit for the parent population of the novae is more likely to be 1 Gyr.

If we consider the most likely limit, then the progenitor age range for the fast and moderately fast novae is likely to be between 3.2 – 1.0 Gyr. The progenitor for the slow novae is likely to originate from the population, in the age range 1 – 10 Gyr.

REFERENCES

1. Duerbeck, H.W., 1984, *Ap&SS*, 99, 363
2. Ciardullo, R., Ford, H.C., Neill, J.D., Jacoby, G.H., Shafter, A.W., 1987, *ApJ*, 318, 520
3. Della Valle, M., Duerbeck, H.W., 1993, *A&A*, 271, 175
4. Della Valle, M., Rosino, L., Bianchini, A., Livio, M., 1994, *A&A*, 287, 403
5. Yungelson, L., Livio, M., Tutukov, A., 1997, *ApJ*, 481, 127
6. Duerbeck, H.W., 1990, in *Physics of Classical Novae*, ed. A. Casatella & R. Viotti (Berlin: Springer), 34
7. Della Valle, M., Bianchini, A., Livio, M., Orio, M., 1992, *A&A*, 266, 232
8. Butcher, H., 1977, *ApJ*, 216, 372
9. Buscombe, W., de Vaucouleurs, G., 1955, *Observatory*, 75, 170
10. van den Bergh, S., 1988, *PASP*, 100, 1486
11. Udalski, A., Szymanski, M., Kubiak, M., Pietrzynski, G., Soszynski, I., Wozniak, P., Zebrun, K., 2000, *Acta Astron.*, 50, 307 (OGLE II data)
12. Subramaniam, A., Anupama, G.C., 2002, *A&A*, (in press)

Novae In External Galaxies From The POINT-AGAPE Survey And The Liverpool Telescope

M. J. Darnley*, M. F. Bode*, E. J. Kerins* and T. J. O'Brien[†]

*Astrophysics Research Institute, Liverpool John Moores University, Birkenhead, CH41 1LD, UK
[†]Jodrell Bank Observatory, The University of Manchester, Macclesfield, SK11 9DL, UK

Abstract. We have recently begun a search for Classical Novae in M31 using three years of multicolour data taken by the POINT-AGAPE microlensing collaboration with the 2.5m Isaac Newton Telescope (INT) on La Palma. This is a pilot program leading to the use of the Liverpool Telescope (LT) to systematically search for and follow novae of all speed classes in external galaxies to distances up to around 5Mpc.

INTRODUCTION

The importance of the study of novae in external galaxies has been recognised since the time of Hubble [1]. However, there are great difficulties in obtaining the frequency and duration of observations with large enough telescopes to determine the peak magnitude and speed class for a meaningful sample of objects (see papers by Shafter, Shara and Della Valle, this volume).

We aim to search for and follow the temporal development of novae across a range of speed classes in selected galaxies over several years. Our goals are to tighten the MMRD, t_{15} and other relationships, which make classical novae potentially very important distance indicators [2]. We hope to help resolve the debate about the dependence of nova rate with galaxy type and stellar population [3, 4]. Finally, we wish to determine whether there are indeed two distinct populations of classical novae [5] and how this might relate to current models of the outburst.

This project is being pushed on two fronts: first with existing data from the POINT-AGAPE gravitational microlensing survey and then with the soon-to-be-commissioned Liverpool Telescope (LT). We plan to automate as far as possible the data reduction and novae selection processes.

Identification of Novae

We are identifying novae using a method similar to that outlined in Shafter et al [6]. The images are first carefully aligned and then have their point-spread functions matched. We identify strongly varying sources by subtracting the aligned and matched

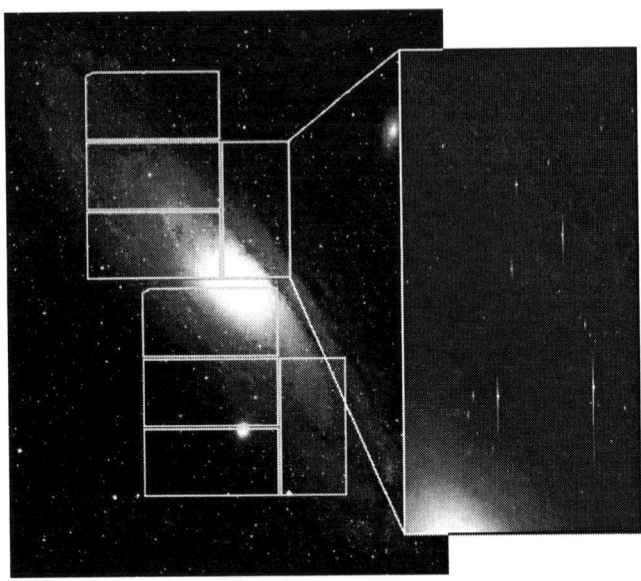

FIGURE 1. The Andromeda Galaxy and the positions of the two POINT-AGAPE fields. (Image adapted from an original image obtained by Bill Schoening, Vanessa Harvey/REU programme/AURA/NOAO/NSF.) Inset shows one of the four chips from one of our INT WFC frames.

images from one another and searching for any residual sources. Photometry is carried out on the data using the DAOPHOT package within the IRAF environment.

In this dataset we are sensitive to objects down to $r' \simeq 23$, though our sensitivity is a strong function of position and decreases towards the bulge. Ultimately we plan to use simulated datasets to correct for this spatial bias.

THE POINT-AGAPE SURVEY

POINT-AGAPE is an Anglo-French collaboration, which is searching for gravitational microlensing events against the mostly unresolved stars in M31 [7]. The survey team is employing a modified detection technique dubbed "pixel lensing" developed to cope with the problems associated with temporal variations in seeing and sky background. The project has evolved from the pilot AGAPE (Andromeda Galaxy Amplified Pixels Experiment) programme conducted at Pic du Midi, which probed the M31 bulge [8]. POINT-AGAPE is using the wide-field camera (WFC) on the INT to survey a much larger area (POINT is an acronym for Pixel-lensing Observations with the INT). Since autumn 1999, two fields covering 0.6 square degrees of the M31 disk have been monitored regularly in three colours (Sloan-like g', r' and i' filters, which correspond to wavelengths of 486, 622 and 767 nm, respectively). The locations of the two fields are shown in Figure 1. To date three seasons of data comprising roughly 180 epochs have been collected.

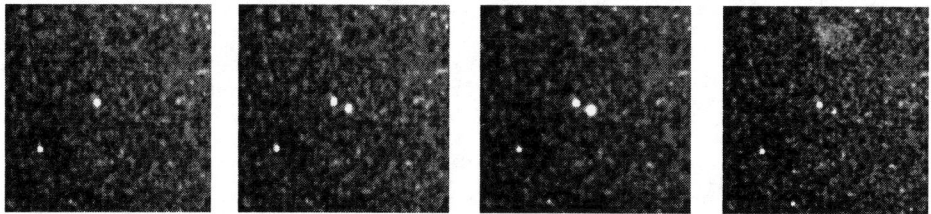

FIGURE 2. An r'-band image sequence of the first nova candidate from the POINT-AGAPE INT survey.

The prerequisites for the detection of microlensing are good temporal sampling and a long baseline of observations and it is these same qualities that make the POINT-AGAPE dataset an excellent repository for novae searches. Since novae are much brighter than typical microlensing events they should, for the most part, show up as resolved sources, negating the need for us to employ the pixel-lensing method.

However, with around 50GB of data to analyse, the task of finding and cataloging novae in M31 remains a challenging one. To meet this challenge we are developing algorithms to automatically reduce the data and search for novae. Parts of this software have already allowed us to discover two probable novae candidates in a small subset of the dataset provided by the POINT-AGAPE collaboration.

First Results

We have identified two nova candidates in the POINT-AGAPE dataset. Both novae were initially discovered in the r'-band data and then located in both the g' and i' bands. The first nova candidate became visible on 7th September 1999, was caught during the initial rise in both r' and g' bands, and was subsequently followed to provide an estimate of $t_2 \simeq 20$ days, classing this candidate as a fast nova [9]. Figure 2 shows a sequence of r'-band images which straddle the outburst. The second nova was discovered near maximum at the beginning of the first observing season on 2nd August 1999. We estimate $t_2 \simeq 30$ days for this event, making it a moderately fast nova [9]. Figures 3 and 4 show the r'-band and i'-band lightcurves for the first and second nova candidates, respectively. The photometric calibration of these lightcurves is still being worked on, but these results illustrate the types of nova observations that we will be able to secure.

The two novae candidates have been found whilst testing our detection methods on a small 1024x1024 pixel section of the data over a period of about three months, amounting to less than 0.3% of the entire dataset. We are now beginning to search for variable objects in the whole dataset and, based on estimates of the M31 novae rate of $29 \pm 4 \, \text{yr}^{-1}$ [10], we are expecting to find 30-50 classical novae. One of the advantages of our dataset is excellent temporal coverage which should allow us to produce reasonably complete lightcurves, hopefully with many sampling the all-important time of maximum light.

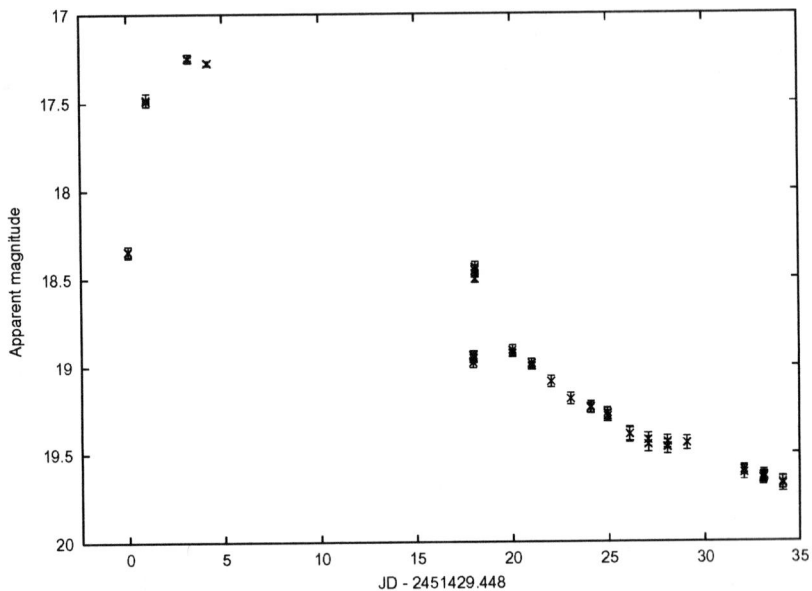

FIGURE 3. r'-band lightcurve of the first nova candidate from the POINT-AGAPE survey data.

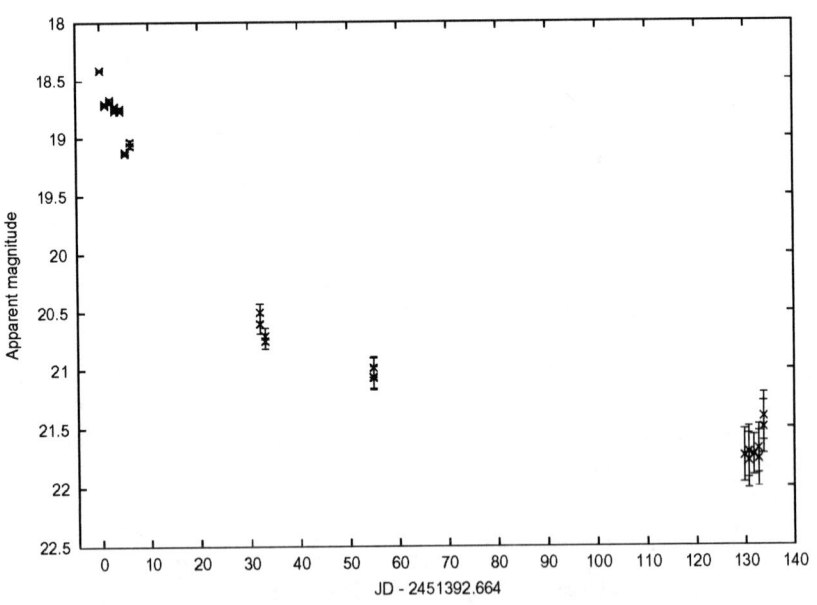

FIGURE 4. i'-band lightcurve of the second nova candidate from the POINT-AGAPE survey data.

THE LIVERPOOL TELESCOPE

The Liverpool Telescope, with a primary mirror diameter of 2m, will be the World's largest fully-robotic telescope. The LT is currently being reassembled at the Observatorio del Roque de Los Muchachos, La Palma. Its primary instruments are a CCD camera with a 12-position filter wheel, a near-IR camera and a fibre-fed spectrograph. The telescope is able to move rapidly to acquire targets and will have arc-second pointing.

The main scientific goals of the LT are: (i) the monitoring of variable objects on all timescales, from seconds to years; (ii) the ability to rapidly react to unpredictable phenomena and their systematic follow up; (iii) simultaneous or coordinated observations with other facilities, both on the ground and in space; and (iv) small-scale surveys and serendipitous source follow-up. Further details about the LT can be found at the telescope's website (http://telescope.livjm.ac.uk).

Guaranteed time has been approved for our novae in external galaxies program and the telescope is expected to see first light in the autumn of 2002. We will use the unique capabilities of the LT to systematically monitor a carefully selected set of external galaxies; M81, M64 and M94, to discover and follow as complete a sample of novae as possible in each system. We shall be able to detect objects down to $m_V \approx 24$. Over a three-year programme, we will discover more classical novae than have been discovered in our own Galaxy to date, and in many cases provide better lightcurve and distance determinations. A by-product of the survey will be the discovery of many other luminous variables in our target galaxies.

ACKNOWLEDGMENTS

We wish to thank the POINT-AGAPE collaboration for use of its data. MJD would like to thank Andy Newsam for help with preparing some of the figures. MJD's research is supported by a PPARC PhD studentship.

REFERENCES

1. Hubble E., 1929, ApJ, 69, 103
2. Warner B., 1989, Classical Novae, J. Wiley & Sons, p8
3. Della Valle, M., Rosino, L., Bianchini, A., Livio, M., 1994, A&A, 287, 403
4. Yungelson, I., Livio, M., Tutukov, A., 1997, ApJ, 481, 127
5. Della Valle, M., Bianchini, A., Livio, M., Orio, M., 1992, A&A, 266, 232
6. Shafter A.W., Ciardullo R., Pritchet C. J., 2000, ApJ, 530, 193
7. Kerins E.J. et al., 2001, MNRAS, 323, 13
8. Ansari R. et al., 1997, A&A, 324, 843
9. Payne-Gaposchkin, C., 1957, The Galactic Novae, North-Holland
10. Capaccioli M., Della Valle M., D'Onofrio M., Rosino L., 1989, AJ, 97, 1622

Early decline spectra of nova SMC 2001 and nova LMC 2002

Elena Mason*, Massimo Della Valle[†], Roberto Gilmozzi*, Robert E. Williams** and Gaspare Lo Curto*

*ESO, Alonso de Cordova 3107, Vitacura, Casilla 19001, Santiago, Chile
[†]Osservatorio Astrofisico di Arcetri, L.go E. Fermi 5, 50125 FI, Italy
**STScI, 3700 San Martin Drive, Baltimore, MD 21218, USA

Abstract. Spectra of two recent novae in the Magellanic Clouds –nova SMC 2001 and nova LMC 2002– are presented and analyzed. The classification of the two novae is then briefly discussed in relation to the nova populations recognized for different Hubble type galaxies.

INTRODUCTION

Spectroscopic observations of extra-galactic novae are relatively rare. We present in this paper the spectra of two recent novae in the Magellanic Clouds, Nova SMC 2001 and Nova LMC 2002. These data are twice important as they provide important pieces of information on both the single nova event (e.g. ejected mass, WD mass and type), and help to constraining the properties of nova populations in different Hubble type galaxies. Indeed the existence of systematic differences affecting the nova populations belonging to different Hubble type galaxies have been recognized only on the basis of the observed nova rate and on the photometric evolution of the novae, never through the spectroscopic investigations [e.g. 2].

DATA

Nova SMC 2001

Nova SMC 2001 was discovered by Liller (IAUC 7738) on Oct 21.09 UT at a magnitude V=12.21. First spectrum was taken on Oct 26.149 UT by Bosch et al. (IAUC 7744), who reported strong low excitation FeII, OI, [OI], and NaI emission lines and broad P-Cyg profiles. They measured an expansion velocity of 1800 and 950 km/sec from the absorption of the Hβ and low ionization lines, respectively. They also measured a FWHM of 1400 km/sec from the Hα line. Similar spectra have been observed by Della Valle et al. and Jensen et al. on Oct 30.25 UT and Nov 2.12 UT, respectively (IAUC 7743). However, Jensen et al. report first evidence of CII emission and point out that the nova strongly resemble Nova LMC 1988 N.1 two weeks after maximum.

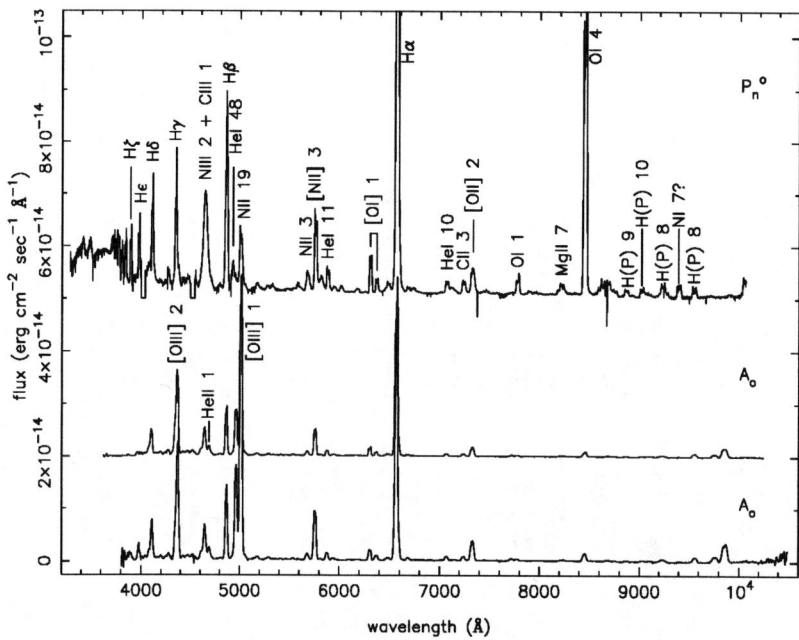

FIGURE 1. Nova SMC 2002 early decline spectra observed at NTT+EMMI. Spectra taken during the same night but with different set up have been combined.

We observed nova SMC 2001 at the NTT+EMMI during three different nights (December 6 2001, February 17 2002, and March 4 2002, figure 1). The December spectrum of Nova SMC 2001 shows a considerable evolution with respect to the near maximum spectra described above. The P-*Cyg* profiles are disappeared and low ionization emission lines (FeII, OI, and NaI) have been replaced by NII, NIII, HeI and CII, as well as forbidden transition from O and N. In particular, the NIII blend at 4640Å is quite strong, though the strongest non Balmer line is still the OI line at 8446Å. The forbidden transition consist on the auroral lines [NII]λ5755 and [OII]$\lambda\lambda$7319,7330, and the nebular lines [OI]$\lambda\lambda$6300,6364. All lines show double peaked saddle shaped profile with FWHM of 1400-1900 km/sec and HWZI between 1200 and 2500 km/sec[1]. We classify this spectrum as P$_{no}$ according to the Cerro Tololo classification [3, 4].

February and March spectra are very similar each other and dominated by strong auroral and nebular lines. The nebular doublet [OIII]$\lambda\lambda$4959,5007 is as strong as the Hα emission line, while [OIII]λ4363 dominates among the auroral lines. Permitted lines have all weakened but the NIIIλ4640 emission is now flanked by the HeIIλ4686 line. Both spectra are classified as A$_o$ according the Cerro Tololo scheme.

---

[1] larger values correspond to the Hα line which is blended with the [NII]

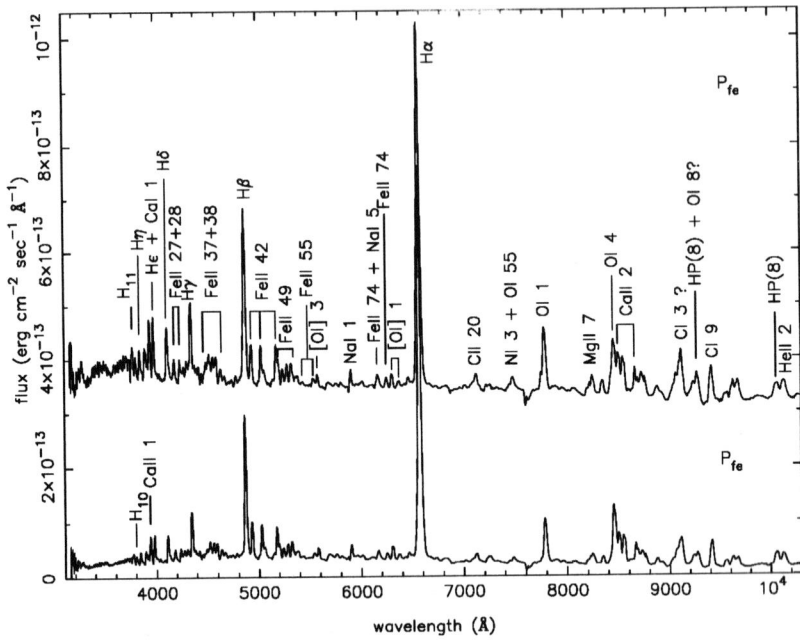

FIGURE 2. Nova LMC 2002 near maximum spectra observed at 3.6+EFOSC. Spectra taken during the same night but with different set up have been combined.

Nova LMC 2002

Nova LMC 2002 outburst maximum occurred close to March 5, 2002 (m_V=10.5, Liller IAUC 7841). We present low resolution spectra (3.6+EFOSC, grisms 9, 12, 14, and 15) on two different nights near maximum (9 and 14 march 2002), when the nova had V magnitude of ∼12. The FWHMs of the Balmer lines provide an average expansion velocity of ∼ 1500 km/sec or less. All lines have single peaked and quite symmetric profile. Spectra from both nights are dominated by Balmer and low ionization emission lines such as the FeII multiplets 27, 28, 37, 38, 42, 49, 74, OI, CaII, MgII, and NaI. Weak emission lines from high ionization elements (HeII, CII and NI) are detected above 7000 Å. Both spectra show also auroral and nebular O lines ([OI]λ5577 and [OI]$\lambda\lambda$6300,6364, respectively). The spectroscopic evolution exhibited by the Nova during the week in between our two observing runs is rather modest and indicates the progressive weakening of the continuum associated to the decrease in intensity of the OIλ7774 line. The spectra near maximum of Nova LMC 2002 are quite similar to those of Nova LMC 1988 N.1 and, Nova Scu 1989, at similar stage, and can be classified as P_{fe}, according to the Cerro Tololo scheme. It seems quite straightforward to conclude that nova LMC 2002 belongs to the *FeII* spectroscopic class identified by Williams [5].

TABLE 1. Flux ratios of the [OI] emission lines $\lambda 6300/\lambda 6364$. We derived also the opacity, τ, and T_e(OI) following Williams [6].

object	obs. date	$r = \frac{F_\lambda(6300)}{F_\lambda(6364)}$	$\tau(6300)$	T_e(K)
nova SMC 2001	06/12/01	2.00	1.44	4600
	07/02/02	2.04	1.36	3400
	04/03/02	2.71	0.32	3700
nova LMC 2002	09/03/02	1.68	2.30	5348
	14/03/02	2.11	1.22	5640

The $\frac{F_\lambda(6300)}{F_\lambda(6364)}$

The spectral range covered by our data allowed us to compute the intensity ratio of [OI] lines which is for most novae below the theoretical value of 3:1 predicted by atomic constants. See Williams [6] for a discussion of this phenomenon.

The absolute magnitude

The well known distances of the Magellanic Clouds [e.g. 7, 8], allow us to determine the absolute maximum magnitude of the novae. We derive $M_V = -6.6$ and $M_V = -7.9$ for nova SMC 2001 and nova LMC 2002, respectively[2]. According to these values, nova SMC 2001 and LMC 2002 fall in the faint tail and middle of the Zwicky's relationship discussed by Della Valle and Livio [1].

ARE NOVA SMC 2001 AND NOVA LMC 2002 PECULIAR?

Our spectroscopic follow up of Nova SMC 2001 is one of the rare observations of novae in the SMC. Our data seem to provide evidence for a *FeII-slow*[3]. The only other spectrum in the literature refers to nova SMC 1994 [12], which was a *slow* nova ($t_2 > 33$) and showed a *standard nebular* spectrum (N_o) at the time of the observation. We can thus assume that nova SMC 1994 was a typical *FeII* nova, too.

The spectrum at maximum of nova LMC 2002 suggests that this object belong to the *FeII* class. Similarly, LMC novae completed of a spectroscopic follow up since their discovery (novae LMC 1988 N.1 and N.2, 1990 N.1 and N.2, 1991, 1992, 1995, and 2000), show low ionization emission lines, often flanked by P-*Cyg* absorptions. Line widths are narrow and provide low expansion velocities, but can evolve to broad while

---

[2] In the computation of the distances we assumed the interstellar extinction, E(B-V), of 0.09 and 0.11 for the SMC and the LMC nova, respectively. The E(B-V) values were derived by the average of values in the literature [8, 9, 10, 11]

[3] Though the high ionization emission lines which are visible in the December spectrum possibly indicate an hybrid nova, the FWHM are typical of a slow nova.

high ionization emission replace the FeII lines within few days (e.g. nova LMC 1991). Thus, among the 8 novae listed above 3 belong to the *hybrid* or *FeII-broad* class, and 5 appear to be *FeII* novae. Whenever reported their photometric evolution appears to be *fast*.

REFERENCES

1. Della Valle, M., Livio, M., 1995, *ApJ*, **452**, 704
2. Della Valle, M., Rosino, L., Bianchini, A., Livio, M., 1994, *A&A*, **287**, 403
3. Williams, R.E., Hamuy, M., Phillips, M., Heatcote, S., Wells, L., 1991, *ApJ*, **376**, 721
4. Williams, R.E., Phillips, M., Hamuy, M., 1994, *ApJS*, **90**, 297
5. Williams, R.E., 1992, *AJ*, **104**, 725
6. Williams, R.E., 1994, *ApJ*, **426**, 279
7. Panagia, N, in Proceedings of IAU Symposium 190, *New views of the Magellanic Clouds*, p.53
8. Massey, P., Lang, C.C., Degioia-Eastwood, K., Germany, C.D., 1995, *ApJ*, **438**, 188
9. Schwering, P.B.W., Israel, F.P., 1991, *A&A*, **246**, 231
10. Capaccioli, M. ... 1990, *ApJ*, **360**, 63
11. Bessel, M.S., 1991, *A&A*, **242**, L17
12. de Laverny, P., et al., 1998, *A&A*, **335**,93
13. Subramaniam, A., Anupama, G.C., 2001, *BASI*, **29**, 379
14. Della Valle, M., Livio, M., 1998, *ApJ*, **506**, 818

A Survey for Novae in M33: Preliminary Results

Stephen J. Williams and Allen W. Shafter

Department of Astronomy, San Diego State University, San Diego, CA 92182-1221, USA

Abstract. An estimate of the nova rate in M33 provides a valuable constraint on the rate of nova production in a disk-dominated, essentially bulgeless galaxy. Presented here are the preliminary results of a new Hα survey for novae in M33. Data analyzed thus far includes the inner $13' \times 13'$ of the galaxy, yielding four novae detected to a limiting magnitude of $m_{H\alpha} \simeq 17.5 \pm 0.3$. After correcting for the fraction of M33 covered in the survey, a preliminary nova rate for M33 of $4 \pm 2 \text{ yr}^{-1}$ is found.

INTRODUCTION

The discovery of novae in external galaxies provides a powerful tool for studying the properties of novae from differing stellar populations. In recent years, surveys of many galaxies spanning a wide range of Hubble types have been undertaken [1, 2, 3]. Despite the wealth of data, there remains no firm consensus as to whether the observed properties of novae (frequency, luminosity, rate of decline) are sensitive to the underlying stellar population. Della Valle et al. [2] have proposed that late-type galaxies with a significant population I component are more prolific nova producers than are their earler type counterparts. However, Ciardullo et al. [1] and more recently Shafter, Ciardullo, & Pritchet [3] argue that the available evidence is not sufficiently compelling to conclude that the nova rate varies systematically with Hubble type.

The nearby spiral galaxy, M31, has been the frequent target of nova surveys beginning with the pioneering study of Hubble [4] and continuing to the present day. A major advantage of targeting M31 is that the absolute nova rate is quite high, with an estimated 30–40 novae erupting per year [5]. A principal drawback of the M31 studies is that the galaxy's spatial orientation (seen at a large inclination angle) makes it difficult to unambiguously separate the galaxy's bulge novae from its disk novae (e.g. see Hatano et al. [6]). On the other hand, despite its relatively low absolute nova rate, the nearby and essentially bulgeless spiral galaxy, M33, provides a good opportunity to determine the nova rate in a population I system. M33 has been studied previously by Della Valle et al. [2] who found a nova rate of $4.6 \pm 0.9 \text{ yr}^{-1}$. Here we present a preliminary analysis of a new nova survey of this important galaxy.

OBSERVATIONS AND DATA ANALYSIS

Observations of M33 were obtained with the Mount Laguna Observatory 1-m reflector. A 2048^2 pixel CCD Camera was used at the f/7.6 Cassegrain focus, resulting in a field

TABLE 1. M33 Novae: Magnitudes at Discovery

Nova	UT Date	Julian Date*	$m_{H\alpha}$	m_B
1995–1	1995 Aug 26	49956	15.4	18.8
1995–2	1995 Oct 28	50012	15.7	19.0
2001–1[†]	2001 Nov 19	52233	15.0	–
2001–2	2001 Nov 19	52233	16.2	–

* 2,400,000+
[†] Discovered independently by O. Trondal [9]

of view of approximately $13' \times 13'$ at a plate scale of $\sim 0.4''$ per pixel. Because novae remain bright in Hα considerably longer than in broad band B [7], we chose to take our images through an Hα filter in addition to broad band B. The Hα filter is centered near 6560 Å and has a FWHM of ~ 70 Å. The complete survey consists of a total of five overlapping fields, with four of the fields forming a 2×2 grid centered on the galaxy, and the fifth, providing considerable overlap with the other four, centered directly on the nucleus. In this preliminary report, we only consider observations from this central field. A series of three 10-min Hα (and occassionally 5-min B) exposures of each field were obtained during 44 nights spanning the period between 1995 July and 2001 November.

Data reduction and analysis was performed with IRAF[1]. After standard processing, which included bias subtraction and flat fielding, individual images from a given night were combined by median-stacking the images with the "imcombine" routine. The median stacking procedure eliminates statistical variations from image-to-image and assures that individual image artifacts, such as cosmic ray events, are suppressed in the final combined image. To find novae candidates two techniques were employed: image subtraction and blinking. In both instances, images from different epochs were aligned and scaled using a sample of field stars common to each image. The resulting images were then either subtracted, and novae searched for on the differenced frame, or blinked, and novae discovered by eye. Confirmation of novae required the appearance of the candidate on successive nights or on successive images from one night, if later nights were not available.

The analysis of the central field has yielded a total of four novae, which are listed in Table 1. Nova Hα magnitudes were estimated by comparison with Hα standards in the field of M31 [5], while B standards were taken from Ivanov, Freedman & Madore [8]. We estimate that we have detected all novae brighter than $m_{H\alpha} = 17.5 \pm 0.3$, which we take to be the limiting magnitude of our survey. We have sufficient temporal coverage to construct a light curve only in the case of nova 1995-1. The light curve, which is shown in Figure 1, exhibits behavior similar to that found for novae in M31 [7]. Specifically, shortly after maximum light (which we probably missed) the Hα and B light curves track each other well, with the nova being ~ 3 magnitudes brighter in Hα.

---

[1] IRAF (Image Reduction and Analysis Facility) is distributed by the National Optical Astronomy Observatories, which are operated by AURA, Inc., under cooperative agreement with the NSF.

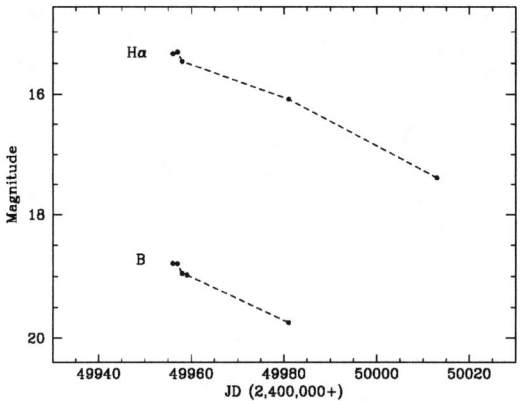

FIGURE 1. The Hα and B light curves of the M33 nova 1995-1.

THE NOVA RATE

To estimate the nova rate in M33 we use the mean nova lifetime approach of Ciardullo et al. [7], which is based on Zwicky's [10] supernova study. In the mean lifetime approach the observed nova rate is corrected for the "effective survey time", which depends on the time that a typical nova is expected to remain brighter than the limiting absolute magnitude of the survey (the mean nova lifetime) and the time sampling of the observations. In effect, it is the total time period sampled by the survey when a typical nova would be detectable. If, following Ciardullo et al. [7], we define τ_c to be the mean nova lifetime, then for multi-epoch observations, the effective survey time is given by

$$T(M < M_c) = \tau_c + \sum_{i=2}^{n} \min(t_i - t_{i-1}, \tau_c). \quad (1)$$

For a total of $N(M < M_c)$ observed novae brighter than the limiting absolute magnitude M_c, the global nova rate in the region of the galaxy surveyed, is then

$$R = \frac{N(M < M_c)}{T(M < M_c)}. \quad (2)$$

If $\tau_c(M_c)$ is known, R can be calculated directly from the nova observations and a knowledge of M33's distance.

Given a distance modulus for M33 of 24.52 ± 0.14, a total reddening of $B - V = 0.2 \pm 0.04$ [11], and our limiting survey magnitude of $m_{H\alpha} = 17.5 \pm 0.3$, we find a limiting magnitude of $M_{H\alpha} = -7.52 \pm 0.34$. If we adopt the most recent second-order $\tau_c(M_c)$ relation from the M31 study of Shafter & Irby [5], we arrive at a mean nova lifetime, $\tau_c = 71 \pm 19$ days. Armed with this mean nova lifetime and our dates of observation, eq (1) yields an effective survey time, $T(M < M_c) = 895 \pm 240$ days. Since a total of four novae were detected over this time period, we obtain a nova rate of 1.6 ± 0.8 yr^{-1} in the surveyed region of M33.

To extrapolate this rate to the entire galaxy we must estimate the fraction of M33's total luminosity covered in our survey. We can estimate the luminosity sampled in our survey by appealing to the H-band aperture photometry of Aaronson, Mould, & Huchra [12]. By interpolating the values from their Table 1, we find that a $13' \times 13'$ field centered on the nucleus of M33 has an integrated magnitude of $H \simeq 4.6$. If we assume $H - K = 0.2$ as representative of Sc galaxies [13], we find that our survey includes an integrated magnitude of $K \simeq 4.4$. According to the data presented in Shafter, Ciardullo, and Pritchet [3], M33 has an integrated magnitude of $K \simeq 3.4$. Thus, we conclude that our central field samples $\sim 40\%$ of M33's total K light. Scaling the nova rate in our central field to the entire galaxy thus yields a global nova rate of $\sim 4 \pm 2$ yr^{-1}.

SUMMARY

The preliminary results of our M33 nova survey suggest a nova rate of 4 ± 2 yr^{-1}, which is consistent with the value of 4.6 ± 0.9 yr^{-1} found by Della Valle et al. [2]. In the near future we hope to strengthen our result with the inclusion of the four outer fields, which were taken as part of the survey but have not yet been analyzed. The inclusion of these outer fields will enable our survey to cover a larger fraction of the total luminosity of M33, and thus minimize the uncertainty resulting from extrapolating the nova rate in the surveyed region to the entire galaxy.

ACKNOWLEDGMENTS

We thank Scott Dahm, Jiyune Lee, & L. Lee Clark for their assistance with observing, and Denise Pollard for help in improving the presentation of this paper. This work has been supported by NSF grant AST 97-31641.

REFERENCES

1. Ciardullo, R., Ford, H. C., Williams, R. E., Tamblyn, P., and Jacoby, G. H., *AJ*, **99**, 1079 (1990).
2. Della Valle, M., Rosino, L., Bianchini, A., and Livio, M., *A&A*, **287**, 403 (1994).
3. Shafter, A. W., Ciardullo, R., and Pritchet, C. J., *ApJ*, **530**, 193 (2000).
4. Hubble, E., *ApJ*, **69**, 103 (1929).
5. Shafter, A. W., and Irby, B. K., *ApJ*, **563**, 749 (2001).
6. Hatano, K., Branch, D., Fisher, A., and Starrfield, S., *ApJL*, **487**, L45 (1997).
7. Ciardullo, R., Shafter, A. W., Ford, H. C., Neill, J. D., Shara, M. M., and Tomaney, A. B., *ApJ*, **356**, 472 (1990).
8. Ivanov, G. R., Freedman, W. L., and Madore, B. F., *ApJS*, **89**, 85 (1993).
9. Trondal, O., , Tech. rep., I.A.U. Circ. 7756 (1988).
10. Zwicky, F., *ApJ*, **96**, 28 (1942).
11. Lee, M. G., Kim, M., Sarajedini, A., Geisler, D., and Gieren, W., *Astrophysical Journal*, **565**, 959 (2002).
12. Aaronson, M., Mould, J., and Huchra, J., *ApJ*, **237**, 655 (1980).
13. Aaronson, M., , Ph.D. thesis, Harvard University, Cambridge, MA 02138 (1977).

OLD NOVA SHELLS AND REMNANTS

The Evolution of Nova Remnants

Michael F. Bode

*Astrophysics Research Institute, Liverpool John Moores University,
Twelve Quays House, Birkenhead, Wirral, CH41 1LD, UK*

Abstract. In this review I concentrate on describing the physical characteristics and evolution of the nebular remnants of classical novae. I also refer as appropriate to the relationship between the central binary and the ejected nebula, particularly in terms of remnant shaping. Evidence for remnant structure in the spectra of unresolved novae is reviewed before moving on to discuss resolved remnants in the radio and optical domains. As cited in the published literature, a total of 5 remnants have now been resolved in the radio and 44 in the optical. This represents a significant increase since the time of the last conference. We have also made great strides in understanding the relationship of remnant shape to the evolution of the outburst and the properties of the central binary. The results of various models are presented. Finally, I briefly describe new results relating to the idiosyncratic remnant of GK Per (1901) which help to explain the apparently unique nature of the evolving ejecta before concluding with a discussion of outstanding problems and prospects for future work.

INTRODUCTION

The observation and modelling of nova remnants are important from several standpoints. For example, imaging and spectroscopy of optically resolved remnants allow us to apply the expansion parallax method of distance determination with greater certainty than any other technique (it should be noted that without knowledge of the three-dimensional shape and inclination of a remnant, this method is prone to significant error). On accurate distances hang most other important physical parameters, including energetics and ejected mass. In addition, remnant morphology (and potentially the distribution of abundances) can give vital clues to the orientation and other parameters of the central binary and the progress of the TNR on the white dwarf surface. Finally, a fuller understanding of nova remnants has implications for models of the shaping of planetary nebulae and (in at least one case) physical processes in supernova remnants.

It is now of course almost thirteen years since the last major conference on Classical Novae, held in Madrid. At that time, a comprehensive review of the optical imagery of novae was given by Richard Wade [1] and of radio remnants by Bob Hjellming [2]. In the abstract of Wade's review, it was stated:

"There is room for much additional work in discovering new remnants and in characterising those that are known", and

"The mechanism that shapes the remnants is not yet known with certainty".

On both these counts, substantial progress has been made, particularly for the optically-resolved remnants.

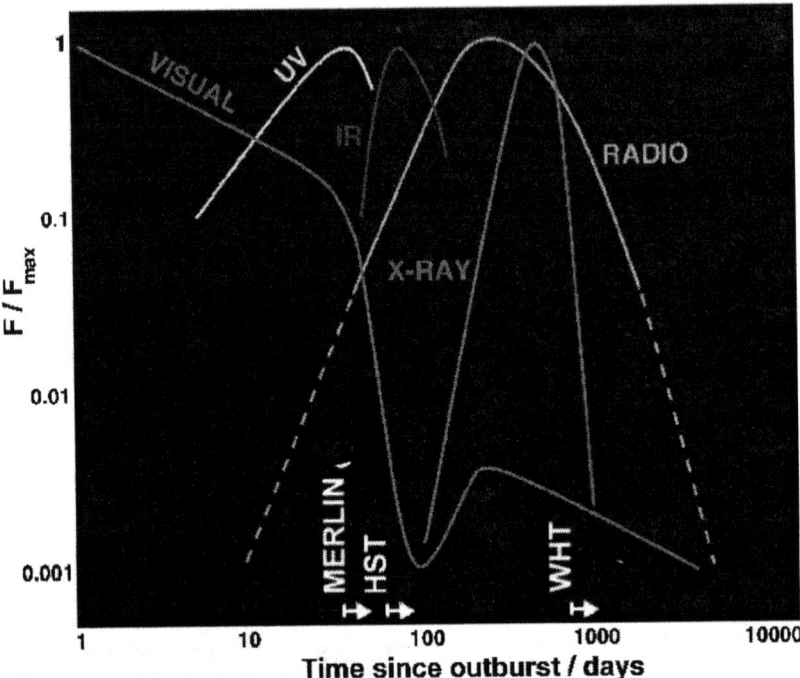

FIGURE 1. Schematic multi-frequency development of a nova outburst with times at which a remnant with v_{exp} = 1000 km s^{-1}, d = 1 kpc becomes spatially resolved in the radio (MERLIN), plus optically from space (HST) and on a conventional ground-based telescope (WHT).

EVIDENCE FOR STRUCTURE IN UNRESOLVED REMNANTS

Intermediate to high-resolution spectroscopy of the spatially unresolved remnants of many novae has long been known to show evidence of organised structure (see e.g. [3] and review by Shore, this volume). This has allowed simple models of remnant geometry to be formulated and the existence of structures such as equatorial and tropical rings and polar caps/blobs to be inferred (see Fig. 2). However, optical depth effects (for the permitted lines) and variations in ionisation/excitation mean such analysis should be treated with caution. It is only when we spatially resolve the shell that we can (with care) become more confident in our conclusions.

In the case of HR Del (1967), Solf [4] performed heroic ground-based spectroscopic observations in exceptional seeing which led him to a more accurate appreciation of the true structure of the remnant (see paper by O'Brien, this volume, for recent optical imagery and spectroscopy of the resolved remnant which yield definitive results).

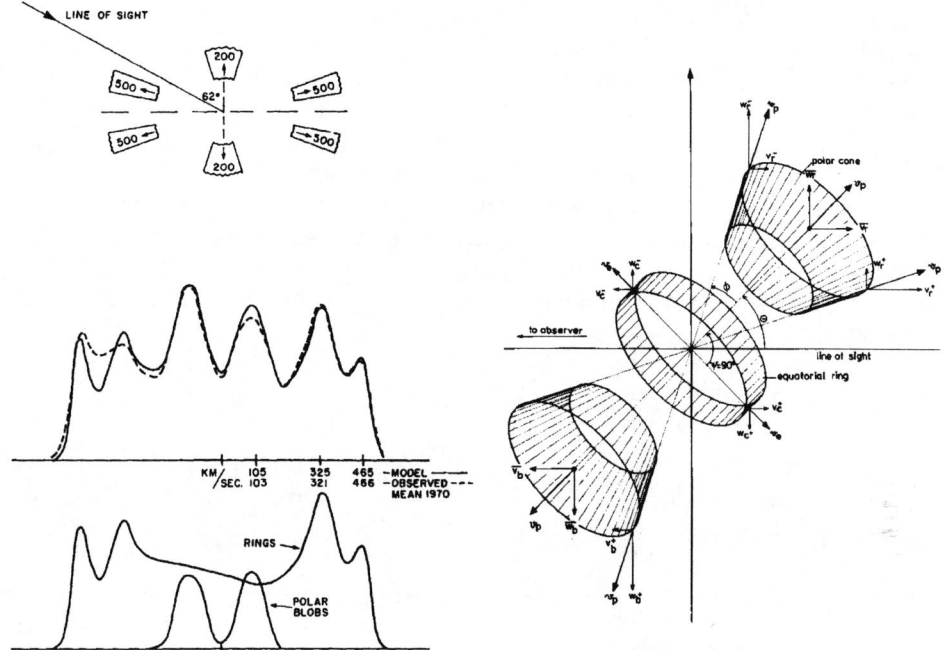

FIGURE 2. Mean Balmer line profiles in 1970 and computed model (left) for HR Del (reprinted from [3]) and (right) model of the same ejecta from high spatial resolution spectroscopy 14-15 years post-outburst (reprinted from [4]). Solf concluded that the remnant was a prolate ellipsoid with polar caps and equatorial ring, with an inclination $i = 38°$.

EVOLUTION OF REMNANTS IN THE RADIO

Gehrz (this volume) gives a comprehensive review of the physical parameters that might be derived from a combination of infrared and radio observations of novae. It is evident from his review, and with reference to Fig. 1, that a typical nova can theoretically be resolved by a radio array such as MERLIN whilst still on the optically thick, rising part of the radio light curve. According to a simple model of radio evolution by Seaquist [5], the peak flux, angular size at peak and time to peak are given approximately by:

$$f_{max} \sim 43 \, (\nu/5 \text{ GHz})^{1.16} (T_e/10^4 \text{K})^{0.46} (M_{ej}/10^{-4} M_\odot)^{0.8} (d/\text{kpc})^{-2} \text{ mJy}$$

$$\theta_{max} \sim 0.6 \, (\nu/5 \text{ GHz})^{-0.42} (T_e/10^4 \text{K})^{-0.27} (M_{ej}/10^{-4} M_\odot)^{0.4} (d/\text{kpc})^{-1} \text{ arcsec}$$

$$t_{max} \sim 1.3 \, (\nu/5 \text{ GHz})^{-0.42} (T_e/10^4 \text{K})^{-0.27} (M_{ej}/10^{-4} M_\odot)^{0.4} (v_{exp}/1000 \text{ km s}^{-1})^{-1} \text{ yr}$$

Thus typically, ~1 yr after outburst, at the time of maximum in the radio light curve, the surface brightness at 5 GHz, $\Sigma_5 \sim 300$ μJy/beam with MERLIN (cf. ~ 50 μJy/beam rms noise) and the expanding remnant should be detectable at least until this time. Thereafter, $f_\nu \propto t^{-3}$ and $\Sigma_5 \propto t^{-5}$; i.e. unless the ejecta are clumped, or (less likely) T_e is increasing, the remnant will rapidly become undetectable.

At the time of the last conference, Hjellming [2] listed only one nova (QU Vul) with a radio-resolved remnant (we do not include GK Per here for reasons we address below). Table 1 shows that in the published literature there are now four novae where the early evolution of the resolved remnant has been followed in some detail. We suspect however that there are more observations of novae in the VLA archive that deserve attention. It should also be noted that the extensive radio observations conducted by the late Bob Hjellming of V1974 Cyg (1992) have so far only been published in conference proceedings [6].

TABLE 1. Radio-resolved Remnants

	Speed Class [7]	Time after outburst first reported resolved (days)	Primary Reference
QU Vul (1984)	F	289	[8]
V1974 Cyg (1992)	F	80	[9]
V705 Cas (1993)	MF	585	[10]
V723 Cas (1995)	VS	840	Heywood (this volume)

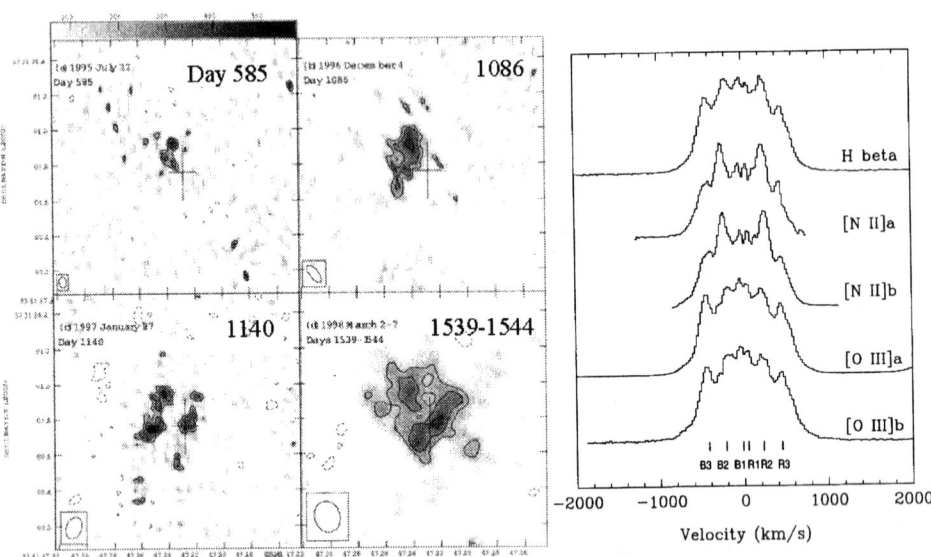

FIGURE 3. MERLIN maps at 5GHz showing the radio evolution of the remnant of V705 Cas (1993 - left, each box ~1.4″ on a side) and WHT optical spectra (right) taken on day 963 (reprinted from [10]).

Hjellming attempted to formulate a unified model of the radio and optical development of V1974 Cyg [6]. This involved a hybrid of the variable wind and Hubble flow models comprising a terminating wind with a linear velocity gradient (see review by Gehrz, this volume). In addition, he required the inner and outer shell boundaries to have different ellipsoidal shapes and an initial temperature rise was included. From this, he derived reasonable fits to the radio behaviour, but the fits to the spectral line shapes from HST observations were obviously too simplistic. Overall, this was a valiant attempt, but the model was largely phenomenological.

Figure 3 shows MERLIN maps of the complex evolution of the remnant of V705 Cas through the optically thick to early optically thin phases [10]. Also shown are WHT spectra from day 963 which imply an ordered structure (consistent with expanding equatorial and tropical rings in a remnant with $i = 60°$ – see also O'Brien, this volume). The optical spectra are difficult to reconcile with the radio observations. Indeed, the radio structure often seems to show drastic changes which make straightforward interpretation very difficult.

However, there are several fundamental complications with radio interferometric observations. First of all, larger-scale, lower surface brightness emission may be "resolved out". More importantly, it is impossible to disentangle the effects of changing optical depth and temperatures without high spatial resolution, simultaneous multi-frequency observations of the resolved shell. Currently with MERLIN for example, this is not possible. We will however address important future developments in the concluding section.

OPTICAL IMAGING AND SPECTROSCOPY

Observations of resolved remnants in the optical are still the most fruitful as far as determining their physical characteristics and gaining insight into shaping mechanisms are concerned. Taking the typical nova expansion velocity and distance (Fig. 1), we would expect to first resolve the remnant after ~ 2 months with HST and ~ 2 years from the ground (without the aid of Adaptive Optics - see below).

TABLE 2. Remnants Resolved Optically Since Wade (1990) Review [1]

	Ground or Space	t_3 (days)	Ref.
DY Pup (1902)	GB	160	[11]
V450 Cyg (1942)	GB	108	[12]
CT Ser (1948)	GB	>100?	[13]
RR Cha (1953)	GB	60	[11]
HS Pup (1963)	GB	65	[11]
QZ Aur (1964)	GB	23-30	Esenoglu (this volume)
V3888 Sgr (1974)	GB	10	[13]
NQ Vul (1976)	GB	65	[12]
PW Vul (1984)	GB	126	[14]
QU Vul (1984)	GB	49	[15]
V1819 Cyg (1986)	HST	89	[13]
V842 Cen (1986)	GB	48	[11]
QV Vul (1987)	HST	53	[13]
V351 Pup (1991)	HST	26	Ringwald (this volume)
V1974 Cyg (1992)	HST	42	[16]
HY Lup (1993)	HST	>25	[13]
V1425 Aql (1995)	HST	~30	Ringwald (this volume)
CP Cru (1996)	HST	4 (t_2)	Ringwald (this volume)

Table 2 gives details of all the optically-resolved nova remnants to appear in published literature since the Wade review [1] which cited 26. The total has substantially increased (a) because older remnants are inevitably becoming resolvable from the ground and (b) since 1989 we have seen the advent of HST.

Ground-based Optical Imagery

The most extensive recent ground-based imaging surveys have been those of Slavin et al [12] for the northern hemisphere and Gill and O'Brien [11] for southern objects. Figure 4 shows a selection of results from these surveys.

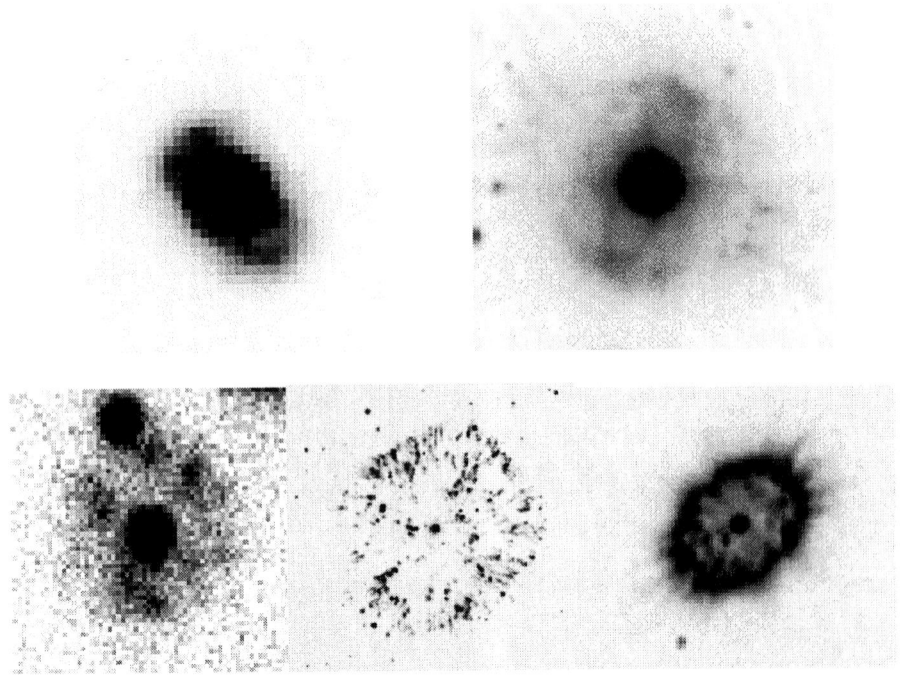

FIGURE 4. Ground-based optical images of novae of varying speed class (from Ref's [11] and [12]). Clockwise from top left: HR Del in [OIII] (outburst 1967, Very Slow, major axis ~12"); RR Pic in Hα/[NII] (1925, S, diameter ~ 21"); DQ Her in Hα (1934, MF, ~23"x17" with halo ~ 47"x29" and "tails" extending ~20" from points of origin); GK Per in [NII] (1901, VF, ~90" diameter), and V1500 Cyg in Hα (1975, VF, diameter ~8").

The optical images tend to confirm the presence of structure as implied from earlier spectroscopy. For example, in RR Pic and DQ Her, equatorial (and for DQ Her, tropical) rings are apparent. In addition, in both these objects, "tails" of emission are evident, streaming away from knots in the shell, suggestive of ablation by faster-moving ejected material (this process is discussed further by O'Brien, this volume). There also appears to be systematically *less* shaping with *increasing* speed class ([12], [13] - see below). Indeed, as pointed out in [12], the very fast nova GK Per, if placed at the distance of V1500 Cyg, would show a very similar, relatively circular, but otherwise amorphous remnant.

HST Imaging and Ground-based Optical Spectroscopy

The combination of HST imaging, long-slit optical spectroscopy and simple modelling has been shown to be a very powerful tool in determining the precise geometry of resolved remnants. It is also potentially important in exploring other physical properties, such as possible abundance gradients. The existence of these would have significant consequences for our understanding of the progress of the TNR across the WD surface. Previous work (e.g. [17]) has not been conclusive in this regard. Enhanced emission in [NII] from equatorial rings (such as that seen for example in Fig. 5, from the work of Gill and O'Brien [18] [19]) may not be due to a simple overabundance of nitrogen. Further discussion on this point is given in O'Brien (this volume), but there is clearly much more work to be done in this regard.

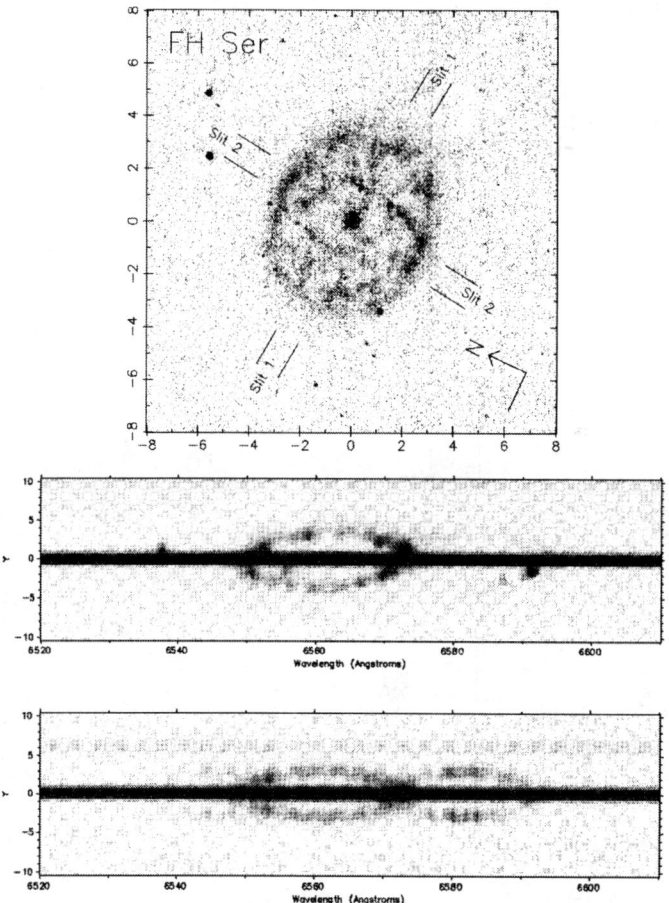

FIGURE 5. HST observations of FH Ser (1970 - top) through the F656N filter (axes in arcsec), taken from [19]. Combined with WHT long-slit spectroscopy (bottom, slit 1 top panel) and modelling, this confirmed that the nova shell is prolate, axial ratio = 1.3 ± 0.1, $i = 62° \pm 4°$, equatorial expansion velocity = 490 ± 20 km s^{-1} and hence $d = 950 \pm 50$ pc [19]. The clear [NII] ring most likely delineates the plane of the orbit of the central binary.

HST results provide further confirmation of a relationship between the degree of shaping and speed class. In Fig. 6 we have included corrections for inclination where possible. This correction appears to strengthen the correlation further.

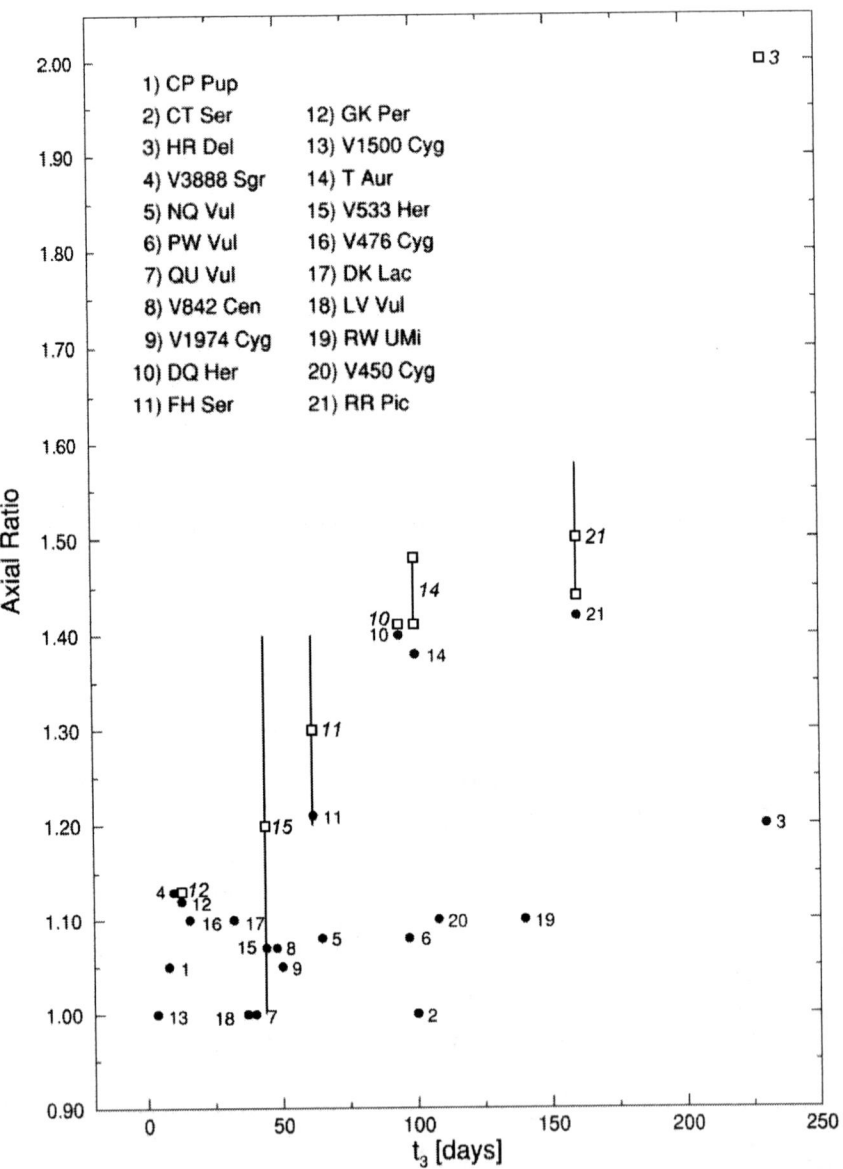

FIGURE 6. Axial ratio of remnants versus speed class. Filled circles from [13]. Open squares show inclination-corrected results (data from [12], [18], [19], plus O'Brien, this volume, for HR Del).

HYDRODYNAMICAL MODELLING

It is easily shown that the ejecta from the WD surface (and indeed the consequent pseudo-photospheric radius) rapidly envelop the secondary star yielding effectively a common-envelope binary. Early results of modelling the effects of the frictional deposition of energy and angular momentum by the secondary on the ejected nova envelope were reported at the Madrid meeting by Shankar et al. [20], [21]. Their 2-D hydrostatic wind models were essentially applicable only to restricted cases of very slow novae.

Subsequently, Lloyd et al. [22] used a 2.5-D hydro-code to investigate remnant shaping for a variety of speed classes. An important ingredient is the observed relation between ejection velocities and speed class [7]. In the case of slower novae for example, lower ejection velocities would lead to longer effective interaction times between the secondary and the ejecta, and hence more shaping might be expected. The basic model involves ejecta in the form of a wind with secularly increasing velocity and decreasing mass-loss rate. This evolved from a model of the early X-ray emission from V838 Her (1991) involving the interaction of ejecta with different expansion velocities early in the outburst [23]. The Lloyd et al. model produces rings, blobs and caps, plus a correlation of speed class to axial ratio in the sense required. However, it also produces *oblate* remnants.

Porter et al. [24] modified this basic model to include the effects of the rotating accreted envelope on the surface of the WD. From consideration of the effective gravity due to envelope rotation, and its effects on local luminous flux driving mass loss at outburst, a mass-loss rate and terminal velocity of ejecta were derived that are dependent on latitude on the WD. Figure 7 shows the comparison of results for a moderately fast nova without and with rotation of the accreted envelope. The Porter et al. models produce prolate shells as required. It should also be noted that a rather more satisfactory fit to the early radio evolution of V1974 Cyg may be provided by such models [25].

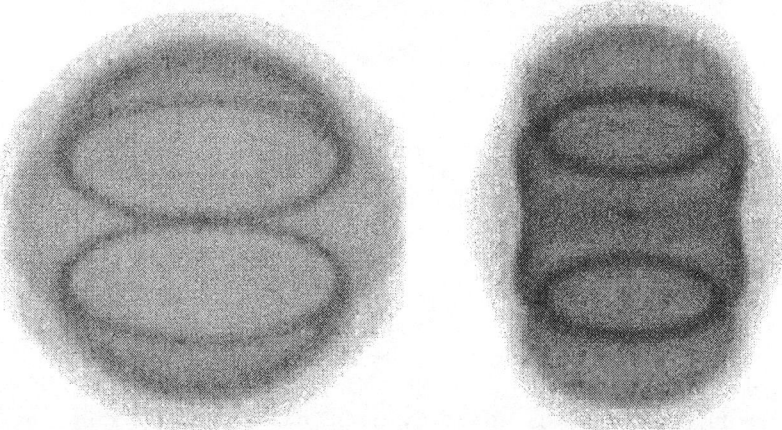

FIGURE 7. Synthetic images of remnants produced for the parameters of the ejecta and central binary of a moderately fast nova. Left - no envelope rotation. Right - accreted envelope rotating at 0.7 of the Keplerian velocity at the WD surface (taken from [24]).

GK PER (1901) - A NOVA SUPER-REMNANT

As reported by Balman in these proceedings, GK Per is an unusual remnant in many respects (see also [26] and references therein). It is for example the only unequivocally non-thermal radio remnant so far discovered among the CNe This, together with the presence of extended X-ray emission and deceleration of the ejecta, has led to the conclusion that the high-velocity material from the 1901 outburst is interacting with a pre-existing circumstellar medium. As shown in Fig. 8, there is evidence in the far-infrared and optical for extended nebulosity that may be associated with a previous phase of the evolution of the central binary (most likely a born-again AGB star [27]).

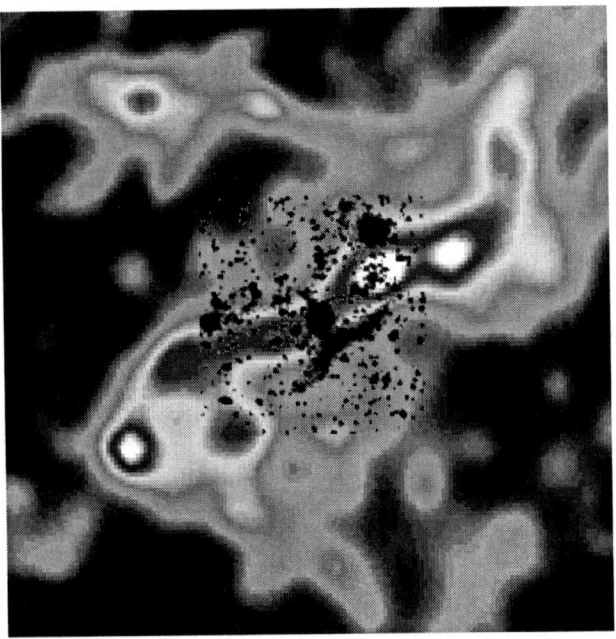

FIGURE 8. 100μm map from IRAS observations of the environs of GK Per [27], superimposed on which is a map of Hα emission [28]. The Hα image is 21' on a side. The remnant of the 1901 nova explosion lies at the centre of the IRAS emission near the cusp of the Hα nebula and is now ~1' in diameter (N at top, W to right). Note that the main interaction region in the nova remnant is to the SW.

One of the major questions has however been why the remnant is so asymmetric with the interaction largely being confined to the SW quadrant. A combination of Isaac Newton Telescope Wide Field Camera imagery and measurement of proper motion from archival plates over a 90-year baseline has confirmed our suspicion that the *whole system* is moving towards the SW. The large-scale "planetary" from the previous evolutionary phase is slowed by its interaction with the ISM. The nova binary travels through this essentially as a bullet. Then in 1901 we saw the very first outburst of the nova. The high velocity nova ejecta naturally impact a higher density of material in the direction of the whole system's motion than elsewhere [29]. Thus, much of the mystery is solved and the important place of GK Per in our understanding of CNe evolution is reinforced. In addition, this remnant, evolving as it does on a time scale two or three orders of magnitude shorter than that of a supernova remnant, and being close enough to resolve on relatively small length scales, is worthy of study in terms of furthering our understanding of the general physical processes occurring in such circumstances.

CONCLUDING REMARKS

We have made major strides in understanding the evolution of nova remnants since the Madrid conference. However, as always, there are still outstanding challenges. These include: the reconciliation of remnant development in the optical and radio (and from which we would particularly aim to determine more accurately the all-important ejecta masses); full 3-D treatment of the shaping of nova remnants and the exploration of a fuller parameter space of models; following the evolution of "clumpiness" and relating this to dust formation (see papers by Evans and Gehrz, this volume), and untangling excitation/ionisation effects from true abundance gradients in resolved remnants.

We are fortunate however to have exceptional observational tools at our disposal, either existing or planned, to tackle these problems head on. In the short term, HST plus STIS will be important in determining whether there really are abundance gradients in the extended nebulae which in turn might be related to variations in the initial conditions of the TNR on the WD surface. In the radio, both the e-MERLIN and EVLA projects will provide superb opportunities for using radio imaging to its full potential. e-MERLIN for example will have full frequency switching and around 30 times the sensitivity of the current array, coupled with resolution as high as 8 m.a.s. This gives the real promise of untangling the effects of temperature and optical depth changes which otherwise lead to misleading interpretations of the evolution of remnants in the radio. Finally, advances in ground-based telescopes, including adaptive optics (see Diaz, this volume) and optical interferometer arrays hold out the prospect of following the evolution of the ejecta from very early in the outburst. The proposed Large Optical Array for example would be able to resolve the remnant of a typical nova within hours of the outburst - a time at which the mechanisms of shaping might still be in full swing.

ACKNOWLEDGMENTS

I would like to thank my many colleagues from whom I have learnt so much over the years. In particular Tim O'Brien, John Porter, Huw Lloyd, Stewart Eyres, Richard Davis and Nye Evans, plus Franz Kahn, who is sorely missed. Thanks also to Andy Newsam and David Hyder who gave me invaluable help in the preparation of the figures for the review. Permission to reproduce figures from previously published works (as cited as appropriate in the text) was gratefully received from Blackwell Publishing, Astronomy and Astrophysics, and the American Astronomical Society.

REFERENCES

1. Wade, R.A., *Physics of Classical Novae*, Berlin, Springer-Verlag, 1990, p. 179
2. Hjellming, R.M., *Physics of Classical Novae*, Berlin, Springer-Verlag, 1990, p.169
3. Hutchings, J.B., *MNRAS*, **158**, 177 (1972)
4. Solf, J., *ApJ*, **273**, 647 (1983)
5. Seaquist, E.R., "Radio Emission from Novae", in: *Classical Novae*, edited by M.F. Bode and A. Evans, Chichester, J. Wiley & Sons, 1989, p.143
6. Hjellming, R.M., *Radio Emission from Stars and Sun*, San Francisco, ASP Conference Series, **93**, 1996, p.174
7. Payne-Gaposchkin, C., *The Galactic Novae*, Amsterdam, North Holland, 1957
8. Taylor, A.R., Hjellming, R.M., Seaquist, E.R., & Gehrz, R.D., *Nature*, **335**, 235 (1988)
9. Eyres, S.P.S., Davis, R.J., & Bode, M.F., *MNRAS*, **279**, 249 (1996)
10. Eyres, S.P.S., Bode, M.F., O'Brien, T.J., Watson, S.K., & Davis, R.J., *MNRAS*, **318**, 1086 (2000)
11. Gill, C.D. & O'Brien, T.J., *MNRAS*, **300**, 221 (1998)
12. Slavin, A.J., O'Brien, T.J., & Dunlop, J.S., *MNRAS*, **276**, 353 (1995)
13. Downes, R.A., & Duerbeck, H.W., *AJ*, **120**, 2007 (2000)
14. Ringwald, F., & Naylor, T., *MNRAS*, **278,** 808 (1996)
15. Della Valle, M., Gilmozzi, R., Bianchini, A., & Esenoglu, H., *A&A*, **325**, 1151 (1997)
16. Paraesce, F., Livio, M., Hack, W., & Korista, K., *A&A*, **299**, 823 (1995)
17. Evans, A., Bode, M.F., Duerbeck, H.W., & Seitter, W.C., *MNRAS*, **258**, 7P (1992)
18. Gill, C.D., & O'Brien, T.J., *MNRAS*, **307**, 677 (1999)
19. Gill, C.D., & O'Brien, T.J., *MNRAS*, **314**, 175 (2000)
20. Shankar, A., Truran, J.W., Burkert, A., & Livio, M., *Physics of Classical Novae*, Berlin, Springer-Verlag, 1990, p.297
21. Livio, M., Shankar, A., Burkert, A., & Truran, J.W., *ApJ*, **356**, 250 (1990)
22. Lloyd, H.M., O,Brien, T.J., & Bode, M.F., *MNRAS*, **284**, 137 (1997)
23. O'Brien, T.J., Lloyd, H.M., & Bode, M.F., *MNRAS*, **271**, 155 (1994)
24. Porter, J.M., O'Brien, T.J., & Bode, M.F., *MNRAS*, **296**, 943 (1998)
25. Lloyd, H.M., O'Brien, T.J., & Bode, M.F., *Radio Emission from Stars and Sun*, San Francisco, ASP Conference Series, **93**, 1996, p.200
26. Seaquist, E.R., Bode, M.F., Frail, D.A., Roberts, J.A., Evans, A., & Albinson, J.S., *ApJ*, **344**, 805 (1989)
27. Dougherty, S.M., Waters, L.B.F.M., Bode, M.F., Lloyd, H.M., Kester, D.J.M., & Bontekoe, Tj. R., *A&A*, **306**, 547 (1996)
28. Tweedy, R.W., *ApJ*, **438**, 917 (1995)
29. Bode, M.F. & O'Brien, T.J., in preparation (2002)

The structure of the shell of HR Del

T.J. O'Brien*, D.J. Harman* and M.F. Bode[†]

*Jodrell Bank Observatory, The University of Manchester, Macclesfield, SK11 9DL, UK
[†]Astrophysics Research Institute, Liverpool John Moores University, Twelve Quays House, Birkenhead, CH41 1LD, UK

Abstract. We present HST/WFPC2 imaging of HR Del in filters centred on the lines of Hα, [NII] and [OIII], with corresponding WHT/ISIS slit spectroscopy for the lines of Hα, Hβ, [NII] and [OIII]. The data show that the ejected shell of HR Del is bipolar in structure with an equatorial ring, and polar rings or caps. The shell is composed of many small, bright knots and diffuse material which flows radially outward from these knots. Three-dimensional models of the shell that reproduce the observed emission-line images and longslit spectra in both the [OIII] and Hα morphologies are presented. A large, bright [OIII] knot is apparent in the imaging that we cannot associate with emission in either Hα or [NII] or the spectroscopy. In addition there are [NII] and [OIII] knots at both higher and lower velocity than that of the main shell which are symmetrically placed to either side of the central system. We estimate an inclination of $35 \pm 3°$, a polar expansion velocity of 560 ± 50 km s^{-1}, an axial ratio of 1.75 ± 0.15 and a distance of 970 ± 70 pc.

INTRODUCTION

HR Del (Nova Delphini 1967) is a very slow nova which outburst in 1967 [1]. The optical lightcurve was punctuated by flaring behaviour similar to that seen in the more recent nova V723 Cas [2]. Optical spectra display 6-peaked profiles which were interpreted as evidence for the ejection of an oblate shell with two equatorial bands and polar blobs [3]. The first resolved photographs of the shell [4] taken in the light of [OIII] revealed an oval brightness distribution of size 3.7×2.5 arcsec. Solf [5] obtained high resolution spatially-resolved spectra of the nebular shell and proposed that the shell was in fact prolate (also see Bode's review in these proceedings). More recent ground-based CCD images [6] resolved structure in the shell confirming the bipolar [OIII] structure suggested by the earlier spectroscopic observations, and a more ring-like structure in the Hα/[NII] images.

Here we report on the first space-based imaging of the nebular remnant of HR Del taken as part of a survey of several nova shells [7] made with WFPC2 on the Hubble Space Telescope (HST). The imaging is combined with ground-based spectroscopy using the ISIS spectrograph on the 4.2m William Herschel Telescope (WHT) to enable us to construct a detailed kinematical model for the structure of the shell. Details of the observations are given in Table 1.

TABLE 1. Summary of the observations of HR Del presented in this paper.

	Date	Wavelength	Exposure time
HST WFPC2 images	1997 May 25	6564 Å (F656N, Hα)	2300 s
	1998 Oct 29	5013 Å (F502N, [OIII])	900 s
	1998 Oct 29	6590 Å (F658N, [NII])	760 s
WHT ISIS spectra	1996 Aug 1 & 3	4740-5160 Å (R1200B, [OIII])	1800 s
	1996 Aug 1 & 3	6390-6810 Å (R1200R, Hα/[NII])	1800 s

OBSERVATIONS AND MODELLING

The HST images are presented in Figure 1. There was noticeable expansion of the shell by about 5% in the 522 days between the two imaging epochs and the Hα image has been magnified by this factor in Fig. 1.

Images in all three wavebands show the shell to be extremely clumpy. The distribution of clumps is more bipolar in [OIII] and [NII] than in Hα. The Hα emission has a significant diffuse component, often appearing as tails beginning on bright clumps and pointing radially away from the central stellar system. The clumps at the southwest pole in the [OIII] and [NII] images take the form of a ring. The northeast pole is more unstructured. There is a bright clump to the north-northwest of the centre of the [OIII] image which breaks the bipolar symmetry of the shell. This feature appears as an arrowhead pointing roughly towards the central star. It is not associated with any emission features in the other two HST images.

The spectra shown in Figure 2 were taken with five separate slit positions aligned along the major axis of the shell as shown in Fig. 1. In each case the central stellar continuum has been subtracted, a residual component is still evident in slit 3. The spectra all show clear 'velocity ellipses' demonstrating that the emission arises from an expanding shell tilted to the line of sight with a red-shifted northeast pole. In fact the ellipses appear to be 'figure-of-eight' shaped indicating that the shell is not a simple ellipsoid but has a constriction at the waist giving it the appearance of a 'closed hourglass'. The shell is clumpy with a significant brightening of clumps in the polar regions and in an equatorial band. This is indicative of a brightening around the waist of the shell in an equatorial ring or disk.

The [OIII] emission is clearly concentrated into clumps at the poles with a small amount at the equator and little emission joining the two. Symmetrically placed knots can be seen in both [OIII] and [NII]. There are a pair of high velocity knots seen most obviously in slit 2 at positions in [OIII] of $(-600$ km s^{-1}, $-3'')$ and $(+650$ km s^{-1}, $+3'')$. These knots have tails of diffuse emission extending outwards and to higher velocities. Only the blue-shifted knot can be readily seen in the [NII] spectra. A knot internal to the main shell profile is also seen in slit positions 1 and 2.

The structure seen in the images and spectra was modelled with a shell in the form of a closed hourglass, expanding radially with the velocity at each point proportional to its distance from the centre. The emission from the shell is enhanced in rings around the equator and towards the poles. Comparisons can be made with the observations by integrating along lines of sight at some inclination angle and hence making synthetic

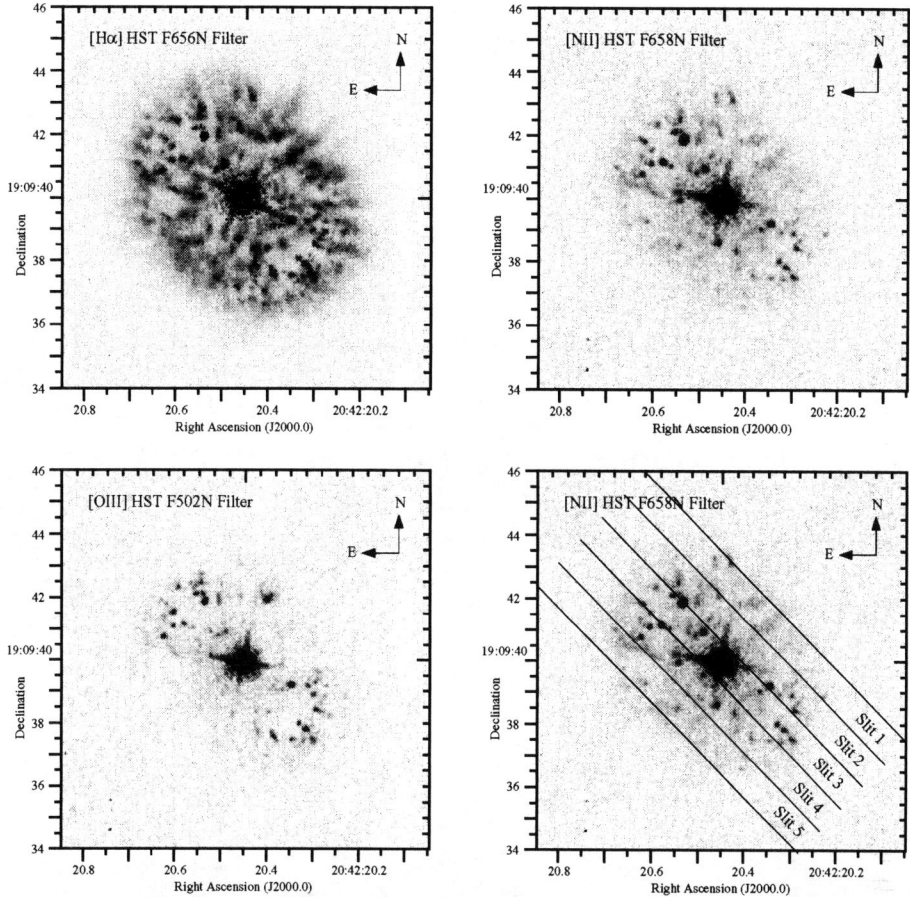

FIGURE 1. HST WFPC2 images of HR Del at Hα, [NII] and [OIII]. The [NII] image is also shown with the slit positions for the ISIS spectroscopy overlaid.

images. It is also possible to generate synthetic spectra by only extracting information corresponding to the spectrometer slits. The results are shown in Figure 3. In order to best match the model with the observations the emission is made clumpy by assuming a gaussian distribution of clump sizes. Artificial 'seeing' is also applied.

DISCUSSION

Several features are worthy of additional discussion. The high-velocity tails seen most obviously in [OIII] are very similar to features seen in the shells of DQ Her and RR Pic [8]. They are most easily interpreted as the result of a fast wind which blows through the

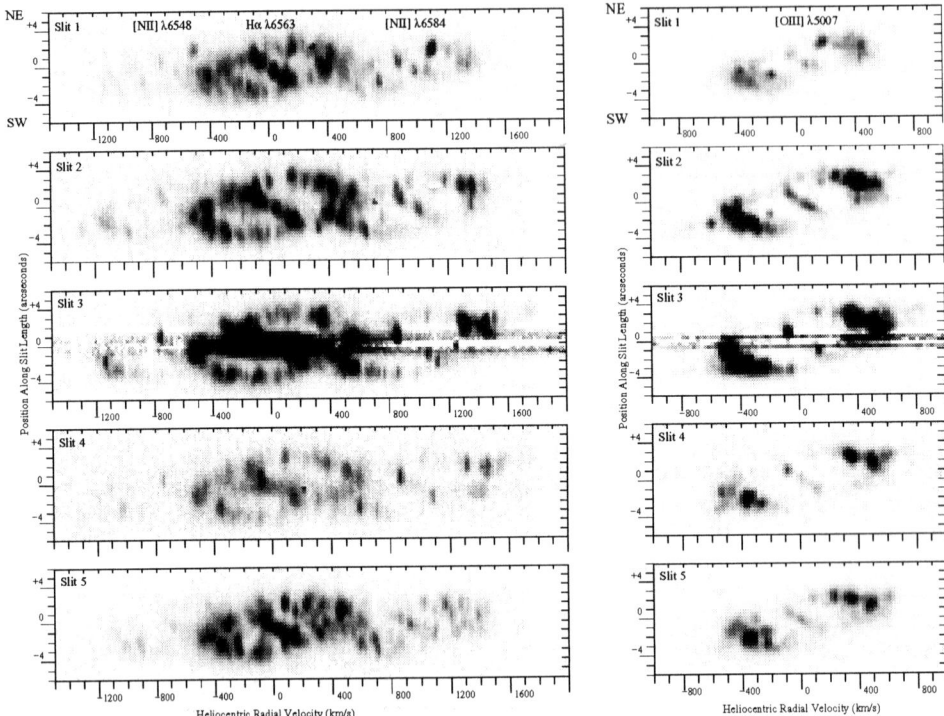

FIGURE 2. The ISIS spectra at all 5 slit positions of Fig. 1 in the regions of Hα/[NII] (left) and [OIII] (right). The latter spectra are labelled with velocities relative to the rest wavelength of Hα rather than either of the two [NII] lines.

clumpy shell, accelerating material ablated from the clumps. Interestingly, evidence for the existence of such a fast wind in HR Del has been provided in these proceedings by Selvelli & Duerbeck. It is not clear however whether these structures are evidence for continuing ablation or are the 'fossil' remains of interactions which took place nearer outburst and which are now expanding self-similarly.

The bright blob visible in the [OIII] HST image has no obvious counterpart in the spectroscopy and remains a perplexing feature. If it is formed as the result of interaction between the fast wind and a clump (suggested by the arrowhead shape of the emission), it might be smeared out in the spectrum and made less obvious. Alternatively it could be a transient feature which was not present at the time the spectra were obtained.

In conclusion, the observations demonstrate that the shell of HR Del is bipolar, resembling a closed hourglass with an equatorial ring. There is also evidence for two polar rings. The modelling allows a number of important parameters to be accurately determined. The shell is inclined at an angle of $35 \pm 3°$ with a polar velocity of 560 ± 50 km s^{-1} and an axial ratio of 1.75 ± 0.15. The distance is calculated to be 970 ± 70 pc.

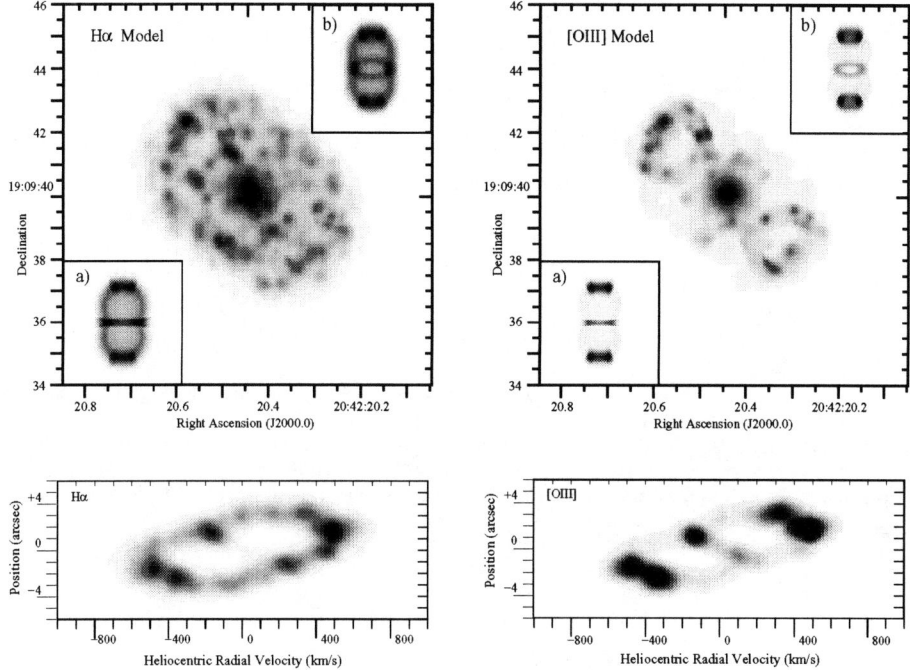

FIGURE 3. Synthetic images (top) and spectra (bottom) in the lines of Hα (left) and [OIII] (right). The insets show inclinations of (a) 90 degrees and (b) 65 degrees. The synthetic spectra are for slit position 3.

Since HR Del is an extremely slow nova accurate knowledge of these values is essential in tying down one extreme in the relationships with speed class of parameters such as shaping, luminosity or ejection velocity.

REFERENCES

1. Candy, M.P., Alcock, G.E.D., and Zissell, R.E., *IAU Circ.* **2022**, (1967).
2. Chochol, D., and Pribulla, T., *Contrib. Astron. Soc. Skalnate Pleso* **27**, 53-69 (1997).
3. Hutchings, J.B., *MNRAS* **158**, 177-198 (1972).
4. Kohoutek, L., *MNRAS* **196**, 87-89 (1981).
5. Solf, J., *ApJ* **273**, 647-659 (1983).
6. Slavin, A.J., O'Brien, T.J., and Dunlop, J.S., *MNRAS* **314**, 175-182 (1995).
7. Gill, C.D., and O'Brien, T.J., *MNRAS* **314**, 175-182 (2000).
8. Gill, C.D., and O'Brien, T.J., *MNRAS* **300**, 221-232 (1998).

Nova outburst luminosities, postnova magnitude behaviour, and long term evolution of nova shell luminosities

Hilmar W. Duerbeck

WE/OBSS, Vrije Universiteit Brussel, Belgium

Abstract. A luminosity calibration of galactic novae indicates that all novae at brightness maximum radiate over-Eddington, and that speed- and light curve classes are intimately related. In later stages, the Balmer and [O III] line fluxes decline in similar ways for novae of all speed classes, except in slow ones where Balmer emission diminishes faster, and [O III] persists for decades. The brightness of the central source declines during the first century after outburst; decline rates for novae with orbital periods above 0.2 days are in good agreement with theoretical predictions, but there are indications that the luminosity will remain constant afterwards. Postnovae with shorter periods appear to decline more rapidly, and they often erupt from low-luminosity stages.

NOVA LUMINOSITIES AT MAXIMUM

The determination of nova shell fluxes requires a knowledge of nova distances and reddenings. A recalibration of nova luminosities based on expansion parallaxes was carried out in collaboration with R.A. Downes (Downes & Duerbeck 2000). These nova luminosities were used to derive new Maximum Magnitude – Rate of Decline relations (MMRD). One of these is shown in Fig. 1. Novae of light curve class A show all speed class VF. Light curve class B,C,D novae belong to speed classes F, MF, S (for a definition of light curve classes see Duerbeck 1981, for a definition of speed classes Payne-Gaposchkin 1957).

It is interesting to note that the luminosity-decline rate dependence is less expressed when applied to two groups of novae separately, the very fast group, and that of all other speed classes. In the "transition phase", when the luminosity has dropped to a value between $M = -8$ and -6, very fast novae sometimes show brightness oscillations (light curve class Ao objects). They are still above the Eddington limit for 1 $M_\odot$, and a "super Eddington wind" model describing this behaviour was developed by Shaviv (2001, 2002). Slower novae with "structured" light curves are also found in this luminosity range, and a similar model may also be applicable to model their light curves.

NOVA SHELL LUMINOSITIES

Nova shell luminosities in later stages of the outburst, months, years and decades after outburst were derived from new narrowband filter imaging, from spectrophotometric observations during the outburst, and from nova shell images in the HST archive.

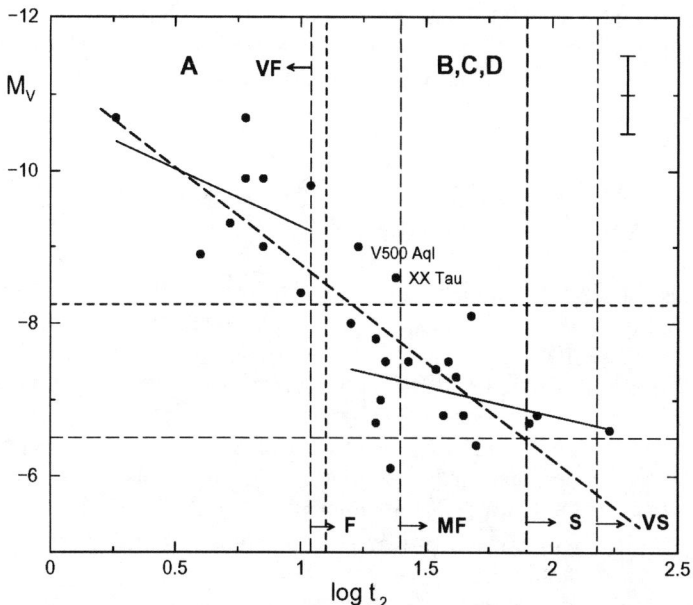

FIGURE 1. An MMRD-relation, adapted from Downes & Duerbeck (2000), with a revised luminosity for HR Del (T. O'Brien, priv. comm.). Two dotted lines define four quadrants, while the speed class regions are separated by vertical dashed lines. In the upper left quadrant, only very fast (VF) novae with smooth light curves (A and Ao, with and without oscillations during the transition stage) are found. In the lower right quadrant, fast (F), moderately fast (MF), slow (S) and very slow (VS) novae with structured light curves (B,C,D) are found. The Eddington limit for a 1 $M_\odot$ object is indicated by a horizontal dashed line at $M_V = -6.5$ (with B.C. = 0.25 of an F5 Ia star, comparable to a slow nova at maximum light). A typical error of a data point is indicated in the upper right-hand corner.

Distances were taken from the above recalibration or from the new MMRD relations. 1200 flux observations of 96 objects are available. Speed class was found to be a useful criterium to arrange shell luminosities of different novae, observed at different times, into groups which contain enough data points. Results are given by Downes, Duerbeck & Delahodde (2001), and are briefly summarized here:

For very fast, fast and moderately fast novae, the slope of the [O III] $\lambda 5007$ decline is steep, 'switchoff' of [O III] $\lambda 5007$ emission occurs after ~ 20 years. The slope of the Balmer luminosity is less steep and similar in all groups. In slow novae, the decline in Balmer luminosity is more rapid than in all types of fast novae. The slope in [O III] $\lambda 5007$ is less steep; slow novae show [O III] $\lambda 5007$ emission after 100 years.

In VF novae, the break in [O III] $\lambda 5007$, which leads to a rapid decrease in line flux, coincides with the nuclear turnoff at $\sim 250^{+300}_{-150}$ days; F and MF novae show a similar break at later times. The slow decline of [O III] $\lambda 5007$ flux in slow novae indicates that nuclear burning on the white dwarf surface continues for a long time.

Hot shells are observed not only around slow novae, but also for the very fast nova GK Per and the recurrent nova T Pyx: both shells interact with circumstellar material.

In (fast) recurrent novae, [O III] $\lambda 5007$ is usually inconspicuous or absent. In objects with giant companions, the Balmer luminosity decreases very slowly because of the contribution from regions the giant wind. Objects with dwarf companions show a very rapid decline in Balmer luminosity.

NOVA LUMINOSITIES AT MINIMUM: INDICATIONS FOR HIBERNATION?

This project is carried out in collaboration with M. Shara, E. Leibowitz, and W. Seitter. Here, only another progress report can be given; it is hoped that the final results will be published within a year's time.

The project consists of (1) CCD BVR photometry of postnovae and nova fields; (2) derivation of postnova magnitudes; (3) calibration of photographic (pg) nova fields; (4) conversion of prenova pg into B magnitudes; (5) conversion of postnova pg and visual into B and V magnitudes, and finally, (6) to look for differences ($m_{post} - m_{pre}$), and for the slope $dm/d\log t$ and their trends with time after outburst.

The most detailed work on nova pre- and postoutburst magnitudes until now is that of Robinson (1975). He found, from a sample of 18 objects, $m_{post} - m_{pre} = -0.06 \pm 0.38$ s.e., and concluded that $m(\text{prenova}) \approx m(\text{postnova})$, where magnitudes were often taken verbatim from the literature. Our new results, based on carefully measured or calibrated pre- and postnova magnitudes of 37 objects, yield $m_{post} - m_{pre} = -0.37 \pm 1.02$ s.e. Note that in spite of the better quality of modern data, the scatter has increased. The average value does not contain any useful information: an analysis reveals that the majority of novae appears slightly *fainter* after outburst, while a minority of novae is much *brighter* after outburst. The following groups can be discriminated:

Normal postnovae: At short times after maximum light, a "normal" postnova is still declining. At about $\log t = 1.3$, i.e. after 20 years, it has settled at a level in which it apparently remains up to 100 years. Only 8 out of 16 novae in the "normal" group have known orbital periods, but *all of them* are longer than 0.2 days.

Unusual postnovae: Another group of novae is, for up to 50 (and likely up to 100) years after outburst *brighter* than before outburst. Orbital periods of these "brighter postnovae", are, *without exception*, below 0.2 days.

It appears that this class can be divided into two subgroups:

Unusual postnovae/extreme group: Examples: V1500 Cyg, CP Pup, RW UMi, GQ Mus, V2214 Oph, V1974 Cyg. These objects clearly erupt from a low-luminosity state, i.e. they should properly be called "faint prenovae".

Unusual postnovae/slowly declining group: Examples: RR Pic, DN Gem, V603 Aql, DQ Her. While the outbursts lie long in the past, the postnovae are still brighter than before outburst. It is not clear whether they erupt from a somewhat lower-luminosity prenova state, as compared with normal postnovae.

If we analyze the present-day absolute visual magnitudes of postnovae, it appears that all of them, irrespective of membership in different groups, have quite similar values. Thus "bright" postnovae are thus not *absolutely* bright, they simply erupted from a low (the "slowly declining group") or a very low (the "extreme group") state, and behave after outburst almost like "normal" postnovae.

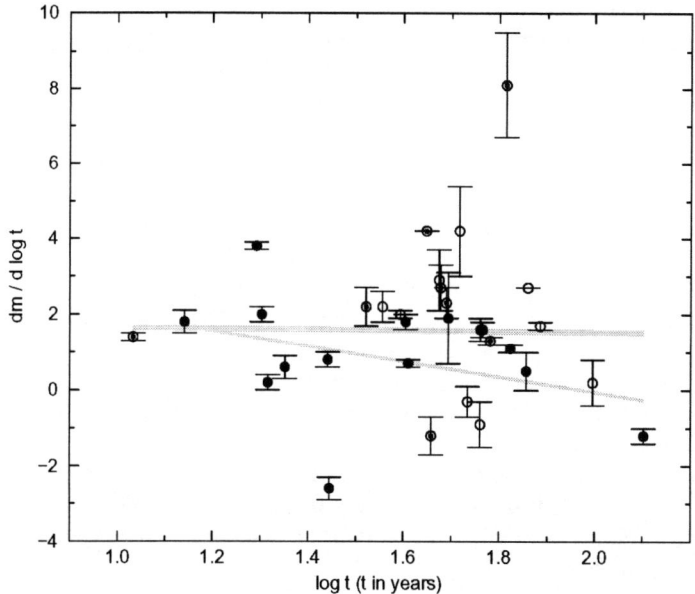

FIGURE 2. The light curve slope $dm/d\log t$ observed in the first 100 years after outburst. A linear regression through all points yields an average slope of $dm/d\log t = 1.6 \pm 0.3$ (thick grey line). If only "normal" postnovae with orbital periods $P > 0.2$ days (filled circles) are considered, the decline rate approaches zero after 100 years (thin grey line).

Kovetz, Prialnik & Shara 1988 gave a theoretical prediction for the decline of a postnova, $dm/d\log t = 0.7$. On the observational side, all novae (on the average 40 years after outburst) show an average slope of $dm/d\log t = 1.6 \pm 0.3$, i.e. they become increasingly fainter (Fig. 2). No obvious trend with time after outburst is visible. Nevertheless, if we isolate the group of "normal" postnovae, their (average) slope is $dm/d\log t = 0.9 \pm 0.4$, and it is indicated that $dm/d\log t \to 0$ after 100 years, i.e. the fading stops after that time. Better and more numerous data are needed to substantiate this finding.

SUMMARY

Activities in nova shells and central stars die off with timescales of 10 years ([O III] λ 5007), 100 years (fading of "normal" postnovae), or possibly much later (fading of "unusual" postnovae) to a level in which they may stay for centuries or millennia.

Do we observe hints of hibernation, i.e. a fading of the postnova to much fainter luminosity levels? And do we need hibernation to explain classical nova eruptions? Hibernation was introduced ~ 20 years ago to keep *average* accretion rates in novae at low values ($\sim 10^{-8}$ to 10^{-10} $M_\odot$/yr) in order to produce TNRs from sufficiently degenerate conditions.

Nowadays, theoreticians can "make" novae with recurrence times of ~ 100 years and accretion rates of 10^{-6} $M_\odot$/yr, however, *without* any mass ejection. Nature, however, can make novae with recurrence times of ≤ 100 years *with* mass ejection (CI Aql, IM Nor) – and these (slow) recurrent novae resemble classical novae in their spectroscopic and photometric appearance. Thus our ideas on recurrence times of classical novae may have to be reconsidered: shorter recurrence times mean smaller space densities, and these is no reason any more to "hide" classical novae between eruptions from the census of novalike objects by camouflaging them as dwarf novae or even detached systems, as the hibernation scenario implies.

Let us imagine that accretion rates of novae after outburst remain high. Recurrence times might be shorter than previously assumed, of the order of 100...10000 years, with an average value of 1000 years; thus *we just may start to witness the short-period ("recurrent") tail of classical novae*. This would solve the nova space density dilemma (Duerbeck 1984, Patterson 1984, Duerbeck and Covarrubias 1995), and would favor the scenario described by Mukai and Naylor (1995) and Naylor (2002): novae are those CVs which keep their high accretion rates. Our studies of the brightness evolution within 100 years after outburst indicates that at least the "normal" novae follow this track.

Another (smaller) group of objects are novae which erupt from faint prenova states. These are novae with orbital periods below 0.2 days, which exhibit clear indications of brightness declines even after 100 years after outburst. Some type of "hibernation" clearly operates in these systems.

ACKNOWLEDGMENTS

This research was supported by the Flemish Ministry for Foreign Policy, European Affairs, Science and Technology. I thank Ron Downes, Mario Livio and Mike Shara (now AMNH) for their hospitality at Space Telescope Science Institute, and Tim O'Brien and Tim Naylor for useful comments.

REFERENCES

1. Downes, R.A., Duerbeck, H.W., 2000, *Astr. J.*, **120**, 2007-2037.
2. Downes, R.A., Duerbeck, H.W., Delahodde, C., 2001, *J. Astr. Data*, **7**, No. 6.
3. Duerbeck, H.W., 1981, *Publ. Astr. Soc. Pacific*, **93**, 165-175.
4. Duerbeck, H.W., 1984, *Astrophys. Space Sci.*, **99**, 363-385.
5. Duerbeck, H.W., Covarrubias, R., 1995, p. 264-267, in *Flares and Flashes* (IAU Coll. No. 151), eds. J. Greiner et al., Springer, Berlin.
6. Kovetz, A., Prialnik, P., Shara, M.M., 1988, *Astrophys. J.*, **385**, 828-836.
7. Mukai, K., Naylor, T., 1995, p. 517-522, in *Cataclysmic Variables*, eds, A. Bianchini et al., Kluwer, Dordrecht.
8. Naylor, T., 2002, these proceedings.
9. Patterson, J., 1984, *Astrophys. J. Suppl.*, **54**, 443-493.
10. Payne-Gaposchkin, C.H., 1957, *The Galactic Novae*, North Holland Publ. Co., Amsterdam.
11. Robinson, E.L., 1975, *Astr. J.*, **80**, 515-524.
12. Shaviv, N., 2001, *Mon. Not. R. Astr. Soc.*, **326**, 126-146.
13. Shaviv, N., 2002, these proceedings.

Multicolor studies of the old novae behavior

Elena P. Pavlenko*, Sergei Yu. Shugarov, Vitaly P. Goranskij,[†] and Nataly V. Primak**

*Nauchny, 19-17, Crimea 98409, Ukraine
[†]Sternberg Astronomical Institute, Moscow University, Moscow, 119992, Russia
**Kiev National University, Ukraine

Abstract. We present a result of our multicolor investigation of the old non-magnetic novae Q Cyg (Nova Cyg 1876), DI Lac (Nova Lac 1910), V446 Her (Nova Her 1960) in 1991 – 2001 and magnetic nova V1500 Cyg (Nova Cyg 1975) in 2000 – 2001. The non-magnetic novae display the outburst-like brightening in a wide region of amplitudes ($0^m.5 - 2^m.5$) and cycle duration ($\sim$ 23-d – $\sim$ 65-d). Similarly to the dwarf novae outbursts, they show a loops on the color – magnitude diagram. A shape of the loop (its width and slope) is different for different novae. In some cases the behavior of ouburts is in agreement with the model of tidal instability of accretion disk. Magnetic nova V1500 Cyg displays a strong brightening every 8.5-d, probably caused by a periodical enhancement of accretion stream and/or accretion ring surrounding the magnetic dwarf in asynchronous binary.

INTRODUCTION

It is known that only a few patterns among known old novae display the quasy-regular brightening which look like dwarf nova outbursts, but with less amplitude. The most famous stars are GK Per [1], V446 Her, DI Lac, V841 Oph [2] . The less famous is Q Cyg [3]. There is still no common meaning on the nature of these outbursts: whether they are caused by dwarf nova-like accretion disk instabilities [4] or mass transfer events [5].

All suggestions were based on studies of the amplitude, shape and spacing of outbursts obtained from V bandpass photometry. Thus the outbursts which amplitude is comparable to those of dwarf novae are suggested to be caused by the accretion disk instability while the small-amplitude outbursts may be due to the mass transfer variations. Meanwile the Schreiber's et al. [6] simulations, taking into account irradiated accretion disk in post novae, shown that the low-amplitude ($0^m.5$!) outbursts could be also produced by accretion disk instability if the white dwarf is hot enough after nova explosion. They also shown that the cooling of white dwarf should promote increase of the outburst amplitude.

Accordingly to observations of some dwarf novae and Smak's models [7], the dwarf novae outbursts show a prominent reddening and loops on the magnitude – color diagram: the star is more red during rising brunch of the outburst and more blue during descending one. The loop is wide if the thermal instability starts in outer parts of accretion disk and propagates inward ("outward – in" instability), while narrow loop corresponds to the "inward – out" instability.

Unlike to several dwarf novae we found no (with exception of GK Per [1]) information on the multicolor behavior of the old nova outbursts.

Here we present the results of multicolor study of non-magnetic old novae Q Cyg, V446 Her and DI Lac as well as the only known asyncronous magnetic nova V1500 Cyg [8], [9]. All the data were obtained in the Crimean Astrophysical Observatory and the Crimean Laboratory of Sternberg Astronomical Institute at 0.6-m, 0.5-m or 0.38-m telescopes with different photometric equipment: UBV-photometer, high-sensitive TV tube or CCD.

Q CYG

Q Cyg displays the outbursts with typical cycle duration $\sim$ 65-d and variable amplitude $0^m.5 - 1^m$ [3]. Here we present observations made in 1995 – 1996 (Fig. 1,a-c). We can't define the shape of outburst because of irregularity of the time series. A behavior at the $V, U - B$ and $V, B - V$ diagrams is shown in Fig. 1, b-c. Data which belong to different outbursts are marked by different symbols. Generally Q Cyg shows a reddening with fading by 1^m within the stripe of $0^m.3$ in $B - V$ and up to 1^m in $U - B$. Note that outburst marked by filled circles shows a broad anti-clockwise loop at $V, U - B$ diagram similar to the Smak's case A "outside – in" thermal instability in accretion disk. However, behavior at $V, B - V$ diagram does not show the loop and is rather complicated within the stripe of scattering.

DI LAC

The outbursts of DI Lac occur every $\sim$ 36-d – $\sim$ 40-d with amplitude $\sim 0^m.5 - 0^m.8$ [2], [10]. The mean profile of some outbursts is asymmetric one with more steep ascending branch [10]. We combined data from several outbursts or their fragments occured in 1991 – 2002 together and placed them on the magnitude – color diagrams. Similarly to Q Cyg, DI Lac displays color variation within the stripe $0^m.3$ at the $V, B - V$ diagram, but does not displays reddening with fading. Some hint on the reddening is visible at $V, U - B$ diagram (Fig. 2, a,b). This effect is more prominent at $V, V - R$ and $V, V - I$ diagrams (Fig. 2, c,d), the star shows narrow clock-wise loops [10].

V446 HER

V446 Her also shows the outbursts at a mean interval of $\sim$ 23-d [2], with variable amplitude. The largest one reaches $1^m.5$, but when corrected for the contribution from two optically close stars, amplitude is increased to $\sim 2^m.5$ [11], consistently with amplitudes of dwarf novae outbursts. Schreiber et al. [6] interpreted the large and small outbursts of V446 Her as events of thermal instability in accretion disk caused by irradiation of disk by a hot white dwarf after the nova explosion. In Fig. 3,a we present the mean profile of the largest outburst obtained by shifting the two original profiles until the best coin-

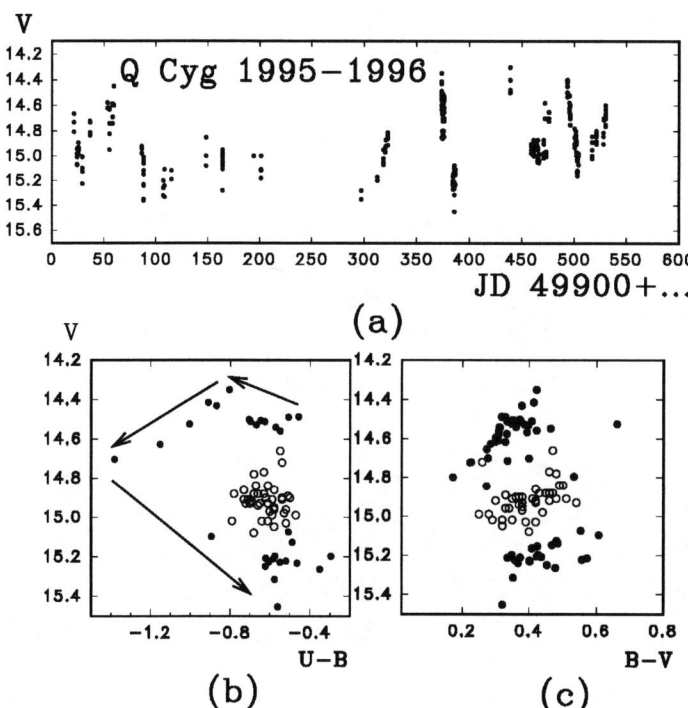

FIGURE 1. Q Cyg in 1995 – 1996. a)The light curve, b)V, $U - B$ and c)V, $B - V$ diagrams. The chronological development of outburst occured at JD ... 50273 – 50287 is shown by arrows.

ciding. The profile is symmetric. These two outbursts display a strong reddening with fading and very broad anti-clockwise loop at the V, $V - R$ diagram (Fig. 3,b). The last circumstance strengthens the idea of the dwarf nova-like outbursts of V446 Her. In the framework of the theory of thermal instability, the symmetric profile implies narrow loop and corresponds to the "inside – out" disk instability, while the "outside – in" instability shows delay of radiation at shortward wavelengths and produce an asymmetric outburst profile with a wide loop at the magnitude – color diagram [7]. The combination of the symmetric outburst profile and wide loop observed in V446 Her is somewhat unusual.

V1500 CYG

The multicolor monitoring of the magnetic nova V1500 Cyg is very important for study the white-dwarf cooling after the nova explosion. The irradiated secondary in this binary dominates the visual flux, producing the orbital light modulation. Cooling of the white dwarf should lead to decrease of orbital amplitude with time. Somers and Naylor [12] first derived the cooling rate of the white dwarf, using B band observations, which coincided with Prialnik's theoretical prediction [13]. In Fig. 4,a we present our data

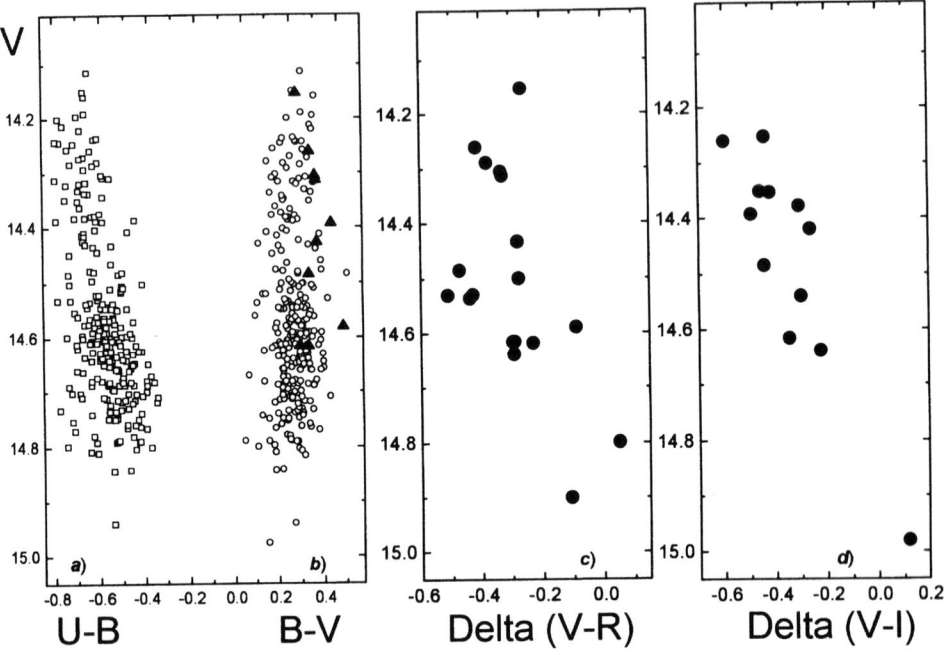

FIGURE 2. Behavior of DI Lac during outbursts at V, $U-B$ (a), V, $B-V$ (b), V, $V-R$ (c) and V, $V-I$ (d) diagrams.

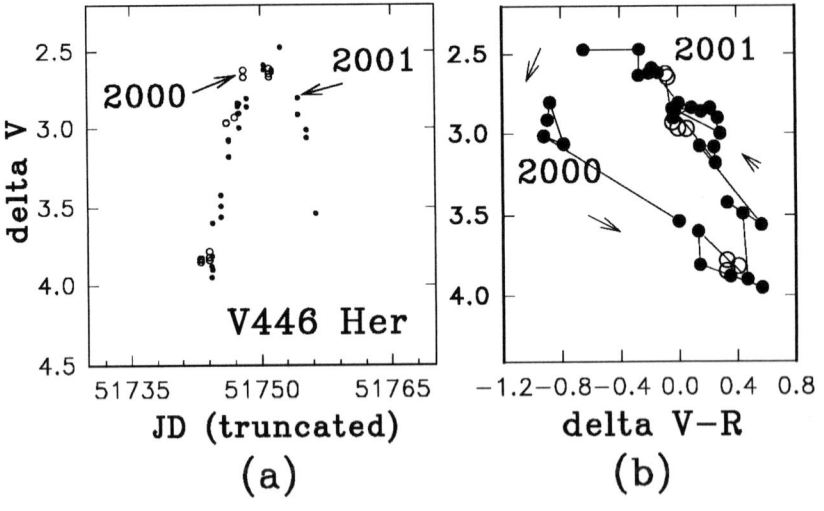

FIGURE 3. The mean profile of the largest outburst (a) and V, $V-R$ diagram (b). Data of the two separate outbursts are marked by solid and open circles.

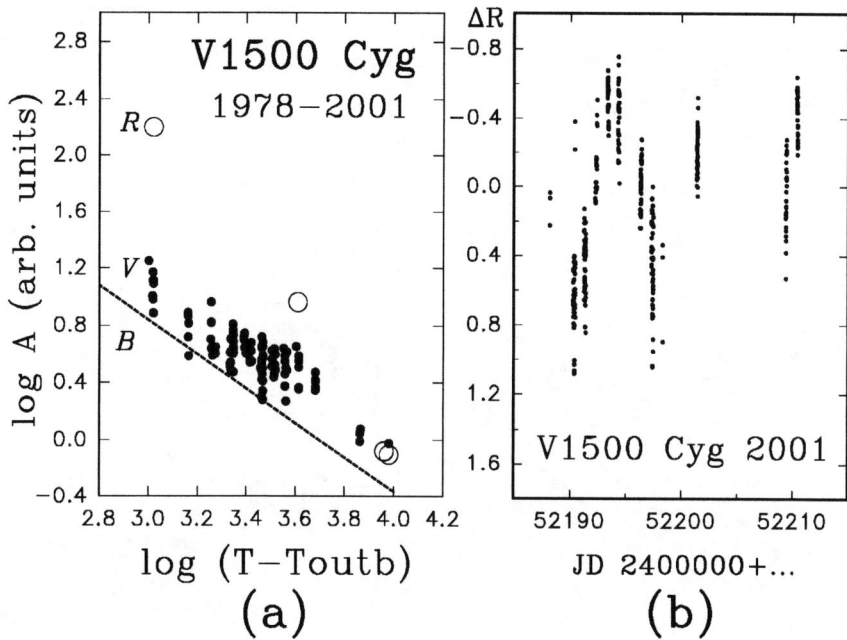

FIGURE 4. (a): Orbital amplitude in B, V and R spectral bands decrease with time passed after nova explosing (a). Data taken from [12], are drawn by line. Our V band data are marked by filled circles and R data – by open circles. All data are expressed in relative flux unites. (b): Example of the 8.5-d light modulation.

of the orbital amplitude change in V (filled circles) and in R (open circles). Note the same rate of amplitude decrease in B and V.

V1500 Cyg displays the prominent 8.5-d spin-orbital beat light variations (Fig. 4,b) caused, probably by a periodical enhancement of accretion stream and/or accretion ring surrounding the magnetic dwarf in asynchronous binary.

ACKNOWLEDGMENTS

This study was partially supported by Ukrainian Fund of Fundamental Researches 02/07/00451, by Russian Foundation of Base Research's grants 00-15-96553, 0202-1642, 02-02-26723 and grant No.1.4.2.2 of Russian scientific and technical program "Astronomy". Authors are grateful to the Conference LOC for supporting their participation in the Conference.

FIGURE 5. Only a few old novae are permitted to carnival.

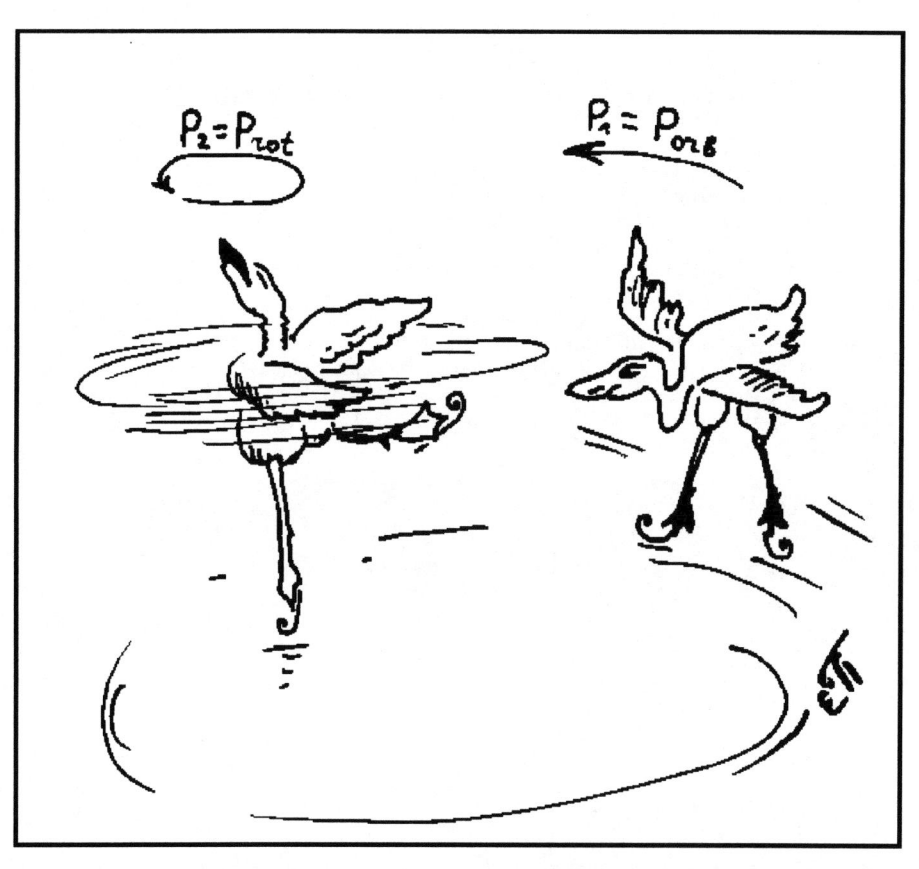

FIGURE 6. Asynchronous dance of V1500 Cyg.

REFERENCES

1. Bianchini, A., Sabbadin, F., Favero, G.C., Dalmeri, I., *Astron. and Astroph.*, **99**, 367-373 (1986).
2. Honeycutt, R.K., Robertson, J.W., Turner, G.W., "Periodic and near-periodic decaday lightcurves in old nova and nova-like CVs", in *Cataclysmic Variables*, edited by A. Bianchini et al., ASC library 205, Kluwer, Dordrecht, 1995, pp. 75-81.
3. Shugarov, S.Yu., *Variable Stars*, **21**, 807-816 (1983).
4. Osaki, Y., *Publ. Astron. Soc. Pacific*, **108**, 39-60 (1996).
5. Bath, G.T., *Nature Phys. Sci.*, **246**, 84 (1973).
6. Schreiber, M.R., Gansicke, B.T., Cannizzo, J.K., *Astron. and Astrop.*, **362**, 268-272 (2000).
7. Smak, J., *Acta Astronomica*, **34**, 161-189 (1984).
8. Schmidt, G.D., Liebert, J., Stockman, H.S., *Astropys. J*, **441**, 414-423 (1995).
9. Pavlenko, E.P. and Pelt, J., *Astrofizika*, **34**, 169-174 (1991).
10. Pavlenko, E.P., Primak, N.V., Shugarov S.Yu., *Astrofizika*, in press (2002).
11. Honeycutt, R.K., Robertson, J.W., Turner, G.W., Henden, A.A., *Astrophys. J.*, **495**, 933 (1998).
12. Somers, M.W., Naylor, T., *Astron. and Astroph.*, **352**, 563-571 (1999).
13. Prialnik, D., *Astroph. J*, **310**, 222-237 (1986).

Photometric and spectroscopic properties of four old novae

L. Schmidtobreick[*], C. Tappert[†], A. Bianchini[**] and R. Mennickent[†]

[*]*European Southern Observatory, Casilla 19001, Santiago 19, Chile*
[†]*Grupo de Astronomía, Universidad de Concepción, Casilla 160–C, Concepción, Chile*
[**]*Dipartimento di Astronomia, Università di Padova, Vicolo dell'Osservatorio 2, I-35122, Padova, Italy*

Abstract. We present optical spectroscopy and multi colour photometry of the old novae V630 Sgr, V693 CrA, V728 Sco, and V840 Oph. We show that novae stand clearly out in colour diagrams as very blue objects with an additional red component. We use the diagrams to select best candidates for the nova identifications in the hitherto uncertain cases of V728 Sco and V840 Oph, and provide spectroscopic confirmation for the latter.

INTRODUCTION

In the course of a long-term project investigating novae with large outburst amplitudes we have observed the four old novae V630 Sgr, V693 CrA, V728 Sco, and V840 Oph, the latter two still counting with ambiguous identification. Due to the presence of the accretion disc (perhaps partially truncated in magnetic systems) and the late-type secondary, the spectral energy distribution of novae does not resemble those of a single black body but of at least two components, the blue disk and the red secondary (the white dwarf being too faint in the optical). Hence, in colour diagrams old novae should stick out as objects being both, more blue and more red than average main sequence stars. Multi-colour photometry should therefore be able to provide a good candidate selection, although follow-up spectroscopy is required for confirmation.

DATA AND REDUCTION

The photometric data were obtained in one night on 2001-07-15 at the 1.54 m Danish telescope at ESO, La Silla, using DFOSC. The spectrum for V630 Sgr was taken in the same night, using DFOSC's grism 4 with a 1.5″ slit, yielding a spectral resolution of 10 Å. V840 Oph was observed spectroscopically on 2002-04-20 using EFOSC2 at the 3.6 m telescope at La Silla. Wavelength calibration gave a resolution of 18 Å.

The basic reductions have been performed with IRAF. Due to the crowding in the field, PSF-photometry has been determined using the standalone version of DAOPHOT and the ALLFRAME package. For the photometric calibration, standard stars selected from Landolt (1992) have been measured.

THE SAMPLE

V 630 Sgr has been detected in outburst in 1936 (Okabayashi 1936). It has been a very fast nova and was well observed until about 1950. Recent photometry by Woudt & Warner (2001) revealed it to be an eclipsing system with an orbital period of 2.83 h.

V 693 CrA is a recent nova which has been discovered in 1981 by M. Honda as a star of V=7.0 mag (Kozai et al. 1981). By systematic observations with the IUE satellite, Williams et al. (1985) analysed the chemical composition of the nova during its nebula phase and found very high elemental abundances in the ejecta.

V 840 Oph has first been visible as a star of 6.5 mag on Harvard photographs taken in May 1917 (Bailey 1920). It has been a fast nova with two very distinct secondary maxima (Shapley 1921). Three stars are close to the position of the nova and hence provide possible candidates for the nova remnant.

V 728 Sco has been visually discovered by J. Tebbutt in 1862 as a star of 5 mag (Tebutt 1878). The given coordinates are very uncertain, the finding chart of Downes et al. (2001) refers to the star closest to the given position.

RESULTS

U−B versus B−V diagrams of the stars in the field of V 630 Sgr, V 840 Oph, and V 728 Sgr are presented in Figures 1 and 3. No U data have been taken for V 693 CrA yet.

On the left side of Figure 1 the colour diagram of V 630 Sgr, a nova with confirmed position, is plotted. V 630 Sgr itself is marked with an asterisk and is distinctively displaced from the main sequence defined by the majority of the field stars. As assumed from theoretical considerations the nova is a blue object with red component and hence lies above the main sequence.

V 840 Oph, which is plotted on the right side of Figure 1, instead is one of the uncertain cases, as three possible positions are given for the nova remnant. These three candidates are highlighted in the diagram. The obvious best candidate indicated by the asterisk in the upper left corner has been confirmed spectroscopically (see Figure 2). In spite of the low S/N, the typical nova emission lines (H, HeII) are clearly visible in the spectrum of our best candidate for V840 Oph, thus confirming its photometric selection. This example shows that old novae with uncertain coordinates can be identified successfully by performing UBV photometry of the stars in the field.

In Figure 3 U−B versus B−V diagrams are plotted for the field of V 728 Sco, the other nova in our sample with uncertain coordinates. The candidate closest to the original position is indicated by the asterisk in the diagram on the left side. However, its position with respect to the main sequence makes it a highly unlikely nova remnant. The diagram on the right shows all stars less than 1' away from the original nova position. The star in the upper left corner is obviously the best nova candidate. However, it has not yet been confirmed spectroscopically.

Coordinates and photometry of all novae of our sample are presented in Table 1. In the

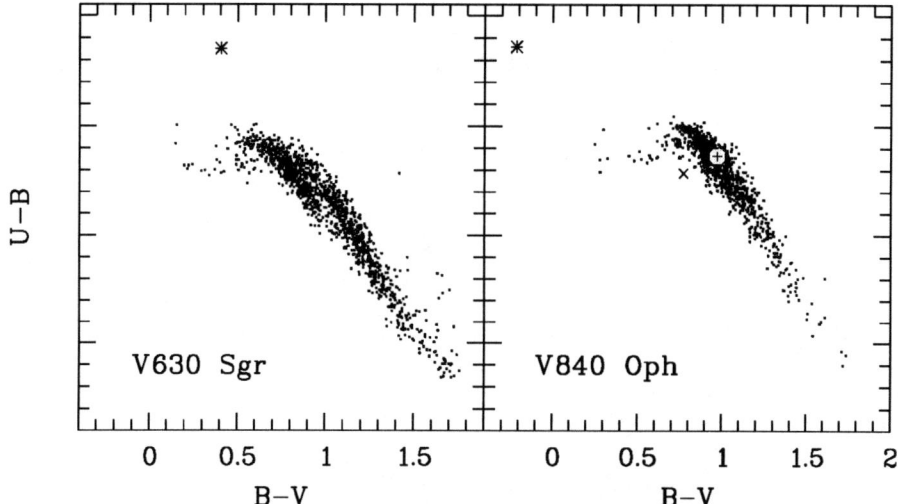

FIGURE 1. The diagram on the left shows U−B versus B−V for the stars in the field of V 630 Sgr. The nova itself is plotted as an asterisk. On the right side, the same plot is presented for V 840 Oph, the three stars inside the error-box of the coordinates are symbolised by +, ×, and ∗ (best candidate).

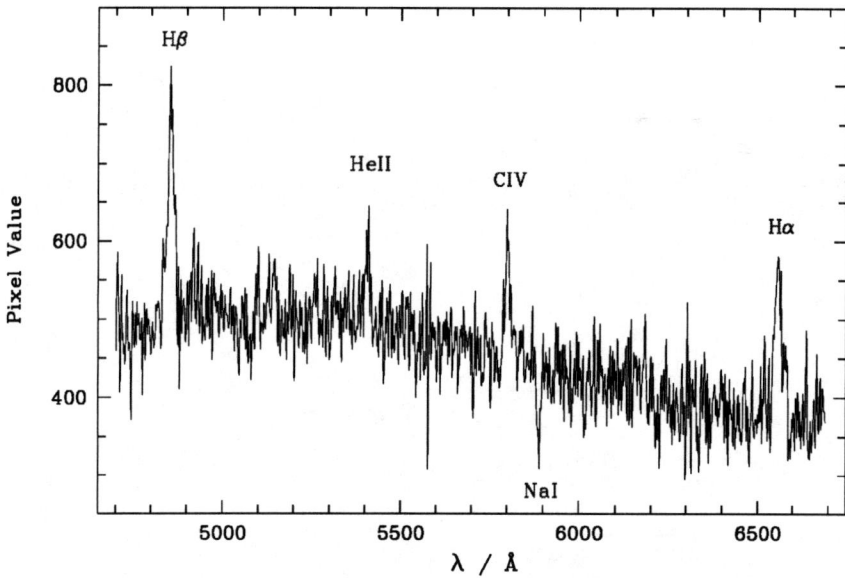

FIGURE 2. Uncalibrated spectrum of our best candidate for V 840 Oph. Its nova character is confirmed by the presence of typical nova emission lines. The feature near $\lambda 5580$ is an artifact.

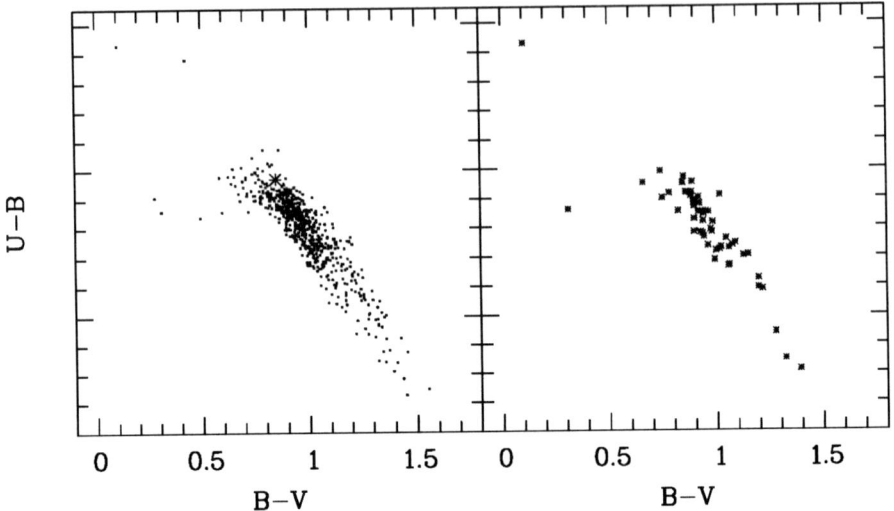

FIGURE 3. U−B versus B−V is plotted for the stars in the field of Nova V 728 Sco. The left diagram shows all stars in the field and the star closest to the original nova position is marked with an asterisk. The diagram on the right shows only the stars within a 1′ vicinity of the nova position. The star in the upper left corner is obviously the best candidate for the nova remnant.

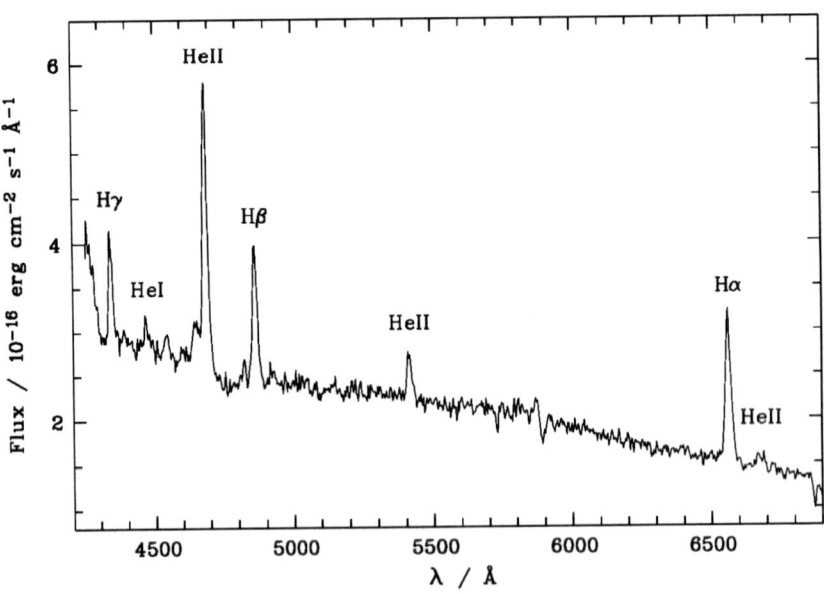

FIGURE 4. Calibrated spectrum of V630 Sgr. Principle emission lines are indicated.

TABLE 1. Coordinates, outburst magnitude (visual or photographic), magnitudes in quiescence, and outburst amplitude (v−V or p−B). In the case of V 728 Sco, the data of the best candidate are listed.

Nova	RA_{2000}	DEC_{2000}	outburst	U	B	V	I	amplitude
V 630 Sgr	18 08 48.3	−34 20 21	4.5v	17.18	17.88	17.48	17.90	13.0
V 693 CrA	18 41 57.8	−37 31 14	7.0v		21.11	21.00	20.78	14.0
V 840 Oph	16 51 33.9	−29 32 40	6.5p	18.39	19.11	19.32	18.33	12.6
V 728 Sco	17 39 07.5	−45 29 50	5.0v	18.58	19.44	19.33	19.07	14.3

TABLE 2. Equivalent widths (in Å) of the principle emission lines in the spectra of V630 Sgr and V840 Oph. Note that the spectral range for the V840 data does not include wavelengths < 4700 Å.

	Hγ 4340	CIII/HeII 4650/4686	Hβ 4861	HeII 5412	CIV 5805	Hα 6563
V630 Sgr	−7	−35	−15	−6	–	−31
V840 Oph			−15	−4	−8	−23

case of V 728 Sco the data for the best candidate according to the colour diagrams are given. The outburst amplitude has been computed as the difference from the outburst magnitude and the most recent quiescence magnitude.

Finally, the spectrum of V630 Sgr is shown in Fig. 4, while Table 2 lists the equivalent widths of its principle emission lines, together with the corresponding data for V840 Oph. The spectroscopy of V630 Sgr represents the first such observation of the quiescent nova. Note that, in spite of the system being seen at high inclination (Woudt & Warner 2001), the emission lines are single-peaked. This might be indicative of the absence of an accretion disc in V630 Sgr, although our resolution (10 Å) is clearly too low to be conclusive in this respect.

ACKNOWLEDGMENTS

We would like to thank George Hau and Michael Sterzik who kindly performed the spectroscopic observations of V840 Oph.

REFERENCES

1. Bailey, S.I., 1920, AN 210, 375
2. Landolt, A.U., 1992, AJ, 104, 340
3. Downes, R., Webbink, R.F, Shara, M.M., et al., 2001, PASP 113, 764
4. Kozai, Y., Kosai, H., Honda, M., Cragg, T., 1981, IAU Circ. 3590
5. Okabayashi, 1936, AN 261, 65
6. Shapley, H., 1921, PASP 33, 189
7. Tebbutt, J., 1878, MNRAS 38, 330
8. Williams, R.E., Ney, E.P., et al., 1985, MNRAS 212, 753
9. Woudt, P.A., Warner, B., 2001, MNRAS 328, 159

The UCT Survey of Old Novae

Patrick A. Woudt* and Brian Warner*

Department of Astronomy, University of Cape Town, Rondebosch 7700, South Africa

Abstract. We present a status report on our high speed photometric survey of faint Cataclysmic Variables, which is concentrating on old novae.

The high speed photometric survey of southern Cataclysmic Variable stars, of which one paper has been published (Woudt & Warner 2001) and another is in press (Woudt & Warner 2002), uses the UCT CCD photometer attached to the 1.0-m and 1.9-m reflectors at the Sutherland site of the South African Astronomical Observatory. We have concentrated on faint old novae in crowded fields; these have generally been neglected in the past as being difficult to observe, but they prove to be a rich source of interesting phenomena, especially relating to orbital modulations and magnetic effects.

From photometric modulations we have determined orbital periods for the old novae RS Car (N1895), V365 Car (N1948), RR Cha (N1953), BY Cir (N1995), DD Cir (N1999), AP Cru (N1936), CP Cru (N1996), V351 Pup (N1991), V630 Sgr (N1936), V697 Sco (N1941) and V992 Sco (N1992). Details of these can be found in Table 1 of the Review paper by Warner in these Proceedings. The Galactic distribution of old novae is shown in Fig. 1, where those with known orbital periods are shown as large filled circles. The southern sky is no longer so undersampled as it was a few years ago.

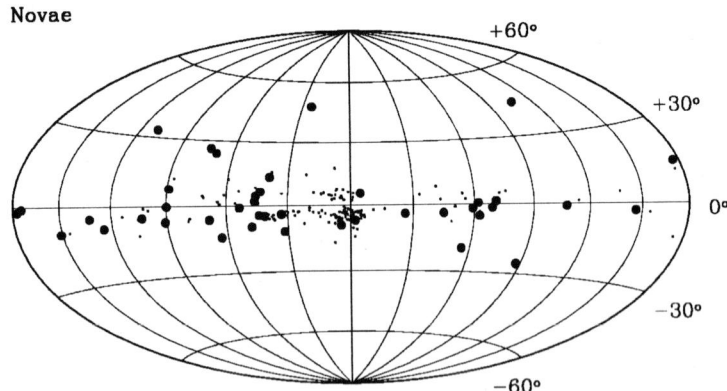

FIGURE 1. The distribution of old novae in Galactic coordinates (Aitoff projection). The Galactic Centre is at the centre of the figure, with increasing longitudes to the left. Novae with known orbital periods are indicated by the large filled circles.

In addition, from high speed flickering activity we have been able to determine the

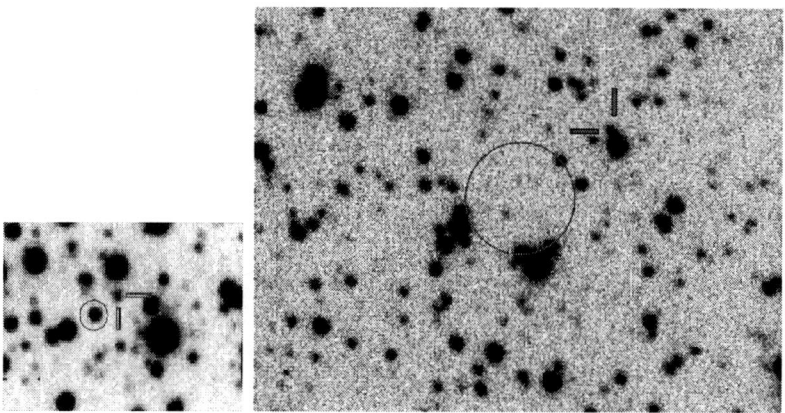

FIGURE 2. Finding charts of MT Cen (left panel) and X Cir (right panel). MT Cen was observed with the 1.9-m telescope (f.o.v. $50'' \times 34''$), X Cir with the 1.0-m telescope (f.o.v. $109'' \times 74''$). Previous positions are indicated by circles, updated indentifications are indicated by markers.

TABLE 1. Old novae with newly detected spin periods.

Star	P_{spin} (sec)	Star	P_{spin} (sec)
RX J1039.7-0507	1444	RR Cha	1950
AP Cru	1850	V697 Sco	~12000

correct identification for 4 previously misidentified or unidentified old novae; MT Cen, X Cir, V552 Sgr and CQ Vel. Finding charts for V552 Sgr and CQ Vel are shown in Woudt & Warner (2001; 2002, respectively). Finding charts for MT Cen and X Cir are shown in Fig. 2.

Apart from the new orbital periods, we have also found periodicities that are ascribed to the rotation periods of the white dwarf components. These are the signatures of intermediate polars, which have magnetic primaries. Table 1 gives details of the spin periods found during our survey.

ACKNOWLEDGMENTS

PAW is funded by the National Research Foundation and by strategic funds made available to BW. BW is funded entirely by the University of Cape Town.

REFERENCES

1. Warner, B., these proceedings (2002).
2. Woudt, P.A., and Warner, B., MNRAS, **328**, 159 (2001).
3. Woudt, P.A., and Warner, B., MNRAS, in press (2002).

Optical Spectroscopy of GK Persei during outburst and quiescence

U.S. Kamath* and G. C. Anupama*

Indian Institue of Astrophysics, Bangalore 560 034, India

Abstract. We present here the optical spectra of GK Persei obtained during 1995 – 2001. This period includes two dwarf-nova outbursts and the quiescence phases.

INTRODUCTION

The old nova GK Persei is well-known for exhibiting dwarf-nova outbursts since 1948. GK Per has a quasiperiodic outburst interval ranging from 900 to 1340 days. Some of the latest events occurred in 1996, 1999 and 2002, taking about a month to reach maximum at mean magnitude of 10.3 – 10.5 from its quiescent level of 13.5 mag. Thermally unstable disc models have generally been used to explain the outburst photometric propeties of GK Per [4]. However, neither standard steady state nor non-steady state models can satisfactorily explain the uv spectra of GK Per at minimum or maximum [5]. The disc-heating mechanism is thus still elusive.

In a previous study [1], we had deduced the parameters of the secondary star, the white dwarf and the accretion disc. One conclusion was that the accretion disc is smaller at quiescence than during outburst. We have continued monitoring GK Per and here we present some spectra obtained during the various phases of its activity.

OBSERVATIONS

CCD spectra were obtained during 1995 – 2001 using the Boller & Chivens and OMR spectrographs at the 2.3m Vainu Bappu Telescope and cover the wavelength range 4400 – 8000 Å, with dispersions of 2.6 or 5.3 Å/pixel. FeAr and HeNe lamps were used for wavelength calibration and different standard stars were used for flux calibration. However, the spectra presented here are on a relative flux scale.

DESCRIPTION OF THE SPECTRA

The spectrum of GK Per is a composite of the secondary absorption and accretion disc emission components. The secondary contributes about 40% of the light during quiescence [1].

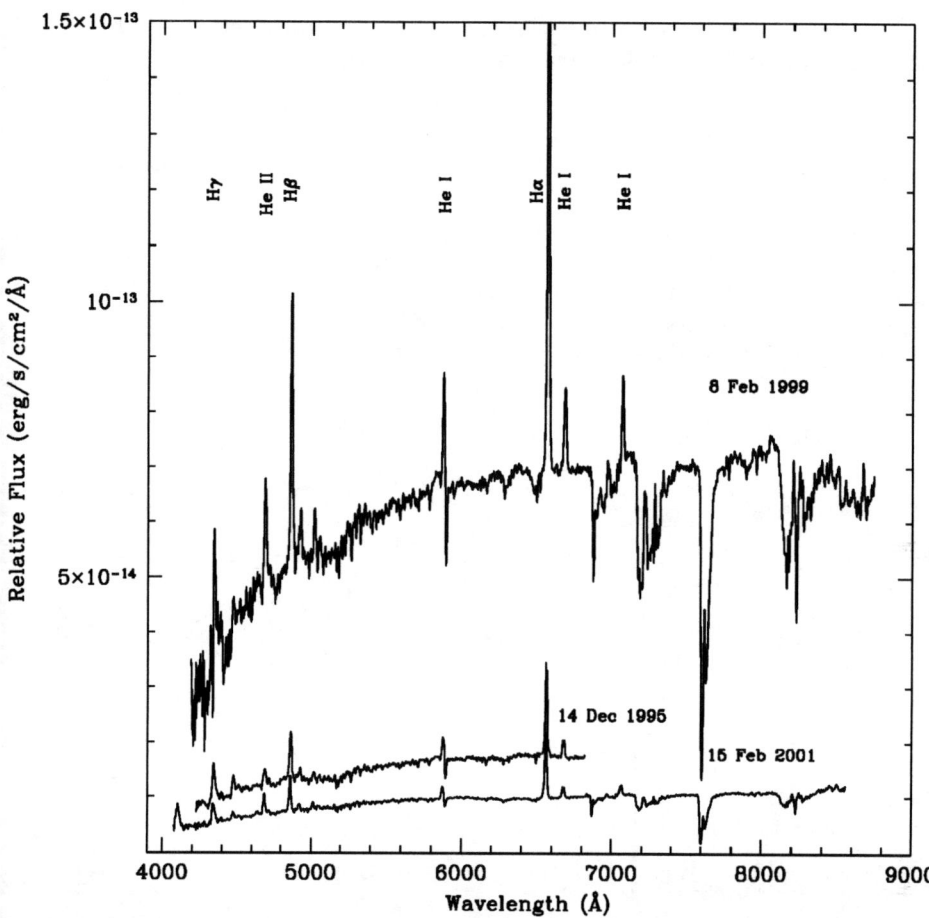

FIGURE 1. Spectra of GK Persei during quiescence.

Spectra at quiescence

- 14 Dec 1995 : An interesting feature of this spectrum is that the Hα and He I 6678 lines are split. This is probably the first time that the Hα line in GK Per has been observed to show some structure.
- 8 Feb 1999 : GK Per rising towards outburst, which occurred in late March 1999. The brightness increase is reflected in the higher continuum level. The equivalent widths of the Balmer lines have also increased and are the highest among these quiescent spectra.

- 15 Feb 2001 : Spectrum similar to Dec 1995 spectrum. However, the He I lines show a flat-topped profile with some structure.

The observed Hα/Hβ ratios are in the range 1.7 to 2.2, similar to that observed in earlier spectra [3]. The departure from the standard Case A or B recombination has been attributed by them to a second emission line region which is the source of the Balmer radiation.

It is seen that the equivalent widths of various helium lines lies in the range of 3 to 7 Å at all epochs. The Balmer lines, however, show considerable variation in their equivalent widths with time. The Hα equivalent width is between 16 to 27 Å. The equivalent width of Hβ is 11 to 16 Å. Thus, as has been observed earlier (e.g., [2]), GK Per shows activity in terms of varying emission line strengths even at quiescence.

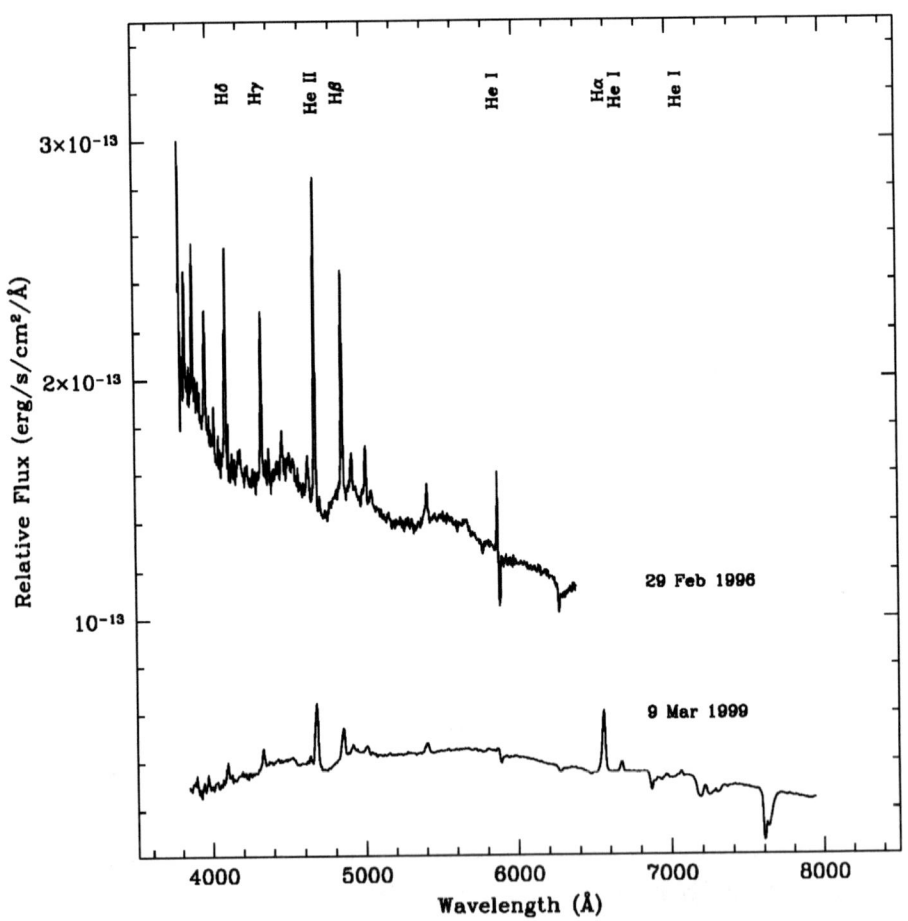

FIGURE 2. Spectra of GK Persei during or rising towards outburst.

Spectra at outburst

We have spectra for the previous two outbursts of GK Per.

- 29 Feb 1996 : GK Per had a minor outburst from late February to April 1996. The rise in the flux levels at the shorter wavelengths and the slope of the continuum clearly indicates that the accretion disc has brightened up.
- 9 Mar 1999 : GK Per had a small outburst starting late February 1999. Since GK Per takes about a month to reach its maximum, the outburst nature is not readily apparent in this spectrum. However, a comparison with the quiescent spectra shows the increased level of the continuum. The $H\alpha/H\beta$ ratio is 2.2. The He II 4686 Å feature is considerably stronger than $H\beta$ during outbursts ; in quiescence it has an equivalent width only about 50% of that of $H\beta$.

CONCLUDING REMARKS

The emission lines show structure which varies from epoch to epoch. This, and the Balmer decrements, can tell us about the nature and location of the emssion regions. The interesting structures shown by various lines needs to be investigated. These spectra will be compared with model spectra in a detailed study in order to better understand the quiescent and outburst nature of GK Per. Such spectroscopic studies have the potential to refine the models of GK Persei's outburst.

REFERENCES

1. Anupama, G.C., Prabhu, T.P., *MNRAS*, **263**, 335 (1992)
2. Crampton, D., Cowley, A.P., Fisher, W.A., *ApJ*, **300**, 788 (1986)
3. Garlick, M.A., Mitlaz, J.P.D., Roisen, S.R., Mason, K.O., *MNRAS*, **269**, 517 (1994)
4. Kim, Soon-Kwok, Wheeler, J.C., Mineshige, S., *ApJ*, **384**, 269 (1992)
5. Yi, I., Kenyon, S.J., *ApJ*, **477**, 379 (1997)

Physical and Chemical Diagnostic of Structured Nova Shells

Marcos P. Diaz

Departamento de Astronomia, IAG, Universidade de São Paulo, 05508-900, São Paulo, Brazil

Abstract. Photoionization models of inhomogeneous nova shells are presented and compared with spherically symmetric calculations. The methods employed for assembling 3D models are briefly discussed. An application of these tools aiming the prediction of the Li^7 optical line intensities in recent nova shells is described.

SPECTRAL ANALYSIS OF STRUCTURED NOVA SHELLS

The evolving photoionization spectra presented by novae during their decay offers a unique opportunity for chemical and physical diagnostic of the central source and the shell itself. However, most of the modeling procedures are not well constrained and numerical solutions are often degenerate. Besides the unknown spectrum of the ionizing source and chemical abundances, the observed emission line spectrum depends strongly on the mass distribution in the shell. In most cases the presence of density condensations and voids remain unknown due to limited imaging resolution. On the other hand, spectral analysis of unresolved shells claimed the presence of such structures in many novae [1]. These studies show the simultaneous occurrence of high and low ionization species during different stages of shell expansion. In fact, these findings represent a major discrepancy between current model spectra of shells and observed line fluxes. During the calculation of models we often overcame the problem of condensations with an extra free parameter in the form of an arbitrary "equivalent" radial density profile, with spherical or cylindrical symmetry.

The presence of condensations provide a way of explaining the presence of strong lines from species which are spread over a wide ionization range, specially in matter bounded shells. The problem of calculating 3D models of nova envelopes can be divided into two parts. The first consists of defining the mass density distribution in the shell and second concern the calculus of thermal and photoionization equilibrium, radiation field and emitted spectrum from the inhomogeneous shell. A general 3D code (RAiny3D) is being built on the basis of the photoionization code CLOUDY 94 [2] with the goal of solving both problems. The effect of neutral gas condensations in the nova envelope was investigated by adding Gaussian globules to a background radial density law. These globules follow power law probability distributions in density and size. They are also randomly positioned within the shell and account for a fraction "f_c" of the total mass. A more detailed description of the 3D photoionization modeling procedures may be found in [3].

APPLICATION: THE LITHIUM PROBLEM IN NOVAE

Over the last two decades theoretical models of thermonuclear runaways (TNR) in novae have predicted a physical scenario where significant amounts of ^{7}Li can be formed. According to recent hydrodynamic calculations, the convective time–scale in the core–envelope interface was found to be shorter than the life–time of ^{7}Be, allowing ^{7}Li enhancements of 10^2 to 10^3 with respect to solar values [4]. Although lithium absorption lines have been detected in the secondary spectrum of a few black hole binaries, there is no observational confirmation of the nova ^{7}Li production mechanism in cataclysmic binaries. Depending on the conditions during outburst such a process may result in a significant contribution to the ISM lithium abundance in the Galaxy [5]. The problem of detecting such a lithium enhancement in optical wavelengths may be addressed by observations of the early pseudo–photosphere absorption spectrum or by measuring the emission line spectrum formed in the ejected envelope. In both cases the line forming regions are evolving fast and therefore require time–resolved descriptions. On the other hand, the intrinsic scatter in nova basic parameters has to be taken into account in the search for possible episodes of lithium line formation.

The nova envelope models presented here share many fixed parameters which are described bellow. Average chemical abundances for non–ONeMg novae [6] were employed with a constant ^{7}Li enhancement of 1000 times the solar value. Typical maximum and minimum nova shell expansion velocities of 600 and 1000 km/s define the shell size and thickness as a function of time in the impulsive ejection approximation. The mass distribution inside the spherically symmetric background component is defined by a radial power density law with index a = −0.9 [7]. An extensive model grid containing more than 13000 points was used to study the complex line flux behavior in the parameter domain given by Table 1. For the sake of generality, no particular prescription for the central source evolution was adopted. Its ionizing spectrum is approximated by NLTE, hot and high–gravity stellar model atmospheres [8]. The line transfer was calculated under the Sobolev approximation for a few control points in the grid, yielding unimportant discrepancies when compared to the static case. Clumps with a maximum density contrast ranging from 1.1 to 12 where simulated.

Sample behavior of the LiI λ6707 line strength as a function of the central source temperature and luminosity is shown in Fig. 1. These results suggest that the line emission is strongly suppressed in luminous and super–Eddington remnants (i.e. in very fast novae). In addition, the line emission in hot sources is restricted to ~18 months after t_0 (Fig. 2). Relatively low temperature sources ($T_{eff} < 130,000$ K) may eventually produce significant flux during an extended period of time (>2.5 yr.) which hold the possibility of spatially resolved diagnostic of nearby novae. The LiI λ6707 line is mainly formed by

TABLE 1. Shell Parameter Ranges

M_{SHELL}	($M_\odot$)	$2.8 \times 10^{-6} - 1.4 \times 10^{-4}$
L_c	(erg.s^{-1})	$10^{35} - 10^{38.5}$
T_{eff}	(K)	$80000 - 350000$
$(t - t_0)$	(days)	$90 - 5000$
$^a f_c$	(%)	$0 - 70$

^{a}Condensed mass fraction (see text).

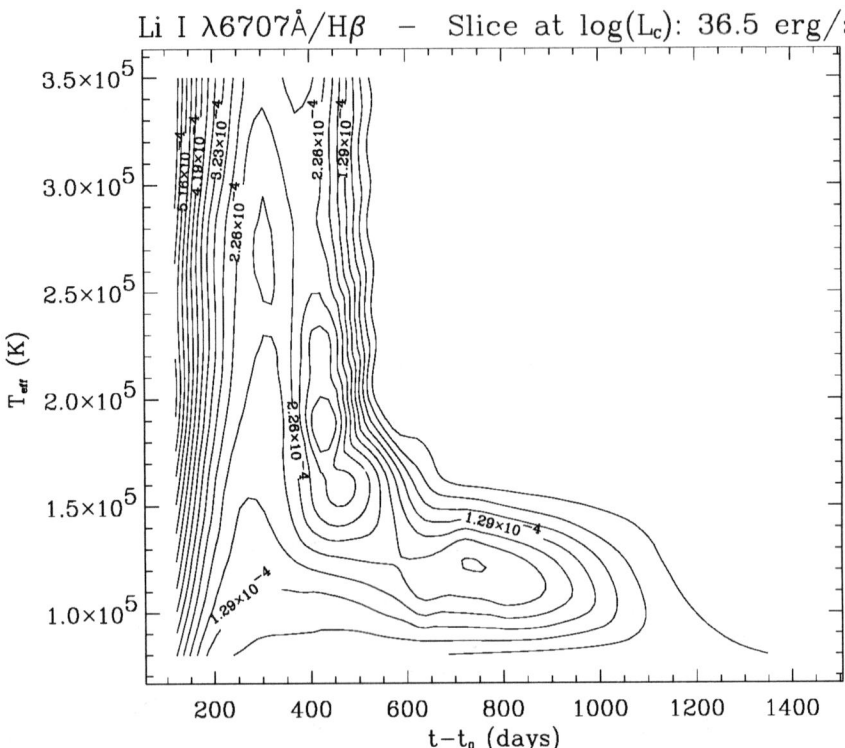

FIGURE 1. Lithium λ6707Å line flux relative to Hβ for a grid of nova shell models with fixed central source luminosity (L_c). The integrated line flux over the ejecta is presented as a function of the ionizing source temperature and time after outburst for a 6×10^{-5} $M_\odot$ shell. Contour lines are linearly spaced in relative flux.

collisional excitation from the ground state in the shell environment. Sources with low ejection efficiency (i.e. having lower nebular densities) are shown in the rising part of the mass dependence curve (Fig. 3). Due to the low first ionization potential of lithium, most of the line production comes from almost neutral gas. This explains the important effect on the expected flux produced by the presence of neutral mass clumps in the envelope (Fig. 4). Facing the current uncertainties in condensation properties (distribution, size and density contrast), the numerical results presented here should be taken as order of magnitude indications only.

The simulations suggest a well defined set of constraints for guiding deep spectroscopic measurements of real nova remnants. In general, the expected 6707 relative flux is a very small quantity reaching, in the most favorable situations, values around $10^{-4} \times$ Hβ. The maximum predicted integrated fluxes in this line are close to 10^{-16} erg.s^{-1}.cm^{-2} for an unabsorbed remnant at 1.0 kpc. The corresponding surface brightness levels predicted for recent novae may be actually observed using high–resolution imaging spectrographs on large telescopes.

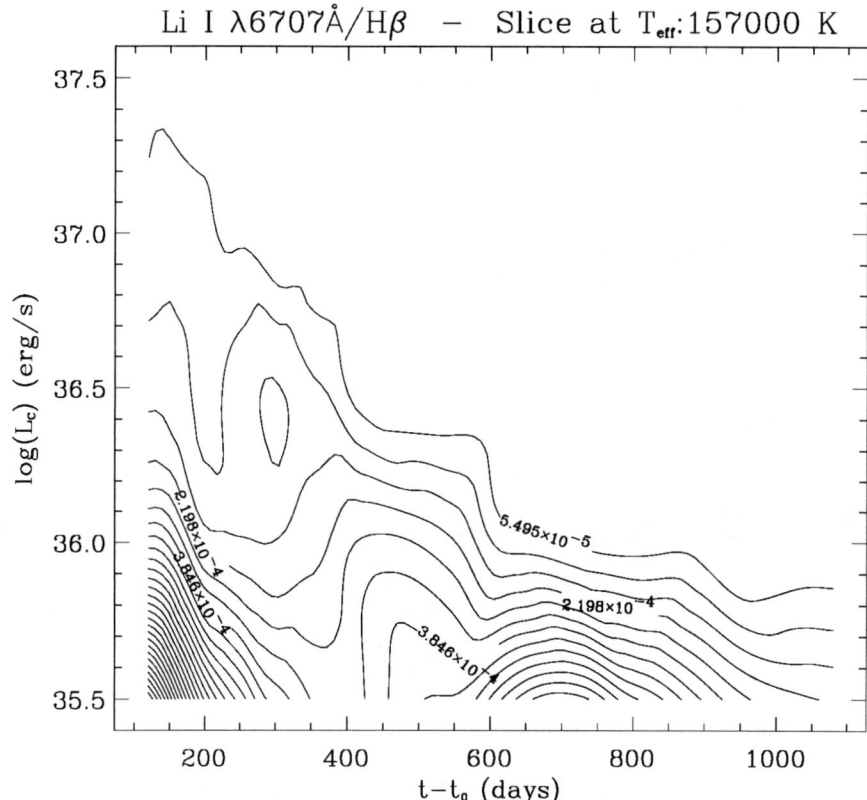

FIGURE 2. Same as Figure 1 for fixed central source temperature T_{eff}. The relative flux integrated over the shell is presented as a function of the ionizing source luminosity and time after outburst.

ACKNOWLEDGMENTS

This work was supported by CNPq (301029) and FAPESP (02/00805-4).

REFERENCES

1. Williams, R.E., *ApJ* **376**, 721–737 (1991).
2. Ferland, G., et al., *PASP* **110**, 761–778 (1998).
3. Diaz, M. P., in "*Spectroscopic Challenges of Photoionized Plasmas*", edited by G. Ferland and D. Savin, ASP Conference Series 247, San Francisco, 2001, pp. 227–230.
4. Hernanz, M., José, J., Coc, A., and Isern, J., *ApJ* **465**, L27–L30 (1996).
5. Romano, D. and F. Matteucci, these proceedings, (2002).
6. Gehrz, R. D., Truran, J. W., Williams, R. E., and Starrfield, S., *PASP* **110**, 3–26 (1998).
7. Diaz, M. P., Williams, R. E., Phillips, M., and Hamuy, M., *MNRAS* **277**, 959–964 (1995).
8. Rauch, T. , *A&A* **320**, 237–248 (1997).

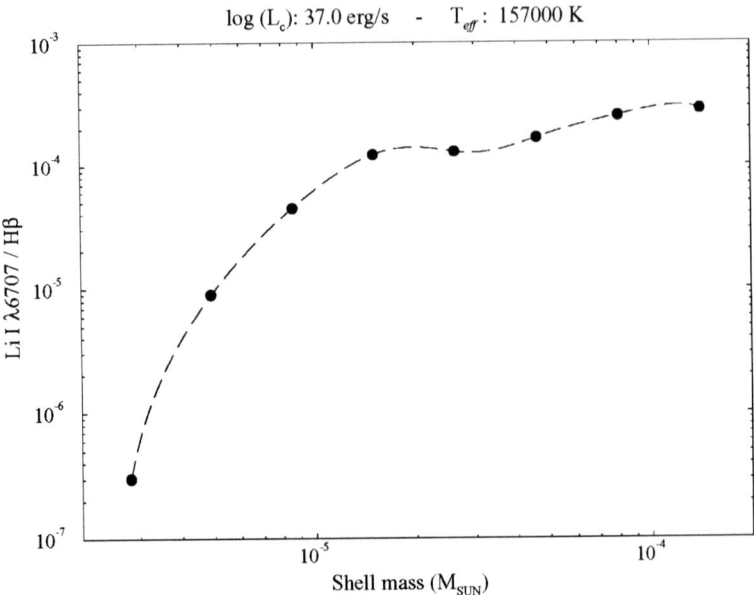

FIGURE 3. Lithium line λ6707Å line flux relative to Hβ for models with different shell masses. The maximum intensity along the shell evolution is plotted for a power law radial density profile.

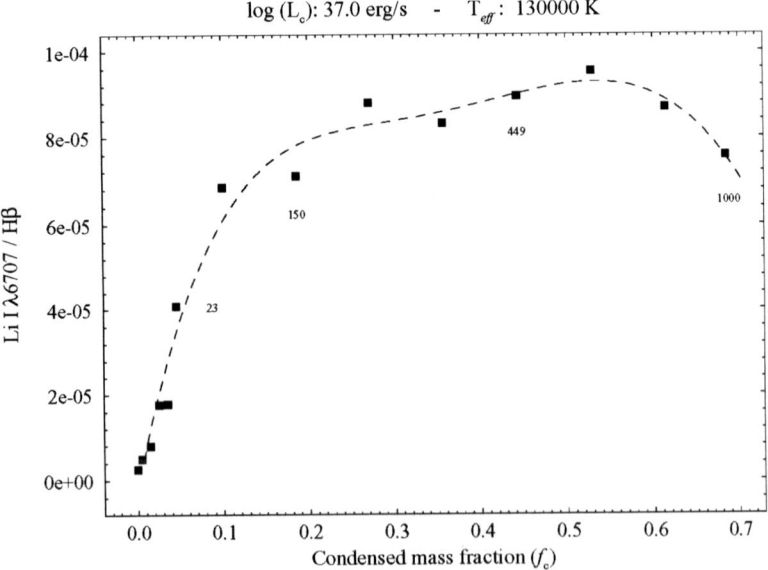

FIGURE 4. Relative line flux for clumpy model shells 21 months after TNR. The values close to the points indicate the number of condensations in a 3D random distribution. $M_{shell} = 5 \times 10^{-5}$ $M_\odot$.

Nova V Persei – a classical nova in the "period gap"

Nataly A. Katysheva[*], Elena P. Pavlenko[†] and Sergei Yu. Shugarov[*]

[*] *Sternberg Astronomical Institute, Universitetskij pr., 13, Moscow V-234, 119992, Russia*
[†] *Crimean Astrophysical Observatory, Nauchny 19-17, Crimea, 98409, Ukraine*

Abstract. We report new CCD-photometric observations of the old nova V Per. The monitoring was carried out in autumn 2001 in *R* and *V*-bands. The eclipsing light curves are presented and analyzed. The value of the orbital period derived by Shafter and Abbott [1] was corrected a little. The period of the system is in the middle of the "period gap". In fact a half of the stars in the "period gap" belongs to superhumpers.

INTRODUCTION

The outburst of a classical nova V Per (1887) was discovered on a Harvard objective prism plate on the 3rd of November 1887. There were the hydrogen emission lines noted in spectra. Later, in 1946, McLaughlin [1], while examining those spectra, indicated that V Per had the nebular lines and so, it was discovered 5–6 month after the eruption. At maximum it possibly reached 4th or 5th magnitude. Shafter & Abbott [2] discovered the eclipses in the light curve of V Per with a period 2.57 h, an eclipse deep of ~1.3 mag and the width of about 15 min:

$$T_{mid-ecl} = HJD\ 2447445.9322(2)+0.10712(1)E. \qquad (1)$$

V Per was suggested to be a magnetic system [2, 3] – intermediate polar, namely. Wood et al. [3] did not detect any signs of the hot spot in the light curves. Shafter [4] gave an estimation of a distance about 1.0 kpc to V Per.

It should be noted that some novae are either intermediate polars (V603 Aql, DQ Her, V533 Her, GK Per) or polars (V1500 Cyg).

CATACLYSMIC VARIABLES IN THE "PERIOD GAP"

This nova is very interesting because its period is in the middle of the "period gap" of cataclysmic variables (CVs) [5]. The "gap" is the major feature of the period distribution and it occurs in the orbital periods of observed CVs between 2 and 3 hours, precisely 2.1 and 2.8 hours. There are some conceptions about the existence of the "period gap". One of them is the sudden enhancement of orbital angular moment [6]; the other (the most popular) – a decreasing of magnetic activity of the secondary when it becomes fully convective and the mass transfer temporarily switches off [7–9]. If the mass loss through point L_1 stops during the fully convective phase any evidences of activity should be absent.

In the early 90s there were only a few cataclysmic variables in the "period gap", polars in general. But there are 35 CVs in the extended "period gap" (2–3 h) and 19 CVs between 2.1 and 2.8 h now [10] (see Table 1).

TABLE 1. Cataclysmic variables into and near the "period gap"

Name	P_{orb}, min 2.0 –2.1 h	Type	Star	P_{orb}, min 2.1 –2.8 h	Type
EF Peg	120.5*	dN SU	UZ For	126.5	NL AM
KK Tel	121.0	dN SU	GX Cas	128.0*	dN SU
TY PsA	121.1	dN SU	UW Pic	133.5	NL AM
ES1113+432	122.4	NL AM	V589 Her	130.9*	dN SU
DV UMa	123.6	dN SU	V419 Lyr	131sp	dN SU
HU Aqr	125.0	NL AM	Var73 Dra	131*	dN SU
YZ Cnc	125.0	dN SU	V725 Aql	138.3	dN SU
IR Com	125.3	dN SU	QS Tel	140.0	NL AM
EU Cnc	125.4	NL AM	NY Ser	144	dN SU
V344 Lyr	126.1*	dN SU	V348 Pup	146.6	NL IP SH
UV Gem	126.*	dN SU	V Per	154.3	Nova
	2.8–3.0 h		V349 Pav	160.0	NL AM
V2214 Oph	169.2	NA IP? SH	QU Vul	160.9	Nova
J0501--03	171	NL AM	J1554.2+2721	165.2	NL AM
LQ Peg	174.2:	NL UX	V592 Cas	165.7*	NL UX SH
DM Gem	176.6:	NA	WX LMi	166.9	NL AM
AH Men	177.71	NL IP? SH	V795 Her	168.0	NL IP SH
V442 Oph	179.1	NL VY	TU Men	168.8	dN SU
UZ Boo	180.0:	dN SU			

Currently about a half of the "gap"–stars are the magnetic stars and the other part (except some ones) – SU UMa-stars (dwarf novae with superhumps): there are seven AM Her-stars and two IPs, seven SU UMa-stars, two novae, one (V592 Cas) is UX UMa-star. We emphasize the fact that a half of the stars in the "period gap" belongs to superhumpers – either to SU UMa or to permanent superhumpers. There are usual signs in Table 1: dN – dwarf nova, SU – SU UMa-type star, NL – nova-like star, IP – intermediate polar, AM – polar, SH – superhumper, UX – UX UMa-type star, VY – VY Scl-type star.

OBSERVATIONS OF V PER

V Per has been observed in 2001. The observations were provided by using the CCD-cameras ST-6 and ST-7 at Crimean Laboratory of Sternberg Astronomical Institute with 0.6 and 1.25-cm telescopes in V and R-bands. Our monitoring (in R) of V Per was carried out in eleven nights in the range from JD 24452171 to 24452257. On September, 24, 2001 (JD 24452177) we observed in V-band. We found the orbital period P=$0.^h1071256$, close to value by Shafter and Abbott [2]:

$$T_{mid-ecl} = HJD\ 2452172.541(2)+0.1071256(1)E. \qquad (2)$$

This period was corrected by using the Shafter and Abbott [2] epoch.

In the Fig.1 the periodograms and data in R-band folded with $0.^d1071256$ are presented.

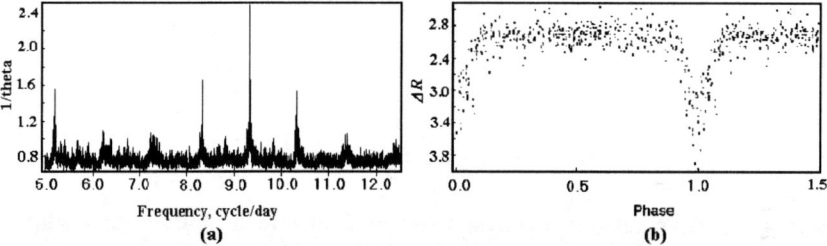

FIGURE 1. The periodogram a) and b) R-data folded with above period.

The light curve in V-band is shown in Fig.2.

V Per is in the middle of "period gap" and challenges all existent models of the "gap". Its behaviour doesn't fit the switching-off of the mass losses. The morphology of the light curve and the spectral features [3] both demonstrate mass transfer high enough. Moreover our investigation shows that the rate of mass transfer is not so stable (see Fig.3). We can see that the light curve is unstable a half-hour before the eclipse: at some nights there is a hump before the eclipse (A hot spot? A hot line [11]? Accretion column?). The other nights it is absent. Shafter and Abbott [2] also observed a hump at one night from the four. We also can see the different depth of eclipses: from 1^m to $1.^m6$.

The photometric variability in the light curves is the sign of magnetic systems, but many non-magnetic systems also demonstrate it.

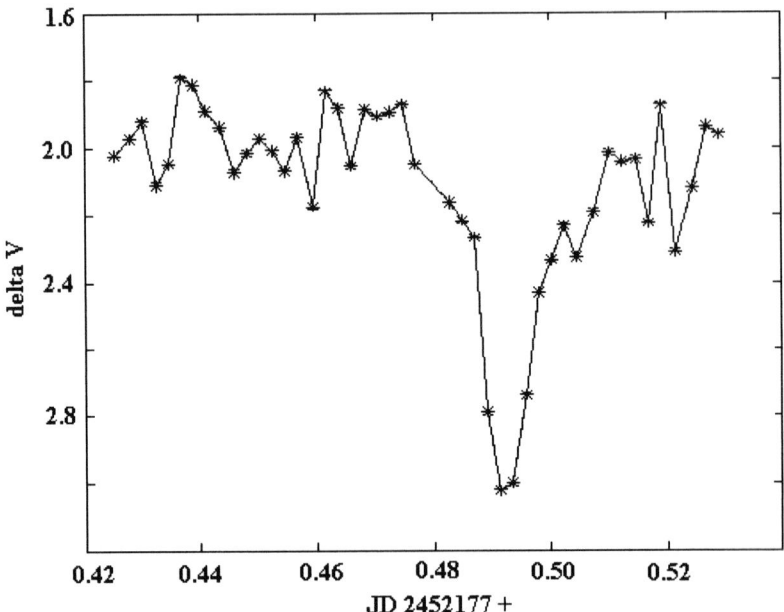

FIGURE 2. The light curve of V Per in *V*-band (24 September 2001).

It is necessary to provide the polarimetric observations to confirm or to refuse a magnetic hypothesis.

ACKNOWLEDGMENTS

This work was supported in part by a Russian Foundation of Base Research's grants 00-15-96553, 0202-16462 and 02-02-26723 (N.K. and S.Sh.) and an Ukranian Fund of Fundamental Researches grant 02/07/00451 (E.P.). Authors also are grateful to the SOC and LOC for hospitality.

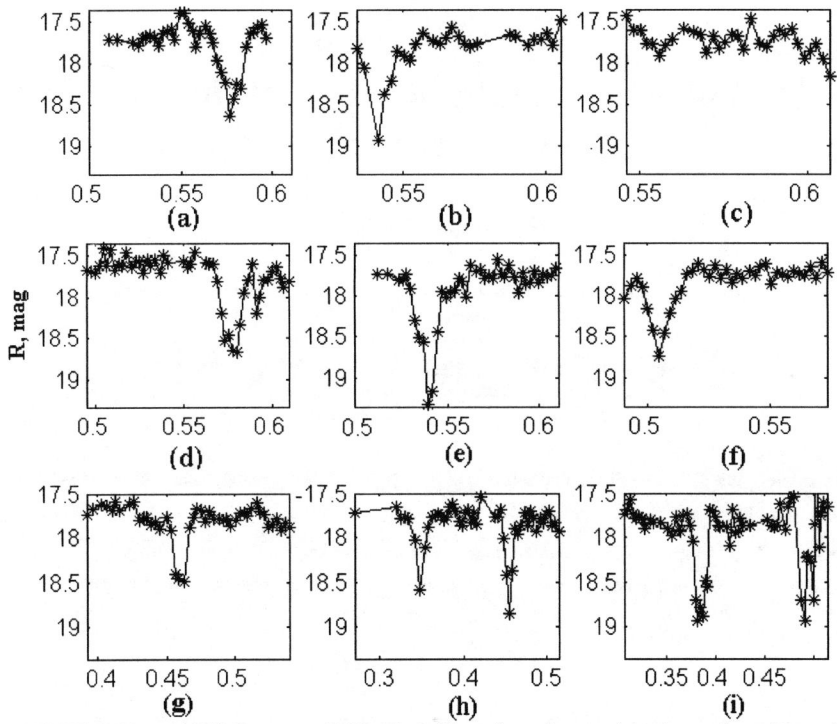

FIGURE 3. The orbital light curves of V Per in *R*. The eclipse shape and depth vary from date to date The abscissa denotes the portion of the Julian day (2452000 + ...; (a) – 171; (b) – 172; etc.). (a – 18.09; b – 19.09; c – 20.09; d – 21.09; e – 22.09; f – 23.09; g – 18.11; h – 08.12 and i – 10.12.02).

REFERENCES

1. McLaughlin, D.B., *Publ. Astron. Soc. Pacific,* **58,** 218–221 (1946).
2. Shafter, A.W., and Abbott, T.M.C., *Astroph. Journal,* **339,** L75–L78 (1989).
3. Wood, J.H., Abbott, T.M.C. and Shafter, A.W., *Astroph. Journal,* **393,** 729–741 (1992).
4. Shafter, A.W., *Astroph. Journal,* **487,** 226–236 (1997).
5. Mestel, L., and Spruit, H.C., *Mon. Notic. Roy. Astron. Soc.,* **226,** 57–66 (1987).
6. Paczynski, B., and Sienkiewich, R., *Astroph. Journal,* **268,** 825–831 (1983).
7. Tutukov, A.V., *Pis'ma Astron. Zh,* **9,** 160–165 (1983); *Sov. Astron Lett.,* **9,** 86–91 (1983).
8. Spruit, H.C., and Ritter, H., *Astron. Ap.,* **124,** 267–272 (1983).
9. Rappaport, S., Verbunt, F., and Joss, P.C., *Astroph. Journal,* **275,** 713–731 (1983).
10. Katysheva, N.A., and Pavlenko, E.P., *Astrofizica*, in press (2002).
11. Khruzina, T.S., *Astron. Report,* **45,** 255–268 (2000).

A Possible Detected Faint Shell of the Classical Nova QZ Aurigae

Hasan H. Esenoglu[1]

University of Istanbul, Faculty of Science, Department of Astronomy and Space Sciences
34452 - University, Istanbul, Turkey
and
Istanbul University Observatory, Research and Application Center
34452 - University, Istanbul, Turkey
(esenoglu@istanbul.edu.tr)

Abstract. We observed the faint shell of QZ Aur (Nova Aurigae 1964) nearly 30 years of its outburst. Despite a diameter of $0.55^{-0.16}_{+0.12}$ arcsec of the intrinsic point source of QZ Aur, a narrow-band Hα image shows that the diameter of the nova shell in 1995 was $0.86^{-0.17}_{+0.14}$ arcsec. This rather faint shell exists for a nova suggests that QZ Aur is an object worthy of further study.

INTRODUCTION

We collected some information from literature on direct QZ Aur under two subtitles the following.

A Review on Basic Parameters

The classical nova QZ Aur (Nova Aurigae 1964) was not identified at quiescence until a decade after its eruption. It was eventually recovered in 1975 on an objective prism plate taken more than a decade earlier, in November 4, 1964 at the Warner and Swasey Observatory (Sanduleak 1975). A subsequent study of archival Sonneberg plates constrained the time of eruption to early February 1964. Maximum light was not covered by observations, the actual maximum brigthness of the nova therefore was unknown, but the nova was relatively fast, characteristic of this light curve by time of decay by 2 and 3 magnitudes $t_2 < 17$ days (Warner 1987) and $t_3 = 23\text{-}30$ days (Gessner 1975; Duerbeck 1987). The following outburst of QZ Aur was observed by Gessner (1975): photographic magnitude of about 17 in January 16-19, 1964 and 6 mag in February 14, 1964. QZ Aur is listed in the Duerbeck atlas (1987) as a fast nova with a smooth light curve, although the decline time could not be quantitatively determined as it was found on plates 8 months after outburst and the early decline was not well

---

[1] Visiting Astronomer, Asiago and Padova Observatories, under scholarship of Italian Government Ministry of Foreign Affairs.

covered. It reached 5th-6th magnitude at outburst. Photometry at the quiescence in 1988 (Szkody 1992) showed QZ Aur at about V = 17.2. Time-resolved photometric observations of the V magnitude of about 18.5 during 1990-1992 revealed QZ Aur to be eclipsing with an orbital period of 8.6 hr (Campbell & Shafter 1995). QZ Aur was established as the longest period eclipsing nova yet discovered. In addition, this long period suggested that QZ Aur would be a likely double-lined system. Prompted by these results, Szkody & Ingram (1994) undertook a preliminary radial velocity study of QZ Aur. Surprisingly, their spectra, which were centered near Hα, did not reveal any obvious evidence for the secondary star. Campbell & Shafter (1995) found that the secondary star in QZ Aur had a temperature of about 5200 K°, suggesting a spectral type near K0 V, and expected flux from this secondary star was used to estimated a distance of about 2 kpc to the nova, adopting an interstellar extinction of A_v = 1.3 mag. The galactic latitude of fast nova QZ Aur is b = - 1°, according to the catalog of Downes et al (1997). The combination of its distance relative to the Sun (2 kpc) and its galactic latitude reveals a distance to the galactic plane of z = 35 pc which places it in the disk of the Galaxy. Szkody & Ingram (1994) estimated a mass of M > 0.65 M$\odot$ for relatively massive white dwarf of QZ Aur as the primary component. This is the standard font and layout for the individual paragraphs.

A Review on Shell

It is known that having a well-found understanding of the distances and luminosities of novae allows them to be studied as astrophysical objects and also to be exploited for other purposes. A distance to a classical nova is best inferred by comparing the angular size of the resolved nova shell with the linear size of the shell which is calculated from the expansion speed of the shell gas and the known time since the outburst. Unfortunately, however, it has not known any accurate value of the expansion velocity of QZ Aur in literature yet, and therefore it could not be estimated the distance of the nova via expansion parallax method in our work. To determine extent of nova shells especially faint shells it has been used a CCD detector with the highest possible efficiently and lowest possible noise and then used an automatic procedure to measure seeing values for a number of stars together with a nova placed in the same CCD frame (Esenoglu et al 2000). QZ Aur has been a very poorly studied nova on its faint shell which there is no any detailed information in the literature. Therefore the work presented in this paper follows by the same methodology mentioned above to apply the faint shell of QZ Aur in order to find the nebular diameter. This includes the two initial frames from Asiago observations at Hα and 6185 Å narrow filters but this work doesn't contain in detailed the image processing technique (see Esenoglu et al 2000 for background information). We aim to study a survey of nova shells (for example RW UMi in the near past and now QZ Aur) by this simple technique with few direct shell observations, and then the technique probably will be turned into a standard method.

This short paper presents an initial effort at analysis of the past few images and attract attention to the shell of QZ Aur, and to give only some the results of the first shell observations together with a brief implication of these results.

OBSERVATIONS

We obtained direct images of the shell of QZ Aur in 1995, February 1 with the 1.82 m telescope at Asiago – Italy. We used a Tektronix TK512M CCD at the f/9 Cassegrain focus, which yielded a pixel scale of 0.3375 arcsec per pixel. The exposure times were 3600 s in 6185 Å band (Hα-off), and 3600 s with an Hα filter (Hα-on). The log of observations is shown in Table 1 and the images are shown in Fig. 1. We took also flat and bias frames, but neither the bias nor the flat fields were subtracted from the images in a motivation to better determine the spatially extended structure of QZ Aur.

TABLE 1. Log of observations of QZ Aur.ype Table Name Here.

Start of JD	Exp (sec)	Filter (Å)	Band Pass (Å)
245 0115.33	3600	Hα	45
245 0115.37	3600	6185	50

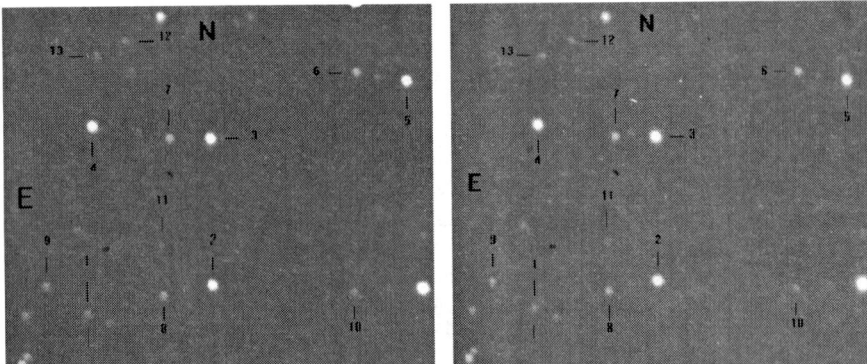

FIGURE 1. Left side: A direct image of QZ Aur obtained in 1995, February, 1st through the Hα filter with an exposure time of 3600 s. Marked numbers 1 and from 2 to 13 in the figures show QZ Aur and nearby stars, respectively. Right side: A direct image through the 6185 Å filter with an exposure of 3600 s.

RESULTS

To determine the extent of the Hα nebulosity we used an automatic procedure with point spread function (PSF) in IRAF environment to FWHM for a number of stars in the field (see Esenoglu et al 2000 for more information). We probably determined the first estimated shell of QZ Aur using the first direct imaging observations with the CCD detector made in 1995. A new radius of the shell of $0.43^{-0.09}_{+0.07}$ arcsec was found. We used the following method: The mean seeing (PSF) which was found from the FWHM Gaussian for the nearby field stars around QZ Aur in the image of the 6185 Å filter was 2.48±0.03 arcsec, compared with 2.54 arcsec for QZ Aur. These values yield a nebular diameter of $0.55^{-0.16}_{+0.12}$ arcsec according to Equation 1

$$D = \sqrt{D_o^2 - PSF^2} \qquad (1)$$

where D and D_o are nebular diameter and seeing value of QZ Aur respectively while is also the average seeing of the twelve nearby field stars. From the same analysis of the Hα image we have measured the FWHM of QZ Aur to be 2.71 arcsec, whereas the mean FWHM of the twelve nearby field stars gives 2.57±0.05. Here, we consider the FWHM of QZ Aur of a point source plus a shell. Therefore, a nebular diameter of $0.86^{-0.17}_{+0.14}$ arcsec from its Hα image for QZ Aur was found. This shell dimension with 31.1 yr after the outburst yields an average expansion rate of $0.014^{-0.003}_{+0.002}$ arcsec per year. The results are listed in Table 2. All seeing values have a maximum error of − 0.17 arcsec, +0.14 arcsec. The intrinsic point source of QZ Aur depends somewhat on whether a standard deviation of the mean value of the nearby stars is obtained or some errors of the FWHM Gaussian in IRAF. Obviously the major error of the shell of QZ Aur comes from the mean seeing rather than the FWHMs, and in this way it was computed the errors on our derived parameters.

TABLE 2. The seeing values for QZ Aur (D_o) and twelve nearby stars in the Hα-on and -off bands.

No	Seeing Hα-on (arcsec)	Seeing Hα-off (arcsec)
1 (D_o)	2.71	2.54
2	2.56	2.48
3	2.53	2.46
4	2.60	2.50
5	2.50	2.40
6	2.51	2.45
7	2.60	2.49
8	2.65	2.52
9	2.57	2.52
10	2.55	2.47
11	2.57	2.50
12	2.65	2.49
13	2.61	2.51
Mean (PSF)	2.57 ± 0.05	2.48 ± 0.03
Diameter (D)	$0.86^{-0.17}_{+0.14}$	$0.55^{-0.16}_{+0.12}$

Unfortunately, we found no one published information on the expansion velocity of QZ Aur, and therefore we could not estimate the distance and an absolute magnitude at maximum of QZ Aur. Consequently, it has not known the accurate distance of QZ Aur from the expansion parallax method in the literature yet. Notably, determination of the expansion velocity of QZ Aur is required.

The main result of this work is as follow:

We observed the shell of QZ Aur nearly 31 years after its outburst and found that the diameter of the nova shell in 1995 was $0.86^{-0.17}_{+0.14}$ arcsec. However, this rather faint shell exists for a nova suggests that QZ Aur should be embarked upon a deep imaging study of the unresolved structure of its nebular remnant using large telescopes with modern detectors.

ACKNOWLEDGMENTS

I would like to thank Şölen Balman for presentation during the meeting of classical nova explosions. I gratefully thanks Padova and Asiago Observatories for hospitalities. And also I specially thanks Atilla Durmaz (ADR@mynet.com.tr) from REEL Company for present of an office furniture. This work was supported by the Research Fund of the University of Istanbul. Project Number: 1506/280700.

REFERENCES

1. Campbell, R.D., and Shafter, W., *ApJ* **440**, 336-344 (1995).
2. Downes, R.A., Webbing, R.F., Shara, M.M., *PASP* **109**, 345-440 (1997).
3. Duerbeck, H.W., A Reference Catalogue and Atlas of Galactic Novae, *Space Sci. Reviews*, Vol.**45**, Nos.1-2 (1987).
4. Esenoglu H.H., Saygac, A.T., Bianchini, A., Retter, A., Özkan, M.T., Altan, M., *A&A* **364**, 191-198 (2000).

The active quiescence of the ex-nova HR Del

Pierluigi Selvelli* and Michael Friedjung[†]

*IASF, Osservatorio Astronomico di Trieste -Italy
[†]Institut d'Astrophysique - Paris

Abstract. We outline here some results of a recent study on HR Del that has included all of the IUE data taken from 1979 to 1992. This has allowed a detailed analysis of both the long-time and the short-time variations in its UV spectrum. Assuming $d = 850$ pc, after correction for reddening ($E_{B-V} = 0.16$), we derive a mean UV luminosity $L_{UV} \sim 56 L_\odot$, the highest value among classical novae at "minimum". Also the optical absolute magnitude ($M_v = +1.35$) indicates a very bright object. The mass accretion rate $\dot{M}$ is close to $1.4 \times 10^{-7} M_\odot yr^{-1}$. The UV continuum has declined by a factor less than 1.2 over the 13 years of the IUE observations, while the UV emission lines have faded by larger factors. The wind components in the P Cyg profiles of the CIV and NV resonance lines are strong and variable on short timescales, with v_{edge} up to -5000 kms^{-1}, a remarkably high value. The modelling of the UV continuum shows that the quiescent optical magnitude at 12^m has origin from a hot component that has a black body temperature close to $\sim 33,900$ K. The fact that the pre-nova was at near the same magnitude and temperature, implies that it was necessarily at the same L_{UV} and $\dot{M}$ values as the ex-nova.

INTRODUCTION

HR Del = Nova Del 1967 brightened in July 1967 as a star of mag 5.5 (Allcock, 1967) from a pre-nova magnitude near $m_v \sim 12$ (Stephenson 1967, Robinson 1975). The extra slow decline in m_v lasted for about 15 years with an asymptotical approach toward the pre-nova value ($m_v = 12.0$) that was reached around 1981-1982. At the present time the visual magnitude of HR Del shows small oscillations around this value.

THE UV SPECTRA AND THE REDDENING

HR Del was the target of several IUE observations centered on three epochs : 1979-1980, 1988, and 1992. However, the UV studies in the literature refer to the 1979-80 data only (Krautter et al. 1981, Rosino et al. 1982, Friedjung et al. 1982).

>From the IUE-INES archive (Rodriguez-Pascual et al., 1999) we have retrieved the whole set of 49 spectra of HR Del. We point out that the IUE data extraction and calibrations methods have undergone several revisions. This has resulted in not negligible changes both in the quality of the line spectrum and in shape of the continuum curve (see Gonzalez-Riestra, Cassatella and Wamsteker, 2001).

The individual spectra are quite similar to each other in the continuum and line features. This justifies the creation of an "average" spectrum by co-adding and merging all SW and LW spectra for the epochs (1979, 1980 and 1988) for which data in both

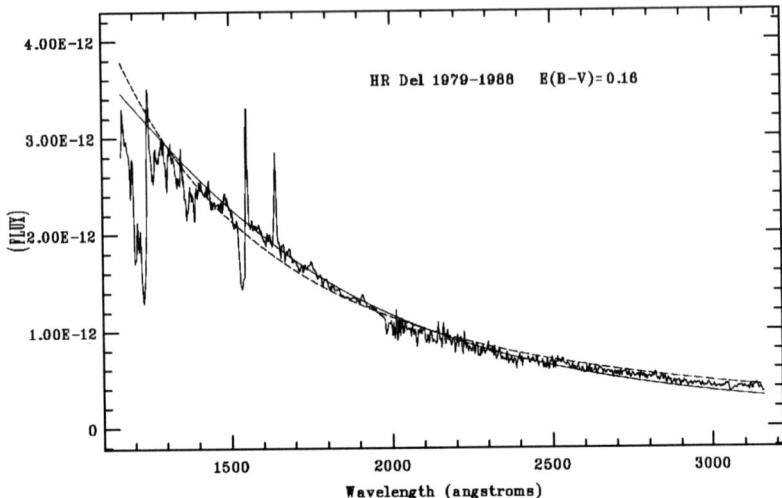

FIGURE 1. The "average" 1980-1988 UV spectrum corrected for E=0.16. A 33,900 K black-body (solid line) and a power-law with $\alpha = -2.2$ (dashed line) are superimposed

spectral ranges are available (Fig. 1). In 1992 only SWP data were taken and have not been included in this "average". The improved S/N in the "average" spectrum has allowed both the detection of weak line features and an accurate determination of the reddening ($E_{B-V} = 0.16 \pm 0.02$) using the well established method of removing the λ 2175 Å bump. In the following we will adopt $A_V = 0.315 \times 0.16 \sim 0.50$.

Prominent spectral features are the P Cyg profiles in the CIV λ1550 Å and NV λ1240 Å resonance lines, and the HeII λ 1640 Å emission. Nebular lines (i.e., NIV λ 1483 Å , OIII λ 1666 Å , NIII λ 1750 Å and CIII λ 1909 Å) are clearly present only in the spectra of 1979-1980 . Un-displaced absorption lines are SiII λ1190 Å , SiII λ1260 Å , OI+SiII λ1303 Å , CII λ1335 Å , and OV (?) λ1370 Å .

THE LONG-TIME VARIATIONS

Fig. 2 is a plot (on an absolute scale) of the mean SWP spectra for the 1980, 1988 and 1992 epochs that shows and almost "gray" decay with time in the continuum, together with a more pronounced decline in the emission line intensities, especially NV λ1240 and HeII λ1640. The short wavelength continuum (SWP region) has declined by a factor ~ 1.19 from 1979-1980 to 1992 (about 1.08 from 1980 to 1988) but the emission lines decline is definitely larger : NV λ1240 is down by ~ 6, CIV λ1640 by ~ 2.3 and He II λ1640 by ~ 1.6. We recall that at the epochs of the first IUE observations the V mag was at about 12.1 and declined to the pre-nova value 12.0 around 1981-1982.

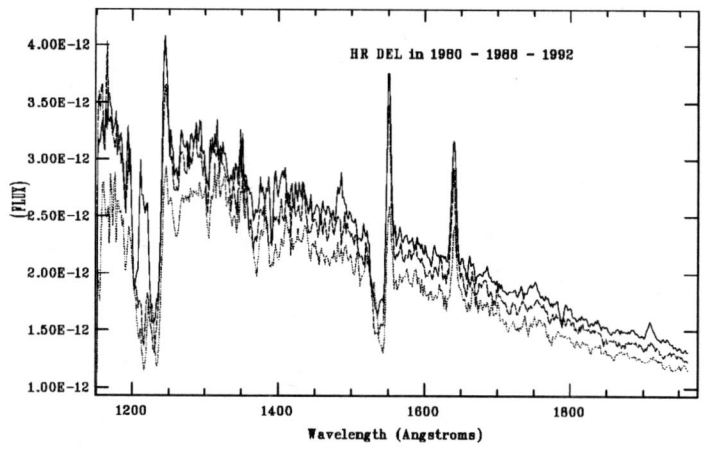

FIGURE 2. The 1980-1988-1992 slow decline in the SWP range

THE SHORT-TIME VARIATIONS AND THE WIND FEATURES

In 1980, 1988 and 1992 several SWP spectra were taken in a strict sequence of exposures, each sequence covering more than one orbital period ($P_{orb} = 5^h6^m24^s$, Bruch, 1982). This has allowed a study of the short-time variations in the absorption component (wind) of the CIV λ1550 Å and NV λ1240 Å resonance lines. The variations appear as a-periodic and suggests the presence of an inhomogeneous outflow with irregular variations on short timescale and the sudden ejection of puffs of thick material.

The absorption component in the CIV λ1550 line reaches $v_{edge} \sim 5000$ km s^{-1} and its shape suggests the presence of two structured components. This surprisingly high v_{edge} value for the ex-nova contrasts with the moderate outflow velocities (~ -1000 km s^{-1}) recorded during the outburst. We recall that in the first IUE spectra, these P Cyg features were interpreted in the framework of a perduring mass outflow in the outburst of a very slow nova (Hutchings 1979). Instead, the persistence of such features in the IUE spectra of 1992, a decade after the return to "quiescence" indicates a different origin, likely a conical-shaped wind region nearly perpendicular to the accretion disk. In turn, the prolateness in the shape of the ejecta, see O'Brien (2002), could be associated with the presence of such a bipolar outflow.

THE $\lambda\lambda$ 1180-3250 Å CONTINUUM DISTRIBUTION

The best single curve fit to the UV continuum distribution of HR Del comes from a black-body distribution with T $\sim 33,900$ K. A power-law distribution with $F_\lambda = 2.0 \cdot 10^{-5} \cdot \lambda^{-2.20}$ ($erg\ cm^{-2}s^{-1}Å^{-1}$) is also a good approximation with a small uncertainty

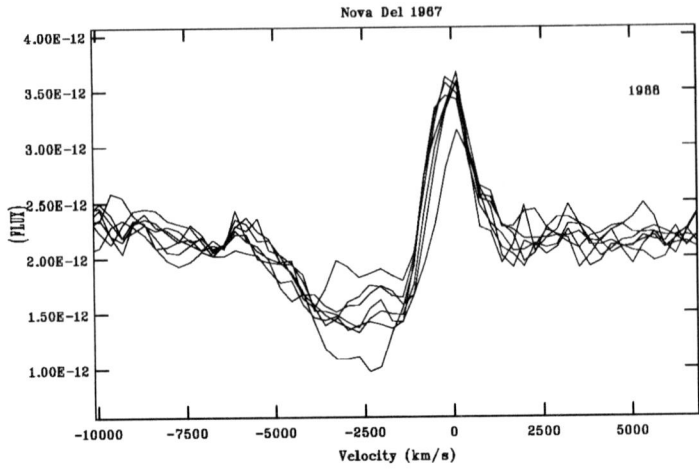

FIGURE 3. The short-time changes in the CIV λ1550 P Cyg profile in the spectra of 1988.

of ±0.05 in the index. The spectral index of the power-law fitting is near that of the Lynden-Bell law for a standard disk ($\alpha = -2.33$), as already found by Friedjung et al (1982) from the early IUE spectra.

It is remarkable that the extrapolation toward λ5450, (the λ_{eff} of the V-band for an hot star) of either the black-body or the power-law distribution yields optical fluxes ($4.7 \times 10^{-14} erg\, cm^{-2} s^{-1} Å^{-1}$ and $11.8 \times 10^{-14} erg\, cm^{-2} s^{-1} Å^{-1}$ respectively) that after conversion to magnitudes and proper "reddening" give $m_v \sim 12.7$ and $m_v \sim 11.7$ respectively. Therefore, the "quiescent" optical magnitude at 12^m comes from the "tail" of the UV continuum. The optical flux ($5 \times 10^{-14} erg\, cm^{-2} s^{-1} Å^{-1}$) found by Ringwald, Naylor, and Mukai (1996) is close to that of the UV-extrapolated values.

THE UV LUMINOSITY AND THE MASS ACCRETION RATE

If the total disk luminosity and the mass of the WD are known $\dot{M}$ can be directly calculated from the relation

$$\dot{M} = \frac{2 \cdot R_1 \cdot L_{disk}}{G \cdot M_1}$$

The observed L_{UV} provides a lower limit to the "intrinsic" (face-on and bolometric) disk luminosity L_{disk}. If the mean accretion disk temperature is close to 33,900 K and the inclination is close to $40°$ it is justified to assume $L_{disk} \sim 2 \times L_{UV}$. Therefore, if $L_{disk} \sim 112 L_\odot$, $M_1 \sim 0.65 \times M_\odot$ and $R_1 \sim 0.0125 \times R_\odot$ we obtain $\dot{M} \sim 1.4 \times 10^{-7} M_\odot yr^{-1}$ as a rough estimate for the mass accretion rate.

An independent determination for $\dot{M}$ can be obtained through the $\dot{M}$ - HeII λ1640 luminosity relation given in Table 2 of Patterson and Raymond (1985). The average (dereddened) flux on earth in the HeII $\lambda 1640$ line is of $7.0 \times 10^{-12} erg\ cm^{-2} s^{-1}$ and the corresponding luminosity is $6.1 \times 10^{32} erg\ s^{-1}$ that, for $M_1 = 0.7 M_\odot$ gives an accretion rate $\dot{M} \sim 10^{19} gr\ s^{-1}$, that is $\dot{M} = 1.4 \times 10^{-7} M_\odot yr^{-1}$, in very good agreement with the estimate based on the UV continuum.

We point out also that from $m_v = 12$, assuming $d = 850\ pc$, and $A_V = 0.50$ we obtain $M_v = +1.85$. After correction for the inclination ($i \sim 40^o$), with a factor $\Delta M_v = 0.50$, as an average from Warner (1986) and Webbink et al (1987) relations, we obtain $M_V = +1.35$, a very bright value that places HR Del in the extreme tail of the frequency distribution of absolute magnitudes for nova remnants whose average value is centered about $M_V = +4.5$ (Warner, 1987).

THE PRE-NOVA AND THE POST-NOVA

We recall that Stephenson (1967) found $m_v \sim 12$ for the pre-nova magnitude. and classified the pre-nova continuum (seven years prior to outburst) to be that of an O or early B star while Hutchings (1968) estimated the temperature of the pre-nova star to be ~ 32000 K. Barnes and Evans (1970) found $(B - V) \approx -0.18$ for the pre-nova in 1951. After correction for $E_{B-V} = 0.16$, this gives $(B - V)_o = -0.34$ in good agreement with the value expected from an object with $T \sim 33,000 K$. Therefore, for many years before the 1967 outburst HR Del was at near the same values of visual magnitude and temperature ($m_v \sim 12$ and $T = 33,900 K$) as in the post-nova stages corresponding to the IUE observations. Recalling that the "quiescent" magnitude at $m_v = 12$ comes from the "tail" of the observed UV continuum, it follows that during the pre-nova stage at 12^m HR Del had necessarily the high T, L_{UV}, M_V and $\dot{M}$ values found in the present study. It is not clear why the pre-nova was so bright, and we wonder whether both the pre-nova and ex-nova "minima" at $m_v = 12$ are real minima.

REFERENCES

1. Allcock, G.E.D., IAU Circ. No. 2022 (1967)
2. Barnes, T. G., Evans, N. R., *PASP* **82**, 889 (1970)
3. Bruch, A., *PASP* **94**, 916 (1982)
4. Friedjung, M., Andrillat, Y., Puget, P., *A&A* **114**, 351 (1982)
5. Gonzalez-Riestra, R., Cassatella, A., and Wamsteker, W., *A&A* **373**, 730 (2001)
6. Hutchings, J.B., *Pub. Dom. Astr. Obs.* Vol XII, No 16, p. 397 (1968)
7. Krautter, J., Klare, G., Wolf, B., Duerbeck, H.W., Rahe, J.W., Vogt, N., *A&A* **102**, 337 (1981)
8. O'Brien, T.J., *These Proceedings*
9. Patterson, J., Raymond, J.C., *Ap.J.* **292**, 550 (1985)
10. Ringwald, F. A., Naylor, T., Mukai, K., *MNRAS* **281**, 192 (1996)
11. Robinson, E. L., *A.J.* **80**, 515 (1975)
12. Rodriguez-Pascual, P.M., Gonzalez-Riestra, R., Schartel, N., Wamsteker, W., *A&AS* **139**, 183 (1999)
13. Rosino, L., Bianchini, A., Rafanelli, P., *A&A* **108**, 243 (1982)
14. Stephenson, C. B., *PASP* **79**, 584 (1967)
15. Warner, B., *MNRAS* **227**, 23 (1987)

Emission line flaring in the SW Sex old nova V533 Herculis

Pablo Rodríguez-Gil[*] and Ignacio G. Martínez-Pais[*†]

[*]*Instituto de Astrofísica de Canarias, Vía Láctea, s/n. La Laguna. E-38200. Santa Cruz de Tenerife. Spain*
[†]*Departamento de Astrofísica. Universidad de La Laguna. Tenerife. Spain*

Abstract. We present high time resolution spectroscopy of the non-eclipsing old nova V533 Herculis (N Her 1963). It is the second nova remnant affected by the 'SW Sex syndrome'. A modulation of the equivalent width of the emission lines with a period of 23.33 min has been detected. This, together with the strong He II λ4686 emission characteristic of magnetic systems, leads us to link this period to the spin of a magnetic white dwarf. Similar flaring activity has been recorded in other SW Sex stars, namely, the old nova BT Mon, LS Peg and DW UMa, supporting the idea of these systems being magnetic accretors. Stationary emission features are also observed in the Balmer lines, which we attribute to the ejected nova shell.

INTRODUCTION

V533 Herculis is the remnant of a bright nova observed in 1963. [1] found its orbital period ($P_{orb} = 0.147$ d). The presence of transient absorption components in the Balmer and He I lines and the delay of the Balmer lines with respect to the white dwarf motion, lead the authors to conclude that V533 Her is a SW Sex star.

The first old nova displaying SW Sex behaviour was BT Mon (Nova Mon 1939, [2]). Its Balmer and He II λ4686 lines showed a quasi-periodic modulation in their fluxes (emission line flaring) similar to the observed in intermediate polar CVs, suggesting that magnetic accretion could be in action. The recently proposed magnetic model by [3] predicted a timescale of ~ 23 min for the possible line flaring in V533 Her. So we decided to perform high time resolution spectroscopy of V533 Her to search for emission line flaring. A positive detection would strongly support the magnetic model for SW Sex stars in general.

OBSERVATIONS

The analysis is based on spectra taken with the 4.2-m William Herschel Telescope (WHT) on La Palma. The data were obtained on the night of 2001 October 29 using the blue arm of the ISIS spectrograph. The useful wavelength range is 3760–5065 Å at 0.9 Å spectral resolution (60 km s^{-1}). We obtained 57 spectra of V533 Her, covering almost one orbital period. The exposure time was fixed at 180 sec, with a dead-time of about 9 sec.

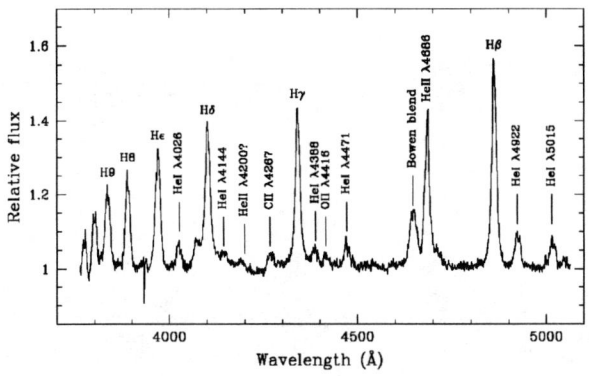

FIGURE 1. Average spectrum of V533 Her

RESULTS

The average spectrum

The average of all 57 spectra, uncorrected for orbital motion, is presented in Figure 1. The spectrum of V533 Her is dominated by Balmer (from Hβ to H11) and He I lines ($\lambda 5015$, $\lambda 4922$, $\lambda 4471$, $\lambda 4026$ and other less intense). High excitation lines as He II $\lambda 4686$, the Bowen blend, C II $\lambda 4267$, and probably He II $\lambda 4200$ are also present.

Trailed spectra

We rebinned the spectra onto a constant velocity interval scale centred on the rest wavelengths of the lines, and then constructed the trailed spectra of the Balmer (from Hβ to H8), He II $\lambda 4686$, He I $\lambda 4922$, and He I $\lambda 4471$ lines. The resulting trailed spectra are shown in Figure 2.

The Balmer lines seem to have two emission components. The most intense has a velocity semi-amplitude of ~ 300 km s^{-1} and reaches its reddest velocity at $\simeq 0.95$, lagging the motion of the white dwarf by $\simeq 0.2$ (a clear sign of SW Sex behaviour). The other emission component has roughly the same velocity semi-amplitude and its maximum excursion to the red occurs at $\simeq 0.6$. The He I $\lambda 4922$ and $\lambda 4471$ lines have the same two components, but now the most diffuse in Balmer seems to dominate. On the other hand, the He II $\lambda 4686$ line has a different structure. It is dominated by a single emission S-wave which follows the motion of the compact object with a velocity semi-amplitude of ~ 100 km s^{-1}. The line displays an enhanced blue wing extending out to ~ -800 km s^{-1}.

The trailed spectra of the Balmer lines show a narrow and stationary feature located at ~ -200 km s^{-1}. We identify the absorption in Hε as the Ca II H line at 3968 Å. The Ca II K absorption line is also stationary. In contrast, the 'tramline' in Balmer is in emission. These spectral features must originate outside the binary. Actually, the Ca II absorptions

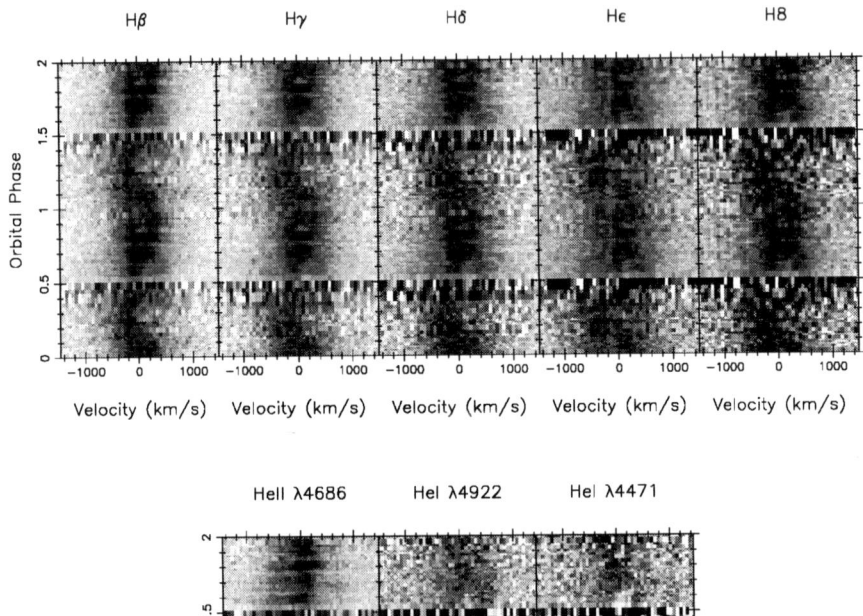

FIGURE 2. The trailed spectra of Balmer, He II $\lambda 4686$, He I $\lambda 4922$, and He I $\lambda 4471$. Dark means emission and a full orbital cycle has been repeated for clarity.

have an interstellar origin. With respect to the Balmer 'tramline' emission (which is not present in He II $\lambda 4686$), we point out that a nova shell has been observed in V533 Her [4] with an expansion velocity of 850 ± 150 km s^{-1} (the maximum line-of-sight velocity). The observed stationary emissions are probably formed in this shell.

The most striking finding is that the fluxes of the Balmer and He II $\lambda 4686$ lines are modulated with a period shorter than the orbital period. This periodic line flaring is clearly patent in He II $\lambda 4686$, visible as horizontal stripes in the trailed spectra. In the Balmer lines the pulses are easily seen between phases 0.6 and 0.9, when they are more intense. The same flaring has been detected in other SW Sex stars, like the old nova BT Mon [2], LS Peg [3], and DW UMa [5].

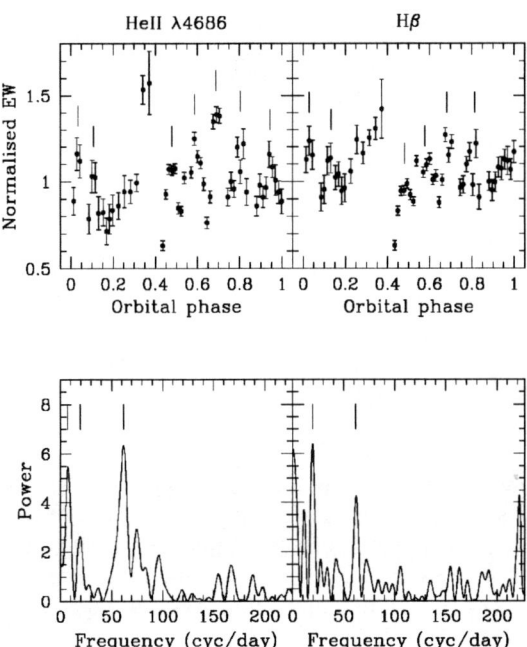

FIGURE 3. *Upper panel*: EW curves of the He II $\lambda 4686$ and Hβ emission lines. The flares are indicated with a vertical mark. *Lower panel*: Scargle periodograms constructed from the HJD–EW curves.

EMISSION LINE FLARING

As we showed above, the He II $\lambda 4686$ trailed spectra of V533 Her display flaring activity similar to the observed in intermediate polar (IP) CVs (e.g. FO Aqr, [6]). Although less pronounced, the Balmer lines also show this flaring. The equivalent width (EW) curves of He II $\lambda 4686$ and Hβ as a function of orbital phase are presented in the upper panel of Figure 3 (the points in the vicinity of phase 0.4 have been excluded, because of their low signal-to-noise ratio). None of the lines exhibits a clear modulation with the orbital period. The most important fact is that the EW of both lines shows pulses, clearly visible in the range $\simeq 0.5 - 1.1$. Although the bulk of Hβ emission displays a phase lag of $\simeq 0.2$ with respect to He II $\lambda 4686$, the phasing of the pulses seems to be the same for both lines.

The periodograms from the He II $\lambda 4686$ and Hβ HJD–EW curves are shown in the lower panel of Figure 3. The most interesting peak is centred around 62 d^{-1} and is present in the periodograms for both He II $\lambda 4686$ and Hβ. We have measured by eye the flare separation in both EW curves, obtaining a mean value of $0.11 P_{\mathrm{orb}} = 0.016$ d, which gives a frequency of $\simeq 62$ d^{-1}. Clearly, this peak represents the frequency of the emission line flaring. After fitting a gaussian function to this peak in the He II $\lambda 4686$ periodogram, we get $v_1 = 61.87 \pm 4.21$ d^{-1} (the error has been chosen as half the FWHM of the peak, so it is highly pesimistic), which means a period of $P_1 = 0.0162 \pm 0.0011$

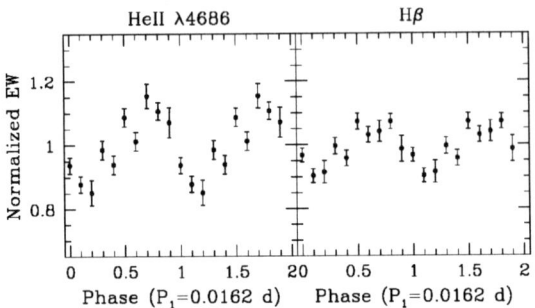

FIGURE 4. Normalized He II λ4686 and Hβ HJD–EW curves folded on the 23.33-min period after averaging the data into 10 phase bins. A complete cycle is repeated for clarity.

d = 23.33 ± 1.58 min. This period coincides with the predicted by the magnetic model [3] for the orbital period of V533 Her. We have folded both EW curves on this period after averaging the data into 10 phase bins, but only considering the orbital phase range $\simeq 0.5$–1.2 to avoid the spectra with low signal-to-noise ratio. The folded EW curves are shown in Figure 4, where the modulation on this period is evident.

DISCUSSION

Emission line flaring has been observed in three SW Sex systems: the old nova BT Mon [2], the non-eclipsing LS Peg [3], and the deeply-eclipsing DW UMa [5]. This is a common feature observed in other CVs with discs and magnetic primaries: the IPs (see e.g. [7]). V533 Her has now to be included in the list of SW Sex stars showing emission line flaring. It is the second nova remnant exhibiting this behaviour in the emission lines. The 23.33-min period we have detected in V533 Her is best defined in the He II λ4686 EW curve. On the other hand, it is well known that the He II λ4686 emission in SW Sex stars forms very close to the white dwarf, so we propose that the emission line flaring in this SW Sex system is related to the rotation of a magnetic white dwarf. But we can not confirm whether the 23.33-min period is either its spin period or the beat between the spin and orbital period.

REFERENCES

1. Thorstensen, J. R., and Taylor, C. J., *Mon. Not. R. Astr. Soc.*, **312**, 629 (2000).
2. Smith, D. A., Dhillon, V. S., and Marsh, T. R., *Mon. Not. R. Astr. Soc.*, **296**, 465 (1998).
3. Rodríguez-Gil, P., Casares, J., Martínez-Pais, I. G., Hakala, P., and Steeghs, D., *Astrophys. J.*, **548**, L49 (2001).
4. Gill, C. D., and O'Brien, T. J., *Mon. Not. R. Astr. Soc.*, **314**, 175 (2000).
5. Smith, D. A., Dhillon, V. S., and Marsh, T. R., *Mon. Not. R. Astr. Soc.*, p. in press (2002).
6. Marsh, T. R., and Duck, S. R., *New Astr.*, **1**, 97 (1996).
7. Warner, B., *Mon. Not. R. Astr. Soc.*, **219**, 347 (1986).

CONCLUDING REMARKS

Summary of the Meeting

Sumner Starrfield

*Department of Physics and Astronomy, Arizona State University,
P. O. Box 871504 Tempe, Arizona 85287-1504
email: starrfield@asu.edu*

The purpose of this meeting was to cover as many topics in the studies of novae, from the outburst to quiescence, that could be fit into a 4.5 day meeting. I have only enough space in this summary to present the highlights, and hope that my remarks are sufficient to interest the reader to examine the papers in these proceedings.

The meeting began on Monday with a review of the general properties of quiescent novae by Brian Warner. Brian, in his usual comprehensive fashion, discussed the orbital period distribution and compared old novae to the magnetic CV systems. It is important to understand the connections between the magnetic CV's and novae. At the present time, only V1500 Cygni is known to be a magnetic system, and there have not been any studies of the impact of magnetic fields on the evolution of the outburst. One of the more interesting ideas about the interoutburst evolution of nova systems is that of "Hibernation" proposed by Mike Shara in the 1980's. Over the past few years, Tim Naylor and his collaborators have been investigating the observational evidence for and against this hypothesis and he gave a review of the current situation which may be that some novae hibernate and some do not (the contribution of Hilmar Duerbeck also relates to this question).

Ed Sion and Paula Szkody then reviewed the studies of the white dwarfs in CV systems. Through hard work, clever ideas, and access to ultraviolet data (IUE and HST), they have been able to study white dwarfs in those systems where the accretion disks provide only a minimal contribution to the total light, so that the white dwarf can be seen and its spectrum analyzed. They have determined temperatures, rotation velocities, and atmospheric abundances for a number of systems. At this meeting they reported on a number of systems where the white dwarfs show evidence for nuclear processed material in their atmospheres, which implies that these systems may have undergone nova outbursts in the past. Clearly, the atmospheric abundances that they are obtaining are providing clues to the past behavior of some interesting CV systems.

We switched topics, on Monday afternoon, and G. C. Anupama gave a review of the observational details of Recurrent Novae (RN). These are systems in which the outburst has been observed to reoccur on astronomer's lifetimes rather than astronomical lifetimes. As shown by Anupama, there are two major classes of RN: the U Sco class (U Sco, V394 CrA, and LMC 1990 #2: orbital period $\sim$day) and the RS Oph class (RS Oph, V745 Sco, T CrB, V3890 Sgr: orbital period $\sim$ year). In both cases the secondary is evolved. In the U Sco class the secondary is compact and the orbital periods are short, while in the RS Oph class the secondary is a giant and the orbital periods are much longer. We have shown that, in order for the outburst to re-occur on such short time scales, the white dwarf must be massive and the mass accretion rates large. Since the systems accrete more mass than is lost in the outburst, the white dwarfs are growing to the Chandrasekhar limit and should become SNe Ia. However, systems with giant companions should show too much hydrogen in the SN explosion spectra to agree with the observations. One concern has been that the number of such systems is thought to be too small. Of importance to this speculation is the discovery, in the past two years, of two new RN: CI Aql and IM Nor. However, in neither case does their outburst resemble the major classes and detailed studies at quiescence are needed.

U. Munari continued with a discussion of the recent "outburst" of V838 Mon. This was a variable that erupted in January 2002, became very red, and then developed a light echo (discovered in late February 2002). At the present time, its outburst is considered to be unique and it is not even clear if it is a binary. It is very cool and a great deal of information on its outburst is presented in the paper by Munari et al. (these proceedings).

On Tuesday, we began the discussion of novae in outburst. I reviewed the evolution of our understanding of the outburst in the 20th Century. Dina Prialnik followed with a discussion of her multicycle evolution models. She and Attay Kovetz have done a large number of studies of thermonuclear runaways on white dwarfs. They include chemical diffusion and vary the initial conditions over a broad range of values. They have found some interesting agreements with observations. Another important paper was presented by one of the organizers: Jordi José. He reviewed his work on nucleosynthesis during the nova outburst. In contrast to the studies of Starrfield and collaborators, who use the ONeMg composition of Arnett and Truran, Jordi and Margarita Hernanz have used a more modern composition that was calculated by Ritossa, Iben (who attended this meeting but was uncharacteristically quiet), and others. Their chosen abundance distribution has important consequences on the production of ^{26}Al and ^{22}Na in the outburst. Their studies also confirm that it is also possible to obtain estimates of the mass of the underlying white dwarf by comparing the observed abundances to the predictions. Unfortunately, as emphasized by Christian Iliadis in the next talk, nuclear reaction rates for charged particle reactions on some of the more important nuclei (for nova nucleosynthesis) are only now being determined and the predictions will have to be updated frequently. Iliadis and his collaborators have been involved in a massive study of the important reactions in nova nucleosynthesis. They have used a Monte Carlo study to determine which reaction rates have the most important effects on the predictions and, thereby, need to be improved. He also pointed out that many of these same rates are also important for light element nucleosynthesis in SNe explosions.

Ami Glasner reviewed the multidimensional studies of the nova outburst that have been done in the past few years. He reported that the results of his group and the German group are in rough agreement, but there are puzzling discrepancies. One of the problems is trying to resolve the multidimensional structure of the Kelvin-Helmholtz instability. Further work is planned with the FLASH code under development at the University of Chicago, and some of the FLASH results were reported on by Dursi and Calder of the Flash Center. The final talk of this session was given by Giora Shaviv on problems with electron screening in dense plasmas. Since thermonuclear runaways occur in material that is partially degenerate and satisfies the conditions for intermediate screening, doing screening correctly may influence the outburst characteristics. This is a difficult problem requiring large scale numerical computations of the interactions between electrons and a few nuclei.

In the late afternoon, we turned to the observational studies of the outburst. Steve Shore led off with a multiwavelength perspective. Since a white dwarf is small and hot most of the energy emitted during the outburst occurs outside the optical. If there is any dust created in the ejected material, it will absorb some or all of the optical and reradiate it in the infrared. He described the physical processes in the expanding shell, the development of the iron curtain, and its disappearance as the material expands and the density drops. Angelo Cassatella continued with a discussion of some of the IUE results. T. Ijima reported on the optical spectroscopy going on at the Asiago Observatory and Massimo Della Valle finished the day with a discussion of V382 Vel, one of the brightest novae in the second half of the 20th Century.

Bob Gehrz led off Wednesday morning with a review of the infrared evolution of novae during their outbursts. He concentrated on the formation of dust, the contributions of novae to the ISM, and determinations of both the ejecta characteristics and the elemental abundances from infrared line features. He emphasized that by obtaining data in a variety of wavelengths those data can be intercompared and a more accurate determination of the evolution will result. Dave Lynch, who had an unfortunate beginning in Barcelona, then reported on his infrared observations.

Peter Hauschildt finished off the early morning session with a discussion of the stellar atmosphere calculations that he has been doing for novae in outburst. Peter and his collaborators have developed PHOENIX, a spherical, line blanketed, expanding, Non-LTE, stellar atmosphere code that includes a large number of species and lines. It has been applied to novae, supernovae, cool stars, brown dwarfs, and virtually everything that glows in the dark. Most recently Jason Aufdenberg has added a stellar wind module and Peter is calculating a new set of Non-LTE atmospheres to be used in studies of the hot photospheres for white dwarfs with ongoing nuclear burning near the surface.

One of the more interesting radiation problems that has arisen from studies of the novae in the LMC is that all fast novae exceed the Eddington Luminosity at maximum light. For example, the extraordinary nova, LMC 1991, was super Eddington for more than two weeks. Nir Shaviv has been working on this problem and may have solved it. As part of this solution, he has developed an effective opacity to measure the flow of radiation through an inhomogeneous medium, and this opacity allows for super-Eddington flows. It should be interesting to use his opacity in hydrodynamic codes and follow the evolution of the expanding shell.

Rawlings and Evans continued with a discussion of grain formation in novae. This is another perplexing problem derived from studies of novae in outburst. How does the same nova form at least four different kinds of dust during its outburst? Those novae that exhibit a deep transition phase, in which sufficient dust forms to block the optical light from the underlying material and reradiate it in the infrared, are observed to form carbon rich grains at first and oxygen rich grains later in the outburst. This problem is exacerbated by the fact that abundance studies of nova ejecta suggest that they are oxygen rich and, therefore, only silicate grains should form. In their studies, Rawlings and Evans find that CO formation does not go to saturation (it is normally assumed that whichever element is left over after the CO molecule is formed then goes to form grains), and so both carbon and oxygen are present during the ejecta expansion. In any case, grain formation must be a non-equilibrium process.

We jumped to higher energy photons on Thursday morning when Joachim Krautter led off with a review of the X-ray evolution of novae. Novae in outburst have now been observed with a variety of X-ray satellites including, EXOSAT, ROSAT, BEPPOSAX, ASCA, XMM, and CHANDRA. Joachim discussed the physical cause for the emission in X-rays, and then reviewed the recent observations of V382 Vel and V1494 Aql. The X-ray spectroscopy of V382 Vel showed only emission lines and analyses of these lines are providing important information about the ejecta. The X-ray spectra of V1494 Aql showed that it had evolved to that of a Super Soft Source.

In addition to the impressive advances that have come from high resolution X-ray spectroscopy, the time tagged photons from CHANDRA data have also provided some surprises. The light curve of V1494 Aql shows that, in October 2000, it was undergoing oscillations with a period of ~2500 seconds. In addition, there was a "burst" of X-rays that lasted for about 10 minutes in the middle of one of the observations. V382 Vel also showed interesting variations in its BEPPOSAX light curve as reported by Marina Orio and her collaborators. One important consequence of the detection of X-ray oscillations and Super Soft spectrum of V1494 Aql is that we are clearly observing the hot photosphere of the rekindled white dwarf 9 months into the outburst. Jochen Greiner reviewed the XMM observations of recent novae and Marina Orio presented the XMM observations of LMC 1995 which was still a Super Soft Source 6 years after outburst. Solen Balman continued with a discussion of her CHANDRA observations of the shell around GK Per, and Koji Mukai ended the X-ray session with a description of his CHANDRA observations of DQ Her and V603 Aql. He finds that V603 Aql is not a magnetic system.

Moving on to even higher energies, Margarita Hernanz reviewed the predictions of γ-ray emission from novae. This area actually has a long history since there have been predictions of γ-rays from novae since the mid - 1970's. However, the predictions have always been slightly below the observational capabilities of the balloon flights or the satellite detectors. Changes in the modeling, the abundances, and nuclear reaction rates have made the task very difficult. Margarita described the various radioactive isotopes that are produced in nova explosions and discussed their importance and probability of detection with INTEGRAL. What is needed is to have a nova go off within 1 kpc at the time that INTEGRAL is flying. The detection of a radioactive signature from a particular nova will give us a tremendous amount of information about the characteristics of the TNR in that particular nova. Finally, two papers ended the session by discussing the reaction rate for $^{18}F(p,\alpha)^{15}O$. ^{18}F is another isotope with a short enough half life (109 min) to be important in the detection of novae with INTEGRAL. Its rate is uncertain and these two groups (Orsay and Oak Ridge) are trying to improve the situation.

The final session on Thursday afternoon began the discussion of nova populations and novae in external galaxies. Massimo Della Valle (who should organize a nova meeting at Arcetri) reviewed our current knowledge on novae in the Local Group and in M81. He has shown that there appear to be two classes of novae: bright and fast novae which originate in the disk and slower and less luminous novae which appear to originate in the bulge. These observations corroborate the original predictions for CO and ONeMg novae. He was followed by Mike Shara who has analyzed archival and new HST observations of M87. Mike has found more than 400 objects with the behavior of novae allowing him to derive a detailed spatial and brightness distribution of novae in this galaxy. Allen Shafter reviewed the data on both Galactic and extragalactic nova rates and has come to the conclusion that novae in M31 are not bulge dominated. He has done a very careful job in this area and identified a number of problems with the previous work. He arrives at a rate for our own Galaxy of about 35 per year. Bill Liller then presented the "outburst" of a peculiar variable star which hopefully is not a nova. It went through an outburst in less than 24 hours.

The last topic of the meeting was devoted to nova shells and Mike Bode reviewed both previous work and his work with MERLIN on resolving nova shells. He discussed the shell of GK Per and those of QU Vul, V1974 Cyg, V705 Cas, and V723 Cas. These were all important novae and resolving the shells provides information on the ejection processes. This area was continued with a talk by Tim O'Brien who has also been doing MERLIN radio imaging of nova shells and is trying to understand the shaping mechanisms. Hilmar Duerbeck described his work on the long term evolution of novae. He has examined the evolutionary behavior of a large number of novae and finds that mass accretion rates appear to decrease with time (at least for the first century) and the Balmer and [OIII] lines decrease in intensity in a similar fashion.

Thanks to the Organizers One of the more pleasant duties of the person giving the meeting summary is to make sure that the organizers and their assistants are thanked for their efforts in organizing the meeting. Therefore, I would like to thank Margarita Hernanz and Jordi José for not only choosing Sitges/Barcelona as the site of the meeting but for all their efforts to make this one of the more exciting and important meetings on Classical Nova outbursts. I would also like Margarita and Jose to please thank their associates for all their help – it was a well run meeting. I would also like to acknowledge partial support for my research from grants to ASU from NSF and NASA.

APPENDICES

Conference Program

Monday, 20 May

- **Registration: 8:30-9:30**
- **Opening 9:30-9:50**
- **SESSION 1: SCENARIO I**
  Chair: M. Shara

    - **9:50-10:20**: "General properties of quiescent novae", *B. Warner*. Invited REVIEW

- **COFFEE BREAK: 10:20-11:00**
- **SESSION 2: SCENARIO II**
  Chair: M. Shara

    - **11:00-11:20**: "Alternatives to Hibernation", *T. Naylor*.
    - **11:20-11:40**: "Spectroscopic evidence from HST Studies of Dwarf Novae as Past Classical Novae", *E. Sion*.
    - **11:40-12:00**: "More on Peculiar Abundances in Dwarf Novae and NLs: Implications for Past Novae Explosions", *P. Szkody*.

- **POSTER SESSION: 12:00-13:00**
- **LUNCH BREAK: 13:00-15:00**
- **SESSION 3: SCENARIO III**
  Chair: E. Sion

    - **15:00-15:30**: "The recurrent novae and their relationship with classical novae", *G.C. Anupama*. Invited REVIEW
    - **15:30-15:50**: "Evolution of the symbiotic nova RX Pup", *J. Mikolajewska*.
    - **15:50-16:10**: "An evolutionary scenario for the U Scorpii", *M. Sarna*.
    - **16:10-16:30**: "V838 Mon and its spectacular light-echo: a nova, a born-again AGB, a M31-RedVar analogue or what else?", *U. Munari*.

- **COFFEE BREAK: 16:30-17:00**
- **SESSION 4: SCENARIO IV**
  Chair: E. Sion

    - **17:00-17:20**: "The maximum mass of CO white dwarfs", *I. Domínguez*.
    - **17:20-17:40**: "The evolution of intermediate mass close binary systems: scenarios leading to novae", *P. Gil-Pons, E. García-Berro*.

- **VISIT and SITGES MAJOR'S WELCOME at PALAU MARICEL: 18:45**

Tuesday, 21 May

- **SESSION 5: EXPLOSION MECHANISM AND MASS LOSS I**
  Chair: G. Shaviv

    - **9:00-9:30**: "Recent Developments in Studies of the Nova Outburst", *S. Starrfield.* Invited REVIEW
    - **9:30-10:00**: "Multicycle Nova Evolution Models", *D. Prialnik, A. Kovetz.* Invited REVIEW
    - **10:00-10:20**: "Evolutions of CO WDs accreting H-rich matter as a function of metallicity", *L. Piersanti, S. Cassisi, I. Iben, A. Tornambe.*

- **COFFEE BREAK: 10:20-11:00**

- **SESSION 6: EXPLOSION MECHANISM AND MASS LOSS II**
  Chair: A. Kovetz

    - **11:00-11:30**: "Nucleosynthesis in Classical Novae", *J. José.* Invited REVIEW
    - **11:30-11:50**: "The Effects of Thermonuclear Reaction Rate Variations on Nova Nucleosynthesis", *C. Iliadis.*
    - **11:50-12:10**: "Nuclear Astrophysics at ISAC with DRAGON: Initial Studies", *A. Olin.*

- **POSTER SESSION: 12:10-13:00**

- **LUNCH BREAK: 13:00-15:00**

- **SESSION 7: EXPLOSION MECHANISM AND MASS LOSS III**
  Chair: W. Sparks

    - **15:00-15:30**: "Multidimensional Studies of Classical Novae Explosions", *A. Glasner.* Invited REVIEW
    - **15:30-15:50**: "Numerical Studies of Pre-runaway Classical Novae", *J. Dursi, A. Alexakis, A.C. Calder, R. Rosner, J.W. Truran, B. Fryxell, P.M. Ricker, K. Olson, F.X. Timmes, H. Tufo, M. Zingale, P. MacNeice.*
    - **15:50-16:10**: "The role of novae in galactic chemical evolution", *D. Romano, F. Matteucci.*
    - **16:10-16:30**: "Screening in dense plasma", *G. Shaviv.*

- **COFFEE BREAK: 16:30-17:00**

- **SESSION 8: OBSERVATIONS: OPTICAL, UV, IR AND RADIO SPECTRA I**
  Chair: R. González-Riestra

    - **17:00-17:30**: "Panchromatic Study of Novae in Outburst: Phenomenology and Physics", *S.Shore.* Invited REVIEW
    - **17:30-17:50**: "Clues to the interpretation of the UV spectra of classical novae in outburst", *A. Cassatella, R. González-Riestra.*
    - **17:50-18:10**: "Spectroscopic observations of classical novae at Asiago Observatory", *T. Iijima.*
    - **18:10-18:30**: "The Evolution of Nova Vel 1999", *M. Della Valle.*

Wednesday, 22 May

- SESSION 9: OBSERVATIONS: OPTICAL, UV, IR AND RADIO SPECTRA II / MODEL ATMOSPHERES AND LIGHT CURVES I
  Chair: A. Evans

    - **9:00-9:30**:"The Determination of the Physical Characteristics of Classical Nova explosions and the abundances in their Ejecta from Infrared and Radio Observations", *B. Gehrz*. Invited REVIEW
    - **9:30-9:50**:"0.8-2.5 micron spectroscopy of novae", *D.K. Lynch, R.J. Rudy, C.C. Venturini, S. Mazuk, W. Dimpfl, J.C. Wilson, N.A. Miller, R.C. Puetter.*
    - **9:50-10:20**:"Nova Model Atmospheres and Spectra", *P. Hauschildt*. Invited REVIEW

- COFFEE BREAK: 10:20-11:00

- SESSION 10: MODEL ATMOSPHERES and LIGHT CURVES II
  Chair: A. Cassatella

    - **11:00-11:20**:"The super Eddington evolution of classical nova eruptions". *N. Shaviv.*
    - **11:20-11:35**:"A few comments on nova models", *M. Friedjung.* (15 min. talk)
    - **11:35-11:55**:"Formation and evolution of dust in novae", *A. Evans, J.M.C. Rawlings*
    - **11:55-12:05**:"The properties of the dust around Nova V705 Cas", *A. Evans, O.Smith, V.H. Tyne, J.M.C. Rawlings* (10 min. talk)
    - **12:05-12:25**:"A solution to the transition phase in classical novae", *A. Retter.*
    - **12:25-12:45**:"Recurrent Novae as a progenitor system of Type Ia supernovae", *I. Hachisu.*

- FREE AFTERNOON

Thursday, 23 May

- **SESSION 11: X-RAYS FROM NOVAE I**
  Chair: M. Bode

    - **9:30-10:00**:"X-ray observations of novae", *J. Krautter*. Invited REVIEW
    - **10:00-10:30**:"XMM/Chandra observations of recent novae", *J. Greiner*.

- **COFFEE BREAK: 10:20-11:00**

- **SESSION 12: X-RAYS FROM NOVAE II**
  Chair: M. Bode

    - **11:00-11:20**:"XMM observation of Nova LMC 1995", *M. Orio*.
    - **11:20-11:30**:"Search for ionization nebulae around post-nova X-ray supersoft white dwarfs", *M. Orio*. (10 min. talk)
    - **11:30-11:50**:"On the Possible X-ray emission from Old Classical Nova Shells (CNRs) and the First Resolved and detected CNR in X-rays; The Shell of Nova Per 1901 (GK Per)", *S. Balman*.
    - **11:50-12:10**:"Chandra observations of old novae", *K. Mukai*.

- **POSTER SESSION: 12:10-13:00**

- **LUNCH BREAK: 13:00-15:00**

- **SESSION 13: GAMMA-RAYS FROM NOVAE**
  Chair: J. Greiner

    - **15:00-15:30**:"Gamma-ray emission from classical novae", *M. Hernanz*. Invited REVIEW
    - **15:30-15:50**:"The contribution of ONeMg novae to the 26Al production in the Galaxy", *M. Politano*.
    - **15:50-16:10**:"Global Galactic Distribution of Classical Novae", *A. Iyudin*
    - **16:10-16:30**:"A new experiment for the determination of the 18F(p,alpha) reaction rate at nova temperatures", *N. de Sereville, A. Coc*.
    - **16:30-16:45**:"Direct Study of the 18F(p,alpha)15O Reaction at Energies Relevant for 18F Nucleosynthesis in Novae", *D. Bardayan*. (15 min. talk)

- **COFFEE BREAK: 16:45-17:15**

- **SESSION 14: NOVA POPULATIONS; NOVAE IN EXTERNAL GALAXIES I**
  Chair: A. Shafter

    - **17:15-17:45**: "Nova Populations", *M. Della Valle*. Invited REVIEW
    - **17:45-18:05**: "HST observations of novae in M87, confrontation with theory", *M. Shara*.

- **CONFERENCE DINNER: 20:30**

Friday, 24 May

- **SESSION 15: NOVA POPULATIONS; NOVAE IN EXTERNAL GALAXIES II/ OLD NOVA SHELLS AND REMNANTS I**
  Chair: H. Duerbeck

    - **9:00-9:30**: "Galactic Nova Rates", *A. Shafter*. Invited REVIEW
    - **9:30-9:50**: "The Mystery of the Eruption of V2434 - LMC", *W. Liller, M. Morel.*

    - **9:50-10:20**:"The evolution of nova remnants", *M.F. Bode*. Invited REVIEW
    - **10:20-10:40**:"Nebular remnants of classical novae", *Tim O'Brien.*

- **COFFEE BREAK: 10:40-11:20**

- **SESSION 16: OLD NOVA SHELLS AND REMNANTS II**
  Chair: B. Warner

    - **11:20-11:40**:"Long term evolution of nova luminosities and shell fluxes", *H.W. Duerbeck.*
    - **11:40-12:00**:"Multicolor studies of the old nova behaviour", *E. Pavlenko.*

- **12:00-12:30: CONCLUDING REMARKS**, *S. Starrfield.*

- **END OF THE WORKSHOP**

POSTERS

SCENARIO

1. "Abnormal CNO abundances of three peculiar magnetic cataclysmic variables", *M. Mouchet.*

2. "Did EY Cyg go through nova explosion?", *G. Tovmassian, M. Orio, S. Zharikov.*

3. "The 2001 superoutburst of WZ Sge: Is there any connection of an 8 mag superoutburst with Nova Outbursts?", *J.M. Echevarria, R. Costero, G. Tovmassian, S. Zharikov, R. Michel, A. Arellano.*

4. "The influence of mass loss on orbital elements of binary systems by periastron effect", *M. Andrade, J.A. Docobo.*

EXPLOSION MECHANISM AND MASS LOSS

5. "Numerical Studies of Pre-runaway Classical Novae" (cont.), *A. Calder.*

6. "Movies of Novae Explosions: Restricted Three-Body Dynamics and Geometry of Novae Shells for Purely Gravitational Development", *D.K. Lynch, E. Campbell, S. Mazuk, C.C. Venturini.*

7. "Three Dimensional Modeling of the Pre-Nova Configuration", *D. Dearborn.*

8. "Nova Nucleosynthesis Calculations: Robust Uncertainties, Sensitivities, and Radioactive Ion Beam Measurements", *M.S. Smith, W.R. Hix, S. Parete-Koon, L. Dessieux, M.W. Guidry, D.W. Bardayan, S. Starrfield, D.L. Smith, A. Mezzacappa.*

9. "The imprint of nova nucleosynthesis in presolar grains", *J. José, M. Hernanz, S. Amari, E. Zinner.*

OBSERVATIONS: OPTICAL, UV, IR AND RADIO SPECTRA

10. "Spectral Evolution of Galactic Novae", *E. Mason.*

11. "Abundance analysis of the ONeMg nova QU Vul", *G. Schwarz.* (presented by S. Starrfield)

12. "Elemental Abundances of Nova Cygni 1992", *K. Vanlandingham.* (presented by S. Starrfield)

13. "The INES Guide for Classical Novae", *R. González-Riestra, A. Cassatella.*

14. "The spectroscopic evolution of Nova Vel 1999 (V382 Vel)", *A. Augusto, M. Diaz.*

15. "WSO/UV, World Space Observatory / Ultraviolet", *W. Wamsteker et al.* (presented by M. Hernanz)

16. "Radio Emission from Nova Cassiopeiae 1995", *I. Heywood.*

MODEL ATMOSPHERES and LIGHT CURVES

17. "Spectra and Light Curves for IM Nor and N Sgr 2002", *W. Liller*.

18. "The recurrent nova IM Nor in a scheme of a novae classification on the shape of a light curve", *A. Rosenbush.* (presented by E. Pavlenko)

19. "Spectroscopic and photometric observations of the recurrent nova IM Normae", *H.W. Duerbeck, R. Baptista, C.M. Dutra, L. Freyhammer, H. Hensberge, A. Jones, C. Sterken.*

20. "Nova Monocerotis 2002 (V838 Mon) in the early stages of its outburst", *E.A. Barsukova, N.V. Borisov, V.P. Goranskij, A.V. Kusakin, N.V. Metlova, S. Yu. Shugarov.*

21. "The discovery of Nova Aql 1985", *S. V. Antipin, S. Yu. Shugarov, P. Kroll.*

22. "V723 Cas a borderline classical nova?", *M. Friedjung, T. Iijima.*

23. "Radial pulsation of the cooling white dwarf in the decay of Nova Cassiopeiae 1995 (V723 Cas)", *V.P. Goranskij.*

24. "UBV photometry of two novae in Cygnus", *I. Voloshina.*

25. "BVRI photometry of extremely slow nova Aql 2001=V1548 Aql", *N. Primak.*

26. "The Photometry of V1974 Cyg = N Cyg 1992", *S. Yu. Shugarov, V.P. Goranskij.*

27. "The Problem of the Flickering Activity of the Recurrent Nova T Corona Borealis", *L. Hric, K. Petrik.*

28. "Activity of the super-soft X-ray source V Sge", *V. Simon, J.A. Mattei.*

29. "The colors and luminosities of the super-soft X-ray sources and classical novae", *V. Simon.*

X-RAYS FROM NOVAE

30. "Observing novae with Chandra", *Y. Butt.*

31. "Nova Velorum 1999: Spectroscopy with Chandra", *V. Burwitz.*

32. "XMM-Newton observations of recent Galactic classical novae", *M. Hernanz, G. Sala.*

33. "Photoionization as a source of X-ray emission from classical novae", *G. Sala, M. Hernanz.*

34. "Supersoft sources in M31: how many are classical or recurrent novae", *M. Orio.*

GAMMA-RAYS FROM NOVAE

35. "The diffuse 1.275 MeV emission from Galactic ONe novae", *P. Jean.*

36. "Future INTEGRAL observations of classical novae", *M. Hernanz.*

NOVA POPULATIONS; NOVAE IN EXTERNAL GALAXIES

37. "The local stellar population of novae regions in the Large Magellanic Cloud", *G.C. Anupama, A. Subramaniam.*

38. "Novae in External Galaxies from the POINT-AGAPE Survey and the Liverpool Telescope", *M. Darnley.*

39. "Two Recent Extra-galactic Novae: Nova SMC 2001 and Nova LMC 2002", *E. Mason.*

40. "A survey for novae in M33", *S.J. Williams, A.W. Shafter.*

OLD NOVA SHELLS AND REMNANTS

41. "An HST Snapshot Survey of Nova Shells", *F. Ringwald.*

42. "Photometric and spectroscopic properties of four old novae", *L. Schmidtobreick, C. Tappert, A. Bianchini, R.E. Mennickent.*

43. "The UCT Survey of Old Novae", *P.A. Woudt, B. Warner.*

44. "Optical spectroscopy of GK Per during outburst and quiescence", *U. S. Kamath, G.C. Anupama.*

45. "Physical and Chemical Diagnostics of Structured Nova Shells", *M. Diaz.*

46. "Observations of the eclipsing classical nova V Per (Nova Persei 1887) in 2001", *N. Katysheva.*

47. "A possible detected faint shell of the classical nova QZ Aurigae", *H. Esenoglu.* (presented by S. Balman)

48. "The active quiescence of HR Del (Nova Del 1967)", *P. Selvelli, M. Friedjung.*

49. "Emission line flaring in the SW Sex old nova V533 Her", *P. Rodriguez-Gil.*

List of Participants

Last name	First Name	Institution	e-mail
Altamore	Aldo	Università Roma Tre	altamore@fis.uniroma3.it
Andrade	Manuel	Observatorio Astronómico R.M. Aller / Univ. Santiago de Compostela	oandrade@usc.es
Anupama	G. C.	Indian Institute of Astrophysics	gca@iiap.ernet.in
Balman	Solen	Middle East Technical University	solen@astroa.physics.metu.edu.tr
Bardayan	Dan	Oak Ridge National Laboratory	bardayan@mail.phy.ornl.gov
Barsukova	Elena	Special Astrophysical Observatory	bars@jet.sao.ru
Bode	Michael F.	Liverpool John Moores University	mfb@astro.livjm.ac.uk
Burwitz	Vadim	MPI für Extraterrestrische Physik	burwitz@mpe.mpg.de
Busegnies	Yves	Université Libre de Bruxelles	yves_busegnies@astro.ulb.ac.be
Butt	Yousaf	Center for Astrophysics	ybutt@head-cfa.harvard.edu
Calder	Alan	University of Chicago	calder@flash.uchicago.edu
Casas	Ricard	Agrupació Astronòmica de Sabadell	ricardcasas@wanadoo.es
Cassatella	Angelo	Università Roma Tre	cassatella@fis.uniroma3.it
Coc	Alain	CSNSM	coc@csnsm.in2p3.fr
Darnley	Matthew	Liverpool John Moores University	mjd@astro.livjm.ac.uk
De Séréville	Nicolas	CSNSM	deserevi@csnsm.in2p3.fr
Dearborn	David	Lawrence Livermore National Laboratory	ddearborn@llnl.gov
Della Valle	Massimo	Arcetri Astrophysical Observatory	massimo@arcetri.astro.it

Last name	First Name	Institution	e-mail
Díaz	Marcos	Universidade de Sao Paulo	marcos@astro.iag.usp.br
Domínguez	Inma	Universidad de Granada	inma@ugr.es
Duerbeck	Hilmar W.	Brussels Free University (VUB)	hduerbeck@vub.ac.be
Dursi	Jonathan	University of Chicago	ljdursi@flash.uchicago.edu
Echevarría	Juan Manuel	Instituto de Astronomía, UNAM	jer@astroscu.unam.mx
Evans	Nye	Keele University	ae@astro.keele.ac.uk
Friedjung	Michael	Institut d'Astrophysique (CNRS)	fried@iap.fr
García-Berro	Enrique	Universitat Politècnica de Catalunya	garcia@fa.upc.es
Gehrz	Robert D.	University of Minnesota	gehrz@astro.umn.edu
Gil-Pons	Pilar	Universitat Politècnica de Catalunya	pilar@fa.upc.es
Glasner	Ami	Racah Institute of Physics	ami@saba.fiz.huji.ac.il
González-Riestra	Rosario	XMM Scientific Operations Centre, VILSPA	rgonzalez@xmm.vilspa.esa.es
Goranskij	Vitaly P.	Special Astrophysical Observatory	goray@sao.ru
Greiner	Jochen	MPI für Extraterrestriche Physik	jcg@mpe.mpg.de
Guerrero	Josep	IEEC	guerrero@ieec.fcr.es
Hachisu	Izumi	University of Tokyo	hachisu@chianti.c.u-tokyo.ac.jp
Hauschildt	Peter	University of Georgia	yeti@hal.physast.uga.edu
Hernanz	Margarida	IEEC/CSIC	hernanz@ieec.fcr.es
Heywood	Ian	The University of Manchester	iheywood@jb.man.ac.uk
Hric	Ladislav	Astronomical Institute of the Slovak Academy of Sciences	hric@ta3.sk
Iben, Jr.	Icko	University of Illinois	icko@astro.uiuc.edu

Last name	First Name	Institution	e-mail
Idan	Irit	Rafael	idan@tx.technion.ac.il
Iijima	Takashi	Osservatorio Astrofisico di Padova	iijima@astras.pd.astro.it
Iliadis	Christian	University of North Carolina	iliadis@unc.edu
Isern	Jordi	IEEC/CSIC	isern@ieec.fcr.es
Iyudin	Anatoli	MPI für Extraterrestrische Physik	ani@mpe.mpg.de
Jean	Pierre	Centre d'Etude Spatiale des Rayonnements	jean@cesr.fr
José	Jordi	IEEC / UPC	jjose@ieec.fcr.es
Kato	Mariko	Keio University	mariko@educ.cc.keio.ac.jp
Katysheva	Nataly	Sternberg Astronomical Institute	nk@sai.msu.ru
Kovetz	Attay	Tel Aviv University	attay@etoile.tau.ac.il
Krautter	Joachim	Landessternwarte	J.Krautter@lsw.uni-heidelberg.de
Lee	Tae Hoon	Soongsil University	thlee@physics.soongsil.ac.kr
Liller	William	Novahound Observatory	wliller@compuserve.com
Lynch	David K.	The Aerospace Corporation	david.k.lynch@aero.org
Marom	Ariel	Rafael	
Mason	Elena	ESO	emason@eso.org
Matteucci	Francesca	Università di Trieste	matteucci@ts.astro.it
Mikolajewska	Joanna	N. Copernicus Astronomical Center	mikolaj@camk.edu.pl
Mouchet	Martine	Observatoire de Paris-Meudon	martine.mouchet@obspm.fr
Mukai	Koji	NASA/GSFC	mukai@milkyway.gsfc.nasa.gov
Munari	Ulisse	Osservatorio Astronomico di Padova	munari@pd.astro.it

Last name	First Name	Institution	e-mail
Naylor	Tim	University of Exeter	timn@astro.ex.ac.uk
O'Brien	Tim	University of Manchester	tob@jb.man.ac.uk
Olin	Art	TRIUMF	olin@triumf.ca
Orio	Marina	University of Wisconsin/INAF Torino	orio@cow.physics.wisc.edu
Pavlenko	Elena	Crimean Astrophysical Observatory	pavlenko@crao.crimea.ua
Petrik	Karol	Trnava University	kpetrik@pobox.sk
Piersanti	Luciano	Osservatorio Astronomico di Teramo	piersanti@astrte.te.astro.it
Politano	Michael	Arizona State University	politano@asu.edu
Prialnik	Dina	Tel Aviv University	dina@planet.tau.ac.il
Primak	Natalia	Kiev National University	prime_str@mail.ru
Retter	Alon	University of Sydney	retter@physics.usyd.edu.au
Ringwald	Frederick	California State University	frederick_ringwald@csufresno.edu
Rodriguez - Gil	Pablo	Instituto de Astrofísica de Canarias	prguez@ll.iac.es
Romano	Donatella	SISSA/ISAS	romano@sissa.it
Rossi	Corinne	Università La Sapienza	corinne.rossi@roma1.infn.it
Sala	Gloria	IEEC/CSIC	sala@ieec.fcr.es
Sarna	Marek J.	N. Copernicus Astronomical Center	sarna@camk.edu.pl
Schmidtobreick	Linda	ESO	lschmidt@eso.org
Seitter	Waltraut C.	Muenster University	
Selvelli	Pierluigi	Osservatorio Astronomico di Trieste	selvelli@ts.astro.it
Shafter	Allen	San Diego State University	shafter@proteus.sdsu.edu

Last name	First Name	Institution	e-mail
Shara	Michael	American Museum of Natural History	mshara@amnh.org
Shaviv	Giora	Technion	gioras@physics.technion.ac.il
Shaviv	Nir	Racah Institute of Physics	shaviv@phys.huji.ac.il
Shore	Steven N.	Indiana University South Bend	sshore@paladin.iusb.edu
Shugarov	Sergei	Sternberg Astronomical Institute	shugarov@sai.msu.ru
Siess	Lionel	Université Libre de Bruxelles	lionel.siess@astro.ulb.ac.be
Simon	Vojtech	Astronomical Institute, Czech Academy of Sciences	simon@asu.cas.cz
Sion	Edward	Villanova University	emsion@ast.villanova.edu
Sparks	Warren	Los Alamos National Laboratory	wms@lanl.gov
Starrfield	Sumner	Arizona State University	sumner.starrfield@asu.edu
Szkody	Paula	University of Washington	szkody@astro.washington.edu
Tappert	Claus	Universidad de Concepción	claus@gemini.cfm.udec.cl
Tovmassian	Gaghik	Instituto de Astronomía, UNAM	gag@astrosen.unam.mx
Voloshina	Irina	Sternberg Astronomical Institute	vib@sai.msu.ru
Warner	Brian	University of Cape Town	warner@physci.uct.ac.za
Williams	Stephen	San Diego State University	williams@bellus.sdsu.edu

Author Index

A

Abada-Simon, M., 67
Alexakis, A., 134, 139
Amari, S., 167
Andrade, M., 82
Angulo, C., 420
Anupama, G. C., 32, 42, 476, 534
Arellano-Ferro, A., 77
Assunção, M., 420
Augusto, A., 233

B

Balman, Ş., 365
Baptista, R., 299
Bardayan, D. W., 161, 425
Baron, E., 249
Barstow, M., 238
Barsukova, E. A., 303
Batchelder, J. C., 425
Beaumel, D., 420
Bennett, K., 415
Beuermann, K., 67
Bianchini, A., 214, 527
Bishop, S., 119
Blackmon, J. C., 425
Bode, M. F., 242, 481, 497, 509
Bonnet-Bidaud, J. M., 67
Borisov, N. V., 303
Bouzid, B., 420
Brandi, E., 42
Brosch, N., 238
Buchmann, L., 119
Burwitz, V., 377

C

Calder, A. C., 134, 139
Campbell, E., 155
Cassatella, A., 188, 228
Cassisi, S., 99
Champagne, A. E., 114, 425
Chatterjee, M. L., 119
Chen, A., 119

Cherubini, S., 420
Coc, A., 420, 435
Corradi, R. M. L., 52
Costero, R., 72, 77
Couder, M., 420

D

Dalmazzo, A., 391
Darnley, M. J., 481
D'Auria, J. M., 119
Davinson, T., 425
Davis, R. J., 242
Della Valle, M., 214, 443, 486
Demaret, P., 420
de Martino, D., 67
Dennefeld, M., 238
de Oliveira Santos, F., 420
de Séréville, N., 420
Dessieux, L., 161
Diaz, M. P., 233, 299, 538
Diehl, R., 415
Dimpfl, W. L., 208
Dobrotka, A., 328
Docobo, J. A., 82
Domínguez, I., 57
Dopita, M., 238
Duerbeck, H. W., 299, 514
Dursi, L. J., 134, 139
Dutra, C. M., 299

E

Echevarría, J., 72, 77
Engel, S., 119
Ergma, E., 47
Esenoglu, H. H., 548
Evans, A., 270, 275
Eyres, S. P. S., 242, 275

F

Ferlet, R., 67
Ferrer, O., 42

Figuera, P., 420
Fitzgerald, R., 425
Fortier, S., 420
Freyhammer, L., 299
Fried, R., 67
Friedjung, M., 266, 308, 553
Fryxell, B., 134, 139
Fu-Zhen, C., 238

G

Gaelens, M., 420
Gális, R., 328
Gänsicke, B., 67
Garcia, L., 42
García-Berro, E., 62
Geballe, T. R., 275
Gehrz, R. D., 198
Gerškevitš, J., 47
Gigliotti, D., 119
Gilmozzi, R., 486
Gil-Pons, P., 62
Giménez, A., 435
Glasner, A., 124
Gómez de Castro, A. I., 238
González-Riestra, R., 188, 228, 238
Goranskij, V. P., 303, 311, 319, 323, 519
Greife, U., 119
Greiner, J., 355
Guidry, M. W., 161

H

Hachisu, I., 284
Hammache, F., 420
Harman, D. J., 509
Hartmann, W., 355
Haubold, H., 238
Hauschildt, P. H., 249
Henden, A., 52
Hensberge, H., 299
Hermsen, W., 415
Hernanz, M., 167, 238, 381, 386, 399, 430, 435
Heywood, I., 242
Hix, W. R., 161, 425
Howell, S., 67
Hric, L., 328

Hunter, D., 119
Hussein, A., 119
Hutcheon, D., 119

I

Iben Jr., I., 99
Iijima, T., 193, 308
Iliadis, C., 114, 425
Isern, J., 57, 435
Iyudin, A. F., 415

J

Jean, P., 430, 435
Jewett, C., 119
Jones, A. F., 299
José, J., 104, 114, 167, 430, 435

K

Kamath, U. S., 534
Kappelmann, N., 238
Kato, M., 284
Katysheva, N. A., 543
Kerins, E. J., 481
Kiener, J., 420
King, J., 119
Kozub, R. L., 425
Krautter, J., 345, 377
Kubono, S., 119
Kusakin, A. V., 303

L

Labar, D., 420
Laird, A. M., 119
Lamey, M., 119
Lecavelier, A., 67
Lefebvre, A., 420
Leleux, P., 420
Lewis, R., 119
Lichti, G. G., 415
Liller, W., 289, 472
Liu, W., 119
Livne, E., 124

Lo Curto, G., 486
Loiselet, M., 420
Lynch, D. K., 155, 208

M

Ma, Z., 425
MacNeice, P., 134
Martínez, P., 238
Martínez-Pais, I. G., 558
Mason, E., 214, 486
Mattei, J. A., 333
Matteucci, F., 144
Mazuk, S., 155, 208
Mennickent, R., 527
Metlova, N. V., 303, 311, 315
Mezzacappa, A., 161
Michel, R., 72, 77
Michimasa, S., 119
Mikołajewska, J., 42
Miller, N. A., 208
Morel, M., 472
Mouchet, M., 67
Mukai, K., 67, 372
Munari, U., 52

N

Naylor, T., 16
Nedialkov, P., 391
Ness, J.-U., 377
Ninane, A., 420

O

O'Brien, T. J., 242, 481, 509
Olin, A., 119
Olson, K., 134, 139
Orio, M., 72, 355, 360, 372, 391
Ottewell, D., 119
Ouichaoui, S., 420

P

Pagano, I., 238
Parete-Koon, S., 161, 425

Parker, P. D., 119, 425
Pavlenko, E. P., 319, 323, 519, 543
Petrík, K., 328
Piersanti, L., 99
Politano, M., 409
Porquet, D., 67
Primak, N. V., 319, 519
Puetter, R., 208

Q

Quiroga, C., 42

R

Rawlings, J. M. C., 270, 275
Retter, A., 279
Richer, M., 77
Ricker, P., 134, 139
Ringwald, F., 372
Rodríguez-Gil, P., 558
Rogers, J., 119
Romano, D., 144
Rosenbush, A. E., 294
Rosner, R., 134
Roueff, E., 67
Rovithis-Livaniou, H., 315
Rudy, R. J., 208
Ryan, J., 415
Ryckewaert, G., 420

S

Sahade, J., 238
Sala, G., 381, 386, 435
Sarna, M. J., 47
Schmidtobreick, L., 527
Schönfelder, V., 415
Schwarz, G. J., 219, 249
Selvelli, P., 553
Shafter, A. W., 462, 491
Shara, M. M., 457
Shaviv, G., 150
Shaviv, N. J., 150, 259
Shore, S. N., 175, 224
Short, C. I., 249
Shu, N., 425

Shugarov, S. Y., 303, 311, 319, 323, 519, 543
Shustov, B., 238
Šimon, V., 333, 338
Sion, E. M., 21
Smirnova, N., 420
Smith, D. L., 161
Smith, M. S., 161, 425
Smith, O., 275
Solheim, J.-E., 238
Starrfield, S., 89, 114, 161, 224, 249, 377, 435, 565
Sterken, C., 299
Still, M., 355, 372
Straniero, O., 57
Strieder, F., 119
Subramaniam, A., 476
Szkody, P., 28, 67

T

Tappert, C., 527
Tatischeff, V., 420
Thibaud, J.-P., 420
Timmes, F. X., 134, 139
Tornambé, A., 57, 99
Tovmassian, G., 72, 77, 360
Truran, J. W., 62, 134, 139, 435
Tupper, P., 114
Tyne, V. H., 275

V

Vanlandingham, K. M., 224
Venturini, C. C., 155, 208
Voloshina, I., 315

W

Wagner, R. M., 224
Wamsteker, W., 238
Warner, B., 3, 532
Wiescher, M., 119
Williams, R. E., 486
Williams, S. J., 491
Wilson, J. C., 208
Woods, P. J., 425
Woudt, P. A., 532
Wrede, C., 119

Z

Zharikov, S., 72, 77
Zingale, M., 134, 139
Zinner, E., 167
Zurek, D. R., 457
Zwitter, T., 52